Labelling Strategies in Environmental Policy

The International Library of Environmental Economics and Policy
General Editor: Tom Tietenberg

Titles in the Series:

Fisheries Economics, Volumes I and II
Lee G. Anderson

Green Accounting
Peter Bartelmus and Eberhard K. Seifert

The Economics of International Environmental Agreements
Amitrajeet A. Batabyal

The Economics of Agri-Environmental Policy
Sandra S. Batie and Richard D. Horan

The Stated Preference Approach to Environmental Evaluation, Volumes I, II and III
Richard T. Carson

International Trade and the Environment
Judith M. Dean

Economics of Water Resources
K. William Easter and Mary E. Renwick

The Economics of Water Quality
K. William Easter and Naomi Zeitouni

Distributional Effects of Environmental Policy
Don Fullerton

Spatial Aspects of Environmental Policy
Jacqueline Geoghegan and Wayne Gray

Economic Costs and Consequences of Environmental Regulation
Wayne B. Gray

Controlling Automobile Air Pollution
Winston Harrington and Virginia McConnell

The Theory and Practice of Command and Control in Environmental Policy
Gloria E. Helfand and Peter Berck

Revealed Preference Approaches to Environmental Valuation
Joseph Herriges and Cathy Kling

The Economics of Residential Solid Waste Management
Thomas C. Kinnaman

Property Rights and Environmental Problems, Volumes I and II
Bruce A. Larson

Environmental Taxation in Practice
Adrian Muller and Thomas Sterner

The Economics of Land Use
Peter J. Parks and Ian W. Hardie

The Economics of Sustainability
John C.V. Pezzey and Michael A. Toman

The Economics of Biodiversity Conservation
Stephen Polasky

The Economics of Environmental Monitoring and Enforcement
Clifford S. Russell

Discounting and Environmental Policy
Joel D. Scheraga

Economics of Forestry
Roger A. Sedjo

Economics and Liability for Environmental Problems
Kathleen Segerson

Experiments in Environmental Economics, Volumes I and II
Jason F. Shogren

Corporate Strategies for Managing Environmental Risk
Bernard Sinclair-Desgagné

Labelling Strategies in Environmental Policy
Mario Teisl

Emissions Trading Programs, Volumes I and II
Tom Tietenberg

Climate Change
Michael A. Toman and Brent Sohngen

The Theory of Environmental Taxation
Roberton Williams

Labelling Strategies in Environmental Policy

Edited by

Mario Teisl

University of Maine, USA

ASHGATE

Wherever possible, these reprints are made from a copy of the original printing, but these can themselves be of very variable quality. Whilst the publisher has made every effort to ensure the quality of the reprint, some variability may inevitably remain.

Published by
Ashgate Publishing Limited
Gower House
Croft Road
Aldershot
Hampshire GU11 3HR
England

Ashgate Publishing Company
Suite 420
101 Cherry Street
Burlington, VT 05401-4405
USA

Ashgate website: http://www.ashgate.com

British Library Cataloguing in Publication Data
Labelling strategies in environmental policy
1. Eco-labeling 2. Eco-labeling – Economic aspects
3. Environmental policy
I. Teisl, Mario F. (Mario Francis), 1960–
658.8′23

Library of Congress Cataloging-in-Publication Data
Labelling strategies in environmental policy / edited by Mario Teisl.
p. cm. – (The international library of environmental economics and policy)
Includes bibliographical references and index.
ISBN: 978-0-7546-2730-2 (alk. paper)
1. Eco-labeling. I. Teisl, Mario F. (Mario Francis), 1960–

HF5413.L33 2007
381.3'4–dc22

2006035274

ISBN: 978-0-7546-2730-2

Printed in Great Britain by TJ International Ltd, Padstow, Cornwall

Contents

PART III FACTORS IMPACTING THE SUCCESS AND DESIGN OF ECO-LABELLING PROGRAMMES

PART IV INTERNATIONAL DIMENSIONS OF ENVIRONMENTAL LABELLING

Acknowledgements

The editor and publishers wish to thank the following for permission to use copyright material.

A B Academic Publishers Douadia Bougherara and Gilles Grolleau (2005), 'Designing Ecolabels in Order to Mitigate Market Failures: An Application to Agrofood Products', *Journal of Interdisciplinary Economics*, **16**, pp. 411–30. Copyright 2005 A B Academic Publishers.

Blackwell Publishing Limited for the essays: Wesley Nimon and John Beghin (1999), 'Are Eco-labels Valuable? Evidence from the Apparel Industry', *American Journal of Agricultural Economics*, **81**, pp. 801–11. Copyright © 1999 American Agricultural Economics Association; Huilan Tian (2003), 'Eco-labelling Scheme, Environmental Protection, and Protectionism', *Canadian Journal of Economics*, **36**, pp. 608–33; Arnab K. Basu, Nancy H. Chau and Ulrike Grote (2004), 'On Export Rivalry and the Greening of Agriculture – The Role of Eco-labels', *Agricultural Economics*, **31**, pp. 135–47. Copyright © 2004 Elsevier; Daniel Melser and Peter E. Robertson (2005), 'Eco-labelling and the Trade–Environment Debate', *The World Economy*, **28**, pp. 49–62. Copyright © 2005 Blackwell Publishing Limited; Arnab K. Basu, Nancy H. Chau and Ulrike Grote (2003), 'Eco-labeling and Stages of Development', *Review of Development Economics*, **7**, pp. 228–47. Copyright © 2003 Blackwell Publishing Limited.

Cambridge University Press for the essay: Bruce A. Larson (2003), 'Eco-labels for Credence Attributes: The Case of Shade-Grown Coffee', *Environment and Development Economics*, **8**, pp. 529–47. Copyright © 2003 Cambridge University Press.

Elsevier for the essays: Timothy J. Brennan (2006), '"Green" Preferences as Regulatory Policy Instrument', *Ecological Economics*, **56**, pp. 144–54. Copyright © 2005 Elsevier BV; Jimin Zhao and Qing Xia (1999), 'China's Environmental Labeling Program', *Environmental Impact Assessment Review*, **19**, pp. 477–97. Copyright © 1999 Elsevier Science Inc.; Mario F. Teisl, Brian Roe and Robert L. Hicks (2002), 'Can Eco-Labels Tune a Market? Evidence from Dolphin-Safe Labeling', *Journal of Environmental Economics and Management*, **43**, pp. 339–59. Copyright © 2002 Elsevier Science Inc.; Thomas Bue Bjørner, Lars Gårn Hansen and Clifford S. Russell (2004), 'Environmental Labeling and Consumers' Choice – An Empirical Analysis of the Effect of the Nordic Swan', *Journal of Environmental Economics and Management*, **47**, pp. 411–34. Copyright © 2003 Elsevier Inc.; Benjamin Cashore, Graeme Auld and Deanna Newsom (2003), 'Forest Certification (Eco-labeling) Programs and their Policy-making Authority: Explaining Divergence among North American and European Case Studies', *Forest Policy and Economics*, **5**, pp. 225–47. Copyright 2002 Elsevier Science BV; Bernhard Truffer, Jochen Markard and Rolf Wüstenhagen (2001), 'Eco-labeling of Electricity – Strategies and Tradeoffs in the Definition of Environmental Standards', *Energy Policy*, **29**, pp. 885–97. Copyright © 2001 Elsevier Science Limited; Helen Nilsson, Burcu Tunçer and Åke Thidell (2004), 'The Use of Eco-labeling like Initiatives on Food Products to Promote Quality Assurance – Is There Enough Credibility?', *Journal of Cleaner Production*, **12**, pp. 517–26. Copyright © 2003 Elsevier Limited; Gregory S. Amacher, Erkki Koskela and Markku Ollikainen (2003), 'Environmental Quality Competition and Eco-labeling', *Journal of Environmental Economics and Management*, **47**, pp. 284–306. Copyright © 2003 Elsevier Inc.; Abhijit Banerjee and Barry D. Solomon (2003), Eco-labeling for Energy Efficiency and Sustainability: A Meta-Evaluation of US Programs', *Energy Policy*, **31**, pp. 109–23. Copyright © 2002 Elsevier Science Limited; Charlotte Leire and Åke Thidell (2005), 'Product-related Environmental Information to Guide Consumer Purchases – A Review and Analysis of Research on Perceptions, Understanding and Use among Nordic Consumers', *Journal of Cleaner Production*, **13**, pp. 1061–70. Copyright © 2004 Elsevier Limited.

Inderscience Enterprises Limited for the essay: Lisette Ibanez and Jacques Laye (forthcoming), 'Ecocertification, Differentiation in Retailing and Upstream Market Power', *International Journal of Agricultural Resources, Governance and Ecology*, pp. 1–16.

Marine Resource Economics for the essay: Eyjólfur Gudmundsson and Cathy R. Wessells (2000), 'Ecolabeling Seafood for Sustainable Production: Implications for Fisheries Management', *Marine Resource Economics*, **15**, pp. 97–113.

Springer BV for the essays: Young-Hwan Ahn and Byong-Hun Ahn (2001), 'Eco-labeling and Non-product-related Process and Production Methods', *Environmental Economics and Policy Studies*, **4**, pp. 167–89. Copyright © 2001 Springer Verlag; Douadia Bougherara, Gilles Grolleau and Luc Thiébaut (2005), 'Can Labelling Policies Do More Harm than Good? An Analysis Applied to Environmental Labelling Schemes', *European Journal of Law and Economics*, **19**, pp. 5–16. Copyright © 2005 Springer Science; Stefanie Engel (2006), 'Overcompliance, Labeling, and Lobbying: The Case of Credence Goods', *Environmental Modeling and Assessment*, **11**, pp. 115–30. Copyright © 2006 Springer; Cesare Dosi and Michele Moretto (2001), 'Is Ecolabelling a Reliable Environmental Policy Measure?', *Environmental and Resource Economics*, **18**, pp. 113–27. Copyright © 2001 Kluwer Academic Publishers; John Thøgersen (2000), 'Psychological Determinants of Paying Attention to Eco-labels in Purchase Decisions: Model Development and Multinational Validation', *Journal of Consumer Policy*, **23**, pp. 285–313. Copyright © 2000 Kluwer Academic Publishers; Gunne Grankvist, Ulf Dahlstrand and Anders Biel (2004), 'The Impact of Environmental Labeling on Consumer Preference: Negative vs. Positive Labels', *Journal of Consumer Policy*, **27**, pp. 213–30. Copyright © 2004 Kluwer Academic Publishers; Paulo A.L.D Nunes and Yohanes E. Riyanto (2005), 'Information as a Regulatory Instrument to Price Biodiversity Benefits: Certification and Ecolabeling Policy Practices', *Biodiversity and Conservation*, **14**, pp. 2009–27. Copyright © 2005 Springer; Robert J. Johnston, Cathy R. Wessells, Holger Donath and Frank Asche (2001), 'Measuring Consumer Preferences for Ecolabeled Seafood: An International Comparison', *Journal of Agricultural and Resource Economics*, **26**, pp. 20–39. Copyright © 2001 Kluwer Academic Publishers.

University of Wisconsin Press for the essay: Roger A. Sedjo and Stephen K. Swallow (2002), 'Voluntary Eco-labeling and the Price Premium', *Land Economics*, **78**, pp. 272–84. Copyright © 2002 Board of Regents of the University of Wisconsin System.

Every effort has been made to trace all the copyright holders, but if any have been inadvertently overlooked the publishers will be pleased to make the necessary arrangement at the first opportunity.

Series Preface

The *International Library of Environmental Economics and Policy* explores the influence of economics on the development of environmental and natural resource policy. In a series of twenty-five volumes, the most significant journal essays in key areas of contemporary environmental and resource policy are collected. Scholars who are recognized for their expertise and contribution to the literature in the various research areas serve as volume editors and write an introductory essay that provides the context for the collection.

Volumes in the series reflect the broad strands of economic research including 1) Natural and Environmental Resources, 2) Policy Instruments and Institutions and 3) Methodology. The editors, in their introduction to each volume, provide a state-of-the-art overview of the topic and explain the influence and relevance of the collected papers on the development of policy. This reference series provides access to the economic literature that has made an enduring contribution to contemporary and natural resource policy.

TOM TIETENBERG
General Editor

Introduction

Is Eco-labelling a 'Fog of Information' or Part of the 'Oxygen of the Modern Age'?[1]

In 1998 I was invited to present a professional paper on the economics of health and environmental labelling.[2] I had recently finished my dissertation on the welfare effects of nutrition labelling and had completed a fellowship at the US Food and Drug Administration, so I felt reasonably comfortable in taking on the 'health labelling' task. However, as someone who had recently been hired as an environmental economist, I wanted to focus the paper more on environmental labelling (hereafter eco-labelling). I was quick to find out that there was only a small amount of work published on eco-labelling, most of it comprising industry and government publications.

This lack of peer-reviewed publications was somewhat surprising in that the environmental characteristics of products had become increasingly important to consumers, producers, government and non-governmental groups. Governments and non-governmental organizations (NGOs) around the world had been organizing, implementing and verifying eco-labelling programmes that covered thousands of products. For example, the US Federal Trade Commission first issued its *Environmental Marketing Guidelines* in 1992, and many eco-labelling programmes had been in existence for a number of years, the oldest being started in 1977 (see Table 1). Firms had also responded by placing information on products that highlighted the product's environmental attributes and by introducing new, or redesigned, 'green' products. Given that economists have long held that the flow of information plays a critical role in the efficient operation of markets, it was rather surprising to find that there has been little in the way of empirical economics research to determine the market and policy effectiveness of such programmes.

Since 1998 interest in eco-labelling has continued to grow; in general, the number of products being covered by some sort of eco-label has more than doubled and the scope of products being awarded an eco-label has also expanded. A decade ago, eco-labels were primarily attached to food products, soaps and detergents, paper and wood products, electricity-drawing appliances and a few speciality products. Now eco-labels are being awarded to everything from buildings, passenger vehicles, floral products, paints, lubricants and even beaches and other tourist destinations.

---

1 'The fog of information can drive out knowledge' (Boorstin, 1983) and 'Information is the oxygen of the modern age' (Reagan, 1989).

2 Labelling is any policy instrument of a government or other third party, which somehow regulates the presentation of product-based information to consumers. Eco-labelling does not include more general environmental disclosure policies, such as the Toxic Release Inventory in the US and PROPER in Indonesia (see Pargal *et al.*, 1997).

Table 1 Growth in the number of products awarded an eco-label, by various programmes

Name of Programme	Year initiated	Number of Awards 1997[a]	2006[b]
Austrian Eco-label	1991	150	337
Chlorine-Free Product Association	1997	2	16
European Union Eco-label	1992	182	1776
Germany Blue Angel	1977	4135	3652
Green Seal	1989	300	567
Marine Stewardship Council	2000	200	300
New Zealand's Environmental Choice	1990	55	223
Sweden – Nordic Swan	1989	350	877
Thailand – Green Label	1993	0	137

a From US Environmental Protection Agency (1998).
b Documentation from individual programmes' websites.

It is reasonable to ask what caused this marked growth in eco-labelling programmes. While it may be partly explained by consumers having increased levels of environmental concern; except for a recent upturn, Gallup's Environmental Poll (see Figure 1) shows a general decline in environmental concern among US consumers (Hayward, 2006). Alternative explanations for the growth in eco-labelling could include the market's increased level of globalization (Cashore *et al.*, Chapter 24) and the fact that eco-labelling is World Trade Organization (WTO)-compliant whereas other environmental regulations may not be. Politically, eco-labelling programmes may be a reaction on the part of increasingly sophisticated and organized international non-governmental groups towards governmental agencies that lack the will to impose stricter environmental regulations. On the supply-side, eco-labelling may have been helped along by the increased opportunities for rent-seeking by retailers of eco-labelled products, by improvements in the techniques and technologies related to supply-chain management or by reduced supply-chain management costs associated with increased levels of consolidation within specific industries. It is likely that all of the above (and more) factors helped contribute to eco-labelling's growth, and, while the essays presented in this volume address some of the above issues, determining the reasons driving this growth may remain elusive.

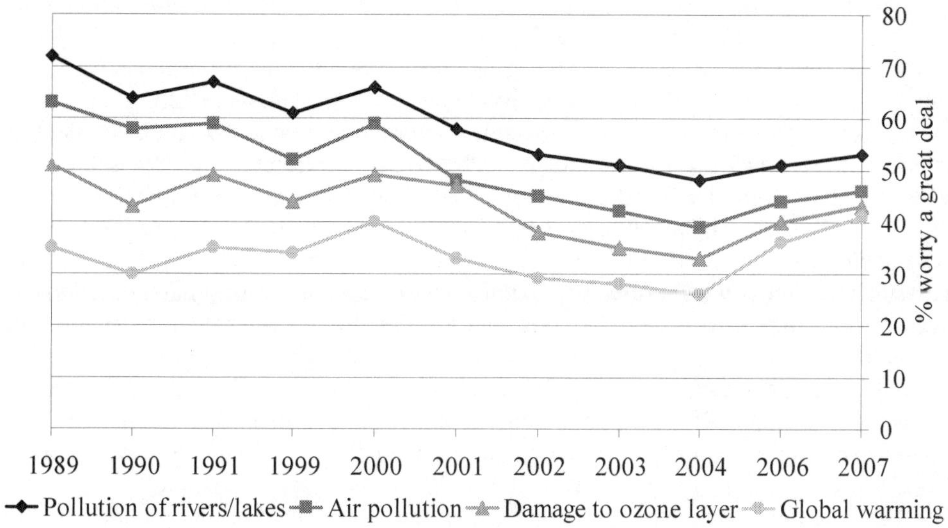

Source: Based on Gallup Poll data, see http://www.galluppoll.com/content/default.aspx?ci=1615&pg=2.

Figure 1 Trends in environmental concern

As the previous paragraph suggests, there are several, sometimes conflicting, objectives to eco-labelling programmes, and these are related to the views of the various stakeholders involved. Producers may use eco-labelling primarily to capture price premiums or increases in market share, while retailers may seek to capture customers by developing consumer goodwill. NGOs may look to improve the local or global environment, or may choose to protect the natural resources of lesser developed countries. Regulators may use eco-labelling to protect the safety of their citizens, while trade officials may promote eco-labelling if they think that it can protect domestic industries. Given these disparate motivations, all of these groups will have different views on what constitutes an effective eco-labelling programme. However, the main idea is that some consumers exposed to eco-labelling will change their information search[3] or product purchase behaviours. Given that firms which can take advantage of a labelling programme could be rewarded with increased profits (Moorman and Slotegraaf, 1999), changes in consumer behaviours may lead to changes in producer behaviours.[4] For example, firms may develop new marketing strategies or target different consumers, develop new products, or alter the attributes of current products (see Moorman, 2001 for a list of studies looking at firms' responses to changes in information policies).

While many groups expect (or hope) that these market changes will lead to reduced environmental impacts, from an economist's point of view this is not the only reason for eco-labelling. For economists, a primary function (benefit) of eco-labelling is to improve the flow

[3] A point made by Larson (Chapter 13), is that an eco-label may create an incentive for individuals to become more informed about the issues surrounding eco-labelled goods.

[4] Importantly, not all consumers in a market need to be affected by the information programme to alter markets; only a subset of consumers need to respond to the information to make an impact on producer behaviors (Moorman, 1998).

of information to consumers (Mitra and Lynch, 1995), for when product information is well understood and credible, then consumers' choices match their preferences and money spent is in line with any underlying willingness to pay for the bundle of attributes received. Thus, labelling provides little benefit for product attributes that can (without labelling) be assessed prior to purchase (for example, colour or size) or after purchase and use (for example, taste, convenience). By contrast, it is usually reserved for *credence* attributes – attributes that cannot be easily verified even after purchase and use but whose value affects utility (for example, nutritional quality, environmental degradation caused by the product's production). Given the above, labels that provide information about a product's credence attributes can lead to an increase in a consumer's welfare, even if it does not change their consumption decisions.[5]

But labelling may also increase a consumer's utility by allowing them to change their consumption decisions. In general, most consumers make choices based on the private benefits they accrue when consuming goods; however, even here, the market dynamics of labelling may lead to additional public benefits. For example, many individuals buy foods labelled as organic because they perceive them to be tastier, more nutritious and safer than non-organic foods (Fein, Levy and Teisl, 2002). Likewise, individuals may purchase tuna labelled as dolphin-safe because they see these products as providing the private benefit of enhanced 'moral quality' (Bockstael and Strand, 2000). Alternatively, eco-labels, like the dolphin-safe tuna label, may allow individuals to altruistically contribute to the provision of enhanced environmental quality (a public good). Whatever the motivation, by providing information about credence attributes, eco-labelling allows for the correction of market failures in product production – namely, the fact that consumers with preferences for (or against) specific production outputs (for example, pesticide use, tuna-related dolphin mortality) lack the information to reward (or punish) a product manufacturer through their market decisions (see Haener and Luckert, 1998 for a nice presentation of this issue).

Although labelling can provide both private and public benefits, it is not a costless activity. Typically, the costs of the actual physical labelling (for example, label design and printing) are a tiny fraction of the costs of compliance and certification (supply chain costs). Other costs may include the costs to public authorities of monitoring and enforcing compliance with labelling regimes, the costs of trade impacts from labelling programmes, costs arising from possible changes in market structures, and costs from the dilution of information already included on labels. As an example of the latter potential cost, labelling may impose cognitive costs on some consumers in that simply increasing the amount of information content on a label may actually decrease their ability to process other, more important, label information (Roe, Levy and Derby, 1999). In addition, requiring specific information to be placed on a label imposes a cost in that the limited space on the product label could have been devoted to other, potentially more useful and important information. Consequently, because information content is competing for valuable space on the label, labelling requirements should be justified in terms of the importance of the required information. As Bougherara and Grolleau (Chapter 16) imply, a policy of 'more information is better' does not necessarily characterize an optimal labelling policy.

[5] For example, information provided by labelling may make an individual happier with the choices they were already making.

Objectives

The market interest in eco-labelling has (fortunately for this book) led to an increased interest among academics (see Figure 2). By the end of 1997 there were approximately seven academic essays related to eco-labelling; currently there are over 100. As with the market-related activity in eco-labelling; the activity in the literature has not only been to increase the number of essays, but also to expand the coverage of topics related to eco-labelling. For example, a decade ago the emphasis of the literature was to develop a basic theoretical model of the consumer/market with eco-labelling and to empirically demonstrate whether or not eco-labelling was effective. Since then there have been several theoretical advancements, especially in terms of understanding producer behaviour related to eco-labelling. In addition, the empirical literature has moved from answering the basic question of effectiveness to examining the factors that influence an eco-label's effectiveness. Most recently there has been increased attention to the effects of eco-labelling on international relationships and how eco-labelling can affect the economic development of lesser developed countries.

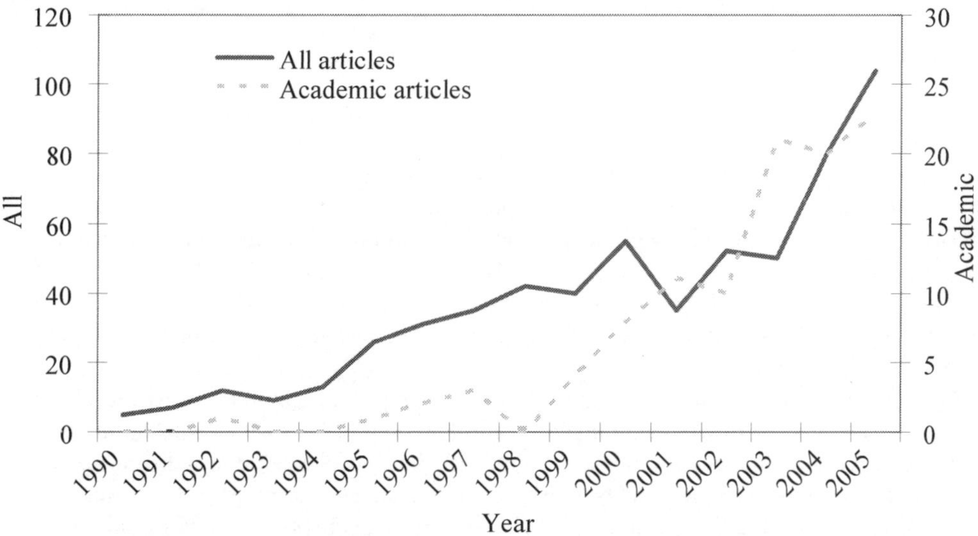

Figure 2 Growth in eco-labelling literature

About this Volume

Given the growth in the size and scope of the eco-labelling literature, the timing of this volume, which presents the current range of theoretical and empirical research on the design and effects of environmental labelling programmes, seems apropos. Eco-labelling is still a rather new and extremely broad topic, affecting not only the environment, but also consumers, producers, regulators and other stakeholders (for example, international development agencies). As a result, this volume is aimed at researchers and decision-makers from various disciplines. Although it concentrates on the economic literature, several of the essays approach the problem from a psychological, sociological or political science perspective. The target audience is primarily environmental economists and market researchers; however, members of environment- or consumer-oriented stakeholder groups (businesses, regulators and members of non-governmental organizations) will also find relevant content that is accessible. Essays were chosen to reflect the best treatment of a range of sub-topics, approaches and applications.

Nevertheless, taking a collection of existing works and attempting to organize them into clusters of similar focus is a difficult task as many essays will cover multiple topic areas. This book consists of four sections, each titled with what I consider to be the main strands of the eco-labelling literature. However, while the principal theme of most essays fits quite closely to the section heading, some, arguably, could have been placed into one of the other sections.

Part I addresses a range of theoretical developments in the eco-labelling literature, while Part II focuses on empirically measuring whether eco-labelling programmes are effective. Part III then takes a more detailed look at those factors that influence the success and design of eco-labelling programmes. Finally, Part IV outlines issues of recent interest – the effects of eco-labelling on international trade relations and development.

Theoretical Developments

Compared to command-and-control options that directly affect the attributes firms produce, clear labelling of eco-attributes may provide a less intrusive and often less expensive method of improving market outcomes. In fact, the labelling of attributes may allow some goods to exist that would otherwise be eliminated under a command-and-control policy. Thus, a cost–benefit analysis may support a labelling policy to promote environmental goals, whereas another type of policy may not pass the cost–benefit test. However, given that it is unlikely that labelling will replace other policy instruments, a potentially rewarding avenue for economic research is to determine the theoretical and empirical interactions between labelling programmes and other policy instruments.

Part I begins with an essay by Timothy Brennan, who provides answers to two fundamental questions. The first is: to what extent should eco-labelling polices act as a substitute to other forms of environmental regulation? He points out that, even when a significant number of consumers buy eco-labelled products, their production and consumption still imposes external costs. As a result, eco-labelling should be viewed not as a substitute, but as a complement to other regulatory methods (for example, taxes, permits).[6] The second question Brennan poses is whether using green preferences as a policy tool is conceptually problematic for economists because policies which allow preferences to change lead to uncertainty as to what the standard should be. Further, changing preferences leads to a valuation problem – for example, what perspective (before or after the labelling programme) should be used for valuation.[7] Because of the ambiguities, he argues that, from a policy development standpoint, eco-labelling only makes sense when the standards used for the eco-labelling programme are not defined by

6 This is not surprising since eco-labelling will only capture the values of individuals who are in the market for the goods being eco-labelled; the utility of individuals with similar environmental preferences but different product preferences will not be included.

7 This problem, and a solution, related to information provision and welfare evaluation was first presented by Foster and Just (1989) and framed in a labelling framework by Teisl et al. (2001).

consumer values but rely on a set of criteria created by others (for example, ecologists). This is a provocative stance that is unlikely to be accepted by many economists or market researchers as he seems to ignore the consumer benefits of eco-labelling. However, because Brennan's argument focuses on the environmental benefit of eco-labelling, his call for non-consumer-based criteria is likely to be accepted by some regulators and environmental advocates.

When eco-labelling was beginning to develop as an accepted strategy, many people assumed that eco-labelled products would necessarily be more expensive than their non-labelled counterparts. This would presumably mean that, under a voluntary system, those firms which were able to obtain an eco-label in a cost-effective manner would increase their profits, while firms which chose not to participate would not necessarily be adversely affected. However, in Chapter 2 Roger Sedjo and Stephen Swallow derive conditions where the market may not lead to any price premium for eco-labelled goods, even if there are a significant number of eco-conscious consumers willing to pay for these goods. They also find conditions where eco-labelling-induced price changes reward both green and conventional (brown) firms and conditions where firms which do not participate in an eco-labelling programme may see their profits decline even when the eco-labelling programme is voluntary.

In a similar vein, eco-labelling has been assumed to necessarily lead to an improved environment. However, Douadia Bougherara, Gilles Grolleau and Luc Thiébaut (Chapter 3) show how eco-labelling can perversely lead to environmental degradation, even when there is an increase in purchases of eco-labelled goods. This is because eco-labels generally display information about the environmental impact per unit, so consumers are not likely to be able to compute the total environmental impact of their consumption. As a result, aggregate consumption may increase, leading to an overall deterioration in the environment. Bougherara *et al.* also show that this perverse effect is more likely to occur as the cost of obtaining an eco-label diminishes. The implication is that policy-makers need to consider these effects when designing eco-labelling programmes.

In most instances, to be awarded an eco-label by a third party requires some sort of certification process, which adds to production costs. When the fixed costs of certification are high, they may limit the ability of smaller producers to participate in the eco-labelling programme. One strategy being used to reduce the fixed cost of obtaining third-party certification of environmentally managed forest lands is to allow groups of smaller individual landowners to obtain a group certification. However, as Lisette Ibanez and Jacques Laye (Chapter 4) point out, the process of group certification may encourage producers to coordinate in socially undesirable ways. Comparing two situations where labelling occurs with and without coordination, they show that eco-labelling is always profitable for the retailer of the eco-labelled goods unless the eco-labelling induces coordination among producers. Under coordination, the retailers and consumers of eco-labelled goods can suffer a loss, while eco-certified producers and the retailers of non-eco-labelled goods benefit from the eco-labelling programme.

The essay by Young-Hwan Ahn and Byong-Hun Ahn (Chapter 5) is framed as an international trade conflict between developed and developing countries when the developed country imposes eco-labelling mandates that include production-related externalities. However, the essay helps one understand more fully the implications of constructing production-related versus consumption-related eco-labelling. Ahn and Ahn compare two eco-labelling schemes – one including and one excluding production-related externalities in the presence of an existing tax on these externalities. They find that having the eco-labelling programme exclude production-related externalities can provide more environmental benefits to the consuming country, especially when the cost of abatement is relatively low for the consumption-related externality. This result is a challenge to individuals who advocate that life-cycle analysis is the only proper framework for eco-labelling programmes.

Although many of the essays on eco-labelling have focused on how eco-labels affect market outcomes, few have explicitly modelled the impact of eco-labelling on environmental outcomes (particularly resource extraction). The problem of how different eco-labelling strategies can affect the management of open-access, limited-access and optimally managed fisheries resources is the subject of an essay by Eyjólfur Gudmundsson and Cathy Wessells (Chapter 6). Using bioeconomic modelling, the authors examine how eco-labelling affects users of fisheries resources. They find that extraction effort depends critically on the shape of the premium function; under reasonable conditions eco-labelling may in fact lead to increases in user effort, leading in turn to higher extraction rates. Alternatively, when a fisheries resource is being optimally managed but is currently below the equilibrium size, then eco-labelling, coupled with an appropriate premium function, can provide a more rapid recovery of the fishery. However, Gudmundsson and Wessells highlight the extreme difficulty in designing an appropriate premium function.

Research indicates that nutrient- and health-related labelling of food products have led to significant changes in the supply of new products by providing firms with an incentive to improve nutritional quality (Ippolito and Mathios, 1996; Frazao and Allshouse, 1996). Similarly, eco-labelling schemes are assumed to stimulate investment in environmental technologies, thereby reducing the supply of brown products. However, Cesare Dosi and Michele Moretto (Chapter 7) point out that the two events are not necessarily linked. They demonstrate that if the awarding of an eco-label causes consumers to view the entire firm in a more positive light (a 'halo' effect), then eco-labelling could in fact lead to increased investments in brown technologies alongside increases in green investments. They show that rationing eco-labels among producers could mitigate or eliminate some of these perverse effects.

In Chapter 8 Gregory Amacher, Erkki Koskela and Markku Ollikainen also examine how eco-labelling affects firms' environmental investment decisions. Using a vertically-differentiated duopoly model they find that firms' relative cost structures affect their investment decisions. When fixed costs are identical across the two firms, both will invest in eco-technologies. However, if one firm is a more efficient eco-investor, then the other may choose not to invest. The traditional duopoly model usually leads to an overinvestment in quality to reduce price competition. However, Amacher *et al.* find that eco-investments will be lower with eco-labelling than in the traditional model, suggesting that eco-labelling might be a way of reducing excessive investment.

Finally, Stefanie Engel, formerly known as Stefanie Kirschoff (Chapter 9), presents conditions where firms may voluntarily overcomply with eco-labelling programmes, even if this implies a loss of short-run profits. Overcompliance is more likely when consumers learn about all monitoring results as opposed to only finding out about cheating firms. Although companies that cheat are generally not likely to lobby for increased monitoring, Engel describes conditions where firms that cheat may have an incentive to lobby for stricter environmental monitoring because the increased monitoring improves the general reputation and profits of all firms.

Empirical Indications of Effectiveness

The widespread use of eco-labels suggests that they are perceived by some as an effective method of altering consumer behaviour. Although some stakeholders have adopted environmental certification and labelling as a business strategy, others remain sceptical about the environmental and economic benefits of these approaches. Early research measured the effectiveness of eco-labelling either by identifying changes in consumer awareness after exposure to label information (see, for example, Hartwell and Bergkamp, 1992; Hashizume, 1992) or by asking consumers whether labelling programmes would affect their purchasing behaviour (see, for example, Chase and Smith, 1992). However, a change in awareness does not necessarily translate into changes in behaviour, and consumers do not necessarily follow their own purchasing assertions (Bailey and Eastlick, 1993). Part II of this volume documents several different attempts at identifying whether eco-labelling programmes impact on market behaviour. In an early essay, Wesley Nimon and John Beghin (Chapter 10), using data from clothing catalogues, estimate a hedonic price function and obtain mixed results; they identify a price premium for organic fibres but none for environment-friendly dyes.

Using aggregate market data, Mario Teisl, Brian Roe and Robert Hicks (Chapter 11) find that negative media activity surrounding dolphin mortality depressed sales of canned tuna and that a subsequent eco-labelling effort by canned tuna producers increased canned tuna market share. The impact of the programme occurred over a period of several months, demonstrating an information diffusion lag. There may be two reasons for this effect. The first is a pure information diffusion process: only a few interested or knowledgeable consumers immediately notice the label change, while others only notice and react to the label after several in-store exposures or after discussions with others who have already noticed and reacted to the label. The second may be a verification effect: some consumers may have noticed the new labels but did not believe the label information, and, as information verifying the label claim is presented, these consumers may have increased their tuna consumption.

In a similar Danish study, Thomas Bjørner, Lars Hansen and Clifford Russell (Chapter 12) use a consumer market panel dataset to quantify the effect of an eco-label on consumers' choices of paper products and detergents. They find that the eco-label has significant effects on consumers' choices and willingness to pay for eco-labelled toilet paper and laundry detergent. Interestingly, the eco-label has much smaller effects on consumer purchases of eco-labelled paper towels, underscoring the importance of understanding potential interactions between typical product consumption and eco-labelling. Danish households tend to use dishcloths in the kitchen rather than paper towels; as a result, the eco-label may have reinforced consumers' desire to use dishcloths as the green alternative.

In Chapter 13 Bruce Larson, using aggregate trade data, investigates the impacts of an eco-label on the coffee market (changes in market shares and in price premiums). He finds significant but minor effects. He also expands earlier models that only allowed two types of consumers: those that only bought eco-labelled products and those that only bought non-eco-labelled products. Larson's expansion allows consumers to balance their desire for product attributes with their desire for variety in consumption. While preferences for eco-friendly goods generate demand for these goods, a preference for variety makes these demands more price-inelastic.

Studies of eco-labelling effectiveness have primarily been undertaken in Western, more developed countries. However, in Chapter 14 Jimin Zhao and Qing Xia examine the effectiveness of China's eco-labelling programme and find that it has raised consumer awareness of eco-labelled products and has encouraged some firms to adopt cleaner technologies. However, the programme's effectiveness is limited because only a small number of product categories are included.

Factors Impacting the Success and Design of Eco-labelling Programmes

Given that environmental certification and labelling of products is not costless, certification and labelling programmes may not achieve their objectives unless consumers are willing to pay for the underlying improvements in the production practices specified by the programme. Studies across several product types (Ozanne and Vlosky, 2003; van Ravenswaay and Blend, 1999; Ethier *et al.*, 2000; Wessells, Johnston and Donath, 1999) indicate that a sizable portion of consumers state that they are willing to pay a premium for environmentally labelled products. However, in addition to being willing to pay for eco-labelled products, consumers must also notice, understand and believe the information presented to them by the product manufacturer.

If it is to make a significant contribution to the design of labelling programmes, research needs to understand the characteristics of the interaction between the eco-attribute, the consumer and the producer and how they affect the impact of eco-label information. The essays by Paulo Nunes and Yohanese Riyanto (Chapter 15) and Douadia Bougherara and Gilles Grolleau (Chapter 16) provide a general framework for understanding the hurdles that eco-labelling programmes try to circumvent. They both highlight the fact that designing an effective eco-labelling programme is difficult because the programme needs to mitigate the public goods problem by providing information to solve the problem of asymmetric information between buyers and sellers. Bougherara and Grolleau expand the discussion by noting that eco-labelling faces the additional problem of needing to avoid information overload. Nunes and Riyanto add to the discussion by focusing on production costs and the interaction between eco-labelling and other policy instruments. Both essays review and critique some current eco-labelling programmes.

Economic theory suggests that demand for a product or service is a function of a number of factors, one of these being the tastes and preferences of consumers. Traditionally, economists have been rather ill-equipped at incorporating tastes and preferences in their models. Yet, social psychologists have a richer literature focusing on what constitutes tastes and preferences, the steps in the process of behavioural change in response to information, and some of the factors that influence the process. In Chapter 17 John Thøgersen develops a psychological model for explaining differences in consumer attention to eco-labels and highlights that the effectiveness of an eco-label is influenced by the capacity of the consumer to absorb the information and act on it. He indicates that a person's attention to eco-labels is influenced by the level and specificity of their environmental involvement, their perception that their eco-behaviours are effective in attaining environmental goals and by their level of trust in the eco-label.

Because it is impossible for most consumers to verify the promise of improved production practices, the success of eco-labelling uniquely hinges on companies being able to credibly communicate to the consumer that production practices have been altered. Mario Teisl (Chapter 18) indicates that relatively detailed labels are more beneficial for consumers (and environmentally sensitive forest product manufacturers) than simpler eco-seals, because eco-seals have lower perceived credibility and do not give consumers enough information to differentiate products. His results also suggest that when firms can voluntarily label their products with eco-seals from a variety of certification organizations then there are several undesirable effects. First, the lack of standardization may increase consumer confusion and erode consumer confidence in eco-labelling. Second, the combination of multiple seals, in a voluntary setting, forces consumers to examine other product attributes (for example, price) to estimate the eco-qualities of the product, and this may allow firms to make vague or misleading marketing claims to misrepresent the eco-friendliness of their products.

In Chapter 19 Charlotte Leire and Åke Thidell review the research on consumer perceptions, understanding and use of eco-label information and find that, although most consumers can recognize, understand and trust eco-labels, many do not use them. They point out that most studies have examined consumer preferences, knowledge and behaviours outside of a market setting and that reactions to eco-labelling are likely to be influenced by factors occurring during the purchasing moment. They conclude that eco-labelling studies should concentrate on the dynamic context of the purchasing decision.

Bernhard Truffer, Jochen Markard and Rolf Wüstenhagen (Chapter 20) examine and compare different voluntary, third-party eco-labelling schemes used in the United States and Europe. They identify six features of labelling programmes and identify two common, but conflicting, strategies used in eco-labelling. In the short run, a simple labelling strategy is preferred as it can generate first-mover advantages. However, in the long run a more detailed labelling scheme is preferred as increased public awareness and potential criticisms may erode the success of these simpler labelling schemes.

Researchers have long questioned whether the presence of many different eco-labelling programmes causes consumer confusion. However, Helen Nilsson, Burcu Tunçer and Åke Thidell. (Chapter 21) also question whether the number of competing eco-labels decreases the perceived credibility of the labels. They examine 58 eco-labelling programmes and find that consumers place little trust in many of them. They then identify factors which would improve the perceived credibility of the programmes. One important factor was the level of openness and stakeholder involvement in the criteria setting process. The strictness of the criteria was not a major factor in improving the labels' perceived credibility. Use of third-parties to provide independent verifications are important even though the results of, or methods used in, the audits are rarely made public. Another key element was providing off-label information and using aggressive marketing techniques to promote and explain the presence of the programme.

Currently, most eco-labels are voluntary and the message on the label is positively, as opposed to negatively, framed – that is, choose this product because it's better for the environment against avoid this product because it's worse for the environment. However, mandatory eco-labels can be framed either or both ways (for example, as a continuum). In a computer-based experiment Gunne Grankvist, Ulf Dahlstrand and Anders Biel. (Chapter 22) investigate how negatively and positively-framed eco-labels interact with consumer characteristics to affect preferences for labelled products. Not surprisingly, individuals who had weak to no concern for the environment were unaffected by either framing. At the other extreme, individuals with robust levels of environmental concern were equally affected by the two types of label. However, consumers with intermediate levels of environmental concern reacted quite differently to the two framings – these individuals were more affected by a negative label. The implication is that negative framings have a greater potential to affect behaviour as they increase the total number of individuals likely to alter their behaviour. However, this type of framing is only likely to occur under some sort of mandatory labelling programme, which is likely to require government intervention.

International seafood eco-labelling programmes have been proposed as a way of promoting sustainable fisheries management. In Chapter 23 Robert Johnston *et al.* investigate international differences in consumer preferences for eco-labelled seafood and find a wide range of factors explaining differences in consumers' likelihood of purchasing eco-labelled seafood products across countries. Given the global nature of the seafood market, however, the favourable reception of eco-labelled seafood in a single nation, or group of nations, may not prevent overfishing, particularly for migratory fish species. Overall, the international heterogeneity in consumer reactions to an eco-labelling programme indicates that these programmes cannot use a 'one-size-fits all' approach, but must be tailored to the particular preferences and constraints within a country.

Many eco-labelling programmes are administered by non-governmental third parties, and in several instances there are multiple programmes available for a product. These rival eco-labelling programmes then end up competing to be seen as the single legitimate program. A question that Benjamin Cashore, Graeme Auld and Deanna Newsom. attempt to answer in Chapter 24 is what factors explain which programme ends up winning this competition. They do this by analysing the supply-side dynamics in the development and use of rival forest eco-labelling programmes in North America and Europe. The authors focus their analysis particularly on the roles played by forest companies and landowners in granting 'legitimacy' to competing programmes and they identify seven factors that may explain the differences in support for alternative eco-labelling programmes.

In Chapter 25 Abhijit Banerjee and Barry Solomon study consumer and manufacturer responses to five energy-efficiency labelling programmes in the United States (two being privately-run, the remaining being government programmes). They find that government programmes are, in general, more successful than privately-run programmes because government programmes have more credibility, financial resources and long-term viability. They also examine several programme-specific factors that have increased the apparent success of the programmes and highlight the positive role of government in terms of imposing mandates or providing incentives (for example, tax deductions or subsidies).

International Dimensions of Environmental Labelling

Although one of the motivating factors leading to the increased use of eco-labelling is its international (WTO) acceptability, it is only recently that the international dimensions of eco-labelling have begun to be explored. Part IV begins with an essay by Daniel Melser and David Robertson (Chapter 26), who evaluate the environmental benefits from the international growth in eco-labelling programmes and show that the level of environmental damage is independent of the number of countries having such programmes. Melser and Roberrtson question the desirability of promoting eco-labelling over the use of multilateral environmental agreements because the use of eco-labelling alone may actually lead to a degraded environment.

The success of eco-labelling is partly measured by how well the programme links eco-conscious buyers and sellers. Arnab Basu, Nancy Chau and Ulrike Grote. (Chapter 27) show that a programme's effectiveness depends on how the level of monitoring of eco-production in a developing country responds to economic growth. They find that eco-labelling with perfect monitoring optimally links consumers and producers and that brown products are seen as inferior. As a result, income gains achieved via international trade lead to environmental improvements as consumer demand shifts towards eco-labelled products. However, in circumstances where enforcement is costly, they show that unless monitoring keeps pace with income, brown products become normal goods as a result of increased amounts of false labelling. In this scenario, monitoring benefits a developing country only when national income exceeds a critical threshold, explaining why eco-labelling programmes are initially less likely to be found in developing countries. Thus, the initial impact of income growth increases the supply of brown products; however, as income continues to increase, the supply of eco-unfriendly products falls as the benefits of increased monitoring outweighs the increased cost. The authors suggest that the promotion of eco-labelling as a device to improve environmental quality should be exercised with caution.

Although eco-labelling has been used in developed countries for a number of years, recently, the practice is spreading to developing countries. Using a general equilibrium model, Arnab Basu, Nancy Chau and Ulrike Grote. (Chapter 28) investigate what factors influence a country's adoption of eco-labelling as a strategy. Their model indicates that the decision to participate in an eco-labelling programme depends on: the size of a 'green premium'; the programme's fixed costs; the industry's comparative cost advantage; the level of output; and whether other countries decide to participate in the programme. A key insight is that countries with a higher trade orientation are more likely to participate in eco-labelling; this suggests that eco-labelling can mitigate any environmental 'race to the bottom'.

The conventional wisdom is that eco-labelling programmes will benefit developed (importing) countries and harm less-developed (exporting) countries on the assumption that more developed countries will have a cost or managerial advantage in meeting the eco-labelling criteria. Using a model of international duopoly where there is competition in both prices and levels of eco-friendliness, Huilan Tian (Chapter 29) identifies instances where a programme that increases the eco-friendliness level of imported goods may actually lead to not only losses for the firm in the importing country, but also increased levels of imports. Further, conditions exist where consumer surplus may also decline. Thus an eco-labelling programme, initially touted as promoting the interests of domestic consumers and producers, may in fact lead to improving the situation for the exporting country's producers.

Conclusions

Ignorance is bliss but knowledge is power.

In the early 1990s eco-labelling programmes were developed under a simplifying 'field of dreams' presumption.[8] The presumption was that if a worthy product was eco-labelled, then eco-conscious consumers would buy it. When consumer response to some of these early programmes proved to be lacklustre, some argued that this lack of response indicated that consumers do not care about, or at least are not willing to pay more for, such products. As a result, it was argued that the development of eco-labelling programmes was a waste of time and effort. We now know that the 'field of dreams' presumption, although politically expedient, was somewhat naïve. However, we also know now that the arguments of eco-labelling's initial critics are not correct. Research has confirmed that well-designed eco-labelling programmes can alter consumer and producer behaviours. However, after reading this book one may become dismayed by the large number of caveats and exceptions that have been uncovered. Despite the fact that eco-labelling is not as simple as we once thought, we now have a clearer idea when eco-labelling is appropriate and the conditions that are likely to make an eco-labelling programme more effective.

Although we have learned a lot in a relatively short time, there is still much to do. For example, the essays in part I point out several complexities that could limit the effectiveness of an eco-labelling programme, although we still need to empirically develop a better understanding of how prevalent these complexities occur, and how large are the effects. Another area of research that could be expanded would be to empirically examine the impacts of eco-labelling on environmental, as opposed to market, outcomes and, also, whether there are spatial relationships evident in these environmental outcomes. For example, an early justification for the eco-certification and labelling of forest products was to protect tropical rainforests. However, due to the costs of certification, almost all of the certified forest acreage in the world is located in developed countries.

The political manoeuvrings of non-governmental organizations and retailers in the development of eco-labelling programmes seems to be another fruitful area of research. That many countries choose not to have a government-run eco-labelling programme seems to have important ramifications; for example, such countries give up possible benefits associated with mandatory labelling approaches (such as increased programme credibility, an ability to use negative labelling), as well as benefits associated with clearer coordination between eco-labelling and other environmental regulations.

As mentioned earlier, from an economic point of view, the ultimate goal of an eco-labelling programme is to increase social welfare. This assumes that the tools and data exist to perform a cost–benefit analysis. A few studies have estimated *some* of the costs or benefits of an eco-labelling programme, but little has been done in terms of a true welfare analysis (see Carambas, Herrmann and Grote, 2003, for movement in this direction). Part of the reason for this is that eco-labelling programs are primarily run by private entities; a benefit of publicly provided labelling is that any costs and benefits of the programme need to be made transparent.

Another avenue of welfare analysis would be to examine the distribution of costs and benefits across a population. For simplicity, and to gain the benefits of standardization, most labelling programmes choose one way of displaying information. Unfortunately, the optimum amount of eco-information presented on a label is likely to be different for different individuals, and for different products. Determining the optimal form for a labelling

[8] 'If you build it, they will come' – from the 1989 movie *Field of Dreams*.

programme is an important welfare question because, given the unequal distribution over the population of cognitive abilities, preferences for eco-attributes and values of time, labelling regulations will have equity and distributional implications. However, no one has examined the welfare implications of alternative eco-labelling approaches. Policy-makers need to know the characteristics of alternative labelling programmes because using a 'one-size-fits all' approach may not be appropriate.

The problems become exacerbated when adding in international trade. The prospect of having alternative eco-labelling programmes in different countries will lead to conflict as policies will serve to advantage or disadvantage different producers in different countries. At a minimum, having different eco-labelling programmes increases multinational producers' costs. It is unclear whether WTO rules will be able to contain this conflict. Economists, using cost–benefit analysis, may be able to provide a framework in which to answer such policy questions.

Eco-certification and labelling can impose significant costs on producers, yet there is a dearth of academic studies focused on identifying and quantifying the supply-side costs of labelling (for example, costs related to standard setting, and private compliance and certification efforts). Similar to the benefits of labelling, the costs of labelling will be highly dependent on the structure of the particular programme. For example, costs will depend on the environmental standards set by the programme and the timeframe needed to comply with the standards. Costs will also be affected by the degree to which the labelling programme is 'mandatory'[9], the range of products included, and the supply-chain quality-assurance system used to comply with the labelling programme. For example, some products (for example, wood products) may require segregation or traceability (for example, chain of custody) approaches to clearly link an end-product to an environmentally managed resource. However, other products (for example, 'green' electricity) may be able to use tagging and trading systems that reduce costs and provide more flexibility in supplying greener electricity.[10] Research that helps identify, quantify and compare the costs of various labelling approaches is needed, yet this type of research is difficult because many labelling programmes are still relatively young (and developing), voluntary and privately administered.

Research on costs is also relevant since the high costs of certification and labelling can erect a barrier to entry for smaller, less efficient producers (Caswell, 1998; Crespi and Marette, 2001). Not only does this potentially increase the market power of larger firms but it could also reduce the environmental effectiveness of the programme. For example, Ávalos-Sartorio, Blackman and Albers (2006) argue that eco-certification can be a more successful land management instrument if it is less costly for smaller landowners, as these landowners are significant contributors to local deforestation. As mentioned earlier, some eco-labelling

---

9 Mandatory is in 'quotes' because even voluntary programmes may perform as if they were mandatory due to market (for example, retailer, or government) pressures.

10 Tagging and trading systems decouple a consumer's support for greener power from the actual electricity they use. For example, in the New England Power Pool, for each unit of electricity generated, a tag that describes the emission and fuel source characteristics is created. These green tags are then bought and sold in a market separate from the generation market. Producers who buy these tags can then use the information to sell their electricity to consumers. This system allows eco-conscious consumers to support greener electricity even if this type of electricity is locally unavailable, while also reducing supply-chain costs.

programmes (for example, the Forest Stewardship Council) have tried to mitigate this problem by allowing smaller producers to obtain a group certification. Another approach (for example, the Sustainable Tourism Eco-certification Program) is to have a progressive set of certification fees based on a firm's gross revenues, allowing smaller operators to participate. Another possibility is to allow taxpayers to subsidize some of the private or public costs associated with eco-labelling as is done for other non-labelling programmes. For example, Crespi and Marette (2003) indicate that only a minority of the US Department of Agriculture's Food Safety and Inspection Service costs are funded through the collection of user fees (the rest being subsidized by taxpayers).

Part II provides several empirical examples documenting the effectiveness of labelling; the studies highlight that eco-labels can be effective but they show that the effects are rather modest and generally lag behind the initiation of the programme. A possible way of enhancing and speeding up the effects of an eco-label would be to administer an informational campaign to educate consumers as to the presence, goals and means of the programme. These so-called 'social marketing' campaigns may be useful because many consumers are unfamiliar with both eco-labelled products and some of the certifying organizations. Raising consumer awareness of the organization might increase the credibility of a simpler label or lead to a desire for more complex labelling. To date, little academic research has been undertaken into the effectiveness of these social marketing programmes in relation to supporting an eco-labelling programme.

Finally there will be instances where eco-labelling programmes will need to change over time to reflect changes in technology or in consumer preferences. However, appropriate changes in the labelling scheme may be greatly lagged due to institutional bureaucracies and coordination difficulties, causing 'excess inertia' or 'lock-in' in the Farrel and Saloner (1985) sense. Examining the costs and benefits of eco-labelling programmes in a dynamic framework may provide a better understanding of the optimal timepath of such schemes.

References

Beatriz Ávalos-Sartorio, Blackman, Allen and Albers, Heidi J. (2006), *Sustainable Coffee Certification as a Forest Conservation Policy in Mexico: Prospects, Constraints, and Policy*, Issue Brief – RFF IB 06-01, Washington, DC: Resources for the Future.

Bailey, W. and Eastlick, M.A. (1993), 'Study Shows Many Claiming to be Green Don't Buy Green', *Northwest Arkansas Times*, 12 November.

Bockstael, N.E. and Strand, I.E. (2000), *Economic Sanctions and Environmental Preferences: Protecting Dolphins and Tuna Markets*, Working Paper No. 2000-05, College Park: Department of Agricultural and Resource Economics, University of Maryland.

Boorstin, D.J. (1983), 'On Computerization of Libraries', *The New York Times*, 8 July.

Carambas, M.C., Herrmann, R. and Grote, U. (2003), 'Cost-Benefit Analysis of Eco-Labeling in the Agriculture Sector of Developing Countries in Asia', Paper presented at *Deutscher Tropentag 2003: International Research on Food Security, Natural Resource Management and Rural Development Georg-August-Universität Göttingen, Germany, 8–10 October.*

Caswell, J.A. (1998), 'How Labeling of Safety and Process Attributes Affects Markets for Food', *Agricultural and Resource Economics Review*, **27**, pp. 15–18.

Chase, D. and Smith, T.K. (1992), 'Consumers Keen on Green but Marketers Don't Deliver', *Advertising Age*, 29 June.

Crespi J.M. and Marette, S. (2001), 'How Should Food Safety Certification Be Financed?', *American Journal of Agricultural Economics*, **83**(4), pp. 852–61.

Crespi, J.M. and Marette. S. (2003), 'Some Economic Implications of Public Labeling', *Journal of Food Distribution Research*, **34**(3), pp. 83–94.

Ethier, G.R., Poe, G.L., Schulze, W.D. and Clark, J. (2000), 'A Comparison of Hypothetical Phone and Mail Contingent Valuation Responses for Green-Pricing Electricity Programs', *Land Economics*, **76**, pp. 54–67.

Farrel, J. and Saloner, G. (1985), 'Standardization, Compatibility, and Innovation', *Rand Journal of Economics*, **16**(Spring), pp. 70–83.

Fein, S., Levy, A.S. and Teisl, M.F. (2002), *American Attitudes Toward Three Alternative Food Technologies. Genetic Modification, Irradiation and Organic*, Papers presented at the annual meeting of the American Public Health Association, Philadelphia, 10–14 November.

Foster, W. and Just, R.E. (1989), 'Measuring Welfare Effects of Product Contamination with Consumer Uncertainty', *Journal of Environmental Economics and Management*, **17**, pp. 266–83.

Frazao, E. and Allshouse, J.E. (1996)', *Size and Growth of the Nutritionally Improved Foods Market*, Agriculture Information Bulletin No. 723, Washington, DC: Economic Research Service, US Department of Agriculture.

Haener, M.K. and Luckert, M.K. (1998)', 'Forest Certification: Economic Issues and Welfare Implications', *Canadian Public Policy*, **24**(Special Supplement 2), pp. S83–S94.

Hartwell, R.V. and Bergkamp. L. (1992), 'Eco-labeling in Europe: New Market-related Environmental Risks?', *BNA International Environmental Daily*, 20 October.

Hashizume, S. (1992)', *Environmental Labeling in Japan: The Eco-mark*, Tokyo: Japan Environment Association.

Hayward, Steven F. (2006), *Index of Leading Environmental Indicators 2006 The Nature and Sources of Ecological Progress in the U.S. and the World* (11th edn), San Francisco: Pacific Research Institute.

Ippolito, P.M. and Mathios, A.D. (1996), '*Information and Advertising Policy: A Study of Fat and Cholesterol Consumption in the United States, 1977–1990*, Bureau of Economics Staff Report, Washington, DC: Federal Trade Commission.

Mitra, A. and Lynch, J.G. Jr (1995)', 'Toward a Reconciliation of Market Power and Information Theories of Advertising Effects on Price Elasticity', *Journal of Consumer Research*, **21**(March), pp. 644–59.

Moorman, C. (1998)', 'The Market-level Impacts of Information: Competitive Responses and Consumer Dynamics', *Journal of Marketing Research*, 34(February), pp. 82–98.

Moorman, C. (2001), 'Firm Responses to Consumer Information Policy', in P.N. Bloom and G.T. Gundlach (eds), *Handbook of Marketing and Society*, London: Sage Publications.

Moorman, C. and Slotegraaf, R.J. (1999), 'The Contingency Value of Complementary Capabilities in Product Development', *Journal of Marketing Research*, **36**(May), pp. 239–57.

Ozanne, L.K. and Vlosky, R.P. (2003)', 'Certification from the U.S. Consumer Perspective: A Comparison of 1995 and 2000', *Forest Products Journal*, **53**(3), pp. 13–21.

Pargal, S., Hettige, H., Singh, M. and Wheeler, D. (1997)', 'Formal and Informal Regulation of Industrial Pollution: Comparative Evidence from Indonesia and the United States', *The World Bank Economic Review*, **11**(3), pp. 433–50.

Reagan, R. (1989), *Guardian*, London, 14 June. Cited in *The Columbia World of Quotations*, New York: Columbia University Press, 1996.

Roe, B., Levy, A.S. and Derby, B.M. (1999), 'The Impact of Health Claims on Consumer Search and Product Evaluation Outcomes: Results from FDA Experimental Data', *Journal of Public Policy & Management*, **18**(Spring), pp. 89–105.

Teisl, M.F., Bockstael, N.E. and Levy, A.S. (2001)', 'Measuring the Welfare Effects of Nutrition Information', *American Journal of Agricultural Economics*, 83(1), pp. 133–49.

US Environmental Protection Agency (1998), *Environmental Labelling Issues, Policies and Practices Worldwide*, Washington, DC: EPA.

van Ravenswaay, E.O. and Blend, J.R. (1999), *Consumer Demand for Ecolabeled Apples: Results from Econometric Estimation*, Staff Paper 99-17, East Lancing, MI: Department of Agricultural Economics, Michigan State University.

Wessells, C.R., Johnston, R.J. and Donath, H. (1999), 'Assessing Consumer Preferences for Ecolabeled Seafood: The Influence of Species, Certifier and Household Attributes', *American Journal of Agricultural Economics*, **81**, pp. 1084–89.

Part I
Theoretical Developments

[1]

"Green" preferences as regulatory policy instrument

Timothy J. Brennan*

Public Policy and Economics, University of Maryland Baltimore County, United States
Resources for the Future, United States

Received 24 July 2003; accepted 17 January 2005
Available online 9 March 2005

Abstract

We examine here the suggestion that if consumers in sufficient numbers are willing to pay the premium to have power generated using low-emission technologies, tax or permit policies become less necessary or stringent. While there are implementation difficulties with this proposal, our purpose is more fundamental: Can economics make sense of using preferences as a regulatory instrument? If "green" preferences are exogenously given, to what extent can or should they be regarded as a substitute for other policies? Even with "green" preferences, production and consumption of polluting goods continue to impose social costs not borne in the market. Moreover, if green preferences are regarded as a policy instrument, the "no policy" baseline would require a problematic specification of counterfactual "non-green" preferences. Viewing green preferences as a regulatory policy instrument is conceptually sensible if the benchmark for optimal emissions is based on value judgments apart from the preferences consumers happen to have. If so, optimal environmental protection would be defined by reference to ethical theory, or, even less favorably, by prescriptions from policy advocates who give their own preferences great weight while giving those of the public at large (and the costs they bear) very little consideration.

Keywords: Green preferences; Regulatory policy; Cost

1. Introduction and summary

Proposals put forth to reduce emissions include the familiar options of taxes, tradable permits, or a blend of the two. An example of a blended proposal would be to reduce risk by setting a ceiling on abatement cost through government sale of permits to meet demand at a fixed price (Pizer, 1998). Although the empirical benefits of these proposals remain a ripe subject for analysis–particularly in a "second-best" framework (Parry and Oates, 2000)–their theoretical properties are well known. Less well understood in economic terms is the suggestion that environmental problems can be mitigated through "buying green." Examples include choosing to purchase energy-efficient appliances, get electricity from renewable fuel generators, and avoid foods or other products whose production puts endangered species at risk.

* Department of Public Policy, UMBC, 1000 Hilltop Circle, Baltimore, MD 21250, United States. Fax: +1 410 455 3229.
E-mail address: brennan@umbc.edu.

Arguably, what we might call "green preferences" could substitute for taxes and permits as environmental policy instruments. If consumers in sufficient numbers are willing to pay the premium to have power generated using low-emission technologies, perhaps tax or permit policies become less necessary or stringent.

The U.S. government, environmental advocacy groups, and commercial marketers of green power have endorsed using green preferences as a tool to achieve policy goals. The Clinton Administration's statement of the benefits of its proposed Comprehensive Electricity Competition Act included the statement that "Competition also provides opportunities for consumers to vote with their wallets for green power and facilitates the marketing of energy efficiency services along with electricity" (Department of Energy, 1999, p. 1). Under the Bush Administration, the Environmental Protection Agency and the Department of Energy continue to promote private purchasing of "green power."[1]

An example from advocacy groups is Global Green USA, founded by Mikhail Gorbachev in 1993, which:

> fosters a global value shift toward a sustainable and secure world through education, advocacy, partnerships, and programs focused on the safe elimination of weapons of mass destruction, stemming climate change, reducing resource use, and preventing conflicts over fresh water.[2]

Firms such as Green Mountain Energy address this "value shift" in the marketplace by advertising to consumers the opportunity to reduce air pollution by purchasing electricity made from renewables and "less-polluting sources."[3]

Reliance on consumer demand for "environmentally friendly" products is subject to criticism on a number of grounds. Technologies that have some environmental benefits, such as reducing greenhouse gas emissions, are sometimes not regarded as "green," most notably nuclear and hydroelectric power. The political factor in the definition of "green" could presumably be addressed through providing better information to people in their roles as citizens and as consumers. It also is less than clear in many contexts whether purchasing green power actually reduces emissions. If a power company produces 20% of its electricity through green sources and 80% in less environmentally benign ways, it could serve up to 20% of its demand by promising to use pollution-free sources to consumers voicing that preference without changing its fuel mix or reducing emissions at all.

Although one may doubt whether many consumers would be willing in great numbers to pay a premium to bring about environmental benefits, some studies do indicate a modicum of market potential for these initiatives. Surveying market research, Farhar (1999) found that 70% of utility customers would be willing to pay US$5 more per month for renewable power sources, and 21% at least US$15. These percentages tend to increase as customers learn more about green power, and if the price difference represents a choice to forego a price reduction as opposed to having to pay a higher price. Swezey and Bird (1999) found some response to this potential demand, listing 13 green power providers in California (despite only 1.1% of residential customers switching from the incumbent) and three green power suppliers in Pennsylvania, with premiums averaging about 2 cents/kW h.

The purpose of this paper is not to examine the empirical issues, but to address more fundamental methodological questions: Can ecological economics make sense of using encouraging particular preferences as a regulatory instrument? To what extent can and should policy makers coherently incorporate encouraging "green preferences" into a portfolio of policy instruments to achieve environmental ends? Conceptual difficulties arise because the economists' measuring stick for policy evaluation, efficiency, presupposes exogenous preferences. If green prefer-

[1] U.S. Environmental Protection Agency, "Green Power Partnership," http://www.epa.gov/greenpower/, accessed July 24, 2003; U.S. Department of Energy, Energy Efficiency and Renewable Energy Green Power Network, "Buying Green Power," http://www.eere.energy.gov/greenpower/consumers.shtml, accessed July 24, 2003.

[2] Global Green USA, "About Global Green: Our Mission," http://www.globalgreen.org/about/about.html, accessed July 24, 2003.

[3] Green Mountain Energy, About Us: Our Energy, http://www.greenmountain.com/about/energy/index.jsp, accessed July 24, 2003.

ences are exogenously given, to what extent can or should they be regarded as a substitute for other policies?[4]

We will show that even with green preferences, production and consumption of polluting goods continue to impose social costs not borne in the market. Conversely, if green preferences should be regarded as a policy instrument, we need a "no policy" baseline. Such a baseline would have to be based on the level of emissions one would see if consumers had some counterfactual "non-green" ("brown?") preferences. Hence, evaluating green preferences as regulatory policy implies that we specify what these "brown" preferences might be. One question such a specification suggests is whether considering external costs using green preference demand curves involves a double counting, in taking into account both B's willingness to pay for cleaner air and A's willingness to pay to give B cleaner air.

Viewing green preferences as a regulatory policy instrument can be conceptually sensible if the benchmark for optimal emissions is fixed (e.g., when ambient levels of a pollutant above a certain level are necessary and sufficient causes of major adverse health effects). But in that case, preference change essentially does not matter if it leaves fixed the optimal level of environmental degradation. If preference change matters in such a way to change that level, then preference change can be coherently evaluated as a policy instrument if environmental outcomes are judged on the basis of values apart from the preferences consumers happen to have. To put it more bluntly, preference change is a sensible policy instrument when what people apparently prefer regarding the environment does not affect what policy should be. This would be when the optimal environmental protection is defined in reference to ethical theory about ecology or, more problematically, by prescriptions from policy advocates who give their own preferences great weight, while giving those of the public at large (and the costs they bear) very little consideration.

2. Literature on preference change

Understandably, economics does not have much to say on policies that involve appeals to change apparent preferences for a particular product. Answering efficiency questions in assessing whether there is a policy-relevant market failure–whether we make anyone better off without making anyone else worse off–presupposes a prior set of preference orderings that give the phrases "better off" and "worse off" their meaning. Without stable preferences, economists may have little to offer as policy evaluators (Rhoads, 1985, pp. 154–155).

Perhaps the most direct economic response to the views of some other social scientists and critics that preference change is an obvious aspect of "real life" was from Becker and Stigler (1977). They argued not merely against preference change but against preference difference, claiming that differences in choices across consumers, and behaviors that seem to involve preference change–addictions, following fashions, acquired tastes–nevertheless are consistent with stable and identical preferences. One can explain differences in revealed preference across persons and over time by appealing to differences in the information they have and the intellectual capital they have amassed as a result of differences in experience.

One of the few discussions by economists that accepts the possibility of preference change is Dixit and Norman (1978). Their focus was advertising, in particular, the question of whether there is, as many noneconomists believe, too much advertising. It is not difficult to show that there can be too much advertising if the effect of the advertising is to reallocate demand among firms in a less-than-perfectly competitive market, when price exceeds marginal cost, but that argument takes preferences as

[4] Peter Schwarz has suggested to me that an alternative policy that would achieve the same end would be to persuade suppliers to internalize external costs. While the analysis here could cover that scenario, we focus on the demand side. First, as discussed below, the environmental policy justifications for opening retail electricity markets hinge on encouraging consumers to choose power according to green preferences. Second, and more fundamentally, if we assume that producers are primarily interested in profits rather than a broader concept of "utility," the scope for encouraging them to take non-market revealed costs into account is smaller than it is for buyers, at least for non-commercial customers.

given.[5] Dixit and Norman looked at this question when the effect of advertising is to change taste. They propose that one can unambiguously test claims about advertising in calculating social welfare at the margin using both pre-advertising and post-advertising preferences. By that rigorous test, advertising can still be excessive. However, as Hahn (1982, pp. 190–195) pointed out, this strict test is unlikely to apply over the broad range of policies and actions for which preferences may be in flux.

Not surprisingly, philosophers, less committed to stable preferences as a tenet of either positive descriptions of human nature or ethical judgments of human behavior, have had more to say about the possibility of preference change. Among the most influential conceptualizations of preference change has involved the idea that people can, do, and should choose and bear responsibility for the preferences they have. The leading modern discussion of this idea is Frankfurt (1971), which, in trying to make sense of the concept of "will," distinguished between first-order behaviorally revealed preferences and higher, second-order preferences regarding what first-order preferences one wants to have. A "willed" action is one in which what one wants to do (first-order preferences) is what wants to want to do (second-order preferences). Jawoska (1999) applied these ideas to understand the ethical consequences of the ability to reconcile first-order and higher-order preferences in assessing the extent to which the former have moral weight.

There have been some connections between these philosophical insights and the analytical techniques of social sciences. George (1993) modeled preference hierarchies, highlighting the possibility that markets may reduce welfare by making it more costly to undertake activities to change preferences and bring them in line with "preferred preferences." Elster (1983) combined philosophy, economics, and social theory to look at preference change as an adaptation to circumstance. Brennan (1994) discussed the possibility that public support of education and the arts could be a response to a hypothetical market failure if current preferences undervalue any change in preferences that these institutions could produce.

This very brief survey can do nothing more than suggest the breadth of positive and normative issues that arise absent a methodological commitment to stable preferences. But, as Becker and Stigler (1977) pointed out, it may be difficult to distinguish activities that change preferences from activities that change behavior by changing the available information that might be used to maximize utility given a set of stable preferences. In the case of green power, the blurry distinction would be between activities that increase one's underlying preference for environmental protection and those that give consumers information to act on environmental preferences they already have.

As we will see below, the benefits of assuming that shifts in demand result from changes in information, rather than underlying preference, come at a nontrivial cost. Opening retail electricity markets to allow green power companies to inform consumers of their supplies and allow them to purchase green power is, to all appearances, a good thing. But, as we show below, the shift in the demand curve for polluting power induced by an information change fails to eliminate the externality and reduces the optimal level of consumption below what we would have seen prior to the new "information." If the preferences people happen to have depend critically on potentially incomplete information, they lose their privileged status as the basis for efficiency tests of market failure and consumer surplus-based measurements of welfare.[6]

[5] The marginal private gain from reallocating demand must exceed the net social gain, if the other firms were earning marginal profits on their sales. But that marginal private gain must equal the marginal cost of advertising, which is a social cost. Putting these together implies that the social gain from advertising at the margin must be less than its social cost. The argument parallels that for why there can be too many firms in an oligopoly market (Mankiw and Whinston, 1986).

[6] A philosophical technique to justify paternalistic interventions is that they can get people to do what they would do if they were "fully informed." The meaning of "fully informed" in practice is problematic. Were someone fully informed about the importance of the environment, caring about others, doing good, or however high up the ethical ladder one wants to go, perhaps that someone would, of course, see things just as I do. It is just that difficulty in sorting out "information" that makes it empirically difficult to distinguish changes in "information" from changes in "taste," hence that looking shifts in demand as only based on changes in information does not erase many of the conceptual difficulties brought out below.

3. Can green power purchases eliminate environmental externalities?

The problem begins with the conceptual difficulty in viewing green preferences as a policy instrument, while retaining the status of preferences as the basis for policy evaluation. Consider the "Econ 101" diagram of the welfare losses created by negative environmental externalities associated with the consumption of a particular good, in this case, fossil fuel-generated electricity. In Fig. 1, PMC, for "private marginal cost," is the supply curve of such electricity. SMC, the social marginal cost of electricity, includes both the PMC and the external cost associated with emissions of air pollutants associated with power generation.[7] DD reflects the given demand for fossil fuel-generated electricity.[8] The market output of electricity, where PMC intersects DD, is Q_M; the efficient level of output, where SMC intersects DD, is Q_E.

Suppose–"imagine" might be a better verb–that demand for fossil fuel electricity can be changed at relatively negligible cost (e.g., through a mailing campaign that leads people to become more altruistic regarding the cleanliness of the air their neighbors breathe). Specifically, let Z be the difference between SMC and PMC at Q_E. Z is the marginal external cost of the pollution from using fossil fuel electricity at the optimal use level. We assume that this campaign to change demand brings about an aggregate reduction of Z in the willingness to pay for electricity, shifting the demand curve down from DD to the altered demand curve DD_{ALT}, as shown in Fig. 2.

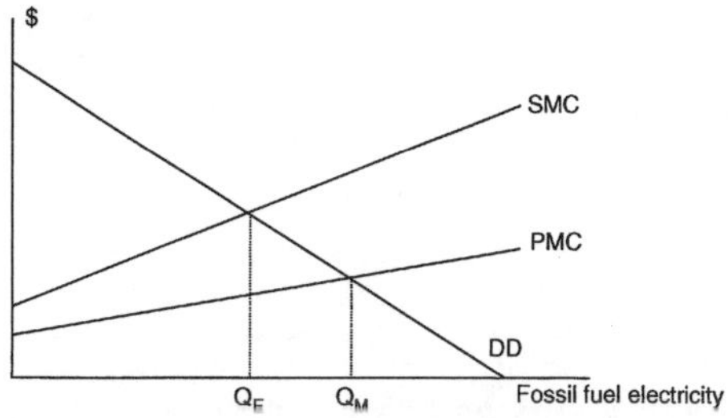

Fig. 1. Basic externality effect.

If we can alter the willingness of people to give up something they would otherwise prefer to buy in the interests of environmental protection to shift the demand for fossil fuel electricity from DD to DD_{ALT}, the new market outcome will be at the intersection of DD_{ALT} and PMC. The output of fossil fuel electricity is Q_E, the policy goal before. In that sense, changing preferences is a policy tool, in this case a "free" tool, to achieve the policy goal.

But if DD_{ALT} is now the demand curve, the optimal amount of electricity falls from Q_E to Q_E^{ALT}, where Q_E^{ALT} is the output level where the new demand DD_{ALT} intersects SMC. Although the quantity demanded shifts back to Q_E after we alter demand, there is still a marginal external harm being imposed by fossil fuel electricity use at that point. Using DD_{ALT} as the reference point, there remains a need for some environmental policy tool–fossil fuel taxes, emissions taxes, or marketable permits–to reduce pollution (via reduced fossil fuel electricity use) to the reduced, more stringent optimum. For example, if the difference between SMC and PMC is constant over the relevant range of fossil fuel electricity consumption, the optimal tax will be the same after the change in preferences as before. In short, changing preferences does not seem able to fulfill a promise to substitute for conventional environmental policies.

---

[7] To use a graph based on production of a good, the level of emissions needs to be uniquely associated with the quantity of fossil fuel electricity produced. Such a diagram posits no possibility of abatement holding output constant, through fuel switching (low-sulfur coal or natural gas for high-sulfur coal), different generation technologies (higher heat rates), or pollution removal devices (scrubbers).

We could incorporate these possibilities easily by using a diagram based on emissions directly rather than electricity. In such a diagram the marginal cost curve reflects the cost imposed by the externalities, and the marginal benefit curve reflects the economic value of being able to pollute by one more unit (from increased output or reduced abatement expense). A change in preferences that makes consumers less willing to buy dirty power would be reflected by a fall in this marginal benefit curve.

Because the market manifestation of a change in preferences is likely to involve demand for the product (substituting renewable power for fossil fuel electricity) rather than the pollutant (for which there is no market), analyzing preference change as a policy instrument seems clarified by looking at the product. However, one could do the analysis using a more general model based on pollution rather than the product, without any change in the conclusions.

[8] A more general analysis could take into account the ability of power sellers to increase demand (i.e., shift DD to the right) through advertising.

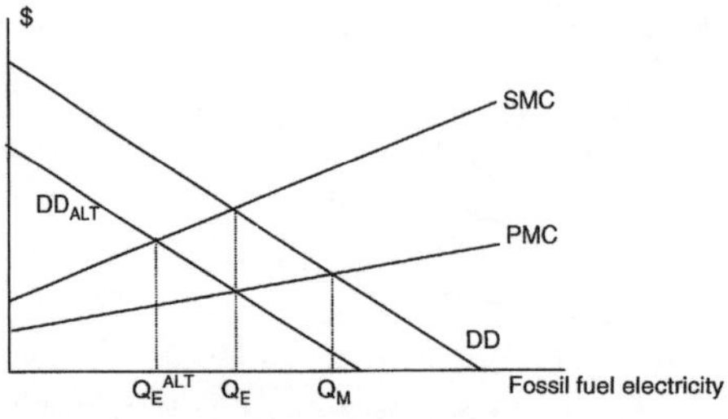

Fig. 2. Altered demand.

Moreover, and as we will see in more detail below, the choice of Q_E^{ALT} itself may be the arbitrary outcome of exactly how DD shifts. To hit the target of Q_E, DD_{ALT} needs to intersect PMS at Q_E. In Fig. 2, DD_{ALT} was constructed to be parallel to DD, reflecting a constant reduction in aggregate consumer willingness to pay for fossil fuel electricity. But that need not be the case. Roughly speaking, if the preference change generates a DD_{ALT} more (less) elastic than that produced by a parallel shift in DD, Q_E^{ALT} will be less (more) than that in Fig. 2.

Can preference change be a policy tool that substitutes for taxes, permits, or command-and-control? Is the right target the original Q_E? Q_E^{ALT}? Which Q_E^{ALT}? To those questions we turn.

4. Counterfactual preferences

One response to this puzzle would be to appeal to a standard independent of fostering green preferences, to get an unambiguous target. There may be some privileged set of preferences that should be the demand curve that counts in deciding the optimal level of pollution, essentially ignoring changes in the calculated optimum that arise with altered preferences. In terms of the above illustrations, one could say that DD is the standard for determining optimal pollution, Q_E. The post-change demand curve DD_{ALT} matters only as a tool for getting the market outcome to be Q_E. That the DD_{ALT} optimum output is different, at Q_E^{ALT}, is irrelevant.

Philosophical thinking on utilitarianism provides some justification for this response to the puzzle, in that in assessing aggregate utility, one ought to disregard contributions to one person's utility based upon the utility levels of others. From an ethical standpoint, counting other regarding preferences leads to a kind of double counting that penalizes the altruistic. We can illustrate this argument formally. Imagine a world with two individuals, A and B, and a fixed quantity of a single good X that can be allocated between them as X_A and X_B, where $X_A+X_B=X$. Let $U_A(X_A)$ and $U_B(X_B)$ be the utility A and B get, respectively, from X_A and X_B. We assume that these functions are identical, increasing ($U'_A, U'_B>0$), with diminishing marginal utility ($U''_A, U''_B<0$). In addition, assume that A is an altruist that also gets utility $kV_A(X_B)$ from knowing how much B is consuming, where k is a variable parameter indicating the strength of the altruistic preference and $V'_A>0$.[9] Hence, A's total utility is $U_A(X_A)+kV_A(X_B)$.

Choosing X_A to maximize the sum of A's and B's total utility, $U_A(X_A)+kV_A(X_B)+U_B(X_B)$, recognizing that $X_B=X-X_A$, gives the equilibrium condition:

$$U'_A - kV'_A = U'_B.$$

A positive k implies that $U'_A>U'_B$; functional identity and diminishing marginal utility from consuming X imply that $X_A<X_B$. A loses because his altruism leads B's interests to be counted more than once.

Some leading moral theorists have based their arguments on the premise that only self-regarding preferences should count in the construction and evaluation of social institutions. To avoid complications and distortions relating to envy, just benevolence, and altruistic preferences, Rawls (1971, esp. Sections 24–26, p. 30) based his principles of justice on hypothetical choices risk-averse individuals, behind a "veil of ignorance" regarding their capabilities and aspirations, would make to promote the "mutually disinterested" well-being of the worst off

9 Non-economists may be more comfortable with having B's utility, rather than consumption, contribute to B's utility function. The advantage of the consumption is that it is easier to observe and measure than utility. However, when taking utilitarianism (rather than only efficiency) seriously, we presume that the ethicist can at least judge utility magnitudes in some rough comparative sense, reducing the conceptual advantage consumption has over utility in this particular setting.

class of persons.[10] Dworkin (1977, pp. 77–83) uses general theory that "external" preferences based on the well-being of others "corrupt" utilitarianism, whether negative (racism) or positive (altruism), to oppose admissions policies that discriminate against minorities (based on the external preference of racism), while supporting redistributive affirmative action policies as increasing aggregate utility (as calculated from "personal" preferences).

Controversial aspects of Rawls's and Dworkin's specific applications aside, the principle that policies be judged according to only those preferences people would have absent any direct consideration for the welfare of others seems a plausible way to resolve the puzzle posed in Section 3: Should one use ex ante or ex post demand curves to determine the optimal intensity of environmental policy? But this potential answer leads to another, empirical question: Is the original, observable demand curve free of such external preferences?

All we know about the original demand curve, DD in Figs. 1 and 2, is that it is the demand curve prior to the implementation of policies designed to encourage persons to take into account the environmental effects of their choices on the well being of others. But nothing precludes the possibility that, to some extent, some people act somewhat altruistically in their energy choices, even absent explicit policies to encourage environmental awareness. Surely it is not extremely unusual (except perhaps among economists) that persons reduce air conditioning use, purchase energy-saving appliances, or turn out the lights partly out of an internalized belief that conserving energy is "politically correct" social behavior. If so, should the pollution policy target, whether achieved through taxes, marketable permits, command-and-control, or preference change, exceed Q_E because revealed preferences understate the purely self-regarding willingness of consumers to pay for fossil fuel electricity?

A thought experiment may illustrate the scope of the potential discrepancy between the observed and hypothetical optimal points when other regarding preferences are neglected. Consider the optimal crime rate given current preferences for law-abiding behavior, balancing the marginal cost of the law enforcement system against the marginal benefit of crime reduction (however construed). Then, imagine what the crime rate would be, holding law enforcement costs constant but assuming that persons lacked an internalized moral code that motivated them to obey the law even when the probability of being caught was negligible.

The two crime rates seem quite different. Is aiming at the second target appropriate, implying that we should be tolerating more crime than we have now? Or should we take advantage of internalized codes of conduct, and, if so, how much? If, as I suspect, most would opt for using the internalized codes of conduct, then at least in that case, "self-regarding" preference is not the appropriate standard for setting policy targets. That standard presents both empirical quandaries in determining what those preferences really are, and normative quandaries regarding the acceptability of the outcomes such a standard may imply.

5. Preference–change equilibria and inelastic external costs

5.1. Definition

One response to the problem presented in Fig. 2, the difference between the pre-change and post-change optimal outcomes, would be to change preferences again to make the post-change outcome the market outcome. Presumably, of course, this could create a new "post-two changes" outcome with even tighter pollution standards. But perhaps we could get to a point where, after enough changes, there is no difference between the pre-change and post-change optimal points, making the ambiguity in policy targets disappear. We call this a "preference change equilibrium," or PCE.

A PCE would exist when the pollution target defined after changing demand is the same as the target aimed at prior to changing demand. Following Figs. 1 and 2, let PMC(Q) be the private marginal cost

[10] Rawls (1971, pp. 147–148) noted that ethical theories should give primacy to direct utility in terms of evaluating that justice is not the same thing as saying that the people should care only about themselves. He appealed to a disposition to act according to principles of justice, supporting the principles that liberty should be maximized and, then, that the well-being of the worst-off class of individuals should be maximized (the "maximin" principle).

In later writings, Rawls (1993, esp. pp. 133–158) attempted to justify these principles as necessary for political stability in a society where people in fact have different but hopefully overlapping comprehensive moral doctrines.

of producing the polluting good–the supply curve, assuming competition in the sale of the polluting good–and let SMC(Q) be the total cost of producing Q, including the external damages. As before, let DD(Q) be the pre-change demand for the good, and let $DD_{ALT}(Q)$ be the post-change demand curve.

The target defined after changing demand will be the value of Q for which:

$$DD_{ALT}(Q) = SMC(Q). \quad (1)$$

If the change in demand was as designed, so that the post-change market equilibrium equals the pre-change target output, then:

$$DD_{ALT}(Q) = PMC(Q). \quad (2)$$

A desirable property of a PCE is that it should meet the pre-change target. If Q is the pre-change target, it was defined by the pre-change social optimum condition:

$$DD(Q) = SMC(Q), \quad (3)$$

and the need for environmental policy at this point implies that:

$$DD(Q) > PMC(Q). \quad (4)$$

At a PCE point meeting the pre-change target, the same output level Q must satisfy all four of these equations. Let Q_{PCE} be such a point. This may occur in one of two ways, requiring perfectly inelastic external cost or perfectly inelastic post-change demand.

5.2. *Inelastic external cost*

The first way these equations can all hold comes from the cost side, assuming that DD_{ALT} is not perfectly inelastic. From Eqs. (1) and (2), $SMC(Q_{PCE})=PMC(Q_{PCE})$, implying that at Q_{PCE}, the target level of output, external costs are zero. However, Eqs. (3) and (4) imply that $SMC(Q_{PCE})>PMC(Q_{PCE})$, since Q_{PCE} was the pre-change target level of output. These two relationships can hold only if SMC-PMC can take both zero and a positive value at Q_{PCE}. Geometrically, this implies a kink in the SMC curve at Q_{PCE}, which implies inelastic external

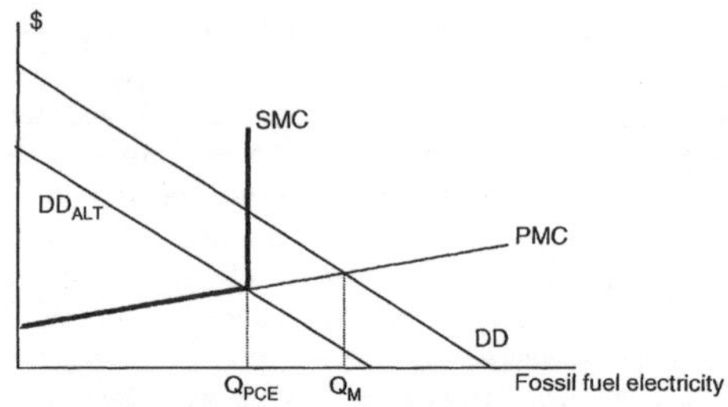

Fig. 3. Inelastic external costs.

cost at Q_{PCE}, as illustrated by Fig. 3, where the heavy line indicates the social marginal cost curve.[11]

If the SMC looks as it does in Fig. 3, the optimal level of polluting output (fossil fuel electricity, here) is largely independent of the level of demand. The harm is negligible below that level of output that is negligible, so pollution does not matter. At that level, the marginal cost is so high that we would never want to exceed it. Hence, the empirical problem of which preferences to use disappears, if the target level of pollution is essentially independent of that value.

Unfortunately, few, if any, pollutants meet this criterion; in particular, there may be no point below which the external costs are zero. Portney (1990, p. 82) cites medical studies suggesting that some people are sufficiently sensitive to some pollutants so that virtually any level can cause some illness or discomfort they would want to avoid.[12] Unless there is a perfectly safe level, Eqs. (1) and (2) cannot hold unless demand is inelastic. Even if there is a perfectly

[11] The SMC curve need not be vertical indefinitely. Nonconvexity (i.e., the fact that beyond some point, the marginal cost of pollution falls because people eventually leave the polluted area) implies that at some point the external marginal costs fall, perhaps back to zero. The optimal response may be to do nothing about pollution at all (e.g., if the potential victims can relocate at a cost less than the cost associated with abatement or foregone output) (Baumol and Oates, 1988, pp. 110–131).

[12] A recent DC Circuit Court of Appeals decision vacated recent particulate matter standards, and remanded ozone standards on the grounds that because there was no perfectly safe level, the Environmental Protection Agency had inadequate delegation from Congress to set any standard. *American Trucking Assn. vs. U.S. Environmental Protection Agency*, D.C. Circ. No. 97-1440, May 14, 1999, available at http://www.ll.georgetown.edu/Fed-Ct/Circuit/dc/opinions/97-1440a.html.

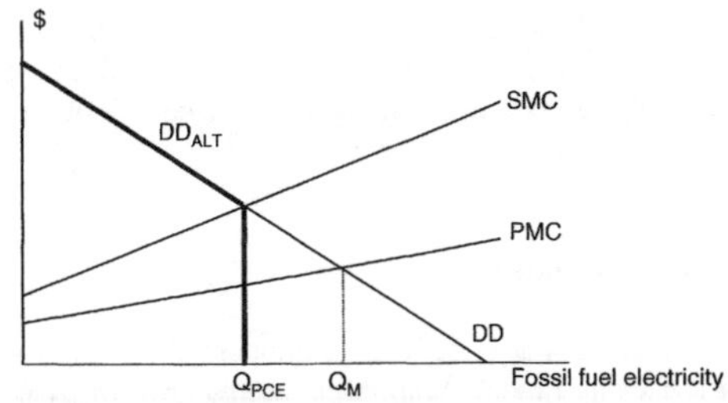

Fig. 4. Inelastic post-change demand.

safe level, Eqs. (3) and (4) will not hold unless the harms are discontinuously large at that crucial juncture. Absent any indication that any pollutant meets these criteria, we should examine the inelastic demand possibility.

5.3. Inelastic "green" demand for polluting goods: do preferences matter?

If there is no value of SMC(Q_{PCE})=PMC(Q_{PCE}), then the only way Eqs. (1) and (2) can hold, with DD_{ALT}(Q_{PCE}) equal to both SMC(Q_{PCE})= PMC(Q_{PCE}), is if the post-change demand curve is perfectly inelastic at Q_{PCE}. But if Q_{PCE} meets the initial target (i.e., if Eq. (3) holds), it and Eq. (1) together imply that:

$$DD_{ALT}(Q_{PCE}) = DD(Q_{PCE})(= SMC(Q_{PCE})).$$

The most obvious post-change demand curve meeting this criterion is where DD_{ALT}=DD for $Q<Q_{PCE}$, and is inelastic at Q_{PCE}, as indicated by the heavy line in Fig. 4.[13]

One can imagine achieving a demand curve of this sort by convincing people that it is ethically mandatory to cut out certain kinds of uses of electricity or to curtail purchases above some given amount. Such a specific change in demand may be as unfeasible as it is unlikely that the harms from a pollutant are sufficiently inelastic to justify a PCE, as described in Section 5.2. More problematic, however, is that any post-change demand curve that is inelastic at some Q between SMC and PMC will be a PCE, although not one that meets the initial target.

Fig. 5 sets out two examples. If post-change preferences were described by the heavy line with the inelastic portion on the left, the PCE would be at a level of output (and pollution) at Q_1, stricter than the target level, Q_E. If those preferences produce a demand curve with the inelastic portion on the right, the PCE will be at Q_2 exceeding Q_E.

If avoiding the need to consider additional policy following the change in preferences is the criterion for setting policy, then the fact that the new equilibria do not agree with the target under the original preferences is of no concern. But this leads to the question of how to choose the target, if any output level can be a PCE. In other words, if preferences (or at least demand curves) are sufficiently malleable, being a PCE is not a constraint. Moreover, we still need to decide which target is the right target.

Malleability to this degree also raises the policy issue of why we focus on changing preferences regarding demand for the polluting product. Why not change the willingness to pay to avoid the externality? If we can eliminate the willingness to pay to avoid the externality, at least up to the market outcome Q_M, there is no need for any policy to reduce pollution. Perhaps we could convince persons that a little coughing (fashionably throaty voices) or shorter life expectancy (reduced chronic illness risk) is not worth paying much to avoid. Maybe an advertising campaign could convince the public that viewing old Grand Canyon pictures over the Internet is as good as being there, to eliminate any willingness to pay to avoid haze.

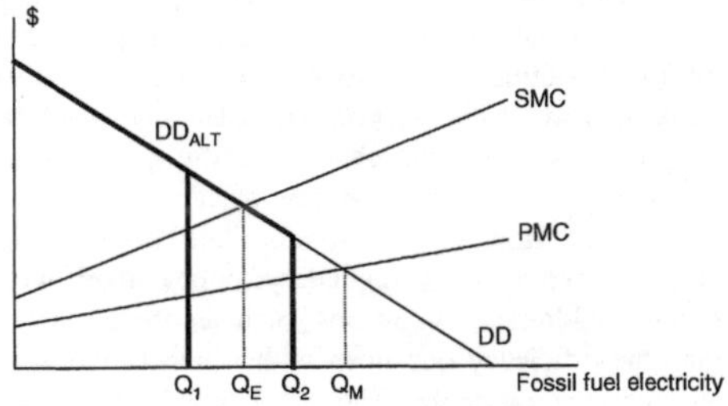

Fig. 5. Multiple post-change equilibria.

[13] More generally, any post-change demand curve will satisfy the equations for a PCE that is inelastic at Q_{PCE} between SMC and PMC, and nonincreasing everywhere else.

6. Bottom line: should preferences matter at all?

The point of these progressively more absurd examples is not to suggest that environmental policy is inherently arbitrary, or that we ought to consider solving the pollution problem by getting people to not mind it. The lesson from the above is that adding preference change to the policy maker's toolkit creates a huge range of ambiguities for the economist's appraisal of policy effectiveness. This should not be surprising; to expect otherwise would be to expect that preferences could be both policy instrument and policy criterion (i.e., both the means and the end of policy).

Widespread advocacy of green preferences among public officials, environmental activists, and renewable energy service companies would makes sense as a policy instrument only if preferences do not count in the evaluation of those policies' outcomes. Making green preferences sensible as a policy instrument may require rejecting the economist's method of basing policies on preferences. Such a view better fits the ecologist's view of environmental policy, where the right level of pollution is defined by criteria drawn from biology and perhaps environmental ethics, rather than from estimated willingness to pay for environmental public goods. It may also fit conceptions of policy choices as being based on deliberations regarding the public welfare rather than mere self-interested, "naked" preferences (Sagoff, 1986; Sunstein, 1990).

Last but not least, the philosophical view that preferences for the goods associated with pollution ought not count for much compared to environmental values is consistent with a view of some kind of ethical objectivity. As most normally regard the truth of a scientific proposition, the truth or validity of an ethical proposition may depend upon "holding up against all rivals through and impartial and informed examination" (Frankena, 1973, p. 109) rather than what people in a particular time or place happen to believe. This paper is hardly the place to recapitulate the debate between objective and relativist ethics. However, the "bottom line" in assessing green preferences as regulatory policy may well involve addressing questions such as these, rather than the efficiency questions with which economists are more comfortable. The institutional means for addressing these questions will inherently rely more on political processes than efficiency criteria or cost–benefit analysis, or other conventional, technical tools for policy evaluation—likely with all of the distortions such processes imply.

Acknowledgment

The author is grateful for comments from Andrew Kleit, Colin Loxley, Ming Miu, Timothy Mount, Peter Schwarz, and other participants in the 19th Annual Rutgers University Advanced Workshop in Regulation and Competition. Errors remain those of the author.

References

Baumol, W., Oates, W., 1988. The Theory of Environmental Policy. Cambridge University Press, New York.

Becker, G., Stigler, G., 1977. De Gustibus Non Est Disputandum. American Economic Review 67, 76–90.

Brennan, T., 1994. Markets, information, and benevolence. Economics and Philosophy 10, 151–168.

Department of Energy, 1999. Comprehensive Electricity Competition Act, Benefits of the Proposal. U.S. Department of Energy, Washington, DC.

Dixit, A., Norman, V., 1978. Advertising and welfare. Bell Journal of Economics 9, 1–17.

Dworkin, R., 1977. DeFunis vs. Sweatt. In: Cohen, M., Nagel, T., Scanlon, T. (Eds.), Equality and Preferential Treatment. Princeton University Press, Princeton, pp. 63–83.

Elster, J., 1983. Sour Grapes: Studies in the Subversion of Rationality. Cambridge University Press, New York.

Farhar, B., 1999. Willingness to Pay for Electricity from Renewable Resources: A Review of Utility Market Research. U.S. Department of Energy National Renewable Energy Laboratory, Golden, CO.

Frankena, W., 1973. Ethics. Prentice Hall, Englewood Cliffs, NJ.

Frankfurt, H., 1971. Freedom of the will and the concept of a person. Journal of Philosophy 68, 5–20.

George, D., 1993. Does the market create preferred preferences? Review of Social Economy 51, 323–346.

Hahn, F., 1982. On some difficulties of the utilitarian economist. In: Sen, A., Williams, B. (Eds.), Utilitarianism and Beyond. Cambridge University Press, New York, pp. 187–198.

Jawoska, A., 1999. Respecting the margins of agency: Alzheimer's patients and the capacity to value. Philosophy and Public Affairs 2, 105–138.

Mankiw, N.G., Whinston, M., 1986. Free entry and social inefficiency. RAND Journal of Economics 17, 48–58.

Parry, I., Oates, W., 2000. Policy analysis in the presence of distorting taxes. Journal of Policy Analysis and Management 19, 603–614.

Pizer, W., 1998. Prices vs. quantities revisited: the case of climate change. Discussion Paper, vol. 98-02. Resources for the Future, Washington, DC.

Portney, P., 1990. Air pollution policy. In: Portney, P., Stavins, R. (Eds.), Public Policies for Environmental Protection. Resources for the Future, Washington, pp. 77–123.

Rawls, J., 1971. A Theory of Justice. Harvard University Press, Cambridge, MA.

Rawls, J., 1993. Political Liberalism. Columbia University Press, New York.

Rhoads, S., 1985. The Economist's View of the World. Cambridge University Press, New York.

Sagoff, M., 1986. Values and preferences. Ethics 96, 301–316.

Sunstein, C., 1990. Political self-interest in constitutional law. In: Mansbridge, J. (Ed.), Beyond Self-interest. University of Chicago Press, Chicago, pp. 209–223.

Swezey, B., Bird, L., 1999. Information Brief on Green Power Marketing, fourth edition. U.S. Department of Energy National Renewable Energy Laboratory, Golden, CO.

[2]

Voluntary Eco-Labeling and the Price Premium

Roger A. Sedjo and Stephen K. Swallow

ABSTRACT. *International environmental organizations propose voluntary eco-labeling as a market incentive to promote industry to operate in an ecologically sustainable and environmentally friendly manner. A microeconomic analysis questions whether voluntary eco-labeling will cause producer profits in a competitive industry to decline and whether eco-labeling will necessarily generate different prices for labeled and unlabeled product. Using wood product as an example, results identify conditions that may exist when firms lose profits, even under a voluntary system, and where existing production constraints may lead to a single price, regardless of labeling.* (JEL Q28)

I. INTRODUCTION

The contradiction between surveys that have found consumers to be willing to pay a premium for certified wood products and the market evidence that certified wood does not trade at a premium is the motivation for this paper. How is this possible? Proponents of eco-labeling have suggested that voluntary labels would allow consumers to reveal their willingness to pay for ecologically-friendly products. Although the issues relate to important attributes of some products, for example, health safety (Wessells and Anderson 1995), for other products the labeling is independent of the safety or physical characteristics of the product, but relates instead to environmentally benign production approaches. For example, eco-labeling is being proposed as a means of encouraging more firms to choose production methods that reduce damage to the ecological quality of forest land (e.g., Upton and Bass 1996). Support for eco-labeling is buttressed by numerous surveys that indicate a significant fraction of consumers would be willing to pay a premium for an eco-labeled product (e.g., WWF 1991; Irland 1993, Winterhalter and Cassens 1994; Forsyth, Haley, and Kozak 1999). Furthermore, most consumers respond that they would choose an eco-labeled product over one that is not labeled (Forsyth, Haley, and Kozak 1999), other things being equal.

On the supply side, an important question is the extent to which eco-labeling would increase production costs. The extent of cost increases would likely vary across products. The removal of dioxin from the wood-pulping process, for example, has involved tens of millions of dollars per plant (Harrison 1998). For the production of certifiable wood, Murray and Abt (1998) have estimated that 50% of the U.S. harvest can be produced at an on-the-ground cost increase of 10% or less; however, the remaining 50% would require cost increases up to 50%. Similar significant cost increases for certified wood are found in a host of other countries (see Sedjo, Goetzl, and Moffitt 1998). In addition, there are costs associated with maintaining the "chain-of-custody" required to ensure that certified commodity flows are not contaminated with non-certified commodities. The existence of a price premium could be important in defraying these costs and could create additional incentives to produce more environmentally friendly eco-labeled goods.

On the empirical side there is some evidence about the market viability of certified products. Not all products are successful in attracting buyers. For example, a program begun in the mid-1990s in Eastern Canada by Scott Paper to introduce a new line of eco-labeled, sanitary-paper products stressed the high-recycled content that competed with its existing non-eco-labeled line. The firm discontinued production of the eco-labeled line after sales declined significantly (Harrison

Land Economics • May 2002 • 78 (2): 272–284
ISSN 0023-7639

The authors are senior fellow and director, Forest Economics and Policy Program, Resources for the Future, Washington, D.C.; and professor, Department of Environmental and Natural Resource Economics, University of Rhode Island.

1999). Additionally, firms in the U.S. that produce wood certified as being grown in "well managed forests" have typically found that a very large portion of the certifiable wood generates no price premium in the marketplace. However, these same firms do find specialty markets in which a price premium for certain products justifies their production.[1] Thus, there are conditions where a price premium will exist for eco-labeled goods even if they are identical in all physical respects to a competing non-eco-labeled product. A purpose of this paper is to identify those conditions.

Thus far, little economic analysis exists to address the issue. Analysts like Kiker and Putz (1997) have focused on practical obstacles in concert with potential environmental advancements. Proponents rely on intuitive persuasion drawn from firms whose experience suggests that choosing environmentally friendly processes results in benefits such as more enthusiastic (ostensibly, more productive) personnel and advantages in retaining market share (e.g., Canadian Chemical Producers Association 1996; Stevens, Ahmad, and Ruddel 1998; Upton and Bass 1996). Mattoo and Singh (1994) provide a basic market analysis, but they focus on a partial-equilibrium model with production constraints that eliminate firms' ability to choose between traditional and environmentally friendly technologies. These assumptions limit the insights that analysts can draw from their model.

This paper focuses on the market implications of labeling consumer products. Although the example addressed in this paper is wood products, the results apply to other products for which a label provides a "stamp of approval," indicating that the raw material is derived from a source that is managed according to independent standards for environmental or ecological purposes. The present example deals with an eco-label based on concerns for on-the-ground forest management consistent with sustainable forestry and ecosystem quality; that is, producers with an appropriate, "certified" management system would qualify their wood for a certified, eco-labeled product. Other motivations include certification that producers have integrated local communities into their management system, or that producers have addressed equity concerns in developing nations (Upton and Bass 1996; Kiker and Putz 1997; Barbier 1998). While our framework may illuminate these other issues, our emphasis concerns whether markets will necessarily generate price signals sufficient to attract firms into certifiable production, whether certification might erode profits of firms choosing not to certify their production, and whether, and under what conditions prices will differ between labeled and unlabeled products.

The main points may be established intuitively, so we place the analytical content in simple graphical tools within a partial-equilibrium analysis, assuming producers are price takers. We assume that producers qualify for a certified eco-label while leaving unharmed the traditional dimensions of product quality.[2] While we begin without assuming that certification generates a premium "willingness to pay" by consumers, much of the analysis relies on the existence of a group of consumers who are willing to pay a price premium for labeled products. Indeed, many of the alleged benefits from eco-labeling, especially under a voluntary system, could not arise without the existence of a group of eco-consumers. Upton and Bass (1996) and Mattoo and Singh (1994) cite evidence that 80% of consumers in U.K. and Canadian markets are willing to pay more for environmentally friendly commodities. However, Stevens, Ahmad, and Ruddell (1998) report that the U.S. industry perceives that only a much smaller share of consumers are willing to pay a price premium. In this paper, the market may not generate a price differential between labeled and unlabeled wood *even if there is a willingness to pay among a significant portion of consumers.*

[1] Conversation with Paul Higby, manager Kane, Pennsylvania plant of Collins Pine, October 8, 1999.

[2] This assumption is not trivial. In addition to concerns that new harvest or management systems may affect fiber quality, informal discussions with producers in Canada, Finland, and the United States quickly reveal concerns that many independently developed labeling schemes will destroy consumer confidence and generate a negative backlash on demand (see also Upton and Bass 1996).

The approach of this paper is to use a static partial equilibrium framework of the entire production sector. Initially, the sector equilibrium price and quantity are examined assuming that certification is mandatory. This assumption is subsequently removed and situations with some portion of consumers willing to pay a premium for eco-labeled products are introduced. The focus is on the relative prices of eco-labeled and non-eco-labeled products and the conditions that must obtain for each situation to occur. One finding is that there are reasonable circumstances in which some portion of consumers are willing to pay a price premium, but a premium will not arise in the market. An important implication of the findings is that even if surveys that indicated that consumers are willing to pay a premium are correct, this is not a sufficient condition to generate a premium in the market.

II. MANDATORY CERTIFICATION WITHOUT ECO-CONSUMERS

To help facilitate the analysis we begin with an economic assessment of mandatory eco-labeling, without a group of eco-consumers, to reveal many of the underlying issues and establish a simple foundation for further analysis. A voluntary approach will follow. Partial equilibrium implications are apparent from Figure 1, which shows the current (non-certified) quantity demanded, $W_D^{NC} = W_D(P_W)$, and quantity supplied, $W_S^{NC} = W_S(P_W, 0)$, of wood products, where P_0 is the pre-certification price of wood products.[3] The second argument in the supply function is C (initially zero), which represents the situation with respect to "certified" eco-labeling. (For now, we assume certification leaves the demand curve unaffected at W_D^{NC}.)

Presently, one may interpret C either as a discrete "with-or-without" condition, or as a continuous variable measuring the stringency of requirements for a producer to gain certification as a practitioner of ecologically sustainable forestry. In either case, C equals zero initially prior to introduction of a certification or eco-labeling program. Henceforth, we will refer to conditions where producers do not meet certification requirements as the case of "not certified" (nc), and to conditions where producers do meet certification requirements as "certified" (c).

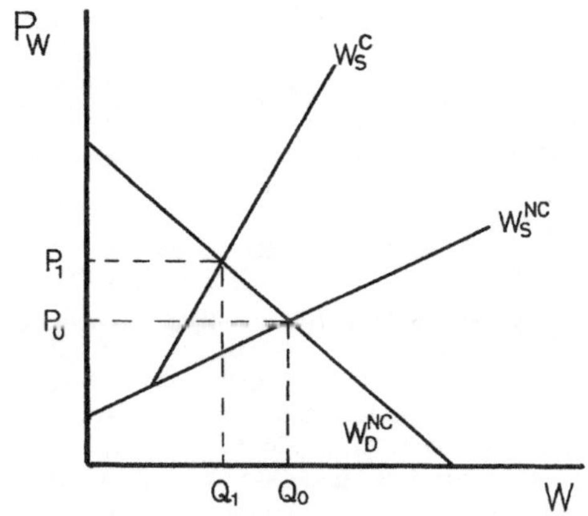

FIGURE 1
MANDATORY CERTIFICATION WITHOUT ECO-CONSUMERS

Under a mandatory certification regulation, all producers must be certified and, presumably, costs will increase in most cases. In Figure 1, the supply curve without certification rises as costs increase with certification to $W_S^C = W_S(P_W, C)$. This supply curve shows that some quantity of wood products could be certified under current conditions, since some forests are already managed for ecological concerns, but most units of output would likely require some degree of higher costs.[4]

[3] Murray and Casey (1998) and Murray and Abt (2001) use this basic framework to motivate some empirical work and welfare considerations, especially regarding the southern United States, but they do not address the equilibrium pricing issues under voluntary eco-labeling.

[4] Here it appears that we assume certification costs increase with the existing marginal costs of production. However, one could interpret the certified marginal cost function as based on an ordering of units of output after certification is in place; this ordering may differ from the *ex ante* ordering. The current paper abstracts from the complications of ecological heterogeneity in determining which lands might improve, ecologically, under certification. Swallow (1996) provides a discussion of analogous issues. Also note that some forests are currently certifiable due to decisions of sympathetic owners or imposition of existing environmental regulations. However, we assume that no producer experiences a decrease in costs because such an outcome suggests the producer is not choosing a currently minimum cost production process.

Accordingly, the price of wood rises from the non-certified level P_0 to the certified level P_1 due to the ordinary equilibrium of initial demand with higher cost supply. While wood prices increase, this situation generates no ''price premium'' because only certified wood is produced.

Proponents of certification may argue that the decline in the quantity supplied, from Q_0 to Q_1, reflects a less intensive use of existing forestland and thereby an increase in ecosystem qualities. Their argument overlooks the possibility that reduced output could stem from a withdrawal of land from forest production. Swallow and Sedjo (2000) address these issues; here we simply note that ecological improvements on land remaining in forest production may be offset partially or entirely by a reduction in the ecological quality of land reallocated to other uses. For present purposes, however, the simple case of mandatory certification produces an increase in the relative price of wood products due to the new equilibrium after the market adjusts to the costs of certification.

Proponents of eco-labeling suggest that some consumers gain more utility from a unit of certified wood than from a unit of non-certified wood. These consumers are unlikely to buy any non-certified wood if certified wood is cheaper or the same price as non-certified wood.[5] If certified wood is more expensive, eco-consumers may choose a mix of certified and non-certified wood, with that mix weighted toward certified wood as the price differential decreases.

In the eco-labeling context, the value of certified wood reveals itself as a rotation of the demand curve for wood. There are two alternative views. Economists commonly represent rotations of demand curves as a pivot around the price intercept in demand, but this view assumes consumers are willing to pay a per-unit premium that increases as their total consumption increases. This view is consistent with non-convexity in the preferences for environmental goods. Following Anderson and Francois (1997), if consumers view the consumption of one unit of certified wood as inconsequential to ecological quality, then they may exhibit a small willingness to pay a premium for the first certified unit, with their willingness to pay on the marginal unit increasing as further consumption increases their ecological impact.

In the second view, certification may be viewed as a ''luxury'' quality for which consumers' marginal willingness to pay a premium diminishes as total consumption increases. This premium may be modeled as a percentage over the willingness to pay for the base commodity. If consumers with lower income enter the market at lower prices, then there would also be a tendency for the premium to diminish as the total quantity consumed increases. We adopt this second view for our analysis.

Figure 2 illustrates the resulting pivot in the demand curve (to W_D^C), assuming that the marginal willingness to pay a premium diminishes. When combined with the certification-induced increase in supply costs (pivot in the supply curve), one observes that a willingness to pay a premium ameliorates the anticipated decline in wood consumption so that, in partial equilibrium, wood output declines from Q_0 to Q_1' rather than the further

---

[5] For these consumers, certified and non-certified wood are imperfect substitutes, and the marginal rate of substitution between certified wood and all other goods (Y) at a particular quantity combination (W, Y) exceeds the marginal rate of substitution between non-certified wood and all-other-goods for an equal quantity combination. Our discussion is consistent with the following model of consumer utility. Let $U^c(\,)$ denote the utility of individuals who prefer certified wood, and $U^{nc}(\cdot)$ denote the utility of consumers who are indifferent between certified and non-certified wood. Define $U^{nc}(W, Y) = U(Y, w_{nc} + w_c)$, so that certified and non-certified wood, w_c and w_{nc} respectively, are perfect substitutes to the consumers who are indifferent to an eco-label; Y is the quantity of ''all other goods consumed.'' Define $U^c(Y, W) = U_1(Y, w_{nc} + w_c) + U_2(w_c, C)$, so that consumers who prefer eco-labeled wood view the certified and non-certified wood as perfect substitutes with respect to their performance for furniture or housing, while, if certification is in place (nonzero C), these consumers gain additional utility from their consumption of w_c. Using Kuhn-Tucker conditions for the maximization of $U^c(\,)$, it is easily shown that the consumers who prefer eco-labeled wood will never consume w_{nc} if the price of w_c is lower, but if the price of w_c is higher than the price of w_{nc}, then these consumers will increase their consumption of w_{nc} as the price differential (or ''price premium'') that they must pay for w_c increases. This utility structure is only adopted for an example discussion. Of course, here, and throughout the paper, prices of the ecologically sensitive product (wood, in this case) are measured relative to the price for a numeraire good, such as Y.

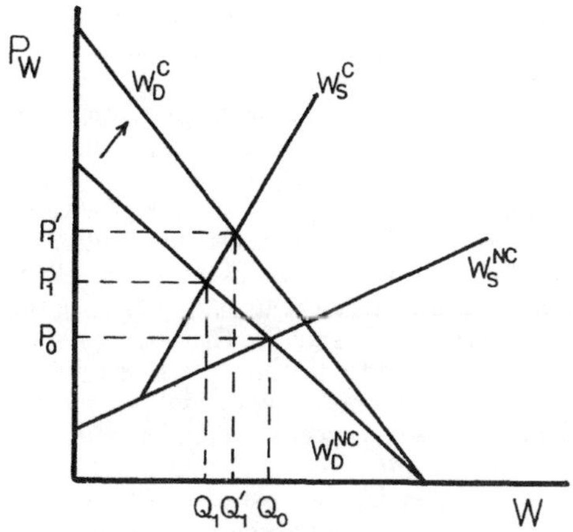

FIGURE 2
MANDATORY CERTIFICATION AND ECO-CONSUMERS' HIGHER WILLINGNESS TO PAY FOR CERTIFIED WOOD

decline to Q_1. Of course, price also rises from P_0 to above P_1 up to P_1'. The willingness to pay a premium encourages producers to increase their output a bit above the post-certification level observed without a premium willingness to pay.

The form of the demand shift carries policy implications. If eco-consumers' marginal willingness to pay diminishes, as in Figure 2, then as certification increases the costs of supply, there is a tendency for the market to provide an increasing level of compensation for producers. Otherwise, if marginal willingness to pay increases with consumption of certified wood (unlike Figure 2), then the market is less willing to compensate producers for certification costs which reduce supply.[6]

Surprisingly, diminishing marginal willingness to pay for certification tends to sustain forest production by providing a greater potential for compensation to producers who invest in certification. Also, while not illustrated here, a sufficiently large willingness to pay for certification could increase demand enough so that the equilibrium quantity of wood products actually increases after certification.

When our analysis considers eco-consumers' willingness to pay a premium, we assume diminishing marginal willingness to pay by eco-consumers. However, our qualitative results depend primarily on the existence of eco-consumers, not on the form of their willingness to pay a premium. Accordingly, one may segregate aggregate demand into the demand from eco-consumers, W_D^C, and the demand from consumers who are indifferent between certified and non-certified wood, W_D^{NC}. From this point forward, our notation for demand curves reflects this segregation of aggregate demand.

III. WILL VOLUNTARY CERTIFICATION GENERATE A PRICE DIFFERENTIAL?

We now turn to our primary question of whether markets will generate different prices for certified and non-certified wood when certification is voluntary, that is, when producers can choose whether to participate in the certification program. We find conditions under which the market generates a price differential between certified and non-certified wood, but we also find a realistic condition under which no price differential exists. The simple analysis of mandatory certification above generates an expectation that general wood prices will increase under certification. However, we find that producers who choose a non-certified process under a voluntary system may still lose, even though the average price of their timber does not decrease.

This analysis separates demand between eco-consumers (W_D^C) and non eco-consumers (W_D^{NC}), but normalizes these demands against the aggregate demand existing prior to certification (W_D^0). Initially we assume eco-consumers simply refuse to purchase a non-certified product. Later we extend our analysis to allow certification to induce a premium willingness to pay by eco-consumers. Unless eco-consumers show some inflexibility in their preference, either by refusing to purchase non-certified products or by being willing to pay a premium for each certified unit, a voluntary certification system could not ex-

[6] If the demand curve pivots around the initial price intercept, unlike Figure 2, then Q_1' and P_1' would be closer to Q_1 and P_1.

ist because competitive producers would always use the least-cost technology, which is presumably the technology they were using prior to certification.

At least three possible price outcomes are possible, and are listed below, where P^C refers to the price of certified product, P^{NC} corresponds to the price of non-certified product, and P^0 represents the equilibrium price prior to (pre-) certification:

1. $P^C > P^{NC} > P^0$;
2. $P^C > P^0 > P^{NC}$;
3. $P^C = P^{NC} > P^0$.

The first outcome is generally expected by labeling advocates, who cite consumer surveys suggesting eco-consumers are willing to pay a percentage above pre-certification prices. This result provides a market incentive in the form of a price premium to produce certified wood. The second outcome also yields a premium for certified products over non-certified products. However, the price of non-certified wood is below the pre-certification equilibrium. The third outcome is the most counterintuitive, with no differential between the price of certified and non-certified products, even while eco-consumers are willing to pay a premium. In this case the common price is, however, above the pre-certification price. These three outcomes are examined in the sections that follow.

Case 1a: Certified Price Higher than Non-Certified Price, and All Gain

Consider Figure 3 which divides demand and supply into two segments, non-certified on the left and certified on the right, with quantities increasing in both directions when moving away from zero along the horizontal axis, and beginning at the intersection with the price axis. The initial demand and supply curves are depicted on the left, respectively labeled W_D^0 and W_S^0, representing conditions prior to the certification program. After a certification system is in place, demand divides into segments labeled W_D^{NC0} and W_D^{C0} (Figure 3). As an expositional convenience, we temporarily assume eco-consumers do not exhibit a willingness to pay a premium, but that eco-consumers are unwilling to consume non-certified wood below the quantities represented along W_D^{C0}. Then the horizontal sum of W_D^{NC0} and W_D^{C0} just equals the initial aggregate demand W_D^0 at each price.

Likewise, we divide aggregate supply by arbitrarily taking the share of supply that was just sufficient to satisfy W_D^{C0} prior to certification and allocating that share to the certified market; $W_S^{C0'}$ represents this portion of supply. This allocation leaves supply W_S^{NC0} that was just sufficient to satisfy W_D^{NC0} at the initial price P^0 (Figure 3). The price question hinges on whether the producers comprising these two segments of supply will voluntarily remain in their respective segments after certification begins.

First, consider the cost of certification. If certification is costly, then the supply curve $W_S^{C0'}$ for certified wood must pivot upward, as represented by the higher curve W_S^{C0}. These cost increases generate a new price for certified wood, as represented by the intersection of W_D^{C0} and W_S^{C0} (above price-line P^C), but the question becomes whether this price increase is sufficient to encourage all producers represented in supply curve W_S^{C0} to remain in the certified market. The outcome could be that even more producers choose to enter the certified market, and this case is represented by Figure 3.

Producers make their choice based on their surplus calculations in comparing the two options. If the higher price in the certified market generates sufficient revenues to cover the higher costs, then producers ''exit'' from the non-certified market and ''enter'' the certified market. This reallocation of suppliers establishes an equilibrium where non-certified supply has decreased to W_S^{NC1} and certified supply has increased to W_S^{C1}. Stated in terms of the marginal tradeoff that producers make, this equilibrium requires the following: a marginal increase in the share of suppliers in the certified market leads to an increase in certified producers' surplus that just offsets the decrease in non-certified producers' surplus that occurs when the share of producers in the non-certified market decreases, at the margin.[7]

[7] Barbier (1998) rightly points out that, in an international trade setting, exporting countries may respond to eco-consumer countries by shifting their exports

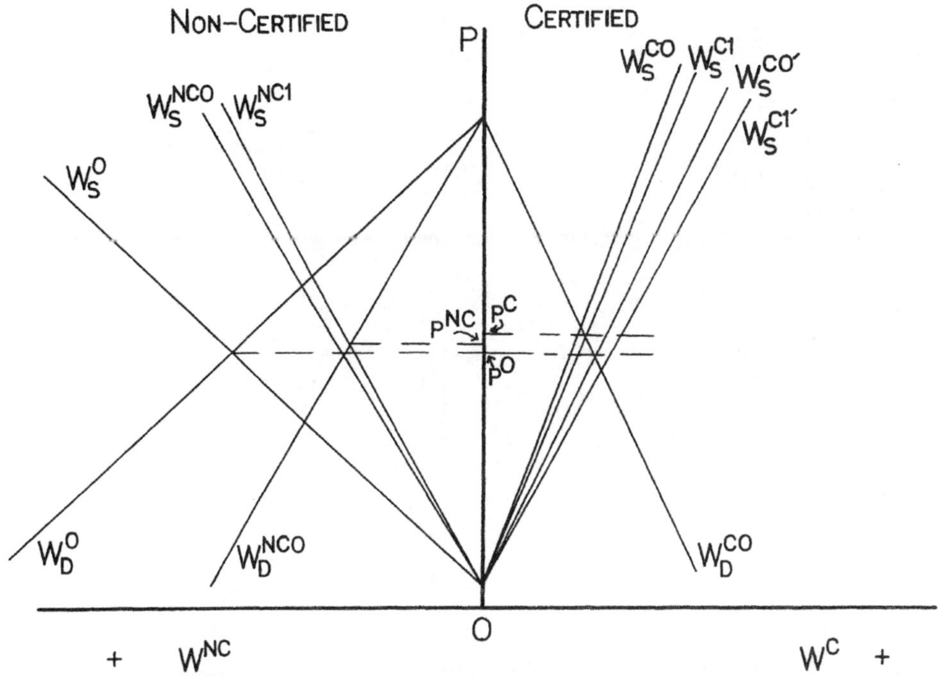

FIGURE 3
PARTIAL EQUILIBRIUM IMPLICATION OF VOLUNTARY CERTIFICATION: CASE 1A, WITH DISTINCT PRICES FOR CERTIFIED (P^C) AND NON-CERTIFIED (P^{NC}) WOOD AND ALL PRODUCERS GAIN

As illustrated in Figure 3, this marginal decision process equilibrates where the certified price, P^C, exceeds the non-certified price, P^{NC}, and both exceed the initial price, P^0. These conditions show that certification may attract a substantial share of production into a sustainable production system. This outcome, of course, captures the intuition of eco-labeling proponents. Moreover, and somewhat unexpectedly, even those producers who choose to remain non-certified will gain from the certification program. The non-certified firms gain because the certification costs have induced an effective supply restriction that raises average price, while non-certified production costs *of the individual suppliers* have remained unchanged.

Case 1b: Certified Price Higher than Non-Certified Price, and Some Lose

Figure 3, however, does not represent the only possible outcome of the marginal conditions for allocating producers between the market segments. Figure 4 represents an alternative case where all the notation is as de-

toward countries generating demand for non-certified products. While the scope of this paper omits the explicit application of our model to international trade issues, the roots of the model in international trade are obvious and one may modify the discussion of suppliers to tailor the analysis for north-south trade. In addition, we note that the partial equilibrium framework here is not sensitive to whether the particular good (wood, in our example) is directed at a specialty or "niche" market.

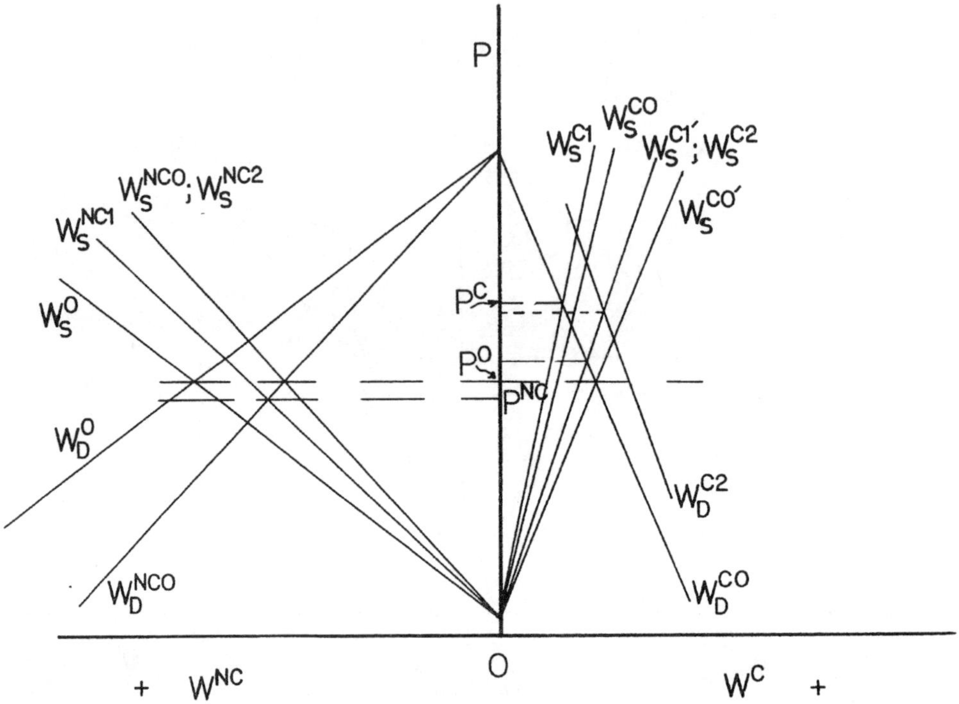

FIGURE 4
PARTIAL EQUILIBRIUM IMPLICATIONS OF VOLUNTARY CERTIFICATION: CASE 1B, WITH DISTINCT PRICES FOR CERTIFIED AND NON-CERTIFIED WOOD, BUT NON-CERTIFIED PRODUCERS LOSE

fined for Figure 3. We again begin after the initial division of aggregate supply W_S^0 between non-certified supply W_S^{NC0} and certified supply W_S^{C0}, which accounts for certification costs. Yet, the scenario in Figure 4 represents a case where the price increase is insufficient to compensate the marginal producer of certified wood. This case, then, represents a case where certification induces some of the producers to "return" to non-certified production, so that certified supply decreases to W_S^{C1} and non-certified supply increases to W_S^{NC1}, as compared to the initial (arbitrary) division of supply. These equilibrium supplies satisfy the eco-consumers' demand W_D^{C0}, establishing a certified price of $P^C > P^0$, and satisfy the non eco-consumers' demand W_D^{NC0}, establishing a non-certified price of $P^{NC} < P^0$.

The case verifies that voluntary certification may yield conditions under which non-certified producers lose, despite their opportunity to avoid certification costs. Their loss arises because demand and supply conditions in the certified market lead to a supply of non-certified product that is large relative to the demand from non eco-consumers remaining in equilibrium, thereby generating a price decrease for non-certified suppliers. While this case leads to two prices in the wood market's (partial) equilibrium, as proponents often argue, it also may lead non-certified producers to lobby against a strong certification program.

Case 1b with Eco-Consumers Willing to Pay a Premium

So far, we have assumed that voluntary certification only induces eco-consumers to establish a separate market share without increasing overall aggregate demand. If eco-consumers are also willing to pay a premium, then their initial demand W_D^{C0} is replaced by the higher demand W_D^{C2}. This certification-induced increase in demand by eco-consumers raises the incentive for producers to become certified. To avoid cluttering Figure 4, we illustrate the implications of the willingness-to-pay premium by assuming that the new equilibrium allocation of suppliers is established by a certified supply curve, W_S^{C2}, that just happens to coincide with curve W_S^{C1} (which, of course, is coincidence and an illustrative convenience). This implies that the corresponding supply of non-certified wood, W_S^{NC2}, happens to coincide with W_S^{NC0}.

Despite this restrictive assumption, Figure 4 now offers two additional insights. First, and somewhat surprisingly, Figure 4 shows that the premium willingness to pay can contribute to a *narrowing* of the price difference between certified and non-certified wood. In this illustration, the equilibrium prices are closer together because non-certified price increases as more suppliers move into (or as more stay in) the certified market and because the relative increase in certified supply causes the certified price to decrease.[8] Second, the premium willingness to pay reduces the chance that voluntary certification imposes losses on producers who choose a non-certified process. In the example illustrated, the equilibrium non-certified price would now occur at the intersection of W_S^{NC2} and W_D^{NC0} which is, by coincidence, the initial price. While this case (1b) is still favorable to eco-labeling proponents, there is no guarantee that eco-consumers will prevent the market from imposing a loss ($P^{NC} < P^0$) for non-certified suppliers.

Case 2: Certified Wood and Non-certified Wood at One Price

If the certified production process adds some cost to all producers who choose to comply with certification, then the foregoing results show that the market *may* generate a price differential between the certified and non-certified products. However, intuitive consideration can convince us that, if certification is always costly to producers, then the market *will* generate a price differential.

This outcome is easily established by consideration of the decision facing the marginal producer of certified wood. For example, suppose initial division of supply and demand establishes that the certified price equals the non-certified price (initial $P^C = P^{NC}$, and both markets clear). This situation cannot represent an equilibrium, because the marginal certified supplier would *refuse to stay* in the certified market. By exiting, this marginal supplier can jettison certification costs and receive the same price from the non-certified market. Of course, as the marginal supplier exits the certified market and enters the non-certified market, that entry-exit decision places an upward pressure on certified price and a downward pressure on non-certified price. Therefore, if certification is always costly to individual producers, the market will *always* produce a price differential with the certified price remaining higher ($P^C > P^{NC}$).

However, certification may *not* always be costly. Some segments of industry argue that they already produce a certifiable product either as a result of ownership philosophy, such as may prevail among some non-industrial forest landowners who compromise timber profits for aesthetic benefits, or as a result of existing government regulations that cause producers to follow strict environmental guidelines, as suggested for example by firms who win timber contracts on Canada's crown lands. If such landowner choices or environmental regulations impose production constraints on enough producers, such that their supply is sufficiently large relative to de-

[8] We note that it is possible that the equilibrium certified-supply, W_S^{C2}, could have ended up intersecting W_D^{C2} at a price *above* P^C. In such a case, which we do not illustrate, the price gap between certified and non-certified may not become narrower, although it is clear that the non-certified price when eco-consumers exhibit a premium willingness to pay will exceed P^{NC} in Figure 4. We do not illustrate this additional subcase here.

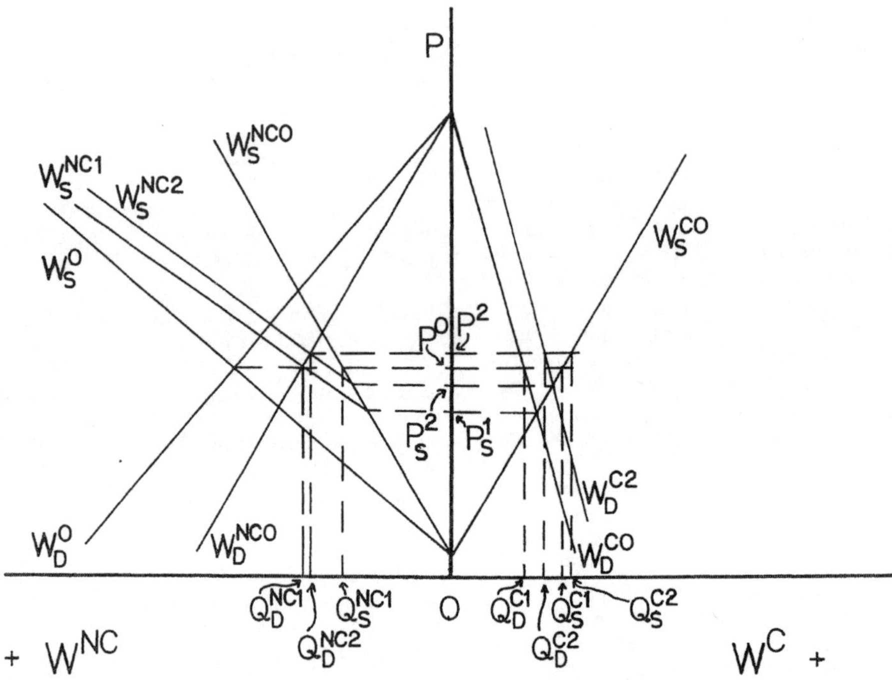

FIGURE 5

PARTIAL EQUILIBRIUM IMPLICATIONS OF VOLUNTARY CERTIFICATION: CASE 2, WITHOUT DISTINCT PRICES FOR CERTIFIED AND NON-CERTIFIED WOOD, AND NO PRODUCERS LOSE

mand of eco-consumers, then certification may fail to generate a price differential. However, in our example, certification may generate an increase in the price of all wood, in which case both certified wood and non-certified wood will sell at the same, higher price.

This case is illustrated in Figure 5, where notation is again as defined above. However, in this case, the initial allocation of supply between the market segments is not arbitrary; the allocation is driven by placing all firms that currently supply by certifiable methods within the supply segment captured by W_S^{C0}. In this case, the demand W_D^{C0} of eco-consumers intersects supply W_S^{C0} at a point below the initial (pre-certification) price P^0, while the intersection of non-certified demand, W_D^{NC0}, and non-certified supply, W_S^{NC0}, occurs above P^0. Since non-eco-consumers are indifferent between certified and non-certified wood, these consumers generate a demand for the excess supply of certified producers, beginning at supply price P_S^1. Defining this excess supply as the horizontal difference between W_S^{C0} and W_D^{C0}, this excess supply is added to the non-certified supply to obtain an aggregate supply available to non eco-consumers, W_S^{NC1}. Under these conditions, W_S^{NC1} constitutes a supply of both certified and non-certified product that is being sold to non-eco-consumers.

So far, however, we are assuming that certification generates a market separation, but W_D^{C0} assumes eco-consumers reveal no increased willingness to pay for certified wood. This process yields an equilibrium at the initial price P^0 where eco-consumers consume Q_D^{C1} from the larger certified quantity Q_S^{C1} and where non-eco-consumers consume Q_D^{NC1}

which is supplied by non-certified production Q_S^{NC1} and excess certified supply given by the difference $Q_S^{C1} - Q_D^{C1}$ (which equals $Q_D^{NC1} - Q_S^{NC1}$ in Figure 5).

If eco-consumers do reveal a willingness to pay a premium, then their demand curve shifts to W_D^{C2} after certification is institutionalized. In this case, the excess supply of certified product available for consumption by non-certified producers would be given by W_S^{NC2}. Accordingly, the common price would become $P^2 > P^0$, with eco-consumers consuming Q_D^{C2} and non eco-consumers consuming Q_D^{NC2}, including an excess of certified product given by the difference $Q_S^{C2} - Q_D^{C2}$. The implication, under partial equilibrium for this case, is a gain for all producers, since no costs have increased but the price received in both markets increases due to the higher demand of eco-consumers.

The condition leading to a single price for the two products requires that supply conditions already lead to a significant certifiable supply at no increase in marginal cost for those units; these conditions represent a type of production constraint. If this supply is insufficient to satisfy eco-consumers' demand at a price below the pre-certification price, then the production constraint would be nonbinding on the equilibrium process, and a version of Case 1 becomes relevant.

Comparing the Cases to Conditions before Voluntary Certification

These market results raise concerns that eco-labeling may reduce the profits of firms that choose to serve the noncertified market. Yet, if eco-consumers comprise a large enough share of the market relative to certification-induced increases in marginal production costs, then, under a voluntary system, the model indicates that even noncertified suppliers may gain. This outcome for producers is even more likely if current regulations have already created a significant supply of certifiable forest land.

IV. CONCLUDING DISCUSSION

The implications of this analysis are significant. A major finding is that there are reasonable circumstances in which some portion of consumers is willing to pay a price premium but a premium (price differential) will not arise in the market. An implication of the findings is that even if surveys that indicate that consumers are willing to pay a premium are correct, this is not a sufficient condition to generate a premium in the market. For a voluntary system, if demand for certified wood is small relative to overall demand, if the costs of certification are insignificant to some producers, and if the amount of new demand created by certification is modest, then the market is less likely to generate a price premium for the certified product, even if there are substantial numbers of consumers willing to pay a premium. However, to the extent that the costs of certification are positive to all or most producers and/or certification with labeling creates significant new demand, the two-price alternative is increasingly likely to be generated by voluntary market activities.

Additionally, the success of eco-labeling depends upon the motivation of forest landowners to favor or oppose eco-labeling. Under a mandatory system of eco-labeling, this issue hinges on whether the net increase in costs due to certification criteria may be offset by an increase in the equilibrium price for producers who retain land in forestry. The chance that producers gain, on net, is certainly greater when some consumers are willing to pay a price premium, especially if eco-consumer's willingness to pay a premium diminishes with the consumption of certified wood (cf. Figures 1 and 2). However, nothing in our analysis indicates that a favorable outcome is certain for producers.

Many proponents have opted to favor a voluntary system of certification, believing that producers who choose not to certify will experience no net loss. However, the outcome of a voluntary system is not unambiguous. The simple existence of a cadre of consumers with a willingness to pay for certified products does *not* guarantee that producers will receive price premiums for these products. Moreover, even in a partial equilibrium analysis, producers who (voluntarily) avoid the certification process can still lose if eco-consumers fail to generate a sufficient incentive to attract a large enough portion of suppliers to volunteer for certified production; the result could be a relative over-supply of non-certified production, leading to a loss for those producers.[9]

These results need not imply that the environmental community should abandon eco-labeling. Rather, the results suggest the need for a thorough empirical analysis of a proposed labeling system. Moreover, the results raise the stakes for finding a pragmatic balance between the costs that a labeling system imposes on producers, even if voluntary, and the benefits that consumers feel they obtain from purchasing certified products.

Finally, our presentation above treats certification as an all-or-nothing proposition, but proponents may evaluate the likely outcome of market feedbacks under various certification standards, each of which would improve ecological conditions "at the forest" to different degrees. These market feedbacks may affect not only producer costs but also both the demand for certified products by bringing in new consumers and the consumer's willingness to pay. These concepts suggest a need for research on consumers' preferences for certification standards and the consequent effects on eco-consumer's willingness to pay.[10] In short, attention to market feedback may clearly affect the degree to which promising, market-oriented tools successfully generate incentives for the conservation objectives of environmental organizations.

References

Anderson, Siwan, and Patrick Francois. 1997. "Environmental Cleanliness as a Public Good: Welfare and Policy Implications of Nonconvex Preferences." *Journal of Environmental Economics and Management* 34 (Nov.): 256–74.

Barbier, E. B. 1998. "The Economics of the Tropical Timber Trade and Sustainable Forest Management." In *Tropical Rain Forest: A Wider Perspective*, ed. F. B. Goldsmith. London: Chapman & Hall.

Canadian Chemical Producers' Association. 1996. *Does Responsible Care® Pay? A Primer on Unexpected Benefits of the Initiative.* Ottawa, Ontario, Canada: CCPA.

Forsyth, Keith, David Haley, and Robert Kozak. 1999. "Will Consumers Pay More for Certified Wood Products?" *Journal of Forestry* 9 (2): 19–22.

Harrison, K. 1998. "Talking with the Donkey: Cooperative Approaches to Environmental Protection." *Journal of Industrial Ecology* 2 (3): 51–72.

———. 1999. "Racing to the Top or the Bottom? Industry Resistance to Ecolabelling of Paper Products in Three Jurisdictions." February. Unpublished.

Holland, Daniel, and Cathy R. Wessells. 1998. "Predicting Consumer Preferences for Fresh Salmon: The Influence of Safety Inspection and Production Method Attributes." *Agricultural and Resource Economics Review* 27 (Apr.): 1–14.

Irland, Lloyd C. 1993. "Wood Products Face Green Marketing Era." *Wood Technology* 120 (2): 34–36.

Kiker, Clyde F., and Francis E. Putz. 1997. "Ecological [sic] Certification of Forest Products: Economic Challenges." *Ecological Economics* 20 (Jan.): 37–51.

Mattoo, Aaditya, and Harsha V. Singh. 1994. "Eco-Labelling: Policy Considerations." *Kyklos* 47 (Fasc. 1): 53–65.

Murray, Brian C., and Robert C. Abt. 1998. "Forest Certification: Timber Market Simulations for the Southeastern U.S.: Draft Working Paper." Working paper prepared for the U.S. Environmental Protection Agency, RTI Project Number 6687-1. Research Triangle Park, N.C.: Research Triangle Institute. May 8.

———. 2001. "Estimating Price Compensation Requirements for Eco-Certified Forestry." *Ecological Economics* 36 (1): 149–63.

Murray, Brian C., and James F. Casey. 1998. "An Economic Framework for Policy Analysis." Working paper prepared for the U.S. Environ-

[9] A less dramatic outcome could be that market equilibrium generates a price premium that is less than the percentage forecast by simple consumer surveys.

[10] Examples for the analogous issues surrounding food safety labels include Wessells and Anderson (1995), Wessells, Kline, and Anderson (1996), Holland and Wessells (1998). Note that while the present analysis focuses on the existence of a price premium for labeled products in partial equilibrium, we acknowledge also that to the extent that the outcome alters the profitability of timber production relative to other goods, the general equilibrium concerns of Swallow and Sedjo (1998) may lead to changes in the allocation and relative intensity of use of key ecological resources, such as forest land.

mental Protection Agency under Contract No. CR 824966-01-0. Research Triangle Park, N.C.: Research Triangle Institute.

Sedjo, Roger A., Alberto Goetzl, and Steverson O. Moffit. 1998. *Sustainability in Temperate Forests – Key Developments and Issues.* Washington, D.C.: Resources for the Future.

Stevens, James, Mubariq Ahmad, and Steve Ruddell. 1998. ''Forest Products Certification: A Survey of Producers and Manufacturers.'' *Forest Products Journal* 48 (June): 43–52.

Swallow, Stephen K. 1996. ''Resource Capital Theory and Ecosystem Economics: Developing Nonrenewable Habitats with Heterogeneous Quality.'' *Southern Economic Journal* 63 (July): 106–23.

Swallow, Stephen K., and Roger A. Sedjo. 2000. ''Eco-Labeling Consequences in General Equilibrium: A Graphical Assessment.'' *Land Economics* 76 (Feb.): 28–36.

Upton, Christopher, and Stephen Bass. 1996. *The Forest Certification Handbook.* Delray Beach, Fl.: St. Lucie Press.

Wessells, Cathy Roheim, and Joan Gray Anderson. 1995. ''Consumer Willingness to Pay for Seafood Safety Assurances.'' *The Journal of Consumer Affairs* 29 (Summer): 85–107.

Wessells, Cathy Roheim, Jeffrey Kline, and Joan Gray Anderson. 1996. ''Seafood Safety Perceptions and Their Effects on Anticipated Consumption under Varying Information Treatments.'' *Agricultural and Resource Economics Review* 25 (Apr.): 12–21.

Winterhalter, D., and D. L. Cassens. 1994. ''Is the Market Ready for Certified Sustainable Forest Products,'' West Lafayette, Ind.: Purdue University, Department of Forestry and Natural Resources.

World Wide Fund for Nature. 1991. ''An Assessment of Claims of 'Sustainability' Applied to Tropical Wood and Timber Retailed in the UK.'' London.

[3]

Can Labelling Policies Do More Harm Than Good? An Analysis Applied to Environmental Labelling Schemes

DOUADIA BOUGHERARA d.bougherara@enesad.inra.fr
Department of Economics and Management, University of Burgundy, UMR INRA – ENESAD CESAER, Agricultural Economics and Sociology Department, 26 Bd Dr Petitjean BP 87999 21079 DIJON Cedex, France

GILLES GROLLEAU g.grolleau@enesad.inra.fr
LUC THIÉBAUT thiebaut@enesad.inra.fr
UMR INRA – ENESAD CESAER, Agricultural Economics and Sociology Department, 26 Bd Dr Petitjean BP 87999 21079 DIJON Cedex, France

Abstract

In certain plausible circumstances, the introduction of labelling schemes can lead to adverse effects. In the case of ecolabelling, the adverse effects are an environmental degradation rather than an environmental improvement. To take into account the environmental sensitiveness or responsiveness of consumers, we introduce the concept of environmental elasticity which enables us to classify goods. In a basic analytical model, we describe the conditions under which different outcomes—overall impacts of change in environmental quality due to environmental labelling—arise after the introduction of an ecolabelling scheme. We show that an ecolabelling scheme can lead to an increase in purchases of environmentally sustainable products. The net effect on the environment can be worse than the initial situation without ecolabelling, because the environmental unit improvement is compensated by an over-consumption. We suggest several tests to detect this potential perverse effect, some policy implications to avoid it and stress the need for further research.

Keywords: labelling, ecolabelling, policy, environmental elasticity

JEL Classification: D11, L15, Q28

"I can sum up my notion of good practice in three injunctions.
1. Keep it simple.
2. Get it right.
3. Make it plausible."

Robert Solow (2001)

1. Introduction and review of the literature

The credible labelling of products is frequently considered as an accurate information-based policy to promote informed choices, notably in the case of credence products[1] such as light products, low-nicotine cigarettes, low-alcohol beers or protective solar filters. Following the same line, the ecolabelling of products is frequently presented as an appropriate, though partial policy tool to regulate environmental problems.[2] It is expected that ecolabels, by providing environmental information, will encourage consumers to discriminate between products and purchase products that have fewer environmental impacts than competing alternatives. Producers will have a greater incentive to manufacture environmentally friendly products. From a public point of view, the net expected effect will be an improvement in overall environmental quality (Morris and Scarlett, 1997).

Similarly to other papers (Mattoo and Singh, 1994; Dosi and Moretto, 2001), we argue that ecolabelling schemes could, in certain plausible circumstances, lead to an adverse effect on the environment. This paper differs from the previous ones by the source of the perverse effect considered and the analytical framework. Mattoo and Singh (1994) distinguish between concerned and unconcerned consumers about environmental problems. Ecolabelling may stimulate concern for the environment and increase the demand for environmentally friendly products. In certain plausible cases, they show that the differentiation of products due to ecolabelling may lead to increased sales of products made by environmentally friendly and environmentally unfriendly methods. Dosi and Moretto (2001) consider the effects of an ecolabelling scheme on the incentives of firms to innovate and to invest in green technology. They argue that an ecolabelling scheme could increase investment in the conventional polluting technology even if green consumption is stimulated by the ecolabelling scheme. It is the case if there are image spillovers, i.e. if the ecolabel projects a positive image not only on the environmentally friendly product but also on the conventional product.

This paper argues that the introduction of an ecolabelling scheme may encourage people to consume more, and in certain cases, the overall effect on the environment may be worse than the initial situation, i.e. the situation without ecolabelling scheme. A simple example will make clear the intuitive rationale for such a result. Say the market for two goods X and Y is in equilibrium respectively at prices $p_x = 10$ and $p_y = 5$ with quantities $x = 10$ and $y = 120$ for an income $R = 700$. For each good, one unit has a negative impact, $i_x = 1$ and $i_y = 0.2$. The overall environmental impact of all the products is then 34 ($10 \times 1 + 120 \times 0.2$). Now ecolabelling is introduced for the good X, which means that this product with fewer environmental impacts is launched at the same price as before. Its environmental impact is only of $i_x = 0.8$. *Let us further assume that the introduction of environmentally friendly products stops consumers feeling guilty and generates an increase in the consumption of the good X.* If this increase exceeds a certain level, in our example 15 units, the overall impact on the environment will be greater after the introduction of the ecolabelling scheme. The increase in the quantity of the good X will lead to a decrease of the quantity of good Y. For a constant income, the quantity of good Y is now 60 ($(700 - 10 \times 40)/5$). Indeed, if people consume now 40 units of good X and 60 units of good Y, then the overall impact is worse, i.e. 44 ($40 \times 0.8 + 60 \times 0.2$) which is superior to the initial level, i.e. 34. The same reasoning applies for an increase in the price of X. Although this example seems rather diagrammatic, these potential perverse effects correspond to a real concern of practitioners and policy-makers (Byron, 2001:6; European Environmental Bureau, 2003:11).[3] In other words, the environmental improvement per unit of product is cancelled by a consumption increase. In our reasoning, consumers care about the environmental impact per product unit and do not compute the overall impact of their consumption.[4] Manufacturers encourage consumers to buy the highest quantity of their products in order to maximise their profits. Consequently, manufacturers emphasise the individual qualities of their products, notably in their advertising. They have little incentives to highlight the potentially adverse effects of the total consumption.[5]

The remainder of the paper is organised as follows: In the next section, the concept of environmental elasticity is introduced and some empirical tests in order to characterise the properties of goods are suggested. In Section 3, we examine the possible outcomes in terms of environmental quality after the introduction of an ecolabelling scheme in a basic model. We identify the conditions under which potentially 'perverse effect' of introducing ecolabelling schemes can arise. Section 4 outlines the relevance of the results for policy makers. Section 5 provides some concluding remarks, notably by stressing numerous potential extensions of this paper.

2. Environmental elasticity of demand and the classification of goods

The environmental elasticity ε of demand for a given product measures the responsiveness of demand for this product to a given change in the environmental impact i of the product, or in the level of pollution, due to the production of the good. Using the conventional equation in the case of a continuous and derivable function, the environmental elasticity ε is given by:

$$\varepsilon = \frac{\frac{dx}{x}}{\frac{di_x}{i_x}} \tag{1}$$

dx corresponds to the variation in the quantity exchanged. For a discrete function, it is the difference between the quantity exchanged of environmentally friendly products (x_e) and the quantity exchanged of conventional products (x_c), before the introduction of the ecolabelling scheme ($\Delta x = x_e - x_c$).

di_x corresponds to a small change in the environmental quality of the products. For a discrete function, it is the difference between the environmental impact (or pollution) generated by the environmentally friendly product (i_x^e) and the environmental impact generated by the conventional product (i_x^c) ($\Delta i_x = i_x^e - i_x^c$). Note that $\Delta i_x < 0$ because we postulate an environmental improvement by diminishing the value of i_x, i.e. $i_x^e < i_x^c$.

In our introductive example, the elasticity is negative and its absolute value greater than one, i.e. -15. In a first approach, we consider that ecolabelling schemes only affect the environmental impacts (and the environmental perception) generated by the product and leaves unchanged the other qualities of the good.

According to the concept of environmental elasticity, we propose an intuitive classification of goods in Table 1. For example, we can intuitively suppose that the demand for clothes is not influenced by the environmental qualities of the clothes ($\varepsilon = 0$) while the demand for paper will increase substantially ($|\varepsilon| < 1$) because of the introduction of an environmentally friendly paper, such as recycled paper.[6]

The classification of goods according to their environmental elasticity is particularly relevant for policy makers who aim at maximising the efficiency of funds invested in ecolabelling schemes. This point is briefly developed in Section 4. An empirical way to determine the environmental elasticity of goods is to realise well-designed surveys to simulate the reaction of consumers to the launch of ecolabeled products. Another way is to exploit data related from products launched for their environmental friendliness. A third way to test this environmental elasticity is to achieve pilot studies in supermarkets by proposing identical products (same price, brand name and other properties) with added environmental qualities.

Table 1. Types of goods according to their environmental elasticity.[a]

Value of the elasticity	Demand	Effect of a decrease in the environmental impact i_x on the demand x
$\varepsilon < 0$		
$\varepsilon < -1$	Elastic	x increases proportionally more than the proportional change in i_x
$\varepsilon = -1$	Unit elasticity	x increases proportionally as much as the proportional change in i_x
$-1 < \varepsilon < 0$	Inelastic	x increases proportionally less than the proportional change in i_x
$\varepsilon = 0$	Perfectly inelastic	No change in the consumed quantity x
$\varepsilon > 0$		
$0 < \varepsilon < 1$	Inelastic	x decreases proportionally less than the proportional change in i_x
$\varepsilon = 1$	Unit elasticity	x decreases proportionally as much as the proportional change in i_x
$\varepsilon > 1$	Elastic	x decreases proportionally more than the proportional change in i_x

[a]In developed countries, people have a high level of satiation with many purchased products and it is possible to argue that we will not really see significant changes in quantities purchased with most products. However, much of the argumentation developed in this article relies on the presence of obstacles leading people to limit their consumption. If these obstacles disappear (e.g. feeling guilty in the case of unsustainable consumption, fear of diseases related to solar exposition or nicotine level in cigarettes), people can naturally choose to increase their consumption.

These different ways are neither exhaustive nor mutually exclusive and can be combined to get better results.

3. The model

The basic results can be demonstrated in a very simple model. To keep the analysis simple and isolate the main effects, we consider, in a first approach, two goods $X$and Y. Each good leads to an environmental impact per unit, or a level of pollution, respectively i_x and i_y. We consider that good X is awarded an ecolabel, after the adoption of environmental criteria leading to a decrease in i_x. We assume several restrictive hypotheses: (1) All consumers are equally concerned about the environment. Their preferences are perfectly homogeneous. (2) The perceived change in the environmental performances of the product shapes the quantities demanded. The perceived change corresponds to the real difference in i_x. We admit that marketing strategies and other parameters can influence the perceived environmental improvement. This effect would be captured by the concept of environmental elasticity. (3) Each consumer perfectly perceives the environmental improvement per unit of product but does not care of the overall environmental impact of his total consumption. This hypothesis means that the consumer does not compare the overall environmental impact or pollution level before and after the introduction of the ecolabelling scheme, but only takes into account the marginal improvement generated by the products. (4) Manufacturing ecofriendly products frequently implies higher production and transaction costs and consequently a price higher than the price of the good before ecolabelling.

Under those hypotheses, we are interested in determining the overall impact I on the environment of the consumption of good X and good Y. I is a function of each good's quantity and each good's environmental impact so that we can write: $I = I(x, i_x, y, i_y)$. In our model, i_y is constant. Thus, $I = I(x, i_x, y)$. The total differential of I is:

$$dI = \frac{\partial I}{\partial x}dx + \frac{\partial I}{\partial i_x}di_x + \frac{\partial I}{\partial y}dy \quad \text{with} \quad \frac{\partial I}{\partial x} > 0, \frac{\partial I}{\partial y} > 0 \quad \text{and} \quad \frac{\partial I}{\partial i_x} > 0 \tag{2}$$

Each demand function x and y is in turn a function of several parameters. As demand functions, x and y are functions of prices and income: p_x, p_y and R. In our model we have an additional variable: the environmental impact per unit, so that not only x and y are a function of respectively i_x and i_y but the price of good X also depends on i_x. Demand functions are then: $x = x(p_x(i_x), p_y, R, i_x)$ and $y = y(p_x(i_x), p_y, R, i_y)$. In our model, the price of the good Y, p_y, the income, R, and the environmental impact of good Y, i_y, are constant.

$$x = x(p_x(i_x), i_x) \tag{3}$$

$$y = y(p_x(i_x)) \tag{4}$$

We rewrite I as $I = I[x(p_x(i_x), i_x), i_x, y(p_x(i_x))]$

$$(3) \Rightarrow \quad dx = \frac{\partial x}{\partial p_x}dp_x + \frac{\partial x}{\partial i_x}di_x \quad \text{with} \quad \frac{\partial x}{\partial p_x} < 0 \quad \text{and} \quad \begin{cases} \dfrac{\partial x}{\partial i_x} > 0 & \text{for } \varepsilon < 0 \\ \dfrac{\partial x}{\partial i_x} < 0 & \text{for } \varepsilon > 0 \end{cases} \tag{5}$$

$$(4) \Rightarrow \quad dy = \frac{\partial y}{\partial p_x}dp_x \quad \text{with} \quad \frac{\partial y}{\partial p_x} < 0 \tag{6}$$

$$p_x = p_x(i_x) \Rightarrow dp_x = \frac{\partial p_x}{\partial i_x}di_x \quad \text{with} \quad \frac{\partial p_x}{\partial i_x} < 0 \tag{7}$$

With (5), (6) and (7) in (2):

$$\begin{aligned} dI &= \frac{\partial I}{\partial x}\left[\frac{\partial x}{\partial p_x}\frac{\partial p_x}{\partial i_x}di_x + \frac{\partial x}{\partial i_x}di_x\right] + \frac{\partial I}{\partial i_x}di_x + \frac{\partial I}{\partial y}\frac{\partial y}{\partial p_x}\frac{\partial p_x}{\partial i_x}di_x \\ dI &= \frac{\partial I}{\partial x}\frac{\partial x}{\partial p_x}\frac{\partial p_x}{\partial i_x}di_x + \frac{\partial I}{\partial x}\frac{\partial x}{\partial i_x}di_x + \frac{\partial I}{\partial i_x}di_x + \frac{\partial I}{\partial y}\frac{\partial y}{\partial p_x}\frac{\partial p_x}{\partial i_x}di_x \end{aligned} \tag{8}$$

Equation (8) shows that the variation in the overall environmental impact due to a decrease in i_x ($di_x < 0$) is an addition of four effects:

A Price Effect. The first term of Eq. (8), $\frac{\partial I}{\partial x}\frac{\partial x}{\partial p_x}\frac{\partial p_x}{\partial i_x}di_x$, describes the effect on the overall impact I of a decrease in the quantity of good X due to an increase in its price p_x due

Table 2. Total effect of a decrease in i_x on the overall environmental impact I according to the environmental elasticity.

	Value of the environmental elasticity		
Effects on I	$\varepsilon < 0$	$\varepsilon = 0$	$\varepsilon > 0$
Price effect	–	–	–
Environmental elasticity effect	+	0	–
Direct effect	–	–	–
Revenue share effect	–	–	–
Total effect	?	–	–

Decrease in the overall impact I, 0: No change in I, +: Increase in I.

itself to a decrease in its environmental impact per unit, i_x. The *Price Effect* is negative, which means a decrease in the overall impact I.

An Environmental Elasticity Effect. The second term of (8), $\frac{\partial I}{\partial x}\frac{\partial x}{\partial i_x}di_x$, corresponds to the effect on I of an increase in the quantity of good X due to a decrease in its environmental impact per unit, i_x ($\varepsilon < 0$). The *Environmental Elasticity Effect* is positive when $\varepsilon < 0$ and is negative when $\varepsilon > 0$.

A Direct Effect. The third term of (8), $\frac{\partial I}{\partial i_x}di_x$, is the direct effect of a decrease in good X's environmental impact on the overall environmental impact I. The *Direct Effect* is negative.

An Income Share Effect. The last term of (8), $\frac{\partial I}{\partial y}\frac{\partial y}{\partial p_x}\frac{\partial p_x}{\partial i_x}di_x$, measures the effect on I of a decrease in the demand for good Y due to an increase in the price of good X, due itself to a decrease in the environmental impact per unit of good X, i_x. The *Income Share Effect* is negative.

A perverse effect on the environment may appear when the *Environmental Elasticity Effect* is positive and overcomes the other effects (Table 2).

An important implication is that an adverse effect is more likely to occur if the ecofriendly product can be produced with low cost technologies and if consumers value the environmental improvement per unit of product. Indeed, if reaching the ecolabel criteria is accessible at low costs, the *Price Effect* and the *Income Share Effect* are low and give an advantage to the *Environmental Elasticity Effect*, i.e. a consumption increase due to the environmental improvement. If this consumption increase exceeds a certain level, it will generate an adverse effect on the environment.

We are able to determine the optimal quantities of good X and good Y as consumers are utility maximizers under a budget constraint:

$$\begin{cases} \text{Max}\,U(x(i_x), y) \\ p_x x + p_y y = R \end{cases}$$

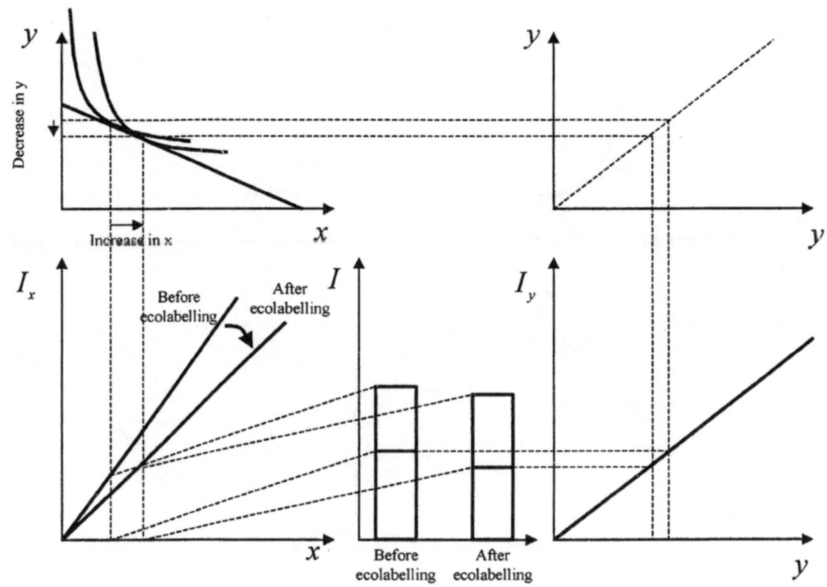

Figure 1. Lower overall environmental impact after ecolabeling for p_x constant.

We write the first order conditions (Lagrangian function):

$$\begin{cases} \dfrac{\partial L}{\partial x} = \dfrac{\partial U}{\partial x} - \lambda p_x = 0 \\ \dfrac{\partial L}{\partial y} = \dfrac{\partial U}{\partial y} - \lambda p_y = 0 \\ \dfrac{\partial L}{\partial x} = -(p_x x + p_y y - R) = 0 \end{cases}$$

According to the chain rule for derivation, $\frac{\partial U}{\partial i_x} = \frac{\partial U}{\partial x} \times \frac{\partial x}{\partial i_x} + \frac{\partial U}{\partial y} \times \frac{\partial y}{\partial i_x}$. Since y is independent of i_x, $\frac{\partial y}{\partial i_x} = 0 => \frac{\partial U}{\partial i_x} = Um_i = Um_x \times \frac{\partial x}{\partial i_x}$.

According to the definition of environmental elasticity, $\dfrac{\partial x}{\partial i_x} = \dfrac{x\varepsilon}{i_x}$.

$$Um_i = Um_x \times \frac{x^*\varepsilon}{i_x} \quad \text{then: } x^* = \frac{Um_i}{Um_x}\frac{i_x}{\varepsilon}$$

The budget constraint gives y^*:$y^* = \frac{R - p_x x^*}{p_y}$ then: $y^* = \frac{R}{p_y} - \frac{p_x}{p_y} \times \frac{Um_i}{Um_x} \times \frac{i_x}{\varepsilon}$ (x^*, y^*) constitute the optimal quantities of goods.

Let us take a simple case: the price of the good X remains constant. Figure 1 shows the move of the equilibrium quantities after ecolabelling and the consequences on the overall impact on the environment I as a sum of I_x, the environmental impact due to good

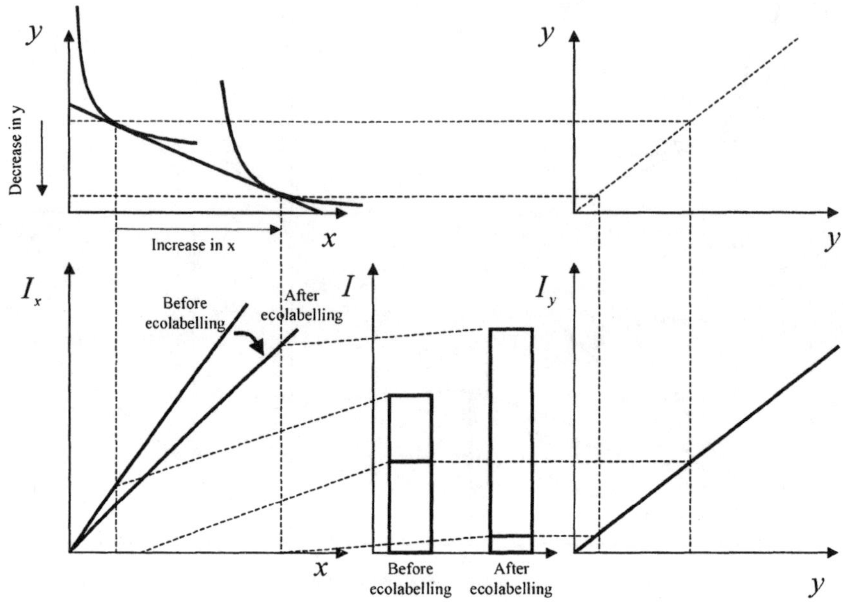

Figure 2. Adverse effect: Higher overall environmental impact after ecolabeling for p_x constant.

X consumption and I_y, the environmental impact due to good Y consumption. Since p_x is constant, the variation in the overall environmental impact is only determined by the *Environmental Elasticity Effect* and the *Direct Effect*. An increase in x leads to a decrease in y. We consider a decrease in i_x after ecolabelling represented by the slope of the curve I_x: $I_x = xi_x$. The unit impact decrease leads to a shift downwards. In figure 1, a decrease in i_x leads to a decrease in I. The introduction of an ecolabel may also lead to an adverse effect if the overall impact after ecolabel introduction is higher than the overall impact before ecolabel introduction (figure 2). The adverse effect, which may appear at the individual level, is still true at an aggregate level. Indeed, there exists a threshold quantity $\tilde{x}$ for which the overall environmental impact is unchanged after the introduction of the ecolabel.

Let us now consider the case where the price of good X varies. Figures 3 and 4 show the change in the equilibrium quantities after ecolabelling due to the decrease in i_x and an increase in p_x. The variation in the overall environmental impact is determined by the four effects identified in Table 2. A decrease in i_x after ecolabelling may lead to a lower overall environmental impact as in figure 3 or to a higher overall impact as in figure 4. In figure 4, the introduction of an ecolabel has lead to an adverse effect: the *Environmental Elasticity Effect* has overcome all the other effects.

4. Policy considerations for introducing ecolabelling schemes

There are many thousands of product categories available in the marketplace. The costs of generating a specific level of environmental improvement are not the same for all product

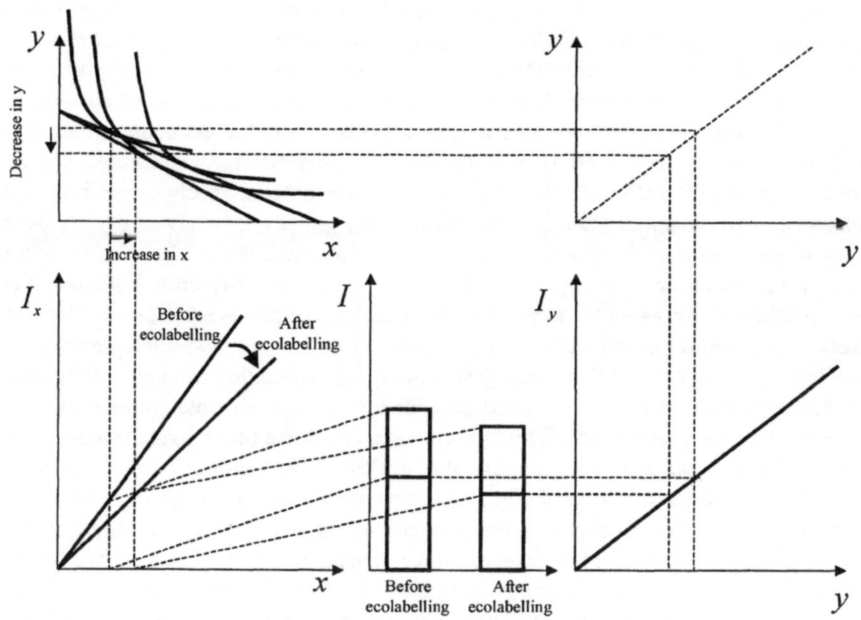

Figure 3. Lower overall environmental impact after ecolabeling when p_x increases.

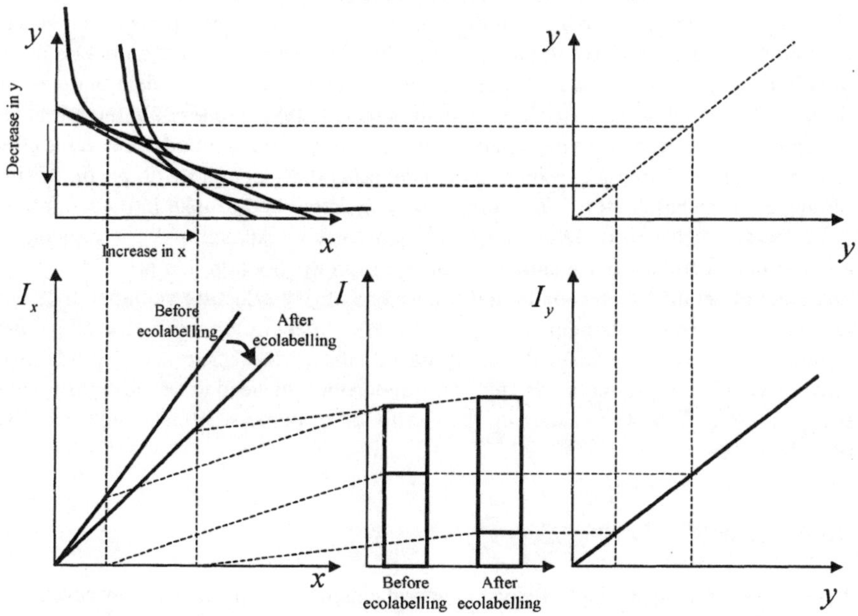

Figure 4. Adverse effect: Higher overall environmental impact after ecolabeling when p_x increases.

categories. If we postulate that consumers want to allocate a fraction of their resources to purchase environmentally friendly goods, they will prefer choosing product categories, which will make the most significant improvement to the environment for a given contribution. Indeed, consumers have limited resources and will prefer the most cost-effective solutions. To maximize the environmental effectiveness and the economic efficiency of ecolabels, policy designers need to select the product categories for which ecolabels would make the most significant improvement to the environment. Conversely, profit maximising firms could be interested in awarding environmentally goods characterised by an elasticity leading to a consumption increase. Policy makers include here various kinds of institutions such as governments or environmental associations, which are frequently initiators of ecolabelling schemes. Knowing the environmental elasticity of goods can be useful for policy makers who can select product categories according to the overall expected environmental net effect. Using the empirical tests suggested in the second section can help policy designers to better allocate scarce resources and generate higher environmental improvement, or at least avoid a hidden perverse effect by introducing an ecolabel for an inappropriate product category. Our analysis does not conclude that policy makers have to ban certain product categories, but stresses the need of complementary measures e.g. educating consumers or taxing products to prevent over-consumption in certain cases. Moreover, there is also a temporal dimension in the policy intervention because ecolabelling criteria are regularly reviewed to strengthen the environmental effectiveness of the scheme (Morris and Scarlett, 1996). In a dynamic perspective, we can suppose that meeting ecolabel criteria will be accessible at low cost in the first period. In the following periods, the costs of meeting the updated criteria will generally increase, leading to a stronger price effect. Consequently an adverse effect is more likely to arise in the starting period, where the price effect is weak. That means that regulators have to take into account this timing. Indeed, an intervention is potentially more effective in the first stages of the ecolabel development than in the following ones. In the same line of reasoning, regulators have to take into account the initial level of consumption, which can provide a partial indication on the likelihood of perverse effects

Note that measures aiming at reducing demand potentially conflict with profit maximising strategies of manufacturers. Of course policy makers need to take into account other related effects, which will together shape the quantities consumed and consequently the net environmental effects. For example, as analysed in the Section 3, a price increase can attenuate the effect due to environmental improvement. By selecting ecolabel criteria accessible at high costs, government can reduce the risk of generating a perverse effect on the environment. However, high costs of complying with the ecolabel criteria may hinder firms' incentives to have their products awarded. This underlines the need to define environmental criteria as a trade-off between lowering the likelihood of perverse effects and maximising firms' incentives.

5. Conclusion

We have introduced the concept of environmental elasticity[7] and showed how ecolabelling schemes can generate an adverse effect on the environment. The considered source of the 'perverse effect' differs from the previous literature, notably in introducing psychological considerations in the decision making process of consumers. The impact of immediate claims—such as 'less pollutant, less toxic per unit'—influences the decision making process more than computed information about the individual or global overall environmental impact. Indeed, this complex computation consumes time, knowledge and expertise abilities. Moreover, from a public point of view, ecolabelling schemes can be more environmentally effective and economically efficient for certain categories than others. In other words, the selection of product categories matters. In some cases, ecolabelling schemes can need complementary devices to achieve their promises such as consumer education. Although we have focused our attention on ecolabelling schemes, the framework and the results are generic and can be easily applied to many other topics, such as light products, low-nicotine cigarettes,[8] low-alcohol beers or protective solar filters.[9] These products also provide a favourable field for empirical investigation, e.g. the evolution of consumption after the introduction of low-nicotine cigarettes and the combined effect of simultaneous quality and price changes. The main results have been derived under very simplifying assumptions. Many extensions can be analysed such as the combination of other related effects. Indeed, this paper has assumed consumers and firms behave somewhat rigidly by postulating identical consumers' preferences and the sole presence of the environmentally friendly product. All consumers do not react in the same way to the introduction of ecolabelling schemes. The co-existence of two identical products differing only by their environmental quality and two kinds of consumers—unconcerned consumers and concerned ones—will make the model closer to the real world. Moreover, our reasoning is based on an informational policy instrument, which is market anchored and do not consider an eventual lifestyle change, e.g. substituting bike trips for car trips. Extending this setting and testing it empirically is a challenging topic for future research.

Acknowledgments

We are indebted to Carl Gaigné for his numerous detailed comments and constant encouragement. We greatly appreciate the comments of the three referees. We are also grateful to Gabriel Lecat, Pascale Lofredi, Julie A. Caswell and Naoufel Mzoughi for stimulating conversations. The usual disclaimer applies.

Notes

1. Credence goods are goods for which consumers cannot determine quality either through search, inspection or experience. Consumers cannot determine quality even after they buy and consume the product. Frequently for these goods, one must rely on a third party (third party certification or governmental regulations) to provide credible information to the consumer whether the product is high quality or not.
2. To make the exposition clear, we deal especially with ecolabelling, but the approach can be easily generalized to numerous other applications. Anyway, we deal with labelling schemes indicating to consumers attributes of the product having less negative impacts on diverse targets like the environment or health.
3. Byron, Neil, 2001, Environmental Certification and Labelling, The Common Property Resource Digest, 56, p.5-6. European Environmental Bureau, 2003, An Environmental NGO Vision Paper Towards a European Integrated Product Policy, 13p, March 2003.

4. Note that we do not consider the overall environmental impact from the consumer viewpoint, but from the state viewpoint, the state wanting to minimize the overall environmental impact. Indeed, we only assume that the consumer is interested by minimizing the impact per product unit consumed, regardless of the individual and global overall impacts. Consequently, the consumer only takes into account the environmental impact per unit and does not compute either the aggregated impact of his total consumption or the overall impact resulting from all consumers' decisions.
5. Several studies stress the potentially perverse effects resulting from diverse labelling schemes. For example, a French report of the National Council of Consumption (Rapport du Conseil national de la Consommation sur les produits cosmétiques de protection solaire, 2000) explains that the labelling of higher solar protection makes "consumers think they are well protected, notably products labeled with high Sunburn Protection Factor (...). Consumers tend to go out in the sun longer and adopt a risky behavior for their health".

   We find the same line of reasoning for nutritional and functional claims, where the labelling of a specific improvement can lead to an over-consumption (See for example: Parliamentary Assembly of the European Union, 2002, Functional food : serving the interests of the consumer on the food industry? and European Commission, 2001, Discussion Paper on Nutrition Claims and Functional Claims). Lastly, according to Dr Pellae, "a study of Pr Apfelbaum has showed that people eat twice more because the products are light. When certain people read a claim "light" on a foodstuff, they perceive it like a miraculous product to lose weight (...)." (www.doctissimo.fr/html/nutrition/mag_2002/mag0607/ nu_5559_faux_sucres_danger_itw.htm).

   Another meaningful and well documented example relates to cigarettes labelled as "safer". Several studies show that these "safer" cigarettes frequently do more harm than good. According to a literature review (The Cigarettes Companies and "Safer" Cigarettes–A Long History of Exploiting Consumers Health Concerns to Keep Them Smoking, 2000, http://tobaccofreekids.org/ research/factsheets/), "soon after the cigarette companies began selling low tar and nicotine cigarettes, it became clear the smokers that switched to these "light" brands were compensating for the lower nicotine levels by adopting new ways of smoking the cigarettes so they would still consume the same amounts of nicotine as they would with regular brands."
6. Marketing literature uses the concept of salient attributes which are attributes taken into account in the purchase of a good. We could say that generally environmental attributes are not as salient for clothes as they are for paper.
7. In many cases related to credence goods, note that the elasticity is based on consumers' perception resulting more from the information tool (i.e. labelling) than the objective perception of the quality change.
8. Despite mandatory warnings, it is broadly admitted that tobacco firms have tested and devised manipulative strategies to lower consumer risk perceptions and increase demand, e.g. by introducing low-nicotine cigarettes.
9. Note that ecolabelling relates to the private provision of a public good, i.e. an environmental quality improvement while other labelling schemes (e.g. low nicotine cigarettes or protective solar filters) mainly relate to the provision of a private good. An intuitive implication is that properties described in the case of ecolabelling will be more emphasized in the case of labelling schemes promoting private benefits.

References

Dosi, C. & Moretto, M. (2001). "Is Eco-labelling a Reliable Environmental Policy Measure?" *Environmental and Resource Economics*. 18(1), 113–127.

Mattoo, A. & Singh, H. (1994). "Eco-Labelling: Policy Considerations." *Kyklos*. 47(1), 53–65.

Morris, J. & Scarlett, L. (1996). "Buying Green: Consumers, Product Labels and the Environment." *Policy Study*, 217, November, Reason Foundation, Los Angeles, CA.

[4]

Ecocertification, differentiation in retailing and upstream market power

Lisette Ibanez* and Jacques Laye

UMR ENGREF-INRA LEF,
14 rue Girardot,
Nancy Cedex, 54042 France
Fax: +33-3-83-37-06-45 E-mail: ibanez@nancy-engref.inra.fr
E-mail: laye@nancy-engref.inra.fr
*Corresponding author

Abstract: The purpose of ecocertification in the forest sector is to guarantee a sustainable management of the forest resources by the producers (forest owners) and define production rules for the transformation industry. We propose an analysis of the ex-post impact of ecocertification, when it is, on the one hand, a way for retailers to differentiate their products, that may lead, on the other hand, to a reinforcement of producers' market power. We build a model of vertical relationships where two supply chains are in competition, one sells ecocertified wood products, whereas the other sells standard wood products. We show that certification is always profitable for the ecocertified retailer and welfare enhancing. However, if ecocertification works as a catalyst for coordination among producers, the ecocertified retailer can be at a disadvantage if the differentiation between ecocertified and standard wood products is not clear enough in the eyes of the consumer. Interestingly, in this case, only the producers and the retailer of standard wood products benefit from the ecocertification initiated by the ecocertified retailer.

Keywords: wood industry; ecocertification; product differentiation; upstream market power.

Reference to this paper should be made as follows: Ibanez, L. and Laye, J. (XXXX) 'Ecocertification, differentiation in retailing and upstream market power', *Int. J. Agricultural Resources, Governance and Ecology*, Vol. X, No. Y, pp.XXX–XXX.

Biographical notes: Lisette Ibanez is a Researcher at the French National Institute for Agricultural Research. She worked on various eco-labelling issues concerning both theoretical analysis through industrial economics as well as applications on agricultural and forest certification.

Jacques Laye received a PhD in Economics of the Laboratory of Econometrics (Ecole Polytechnique, Paris, France). He is a Researcher at the Laboratory of Forest Economics (LEF-Inra, Nancy, France). He works on industrial organisation and computational economics (oligopolistic competition, vertical relationships, product differentiation and coalition theory) applied to sectors like the wine industry, the paper industry or internet retailing. Other fields of research concern operations research (stochastic control applied to the management of renewable resources) and agent-based computational economics.

1 Introduction

Since the Earth Summit of Rio in 1992, a growing awareness of the over-exploitation of forests and its consequences on the environment led to the emergence of ecocertification labels.[1] The purpose of this ecocertification is designed to guarantee the sustainable management of the forest resources which takes into account the environmental and social functions of forests, along with the economical ones. So far, 40% of the forests in Europe have been ecocertified (FAO, 2003). The certification process has provided access to a larger database and a better information on the condition of forests in the world (Siry et al., 2005). Using the available data and a partial equilibrium model, Schwarzbauer and Ramesteiner (2001) examine the economical consequences of forest ecocertification on Europe's wood markets. Their model brings to the fore contrasting effects. They argue that forest certification leads either to a production cost increase for intermediary products (in order to take into account extra expenses related to traceability) or to a smaller wood supply. The authors consider a single homogeneous wood product, while, in our model, we distinguish between certified wood and standard wood.

According to Roda (2001), there are two main aspects in the concept of ecocertification of wood products. The first one is related to the environmental quality. The second has to do with tracing the origins of products and controlling the industrial process. In our model, we focus on the first aspect and suppose perfect certification.[2] Various works have also examined the impact of wood ecocertification from the perspective of vertical differentiation. It should be noticed that neither there is definite conclusion on the willingness to pay for the sustainable management of forests nor on the assessment of this extra quality. In Ibanez and Pourcher (2001), the increase in the consumers' willingness to pay for an ecocertified product is estimated to be not more than 5%. Nevertheless, Cason and Gangadharan (2002) give empirical evidence that consumers are willing to pay higher prices for products that cause less damage on the environment. Although this market is rather limited, the rapid development of ecocertified forest areas seems to indicate that the actors of the wood supply chain are ready to respond to this potential market.

Other Papers study the motivations for the actors of the wood industry to willingly adopt sustainable management, focusing on the expenses and profits related to certification (Haener and Luckert, 1998; Takahashi et al., 2003). Sedjo and Swallow (2000) suggest that the demand for certified and non-certified wood will be affected by how consumers respond to non-timber substitutes. Costa and Ibanez (2002) confirm that the introduction of sustainable forest management may well decrease the demand for timber. However, none of those studies look into the indirect effects on the supply chain. Indeed, ecocertification affects the wood sector either at the resource level (species and qualities of wood available and in demand, etc.) or at the industry level (inter- and intra-market competition, organisation among actors, comparative advantage, etc.).

Because of the management, information, planning and monitoring systems generally required to achieve certification, larger organisational units benefit from economies of scale over smaller units (Meidinger et al., 2003). Therefore, individual actors (i.e. private forest owners) with well-defined and identical interests in terms of feedback on sustainable forest management have implemented group certification. Group certification is the process of assessing and certifying groups of forest owners under a single certificate. It allows individual forest owners to benefit from economies of scale as a consequence of being part of a larger organisation. The advantages of group certification include the sharing of information and the mutual support offered among its members as well as increased efficiency in implementing the standard and achieving certification by an independent auditor. We can thus suppose that the implementation of group certification allows coordination among private forest owners, which is one of the features of our model.[3]

An empirical study (Ibanez and Laye, 2005) showed that 53% of the certified firms of the French wood supply chain were actually urged to move on to ecocertification by their customers, for fear of boycotting. Whatever the motivation of these customers (some major retailers) to differentiate their products is, it is their relatively high market power that enables them to put pressure on producers, still very poorly organised and competing with each other. We may then wonder about the ex-post impact of ecocertification on the market power of the actors of an ecocertified supply chain. More precisely, if ecocertification becomes a kind of catalyst for a coordination that would reinforce the market power of wood producers and not only a way for retailers to differentiate their products, how will profits then be shared out in the wood supply chain?

In order to address those issues, we build a model of vertical relationships with two independent supply chains, each one consisting of upstream producers (forest owners) and a single retailer. One of the supply chains sells ecocertified wood products, while the other sells standard wood products. The consumers in the final market are considered to be heterogeneous in their taste for environmental quality. Forest owners (producers) have to decide whether to supply standard or ecocertified wood products. The decision will always depend on the degree of differentiation and, eventually on the possibility of horizontal merger upstream the ecocertified retailer. In a first scenario, we consider a game in which the standard wood producers decide whether or not to enter the 'group' of ecocertified producers and compete with them. In a second scenario we add, for the ecocertified producers, the possibility to coalesce and behave as an upstream monopoly. We take as a benchmark a situation in which the two supply chains produce standard wood products. After computing the optimal proportion of ecocertified producers, we compare the profits and the consumers' surplus in the two scenarios described above (competition or coordination among ecocertified producers) with the benchmark situation. We show that the coordination scenario can turn to the disadvantage of the ecocertified retailer and the consumers (compared to the benchmark situation), if the latter do not perceive enough differentiation between the two types of products. Interestingly, in this case, only producers and the standard retailer benefit from the ecocertification initiated by the ecocertified retailers.

Other contributions use similar frameworks and study concentration phenomena in vertical structures. Gaudet and Van Long (1996) study vertical mergers in a two-level oligopoly. The incentives for vertical integrations are found to depend on the number of firms in the up- and downstream oligopolies. They also focus on foreclosure issues. Using the same framework, Abiru et al. (1998), focus more on the impact of the unequal number case. The main differences with our model is that we work on a differentiated market and we do not consider vertical mergers: our coalitions take place among producers. Considering such horizontal concentrations relates more our work to that of Allain and Souam (2005) who also analyse the effects of horizontal mergers in vertically-related industries. In a successive oligopoly model with an intermediate market, they show that horizontal mergers (leading to size effects) are more profitable downstream than upstream. Furthermore, they show that concentration reduces the incentives to merge in the vertically-related industries. Their model is more general than ours in studying horizontal mergers, since we study only extreme scenarios, but applies for a single product. Our focus is more on the interaction between horizontal concentration and product differentiation of two independent supply chains.

2 The model

Vertical structure: as a starting point, we assume that some actors of the wood sector decide to ecocertify their products in order to differentiate them from the non-ecocertified ones. This results in the emergence of a new distribution channel for ecocertified products (see Figure 1). We consider the competition between only two supply chains, each one offering a homogeneous wood product: one of them sells ecocertified wood products (indexed by e) and the other sells standard wood products (indexed by s). The ecocertified supply chain consists of n_e upstream producers (owners of ecocertified forests), either coalesced in a coalition C (modelled as a monopoly) or independent in their pricing policy (imperfect competition among producers). They sell their raw material at price w_e to a single ecocertified retailer R_e. Likewise, the standard wood supply chain consists of n_s producers in competition, selling at price w_s to a single standard retailer R_s. Let $n = n_s + n_e$ denote the total number of upstream wood producers. As we see, our model is built for large values of n, which correspond to empirical data.[5]

Figure 1 Standard and ecocertified supply chains

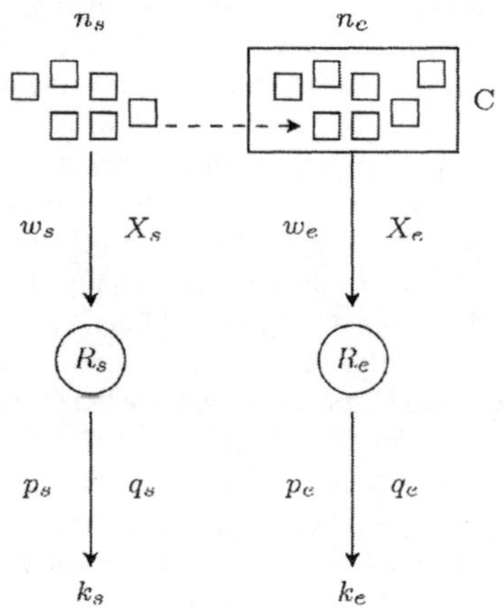

Cost structure: we normalise production costs for standard wood to zero. We consider that ecocertified producers incur a cost $f \geqslant 0$ that is proportional to the quantity produced.[5] As for the transformation and the retailing costs, we assume that they do not depend on whether the wood is ecocertified or not and we normalise them to zero as well.

Final market: we suppose that there is no physical difference between the two types of wood products and that the differentiation in perceived quality levels results only from the fact that ecocertified products respect environmental and social constraints and are therefore perceived by consumers as products of higher quality. k_e denotes the perceived quality of the ecocertified product and k_s is the perceived quality of the standard one, with $k_e \geqslant k_s > 0$. We can suppose, without loss of generality, that $k_e = 1$. The ecocertified product is sold on the final market at price p_e and the standard one at price p_s. We model consumers' demand using the classical model of Mussa and Rosen (1978). More precisely, the two supply chains face a demand from consumers that have a sensitivity towards forest ecocertification θ, which we assume to be uniformly distributed on the segment [0, 1]. The utility derived from buying one unit of standard product is given by: $u_s(\theta) = \theta k_s - p_s$ and $u_e(\theta) = \theta k_e - p_e$ for one unit of ecocertified product. Consumers who do not buy any wood product get zero utility. In other words, environmental quality is seen as a parameter that affects positively the consumers' willingness to pay, in proportion to their taste for quality.

Games: the two scenarios (competition or coordination among ecocertified producers) are formalised by two games with the following general structure:

Stage 1: choice of supply chain by the producers.

Stage 2 or 2': choice of quantities to be sold by the producers.

Stage 3: retailers' demand in the intermediate market.

Stage 4: choice of quantities to be sold by the retailers.

Stage 5: consumers' demand in the final market.

Let r be the ratio of qualities produced by each supply chain with $r = k_s/k_e \in [0, 1]$. The game $G(r, n)$ for the competition case is made of stages 1–5 and the game $G'(r, n)$, for the coordination case, is obtained by replacing the stage 2 by stage 2' in the game $G(r, n)$. Let us now define more precisely the stages of the games.

In stage 1 (*choice of supply chain by the producers*), n homogeneous producers choose whether to supply the standard or the ecocertified supply chain. Let n_s be the resulting number of standard wood producers and n_e the number of ecocertified wood producers, with $n = n_s + n_e$.

The external and internal stability criteria for the group of ecocertified producers respectively state that $\pi_s^i(n_e) \geqslant \pi_e^i(n_e + 1)$ and that $\pi_e^i(n_e) \geqslant \pi_s^i(n_e - 1)$. The first inequality means that no standard producer has an interest to join the group of ecocertified producers. The second inequality means that no ecocertified producer has an interest to join the group of standard producers. As we consider large values of n, our *stability criterium* becomes $\pi_s^i(n_e) = \pi_e^i(n_e)$. Solving this equation allows us to endogenise the equilibrium proportion of ecocertified producers $a = n_e/n$. The optimal proportion a^* (or a_c^* for the coordination case) of ecocertified producers will depend on n and r, the two exogenous variables of the model.

In stage 2 (*choice of quantities to be sold by the producers*), the n_s producers of standard wood are in competition to sell their woods to the retailer R_s. Let X_s be the demand of the retailer R_s. Standard producers set their quantity x_s^i in order to maximise their profit given by $\pi_s^i = w_s(X_e, X_s)x_s^i$. Similarly, ecocertified producers are in competition to sell their wood to the retailer R_e. Let X_e be the demand of the retailer R_e. Ecocertified producers choose x_e^i in order to maximise their profit $\pi_e^i = [w_e(X_e, X_s) - f]x_e^i$. Moreover, we suppose that the demands X_s and X_e are satisfied by the producers: $X_s = \sum_{i=1}^{n_s} x_s^i$ and $X_e = \sum_{i=1}^{n_e} x_e^i$.

Stage 2' is similar to stage 2 but we consider that ecocertified producers form a coalition C and coordinate their choice of quantity as an upstream monopoly. The quantity X_e^c is set by the coalition in order to maximise the profit $\pi_e^c = [w_e(X_e^c, X_s) - f]X_e^c$ that is supposed to be shared equally among the members of the coalition, that is to say $\pi_e^{c,i} = \pi_e^c/n_e$.

In stage 3 (*retailers' demand in the intermediate market*), we assume that the retailers buy each unit on the upstream market to resell it in the downstream market. Thus, we have $X_s = q_s$ and $X_e = q_e$ in the intermediate market, where q_s and q_e are the quantities that the retailers expect to sell in the final market.

In stage 4 (*choice of quantities to be sold by the retailers*), the profits of standard and ecocertified wood retailers are written $\Pi_s = (p_s - w_s)q_s$ and $\Pi_e = (p_e - w_e)q_e$. Retailers set the quantities they sell on the final market in order to maximise their profits. We suppose that these quantities meet the demand functions of the consumers in the final market ($q_s = D_s$ and $q_e = D_e$).

In stage 5 (*consumers' demand in the final market*), the consumers demand each type of product (D_s and D_e) depending on their taste for quality. Consumers do not buy any wood product if $u_s(\theta, k_s) \leqslant 0 \Leftrightarrow \theta \in [0, \underline{\theta}]$, where $\underline{\theta} = p_s/k_s$ corresponds to a consumer who is indifferent between buying standard wood and not buying at all. Consumers buy the standard product if $u_s(\theta, k_s) \geqslant 0$ and $u_s(\theta, k_s) \geqslant u_e(\theta, k_e)$ $\Leftrightarrow \theta \in [\underline{\theta}, \widetilde{\theta}]$, where $\widetilde{\theta} = (p_e - p_s)/(k_e - k_s)$ corresponds to a consumer who is indifferent between standard and ecocertified products. Finally, consumers buy the ecocertified product if $u_e(\theta, k_e) \geqslant u_s(\theta, k_s) \Leftrightarrow \theta \in [\widetilde{\theta}, 1]$. As a result, we find that $D_s = \widetilde{\theta} - \underline{\theta}$ and $D_e = 1 - \widetilde{\theta}$. The consumers' surplus is given by $S = (k_e D_e^2 + k_s D_s^2 + 2k_s D_e D_s)/2$.

Benchmark: we take as a benchmark the situation in which the two supply chains are identical and sell standard wood products ($k_e = k_s \leqslant 1$). The corresponding game is denoted by $G_0(r, n)$. In this case, we can show (see the Appendix) that, $n/2$ producers supply each of the two retailers ($a_0^* = 1/2$). As for the notations at equilibrium, producers sell a quantity x_0 at price w_0 in the intermediate markets. The retailers sell a quantity q_0 at price p_0 in the final market. This leads to profits π_0 for the producers and Π_0 for the retailers. The consumers get a surplus denoted by S_0.

Total welfare: the general formula for the total welfare is $W = \Pi_s + \Pi_e + (1-a)n\pi_s^i + an\pi_e^i + S$, that is to say, the sum of the two retailers' and the n producers' profits and the consumers' surplus.

3 Results

Let us first suppose that there is no certification cost ($f = 0$). By making the assumption of rational expectations, it is possible to solve the subgame perfect equilibria for $G(r, n)$ and $G'(r, n)$ by backward induction. This allows us to determine the equilibrium quantities and prices in the intermediate and final markets, and from there, the profits of the producers and the retailers as well as the consumers' surplus (stages 5 – 2). Stage 1 is then solved thanks to the hypothesis stating that n is large enough for the *stability criterium* $\pi_s^i(n_e) = \pi_e^i(n_e)$ to be applicable.

Proposition 1: *There exist unique optimal proportions a^* (in the competition scenario) and a_c^* (in the coordination scenario). These proportions are interior solutions (belong to* [0, 1]*) under assumption (H): $n > 1/2r$. Furthermore, if $n > 8$, $a^* \leqslant a_c^*$ is always satisfied.*

Proof: See the appendix.

Assumption[6] (H) ensures that none of the supply chains disappear at equilibrium and we consider that this is the case in the rest of the model. We obtain that the proportion of ecocertified producers is higher when there is coordination among them. Indeed, in the coalition, producers have higher market power than when they compete with each other, making ecocertification more profitable for them.

If we replace the proportion of ecocertified producers a^* and a_c^* in all the economic variables (prices, quantities, profits and surplus), they can be written only in function of n and r. However, in the rest of this paper, we focus more on the impact that the ratio of qualities r has on the variation of the profits and surplus since n (the total number of producers) is an objective reality of the market that can be easily known. Indeed, we will suppose that n is large[7] so that the variation of n is lesser crucial than the variation of $r \in [0, 1]$.

Proposition 2: *There exists $\widehat{n} > 1/2r$ such that the coordination among ecocertified producers has a positive impact on the producers' and the standard retailer's profits and a negative impact on the ecocertified retailer's profit, the consumers' surplus and the total welfare, if $n > \widehat{n}$.*

Proof: See the appendix.

When there is no coordination of ecocertified producers, ecocertification is always profitable to the ecocertified retailer. The reason is that consumers perceive the difference between the two types of goods ($k_s < k_e$) and are willing to pay for it. It is then obviously optimal to segment the offer. All producers and final consumers gain from this segmentation as well. Only the standard retailer gets lower profit compared to the benchmark situation.

However, when ecocertified producers coordinate, these results might no longer be true. Smaller quantities of ecocertified products are sold when producers coordinate. This restriction of the offer made by the coalition coincides with the setting of a higher price for the ecocertified product in the intermediate and final markets. At the same time, the coordination of ecocertified producers leads to an increase in the quantities sold in the standard supply chain, but also to an increase in the price of standard products in the final market.

Let us now analyse in more detail the role of the coordination among producers, on producers' and retailers' profits and on consumers' surplus. In order to do so, we study profits and surplus compared to the benchmark situation. The graphs of the functions are given for $k_e = 1$, $f = 0$ and $n \to \infty$ so that the economic variables[8] are only functions of $r \in [0, 1]$.

Producers' profit: the profits at equilibrium are equal for standard and ecocertified producers (stability criteria) and increased by the segmentation of the market through ecocertification (compared to the benchmark situation). These profits significantly increase when there is a coordination of ecocertified producers.

Retailers' profit: when ecocertified producers coordinate, the profit of the ecocertified retailer can be lower than the one gained in the benchmark situation (selling standard wood). Figure 2 shows the impact of a coalition of ecocertified producers on the profit of the ecocertified retailer. This profit is a function of the ratio of qualities r (in $r = 1, k_s = k_e$).

Figure 2 Ecocertified retailer's profits compared to benchmark profit

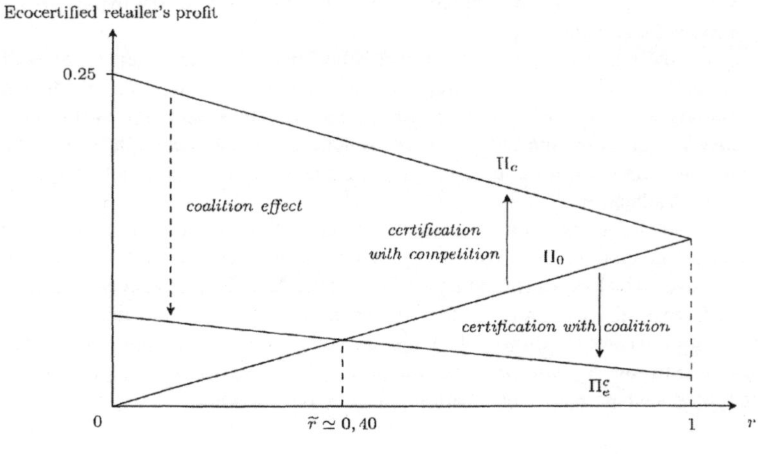

In the benchmark situation, the retailer gets a profit Π_0. When there is a segmentation of the market, the profit gained by an ecocertified retailer is Π_e. We can see that when ecocertified producers coalesce, the profit of the ecocertified retailer Π_e^c is smaller. As a result, when the two types of products are not differentiated enough ($r > \widetilde{r}$), the higher price of the ecocertified product does not counterbalance the market power loss. Interestingly, in this case, the ecocertified retailer, initiator of the certification process, gets a lower profit than before the certification move. If the retailer is able to anticipate the degree of differentiation and the ability of the producers to coordinate in a previous step of the game, the ecocertified supply chain should not exist at equilibrium for $r > \widetilde{r}$.

We observe opposite effects for the standard retailer, as shown in Figure 3.

Figure 3 Standard retailer's profits compared to benchmark profit

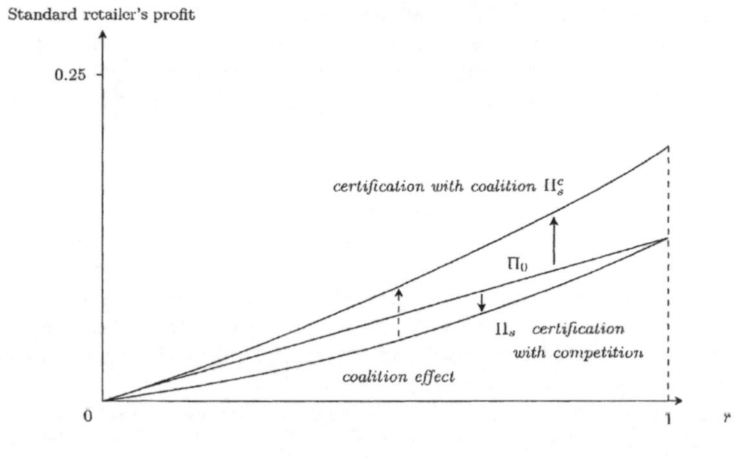

Even though the segmentation of the market lowers the profit of the standard retailer compared to the benchmark situation ($\Pi_s < \Pi_0$), the fact that ecocertified producers coordinate increases this profit (Π_s^c) since it reduces the competition with the ecocertified retailer, which is handicapped from being supplied by an upstream monopoly.

Consumers' surplus: regarding the consumers, the analysis is similar to the one we made for the ecocertified retailer (Figure 4).

Of course the segmentation of the market is beneficial for consumers ($S > S_0$, for all r), if ecocertified producers do not coordinate. When there is coordination, the final prices' increase along with the reduction of the total offer leads to a reduction of the consumers surplus. As a result, when the two products are not differentiated enough ($r > \widehat{r}$), the gain from segmentation does not counterbalance the change in market structure leading to a lower total consumption.

We give details in the Annex about the impact of the coordination of ecocertified producers on the total welfare. We show in particular that there is also a switching value $\underline{r} \simeq 0.87$, over which ecocertification is not socially beneficial if ecocertified producers coalesce. Conversely, if enough differentiation is perceived by consumers ($r < \underline{r}$), an ecocertification scheme is welfare enhancing: the benefit of ecocertification counterbalances the cost of allowing a coalition. An evaluation of the degree of differentiation should therefore be done before any certification policy is to be adopted.

Figure 4 Consumers' surplus compared to benchmark situation

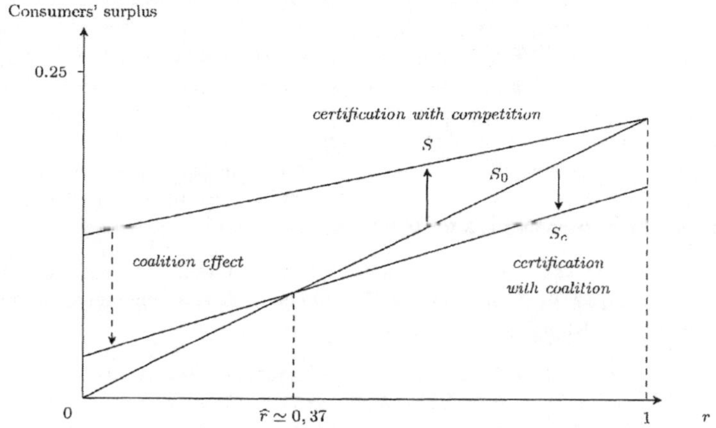

To sum up, the coordination of ecocertified producers can lead to a loss of profit for the ecocertified retailer and a loss of consumers surplus, if consumers do not perceive enough differentiation between standard and ecocertified products. However, we considered that $f = 0$ (no certification costs). We may then wonder how certification costs will affect these results.

Proposition 3: *The certification cost*[9] $f \in [0, 1/2]$ *incurred by the ecocertified producers has:*

- *a positive impact on the standard retailer's profits and*
- *a negative impact on the producers' and the ecocertified retailer's profits, on the consumers' surplus and on the total welfare.*

Proof: See the appendix.

The certification cost is an added handicap for the ecocertified retailer, which benefits to the standard retailer. As a result, the impact for the retailers and the consumers, of the certification cost incurred by the ecocertified producers goes in the same direction as the impact of the coordination among them. On the contrary, for producers, ecocertification becomes less attractive when it is costly.

4 Concluding remarks

Our analysis shows that ecocertification may have pitfalls for a high quality product retailer. The most evident one is related to the existence of certification costs that increase the raw material price in the intermediate market. The perception of the quality differentiation must be sufficient to make the segmentation profitable for the retailer. A more original side effect of ecocertification is related to the possibility of a reorganisation of the market structure . Indeed, if ecocertification leads to a coalition among certified producers, which in practice seems to be more likely in group certification processes, the high quality product retailer is, then, again penalised unless there is enough differentiation between standard and certified products. Both effects turn to the advantage of the standard retailer. Furthermore, the price increase results in a loss of market shares for the wood sector since more consumers are priced out and the consumers may see their surplus diminished. Conversely, if the consumers perceive enough product differentiation, an ecocertification policy is welfare enhancing.

The importance of the differentiation between ecocertified and standard wood product is particularly highlighted by our model, since it can completely change the economic results. Since the differentiation put in evidence in empirical studies (Ibanez and Pourcher, 2001) is estimated to be relatively small, we should wonder about the motivations of the retailer that first initiated and encouraged the certification of all the supply chains. Many reasons can explain the ecocertification strategy:

- The retailer might fear consumers or non-governmental pressures that could lead to a consumers' boycott of its product, which could be studied through an uncertain demand for the standard supply chain.
- The retailer expects the differentiation between standard and ecocertified wood products to increase with time. This emphasises the necessity to evaluate over time the degree of differentiation perceived by the consumers. Moreover, if the ecocertified retailer is able to influence the perceived quality through advertising campaigns for ecocertified products, product differentiation can be strategically fixed within an interval leading to a profit improvement.
- By offering 'green' products, the retailer wants to attract new consumers with increasing ecological concerns, that is to say, increase the demand for wood products. Our model considers that the total demand is fixed. Consumers who did not buy any wood product might be willing to turn to wood for its environmental characteristics. It would then be interesting to study the impact of this 'new' demand on our results.
- The retailer seeks for a general image improvement. Offering ecocertified wood products can attract consumers to the retailer's store and increase its sales on other non-wood products. An extension of our model with a multiproduct retailer could capture this phenomenon.

Another kind of explanation is that the high quality retailer does not believe that producers are able to strongly coordinate their pricing strategy. The results we obtained in our model are based on the comparison of two extreme situations: either the producers are price takers (when they compete) or price makers (when they coalesce and behave as an upstream monopoly). Considering a Nash negotiation which introduces a weight in the market power of the ecocertified producers' coalition, on the one hand and the ecocertified retailer on the other hand, would lead to an intermediary profit sharing between the two extreme outcomes. Even though we can expect the same qualitative results to hold, the range of product differentiation for which the segmentation is profitable for the retailer would be enlarged.

References

Abiru, M., Nahata, B., Raychaudhuri, S. and Waterson, M. (1998) 'Equilibrium structures in vertical oligopoly', *Journal of Economic Behavior and Organization*, Vol. 37, No. 4, pp.463–480.

Allain, M-L. and Souam, S. (Forthcoming) 'Horizontal mergers and vertical relationships', *Annales d'Economie et Statistiques*.

Cason, T. and Gangadharan, L. (2002) 'Environmental labeling and incomplete consumer information in laboratory markets', *Journal of Environmental Economics and Management*, Vol. 43, No. 1, pp.113–134.

Costa, S. and Ibanez, L. (2002) 'Who benefits from new certified products?', *Scandinavian Forest Economics*, Vol. 39, pp.33–41.

FAO (2003) 'Forest products: annual market analysis 2002–2004', *Timber Bulletin*, Vol. 56, No. 3.

Gaudet, G. and Van Long, N. (1996) 'Vertical integration, forclosure and profit in the presence of double marginalization', *Journal of Economics and Management Strategy*, Vol. 5, pp.409–432.

Haener, M.K. and Luckert, M.K. (1998) 'Forest certification: economic issues and welfare implications', *Canadian Public Policy*, Vol. 24, No. 2, pp.83–94.

Ibanez, L. and Laye, J. (2005) 'Stratégies d'écocertification, ou comment le rêve des uns peut faire le bonheur des autres', *Les Cahiers Scientifiques du Bois*, Vol. 3, pp.325–332.

Ibanez, L. and Pourcher, E. (2001) 'Existe-t-il un marché durable pour les bois écocertifiés?', *Revue Forestière Française*, Vol. 53, No. 6, pp.715–722.

Marette, S. and Crespi, J. (2003) 'Can quality certification lead to stable cartels?', *Review of Industrial Organization*, Vol. 23, pp.43–64.

Meidinger, E., Elliot, C. and Oesten, G. (2003) *Social and Political Dimensions in Forest Certification*, Verlag, ISBN: 3-935638-23.

Mussa, M. and Rosen, S. (1978) 'Monopoly and product quality', *Journal of Economic Theory*, Vol. 18, pp.301–317.

Roda, J-M. (2001) 'Ecocertification tropicale et idées préconçues', *Bois et Forêts des Tropiques*, Vol. 270, No. 4.

Schwarzbauer, P. and Rametsteiner, E. (2001) 'The impact of SFM-certification on forest product markets in Western Europe – an analysis using a forest sector simulation model', *Forest Policy and Economics*, Vol. 2, pp.241–256.

Sedjo, R. and Swallow, S. (2000) 'Eco-labeling consequences in general equilibrium: a graphical assessment', *Land Economics*, Vol. 76, No. 1, pp.28–36.

Siry, J.P., Cubbage, F.W. and Ahmed, M.R. (2005) 'Sustainable forest management: global trends and opportunities', *Forest Policy and Economics*, Vol. 7, pp.551–561.

Takahashi, T., Van Kooten, G.C. and Vertinsky, I. (2003) 'Why might forest companies certify? Results from a Canadian survey', *International Forestry Review*, Vol. 5, pp.329–337.

Appendix

Proof of Proposition 1: We compute the subgame perfect equilibria by backward induction for the games $G(r, n)$ and $G'(r, n)$. We give, in footnotes, the steps of the game $G_0(r, n)$ corresponding to the benchmark situation.

In stage 5 *(consumers' demand in the final market)*, the retailers R_s and R_e have the following demand functions[10] for standard and ecocertified products:

$$D_s(p_s, p_e) = \frac{p_e - p_s}{k_e - k_s} - \frac{p_s}{k_s} \text{ and } D_e(p_s, p_e) = 1 - \frac{p_e - p_s}{k_e - k_s}.$$

In stage 4 *(choice of quantities to be sold by the retailers)*, we first isolate the prices[11] in the demand functions. We find $p_s = k_s(1 - D_e - D_s)$ and $p_e = k_e - k_e D_e - k_s D_s$. The retailers' offers are equal to the consumers' demands ($Q_s = D_s$ and $Q_e = D_e$) and the profits of the retailers are written $\Pi_s = (p_s - w_s)Q_s$ and $\Pi_e = (p_e - w_e)Q_e$. The Nash equilibrium in quantities is obtained by solving the system of equations $\{d\Pi_s/dQ_s = 0, d\Pi_e/dQ_e = 0\}$. We obtain the quantities[12] in function of k_s, k_e, w_s, w_e:

$$\left| \begin{array}{l} Q_s = \frac{k_e k_s + k_s w_e - 2k_e w_s}{k_s(4k_e - k_s)} \\ Q_e = \frac{2k_e - k_s - 2w_e + w_s}{4k_e - k_s} \end{array} \right. \qquad \text{(A1)}$$

In stage 3 *(retailers' demand in the intermediate market)*, the quantities that retailers want to sell in the final market are demanded to the producers:[13] $X_s = Q_s$ and $X_e = Q_e$.

In stage 2 *(choice of quantities to be sold by the producers)*, we first isolate the prices in the demand functions of the retailers in order to obtain the prices in the intermediate market. This leads[14] to $w_s = k_s(1 - X_e - 2X_s)$ and $w_e = k_e - 2k_e X_e - k_s X_s$. Next, the producers have to maximise their profits $\pi_s^i = w_s(X_e, X_s)x_s^i$ and $\pi_e^i = [w_e(X_e, X_s)]x_e^i$, where $X_s = \sum_{i=1}^{n_s} x_s^i$ and $X_e = \sum_{i=1}^{n_e} x_e^i$. We solve the system of equations $\{d\pi_s^i/dx_s^i = 0, d\pi_e^i/dx_e^i = 0\}$ and we search for symmetric Nash equilibria, thus dropping the superscript i in the expressions. The equilibrium quantity for each producer is:[15]

$$\left| \begin{array}{l} x_s = \frac{k_e(2+n_e)}{4k_e(1+n_e)(1+n_s) - k_s n_e n_s} \\ x_e = \frac{2k_e(1+n_s) - k_s n_s}{4k_e(1+n_e)(1+n_s) - k_s n_e n_s} \end{array} \right. \qquad \text{(A2)}$$

At this point, let us solve simultaneously the game $G'(r, n)$, since it is only from stage 2' that it differs from $G(r, n)$. Stage 2 is replaced by stage 2'. Standard producers maximise their profit $\pi_s = w_s(X_e^c, X_s)x_s$ while the coalition of ecocertified producers sets the quantity X_e^c in order to maximise $\pi_e^c = [w_e(X_e^c, X_s)]X_e^c$. We obtain the symmetric equilibrium quantities:

$$\left| \begin{array}{l} x_s^c = \frac{3k_e}{8k_e(1+n_s) - k_s n_s} \\ X_e^c = \frac{2k_e(1+n_s) - k_s n_s}{8k_e(1+n_s) - k_s n_s} \end{array} \right. \qquad \text{(A3)}$$

If we replace the expressions of the quantities sold by the producers at equilibrium and we consider that $k_s = rk_e$, that $k_e = 1$, that $n_s = n - n_e$ and that $n_e = an$, we obtain the expressions of all economic variables at equilibrium, only in function of a and n. In particular, we obtain the profit of each standard producer[16] $\pi_s(a, n)$ and the profit of each ecocertified producer in competition $\pi_e(a, n)$:

$$\left| \begin{array}{l} \pi_s(a, n) = 2(2 + an)^2 r/\delta(a, n) \\ \pi_e(a, n) = 2[2 + n(1 - a)(2 - r)]^2/\delta(a, n) \end{array} \right. \qquad \text{(A4)}$$

with $\delta(a, n) = [4 + 4n + (1 - a)(4 - r)an^2]^2 > 0$ for all $r \in [0, 1]$.

In the coordination case (index c), the profit of the standard producers $\pi_s^c(a, n)$ and the profit of the coalition of ecocertified producers $\pi_e^c(a, n)$ are written:[17]

$$\left| \begin{array}{l} \pi_s^c(a, n) = 18r/\delta_c(a, n) \\ \pi_e^c(a, n) = 2[2 + n(1 - a)(2 - r)]^2/an\delta_c(a, n) \end{array} \right. \tag{A5}$$

with $\delta_c(a, n) = [8 + n(1 - a)(8 - r)]^2 > 0$ for all $r \in [0, 1]$.

Stage 1 consists in solving in a the equation $\pi_s(a, n) = \pi_e(a, n)$. Since a and r belong to $[0, 1]$, this equation is equivalent to solve[18] in a: $\Phi(a) = 2+(1-a)(2-r)n-(2+an)\sqrt{r} = 0$. We have that $\Phi'(a) = n[r - \sqrt{r} - 2] < 0$ for all $r \in [0, 1]$. Hence, Φ strictly decreases with a. We have $\Phi(0) > 0$ since this inequality is equivalent to $n+2(1-\sqrt{r})/(2-r) > 0$, which is always verified. $\Phi(1) < 0 \Leftrightarrow n > \underline{n} = 2/\sqrt{r} - 2$. As a result, if $n \leqslant \underline{n}$, $a^* = 1$ (all the producers choose to supply the ecocertified supply chain and the standard one disappears). If $n > \underline{n}$ then $a^* < 1$. We solve $\Phi(a) = 0$ and we obtain:[19]

$$a^* = \frac{n(2 - r) + 2 - 2\sqrt{r}}{n(2 - r + \sqrt{r})} \tag{A6}$$

which is a decreasing function of n that tends to $(2-r)/(2-r+\sqrt{r}) > 1/2$. If ecocertified producers coordinate, Stage 1 is slightly modified. The equation that we must solve in a is $\Pi_s^c(a, n) = \Pi_e^c(a, n)$. Since $\delta_c(a, n) > 0$ for all $r \in [0, 1]$, this problem is equivalent to finding the root of the function $\Psi(a) = [2 + n(1 - a)(2 - r)]^2 - 9anr$. We have that $\Psi'(a) = -2[2 + n(1 - a)(2 - r)](2 - r)n - 9nr < 0$ for all $r \in [0, 1]$, so Ψ is a strictly decreasing function of a. $\Psi(0) = [2 + n(2 - r)]^2 > 0$. $\Psi(1) = 4 - 9nr$, thus $\Psi(1) < 0 \Leftrightarrow n > 4/9r$. If $n \leqslant 4/9r$, then $a_c^* = 1$ (all the producers choose to supply the ecocertified supply chain and the standard one disappears). If $n > 4/9r$, $a_c^* < 1$. We obtain (by solving $\Psi(a) = 0$):

$$a_c^* = \frac{\alpha + 9r/2 - 3\sqrt{r}\sqrt{2\alpha}}{2n(2 - r)^2} \tag{A7}$$

where $\alpha = 8 + r/2 + 2n(2 - r)^2$.

a^* and a_c^* belong to $[0, 1]$ as long as $n > \max\left(4/9r, 2/\sqrt{r} - 2\right)$. Assumption (H) $n > 1/2r$ is a sufficient condition for this inequality to be verified.

We see that $\lim_{n\to\infty} a^* = (2 - r)/(2 - r + \sqrt{r}) < 1 = \lim_{n\to\infty} a_c^*$. Hence, there exist $\widetilde{n}$ such that $a^* < a_c^*$ for $r \in [0, 1]$ and $n > \widetilde{n}$. Let us now find $\widetilde{n}$. The equation $a^*(r) < a_c^*(r)$ has four solutions:

$$r_1 = 2(21 - 7n + n^2 - 3\sqrt{33 - 14n + 2n^2})/(n - 4)^2$$

$$r_2 = 2(21 - 7n + n^2 + 3\sqrt{33 - 14n + 2n^2})/(n - 4)^2$$

$$r_3 = (9 - 4n + n^2 - 3\sqrt{9 - 8n + 8n^2})/2(n - 1)^2$$

$$r_4 = (9 - 4n + n^2 + 3\sqrt{9 - 8n + 8n^2})/2(n - 1)^2$$

Let us now reject those that are always greater than 1. Indeed, $r_2 - 1 = (26 - 6n + n^2 + 6\sqrt{33 - 14n + 2n^2})/(n - 4)^2$. Since $26 - 6n + n^2$ is always positive, $r_2 > 1$. We can reject in a similar way r_3 and r_4. Let us now study r_1. We have that $r_1(5) \simeq 0.37$ and $\lim_{n\to\infty} r_1 = 2$. Since $dr_1(n)/dn > 0$ for all n, $r_1(n)$ is a strictly increasing function. The equation $r_1(n) = 1$ gives $\widetilde{n} = 8$. As a result, for $\widetilde{n} > 8$, $a^* - a_c^*$ has no root in $[0, 1]$, and has, thus, a constant sign in this interval. For $n = 9$ we verify that $a^* < a^*_c$. For a given r, $a^* - a^*_c$ is a continuous function of n that is different to 0 in the interval $[9, +\infty]$. Hence, the sign of $a^* - a^*_c$ is conserved and we have $a^* < a^*_c$ for all r in $[0, 1]$. What happens for $n < 8$ is irrelevant in our model since in this case, the stability criterium we use is not appropriate.

Proof of Proposition 2: Due to the stability criteria, the profit of a standard producer is equal, at equilibrium, to that of an ecocertified producer. It is evident that the segmentation of the market is beneficial to the producers, so $\pi_s = \pi_e \geqslant \pi_0$. It is also evident that the possibility to coordinate their strategy increases even more the profit of the ecocertified producers. So the coordination among ecocertified producers has a positive impact on all producers' profits.

Taking into account that in our model, we consider large values of n, we can compute the limit of the economic variables at equilibrium when n tends to infinity and for r fixed (the symbol ∞ is added to the notations in this case), and compare these limits. We can then say that, for a given r, there exists $\widehat{n}$ such that the relationships among the economic variables that we obtain when n tends to infinity are true for $n > \widehat{n}$.

We prove in this way that the standard retailer also benefits from this coordination. We have the general expression $\Pi_s = (p_s - w_s)Q_s$. If we replace the prices and the quantity in the expression and we consider that $k_s = rk_e$, that $k_e = 1$, that $n_s = n - n_e$ and that $n_e = an$, we obtain $\Pi_s(a, n) = r[(1-a)(2+an)n]^2/\delta(a, n)$. At equilibrium $\Pi_s(a^*, n) = [((n+2)\sqrt{r}-2)/(4(2-\sqrt{r})+n(4-r))]^2$. When n tends to infinity, this profit tends to $\Pi_s^\infty(a^*) = r/(4-r)^2 \leqslant r/9 = \lim_{n\to\infty} n^2 r/(4+3n)^2 = \Pi_0^\infty$. Thus, the profit of the standard retailer is lower than in the benchmark situation. When ecocertified producers coordinate, $\Pi_s^c(a, n) = 9r[(1-a)n]^2/\delta_c(a, n)$. We replace a by a^*_c and we compute the limit when n tends to infinity. Finally, $\lim_{n\to\infty}\Pi_s^c(a^*_c) = 9r/(8-r)^2 \geqslant \Pi_0^\infty \geqslant \Pi_s^\infty(a^*)$.

Let us now prove, using the same approach, that the coordination has a negative impact on the profit of the ecocertified retailer, whose general expression is $\Pi_e = (p_e - w_e)Q_e$. We compute the profit depending on whether there is coordination or not, and we obtain $\Pi_e(a, n) = [an(2 + (1-a)(2-r)n]^2/\delta(a, n)$ and $\Pi_{e,c}(a, n) = [2 + (1-a)(2-r)n]^2/\delta_c(a, n)$. We replace a by a^* in the first expression, by a^*_c in the second one and we compute the limit of the resulting expressions when n tends to infinity. We find that $\Pi_e^\infty(a^*) = [(2-r)/(4-r)]^2$ and $\lim_{n\to\infty}\Pi_e^c(a^*_c) = [(2-r)/(8-r)]^2 < \Pi_e^\infty(a^*)$. $\Pi_e^\infty(a^*) \geqslant \Pi_0^\infty$ leads to $(1-r)(36-16r+r^2) \geqslant 0$ which is always true for $r \in [0, 1]$, and the equality $\Pi_e^\infty(a^*) = \Pi_0^\infty$ is verified for $r = 1$. Solving $lim_{n\to\infty}\Pi_e^c(a^*_c) = \Pi_0^\infty$ is equivalent to solve the equation of the third degree $\varphi(r) = r^3 - 25r^2 + 100r - 36 = 0$. Only $\widetilde{r}_\infty$, one of the three roots of φ belongs to $[0, 1]$, and $\widetilde{r}_\infty \simeq 0,40$. As a result, if $r < \widetilde{r}$, the profit of the ecocertified retailer remains higher than in the benchmark situation, but if $r \geqslant \widetilde{r}$, that is to say, if differentiation between standard and ecocertified woods is not clear enough, then the ecocertified retailer wins lesser than in the benchmark situation.

Regarding the consumers surplus, we find $S^\infty = (4+r-r^2)/2(4-r)^2$ (when there is no coordination), $S_c^\infty = (4+17-5r^2)/2(8-r)^2$ (in the coordination case) and $S_0^\infty = 2r/9$ (in the benchmark situation). $S^\infty \geqslant S_0^\infty$ is equivalent to $(1-r)(4r^2 - 19r + 36) \geqslant 0$, which is always true for r in $[0, 1]$. Solving $S_c^\infty = S_0^\infty$ is equivalent to solve $\psi(r) = 4r^3 - 19r^2 + 103r - 36 = 0$. The only of the three roots of ψ that belongs to $[0, 1]$ is $\widehat{r} \simeq 0.37$. If $r > \widehat{r}$, then the consumers get a lower surplus than in the benchmark situation.

Finally, the total welfare is given by the general formula $W = \Pi_s + \Pi_e + (1-a)n\pi_s + an\pi_e + S$. Depending on the situation, we obtain the functions when n tends to infinity: $W^\infty = (12 - 5r + r^2)/2(4-r)^2$, $W_c^\infty = (4+r)(7+r)/2(8-r)^2$ and $W_0^\infty = 4r/9$. $W^\infty \geqslant W_0^\infty$ leads to $(1-r)(8r^2 - 65r + 108) \geqslant 0$, which is always true for r in [0, 1]. In the coordination case, $W_c^{\infty} = W_0^{\infty}$ is equivalent to solving the equation $\xi(r) = 8r^3 - 137r^2 + 403r - 252 = 0$. Only $\underline{r}$, one of the three roots of ξ, belongs to [0, 1]. If $r > \underline{r} \simeq 0,87$, the total welfare is lower than in the benchmark situation, due to the coordination of ecocertified producers.

Proof of Proposition 3: The fact of considering a positive certification cost f only changes the profit of the ecocertified producers (stage 2 and 2' of the games). If they compete we have $\pi_e^i = [w_e(X_e, X_s) - f]x_e^i$, if they coordinate we have $\pi_e^c = [w_e(X_e^c, X_s) - f]X_e^c$. We solve the game with $f > 0$ as we did in the proofs of Proposition 1. All prices must be greater than the certification cost and all quantities must be positive when $f > 0$, for all n and for all r in [0, 1]. A necessary condition is that $f < 1/2$.

It is evident that the profit of the standard producers (that is equal to that of the ecocertified ones) is lowered by the certification cost f incurred by the ecocertified producers.

We compute the limit of the retailers profit functions, the consumers' surplus and the total welfare, as in the proof of Proposition 2. As a result, we obtain:

- The standard retailer's profits $\Pi_s^{\infty} = (1+f)^2 r/(4-r)^2$ and $\lim_{n\to\infty}\Pi_s^c = (3+f)^2 r/(8-r)^2$, that increase with f.
- The ecocertified retailer's profits $\Pi_e^{\infty} = (2-2f-r)^2/(4-r)^2$ and $\lim_{n\to\infty}\Pi_e^c = (2-2f-r)^2 r/(8-r)^2$, that decrease with f.
- The consumers surplus $S^{\infty} = (4+r-r^2-(2f-f^2)(4-3r)-2fr^2)/2(4-r)^2$ and $S_c^{\infty} = (4+17-5r^2-(2f-f^2)(4-3r)-2f(2r+r^2)/2(8-r)^2$, that decrease with f since $4-3r > 0$ and $2f - f^2 > 0$ if $f < 2$ (which is true since we supposed $f < 1/2$).
- The comparison of the total welfare obtained with or without certification costs shows that it is also lowered if $f < 8/11$ (competition case) or if $f < 10/11$ (coordination case). Both are verified since $f < 1/2$.

Notes

[1]Such as Forest Stewardship Council (FSC) and Pan European Forest Certification (PEFC).

[2]Consumers perceive perfectly the environmental quality of the products.

[3]Our analysis is based on the forest certification processes but it should be underlined that cartel formation under common labelling is also relevant in agricultural and food processing markets. For example, Marette and Crespi (2003) look at the cartel formation of producers under quality certification. They show that a cartel may lead to welfare improvement by providing information and sharing certification costs even though producers collude and thereby reduce competition.

[4]For instance, in France, 74% of the forest belongs to private owners. Over the 3.5 millions of private forest owners, 1.1 millions own more than 1 hectare. In this context, we can suppose that a major wood product retailer is supplied by a large number of producers.

[5]The costs actually incurred by producers who adopt an ecocertification system are mainly variable costs, due to audits, new forest management rules, etc. For example, the fixed cost associated to a PEFC certification scheme (around 10 Euros and 0.1 Euro per hectare and per year) is insignificant compared with the costs that depend on the quantity produced.

[6]The mathematical condition $n > 1/2r$ is always satisfied if we consider large values of n and r that is not in the neighbourhood of 0. Such a framework is economically relevant since consumers do not perceive large differentiation between ecocertified and standard wood products.

[7]In France, for example, there is around 4 millions private forest owners.

[8]The speed of convergence of the functions towards the limit functions when n tends to infinity (and for a given r) is high. For $n > 20$ and outside the immediate neighbourhood of $r = 0$, which is economically irrelevant, all the properties we can observe on the limit functions are already true. For big values of n, for which our model is built ($n = 1000$ for example), the difference between the graph of a given function of n and the graph of its limit function is insignificant.

[9]If $f < 1/2$, all prices are greater than the certification cost and all quantities are positive.

[10]*Benchmark:* $D_0(p_0) = 1 - p_0/k_s$.

[11]*Benchmark:* $p_0 = k_s - k_s D_0$. In order to differentiate the two supply chains offering standard products we name them A and B.

[12]*Benchmark:* $Q_A = (k_s - 2w_A + w_B)/3k_s$ and $Q_B = (k_s - 2w_B + w_A)/3k_s$.

[13]*Benchmark:* $X_A = Q_A$ and $X_B = Q_B$.

[14]*Benchmark:* $w_A = k_s(1 - 2X_A - X_B)$ and $w_B = k_s(1 - X_A - 2X_B)$.

[15]*Benchmark:* $x_A = (2+n_B)/(4+4n_A+4n_B+3n_An_B)$ and $x_B = (2+n_A)/(4+4n_A+4n_B+3n_An_B)$.

[16]*Benchmark:* $\pi_A(a,n) = 2(2+an)^2 r/\delta_0(a,n)$ and $\pi_B(a,n) = 2r[2+(1-a)n]^2/\delta_0(a,n)$ with $\delta_0(a,n) = [4+4n+3(1-a)an^2]^2 > 0$ for all $r \in [0,1]$.

[17]We suppose that the profit is equally shared among coalesced ecocertified producers.

[18]*Benchmark:* $\Phi_0(a) = (2+an)^2 - (2+(1-a)n)^2 = 0$

[19]*Benchmark:* $a_0^* = 1/2$.

[5]

Eco-labeling and non-product-related process and production methods

Young-Hwan Ahn and Byong-Hun Ahn
Graduate School of Management, Korea Advanced Institute of Science and Technology (KAIST), 207-43 Cheongryang, Seoul 130-012, Korea

Received: November 27, 2000 / Accepted: November 26, 2001

Abstract Eco-labeling has caused some concerns about international trade, particularly between developed and developing countries, owing to the incorporation of non-product-related process and production methods (PPMs) into eco-labeling criteria. Most countries prefer inclusion of non-product-related PPMs for effectiveness and efficiency of eco-labeling. Using a simple duopoly model, this article compares two eco-labeling schemes: one with inclusion of non-product-related PPMs and the other without. This is done for the case where there has already existed a regulation on the production phase in the form of an emission tax. Counterintuitively, the analysis shows that the exclusion scheme is socially desirable under certain conditions. It reminds us of the importance of the policy mix when introducing eco-labeling. The paper concludes with a discussion of future research directions.

Key words Eco-labeling criteria · Non-product-related PPMs · Trade and environment Policy mix · Game theory

Introduction

Eco-labeling has been receiving growing attention in modern environmental policies. Since Germany introduced the "Blue Angel" program in 1977, many countries have introduced their own eco-labeling programs. The objective of these programs has been to improve domestic environments by providing firms with incentives to produce more environmentally friendly products. These are usually set up based on nation-specific situations in favor of their economic, environmental, and technology infrastructure (Salzman 1997; Vossenaar 1997). Such nation specificity in these programs could cause concerns with regard to trade, which arise mainly from two sources: transparency of the schemes and inclusion of non-product-related process and production methods (PPMs) in the eco-labeling criteria. We examine the latter of the two because it presents more of a structural problem.

There are three types of eco-labeling schemes according to the International Organization for Standardization (ISO) classification. Type I and III schemes are certified by third parties, whereas type II schemes are self-declaration schemes.

Type I and III schemes are different in regard to the information provided. Type I labels provide information concerning whether a product meets preset environmental standards established by third parties. Type III labels deliver quantitative environmental information resulting from an independent scientific assessment. Type II schemes are labels based on manufacturers' claims concerning individual environmental characteristics of their products. Among them, this research focuses on type I eco-labeling schemes evaluating products' environmental characteristics throughout their life cycles.

Generally, the environmental impact of PPMs are of two types. One is product-related PPMs and the other is non-product-related PPMs. Product-related PPMs (e.g., product design) determine the environmental characteristics of a product itself when it is used or disposed of, whereas non-product-related PPMs refer to a production process design that damages environmental quality during production. Often an importing country imposes product-related PPM requirements to protect its domestic environment from the use and disposal phase of imported products. The Agreement on Technical Barriers to Trade (TBT), in Article 2.2, also allows making product-related PPMs requirements to avoid consumption externalities. However, the issue becomes difficult when inclusion of non-product-related PPMs (i.e., the production phase) is also included in eco-labeling criteria (OECD 1997b; WTO 1999b).

If a product is traded internationally, its production and consumption take place in different countries. If the importing country includes non-product-related PPMs in its eco-labeling criteria, it affects the production process in the exporting country as well. The exporting country might argue that it is absurd for the importing country to impose environmental restrictions on exporters' non-product-related PPMs that affect the environment of the exporting country only. Exporting countries might even suspect the eco-labels to be de facto nontariff barriers. This concern could be more serious among developing countries exporting their goods to developed (and thus environmentally strict) countries (Anderson 1996; Appleton 1997; United Nations 1997; WTO 1999a).

Despite such concerns, a growing number of eco-labeling schemes are employing the life cycle approach and consequently incorporate non-product-related PPMs (Neitzel 1997; OECD 1997a). One possible justification, aside from myopic protection of the domestic firms' competitiveness, could be to avoid the "ping-pong" of environmental burdens among the various phases of the life cycle (Giezeman and Verhees 1997). If an eco-labeling scheme does not incorporate non-product-related PPMs, firms would pay more attention to emissions from other phases. Some firms would even try to reduce emissions during other phases by sacrificing reductions in the production phase. Because this practice can generate inefficiency when allocating cost of emission reduction among various phases, the inclusion of non-product-related PPMs in eco-labeling criteria can be justified.

In reality, however, we must remind ourselves that most countries, particularly developed countries, have already been employing some form of environmental regulation, notably in the form of command-and-control, emission tax, and a tradable permit system on the production phase (non-product-related PPMs). Thus, when we introduce the eco-labeling scheme, its effects should be judged considering how this is combined with already existing regulations.

This paper compares two eco-labeling schemes: one with non-product-related PPMs included and the other without. This is done for the case where there is already a regulation on the production phase in the form of an emission tax. If the tax level is already quite strict, the additional eco-labeling burden on the production phase might not serve the purpose effectively. To address this question we set up a simple duopoly model (see Section 2). In Section 3 we determine an equilibrium for each scheme and compare the outcomes when a tax is imposed on emissions from the production phase. Section 4 discusses policy implications of the analysis. Finally, Section 5 outlines the conclusions of this study and notes the possible directions for future research.

Model

The life cycle of a product mainly consists of five phases: raw material, production, distribution, use, disposal. Environmental pollutants (e.g., air emissions, water emissions, solid wastes) are generated during each phase. For more comprehensive understanding of the key idea, we assume that the amounts of pollutants from the raw material and the distribution phases are fixed, and the use and disposal phase together are integrated as the consumption phase; then non-product-related PPMs and product-related PPMs determine emissions from the production and consumption phases, respectively.

Structure

We use a simple partial equilibrium model of a single industry. Suppose there are two countries $i = 1, 2$ of which country 1 is developed and country 2 is developing. Each country i has a home firm i selling a product that is identical in all aspects except environmental characteristics. Assume that a market for the product exists only in country 1. Thus firm 2 exports all its products to the market in country 1 and competes with firm 1. Each product of both firms emits pollutants, as much as $\bar{x}$ during production and $\bar{y}$ during consumption, when the firms do not make any effort to reduce emissions. It is assumed that the pollutants have a local effect only. Let x_i and y_i denote emission reductions per product in the production and the consumption phases of firm i, respectively. The firms choose emission reduction levels, x_i and y_i, by choosing proper PPMs, considering the emission tax and an eco-labeling scheme.

Assume that emission tax t_i is imposed on the unit emission from the production phase of a firm in country i; and $t_2 < t_1$, which reflects that the developed country usually has stronger environmental regulations than the developing country.[1] Setting eco-labeling criteria in country 1, the eco-labeling competent body has two alternatives: (1) the inclusion scheme incorporating both the PPMs; or (2) the exclusion scheme incorporating only product-related PPMs. Under the inclusion scheme the total reduction per product, $x_i + y_i$, matters in eco-labeling. Meanwhile, the product with more reduction per product during the consumption phase, y_i, would be labeled under the exclusion scheme. Assume that if the reduction levels of both firms are equal firm 1 would be labeled because of the domestic firm's home turf advantage. Let us now consider the following three-stage noncooperative game.

Stage 1: Firm 1 decides on the reduction levels per product, x_1 and y_1.

Stage 2: Firm 2 observes x_1 and y_1 and decides on x_2 and y_2. Then, according to the labeling scheme, the more environmentally friendly product is granted the *g* (green) label. The other is recognized as *n* (nongreen).

Stage 3: Taking as given the labels the firms have been granted, they choose the prices of their products to maximize profits.

With type I schemes the criteria a product should meet to be labeled are preset by eco-labeling bodies. The model in this paper assumes that the eco-label is awarded to the relatively environmentally friendly product compared with the other product. It might seem that this model does not reflect the actual world. However, what matters is how the criteria are preset. Firms can influence the criteria during the criteria-setting process owing to information asymmetry.

The emission reduction cost of a firm is the information that is not revealed to eco-labeling bodies. A firm might conjecture about other firms' reduction cost on the basis of its own cost with some accuracy. The eco-labeling bodies, however, lack information. Even though the bodies perform life cycle assessments to set the criteria, life cycle assessment usually does not give full information on reduction cost, which is the reason most eco-labeling schemes require stakeholders, including producers, to participate in the criteria-setting process. The main task of the process is to induce the firms to reveal their willingness to reduce emissions to obtain the eco-label. During the process firms have a chance to influence the criteria based on their willingness to reduce emissions. In short, the model can be considered to combine the criteria setting and the application process by firms in type I schemes.[2]

Additionally, the eco-labeling bodies usually set the criteria such that about 20%–30% of the products of a single product category can meet them. The criteria are revised every 2 to 3 years. Therefore, the schemes can be said to perform relative evaluation in terms that there are always some winning products

[1] If we assume $t_1 = t_2$, we can think of both firms as domestic firms. Under this assumption, the firm with more advanced abatement technology would be labeled at equilibrium. Even in this case, we also cannot say that the inclusion scheme is superior to the exclusion scheme at equilibrium because of double incentives to reduce emissions during the production phase.

[2] The model could be applied to type III schemes with some modification of the consumer's preference.

and some losing products. In these respects, the model captures the relevant aspects of the practice.

In the game, firm 1 plays the leader when choosing reduction levels. While setting eco-labeling criteria, eco-labeling bodies have consultations with producers as stakeholders, and usually domestic producers find more and closer consultation opportunities. For example, the European Union (EU) eco-labeling body organizes a Consultation Forum that reflects its opinions when setting the criteria. The members of the Consultation Forum come from domestic industry (retailers, importers), environmental protection groups, and consumer organizations. Domestic firms state their opinions directly, and foreign firms express their opinions indirectly through importers. Even though EU eco-labeling is one of the most transparent schemes, domestic firms have the leading power when setting criteria. Thus, the model takes the form of a Stackelberg game.[3]

Consumers

We base our model on that of Kuhn (1998) when describing the environmentally differentiated market. There are M consumers. Each consumer has unit demand (i.e., each consumes zero or one unit of the product). Of the total M consumers, there is a fraction $1 - r$ of nongreen consumers and a fraction r of green consumers. All consumers have a common willingness to pay, w, for the product without regard to its environmental characteristics. A green consumer is willing to pay a premium of $\alpha(0 < \alpha < w)$ for the eco-label, whereas a nongreen consumer is indifferent to the eco-label. As a result, a green consumer is willing to pay w for a nongreen product and $w + \alpha$ for a green product. A nongreen consumer is willing to pay w for both products. A green consumer's payoff, U_g and a nongreen consumer's payoff, U_g, are

$$U_g = \begin{cases} w + \alpha - p_i^g & \text{if he/she buys a green product} \\ w - p_j^n & \text{if he/she buys a nongreen product} \\ 0 & \text{if he/she buys neither} \end{cases}$$

$$U_n = \begin{cases} w - p_i^g & \text{if he/she buys a green product} \\ w - p_j^n & \text{if he/she buys a nongreen product} \\ 0 & \text{if he/she buys neither} \end{cases}$$

where p_i^g and p_j^n are prices of green and nongreen products and $i, j = 1, 2$, $i \neq j$. Then the demand for the green product and the nongreen product are as follows.

[3] Even if the firms are assumed to choose the emission reductions simultaneously, we can derive the same result. See Appendix 2.

$$q_i^g = \begin{cases} 0 & \text{if } p_i^g > w + \alpha \text{ or } p_i^g > p_j^n + \alpha \\ rM & \text{if } p_j^n < p_i^g \le p_j^n + \alpha \text{ and } p_i^g \le w + \alpha \\ M & \text{if } p_i^g \le p_j^n \text{ and } p_i^g \le w \end{cases}$$

$$q_j^n = \begin{cases} 0 & \text{if } p_j^n > w \text{ or } p_j^n \ge p_i^g \\ (1-r)M & \text{if } p_i^g - \alpha \le p_j^n \le w \le p_i^g \le w + \alpha \\ M & \text{if } p_j^n < p_i^g - \alpha \le w \end{cases}$$

Firms

The demand for firm i's product, q_i^h, is determined by its prices and label, which is dependent on emission reductions.

$$q_i^h = q_i^h\left(p_i^h, p_j^l, x_i, y_i, x_j, y_j\right), \quad \text{for} \quad i,j = 1,2, \quad i \neq j, \quad \text{and} \quad h,l = g,n, \quad h \neq l$$

Production and emission reduction technology of each firm is characterized by the following cost function:

$$C_i\left(q_i^h, x_i, y_i\right) = \left(q_i^h\right)^2 + C_{iP}\left(q_i^h x_i\right) + C_{iC}\left(q_i^h y_i\right)$$

where $C_{iP}(q_i^h x_i)$, and $C_{iC}(q_i^h y_i)$ are reduction costs during the production and the consumption phases of firm i, respectively.[4] The variable $q_i^h x_i$ is the total emission reduction during the production phase of firm i, and $q_i^h y_i$ is that during the consumption phase. The variable cost, $(q_i^h)^2$, is assumed to be quadratic in quantity.

Assuming quadratic reduction costs, components of the cost function of each firm are

$$C_{1P}\left(q_1^h x_1\right) = \frac{1}{2}\beta\left(q_1^h x_1\right)^2, \qquad C_{1C}\left(q_1^h y_1\right) = \frac{1}{2}\beta k\left(q_1^h y_1\right)^2$$

$$C_{2P}\left(q_2^l x_2\right) = \frac{1}{2}\left(q_2^l x_2\right)^2, \qquad C_{2C}\left(q_2^l x_2\right) = \frac{1}{2}k\left(q_2^l y_2\right)^2$$

where β is a parameter representing the technology gap between the countries, and k reflects the reduction cost structure of the product. By assuming $0 < \beta < 1$, firm 1 has more advanced technology for emission reduction than firm 2. If all the parts of the reduction technology can be acquired through outsourcing, there would be no reason for firm 1 to have more advanced technology. In most cases,

[4] Even if the reduction costs depend not on $q_i x_i$ and $q_i y_i$ but on x_i and y_i, we derive similar results. In other words, the results of this analysis are not much affected by the form of the reduction cost functions.

however, some parts of the abatement technology are embedded in the firm itself. In this case, the levels of technology are different from firm to firm. Developed countries have had a longer history of environmental regulation than developing ones. The firms in developed countries have had more chance to advance their abatement technology through "learning by doing" and therefore can be assumed to have more advanced technology.[5] If $k > 1$, emissions can be reduced more easily during the production phase than during the consumption phase. However, if $0 < k < 1$, it would be the opposite. When the emission tax t_i is imposed on unit emissions from the production phase of firm i, firm i's profit is denoted as

$$\Pi_i^h(p_i^h, x_i, y_i) = p_i^h q_i^h - (q_i^h)^2 - C_{iP}(q_i^h x_i) - C_{iC}(q_i^h y_i) - t_i q_i^h(\bar{x} - x_i) \quad (1)$$

for $i = 1,2$, and $h = g,n$

Analysis

Inclusion scheme

We can normalize $w = 1$ and $M = 1$ without a significant loss of generality.[6] Because the model is a three-stage game, we adopt backward induction to determine an equilibrium. At the third stage each firm takes its emission reductions and labeling outcome as given and chooses the price of its product to maximize its profit. The unique Bertrand-Nash equilibrium at this stage is given by

$$p_i^{g*} = 1 + \alpha$$
$$p_j^{n*} = 1$$

which is proved in the Appendix. Then the profits of the firms in the third stage are $\Pi_i^g(1 + \alpha, x_i, y_i)$ and $\Pi_j^n(1, x_i, y_i)$, for $i, j = 1, 2, i \neq j$.

---

[5] It is common that developed countries have higher marginal abatement cost than developing ones. That is the reason, for example, that the Clean Development Mechanism in the Kyoto Protocol is considered one of the good programs for tackling the global warming problem. We need to determine if the marginal abatement cost is higher in firm 1 before introducing eco-labeling. The marginal abatement cost of firm 1 is $MC_{1P} = \partial C_{1P}/\partial x_1 = \beta(q_1)^2 x_1{*}$. That of firm 2 is $MC_{2P} = \partial C_{2P}/\partial x_2 = (q_2)^2 x_2{*}$. The superscripts h and l are omitted because eco-labeling is not introduced. We obtain the optimal reduction levels $x_1{*} = t_1/\beta q_1$ and $x_2{*} = t_2/q_2$ by solving the problem $\max_{x_i y_i} \Pi_i(p_i, x_i, y_i)$. Thus, the marginal abatement costs of the firms are $MC_{1P} = q_1 t_1$ and $MC_{2P} = q_2 t_2$, respectively. We find out that the marginal abatement cost is directly related to the level of regulation, t_i. If there is no special reason to justify that their production levels are different before introducing eco-labeling, the marginal abatement cost is higher in firm 1 because of the assumption $t_1 > t_2$. Although we assumed that firm 1 has more advanced abatement technology, the marginal abatement cost is higher in firm 1, located in the developed country, because of stronger regulation.

[6] If we normalize $M = 1$, all expressions containing quantities can be considered to be given in "per capita" terms, as in Kuhn (1998).

Next we analyze the firms' behavior at the second and first stages. We can consider two kinds of Stackelberg equilibrium. The first is the equilibrium at which firm 1 obtains the eco-label. For the second, firm 2 obtains the eco-label. What determines the kind of equilibrium is the maximum willingness of the firm to reduce emissions, thereby earning the eco-label, but still maintaining their profits at a level that is not less than the profits without the eco-label. If firm 2 can reduce emissions more, it would obtain the eco-label; otherwise, firm 1 would obtain it.

Let us have more explanation. At the second stage, given x_1 and y_1, firm 2 roughly considers two choices. Firm 2 either obtains the eco-label (in which case, $x_2 + y_2$ should be greater than $x_1 + y_1$) or abandons the eco-label. Firm 2's decision is dependent on $x_1 + y_1$ and its maximum willingness to reduce emissions. If $x_1 + y_1$ is lower than firm 2's maximum willingness to reduce emissions, firm 2 would reduce emissions slightly more than $x_1 + y_1$ and acquire the eco-label. On the other hand, if $x_1 + y_1$ is higher than the maximum and it is not profitable for firm 2 to reduce emissions more than $x_1 + y_1$ to be eco-labeled, it would give up the eco-label. At the first stage, firm 1 can anticipate firm 2's reaction. Therefore, if firm 1's maximum willingness to reduce emissions is not less than that of firm 2, firm 1 could obtain the eco-label; otherwise, firm 2 would get it. Thus, we need to know the firms' maximum willingness to reduce emissions.

To ascertain firm 2's maximum willingness to reduce emissions, we first, should know how much profit firm 2 could earn if it chooses not to gain the eco-label. The optimal reduction levels of firm 2, (x_2^*, y_2^*), can solve the following problem:

$$\max_{x_2,y_2} \Pi_2^n(1, x_2, y_2) \tag{P-1}$$

which yields

$$x_2^* = \frac{t_2}{1-r}$$
$$y_2^* = 0$$

As the level of the emission tax t_2 increases, firm 2 reduces more emission per product during the production phase. Meanwhile, the firm pays no attention to the consumption phase because it expects no reward for emission reduction during that phase. This behavior is the same as that before we introduced eco-labeling. Firm 2's profit without the eco-label is $\Pi_2^n[1, t_2/(1-r), 0]$.

Assume that being labeled provides more profits to firms than not being labeled, ceteris paribus both firms want to acquire the eco-label. Otherwise, eco-labeling would have no effect. The condition is described as[7]

[7] We can obtain $(x_2, y_2) = (t_2/r, 0)$ by solving the problem $\max_{x_2,y_2} \Pi_2^g(1+\alpha, x_2, y_2)$. We can also obtain $(x_1, y_1) = (t_1/\beta r, 0)$ and $(x_1, y_1) = [t_1/\beta(1-r), 0]$ by solving the problems, $\max_{x_1,y_1} \Pi_1^g(1+\alpha, x_1, y_1)$ and $\max_{x_1,y_1} \Pi_1^n(1, x_1, y_1)$, respectively.

$$\Pi_2^g(1+\alpha,t_2/r,0) - \Pi_2^n[1,t_2/(1-r),0] \geq 0 \quad (2)$$

$$\Pi_1^g(1+\alpha,t_1/\beta r,0) - \Pi_1^n[1,t_1/\beta(1-r),0] \geq 0 \quad (3)$$

The condition satisfies if and only if

$$r\alpha + (1-2r)t_2\bar{x} \geq 0 \quad (4)$$

$$r\alpha + (1-2r)t_1\bar{x} \geq 0 \quad (5)$$

The left-hand sides of Eqs. (2) and (3) [i.e., Eqs. (4) and (5)] can be interpreted as the maximum benefits the firms can earn by acquiring the eco-label. We assume $r \leq 1/2$ because this assumption satisfies Eqs. (4) and (5) and is close to reality. The benefits then rise in proportion to the premium α and the emission tax level t_i. Maximum benefits could be realized if there was no competition. It can be expected that the more the profit increases the more reduction efforts the firms exert.

We can then draw the maximum willingness of firm 2 to reduce emissions to acquire the eco-label. When firm 2 obtains the eco-label, its profit should not be less than $\Pi_2^n[1, t_2/(1-r), 0]$, as firm 2 could earn that profit even if it did not obtain the eco-label. Therefore, the profile that achieves the maximum willingness to reduce (x_2^I, y_2^I),[8] solves

$$\max_{x_2,y_2} x_2 + y_2 \quad \text{(P-2)}$$

$$\text{subject to } \Pi_2^g(1+\alpha,x_2,y_2) \geq \Pi_2^n[1,t_2/(1-r),0]$$

Using Kuhn-Tucker conditions, we get

$$r^2x_2^I - t_2r = kr^2y_2^I \quad (6)$$

$$\Pi_2^n[1,t_2/(1-r),0] - \Pi_2^g(1+\alpha,x_2^I,y_2^I) = 0 \quad (7)$$

Equation (6) describes the condition of firm 2's reduction allocation between the two phases. It denotes that the net marginal costs of reduction during both the production and consumption phases are equalized at (x_2^I,y_2^I). If the firm reduces one unit of emission per product during the production phase, it costs x_2r^2 additionally but saves on the emission tax as much as t_2r for the firm. Thus, the net reduction cost during the production phase for the firm is $x_2r^2 - t_2r$. Reduction of one unit of emission during the consumption phase costs an additional kr^2y_2. Therefore, the firm makes the net marginal costs equal at (x_2^I,y_2^I) to maximize the reduction sum. It is notable that the marginal reduction cost during the production phase, $x_2^I r^2$, is always greater than the marginal reduction cost during the consumption phase, $kr^2y_2^I$. This is because there are double incentives for the production phase. If the firm reduces emissions during the production phase, it could not only save on the

[8] The superscript I represents the inclusion scheme.

emission tax but also increase chances for the firm to obtain the eco-label. Meanwhile, reduction during the consumption phase works only on the latter. This induces the firm to pay more attention to the production phase.

Equation (7) indicates that firm 2's profit with the eco-label is equal to its profit without the eco-label at the solution, (x_2^I, y_2^I). From Eqs. (6) and (7), we obtain

$$x_2^I = \frac{t_2}{r} + \sqrt{2}k\sqrt{\frac{r\alpha + (1-2r)t_2\bar{x}}{(k+k^2)r^2}}$$

$$y_2^I = \sqrt{2}\sqrt{\frac{r\alpha + (1-2r)t_2\bar{x}}{(k+k^2)r^2}}$$

Remember that we assumed $r\alpha + (1 - 2r)t_2\bar{x} \geq 0$ in Eq. (4). Therefore, firm 2's maximum willingness to reduce is

$$x_2^I + y_2^I = \frac{t_2}{r} + \sqrt{\frac{2(k+1)}{k}}\sqrt{\frac{r\alpha + (1-2r)t_2\bar{x}}{r^2}} \tag{8}$$

With the same logic, we can formulate the problem to ascertain firm 1's maximum willingness to reduce as follows:

$$\max_{x_1,y_1} x_1 + y_1 \tag{P-3}$$

$$\text{subject to } \Pi_1^g(1+\alpha, x_1, y_1) \geq \Pi_1^n[1, t_1/(1-r), 0]$$

The following solution is derived from Kuhn-Tucker conditions.

$$x_1^I = \frac{t_1}{\beta r} + k\sqrt{\frac{2}{\beta}}\sqrt{\frac{r\alpha + (1-2r)t_1\bar{x}}{(k+k^2)r^2}}$$

$$y_1^I = \sqrt{\frac{2}{\beta}}\sqrt{\frac{r\alpha + (1-2r)t_1\bar{x}}{(k+k^2)r^2}}$$

Then, firm 1's maximum willingness to reduce emissions is

$$x_1^I + y_1^I = \frac{t_1}{\beta r} + \sqrt{\frac{2(k+1)}{\beta k}}\sqrt{\frac{r\alpha + (1-2r)t_1\bar{x}}{r^2}} \tag{9}$$

Of the two kinds of equilibrium mentioned previously, the more interesting is the equilibrium where firm 1 obtains the eco-label.[9] For firm 1 to obtain the eco-

[9] In practice, most firms that earn eco-labels are in developed countries.

label at equilibrium, firm 1's maximum willingness to reduce should not be less than that of firm 2. The condition is as follows:

$$\frac{t_1}{\beta r} + \sqrt{\frac{2(k+1)}{\beta k}}\sqrt{\frac{r\alpha + (1-2r)t_1\bar{x}}{r^2}} \geq \frac{t_2}{r} + \sqrt{\frac{2(k+1)}{k}}\sqrt{\frac{r\alpha + (1-2r)t_2\bar{x}}{r^2}} \tag{10}$$

The condition satisfies because $t_2 < t_1$, $0 < \beta < 1$, and $r \leq 1/2$. Therefore, at equilibrium firm 1 would choose to obtain the eco-label and firm 2 would not.

Then, at equilibrium, if $x_1 + y_1 \geq x_2^I + y_2^I$, firm 2 chooses the emission reduction levels $x_2^* = t_2/(1-r)$ and $y_2^* = 0$. Firm 1 tries to maximize its profit subject to the constraint that its total reduction level per product is not less than firm 2's maximum willingness to reduce, $x_2^I + y_2^I$. Firm 1's optimal reduction level $(x_1^I{*}, y_1^I{*})$ is derived from the problem

$$\begin{aligned} &\max_{x_1,y_1} \Pi_1^g(1+\alpha, x_1, y_1) \\ &\text{subject to } x_1 + y_1 \geq x_2^I + y_2^I \end{aligned} \tag{P-4}$$

The Kuhn-Tucker conditions generate[10]

$$\beta r^2 x_1^I{*} - t_1 r = \beta k r^2 y_1^I{*} \tag{11}$$

$$x_1^I{*} - y_1^I{*} = x_2^I + y_2^I \tag{12}$$

Equation (11) denotes reduction allocation between phases. As we have seen in Eq. (6), firm 1 also makes more reduction efforts during the production phase because of the emission tax on that phase. Equation (12) expresses the total reduction per product of firm 1 that is necessary to obtain the eco-label. Solving the equations, we get

$$x_1^I{*} = \frac{t_1}{\beta(1+k)r} + \frac{k}{(1+k)}\left(x_2^I + y_2^I\right)$$

$$y_1^I{*} = \frac{-t_1}{\beta(1+k)r} + \frac{1}{(1+k)}\left(x_2^I + y_2^I\right)$$

We conclude the subgame perfect outcome under the inclusion scheme[11]

---

[10] We consider the case in which the constraint is binding. If the constraint is not binding, the analysis becomes trivial. The constraint is binding when the following conditions satisfy (see Appendix).

$$\frac{t_2}{r} + \sqrt{\frac{2(k+1)}{k}}\sqrt{\frac{r\alpha + (1-2r)t_2\bar{x}}{r^2}} > \frac{t_1}{\beta r}$$

The condition means that firm 2's maximum willingness to reduce should be greater than the optimal reduction level of firm 1 with no competition. The condition makes competition work between the firms.

[11] It is notable that the subgame perfect outcome differs in concept from the subgame perfect equilibrium. The difference is that an equilibrium is a collection of strategies (and a strategy is a complete plan of action), whereas an outcome describes what will happen only in the contingency that are expected to arise, not in every contingency that might arise (Gibbons 1992).

$$\left(x_1^{I}*,y_1^{I}*,p_1^{g}*\right)=\left(\frac{t_1+\beta k t_2}{\beta(1+k)r}+\sqrt{2}k\sqrt{\frac{r\alpha+(1-2r)t_2\bar{x}}{(k+k^2)r^2}}\frac{-t_1+\beta t_2}{\beta(1+k)r}\right.$$
$$\left.+\sqrt{2}\sqrt{\frac{r\alpha+(1-2r)t_2\bar{x}}{(k+k^2)r^2}},1+\alpha\right)$$

$$\left(x_2*,y_2*,p_2^{n}*\right)=\left(\frac{t_2}{1-r},0,1\right)$$

Exclusion scheme

Even if country 1 uses the exclusion scheme, the behaviors of the firms in the third stage do not change. Employing the logic used in the inclusion scheme, we can also determine the maximum willingness to reduce emissions under the exclusion scheme. Note that the maximum willingness to reduce emissions in the exclusion scheme is not total reduction per product but the reduction per product during the consumption phase because the scheme cares only about the emissions from the consumption phase.

Firm 2's profit without the eco-label is $\Pi_2^n[1,t_2/(1-r),0]$, as in the previous analysis. Firm 2's maximum willingness to reduce emissions is derived from the following problem:

$$\max_{x_2,y_2} y_2 \tag{P-5}$$
$$\text{subject to } \Pi_2^g(1+\alpha,x_2,y_2)\geq\Pi_2^n\left[1,t_2/(1-r),0\right]$$

which yields

$$x_2^E=\frac{t_2}{r} \tag{13}$$

$$\Pi_2^n\left(1,t_2/(1-r),0\right)-\Pi_2^g\left(1+\alpha,x_2^E,y_2^E\right)=0 \tag{14}$$

Equation (13) states that the reduction during the production phase is equal to the reduction by the firm without the eco-label. This is because the emission tax is the only incentive to reduce emissions during the production phase. We know from Eq. (14) that firm 2 would reduce emissions during the consumption phase to the point where its profit with the eco-label is equal to that without the eco-label. Solving the equations, we obtain

$$x_2^E=\frac{t_2}{r}$$

$$y_2^E = \sqrt{2}\sqrt{\frac{r\alpha + (1-2r)t_2\bar{x}}{kr^2}}$$

Firm 2's maximum willingness to reduce emissions is y_2^E. Firm 1's maximum willingness to reduce emissions, y_1^E, is derived from

$$\max_{x_1,y_1} y_1 \qquad \text{(P-6)}$$
$$\text{subject to } \Pi_1^g(1+\alpha, x_1, y_1) \geq \Pi_1^n[1, t_1/(1-r), 0]$$

Solving the problem, we obtain

$$x_1^E = \frac{t_1}{\beta r}$$

$$y_1^E = \sqrt{\frac{2}{\beta}}\sqrt{\frac{r\alpha + (1-2r)t_1\bar{x}}{kr^2}}$$

It is clear that y_1^E is greater than y_2^E because $0 < \beta < 1$, and $r \leq 1/2$.

Then, at equilibrium firm 2 chooses the emission reduction levels $x_2^* = t_2/(1 - r)$ and $y_2^* = 0$. Firm 1's optimal reduction level to obtain the eco-label can be obtained by solving the following problem.

$$\max_{x_1,y_1} \Pi_1^g(1+\alpha, x_1, y_1) \qquad \text{(P-7)}$$
$$\text{subject to } y_1 \geq y_2^E$$

which yields[12]

$$x_1^{E*} = \frac{t_1}{\beta r}$$

$$y_1^{E*} = y_2^E$$

Then the subgame perfect outcome in the exclusion scheme is

$$\left(x_1^{E*}, y_1^{E*}, p_1^{g*}\right) = \left(\frac{t_1}{\beta r}, \sqrt{2}\sqrt{\frac{r\alpha + (1-2r)t_2\bar{x}}{kr^2}}, 1+\alpha\right)$$

$$\left(x_2^*, y_2^*, p_2^{n*}\right) = \left(\frac{t_2}{1-r}, 0, 1\right)$$

[12] The constraint is always binding. See Appendix.

Comparison

The outcomes of the two schemes are the same except for firm 1's emission reduction. We therefore compare firm 1's emission reduction for the two schemes in terms of effectiveness. Effectiveness implies the success scheme that induces more total emission reduction per product of firm 1. The following are firm 1's total reductions of the schemes.

$$x_1^I* + y_1^I* = \frac{t_2}{r} + \sqrt{\frac{2(k+1)}{k}}\sqrt{\frac{r\alpha + (1-2r)t_2\bar{x}}{r^2}} \tag{15}$$

$$x_1^E* + y_1^E* = \frac{t_1}{\beta r} + \sqrt{\frac{2}{k}}\sqrt{\frac{r\alpha + (1-2r)t_2\bar{x}}{r^2}} \tag{16}$$

The t_2/r is less than $t_1/\beta r$ because $t_2 < t_1$ and $\beta < 1$. The second term in Eq. (15) is greater than that in Eq. (16) because $\sqrt{k+1} > 1$. Therefore, the effectiveness depends on the values of the coefficients. Expressions t_2/r and $t_1/\beta r$ are the reduction levels of firms 2 and 1, respectively, before introducing eco-labeling. Therefore, we find that if the firms' reduction levels are greatly different before introducing eco-labeling the exclusion would be more effective. Under the inclusion scheme firm 1 can leverage the reduction level $t_1/\beta r$ that it accomplished before introducing eco-labeling to obtain the reduction level of Eq. (15) (i.e., to obtain the eco-label). Under the exclusion scheme, however, firm 1 cannot use the reduction level $t_1/\beta r$ to obtain the eco-label. This is the reason the exclusion scheme can induce more reduction.

We also find that the smaller k is, the smaller is the difference between the second terms. It can make $x_1^E* + y_1^E*$ greater than $x_1^I* + y_1^I*$ because $t_1/\beta r$ is greater than t_2/r. This means that the exclusion scheme is more effective for the products that have a relatively low abatement cost during the consumption phase. Under the inclusion scheme, firm 1 pays more attention to the production phase, as we saw in Eq. (11), and it reduces emissions during the consumption phase as much as possible, as we saw in Eq. (14). Therefore, when the consumption phase does not have much potential for reducing emissions, the reduction levels during the consumption phases are not much different from scheme to scheme, and hence the inclusion scheme is better. However, if the consumption phase has a relatively low abatement cost, during the consumption phase the exclusion scheme could lead to much more reduction than the inclusion scheme.

This result might seem trivial because the exclusion scheme considers only the reduction during the consumption phase; but from the perspective of the advocates of life cycle assessment (LCA) the inclusion scheme is better for all products. They think that even though the reduction cost is different from phase to phase under the inclusion scheme the firm would make the marginal reduction cost equal during both phases to reduce emissions even more. However, the inclusion scheme gives more incentive to reduce emissions during the production phase than during the consumption phase because of the emission tax for the production phase. This makes the inclusion scheme less effective when k is low. The result can be a counter-example to the advocates of LCA.

Among the other parameters, r and $\bar{x}$ do not often change in practice. The parameter α, which is a premium for the eco-label, is worth considering. We find that the smaller α is, the lower are the effects of the second terms in Eqs. (15) and (16). This increases the possibility that Eq. (16) is greater than Eq. (15). This result implies that if the consumers' willingness to pay for the green product is not high the exclusion scheme could be more effective. We think that the exclusion scheme is more effective at the beginning of eco-labeling when the premium for the eco-label is low.

We should also compare the schemes in terms of efficiency. Efficiency denotes which scheme leads to lower reduction cost when the reduction levels are equal. We cannot compare the reduction cost directly because the reduction levels are different in the two schemes. Therefore, we check the efficiency of the schemes indirectly by comparing the differences of the marginal reduction costs under the two schemes. For each scheme, the differences of the marginal reduction costs are

$$MRC_{1P}\left(rx_1^{I}*\right) - MRC_{1C}\left(ry_1^{I}*\right) = rt_1 \tag{17}$$

$$MRC_{1P}\left(rx_1^{E}*\right) - MRC_{1C}\left(ry_1^{E}*\right) = rt_1 - \beta\sqrt{2kr^2}\sqrt{r\alpha + (1 - 2r)t_2\bar{x}} \tag{18}$$

where $MRC_{1P}(q_1x_1)$ and $MRC_{1C}(q_1y_1)$ are the marginal reduction costs during the production phase and the consumption phase, respectively. They are derived by differentiating $C_{1P}(q_1x_1)$ and $C_{1C}(q_1y_1)$ with respect to x_1 and y_1. The ideal situation is that the marginal reduction costs are equal. The greater the difference, the more inefficiently is firm 1 reducing emissions. So long as $t_1 > 0$, neither scheme can reach the ideal situation. There is less difference in the marginal reduction cost with the exclusion scheme when rt_1 is not particularly small. This means that firm 1 reduces emissions more efficiently under the exclusion scheme. Even if the exclusion scheme generates more reductions, it might cost less, as the double incentives during the production phase of the inclusion scheme could lead to excessive reduction during the production phase and too little reduction during the consumption phase.

To complete the comparison, it is necessary to consider both emission reduction and the reduction cost from the social perspective simultaneously. The social welfare function of country 1 is given by

$$SW_1 = CS + \Pi_1 + t_1q_1\left(\bar{x} - x_1\right) - de_1^2$$

where e_1 is the total amount of pollutants emitted in country 1, and d is a damage parameter. The pollutants are composed of emissions from domestic production and consumption as follows:

$$e_1 = q_1\left(\bar{x} - x_1 + \bar{y} - y_1\right) + q_2\left(\bar{y} - y_2\right)$$

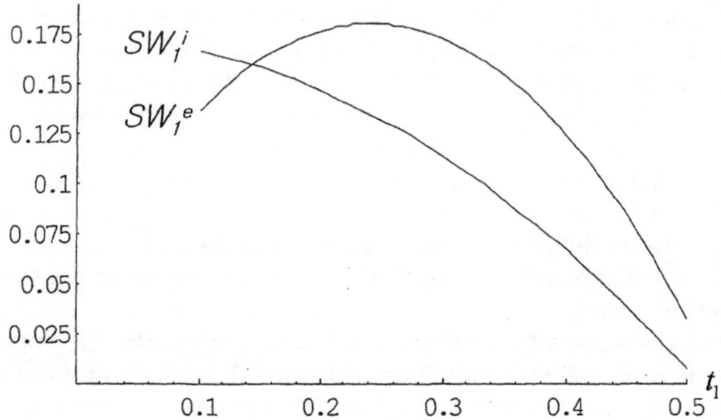

Fig. 1. Social welfare of the schemes on t_1 when $\bar{x} = 1$, $\bar{y} = 1$, $\beta = 0.5$, $r = 0.4$, $\alpha = 0.2$, $k = 0.5$, $t_2 = 0.1$, and $d = 0.3$

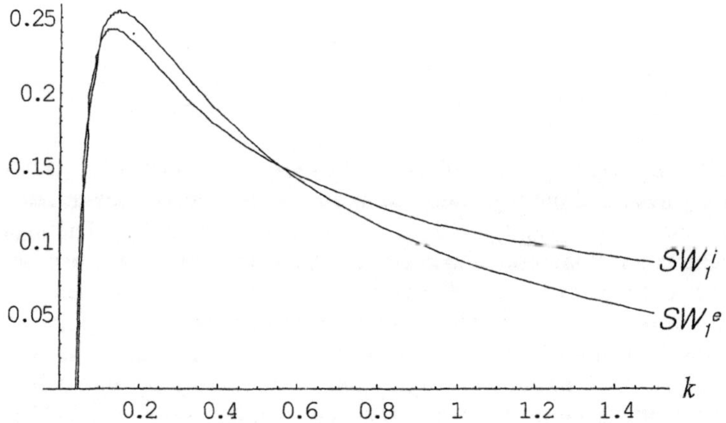

Fig. 2. Social welfare of the schemes on k when $\bar{x} = 1$, $\bar{y} = 1$, $\beta = 0.5$, $r = 0.4$, $\alpha = 0.2$, $t_1 = 0.15$, $t_2 = 0.1$, and $d = 0.3$

It is difficult to clarify in parameters the conditions under which one scheme is better than the other because there are several parameters and the equations are complicated. Instead, we present a numerical example that may be representative.

Figure 1 depicts the social welfare under the schemes in the case that $\bar{x} = 1$, $\bar{y} = 1$, $\beta = 0.5$, $r = 0.4$, $\alpha = 0.2$, $k = 0.5$, $t_2 = 0.1$, $d = 0.3$, and the horizontal axis is t_1. SW_1^i denotes country 1's social welfare under the inclusion scheme and the SW_1^e its social welfare under the exclusion scheme. Figure 2 depicts the social welfare when the condition is the same as in Fig. 1 except that $t_1 = 0.15$ and the horizontal axis is k.

The figures state that the exclusion scheme is socially desirable when the emission tax level of country 1 is relatively high and the reduction cost during the consumption phase is low compared to that for the production phase. This means that even though the exclusion scheme induces more emission reductions from firm 1, the reduction cost is not much more with the exclusion scheme than with the inclusion scheme.

Policy implications

The ISO subcommittee on environmental labeling, ISO/TC 207/SC3, has been attempting to agree on the procedures and principles to underpin all current and future eco-labeling schemes in the ISO 14020 series. ISO 14020 in Principle 5 requires inclusion of non-product-related PPMs in eco-labeling criteria by mentioning that "the development of environmental labels and declarations shall take into consideration all relevant aspects of the life cycle of the product." Article 5.6 in ISO 14024 also demands "life cycle consideration" to fulfill the objective of reducing the environmental impact and to avoid transfer of the effects across media or phases of the product life cycle. Other international organizations, such as WTO/CTE, OECD, and UNCTAD, are fully aware that non-product-related PPMs in eco-labeling may result in trade conflicts, and so guidance is needed on how to deal with them. Nevertheless, they have not yet introduced clear guidelines (Henry 1997; WTO 1999b).

Even though this research does not give a complete guideline for non-product-related PPMs, there are some new results. When the regulation gap between the countries is large, the exclusion scheme is more effective. For the products that have much potential for reducing emissions during the consumption phase, the exclusion scheme is more effective. Moreover, at the beginning of eco-labeling when consumers have low willingness to pay for green products the exclusion scheme is better. This research can contribute more or less to the future research on making full guidelines for non-product-related PPMs.

Non-product-related PPMs could be classified into two types depending on whether the resulting environmental effects are global or local. This paper has analyzed the case of local effects. It has shown that it is socially desirable to exclude non-product-related PPMs that have local environmental effects from eco-labeling criteria, provided the emission tax in the importing country is relatively higher than that in the exporting country and the product has low reduction cost during the consumption phase compared to the production phase.

Emission tax levels in most countries might be substantially lower than the socially optimal level. Nevertheless, there is some difference between the developed and developing countries. Most of the developed countries, especially in Europe, have higher emission taxes. In addition, the results of LCA studies of many consumer products (e.g., household equipment, detergents, heaters, paints) have often concluded that the emission proportion of the production phase is much less than that of the consumption phase (Neitzel 1998). It is clear that the more emissions produced during the consumption phase the more potential there is to reduce these emissions and therefore the less reduction cost there is during that phase. Hence, it is desirable for developed countries to adapt the exclusion schemes for those product groups. Some products (e.g., textiles) emit the most pollutants during the production phase. For these product groups the inclusion scheme might be proper.

The analysis thus far is from the perspective of the developed country employing eco-labeling. Is the exclusion scheme also desirable from the global viewpoint? The exclusion scheme has an advantage that it can mitigate some trade concerns between developed and developing countries. It could cause concern, however, about environmental deterioration in developing countries. It depends on whether the developing countries manage their environment properly or try to do eco-dumping for their industrial competitiveness, when non-product-related PPMs are excluded from the eco-labeling criteria. This is a delicate problem that is not yet clearly proven.

One of the main implications of this analysis is that the appropriate mix of existing regulations and eco-labeling is important. Each nation has several existing tools for each phase of the product life cycle (from raw materials to disposal) to regulate pollution. Even if each tool is designed to be efficient alone, the mix of the tools might be inefficient as a whole. Therefore, what matters is how to mix several tools. In many countries the emission tax is designed by an environmental agency, and eco-labeling is designed by an eco-labeling body that might be independent of the agency. In such a situation, coordination between them is important to prevent duplication of regulation.

This study has considered the case in which the existing regulation on the production phase is in the form of an emission tax. If the existing regulation is in the form of a tradable permit system or an emission standard, we might need additional analyses. However, we can conjecture as follows. If the regulatory tool is a tradable permit system and the market for the tradable permits is competitive, the result would be the same as in the previous analysis because the firms consider the price of the permits as a kind of emission tax. If the regulatory tool is an emission standard, however, the effect of eco-labeling would be better in the inclusion scheme. This is because the emission standard does not give double incentives to emission reduction during the production phase.

Conclusions

The question in this research is whether to include or exclude non-product-related PPMs in eco-labeling criteria in the presence of traditional regulations. We have shown that the exclusion scheme would be socially desirable when the emission tax is high, and it is relatively easy to reduce emissions during the consumption phase. Based on these results, it may be that for some consumer product groups we need a policy mix that applies eco-labeling schemes to internalize externalities generated from all the phases of the product life cycle except the production phase and uses other regulatory instruments to internalize externalities during production. In this way, without including non-product-related PPMs in eco-labeling criteria, this policy mix can cover the whole life cycle of products. This study has shown the importance of policy mix, while introducing eco-labeling, which we may overlook.

At this point we propose four topics for further research. First, it is necessary to investigate what to do with non-product-related PPMs that have global effects. Second, we can consider examining the supply chain including the production phase. It was assumed in the model that only one firm exists during the production phase. However, production phases of some products contain chains of several firms called a supply chain. The effect of eco-labeling criteria may rely on the structure of the supply chain. Third, some industries such as the textile industry have concentrated emissions during the production phase only. Another analysis is necessary for such industries because this study assumes that both the production and consumption phases emit pollutants. Finally, an interesting issue for eco-labeling criteria is to analyze under which scheme consumers are more willing to buy eco-labeled products: the inclusion scheme or the exclusion scheme. According to Osterhus (1997), the more responsible consumers feel, the more environmentally consciously they behave. Do consumers feel responsible for the pollution emitted during the production phase of products? If they do not feel responsible for this pollution the exclusion scheme might be more acceptable to consumers than the inclusion scheme.

Appendix

1 Proof of the Bertrand-Nash equilibrium at the third stage

(a) *Proof of the Bertrand-Nash equilibrium*

At ($p_g^* = 1 + \alpha$, $p_n^* = 1$), the profits of the firms are

$$\Pi_i^g(1+\alpha, x_i, y_i) = r - r^2 + \alpha r - C_{iP}(rx_i) - C_{iC}(ry_i) - t_i r(\bar{x} - x_i)$$

$$\Pi_j^n(1, x_j, y_j) = r - r^2 - C_{jP}[(1-r)x_j] - C_{jC}[(1-r)y_j] - t_j(1-r)(\bar{x} - x_j)$$

To meet individual rationality, assume that

$$\Pi_i^g(1+\alpha, x_i, y_i) \geq 0, \qquad \Pi_j^n(1, x_j, y_j) \geq 0$$
$$\text{for all } x_i, y_i, x_j, y_j, \qquad i, j = 1, 2, \qquad i \neq j$$

If (p_g^*, p_n^*) is not a Nash equilibrium, either firm would have an incentive to deviate.

(i) No incentive for firm g to deviate

Suppose that there is $\tilde{p}_g$ ($\neq 1 + \alpha$) where firm g can make more profit given $p_n = 1$. If $\tilde{p}_g > 1 + \alpha$, $q_g = 0$; then $\Pi_i^g(\tilde{p}_g, x_i, y_i) = 0 \leq \Pi_i^g(1 + \alpha, x_i, y_i)$. No incentive to deviate. If $1 < \tilde{p}_g < 1 + \alpha$, $q_g = r$. $\Pi_i^g(\tilde{p}_g, x_i, y_i) = \tilde{p}_g r - r^2 - C_{iP}(rx_i) - C_{iC}(ry_i) - t_i r(\bar{x} - x_i) < \Pi_i^g(1 + \alpha, x_i, y_i)$. No incentive to deviate. If $\tilde{p}_g \leq 1$, $q_g = 1$. $\Pi_i^g(\tilde{p}_g, x_i,$

$y_i) = \tilde{p}_g - 1^2 - C_{iP}(x_i) - C_{iC}(y_i) - t_i(\bar{x} - x_i) < \Pi_i^g(1 + \alpha, x_i, y_i)$ because $\tilde{p}_g - 1^2 \leq 0 < r - r^2 + \alpha r$. No incentive to deviate. A contradiction.

(ii) No incentive for firm n to deviate

Suppose there is $\tilde{p}_n(\neq 1)$ where firm n can make more profit given $p_g = 1 + \alpha$. If $\tilde{p}_n > 1$, $q_n = 0$; then $\Pi_j^n(\tilde{p}_n, x_j, y_j) = 0 \leq \Pi_j^n(1, x_j, y_j)$. No incentive to deviate. If $\tilde{p}_n < 1$, $q_n = 1$. Then $\Pi_j^n(\tilde{p}_n, x_j, y_j) = \tilde{p}_n - 1^2 - C_{jP}(x_j) - C_{jC}(y_j) - t_j(\bar{x} - x_j) < \Pi_j^n(1, x_j, y_j)$ because $\tilde{p}_n - 1^2 < r - r^2$. No incentive to deviate.

(b) *Proof of uniqueness*

It is easy to check that given $p_n = h(\leq 1)$ the best reaction of firm g is $p_g = h + \alpha$. Meanwhile, the firm n's best reaction to $p_g = h + \alpha$ is $p_n = \text{argmin}\{h + \alpha - \varepsilon, 1\}$. Hence, if $h \neq 1$, there is no Nash equilibrium. ■

2 Two-stage game: simultaneous move

If it is assumed that the firms choose to reduce emissions simultaneously, the structure of the game is changed to a two-stage game.

Stage 1: Each firm decides on the emission levels of its own product during the production and consumption phases. According to the labeling criteria, the more environmentally friendly product of the two firms' products is granted the g (green) label. The other product is recognized as n (nongreen).

Stage 2: Taking as given the labels the firms have been granted, they establish the prices of their products to maximize profits.

We analyze the model in the inclusion scheme. The equilibrium at the second stage of the two-stage game is the same as the equilibrium at the third stage of the three-stage game (Stackelberg game). However, there is no Nash equilibrium for emission reduction during the first stage of the two-stage game.

(a) *Proof of nonexistence of Nash equilibrium during the first stage*

By way of contradiction, suppose that $\{(x_i^N, y_i^N), (x_j^N, y_k^N)\}$ constitutes a Nash equilibrium. A Nash equilibrium is an emission profile, $\{(x_1^N, y_1^N), (x_2^N, y_2^N)\}$, such that for a given (x_2^N, y_2^N), firm 1 chooses (x_1^N, y_1^N) to maximize Π_i^g and for a given (x_1^N, y_1^N), firm 2 chooses (x_2^N, y_2^N) to maximize Π_2^n. Then, there are two cases: (i) $x_1^N + y_1^N = x_2^N + y_2^N$; and (ii) $x_i^N + y_i^N > x_j^N + y_j^N$.

(i) Suppose that $x_1^N + y_1^N = x_2^N + y_2^N$. In this case, firm 1 would obtain the eco-label because of home turf advantage. Then firm 2 would deviate by slightly decreasing its emission reductions to $(\tilde{x}_2, \tilde{y}_2)$, such that $\tilde{x}_2 + \tilde{y}_2 < x_1^N + y_1^N$, and obtain the eco-label from firm 1. This would increase firm 2's profit from $\Pi_2^n(1, x_2^N, y_2^N)$ to $\Pi_2^g(1 + \alpha, \tilde{x}_i, \tilde{y}_i)$. A contradiction.

(ii) Suppose that $x_i^N + y_i^N > x_j^N + y_j^N$. Then, firm i would deviate by decreasing its emission reductions to $(\tilde{x}_i, \tilde{y}_i)$ such that $x_j^N + y_j^N < \tilde{x}_i + \tilde{y}_i < x_i^N + y_i^N$, $\Pi_i^g(1 + \alpha, \tilde{x}_i, \tilde{y}_i) > \Pi_i^g(1 + \alpha, x_i^N, y_i^N)$ and then increase its profit to $\Pi_i^g(1 + \alpha, \tilde{x}_i, \tilde{y}_i)$. A contradiction. ■

Therefore, we need to introduce another equilibrium concept. We employ an overcut-proof equilibrium. In an overcut-proof equilibrium, a labeled firm chooses the emission reduction levels to maximize its profit, subject to the constraint that the reduction levels are sufficiently high that the unlabeled firm would not find it more profitable to set sufficiently higher reduction levels to obtain the eco-label.[13]

To determine an overcut-proof equilibrium, we need to know how much emission firm 2 can reduce in total to obtain the eco-label. When firm 2 obtains the eco-label, its profit should not be less than $\Pi_2^n[1, t_2/(1-r), 0]$. This is because firm 2 could earn profit as much as $\Pi_2^n[1, t_2/(1-r), 0]$ even without the eco-label. Therefore, the problem is

$$\max_{x_2, y_2} x_2 + y_2 \qquad \text{(P-2')}$$

$$\text{subject to } \Pi_2^g(1+\alpha, x_2, y_2) \geq \Pi_2^n\left[1, t_2/(1-r), 0\right]$$

The problem (P-2′) is the same as Problem (P-2). Thus, the solution of the problem is (x_2^I, y_2^I), as in Problem (P-2).

Firm 1 would guess these reduction levels. Then firm 1 would try to maximize its profit subject to the constraint that its total reduction level is not less than firm 2's maximum $x_2^I + y_2^I$. Firm 1's optimal reduction levels solve

$$\max_{x_1, y_1} \Pi_1^g(1+\alpha, x_1, y_1) \qquad \text{(P-4')}$$

$$\text{subject to } x_1 + y_1 \geq x_2^I + y_2^I$$

Problem (P-4′) is the same as Problem (P-4). Therefore, the undercut-proof equilibrium is identical to the subgame perfect outcome of the three-stage game. The same logic can be applied to the exclusion scheme.

3 Solution of problem (P-4) and the condition in which the constraint is binding

The objective function is concave and the constraint is linear—hence convex. Thus, the Kuhn-Tucker conditions are both necessary and sufficient conditions for optimality. The Largrangian is

$$L(x_1, y_1, \lambda) = \Pi_1^g(1+\alpha, x_1, y_1) + \lambda\left(x_2^I + y_2^I - x_1 - y_1\right)$$

The Kuhn-Tucker conditions are

$$\frac{\partial L}{\partial x_1} = -\beta r^2 x_1^{I*} + t_1 r + \lambda = 0 \qquad \text{(1-1)}$$

$$\frac{\partial L}{\partial y_1} = -\beta k r^2 y_1^{I*} + \lambda = 0 \qquad \text{(2-1)}$$

$$x_2^I + y_2^I - x_1^{I*} - y_1^{I*} \leq 0 \qquad \text{(3-1)}$$

[13] We borrowed the concept of the overcut-proof equilibrium from the undercut-proof equilibrium of Shy (1995). For a rigorous definition of an undercut-proof equilibrium, see page 160 in Shy (1995).

$$\lambda \geq 0 \tag{3-2}$$

$$\lambda\left(x_2^I + y_2^I - x_1^{I*} - y_1^{I*}\right) = 0 \tag{3-3}$$

If $\lambda = 0$ (i.e., the constraint is not binding), $x_2^I + y_2^I - x_1^{I*} - y_1^{I*} \leq 0$ and Condition (3-3) satisfies. We obtain $x_1^{I*} = t_1/\beta r$ and $y_1^{I*} = 0$ from Conditions (1-1) and (2-1), respectively. For the solution not to violate Condition (3.1), the following should satisfy:

$$\frac{t_2}{r} + \sqrt{\frac{2(k+1)}{k}}\sqrt{\frac{r\alpha + (1-2r)t_2\bar{x}}{r^2}} \leq \frac{t_1 r}{\beta}$$

Therefore, the constraint is binding if

$$\frac{t_2}{r} + \sqrt{\frac{2(k+1)}{k}}\sqrt{\frac{r\alpha + (1-2r)t_2\bar{x}}{r^2}} > \frac{t_1 r}{\beta}$$

If $\lambda > 0$ (the constraint is binding), $x_2^I + y_2^I - x_1^{I*} - y_1^{I*} = 0$ from Condition (3-3). From Conditions (1-1) and (2-1) we get

$$\beta r^2 x_1^{I*} - t_1 r = \beta k r^2 y_1^{I*}$$

Solving the equations, we obtain

$$x_1^{I*} = \frac{rt_1}{\beta(1+k)} + \frac{k}{1+k}\left(x_2^I + y_2^I\right), \qquad y_1^{I*} = \frac{-rt_1}{\beta(1+k)} + \frac{1}{1+k}\left(x_2^I + y_2^I\right)$$

4 Solution of problem (P-7) and proof that the constraint is binding

The Lagrangian is

$$L(x_1, y_1, \lambda) = \Pi_1^g(1+\alpha, x_1, y_1) - \lambda\left(y_2^E - y_1\right)$$

The Kuhn-Tucker conditions are

$$\frac{\partial L}{\partial x_1} = -\beta r^2 x_1^{E*} + t_1 r = 0 \tag{1-1}$$

$$\frac{\partial L}{\partial y_1} = -\beta k r^2 y_1^{E*} + \lambda = 0 \tag{2-1}$$

$$y_2^E - y_1^{E*} \leq 0 \tag{3-1}$$

$$\lambda \geq 0 \tag{3-2}$$

$$\lambda\left(y_2^E - y_1^{E*}\right) = 0 \tag{3-3}$$

If $\lambda = 0$, then $y_2^E* = 0$ from Condition (2-1). This violates Condition (3-1), which means that the constraint is binding. Thus $\lambda > 0$. Then from Conditions (1-1) and (3-3) we get

$$x_1^E* = \frac{t_1}{\beta}r, \qquad y_1^E* = y_2^E = \sqrt{2}\sqrt{\frac{r\alpha + (1 - 2r)t_2\bar{x}}{kr^2}}$$

Acknowledgments. We would like to thank two anonymous referees and Tsuneyuki Morita at National Institute of Environmental Studies and Toshihiro Matsumura at Tokyo University, Japan, for their helpful comments and suggestions.

References

Anderson K (1996) The multilateral trading system and sustainable development. United Nations University/Institute of Advanced Studies (UNU/IAS) working paper no. 7

Appleton AE (1997) Environmental labeling programmes: international trade law implications Kluwer Law International, London

Giezeman I, Verhees F (1997) Eco-labeling: practical use of the cradle-to-grave approach. In: Zarrilli S, Jha V, Vossenaar R (eds) Eco-labeling and international trade. Macmillan, London, pp 195–205

Henry J (1997) ISO and eco-labeling, In: Zarrilli S, Jha V, Vossenaar R (eds) Eco-labeling and international trade. Macmillan, London, pp 272–276

Kuhn M (1998) Going green or going abroad? Environmental policy, firm location and green consumerism. In: Hanley N, Folmer H (eds) Game theory and the environment. Edward Elgar, Cheltenham, UK pp 341–391

Neitzel H (1997) Application of life cycle assessment in environmental labeling: German experiences. International Journal of LCA 2:241–249

Neitzel H (1998) Applying non-product-related criteria in eco-labeling: some controversies and experiences. Gate Magazine 1998, 2nd issue

OECD (1997a) Eco-labeling: actual effects of selected programmes. OECD/GD (97) 105

OECD (1997b) Process and production methods (PPMs): conceptual framework and considerations on use of PPM-based trade measures. OECD/GD (97) 137

Osterhus TL (1997) Pro-social consumer influence strategies: when and how do they work?, Journal of Marketing 61:16–29

Salzman J (1997) Informing the green consumer: the debate over the use and abuse of environmental labels. Journal of Industrial Ecology 1(2):11–21

Shy O (1995) Industrial organization: theory and application. MIT Press, Cambridge, MA

United Nations (1997) Trade effects of eco-labeling: proceedings of a seminar held in Bangkok. United Nations, New York

Vossenaar R (1997) Eco-labeling and international trade: the main issues. In: Zarrilli S, Jha V, Vossenaar R (eds) Eco-labeling and international trade. Macmillan, London, pp 21–36

WTO Trade and Environment Division (1999a) Developing countries and the multilateral trading system: past and present. Background document of high level symposium on trade and environment. WTO, Geneva

WTO Trade and Environment Division (1999b) Trade and environment in the GATT/WTO. Background document of high level symposium on trade and environment. WTO, Geneva

[6]

Ecolabeling Seafood for Sustainable Production: Implications for Fisheries Management

EYJÓLFUR GUÐMUNDSSON
University of Akureyri
CATHY R. WESSELLS
University of Rhode Island

Abstract *Demand for environmental quality of a fishery product and its effect on the incentive to manage fisheries sustainably is analyzed using bioeconomic modeling. The paper examines the effectiveness of an ecolabel in achieving sustainable fisheries production under open-access, limited-access and optimally managed fisheries with price premiums for ecolabeled products that are either constant or a function of sustainability criteria. In addition, the paper investigates the effect of placing an ecolabel on the rate of recovery of an overfished stock.*

Key words Ecolabeling, fisheries management, seafood, sustainable production.

Introduction

Traditional bioeconomic modeling focuses on the production side of fisheries management, creating mechanisms to avoid or reduce externalities under open-access situations. Over the past two decades, economists have increasingly looked at the effects of fisheries management on demand and supply of fish products. Gates (1974) showed that if fish size affects ex-vessel prices, there are unexpected effects on the outcome of fisheries management. McConnell and Strand (1989) looked at how water quality changes the supply and demand for a fish stock. Homans and Wilen (1997) discussed how, under regulated open-access fisheries management, rent is not only dissipated through harvesting practices, but also by reducing revenues through lower-quality products and higher processing costs.

Increasing demand for seafood products has contributed to the overexploitation of fisheries by increasing prices when harvest is declining, allowing the fishery to become economically viable in the short run, even at low stock sizes.[1] The implication is that demand for fisheries products is an important factor that must be accounted for when designing a fisheries management system. Recently, there have been several proposed programs for ecolabeling seafood products in an effort to provide incentives to fisheries managers to create sustainable fisheries. For example, in

---

Eyjólfur Guðmundsson is lecturer in the Faculty of Fisheries Studies at the University of Akureyri, Iceland, e-mail: eyjolfur@unak.is, and a graduate student in the Department of Environmental and Natural Resource Economics, University of Rhode Island, Kingston, RI 02881. Cathy R. Wessells is a professor in the Department of Environmental and Natural Resource Economics at the University of Rhode Island, Kingston, RI 02881, e-mail: crw@uri.edu.

The authors would like to acknowledge funding for this paper from the Rhode Island Agricultural Experiment Station, the USDA National Research Initiative Competitive Grants Program under award No. 96-35400-3604, and the Rhode Island Sea Grant Program.

[1] Subsidies, of course, can compound that effect.

1997, the Marine Stewardship Council (MSC) was formed via collaboration between the World Wildlife Fund, an environmental organization, and Unilever, an international corporation that markets various international brand names, including frozen seafood under the Bird's Eye and Gorton labels (McHale 1997). Several other initiatives to create ecolabels for seafood have emerged, including efforts by the Global Aquaculture Alliance (Boyd 1998) and the Nordic Council (Nordic Council 1998). The purpose of these initiatives is to provide a market-based incentive for sustainable fisheries management. The first step of these programs is to accredit sustainable fisheries worldwide. Processors, wholesalers, and retailers who purchase fisheries products from these accredited fisheries may purchase the right to place an ecolabel on their seafood products, informing consumers that the product was harvested from a sustainable fishery. Presumably, if there is a demand for environmental quality, consumers will respond by purchasing predominantly ecolabeled products, thus reducing demand for unlabeled products and lowering unlabeled products' prices. This may cause fishermen to put pressure on fisheries managers to manage the fisheries such that they also can receive accreditation as sustainable fisheries, and, hence, receive the price premium. The fishermen themselves may not bear any increased costs from harvesting a certified fishery. The conception of most current ecolabeling programs would place most of the cost on the upper levels of the market, and probably on the costs of fisheries management, to attain the certified status.

These market initiatives raise several questions worthy of research from the perspectives of both consumer demand for ecolabeled seafood and the likelihood of ecolabeling programs achieving their goals. In this paper, we will focus on the latter. Will ecolabels achieve the goal of creating an incentive for managers to manage fisheries on a sustainable basis? What are the conditions under which ecolabeling programs are likely to be most effective? Is the effectiveness of the ecolabel program conditional on the management regime currently in place for these fisheries? Since there are no ecolabeling programs currently in place, we cannot provide empirical answers to these questions.[2] Instead, we must make assumptions about fisheries, and based on those assumptions, provide some hypotheses about the impact of ecolabeling programs on the sustainability of fisheries resources.

We attempt to address these questions first by specifying a theoretically consistent utility framework within which there is some utility derived by the consumer from the information an ecolabel contains, and then by introducing that framework into a bioeconomic model in which output price is endogenous and related to stock (biomass) size. The paper concludes with discussion and implications of the findings regarding the effectiveness of ecolabels in achieving their goals.

Sustainable Fisheries Management

The criteria upon which the ecolabel is based may determine the effectiveness of the label in creating incentives for sustainably managed fisheries. Before determining the criteria for an ecolabel, one must first assess the characteristics of a sustainable fishery. The renewable character of fish stocks creates multiple solutions for sustainable harvesting of fisheries resources. In the Magnuson-Stevens Fisheries Manage-

[2] There is dolphin-safe tuna. However, the purpose of ecolabeling tuna is to protect dolphins from mortality or harm as a result of harvesting tuna. It does nothing to prevent overfishing of tuna. The type of ecolabeling we focus on in this paper is that to promote sustainable fisheries harvest. Since there are no such programs currently in place, we have no empirical evidence to answer the questions posed in this paper.

ment Act, sustainable fisheries are those with stock levels at or above maximum sustainable yield (MSY). However, the concept of MSY has proven inappropriate with regard to optimal utilization of fish stocks (Larkin 1979). Optimum yield is the level of harvesting that is consistent both with biological and economic factors of harvesting fish stocks, and is defined as the level of stock size that maximizes profits. This concept has proved to be difficult to apply in practice, and is sometimes misinterpreted in favor of short-term economic benefits for the harvesting sector (Nixon 1996).

Often there are non-biological factors that are also included in a definition of sustainability. For example, the Food and Agriculture Organization (FAO) of the United Nations has created a Code of Conduct for Responsible Fisheries, which includes provisions to ensure effective conservation and management of aquatic resources, on biodiversity, habitats, harvesting practices, nutrition, waste, and social objectives such as transparent decision making, allowing all interested groups to participate in the decision process and protecting the rights of fishers and those employed by the fishing industry (FAO 1995). The MSC has created principles and detailed criteria that a fishery must meet to be certified as sustainable, including criteria on stock sizes, habitat, diversity, and management methods, as well as social objectives (MSC 1998).

We suggest that the sustainability criteria of the Magnuson-Stevens Fisheries Management Act and the FAO can be split into three categories aimed at different functions of the fishery. The first category covers harvesting methods. Sustainability criteria would perhaps include constraints on use of inputs and levels of output (landings, as in an allocation of total allowable catch). The second category covers biological objectives, such as minimum stock size, age, and size of capture. The third category covers socioeconomic issues, such as income distribution, institutional setting of management regulations, equity, and efficiency. This paper defines sustainability as maintaining a stock size above some minimum safe level.

Whatever the criteria for certification, an important determinant of the ecolabel program's success will be whether the consumer understands the meaning of the label or has some perception of it. The consumer may not have perfect information on the meaning of the label. Therefore, the willingness to pay for the labeled product, as opposed to the unlabeled product, may not reflect their true willingness to pay for the knowledge that purchasing seafood has no detrimental effect on the health of the fisheries. We will continue on the assumption that consumers have perfect information about the label.

Demand for Environmental Quality

Ecolabeling of products is one way to alert consumers that specific measures have been taken by a producer to avoid or reduce undesirable environmental production externalities. This emerging practice has been recognized under the General Agreement on Tariffs and Trade (GATT) administered by the World Trade Organization (WTO) as an acceptable form of product differentiation based on production process rather than innate product characteristics. Recent surveys have shown that many consumers are likely to choose one brand or product over another if they believe that it will help the environment. Jha (1993) quotes survey evidence that suggests slightly over half of the consumers in North America had purchased a product that they felt was better for the environment, boycotted a specific product that they felt was bad for the environment, or boycotted products made by a company that they felt was damaging the environment. Wessells, Johnston, and Donath (1999), in a national survey of U.S. seafood consumers, found that there is at least a hypothetical demand for ecolabeled salmon, cod, and shrimp, if the ecolabel implies no overfishing, and a willingness to pay a premium for these products.

To endogenize the effect of ecolabels in a bioeconomic model of fisheries, consumer demand for ecolabeled fish must first be specified. Assuming that utility from fish consumption is separable from consumption of other goods, and the utility from all fish consumption is represented by a concave function, the utility function can be represented as:

$$U(h, h_L, C) = U(h, h_L) + B(h_L|C) \tag{1}$$

where h represents quantity consumed of non-labeled fish, h_L represents quantity consumed of fish that carries an ecolabel, and $B(h_L)$ represents the utility from consuming fish products from a sustainable fishery. C are the criteria under which the ecolabel is attached to the product. The criteria could be based on stock size, harvesting methods, a given set of fisheries regulations, level of community involvement in the fishery, or a combination of all these factors.

By using this specification of the utility function, we are assuming that the utility from consuming fish and the utility from knowing that the fish come from a sustainable fishery are additively separable. The additional utility from consuming labeled fish does not represent existence value, but rather the marginal willingness to pay for knowing that one is not contributing to unsustainable fisheries by purchasing this particular product. In that sense, we follow Lancaster's (1966) model that consumers receive utility from goods according to their attributes. The consumer is willing to pay more for goods that give them more satisfaction due to certain characteristics. In this case, the characteristic is the information on the sustainability of the resource in question. The unlabeled fish carries no information as to the sustainability of the stock; thus, we do not specify B as a function of h. This product carries imperfect information, because if the product were labeled as being from unsustainable harvest practices, there might be disutility associated with it.

Mattoo and Singh (1994) and Swallow and Sedjo (2000) show the effect on prices can be different depending on the structure of the market, degree of homogeneity between labeled and non-labeled product, and different cost structures. We model the problem as one where there are two prices for seafood products, one for unlabeled fish, p, and another for labeled fish, p_L, but the two products are otherwise identical.

Maximizing the utility function subject to the budget constraint $M = p \cdot h + p_L \cdot h_L$ gives the following first order conditions:

$$U_h - \lambda \cdot p = 0 \tag{2a}$$

$$U_{h_L} + B_{h_L} - \lambda \cdot p_L = 0 \tag{2b}$$

$$M - p \cdot h - p_L \cdot h_L = 0 \tag{2c}$$

where U_h is the marginal utility of the unlabeled fish, and U_{h_L} is the marginal utility of the labeled fish. Both U_h and U_{h_L} are functions only of the physical attributes of the fish. B_{h_L} is the marginal utility of the sustainability information from consuming a product with the ecolabel. This captures the full psychological effect of ecolabeling. M denotes income.

Concavity of the utility function requires:

$$U_h > 0,\ U_{hh} < 0$$

$$U_{h_L} > 0,\ U_{h_L h_L} < 0$$

$$B_{h_L} > 0,\ B_{h_L h_L} < 0$$

Solving equations (2a) and (2b) together, assuming $U_h = U_{h_L}$, gives the marginal conditions in equilibrium:

$$U_h \cdot \left[\frac{p_L}{p} - 1\right] = B_{h_L} \tag{3}$$

A critical assumption is that the marginal utility from the non-labeled fish is identical to the marginal utility from labeled fish, or that $U_h = U_{h_L}$. This is a reasonable assumption, since the good itself will likely not necessarily differ in quality or other sensory attributes, such as taste. There are several real-world examples where this may be true. One might argue that farmed salmon and wild salmon have the same quality attributes, as well as Canadian cod and American cod, Atlantic skipjack tuna and Pacific skipjack tuna, and so on.

Since U_h, p, p_L are positive, if $B_{h_L} \geq 0$, then from equation (3), $p_L \geq p$. In other words, if the consumer receives positive utility from consuming the labeled fish, then the consumer is willing to pay at least the same price for labeled versus non-labeled fish. If the consumer is unwilling to pay a premium, Pr, for the certification, the marginal utility of the label is zero.

Given that the second-order conditions hold, we can derive the Marshallian demand functions for non-labeled and labeled fish as functions of p, p_L, M, C:

$$h^M(p, p_L, M, C) \tag{4a}$$

$$h_L^M(p, p_L, M, C) \tag{4b}$$

By inverting functions (4a) and (4b), assuming there exists such single-valued functions and income is constant, we can derive the inverse demand functions for each product:

$P(h, h_L, C)$ Demand for non-labeled product (5a)

$P_L(h, h_L, C)$ Demand for labeled product (5b)

We assume that labeled and non-labeled products are perfect substitutes based on their non-environmental attributes (quality, freshness, *etc.*). The demand function, D, will be higher for labeled fish, since the utility that the consumer receives from the label information is always positive (by assumption). The actual shift of the demand function depends upon how the per-unit premium changes as h_L increases and decreases.

Using Ecolabels to Obtain Sustainable Use of Fishery Resources

The criteria for sustainable fisheries are by no means obvious. As we saw earlier, the criteria defined by the FAO and the MSC are fairly detailed and complicated. However, many of the biological characteristics, such as saving habitat and diversity, can be indicated by stock size. That is, a large stock size requires food and habitat to thrive. The same can be said about harvesting methods. If the harvesting methods are destructive, the fishery cannot maintain large stock sizes. Hence, we can assume if stock size is known, it can be used as an indicator for the sustainability of the fish stock in question and replace C in the consumer utility function with a variable for stock size, x. The utility function is then:

$$U(h, h_L, x) = U(h, h_L) + B(h_L|x) \quad \text{for } x > x^* \qquad (6)$$

We assume that the consumer receives additional utility from consuming fish products from fisheries that are above minimum stock size, x^*, and is willing to pay a premium per unit of product (per kilogram) for labeled fish products.[3] There are several possibilities for premium functions. First, the consumer may be willing to pay a fixed premium (figure 1a). Second, the consumer may be willing to pay an increasing premium for increased levels of stock size (figure 1b). Third, the consumer may be willing to pay a premium for fish from a fish stock which is recovering, but is not willing to pay a premium for fish from fish stock stocks that are at high levels (figure 1c).

Changes in stock size cause a shift in demand (either a downward shift for non-labeled or upward shift for labeled product), for each level of stock size above the minimum level, x^*. This static result is illustrated in figure 2. The figures are four quadrant graphs with a backward bending supply curve and a demand curve in the northeast quadrant. The southeast quadrant shows the sustainable harvest function. The southwest quadrant is a simple 45 degree line, and the northwest quadrant shows the premium function in relation to stock size. In the northeast quadrant, quantity harvested is on the horizontal axis. On the vertical axis in the northwest quadrant is the premium, represented in dollars per unit of harvest. For example, in figure 2a the premium is constant regardless of stock size. As a result, demand shifts from D to D′, reflecting the additional amount a consumer is willing to pay for the labeled product. In figure 2b, the premium function shows increasing consumer willingness to pay as the stock size grows. Point A (*i.e.*, x^*) represents the minimum stock size necessary to receive a premium.[4] Point B represents the stock size for which the consumer will pay the maximum premium. These two points, A and B, are reflected off the 45 degree line in the southwest quadrant onto the harvest function in the southeast quadrant, showing two different levels of sustainable harvest. These two points are further reflected to the northeast quadrant onto the backward bending supply curve, showing the maximum shift in the demand function. If this exercise continues for all points between A and B, one obtains a locus of new equilibria between G and H.

In figure 2c, for each level of the stock size between points A and B the consumer is willing to pay a premium, but that premium decreases as stock size increases. With the upward shift in the demand function, and *without* a proper manage-

[3] The authors do not believe that the average consumer is going to be aware of the stock size of the fishery. Nor are consumers likely to be knowledgeable about the effects of gear types on the habitat and ecosystem. However, we choose this as a measure of sustainability because it best represents a "sustainable fishery" under most of the definitions of sustainability, and serves the intended purpose of illustrating the effects of ecolabeling on fisheries management.

[4] Note that x^* can be any stock size and is not limited to x_{MSY}.

ment regime in place, stock size will decrease. Due to the backward bending nature of the fisheries supply function, the new equilibrium is reached between points G and H in the northeast quadrant. Only when the consumer is willing to pay a fixed premium for labeled fish will the premium cause a single shift in demand (figure 2a).

The effects of a constant increase in price (*i.e.*, premium regardless of stock size) on fisheries under different management systems is well known (Clark 1990). Figure 3 shows what happens within a fishery when a price increase is constant (corresponding to figure 1a). The x-axis shows level of effort (an index of inputs used in

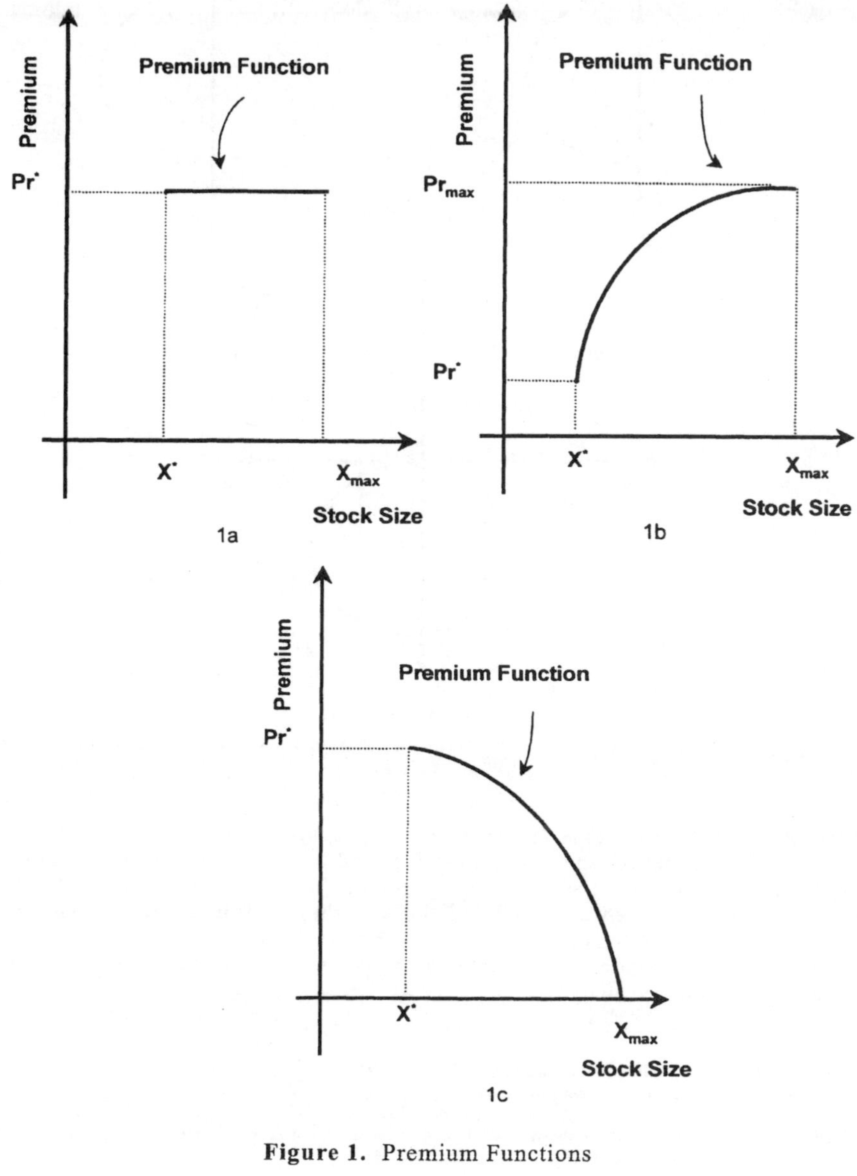

Figure 1. Premium Functions

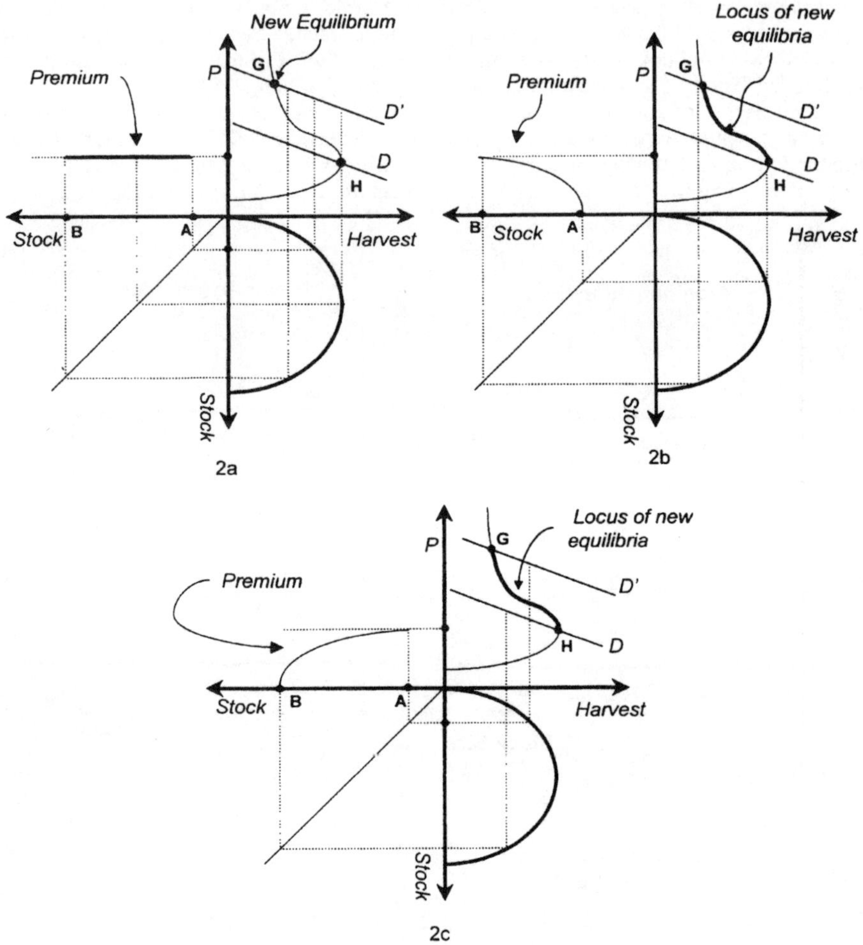

Figure 2. The Effect of an Ecolabel Premium on Demand and Supply

harvesting), and the y-axis shows value. The graph shows two total revenue (TR) functions, one for the labeled product and one for the unlabeled product. On the graph, there are also three equilibrium solutions: an open-access fishery (effort level D), limited or controlled access fishery (effort level B), and optimally controlled fishery (effort level A).[5] The graph also has several cost/price ratio lines. For example, a price increase will lower C/P to C/P′, holding costs constant.[6] A constant price increase for all harvesting levels will increase the

---

[5] For derivation of each of these equilibria, see Clark (1990).

[6] Since there are no functioning ecolabel systems yet in place, we cannot be sure whether there will be any increase in costs to the industry from ecolabels. Most likely there will be some increased costs, but whether those are borne by the fishermen or at higher levels (such as wholesalers and retailers) is not yet known. For simplicity, we assume costs are unchanged.

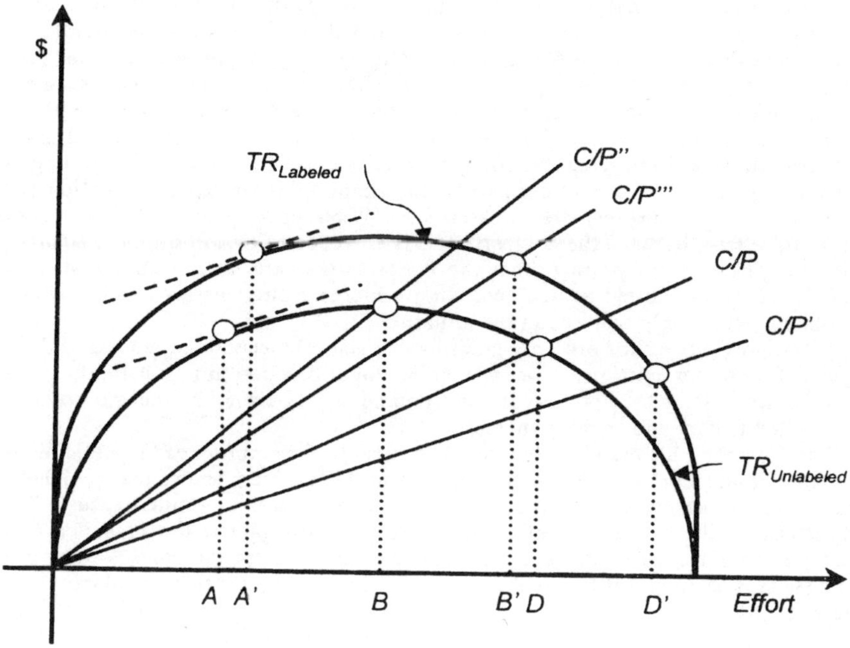

Figure 3. Constant Price Increase and the Static Bioeconomic Model

effort level in the open-access fishery from D to D′ simply because it would be profitable for new effort to enter the fishery. The same is most likely to happen for the limited entry management systems. If the fisheries managers were successful at controlling effort, the fishery would simply stay at point B, and some resource rents would be earned.[7] The more likely situation is increased input stuffing (Holland, Guðmundsson, and Gates 1999) leading to higher effort levels in the fishery. This result is exactly what one obtains when considering a simple exogenous price increase in the classical fisheries bioeconomic model. Even in the case of the optimally controlled fishery, the effort level is increased from A to A′, leading to higher exploitation rates and lower stock size. A key point, then, is that if the premium for the ecolabel is constant, the ecolabeling program will not be effective in providing fishermen or fisheries managers with a greater level of resource sustainability. Increasing effort levels are only a concern if they lead to a stock size below the minimum safe levels. Generally speaking, open-access and limited-access fisheries harvest at levels greater than appropriate for sustainability.

Changing the shape of the TR function by designing a premium schedule (*i.e.*, either figure 1b or 1c) can have significant effects on the static solution under each of the three management regimes. Figure 4 shows the effect of shift-

[7] Townsend (1990) and Holland, Guðmundsson, and Gates (1998) provide evidence that input control or license limitation have not been successful fisheries management tools. Examples can also be found in Hannesson (1996).

ing maximum sustainable revenue to the left by paying a premium for harvest from larger stocks (at lower effort levels) and then decreasing the premium as stock size decreases and effort increases (figure 1b). The premium is cut off at some level of effort corresponding to x^*, such that an open-access fishery would not receive the premium. As seen from figure 4, the optimally controlled fishery would decrease its effort levels from A to A′, while the limited-access fishery would either increase effort or increase costs by input stuffing. The position of A′ relative to A is defined by the shape of the premium function (figure 1b or 1c) and the specific stock size, x^*. Whether A′ is to the left or right of A is irrelevant, because these efforts levels correspond to an optimally managed fishery. The primary point is that the fisheries that are not optimally managed (open-access or limited-access) will likely increase their effort levels given an ecolabel, effectively worsening the situation.

The previous results are only good for the static bioeconomic models with two types of premium functions. If we use the premium function in figure 1c, the results will be similar for the open-access and limited-access fisheries, but the optimally controlled fishery will actually increase effort.

In summary, the premium function is critical to the solution of the static model. In fact, classical bioeconomic theory applied to this problem shows that ecolabeling can actually exacerbate the problem when in open-access or limited-access. What the static analysis does not allow us to do is investigate if the presence of an ecolabel on fisheries products can provide an incentive for the stock size to grow any faster, assuming it is below the sustainable level. It is this that we address in the next section.

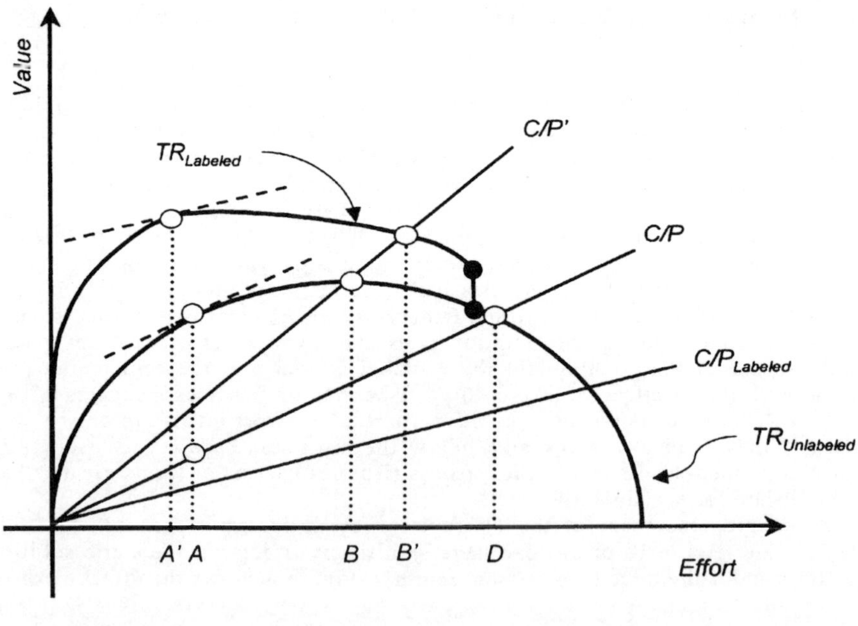

Figure 4. Variable Price Increase and the Static Bioeconomic Model

Ecolabels and Dynamic Bioeconomic Models

We reexamine these results by showing the effect of ecolabels on a dynamic bioeconomic model. We assume that the objective of the fisheries manager is to maximize social welfare, including consumers' utility (as measured by consumer surplus) and producer surplus from harvesting and consuming the fisheries resource. The fisheries manager can either have sole property rights over the resource and lease it to fishermen, or assign the property right to individual fishermen, such as is done under Individual Transferable Quota systems (ITQs).[8]

Given the available fish stock and the known demand function for the final product from the fish stock, the net benefit (NB) function[9] for the labeled (sustainable) fishery is defined as:

$$NB = \int_0^{\infty} e^{-r \cdot t} \left[\int_0^{h_{Lt}} P(z_t, x_t) dz - c(x_t) \cdot h_{Lt} \right] dt$$

where $h_{L,t}$ is the harvest and x_t is the stock size at time t, r is the discount rate, $c(x)$ is harvest cost per unit of harvest (assumed to be linear in h_L), z_t is the integration factor for h_L, $P(h_L,x)$ is consumer demand, and $\int_0^{h_{Lt}} P(z_t, x_t)dz$ represents total revenue, holding all other arguments in the utility function constant. Utility is assumed fully separable in harvest and stock size. Changes in stock size are defined by the stock equation $\dot{x} = F(x) - h_L(t)$, where $F(x)$ represents the growth function. The growth function is assumed to be logistic.

The optimal solution is found using a current value Hamiltonian function and applying Pontryagin's maximum principle:

$$\max H^{cv} = \int_0^{h_L} P(z, x)dz - c(x) \cdot h_L + \mu \cdot \left[F(x) - h_L \right] \tag{7}$$

The necessary conditions for a maximum include (assuming costs are linear in h_L and that there exists a solution to the problem):

$$\mu = P(h_L, x) - c(x) \tag{7a}$$

$$\dot{\mu} = r\mu - \int_0^{h_L} P_x(z, x)dz + c_x \cdot h_L - \mu \cdot F_x \tag{7b}$$

$$\dot{x} = F(x) - h_L \tag{7c}$$

Differentiating equation (7a) with respect to time, and equating the result with equation (7b) yields:

$$\dot{h}_L = \frac{1}{P_{hL}} \cdot \left[r \cdot \mu - F_x \cdot \mu - \int_0^{h_L} P_x(z, x)dz + c_x \cdot F(x) - P_x \cdot \dot{x} \right] \tag{8}$$

[8] See Arnason (1990).

[9] To simplify mathematical notation, time subscripts are omitted from all equations from here onward.

Equilibrium conditions are $\dot{h}_L = \dot{x} = 0$. Defining the total ecolabel premium as $\int_0^{h_L} P_x(z, x)dz \equiv \text{Pr}$, and solving for r yields a modified golden rule, shown as equation (9):

$$r = F_x - \frac{c_x F(x)}{P(h_L, x) - c(x)} + \frac{\text{Pr}}{P(h_L, x) - c(x)} \tag{9}$$

Pr represents the total premium of labeled fish over unlabeled fish (*i.e.*, for the total catch), where the label represents a product harvested from a fishery of sustainable stock. The equality in equation (9) shows that, in order for there to be equilibrium, any premium will have the same effect on the optimal level of harvest as if the discount rate was raised or lowered. If the profit margin is low and the premium high, the incentive to respond to the premium is great. If, however, the fishery is already profitable and the premium is relatively low, there is less incentive for the owner of the fish stock to respond to the premium (or the ecolabel).

To compare the optimal stock size for the labeled fishery to the non-labeled fishery, we follow Clark (1990, p. 42). Equation (10a) repeats Clark's equation 2.18 in our notation.

$$r[P - c(x)] = [P - c(x)] \cdot F_x - c_x F(x) \tag{10a}$$

Rewriting equation (9) to resemble equation (10a):

$$r\left[P(h_L, x) - c(x)\right] = \left[P(h_L, x) - c(x)\right] \cdot F_x - c_x F(x) + \text{Pr} \tag{10b}$$

Let $\rho(x) = [P(h_L, x) - c(x)] \cdot F(x)$, which is the sustainable social economic rent from harvesting the fish stock when the consumers receive additional utility of consuming fish from a fishery with stock size above x^*. Totally differentiating $\rho(x)$ with respect to the stock size, x, we obtain:

$$\frac{d\rho}{dx} = \left[P(h_L, x) - c(x)\right] \cdot F_x - c_x F(x) + P_x \cdot F(x) \tag{11}$$

Comparing the right-hand side (RHS) of equation (11) to the RHS of equation (10b) reveals that the two equations are identical, except that the term $P_x \cdot F(x)$ replaces Pr.

In equilibrium $F(x) = h_L$ we harvest exactly the growth of the resource; hence, it must hold in equilibrium that:

$$P_x \cdot h_L = \int_0^{h_L} P_x(z, x)dz \equiv \text{Pr} \tag{12}$$

We can, therefore, write equation (11) as:

$$\frac{d\rho}{dx} = \left[P(h_L, x) - c(x)\right] \cdot F_x - c_x F(x) + \text{Pr} \tag{13}$$

Equation (13) is the change in the rent from a sustainable fishery given a change in stock level, x, and is the same as the RHS of equation (10b). Substituting equation (13) into equation (10b) results in:

$$r[P(h_L, x) - c(x)] = \frac{d\rho}{dx} \tag{14}$$

Equation (14) looks identical to equation 2.19 in Clark (1990, p. 42), with the same interpretation that when x is optimal, the marginal immediate gain must equal the present value of marginal future loss. The difference between equation (14) and Clark's equation 2.19 is in the specification of the right-hand side. In other words, equation (14) takes into account the willingness to pay for the sustainable attribute of the resource. Figure 5 shows the difference between the two sustainable revenue curves for each case. The upward sloping lines indicate the immediate gain in harvesting additional units. The shift in the gain from harvesting additional units is dependent on the premium function (the left-hand side of equation 14). The downward sloping curves represent the future change in increasing or decreasing the stock size. The shift of these curves is also dependent on the premium function (equation 13, or the right-hand side of equation 14). Depending on assumptions of the shape of the premium functions, the difference between the sustainable revenue functions from Clark (1990) and equation (14) will either be constant (figure 5a), decreasing (figure 5b), or increasing (figure 5c).

Figure 5 also shows that a profit-maximizing fisheries management regime will not necessarily translate to a fishery that will be utilized at higher stock levels. The figure shows that x_L, the optimal stock size when the product is labeled, is either less than, equal to, or greater than x_n, the optimal stock size for non-labeled product. The magnitude of the effect, and whether the difference between the two rent functions is increasing or decreasing, depends on the premium function. So here again, as in our previous static model, consumers' preferences for ecolabeled fish will determine the shape of the premium function, as shown in equations (3) and (6).

Without specific knowledge of consumer preferences, little can be said about what the effects of an ecolabel will be. One of the reasons that ecolabels are being created is to help *restore* overexploited fisheries by rewarding management schemes that aim at rebuilding fish stocks. Countries around the world that have rebuilt their fisheries (New Zealand, Iceland, some fisheries in the United States, and Australia) have some form of private property rights. It is, therefore, interesting to examine what happens if an ecolabel is awarded to a fishery with stock size below that associated with MSY, but at levels high enough to be paid a premium.

Recall that costs are linear in h_L, and assuming that the maximum effort level is less than the optimal effort level ($h_{L,\max} < h_L^*$), harvest will begin before optimal stock size is reached. Under linear control problems, the level of harvest is selected at the beginning of the path and is fixed from that time forward, *i.e.*, $\dot{h}_L = 0$. This is not in equilibrium, but rather in a situation where the fishery manager has decided to increase the stock size in order to receive a premium through the ecolabel. The stock is, therefore, increasing, or $\dot{x} > 0$. Hence, we can rewrite equation (8) as:

$$F(x) = \frac{[P(h_L, x) - c(x)] \cdot (r - F_x) - \int_0^{h_L} P_x(z, x)dz + P_x \cdot h_L}{P_x - c_x} \tag{15}$$

We are assuming separability between harvest and stock size in the consumer utility function $\int_0^{h_L} P_x(z, x)dz = P_x \cdot h_L$. This is not equal to Pr, since in this case the growth is greater than the harvest $[F(x) > h_L]$. $F(x)$ then reduces to:

$$F(x) = \frac{[P(h_L, x) - c(x)] \cdot (r - F_x)}{P_x - c_x} \tag{16}$$

Substituting $F(x)$ into the stock equation:

$$\dot{x} = \frac{[P(h_L, x) - c(x)](r - F_x)}{P_x - c_x} - h_L \tag{17}$$

Since $\dot{h}_L = 0$, this is a linear control problem, and assuming infinitely small changes such that $\lim_{t \to 0} \dot{x} \to 0$, the harvest rate is set such that:

$$h_L = \frac{[P(h_L, x) - c(x)] \cdot (r - F_x)}{P_x - c_x} \tag{18a}$$

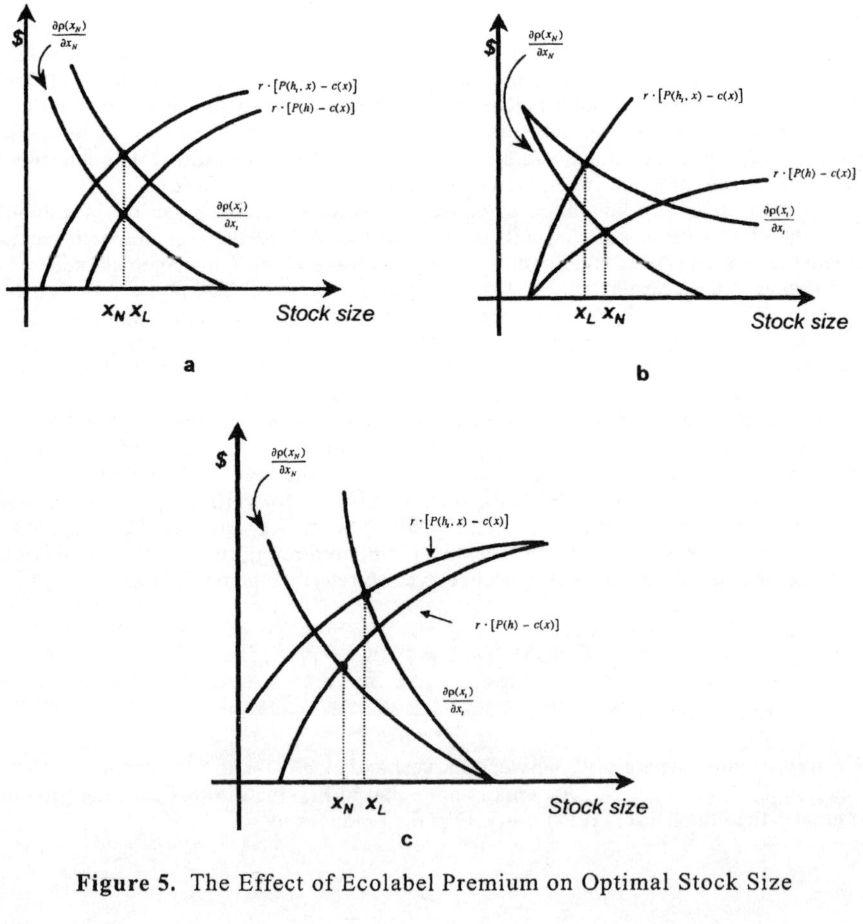

Figure 5. The Effect of Ecolabel Premium on Optimal Stock Size

$$h = \frac{\left[P(h_L, x) - c(x)\right] \cdot (r - F_x)}{-c_x} \tag{18b}$$

Remembering that $-c_x > 0$ and $F_x^- > 0, F_x^{MSY} = 0, F_x^+ < 0$, where the superscripts stand for stock size below (–) MSY and above (+) MSY. For equations (18a) and (18b) the term on the RHS is positive if the stock size is below MSY and $F_x^- < r$, or when stock size is above MSY. Therefore, it holds that $h_L < h$ if and only if $P_x > 0$.

Figure 6 shows the phase diagram for both labeled and non-labeled fisheries. Both fisheries are assumed to have the same biological and cost characteristics (in fact, this is the same fishery with or without the ecolabel). Therefore, the stock equation $\dot{x} = F(x) - h$ is the same in both cases. The harvest function is higher for unlabeled fish (equation 18b), and lower for labeled fish (equation 18a). There are four quadrants on the graph, each with a unique optimal path that will either lead to, or from, the corresponding equilibrium point. All other paths will lead to a disequilibrium situation. At time $x(0)$, the fisheries manager decides that stocks must be rebuilt. The optimal path for the non-labeled fishery will set harvest at point A, and the labeled fishery at point B. The optimal path for the labeled fishery leads to an equilibrium with lower sustainable harvest and higher stock level than in the non-labeled case. However, even for a fishery with less than sustainable stock size, the ecolabel will be beneficial in that the harvest levels will be set lower, allowing for a faster recovery period.

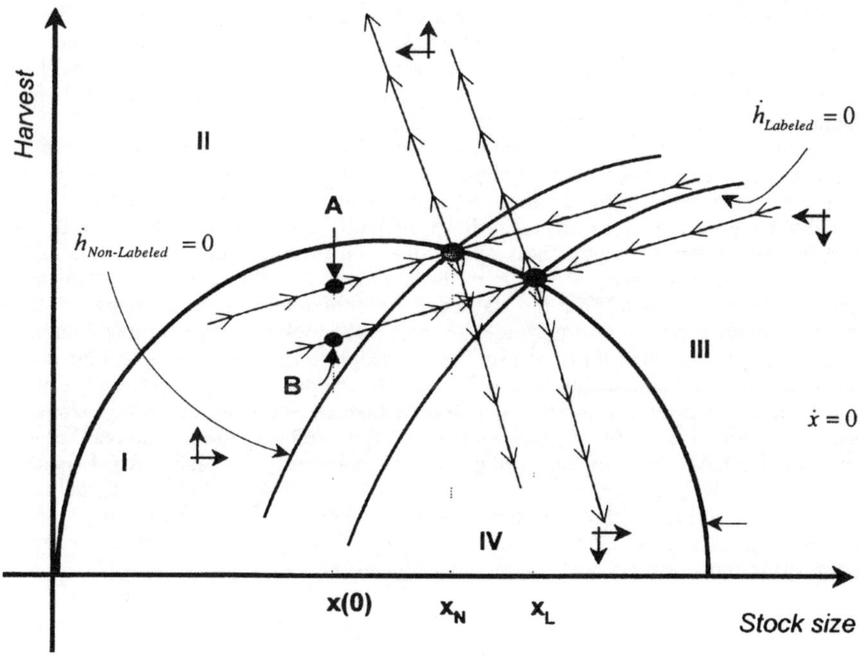

Figure 6. Phase Diagram for Labeled and Non-labeled Fisheries

Discussion

Given the restrictive set of assumptions applied above, are the results applicable in reality? In fact, there are only two critical assumptions that must be met for these results to be valid. First, it is assumed that there will be a price premium paid for the ecolabeled product over the unlabeled product. Using organic produce as an example, where the premium can be as high as 30–40%, this assumption is realistic (Lohr 1998). Second, the labeled product must be differentiable from the unlabeled product; *i.e.*, the label must be obvious to and trusted by the consumer. This can be accomplished with a well-functioning labeling mechanism and through consumer education.

Results of this paper can be presented as dependent on the shape of the premium function. If there is a constant premium (figure 1a), then the static model shows an ecolabel will cause effort to increase, regardless of management regime (figure 3), resulting in a smaller stock size. In the dynamic model, a constant premium may cause effort to increase, decrease, or remain the same, but as the model is set up for a sole-ownership fishery, whatever decision is made will be optimal.

If the premium is not a constant (figures 1b and 1c), then the static model shows effort will always increase in open-access and limited-access fisheries (figure 4). However, effort may increase, decrease, or remain the same in the optimally managed fishery. The dynamic model of an optimally controlled fishery with a premium that is a function of stock size shows that, if the fishery is below the equilibrium stock size, the resource will recover more quickly with an ecolabel.

The recommendation from this analysis for institutions awarding an ecolabel, given the results of the static analysis, is to award it only to an optimally managed fishery. The fishery must be able to adjust for additional effort by having regimes in place to control for input stuffing. Given a well-managed fishery, the ecolabeling institution may award a label to a fishery whose stock size is not currently at an optimum level, because the stock will recover faster with the ecolabel. Thus, an ecolabel sends the message that if the manager stays on the optimal path, the premium from the labeled product will pay him to stay on that path.

Given that our results are dependent on the design of the premium function and the management regime, finding a method of designing a premium that considers the economic incentives within the management scheme is critical for the success of an ecolabeling program. In reality, it will be difficult, if not impossible, to design an ecolabeling system in which the result is that the premium is a function of stock size, or any other criteria. Most likely the premium will be constant, regardless of stock size. The results of both the static and dynamic analyses in this paper indicate that a constant premium will not yield an increased stock size (see figures 3 and 5a). Hence, such an ecolabel will not provide an incentive to build stocks to higher levels. Thus, the final conclusion is that the tragedy of open-access to natural resources is not solved by awarding a label to a fishery. Instead, the label needs to change the economic incentives under the management system, and provide incentives that generate a sustainable level of harvesting. The relevance of these findings for fisheries policy is that the criteria for any sustainability label for fisheries should focus on the economic incentives within the management method in combination with identifiable biological factors, such as stock size, rather than focusing on detailed ecological information or general socio-economic objectives.

References

Arnason, R. 1990. Minimum Information Management in Fisheries. *Canadian Journal of Economics* 23(3):630–53.

Boyd, C. 1998. *Codes of Practice for Responsible Shrimp Farming*. Global Aquaculture Alliance, St. Louis, MO, USA.

Clark, C.W. 1990. *Mathematical Bioeconomics: The Optimal Management of Renewable Resources,* 2nd ed. New York: John Wiley & Sons Inc.

Food and Agriculture Organization of the United Nations (FAO). 1995. *Code of Conduct for Responsible Fisheries*. Rome, Italy.

Gates, J.M. 1974. Demand Price, Fish Size and the Price of Fish. *Canadian Journal of Agricultural Economics* 22(3):1–12.

Hannesson, R. 1996. *Fisheries Mismanagement*. Oxford, England: Fishing News Books.

Holland, D., E. Guðmundsson, and J. Gates. 1999. Do Fishing Vessel Buyback Programs Work: A Survey of the Evidence. *Marine Policy* 23(1):47–69.

Homans, F.R., and Wilen, J.E. 1997. A Model of Regulated Open Access Resource Use. *Journal of Environmental Economics and Management* 32:1–21.

Jha, V. 1993. Green Consumerism, Eco-labeling and Trade. Mimeo, UNCTAD.

Lancaster, K.J. 1966. A New Approach to Consumer Theory. *The Journal of Political Economy* 74:132–57.

Larkin, P.A. 1979. An Epitaph for the Concept of Maximum Sustained Yield. *Transactions of the American Fisheries Society* 106(1):1–11.

Lohr, L. 1998. Implication of Organic Certification for Market Structure and Trade. *American Journal of Agricultural Economics* 80(5):1125–29.

Marine Stewardship Council. 1998. *Principles and Criteria for Sustainable Fishing: Airlie House Draft*. United Kingdom, www.msc.org.

Mattoo, A., and H.V. Singh. 1994. Eco Labelling: Policy Consideration. *Kyklos* 47:53–65.

McConnell, K.E., and I.E. Strand. 1989. Benefits from Commercial Fisheries When Demand and Supply Depend on Water Quality. *Journal of Environmental Economics and Management* 17:284–92.

McHale, T. 1997. What's in a Label. *Pacific Fishing*. February, pp. 36–39.

Nixon, D.W. 1996. Easing Pressures on U.S. Fisheries. *Forum for Applied Research and Public Policy*. Summer, pp. 24–28.

Nordic Council of Ministers. 1998. The Nordic Fisheries in the New Consumer Era: Challenges Ahead for the Nordic Fisheries Sector. *Temanord* 1998:526, Copenhagen, Denmark.

Swallow, S.K., and R.A. Sedjo. 2000. Eco-Labeling Consequences in General Equilibrium: A Graphical Assessment. *Land Economics* 76(1):28–36.

Townsend, R.E. 1990. Entry Restrictions in the Fishery: A Survey of the Evidence. *Land Economics* 66(4):359–79.

Wessells, C.R., R.J. Johnston, and H. Donath. 1999. Assessing Consumer Preferences for Ecolabeled Seafood: The Influence of Species, Certifier, and Household Attributes. *American Journal of Agricultural Economics* 81(5):1084–89.

[7]

Is Ecolabelling a Reliable Environmental Policy Measure?

CESARE DOSI and MICHELE MORETTO
Department of Economics, University of Padova, Via del Santo, 33, I35123 Padova, Italy
(e-mail: dosi@decon.unipd.it; moretto@decon.unipd.it)

Accepted 22 March 2000

Abstract. The rationale of ecolabelling is to enable firms to reap the willingness-to-pay for the environmental attributes of goods by helping consumers to identify "green" products. By so doing, ecolabelling is expected to stimulate spontaneous environmental innovation and to reduce aggregated pollution. Our analysis however outlines situations under which ecolabelling could induce perverse effects, namely increased investment in conventional technologies before the labels are awarded, and examines whether restricting the issue of labels could constitute an antidote.

Key words: ecolabelling, environmental policy, investment decisions, irreversibility, real option theory

JEL classification: Q28, O38

1. Introduction

Similarly to other "voluntary approches", ecolabelling has become increasingly popular among the business community and policy-makers.[1,2] As far as the latter are concerned, the prevailing view is that, though it cannot be considered a panacea, ecolabelling would be a useful accessory in environmental policy, able to encourage spontaneous innovation and a virtuous "environmental competition" among firms. For instance, given the latent (increasing) demand (willingness-to-pay) for "green" (environmental attributes of) products, the issue of labels would serve two purposes at once. On the one hand, labels would allow for consumers' recognition of these products: something which would otherwise be difficult, if not impossible, given the complex nature of the environmental impact of products along their life cycle. On the other hand, as certification highlights more environmentally benign alternatives and provides consumers with guidance, firms with a greater propensity to innovate would be able to reap the benefits deriving from the transformation of their productive processes, and at the same time "penalize" competitors that try to acquire a green reputation by merely making superficial or cosmetic changes to their products.

While we agree with this view of these potential advantages, we feel it is worthwhile to deepen the analysis about the *conditions which may affect the effectiveness of ecolabelling programs* in terms of pollution abatement (reduced supply/consumption of polluting products). For instance, in one of the rare attempts at theoretical analysis of the effectiveness of ecolabelling, Mattoo and Singh (1994) have pointed out the risk that it might lead to an increase in the supply of polluting products, due to possible increases in their relative prices induced by the greater supply of green products, following the establishment of an ecolabelling programme. More recently, Kuhn (1999) – through a more sophisticated model which accounts for market structure and competitive interactions – illustrates situations where ecolabelling may produce a reverse, i.e. the improvement of average environmental performance, brought about by an increase in green producers' market share, is more than offset by the increase in pollution from an expanding market.

While also outlining potential reverse effects of ecolabelling, our analysis is based on a different approach. By drawing upon the theory of irreversible investment decisions under uncertainty,[3] we propose a two-period model[4] which allows for a stylised analysis of the *overall impact* of ecolabelling upon firms' investment decisions, that is, both before and after the label (the "green technology" required to qualify for the ecolabel) is awarded (is adopted).

The paper is structured as follows. The following section presents the basic model which is characterized by the hypothesis that voluntary adoption[5] of a technology which implies the supply of an environment friendly product (using "green capital") is a necessary and *sufficient* condition for a firm to be awarded with an ecolabel, even though it adopts conventional technologies (uses "polluting capital") at the same time. Our analysis shows that if a complementarity relationship between the polluting (unlabelled) and the green (labelled) production lines is expected to emerge – e.g. if the ecolabel is expected to project a positive image over the entire firm, and not merely on the certified product(s) – ecolabelling may lead to an expansion of investment in "polluting capital" *before* the label itself is awarded.

In section 3 we examine the consequences of restricting the issue of labels, by assuming that they are awarded with some positive probability less than one, and that candidates – i.e. firms able to submit products which meet the standards a product must meet to qualify for the ecolabel- are also evaluated by somehow reviewing their "historical" environmental records. In the conclusion we make some remarks on the policy implications deriving from the analysis.

2. The Model

Consider a good, with a stochastically evolving demand, that may be produced by two different processes, i.e. by using two different types of capital, K^p and K^g, with unitary *sunk costs* equal to r and $r + b(b \geq 0)$, respectively.

Adoption of the first type (K^p, the "polluting capital") translates into supply of a polluting product (in the production and/or consumption phase), while the second

(K^g, the "green capital") allows for production of a "green" one. The firm may adopt both technologies simultaneously, setting up two distinct production lines, but is unable to adapt and reconvert the polluting capital. In other words, K^p and K^g are fixed factors and the investment decisions are irreversible.

Let us assume that the relatively more costly green technology only becomes available in the second period. Moreover, we assume that investment in green capital is only justified by the label, without which the firm's innovative effort would not be recognised and rewarded by consumers (willing-to-pay more for green products).

In the first period the firm decides how much polluting capital to invest (K_1^p), knowing that in the second period it may either increase the stock ($K_2^p > K_1^p$) or else abandon it, i.e. by adopting only non-polluting processes (using K^g alone), or maintaining both production lines.

As both types of capital are fixed and their earnings are uncertain, the possibility of varying K^p in the second period may be interpreted as a *call option*, while the possibility of buying green capital may be seen as a *flexible production option*. The optimal choice of K_1^p is therefore the result of *ex ante* maximization of the firm's value, a value constituted by both the discounted flow of expected profits and the above-mentioned option values.

Firms' profits in the first and second period are defined respectively by the following functions: $R^p(K_1^p)$ and $R(K_2^p, K_2^g; \theta)$, where θ – by hypothesis, common to both productive processes and hence independent of the technology adopted – is a stochastic variable, with a differentiable accumulated distribution $\Phi(\theta), \theta \in \Theta \subseteq \mathbf{R}$, representing a profitability index capturing the effects of possible demand (and/or input productivity) shocks.

Assumption 1. $R^p(K_1^p)$ is a continuous, strictly increasing and concave function in K_1^p. Moreover, Inada's conditions hold, i.e.: $\lim_{K^p \to 0} R_K^p = +\infty$, $\lim_{K^p \to \infty} R_K^p = 0$.

Assumption 2. $R(K^p, K^g; \theta)$ is a continuous, strictly increasing function in the set of variables (K^p, K^g, θ). Yet $R_{K^p}(K^p, K^g; \theta) \geq 0$ and $R_{K^g}(K^p, K^g; \theta) \geq 0$ are continuous and strictly decreasing in K^p and K^g respectively, continuous and strictly increasing in θ and satisifies $R_{K^g\theta} R_{K^pK^g} - R_{K^p\theta} R_{K^gK^g} > 0$. Finally, $R(.)$ meets Inada's conditions: $\lim_{K^i \to 0} R_{K^i} = +\infty$, $\lim_{K^i \to \infty} R_{K^i} = 0$, $i = p, g$.

Assumption 3. $R(K^p, 0, \theta) = R^p(K^p, \theta)$, while $R(0, K^g, \theta) = R^g(K^g, \theta)$. Moreover, there is a (small) value $\theta' \geq 0$ beyond which $R_K^p < R_K^g$.

The first and second assumptions are standard, except $R_{K^g\theta} R_{K^pK^g} - R_{K^p\theta} R_{K^gK^g} > 0$. This condition is always satisfied if $R_{K^pK^g} > 0$, otherwise it requires that investment in green capital is not (significantly) affected by demand shocks (see Appendix A, equation 11).[6] Finally, the third assumption implies that, *coeteris paribus*, i.e. for whatever realisation of $\theta > \theta'$, the marginal profitability of the green capital, due to the ecolabel, is higher than that of the polluting capital.[7]

In the second period, once the realization of the profitability index θ is known, the firm maximizes its (short term) profits by choosing, for any given values of r and b, an optimal mix between the two types of capital as solution of:

$$R_{K^p}(K_2^p, K_2^g; \theta) = r \tag{1}$$

$$R_{K^g}(K_2^p, K_2^g; \theta) = r + b \tag{2}$$

Indicating with $K_2^g(\theta)$ the optimal green capital as solution of the system 1–2, by Assumption 2 it is always possible to identify, for any given inherited stock of polluting capital K_1^p, a *critical value for the profitability index*, $\hat{\theta}$, such that:

$$R_{K^p}(K_1^p, K_2^p(\hat{\theta}); \hat{\theta}) = r \tag{3}$$

Since for every optimal stock $K_2^g > 0$, in the interval $S = \{\theta{:}\theta \geq \hat{\theta}\}$ the value of a marginal unit of pollution capital is equal to its installation cost, S may be interpreted as the firm's instantaneous investment in additional units of polluting capital. Similarly, the complement $S^c = \{\theta{:}\theta < \hat{\theta}\}$ indicates the interval in which the firm finds it optimal to invest in green capital. In this case, $K_2^p = K_1^p$, since $R_{Kp}(K_1^p, K_2^p(\theta); \theta) < r$.

Given these hypotheses, and assuming a constant discount rate, $\gamma > 0$, the following proposition may be proved:

PROPOSITION 1: *If, despite contemporaneous use of polluting capital ($K_2^p > 0$), investment in green capital ($K_2^g > 0$) is a necessary and sufficient condition for a firm to be awarded with the ecolabel, the firm's ex ante value is defined by the following expression*:

$$\begin{aligned} V(K_1^p) &= R^p(K_1^p) + \gamma \int_{-\infty}^{\hat{\theta}} \{R(K_1^p, K_2^g(\theta), \theta) - (r+b)K_2^g(\theta)\} d\Phi(\theta) + \\ &\quad \gamma \int_{\hat{\theta}}^{+\infty} \{R(K_2^p(\theta), K_2^g(\theta), \theta) - r[K_2^p(\theta) - K_1^p] \\ &\quad -(r+b)K_2^g(\theta)]\} d\Phi(\theta) \end{aligned}$$

or rearranging,

$$\begin{aligned} V(K_1^p) &= V_{NL}(K_1^p) + \gamma [\int_{-\infty}^{\hat{\theta}} \{R(K_1^p, K_2^g(\theta), \theta) - (r+b)K_2^g(\theta)\} d\Phi(\theta) \\ &\quad - \int_{-\infty}^{\hat{\theta}_0} R^p(K_1^p, \theta) d\Phi(\theta)] + \gamma [\int_{\hat{\theta}}^{+\infty} \{R(K_2^p(\theta), K_2^g(\theta), \theta) \\ &\quad -r[K_2^p(\theta) - K_1^p] - (r+b)K_2^g(\theta)]\} d\Phi(\theta) \\ &\quad - \int_{\hat{\theta}_0}^{+\infty} \{R^p(K_2^p(\theta), \theta) - r[K_2^p(\theta) - K_1^p]\} d\Phi(\theta)] \end{aligned} \tag{4}$$

Proof: see Appendix A.

$V_{NL}(K_1^p)$ represents the value of the firm if, in the second period, there was no "technological flexibility", i.e., under our assumptions, if the award of an ecolabel was not foreseen. That is,

$$V_{NL}(K_1^p) \equiv R^p(K_1^p) + \gamma \int_{-\infty}^{\hat{\theta}_0} R^p(K_1^p, \theta), d\Phi(\theta) + \gamma \int_{\hat{\theta}_0}^{+\infty} \{R^p(K_2^p(\theta), \theta) - r[K_2^p(\theta) - K_1^p]\} d\Phi(\theta).$$

where $\hat{\theta}_0$ indicates the critical value ("trigger value") of the profitability index deriving from condition $R_K^p(K_1^p; \hat{\theta}_0) = r$.

Hence, analysing (4), it is possible to interpret the second and the third term on the r.h.s. as a "*quasi-rent*" which the firm may gain thanks to (the possibility of adopting the green technology, made profitable by) the ecolabel award scheme. The problem of optimization in the first period may be represented as follows:

$$K_1^p = \arg\max[V(K_1^p) - rK_1^p]. \tag{5}$$

Maximization of (5) allows us to determine the optimal stock of polluting capital in the first period. The first order condition is defined by the following expression:[8]

$$V'(K_1^p) \equiv V'_{NL}(K_1^p) + \gamma \int_{-\infty}^{\hat{\theta}} [R_{K^p}(K_1^p, K_2^g(\theta), \theta) - r] d\Phi(\theta) - \gamma \int_{-\infty}^{\hat{\theta}_0} [R_K^p(K_1^p, \theta) - r] d\Phi(\theta) = r. \tag{6}$$

Equation (6) expresses the marginal value of the investment in polluting capital in the first period with the sum of its marginal value, if ecolabelling were not foreseen, $V'_{NL}(K_1^p)$, the value reached – in the interval of "non investment" – $(-\infty, \hat{\theta})$ – by its net expected marginal earnings in the second period when the programme is in force, $\int_{-\infty}^{\hat{\theta}} [R_{K^p}(K_1^p, K_2^g(\theta), \theta) - r] d\Phi(\theta) \leq 0$, net of the value reached (regarding the non-investment interval – $(-\infty, \hat{\theta}_0]$ – by the net expected marginal earnings with no ecolabelling, $\int_{-\infty}^{\hat{\theta}_0} [R_K^p(K_1^p, \theta) - r] d\Phi(\theta) \leq 0$.[9] Indicating as $K_{NL,1}^p$ the stock of polluting capital which the firm would acquire in the first period if there were no ecolabelling,[10] the following proposition may be proved.

PROPOSITION 2: *If, despite contemporaneous use of polluting capital ($K_2^p > 0$), investment in green capital ($K_2^g > 0$) is a necessary and sufficient condition for a firm to be awarded with the label, ecolabelling induces*

i) a reduction, in the period before the label is granted, in investment in polluting capital, if $R_{K^pK^g} < 0$. That is:

$$K_1^p < K_{NL,1}^p$$

ii) an increase, in the period before the label is granted, in investment in polluting capital, if $R_{K^pK^g} > 0$*. That is*:

$$K_1^p > K_{NL,1}^p$$

Proof: see Appendix B.

Proposition 2 thus shows an interesting result. In particular, the possibility that the establishment of an ecolabel award scheme, and the consequent expectation of receiving the label once the firm has adopted the technology required to qualify for the certification ($K_2^g > 0$), might induce perverse effects, i.e. an expansion of the stock of polluting capital *before* the label is awarded, with permanent consequences due to the irreversibility of the investment. This possibility is linked to the existence of a "complementary" among the conventional and the green (labelled) production lines.

If a "complementary relationship" is expected to emerge ($R_{K^pK^g} > 0$), in particular if the ecolabel is expected to project a positive image onto the entire firm, increasing the profitability of all the capital invested, then the firm will find it profitable to expand its stock of polluting capital before adopting the green technology and receiving the label ($K_1^p > K_{NL,1}^p$).[11] In other words, the rationale behind the investment in green capital is to enhance the overall firm's rentability, by exploiting the "leverage effect" associated with the ability to exhibit a labelled product in the firm's "catalogue".

On the other hand, if a "substitutability relationship" is expected to emerge ($R_{K^pK^g} < 0$), that is if the label does not have a positive effect on the profitability of the conventional (polluting) production line, and the supply of a green output implies a decline in the rentability of the polluting one(s), the ecolabel award scheme will induce the firm to revise downward investment in polluting capital before the investment in green capital ($K_1^p < K_{NL,1}^p$).[12]

The risk of undesirable consequences of ecolabel award schemes, namely those due to misleading advertising, has been noticed by legislators. EEC Council Regulation n.880/92, for example, allows firms to refer to the ecolabel in advertising, but "only in relation to the specific product for which it was awarded" (art. 16, 1).

Legal provisions of this kind attenuate, but do not eliminate, the risk or perverse effects of ecolabelling, in that such a risk is connected with behavioural patterns by firms and consumers' misperception alike which are difficult to foresee and not always controllable ex post, as such behaviour is not easily covered by specific regulations on the use of ecolabels and/or general rules on misleading advertising.

Hence there is a need to apply some remedies that are more effective than formal restrictions, such as the prohibition in the above-mentioned Article 5 of the Community Regulation, in order to counter, or at least attenuate, the risk that a measure, aimed not only at stimulating environmental innovation but also at reducing the supply of polluting products, should produce undesirable consequences.

The most radical solution would clearly be to award labels only to firms able to demonstrate the environmental compatibility of all their activities, and not merely the eco-compatibility of a subset of productions lines (Dosi and Moretto 1997b).

A less radical solution would be to adopt selection procedures for candidates which, though not imposing an absolute restriction on the use of polluting technologies, would allow labels to be awarded after controls not only on the specific features of the product(s) proposed for certification, but also an evaluation of the firm's general behaviour, rewarding those firms that have shown a greater ("historical") propensity to adopt environmentally-friendly behaviour. The implications of this approach will be formally explored in the following paragraph.

3. Restrictions on the Issue of Labels

One of the solutions foreseen in the previous paragraph, that regarding the obligation for firms seeking to obtain the ecolabel to demonstrate the "environmental compatibility" of all of their activities ($K_2^p = 0$), and, therefore, not only eco-compatibility of a subset of their products, is rather unrealistic or, at least, does not correspond, to our knowledge, with the current official ecolabel award schemes. In fact, the current schemes base the issue of labels on an evaluation of the specific characteristics of the product put forward for certification ($K_2^g > 0$) by the firm and not on the overall environmental performance of the firm itself.

However, it should be noted, that not only are those characteristics predefined by the legislator but that the final decision to grant the label is to some extent left to the subjective appraisal of the certifiers. Therefore, considering the relative flexibility and discretionary nature of the evaluation procedure, it is not implausible to believe that this decision could also be influenced by the firms' general behaviour, by awarding with the label those firms which have also shown a greater propensity for environment friendly behaviour in the past.

The following analyses the consequences of a certification system which foresees an evaluation procedure which is stricter and gives uncertain outcomes with respect to that which is hypothesised in the basic model. More precisely, a procedure which does not only base the awarding of the label on a check of the specific characteristics of the product put forward for certification, but also on an examination of firm's propensity for eco-compatibility, by taking its past behaviour into consideration.

By adopting our general framework, the implications of this ecolabel award scheme can be formally explored by assuming that a firm able to submit products which meet the standard a product must meet to qualify for the ecolabel ($K_2^g > 0$) will be awarded with the label with some positive probability less than one ($0 < \mathrm{Pr} < 1$), and that Pr is negatively correlated to the stock of polluting capital employed by the firm in the past ($\mathrm{Pr}'(K_1^p) < 0$).

PROPOSITION 3: *If* $R_{K^pK^g} > 0$ *("complementarity relationship"), an ecolabelling programme which foresees that a firm able to submit a certifiable product* ($K_2^g > 0$) *will be awarded with some positive probability (Pr), and* $Pr'(K_1^p) < 0$, *would imply an investment in polluting capital* ($K_{Pr,1}^p$) *lower than the investment* (K_1^p) *induced by programme which forsees that* $K_2^g > 0$ *entitles the firm to be awarded with the label. That is:*

$$K_{\mathrm{Pr},1}^p < K_1^p$$

Proof: see Appendix C.

PROPOSITION 4: *If* $R_{K^pK^g} < 0$ *("subsitutability relationship"), an ecolabelling programme which foresees that a firm able to submit a certifiable product* ($K_2^g > 0$) *will be awarded with some positive probability (Pr), and* $Pr'(K_1^p) < 0$, *would imply an investment in polluting capital* ($K_{Pr,1}^p$)*:*

(i) higher than the investment (K_1^p) *induced by a programme which foresees that* $K_2^g > 0$ *entitles the firm to be awarded with the label, if:*

$$V'(K_1^p) - V'_{NL}(K_1^p) - \frac{\mathrm{Pr}'(K_1^p)}{1-\mathrm{Pr}(K_1^p)}\{V(K_1^p) - V_{NL}(K_1^p)\} < 0$$

(ii) lower than the investment (K_1^p) *induced by a programme which foresees that* $K_2^g > 0$ *entitles the firm to be awarded with the label, if:*

$$V'(K_1^p) - V'_{NL}(K_1^p) - \frac{\mathrm{Pr}'(K_1^p)}{1-\mathrm{Pr}(K_1^p)}\{V(K_1^p) - V_{NL}(K_1^p)\} > 0$$

Proof: see Appendix C.

To summarize, the restrictions on the issue of labels would attenuate the potential perverse ecolabelling's effects stemming from the existence of a *complementarity relationship* between the polluting and the green production lines. On the other hand, in the presence of a *substitutability relationship* the announcement that labels will be awarded with a probability less than one could induce higher investment in polluting capital with respect to an ecolabelling programme which entitles all firms able to submit a certifiable product (i.e., in our simplified model, all firms satisfying the condition $K_2^g > 0$) to be awarded with the label.

4. Final Remarks

Over the last two decades ecolabel award schemes, aimed at encouraging the supply of green products, have become widespread. The popularity of these initiatives depends in general on the implicit assumption that adoption of clean(er) technologies – encouraged by consumers' willingness-to-pay for (certified) green products – is necessarily accompanied by a reduction in investment in conventional production processes and hence a reduction in aggregate pollution.

Our analysis, however, points out that the two phenomena are not necessarily correlated. In other words, the adoption of a green production process and the supply of more environmentally benign products may be accompanied not only by conservation of conventional production lines – a phenomenon generally accepted by the current ecolabelling programs – but also by an increase in investment in "polluting capital" before the adoption of the technology required to submit products which qualify for the label.

The occurrence of such perverse effect depends on whether a "complementarity relationship" between the different production lines is expected to emerge. In particular, if firms expect that the label – obtained for a specific product – will project a positive image over the entire firm, then ecoloabelling, while encouraging the supply of green products, could at the same time induce increased investment in conventional technologies.

Although the risk of a distorted use of ecolabels could be attenuated through legal provisions aimed at avoiding misleading advertising, the occurence of a perverse effect is connected with behavioural patterns by firms and consumers' misperceptions that are difficult to prevent ex post and are not always easy to account for in the specific rules on using the labels and the more geneal legislation on misleading advertising. Hence there is a need to take more effective measures, in particular through the stricter selection of candidates.

The most radical solution would clearly be to award labels only to firms able to demonstrate the environmental compatibility of all their activities, and not merely the eco-compatibility of a subset of productions lines. A less radical solution would be to somehow restrict the awards of labels to a sub-set of firms able to submit products which meet the standards required to qualify for certification. In particular, our analysis shows that if labels are awarded with some positive probability less than one, negatively correlated to past investment in polluting capital, the potential perverse ecolabelling's effects stemming from the existence of a complementarity relationship between the polluting and the green production lines would be attenuated. On the other hand, restricting the issue of labels would reduce the effectiveness of ecolabelling programmes in the presence of a "substitutability relationship" between the two production lines.

The main policy implications can be summarized as follows. First, the design of an ecolabelling program should properly account for the potential impact exerted by the label on the rentability of investment in conventional technologies. Secondly, a precautionary approach, that is, restrictions on the issue of labels, is highly recommended in those contexts where firms' are more prone to abuses of the label and/or where consumers are more likely to be affected by misperceptions about firms' overall green performance.

Acknowledgements

A preliminary version was presented at the First World Congress of Environmental and Natural Resource Economists, Venice 25–27 June, 1998. We are indebted to the Referees, Roberto Tamborini and Jane Wallace for valuable comments and suggestions.

Notes

1. The first public ecolabel award scheme was set up in Germany in 1978; similar initiatives were later undertaken by other governments. In 1992 the Council of European Communities, through Regulation n.880/92, set up a Community eco-label award scheme, with the aim of creating the conditons for ultimately establishing an effective single environmental label in the Community. In many countries, there are also private ecolabelling programs, usually managed by nonprofit organizations, like the Green Seal, operating in the United States, which has developed standards for a number of product categories, such as paper, fluorescent lamps, paint.
2. The programs considered in this paper are only one form of what is sometimes referred to as environmental labelling which also includes government-mandated labels and firm's self-declarations (Van Ravenswaay and Blend, 1997). For instance, although our analysis could also have some implications for the evaluation of the overall effectiveness of mandatory programs or manufacturers' claims about the environmental attributes of certain products, in this paper we only consider voluntary programs which involve certification by a third party.
3. The relationship between irreversible investment decisions and option theory has been developed in the works of Bertola (1988) and Dixit (1989), and by He and Pindyck (1992) with regard to choices of technological flexibility. For an extension of this literature to investment decisions regarding the transition from a polluting technology to a green one, see Dosi and Moretto (1997a, 1998).
4. In particular, as a conceptual scheme for analysising firm's investment decisions, we refer to the two-period model proposed by Abel et al. (1996), which points out the connections between the literature on irreversible investment decisions under uncertainty and the theory of real options.
5. By adopting the taxonomy proposed by Brau and Carraro (1999), the programs considered in this paper falls within the firms' *green reputation enhancing voluntary approaches*, which hold the feature of contributing to a product differentiation that "helps creating niche markets and/or help identifying a firm's products" (p. 6).
6. We are indebted to a referee for having made evident this condition to guarantee the monotonicity property of the marginal profit of polluting capital with respect to θ.
7. We assume that, besides the capital cost, production costs are not affected by the production process (i.e. by the use of 'polluting' or 'green capital'), so that the higher profitability of the labelled output only results from the higher consumers' willingness-to-pay for green products; in other words, the adoption of green capital only implies higher revenues. However, differences between production costs would not affect our results, provided that higher revenues more than compensate higher green technology's production costs.
8. Notice that a change in K_1^p will affect the threshold value $\hat{\theta}(K_1^p)$ (as well as $\hat{\theta}_0(K_1^p)$). Taking account of these changes, the first derivative of (4) with respect to K_1^p becomes:

$$V'(K_1^p) + \gamma[R(K_1^p, K_2^g(\hat{\theta}), \hat{\theta}) - (r+b)K_2^g(\hat{\theta})]\frac{d\hat{\theta}(K_1^p)}{dK_1^p}$$

$$-\gamma[R(K_2^p(\hat{\theta}), K_2^g(\hat{\theta}), \hat{\theta}) - r[K_2^p(\hat{\theta}) - K_1^p] - (r+b)K_2^g(\hat{\theta})]\frac{d\hat{\theta}(K_1^p)}{dK_1^p}$$

where $V'(K_1^p)$ is as in (6). However, by definition we know that $R_{K^p}(K_1^p, K_2^g(\hat{\theta}); \hat{\theta}) = r$ which implies that the optimal second period stock of capital when $\theta = \hat{\theta}$ is equal to K_1^p, so that $K_2^p(\hat{\theta}) \equiv K_1^p$. This makes the sum of the second and third term always null. The same applies for $\hat{\theta}_0$. It is worthwhile to stress that also the second period optimal stocks of capital should depend on K_1^p. For instance, the correct notation would be $K_2^p(K_1^p, \theta)$ and $K_2^g(K_1^p, \theta)$. However, since by the Envelope theorem K_1^p does not play any rule, we have suppressed the notation to avoid any potential confusion.

9. Following Assumption 2 it is easy to check that:

$$V''(K_1^p) = R''^p(K_1^p) + \gamma \int_{-\infty}^{\hat{\theta}} R_{K^pK^p}(K_1^p, K_2^g(\theta), \theta) d\Phi(\theta) < 0,$$

That is, for whatever $r > 0$, there is always a single value of K_1^p which satisfies condition (6).

10. That is, the stock of polluting capital deriving from condition $V'_{NL}(K_{NL,1}^p) = r$.
11. Besides consumers' misperceptions, a complementarity relationship between the two production lines could emerge as a result of intra-firm's technological spillovers stemming from the simultaneous adoption of the two production processes.
12. As the label transfers process innovation into product innovation, the positivity of the "*quasi-rent*" $V(K_1^p) - V_{NL}(K_1^p)$ follows a similar path to equation (1) in Athey and Schmutzler (1995), where, for a given quantity, implementing a product design innovation is beneficial if it raises the demand curve, and a process innovation is beneficial if it lowers the average cost curve. Therefore we can interpret the difference $V'(K_1^p) - V'_{NL}(K_1^p)$ as the marginal effect of the label on the returns to K_1^p. That is, the effect of implementing an innovative technology (product and/or process innovation) on marginal revenue as in equation (3) in Athey and Schmutzler (1995).

Appendix A: Proof of Proposition 1

To identify the frm's ex ante value, let us start by analysing investment decisions in the second period, for a given value of stock of polluting capital inherited from the first period.

Second Period

In the second period, once the realization of the profitability index θ is known, the firm maximizes its profits by choosing an optimal mix between the two types of capital. The conditions of optimality are:

$$R_{K^p}(K_2^p, K_2^g, \theta) = r \quad (7)$$

$$R_{K^g}(K_2^p, K_2^g, \theta) = r + b \quad (8)$$

By virtue of Assumption 2, for any given r and b, the solution of the above system gives rise to a couple of well defined optimal, second period, demands for polluting and green capital $K_2^p(\theta)$ and $K_2^g(\theta)$ respectively. To analyse the property of $K_2^p(\theta)$ we can proceed by steps. First, from (8) there exists a function $K_2^g = K_2^g(K_2^p; \theta)$ such that, for any given values of θ and the polluting capital K_2^p, $R_{K^g}(K_2^p, K_2^g(K_2^p, \theta), \theta) \equiv r + b$, with the properties that $\frac{\partial K_2^g}{\partial \theta} = -\frac{R_{K^g\theta}}{R_{K^gK^g}} > 0$, and $\frac{\partial K_2^g}{\partial K_2^p} = -\frac{R_{K^gK^p}}{R_{K^gK^g}} \begin{cases} > 0 \text{ if } R_{K^pK^g} > 0 \\ < 0 \text{ if } R_{K^pK^g} < 0 \end{cases}$. Second, substituting $K_2^g(K_2^p, \theta)$ in the marginal productivity of K^p we can write:

$$P(K_2^p, \theta) \equiv R_{K^p}(K_2^p, K_2^g(K_2^p, \theta), \theta) \quad (9)$$

Finally, following Assumption 2, it is easy to show that:

$$P_{K^p}(K_2^p, \theta) = \frac{R_{K^pK^p}R_{K^gK^g} - (R_{K^pK^g})^2}{R_{K^gK^g}} < 0 \tag{10}$$

and

$$\begin{aligned} P_\theta(K_2^p, \theta) &= R_{K^p\theta} + R_{K^pK^g}\frac{\partial K_2^g}{\partial \theta} \\ &= \frac{R_{K^p\theta}R_{K^gK^g} - R_{K^g\theta}R_{K^pK^g}}{R_{K^gK^g}} > 0 \end{aligned} \tag{11}$$

Therefore, by (10) and (11), given the stock of inherited, capital K_1^p, we are able to identify a critical value for the profitability index $\hat{\theta}$, which satisfies:

$$P(K_1^p, \hat{\theta}) \equiv R_{K^p}(K_1^p, K_2^g(K_1^p, \hat{\theta}), \hat{\theta}) = r \tag{12}$$

That is, depending on K_1^p and for every optimal value of $K_2^g > 0$, equation (12) identifies two intervals of investment. In the first one, defined by the inequality $\theta > \hat{\theta}$, it is optimal for the firm to acquire new units of both capital typologies up to the point where the marginal profits are equal to the marginal costs as in (7). In this case $K_2^p > K_1^p$. In the second case, expressed by $\theta < \hat{\theta}$, the firm finds it optimal to invest in green capital alone. In this case, $K_2^p = K_1^p$, since $R_{K^p}(K_1^p, K_2^g\,(K_1^p, \theta), \theta) < r$.

Furthermore, since Assumption 3 implies that:

$$R(K^p, \delta K^g, \theta) = \begin{cases} R(K^p, K^g, \theta) & \textit{if}\ \delta = 1, \\ R^p(K^p, \theta) & \textit{if}\ \delta = 0, \end{cases}$$

where δ is a dummy variable indicating the existence of an ecolabel award scheme ($\delta = 1$) or not ($\delta = 0$), applying the theorem of functions implicit to (12), introducing δ and using Assumption 2, we can prove:

LEMMA 1 *(i)* $\frac{d\hat{\theta}}{d\delta} = -\frac{R_{K^p\delta K^g}K_2^g}{R_{K^p\hat{\theta}} + R_{K^p\delta K^g}\frac{\partial K_2^g}{\partial \hat{\theta}}}$, *so that* $\frac{d\hat{\theta}}{d\delta}|_{\delta=0} = 0$ *and* $\frac{d\hat{\theta}}{d\delta}|_{\delta=1}$

$$\begin{cases} > 0 \ \textit{if}\ R_{K^pK^g} < 0 \\ < 0 \ \textit{if}\ R_{K^pK^g} > 0 \end{cases}$$

(ii) With $\hat{\theta}(\delta = 0) = \hat{\theta}_0$, *where* $\hat{\theta}_0$ *is given by* $R_K^p(K_1^p, \hat{\theta}_0) = r$

The Lemma shows that the interval where it is optimal to invest in both types of capital, $\theta > \hat{\theta}$, is reduced with increases in investment in green capital when a substitutability relationship prevails. On the other hand, this relation expands when the two product lines are complementary to one another.

FIRST PERIOD

The value of the firm in period 1, $V(K_1^p)$, is defined as the expected discounted value of current profits in the two periods when the stock of capital in period 1 is given by K_1^p. Recalling (12), this value may be expressed as follows:

$$\begin{aligned} V(K_1^p) &= R^p(K_1^p) + \gamma \int_{-\infty}^{\hat{\theta}} \{R(K_1^p, K_2^g(\theta), \theta) - (r+b)K_2^g(\theta)\} d\Phi(\theta) + \\ &\gamma \int_{\hat{\theta}}^{+\infty} \{R(K_2^p(\theta), K_2^g(\theta), \theta) - r[K_2^p(\theta) - K_1^p] \\ &-(r+b)K_2^g(\theta)]\} d\Phi(\theta), \end{aligned}$$

or rearranging and taking account of Lemma 1, we obtain the expression reported in the text:

$$\begin{aligned} V(K_1^p) &= V_{NL}(K_1^p) + \gamma [\int_{-\infty}^{\hat{\theta}} [R(K_1^p, K_2^g(\theta), \theta) - (r+b)K_2^g(\theta)] d\Phi(\theta) \\ &- \int_{-\infty}^{\hat{\theta}_0} R^p(K_1^p, \theta) d\Phi(\theta)] + \gamma [\int_{\hat{\theta}}^{+\infty} \{R(K_2^p(\theta), K_2^g(\theta), \theta) \\ &-r[K_2^p(\theta) - K_1^p] - (r+b)K_2^g(\theta)]\} d\Phi(\theta) \\ &- \int_{\hat{\theta}_0}^{+\infty} \{R^p(K_2^p(\theta), \theta) - r[K_2^p(\theta) - K_1^p]\} d\Phi(\theta)], \end{aligned} \tag{13}$$

where V_{NL} (K_1^p) indicates the value of the firm if, in the second period, there was no possibility of switching to the green technology, i.e., under our assumptions, if the award of an ecolabel was not forseen. That is:

$$\begin{aligned} V_{NL}(K_1^p) &\equiv R^p(K_1^p) + \gamma \int_{-\infty}^{\hat{\theta}_0} R^p(K_1^p, \theta) d\Phi(\theta) \\ &\gamma \int_{\hat{\theta}}^{+\infty} \{R^p(K_2^p(\theta), \theta) - r[K_2^p(\theta) - K_1^p]\} d\Phi(\theta). \end{aligned}$$

Appendix B: Proof of Proposition 2

For convenience we rewrite the optimal conditon for K_1^p:

$$\begin{aligned} V'(K_1^p) &\equiv V'_{NL}(K_1^p) + \gamma \int_{-\infty}^{\hat{\theta}} [R_{K^p}(K_1^p, K_2^g(\theta), \theta) - r] d\Phi(\theta) \\ &-\gamma \int_{-\infty}^{\hat{\theta}_0} [R_K^p(K_1^p, \theta) - r] d\Phi(\theta) = r. \end{aligned} \tag{14}$$

Equation (14) expresses the marginal value of the polluting factor in period 1 as the sum of the marginal value of the polluting factor should the label not be granted, $V'_{NL}(K_1^p)$, plus the differences between the expected value of marginal profits in period 2 when the label is granted (regarding the non-investment interval $(-\infty, \hat{\theta})$, $\int_{-\infty}^{\hat{\theta}} [R_{K^p}(K_1^p, K_2^g(\theta), \theta) - r] d\Phi(\theta \leq 0)$, and the expected value of marginal profits in period 2 without the label (regarding the non investment interval $(-\infty, \hat{\theta}_0)$, $\int_{-\infty}^{\hat{\theta}_0} [R_K^p(K_1^p, \theta) - r] d\Phi(\theta) \leq 0)$. As in both cases, in the non investment interval the marginal profits are lower than r, the expected value in turn is negative, so that:

$$\int_{-\infty}^{\hat{\theta}}[R_{K^p}(K_1^p, K_2^g(\theta), \theta) - r]d\Phi(\theta) - \int_{-\infty}^{\hat{\theta}}[R_K^p(K_1^p, \theta) - r]d\Phi(\theta) < 0 \text{ if } \hat{\theta} > \hat{\theta}_0,$$

$$\int_{-\infty}^{\hat{\theta}}[R_{K^p}(K_1^p, K_2^g(\theta), \theta) - r]d\Phi(\theta) - \int_{-\infty}^{\hat{\theta}}[R_K^p(K_1^p, \theta) - r]d\Phi(\theta) > 0 \text{ if } \hat{\theta} < \hat{\theta}_0,$$

Then, application of Lemma 1 proves proposition 2 in the text.

Appendix C: Proof of Propositions 3 and 4

The firm's *ex ante* value is defined by the followint expression:

$$\begin{aligned}
V_{\Pr}(K_1^p) &= V(K_1^p) - \gamma(1 - \Pr(K_1^p))\{[\int_{-\infty}^{\hat{\theta}}[R(K_1^p, K_2^g(\theta), \theta) \\
&\quad -(r+b)K_2^g(\theta)]d\Phi(\theta) - \int_{-\infty}^{\hat{\theta}_0} R^p(K_1^p, \theta)d\Phi(\theta)] + [\int_{\hat{\theta}}^{+\infty} \\
&\quad \{R(K_2^p(\theta), K_2^g(\theta), \theta) - r[K_2^p(\theta) - K_1^p] - (r+b)K_2^g(\theta)]\}d\Phi(\theta) \\
&\quad - \int_{\hat{\theta}_0}^{+\infty}\{R^p(K_2^p(\theta), \theta) - r[K_2^p(\theta) - K_1^p]\}d\Phi(\theta)]\}. \qquad (15)
\end{aligned}$$

where the scond term on the r.hs. of (15) indicates the expected loss due to the possibility of not obtaining the label (in the second period) in spite of the investment in green capital ($K_2^g > 0$). Thus, the f.o.c. to determine the optimal stock of polluting capital in the first period becomes:

$$\begin{aligned}
V'_{\Pr}(K_1^p) &\equiv V'(K_1^p) - (1 - \Pr(K_1^p))\{V'(K_1^p) - V'_{NL}(K_1^p)\} - \\
&\quad +\Pr'(K_1^p)\{V(K_1^p) - V_{NL}(K_1^p)\} = r. \qquad (16)
\end{aligned}$$

From (16), considering that the issue of the label implies that the *quasirent* is positive, $V(K_1^p) - V_{NL}(K_1^p) > 0$, and recalling proposition 2, the restriction on the issue of labels implies:

a) a reduction in the investment in polluting capital in period 1 if $R_{K^pK^g} > 0$; or if $R_{K^pK^g} < 0$ and

$$V'(K_1^p) - V'_{NL}(K_1^p) - \frac{\Pr'(K_1^p)}{1 - \Pr(K_1^p)}\{V(K_1^p) - V_{NL}(K_1^p)\} > 0.$$

That is:

$$V'_{\Pr}(K_1^p) < V'(K_1^p) \Rightarrow K_{\Pr,1}^p < K_1^p$$

b) an increase in the investment in polluting capital in period 1 if $R_{K^pK^g} < 0$ and

$$V'(K_1^p) - V'_{NL}(K_1^p) - \frac{\Pr'(K_1^p)}{1 - \Pr(K_1^p)}\{V(K_1^p) - V_{NL}(K_1^p)\} < 0.$$

That is:

$$V'_{\Pr}(K_1^p) > V'(K_1^p) \Rightarrow K_{\Pr,1}^p > K_1^p$$

References

Abel, A., A. Dixit, J. Eberly and R. Pindyck (1969), 'Options, the Value of Capital and Investment', *The Quarterly Journal of Economics* **61**, 753–777.

Athey, S. and A. Schmutzler (1995), 'Product and Process Flexibility in an Innovative Environment', *RAND Journal of Economics* **26**, 557–574.

Bertola, G. (1988), 'Irreversible Investment', Working Paper, Princeton University.

Brau, R. and C. Carraro (1999), 'Voluntary Approaches, Market Structure and Competition', *Note di Lavoro* 53.99, Fondazione Eni E. Mattei, Milan.

Dixit, A. (1989), 'Entry and Exit Decisions Under Uncertainty', *Journal of Political Economy* **97**, 620–638.

Dosi, C. and M. Moretto (1997a), 'Pollution Accumulation and Firm Incentives to Accelerate Irreversible Technological Change under Uncertain Private Benefits', *Environmental and Resource Economics* **10**, 285–300.

Dosi, C. and M. Moretto (1997b), *Ecolabelling e decisioni di investimento: un'analisi teorica.* Paper presented at the XXXVIII Annual Meeting of the Società italiana degli Economisti, Rome 17–8 October, 1997.

Dosi, C. and M. Moretto (1998), 'Auctioning Green Investment Grants as a Means to Accelerate Environmental Innovation', *Revue d'Economie Industrielle* **83**, 99–110.

He, H. and R. S. Pindyck (1992), 'Investment in Flexible Production Capacity', *Journal of Economic Dynamic and Control* **16**, 575–599.

Kuhn, M. (1999), 'Green Lemons. Environmental Labels and Entry into an Environmentally Differentiated Market under Asymmetric Information', *Working Paper* 20, Thunen Series of Applied Economic Theory, Universitat Rostock.

Mattoo, A. and H. V. Singh (1994), 'Eco-Labelling: Policy Considerations', *Kyklos* **47**, 53–65.

Van Ravenswaay, E. O. and J. R. Blend (1997), 'Using Ecolabelling to Encourage Adoption of Innovative Environmental Technologies in Agriculture', *Staff Paper* 97–19, Department of Agricultural Economics, Michigan State University.

[8]

Environmental quality competition and eco-labeling

Gregory S. Amacher,[a,b,*] Erkki Koskela,[c] and Markku Ollikainen[d]

[a] *College of Natural Resources, Virginia Polytechnic Institute and State University, 307 C Cheatham Hall, Blacksburg, VA 24601-0324, USA*

[b] *Department of Economics, Virginia Commonwealth University, Richmond, VA, USA*

[c] *Department of Economics, P.O. Box 17 (Arkadiankatu 7), University of Helsinki, FIN-00014, Finland*

[d] *Department of Economics and Management, P.O. Box 27, University of Helsinki, FIN-00014, Finland*

Received 1 July 2002; revised 20 November 2002

Abstract

A three-stage game of investment, environmental quality provision and price competition is developed to study the impact of green technology investment (eco-labeling), in a duopoly model of vertical product differentiation. The firms' incentives to invest in green technologies depend on their relative cost structure. When firms are identical with respect to fixed costs, both firms will always invest, but if one firm is more efficient in investing, then the other firm may or may not invest depending on the level of unit cost of investment. Quality competition will be tighter when the low-quality firm is more efficient, and looser when the high-quality firm is more efficient in investing. Socially optimal investment for both firms is always positive, but lower than in the duopoly solution. In the absence of environmental externalities, the quality dispersion chosen by profit maximizing firms may be too high or too low, while environmental externalities increase the possibility low-quality dispersion that is too low within the market solution. Finally, and importantly, ecolabeling can be used as a means of reducing excessive investment and increasing environmental quality that is too low.

JEL classification: L13; H23; Q20

Keywords: Product differentiation; Technology investment; Socially optimal quality and investments

*Corresponding author. College of Natural Resources, Virginia Polytechnic Institute and State University, 307 C Cheatham Hall, Blacksburg, VA 24601-0324, USA. Fax: +1-540-231-3698.

E-mail addresses: gamacher@vt.edu (G.S. Amacher), erkki.koskela@helsinki.fi (E. Koskela), markku.ollikainen@helsinki.fi (M. Ollikainen).

Nomenclature

$k=$ H, L	index of technology for high-quality (H) and low-quality (L) cost firms
b	marginal cost parameter for provision of quality by firm $k=$ H, L
$\alpha_k(I_k)$	fixed costs of quality provision depending on investment
s_k	quality of goods provided in second stage
$c_k(s_k, I_k)$	total cost of providing quality for the firm $k =$ H, L
p_k^*	optimal (Bertrand) prices from third-stage price competition for firm $k=$ H, L
u	utility of consumer
π^k	profit function for low-quality ($k =$ L) and high-quality ($k =$ H) firms
s_k^*	quality level of goods provided by duopoly for low ($k =$ L) and high ($k =$ H) quality
s_k^{w}	socially optimal quality level of goods for low ($k =$ L) and high ($k =$ H) quality
s_k^{we}	quality level of goods provided in market under the externality case for low ($k =$ L) and high ($k =$ H) quality
x	difference in fixed costs, $x = x(I_{\mathrm{H}}, I_{\mathrm{L}}) = \alpha_{\mathrm{H}}(I_{\mathrm{H}}) - \alpha_{\mathrm{L}}(I_{\mathrm{L}})$
θ	consumer i's taste parameter, i.e., marginal willingness to pay for good
$\hat{\theta}$	threshold taste parameter for consumer who is indifferent between consumption of high and low environmental quality goods
$\bar{\theta}$	upper bound on consumers' marginal willingness to pay for environmental quality
$\underline{\theta}$	lower bound on consumers' marginal willingness to pay for environmental quality
$\tilde{\theta}$	index of marginal consumer when both high- and low-quality goods are sold at marginal cost
I^k	first-stage investment in technology for firm $k =$ H, L.
$I_{\mathrm{W}k}$	social welfare maximizing level of investment in technology ($k =$ H, L)
d_k	demand (number of consumers) purchasing high- ($k =$ H) and low- ($k =$ L) quality goods
$\tilde{\pi}^k$	profits of high- and low-quality firms ($k =$ (H, L)), gross of investment costs, when prices and qualities have been chosen optimally
V^k	profit function corresponding to first stage, where firms choose their production technology
u^{e}	utility of consumer in the presence of an externality
γ	weight reflecting the magnitude of the externality
s_{a}	average level of environmental quality across the firms
SW	social welfare function in the absence of environmental-quality-related externalities
$\bar{\theta}_k$	partial derivative of upper bound on marginal willingness to pay wrt quality $k =$ H, L
$\underline{\theta}_k$	partial derivative of lower bound on marginal willingness to pay wrt quality $k =$ H, L

286 G.S. Amacher et al. / Journal of Environmental Economics and Management 47 (2004) 284–306

$\hat{v}$	Unit cost of investment
τ	Auditing cost of applying eco-label
SW^e	social welfare function in the presence of environmental-quality-related externalities

1. Introduction

Consumer preference to purchase from "green" firms is well established and often revealed through increased willingness to pay for products viewed as "clean," i.e., produced with environmentally friendly production or abatement technologies such as recycling and use of less polluting inputs. From a polluting firm's perspective, there may be strong incentives to invest in these technologies if public opinion becomes more favorable toward the firm, or if the firm can use its investment as a way of differentiating its product.

Investment in environmentally clean technologies has recently given firms the right to attach a specific "ecolabel" to their product. Well-known examples include dolphin-safe tuna, green electricity, recycling in production processes, organic food production, and production of biodegradable washing powders (e.g., see [26,34, Chapter 2] for a review of ecolabeling and related green technology investments). Such ecolabels are potentially important strategic variables for firms, serving to differentiate a firm's product from those produced by firms that do not make the necessary green investments. This suggests that environmental quality competition might be studied within the framework of product quality models. These models usually follow a duopoly framework under the assumption of vertical product differentiation. Another common assumption is that firms compete in two stages, first by choosing product quality, and then by choosing prices for their (quality-differentiated) products.

The quality competition literature omits one crucial component of existing eco-labeling systems, namely investment.[1] Current ecolabeling systems, such as Green Seal in the United States, the Nordic Swan in Scandinavia, the European Union Eco-Label Award Scheme, the Blue Angel in Germany or the Japanese Eco-Mark, are usually designed to cover from 5 to 20% of the market. The license to use the ecolabel is quite often limited to a relatively short period of time. What is most important, however, is the fact that the criteria for ecolabeling rights are revised, i.e. tightened, on average every 3 or 4 years (for a representation of these and other eco-labeling schemes, see [27]). This feature of ecolabeling systems implies that any firm, wishing to provide high enough environmental quality to secure an ecolabel, is forced to make investments to improve quality (and to reduce the costs of quality production). High provision of environmental quality therefore requires some combination of new advanced abatement technologies and increased abatement efforts. These investments are typically costly in terms of either the capital

[1] Crampes and Hollander [12], Lehmann-Grube [20], and Ronnen [30] provide studies of general quality models, and see [4,5] for application to environmental quality. There are only a few analytical treatments on eco-labeling, despite that it is becoming more and more common among industries. Matto and Singh [21] discuss circumstances under which eco-labels are profitable to firms, and Swallow and Sedjo [35] provide a graphical analysis of mandatory eco-labeling. There is also a growing non-analytical discussion concerning "green labeling" in natural resource markets [18,28], and empirical analysis of the values of the ecolabel [25].

outlays required or the auditing and license costs incurred by the firms. Understanding the inherent dynamics of ecolabeling schemes requires an investigation of firms' investment decisions, and the role of this investment to competition and product differentiation.

The purpose of this paper is to examine eco-labeling within a vertically differentiated market model. To capture features of ecolabeling discussed above, we extend the usual duopoly model of vertical product differentiation with variable costs and full market coverage by including an initial technology investment stage. This allows us to compare socially optimal levels of investments and environmental qualities with those that result from profit maximization.[2] In addition, we study how externalities, associated with inefficient average environmental quality, affect socially optimal provision of environmental quality. Finally, utilizing an approach used to study the role of advertising in the industrial organization literature [15], we extend the model by allowing consumer willingness to pay to depend on investments made by firms.

We show that firms' incentives to invest in technologies (and obtain ecolabels) depend crucially on the differences in cost structures between firms. When firms are identical with respect to fixed costs, both firms will always invest in the green technology. If the high-quality firm is more efficient at the margin in investing, then it invests, but the low-quality firm may or may not invest depending on the unit cost of investment. The opposite holds if the low-quality firm is more efficient at investing.

These different incentives to invest lead to strikingly different outcomes for the provision of environmental quality in markets. In fact, quality competition will be more intense when the low-quality firm is more efficient at investment, and less intense when the high-quality firm is more efficient at investment. Between these extremes, i.e., when the firms are equally efficient in investing, we arrive at results obtained in the basic two-stage model of quality competition by Crampes and Hollander [12].[3] Thus, the two-stage model of price and quality competition between identical firms is simply a special case of our three-stage model.

Relative to the social optimum, previous work demonstrates that duopolists usually over-invest in order to mitigate price competition. In the absence of externalities, if the firms have identical

[2] There has been some work in product differentiation that considers a three-stage game, but in these models firms' entry decisions represent the first stage, followed by quality and price competition [13,19,31]. Our model is loosely related to Zhou et al. [37], who model investment within the traditional two-stage game of product differentiation. In their model, however, investment is not observationally defined and is lumped together with quality into a single composite variable. Unlike Zhou et al. [37], our approach allows an explicit examination of the impact of firms' incentives to invest in technologies used to produce quality. Our model also allows a comparison of incentives to invest between profit maximizing and socially optimal outcomes. These new extensions are important, yet unexplored, issues from the perspective of eco-labeling systems, as they provide insight into the welfare effects of eco-labeling schemes.

[3] More precisely, in Crampes and Hollander [12] both the framework and the focus of analysis are different from our paper. First, in their duopoly model both firms share a common cost function, which is linear in terms of quantity and convex in terms of quality. We do not restrict the cost function to be equivalent among firms. In our case, the Crampes-Hollander assumption holds when firms are equally efficient in investment, measured in terms of the cost reduction that follows from investing. Second, Crampes and Hollander explore the implications of a minimum quality standard to social welfare and the behavior of high-quality and low-quality producers. They show, under a binding minimum quality requirement, that the high-quality producer is worse-off and the low-quality producer is better-off than in an unregulated equilibrium. Moreover, the social welfare effect of the minimum quality standard is positive if the required quality standard is "sufficiently close" to the one chosen by the low-quality producer in the unregulated equilibrium.

cost structures, we find that quality dispersion between products chosen by profit maximizing firms is too high, but average quality is too low from the viewpoint of social welfare. But if the high-quality firm is more efficient in investment, then quality dispersion and average quality are both too low. This is because the high-quality firm may underprovide environmental quality, while the low-quality firm always under-provides environmental quality. This possibility for underproviding high quality is increased by the presence of environmental externalities. These findings represent an important departure from the two-stage quality competition literature. For our problem here, they imply that eco-labeling can be used as a means of reducing excessive investment and increasing environmental quality.

The rest of the paper is structured as follows. In Section 2 we develop a duopoly model of vertical product differentiation to analyze price, quality and investment decisions. In Section 3 we examine the social welfare optimum and compare it with the profit-maximizing solution, and in Section 4 we introduce the presence of a quality externality into the social welfare function. Finally, in Section 5 we offer our conclusions.

2. An extended model with price, quality and investment competition

In this section we develop and analyze a duopoly model of vertical product differentiation with endogenous investments.

2.1. Basic model

We assume that consumers can observe the quality choices of firms and have the following utility function:

$$u = \theta s_k - p_k, \tag{1}$$

where s_k and p_k are, respectively, the (environmental) quality and price of a good of quality k and the term θ represents consumer's taste parameter.[4]

There are two qualities of goods corresponding to two types of firms, $k = \mathrm{H}$ (high quality) and $k = \mathrm{L}$ (low quality). θ can be interpreted as each consumer's marginal willingness to pay for environmental quality, which is *uniformly* distributed on the interval $\theta \in [\underline{\theta}, \bar{\theta}]$. Each consumer purchases one unit of the good. The indifferent consumer has a threshold taste parameter $\hat{\theta} = \frac{p_\mathrm{H} - p_\mathrm{L}}{s_\mathrm{H} - s_\mathrm{L}}$. When the market is fully covered, the demands for high- and low-quality goods are defined by $d_\mathrm{H} = \bar{\theta} - \hat{\theta}$ and $d_\mathrm{L} = \hat{\theta} - \underline{\theta}$, respectively.[5]

Environmental quality refers here to the cleanness of production, which can be increased by abating pollution resulting from production. Abatement of pollution can entail variable and fixed costs. Variable costs reflect ongoing abatement effort by a firm, while fixed costs could be thought

[4] The nomenclature list provides a definition of all symbols. In what follows, derivatives will be noted by primes for functions with one argument, and partial derivatives will be denoted by subscripts for functions with many arguments.

[5] Under full market coverage, production does not change while quality may change. This allows us to focus sharply on quality competition and investments.

of as abatement technology investment.[6] (e.g., [1,2,7,22,23]). With this in mind, we use the following assumption regarding the costs of production:

Assumption 1. Let $C^k(s_k, I_k, d_k)$ be the total costs of output production and $c^k(s_k, I_k, d_k)$ the respective marginal costs with the following properties

(i) $\frac{\partial C^k(s_k,I_k,d_k)}{\partial d_k}>0$; $\frac{\partial c^k(s_k,I_k,d_k)}{\partial d_k}\geqslant 0$,
(ii) $\frac{\partial c^k(s_k,I_k,d_k)}{\partial s_k}>0$; $\frac{\partial^2 c^k(s_k,I_k,d_k)}{\partial s_k^2}>0$,
(iii) $\frac{\partial c^k(s_k,I_k,d_k)}{\partial I_k}<0$; $\frac{\partial^2 c^k(s_k,I_k,d_k)}{\partial I_k^2}>0$,

where d_k is the demand by a firm of type k (= H, L), s_k is the quality choice, and I_k is investment in a technology.

According to Assumption 1 the marginal cost of producing quality is convex and is decreased by investment at a decreasing rate. We use a specific form for these costs, derived by adding the fixed cost component of investment to the quadratic form common in the vertical product differentiation literature,

$$c_k(s_k, I_k) = \tfrac{1}{2}bs_k^2 + \alpha_k(I_k) \quad \text{for } k = \text{H, L}, \quad \text{with } \alpha_k'(I_k)<0 \text{ and } \alpha_k''(I_k)>0. \tag{2}$$

The cost of quality provision features both a variable and fixed component. This is consistent with capacity investment in the Dixit–Spence sense [14,32,33].

Assuming variable linear costs of output production, the profit functions of both firms can now be written,

$$\pi^k = [p_k - c_k(s_k, I_k)]d_k, \quad \text{for } k = \text{H, L}. \tag{3}$$

Next we analyze our three-stage game. In the first stage, firms decide whether and how much to invest in the technology. In the second stage, firms compete by choosing the level of environmental quality for their goods. The third stage has firms competing by choosing prices for their (quality) differentiated product. This sub-game perfect equilibrium relies on commitment by each firm regarding its investment and quality choices.

[6] Investment in technologies is ubiquitous among eco-labeling participants. Many eco-labeling schemes often encourage participating firms to use only the "best available technologies" to achieve required environmental quality targets (e.g., [17,27]). Some schemes in both agricultural and non-agricultural contexts tie eco-labeling specifically to technology adoption [6,29]. A common assumption is that firms must consider either modifications to existing technologies or new technologies in order to satisfy a third-party auditor. This is the case even in de facto labeling schemes, such as green electricity in the US. Here, with power plants, the required investments to achieve the "green" distinction are costly and require large capital outlays (e.g., see [24]). It is also the case with "green" wood manufacturing programs [36]. Indeed, it is not surprising that in practice most eco-labeling systems have given rise to arbitragers who provide information to companies on technologies needed to meet specific label environmental quality standards.

2.2. Price and quality competition

Using backward induction, we first turn to the *third* stage, in which firms choose prices subject to their previous choices for technology investment and level of environmental quality provided for their product. It is straightforward to solve for the Bertrand optimal prices and obtain the following price difference:

$$p_H^* - p_L^* = \tfrac{1}{3}\left[\tfrac{1}{2}b(s_H^2 - s_L^2) + (\bar{\theta} + \underline{\theta})(s_H - s_L) + x\right], \tag{4}$$

where $x = \alpha_H(I_H) - \alpha_L(I_L)$.[7] This term x measures the relative efficiency of each firm, in terms of the difference in their fixed cost components of providing quality. We will refer to this as the difference in cost structures between the firms.

We can see from (4) that, in addition to conventional parameters (market size, quality difference and the cost of providing quality), the price difference depends on x, i.e. on the difference in the firms' fixed costs as determined by the investment stage. If x is negative (positive), meaning that the fixed cost component of the low-quality firm is higher (lower) than that of the high-quality firm, then the price difference is smaller (larger). This will be important later when we interpret the investment results.

The Nash equilibrium qualities for each firm in the *second* stage can be shown to equal,[8]

$$s_H^* = \frac{5\bar{\theta} - \underline{\theta}}{4b} + \frac{2x(I_H, I_L)}{3(\bar{\theta} - \underline{\theta})}, \quad s_L^* = \frac{5\underline{\theta} - \bar{\theta}}{4b} + \frac{2x(I_H, I_L)}{3(\bar{\theta} - \underline{\theta})}, \tag{5}$$

so that,

$$s_H^* - s_L^* = \frac{3(\bar{\theta} - \underline{\theta})}{2b}. \tag{6}$$

The first term of Eq. (5) is conventional in product differentiation results. The second term in (5) is unique to our investment case; this shows that the difference in fixed costs of quality provision will affect firms' optimal level of qualities, but it does not matter for quality dispersion in the market (Eq. (6)). Referring to (5), if the difference in fixed costs is negative (positive), indicating that the fixed cost component of the low-quality firm is higher (lower) compared to the high-quality firm, then the quality levels for both firms will be lower (higher). After examining the investment stage, we will discuss the quality game in more detail.

Substituting the optimal quality levels (5) into the profit functions for each firm and rearranging yields the corresponding profits of the two firms when prices and qualities have been chosen optimally,

$$\tilde{\pi}^H = \frac{1}{6b}(\bar{\theta} - \underline{\theta})\left[\frac{3}{2}(\bar{\theta} - \underline{\theta}) - \frac{4x(I_H, I_L)b}{3(\bar{\theta} - \underline{\theta})}\right]^2, \tag{7a}$$

[7] It is also straightforward to show under the assumptions made that the Bertrand equilibrium is unique and stable (Crampes and Hollander [12, Appendix B] provide a proof, which works in our case as well).

[8] It can also be shown that the resulting Nash equilibrium in qualities is unique and stable, and that product quality choices are strategic complements (see [10] for a discussion of this concept in a different framework).

$$\tilde{\pi}^{\mathrm{L}} = \frac{1}{6b}(\bar{\theta} - \underline{\theta})\left[\frac{3}{2}(\bar{\theta} - \underline{\theta}) + \frac{4x(I_{\mathrm{H}}, I_{\mathrm{L}})b}{3(\bar{\theta} - \underline{\theta})}\right]^2. \tag{7b}$$

These profits depend on market size, the marginal cost of producing quality, and on investment levels I_{H} and I_{L} via the term $x(I_{\mathrm{H}}, I_{\mathrm{L}}^{(\bar{\theta}-\underline{\theta})})$. Firms' profits in (7a) and (7b) are equal if x is zero, like in [12]; otherwise they will differ.

2.3. Technology investment

Now we turn to the *first* stage, where firms determine their levels of investment. Under competitive capital markets the unit cost of investment, $\hat{v}$, is equal for both firms. The *total* unit cost of investment also includes auditing costs paid by the firm, τ, required to obtain an ecolabel.[9] The total unit cost is therefore written as $v = \hat{v} + \tau$. The profit functions corresponding to the maximization problem in the first stage can now be expressed with the help of (7a) and (7b) as

$$V^{\mathrm{H}} = \tilde{\pi}^{\mathrm{H}} - vI_{\mathrm{H}}, \quad V^{\mathrm{L}} = \tilde{\pi}^{\mathrm{L}} - vI_{\mathrm{L}}. \tag{8}$$

Investment in the technology is then determined via the following first-order conditions:

$$V^{\mathrm{H}}_{I_{\mathrm{H}}} = [-A + Bx]\alpha'_{\mathrm{H}}(I_{\mathrm{H}}) - v = 0, \tag{9a}$$

$$V^{\mathrm{L}}_{I_{\mathrm{L}}} = [-A - Bx]\alpha'_{\mathrm{L}}(I_{\mathrm{L}}) - v = 0, \tag{9b}$$

where $A = \frac{2}{3}(\bar{\theta} - \underline{\theta})$, $B = \frac{16b}{27(\bar{\theta}-\underline{\theta})}$ and $x = \alpha_{\mathrm{H}}(I_{\mathrm{H}}) - \alpha_{\mathrm{L}}(I_{\mathrm{L}}) = a_{\mathrm{H}}I_{\mathrm{H}} - a_{\mathrm{L}}I_{\mathrm{L}}$. We show in Appendix A that this equilibrium is unique and stable at the interior solutions, and that investments are strategic substitutes.

Eqs. (9a) and (9b) show that each firm invests to equate the marginal benefit of investment, in terms of reduced costs of providing quality, to the effective unit cost of investment v. The sign of x is again important, this time to the investment choices made by the duopolists. If x *is zero*, then both firms invest. If x is *negative*, then the high-quality firm always has positive technology investment in equilibrium, because the term in brackets of (9a) is negative. The low-quality firm invests only if the bracket term in (9b) is negative. Finally, a strictly *positive* x implies that the low-quality firm invests, but we now have the possibility of a corner solution for the high-quality firm's investment.

To obtain a closed form solution for investment, we further specify the $\alpha_k(I_k)$ term as follows:

$$\alpha_k(I_k) = e^{a_k I_k}, \tag{9c}$$

[9] Costs of eco-labeling can be specifically linked to firms' technologies. For example, the costs of eco-labeling to a firm consist of auditing costs and annual license costs. While the latter is most often defined as a percentage of the annual turnover of sales (usually with an upper limit), the former is a fixed cost closely tied with the production technology used by the firm. The auditing costs vary considerably across ecolabel systems. For instance, in the European Union the Eco-Label Award system the auditing fee is about USD 500. The Nordic Swan auditing fee is USD 2000 and with the Blue Angel it lies between about USD 178-2034 depending on the technology in question [27].

where $a_k<0$. We approximate this part of the cost function log-linearly using $e^{a_k I_k} \approx 1 + a_k I_k$, so that $\alpha'_k(I_k) = a_k(1 + a_k I_k) < 0$ and $\alpha''_k(I_k) = a_k^2 > 0$. The log-linear approximation is commonly used in many other contexts.[10]

Using (9c) for the $\alpha_k(I_k)$ terms and abstracting from the cross and squared terms we can rewrite the first-order conditions (9a) and (9b) for investments as follows:

$$I_{\mathrm{H}} : -Aa_{\mathrm{H}}(1 + a_{\mathrm{H}} I_{\mathrm{H}}) + Ba_{\mathrm{H}}(a_{\mathrm{H}} I_{\mathrm{H}} - a_{\mathrm{L}} I_{\mathrm{L}}) - v = 0, \tag{9a$'$}$$

$$I_{\mathrm{L}} : -Aa_{\mathrm{L}}(1 + a_{\mathrm{L}} I_{\mathrm{L}}) - Ba_{\mathrm{L}}(a_{\mathrm{H}} I_{\mathrm{H}} - a_{\mathrm{L}} I_{\mathrm{L}}) - v = 0, \tag{9b$'$}$$

where $a_{\mathrm{H}}, a_{\mathrm{L}} < 0$. If the fixed cost terms are identical between firms, i.e., $x = 0$, then the optimal investments and their difference are given by

$$I^*_{\mathrm{H}}\big|_{x=0} = -\frac{1}{a_{\mathrm{H}}} - \frac{v}{Aa^2_{\mathrm{H}}}, \quad I^*_{\mathrm{L}}\big|_{x=0} = -\frac{1}{a_{\mathrm{L}}} - \frac{v}{Aa^2_{\mathrm{L}}}, \tag{10a}$$

$$[I^*_{\mathrm{H}} - I^*_{\mathrm{L}}]_{x=0} = \frac{a_{\mathrm{H}} - a_{\mathrm{L}}}{a_{\mathrm{H}} a_{\mathrm{L}}}\left[1 + \frac{v(a_{\mathrm{H}} + a_{\mathrm{L}})}{Aa_{\mathrm{H}} a_{\mathrm{L}}}\right] > (\leqslant) 0 \quad \text{as } a_{\mathrm{H}} < (\geqslant) a_{\mathrm{L}}, \tag{10b}$$

where the term in the brackets is negative.[11]

When $x \neq 0$, the optimal investments and the resulting difference in the firms' cost structure can be characterized as follows:

$$I^*_{\mathrm{H}}\big|_{x\neq 0} = -\frac{1}{a_{\mathrm{H}}}\left[1 + \frac{v[Aa_{\mathrm{L}} - B(a_{\mathrm{H}} + a_{\mathrm{L}})]}{Aa_{\mathrm{H}} a_{\mathrm{L}}(A - 2B)}\right], \quad I^*_{\mathrm{L}}\big|_{x\neq 0} = -\frac{1}{a_{\mathrm{L}}}\left[1 + \frac{v[Aa_{\mathrm{H}} - B(a_{\mathrm{H}} + a_{\mathrm{L}})]}{Aa_{\mathrm{H}} a_{\mathrm{L}}(A - 2B)}\right], \tag{10c}$$

$$x^* = \frac{v(a_{\mathrm{H}} - a_{\mathrm{L}})}{a_{\mathrm{H}} a_{\mathrm{L}}(A - 2B)}\left\{\begin{matrix} < \\ > \end{matrix}\right\} 0 \quad \text{as } a_{\mathrm{H}} \left\{\begin{matrix} < \\ > \end{matrix}\right\} a_{\mathrm{L}}, \tag{10d}$$

where $A - 2B > 0$ at the interior solution for the unique and stable duopoly investment game. The difference between the optimal investment levels of the firms can now be expressed as

$$[I^*_{\mathrm{H}} - I^*_{\mathrm{L}}]_{x\neq 0} = \frac{a_{\mathrm{H}} - a_{\mathrm{L}}}{a_{\mathrm{H}} a_{\mathrm{L}}}\left\{1 + \frac{v[(a_{\mathrm{H}} + a_{\mathrm{L}})]}{Aa_{\mathrm{H}} a_{\mathrm{L}}} \frac{(A - B)}{(A - 2B)}\right\}, \tag{11}$$

where $(A - B)/(A - 2B) > 1$. Thus when $x < 0$ (equivalent to $a_{\mathrm{H}} < a_{\mathrm{L}}$), the difference in (11) is positive. If $x > 0$, an interior solution for both firms requires $a_{\mathrm{H}} > a_{\mathrm{L}}$, then the sign of (11) is reversed. We summarize our findings in

Proposition 1. *At the interior solution for investments, the equilibrium is unique and stable, and investments of the firms are strategic substitutes. The outcome of the investment game depends on the relative efficiency of firms' in terms of their difference in the fixed costs of providing quality, x, as follows:*

(a) *If* $x < 0$, *then* $I^*_{\mathrm{H}} > 0$ *but* $I^*_{\mathrm{L}} = 0$ *is possible if the unit cost of investment is "high enough".*

[10] A log-linear approximation makes the analysis more tractable and transparent; and in our case it makes it possible to provide a closed-form analytical solution for the investment stage. See [11] for examples of the use of this log-linear approximation in the context of growth models and Bovenberg and van der Ploeg [9] for public finance contexts.

[11] Given that the efficiency of investment in reducing fixed costs is expected to be at most about 50%, then $(a_{\mathrm{H}} + a_{\mathrm{L}})/a_{\mathrm{H}} a_{\mathrm{L}} \leqslant -4$. Normalizing $\bar{\theta} - \underline{\theta} = 1$ in the A term makes the negativity clear.

(b) *If* $x = 0$, *then* $I_H^* > 0$, *and* $I_L^* > 0$.
(c) *If* $x > 0$, *then* $I_L^* > 0$ *but* $I_H^* = 0$ *is possible if the unit cost of investment is "high enough"*.

Thus, naturally the more efficient firm invests more. In the case of equal efficiency of investing, both firms invest and their fixed costs become identical ($x = 0$). Note also that the higher is the effective unit cost of investment (i.e., the cost of auditing for eco-labeling), the greater is the possibility of a corner solution with one firm not investing. Later we show that the resulting cost structure has crucial implications for quality competition.

In Appendix B we study the implications of the case where firms can influence the marginal willingness to pay of consumers by investments. Following studies of advertising in the industrial organization literature [15], we can express marginal willingness to pay as a function of investments. This assumes that investment is a product characteristic which increases utilities of consumers [8]. In this case, it turns out that the possibility for $I_L^* = 0$ increases when $x < 0$, but the possibility of $I_H^* = 0$ decreases when $x > 0$. The economic intuition is as follows: Investment by the low-quality firm would increase price competition with the high-quality firm, because willingness to pay for low-quality consumers would increase, and thus there would be less marginal gains from differentiating through investment. Similarly, increased willingness to pay for high-quality strengthens the high-quality firm's incentives to invest, even though it is less efficient in investing. Hence, when marginal willingness to pay depends on investment, this serves to reinforce the incentive for the high-quality firm to obtain eco-labeling rights via technology investment.

2.4. Closing the duopoly competition: environmental quality and price competition

Using the solutions for optimal investments, we can now characterize the quality competition stage under *endogenous* investments. According to Eq. (10c) we have $x^* = \frac{v(a_H - a_L)}{a_H a_L (A-2B)} < (\geqslant) 0$ as $a_H - a_L < (\geqslant) 0$. Substituting this expression for x in (5) we obtain the following profit-maximizing qualities, written in terms of exogenous variables:

$$s_H^* = \frac{5\bar{\theta} - \underline{\theta}}{4b} + \frac{2v(a_H - a_L)}{3(\bar{\theta} - \underline{\theta})(A - 2B)a_H a_L}, \quad s_L^* = \frac{5\underline{\theta} - \bar{\theta}}{4b} + \frac{2v(a_H - a_L)}{3(\bar{\theta} - \underline{\theta})(A - 2B)a_H a_L}. \qquad (5')$$

We can distinguish several alternative outcomes for the quality game. Here we focus solely on interior solutions where both firms invest, because corner solutions with one firm not investing leads to qualitatively similar outcomes as interior solutions.[12]

We start with the benchmark case where the firms' cost structures are identical with respect to investment, $x = 0$ (i.e., $a_H - a_L = 0$). In this case, the last terms in (5′) are zero like in the case analyzed in [12] without investment decisions. The interpretation of their result in our model is different, however. Comparing (5′) with (5) reveals that, in our model, the firms provide *either* higher or lower quality than in the absence of investments. Specifically, if the difference between the initial fixed costs is negative (positive), then quality provision after the investment stage is higher (lower). Thus, the result in [12] is a special case of our investment-based model with identical cost structures for firms.

[12] This is because in a corner solution with one firm not investing x must be either positive or negative. Recall that if $x < 0$ then $I_H > 0, I_L = 0$ and if $x > 0$ then $I_H = 0, I_L > 0$ in the corner solutions.

At the interior solution where both firms invest, the resulting cost structure may differ, i.e., we may have either $x<0$, or $x>0$. If $x<0$, so that the high-quality firm is more efficient in investing, then the resulting qualities provided in the market are lower than in the benchmark case of $x=0$. This is because, under $x<0$, the high-quality firm has lower fixed costs compared to the low-quality firm. Thus, the high-quality firm performs better in price competition, and it does not need to produce very high quality in order to mitigate price competition with the low-quality firm. If instead $x>0$, so that the low-quality firm is more efficient in investing, then the last terms in (5′) are positive and higher quality would be produced than when firms have an identical cost structure ($x=0$). In this case, the high-quality firm is less efficient; it therefore provides higher quality to mitigate price competition.[13] These findings are summarized in

Proposition 2. *The possibility that investment affects fixed costs of providing quality leaves quality dispersion unchanged, but it affects duopolists' incentives for providing (environmental) quality. The quality provided by the market is highest when the low-quality firm is more efficient at investment and lowest when the high-quality firm is more efficient at investment.*

Investment changes the firms' relative profitability and thus conditions of price competition. Mitigation of price competition in turn affects the choice of qualities. This explains our finding of environmental quality being lowest (highest) when the high-quality firm is more (less) efficient at investing. One may ask why the *difference* in quality among firms is equal over all cases. This is because, in our model, investments affect the fixed cost but not the variable cost of quality provision. The marginal cost of quality provision remains the same in all cases and, consequently, the quality dispersion does not change.

3. Socially optimal quality and investment decisions

The equilibrium characterized above results from profit maximizing decisions of firms. We now examine the *first best* levels of qualities and investment as a benchmark. Note first that it is socially optimal to produce both high- and low-quality goods, because variable costs of producing quality are non-zero (Ecchia et al. [16] provides a proof of this and further discussion in a different model).

We consider a utilitarian social welfare function, which is the sum of consumer and producer surplus,

$$SW = (\bar{\theta} - \tilde{\theta})\left[\frac{\bar{\theta}+\tilde{\theta}}{2}s_{\mathrm{H}} - c_{\mathrm{H}}(s_{\mathrm{H}})\right] + (\tilde{\theta} - \underline{\theta})\left[\frac{\underline{\theta}+\tilde{\theta}}{2}s_{\mathrm{L}} - c_{\mathrm{L}}(s_{\mathrm{L}})\right] - v(I_{\mathrm{H}} + I_{\mathrm{L}}). \tag{12}$$

[13] When both firms invest, the duopoly equilibrium with a positive price difference exists for $x>(<)0$, and the profits of the high-quality firm are greater (smaller) than the profits of the low-quality firm when $a_{\mathrm{H}}<(>)a_{\mathrm{L}}$. All consumers are consuming one unit of the commodity when $\underline{\theta}>\frac{p_{\mathrm{L}}}{s_{\mathrm{L}}}$.

In (12) both variants of quality are sold at marginal cost, so that the index of the marginal consumer indifferent between consuming high- and low-quality goods is modified accordingly, as $\tilde{\theta} = \frac{\bar{\theta}+\underline{\theta}}{2}$. The high- and low-quality firms also split the market into two halves of equal size: $\bar{\theta} - \tilde{\theta} = \tilde{\theta} - \underline{\theta}$.

3.1. Environmental quality choice

We first examine differences in quality provided by the two firms under the social welfare maximizing case, and then compare this with the profit maximizing case. First, we differentiate (12) with respect to s_{H} and s_{L} and solve for optimal qualities, obtaining $s_{\mathrm{H}}^{\mathrm{W}} = \frac{3\bar{\theta}+\underline{\theta}}{4b}$ and $s_{\mathrm{L}}^{\mathrm{W}} = \frac{\bar{\theta}+3\underline{\theta}}{4b}$, where the superscript "W" denotes the social welfare optimum, so that $s_{\mathrm{H}}^{\mathrm{W}} - s_{\mathrm{L}}^{\mathrm{W}} = \frac{\bar{\theta}-\underline{\theta}}{2b}$. Due to the absence of strategic interaction of firms in the social optimum, these quality choices are independent of each other and of investments in the (green) technology. Consequently, the optimal qualities and quality difference under the social welfare optimum are both different from the ones obtained under profit maximization.

The following relationship holds comparing the social welfare maximizing qualities with profit maximizing ones (obtained from (5)),

$$s_{\mathrm{H}}^{*} - s_{\mathrm{H}}^{\mathrm{W}} = \left[\frac{\bar{\theta} - \underline{\theta}}{2b}\right] + \frac{2v(a_{\mathrm{H}} - a_{\mathrm{L}})}{3(\bar{\theta} - \underline{\theta})(A - 2B)a_{\mathrm{H}}a_{\mathrm{L}}}, \tag{13a}$$

$$s_{\mathrm{L}}^{*} - s_{\mathrm{L}}^{\mathrm{W}} = -\left[\frac{\bar{\theta} - \underline{\theta}}{2b}\right] + \frac{2v(a_{\mathrm{H}} - a_{\mathrm{L}})}{3(\bar{\theta} - \underline{\theta})(A - 2B)a_{\mathrm{H}}a_{\mathrm{L}}}, \tag{13b}$$

where "*" continues to denote each firm's profit maximizing choices. The quality difference depends on two terms, on the spread of consumer tastes, and the firms' relative efficiency in investment $(a_{\mathrm{H}} - a_{\mathrm{L}})$. The spread term is positive for the high-quality firm and negative for low-quality firm by definition. If the low-quality firm is more efficient in investing $((a_{\mathrm{H}} - a_{\mathrm{L}})>0)$ so that quality competition is more intense, then the latter terms are positive. The high-quality firm overprovides quality, but the low-quality firm's behavior is ambiguous. If the high-quality firm is more efficient in investing $((a_{\mathrm{H}} - a_{\mathrm{L}})<0)$, so that quality competition relaxes, then (13a) is ambiguous while (13b) is negative. Now, the low-quality firm *underprovides*. It may therefore be possible that both firms underprovide qualities. Finally, when $(a_{\mathrm{H}} - a_{\mathrm{L}}) = 0$, we have the special case of [12]. Here, the spread of quality among firms in the profit maximizing duopoly is too large compared to the social optimum.

In sum, profit maximization leads to incentives for the firms to increase quality differences among them in order to relax price competition. The severity of price competition depends on the firms' relative efficiency of investing, and on the resulting cost structure. Therefore, while the social welfare optimum features marginal cost pricing, the market provision of qualities may be lower or higher depending on how investments affect the relative cost structure of firms.

3.2. Technology investment choice

Next we consider the socially optimal level of technology investment, where investment can affect the fixed costs of providing quality. From the necessary conditions for an interior solution, we have

$$I_{\mathrm{H}}^{\mathrm{W}} = -\frac{1}{a_{\mathrm{H}}} - \frac{v}{\frac{1}{2}(\bar{\theta} - \underline{\theta})a_{\mathrm{H}}^{2}}, \tag{14a}$$

$$I_{\mathrm{L}}^{\mathrm{W}} = -\frac{1}{a_{\mathrm{L}}} - \frac{v}{\frac{1}{2}(\bar{\theta} - \underline{\theta})a_{\mathrm{L}}^{2}}. \tag{14b}$$

In (14a) and (14b), the first RHS terms dominate. Like in the case of quality provision, high- and low-quality investments are independent of each other, because socially optimal investment is not determined as a result of firms' interaction in the investment game. The difference between the socially optimal high- and low-quality firm investments is given by

$$I_{\mathrm{H}}^{\mathrm{W}} - I_{\mathrm{L}}^{\mathrm{W}} = \frac{a_{\mathrm{H}} - a_{\mathrm{L}}}{a_{\mathrm{H}}a_{\mathrm{L}}}\left[1 + \frac{v(a_{\mathrm{H}} + a_{\mathrm{L}})}{\frac{1}{2}(\bar{\theta} - \underline{\theta})a_{\mathrm{H}}a_{\mathrm{L}}}\right], \tag{14c}$$

where we assume that the braced term is negative, i.e., the last term is greater than one in absolute value. The sign of the first term before the brackets depends on firms' relative efficiency in investing. Thus, the whole expression is positive (negative) if the high-quality firm is more (less) efficient in investing.

Now consider how the socially optimal investment levels relate to the profit maximizing duopoly investment levels. Using (14a), (14b), (10a) and (10b) for the case $x = 0$ yields

$$I_{\mathrm{H}}^{*} - I_{\mathrm{H}}^{\mathrm{W}} = \frac{v}{3a_{\mathrm{H}}^{2}A} > 0, \tag{15a}$$

$$I_{\mathrm{L}}^{*} - I_{\mathrm{L}}^{\mathrm{W}} = \frac{v}{3a_{\mathrm{L}}^{2}A} > 0. \tag{15b}$$

Thus, from society's viewpoint both firms invest too much, because they use investing in the green technology as a means of mitigating price competition. For the case $x \neq 0$ we obtain from (14a), (14b) and (10c), (10d),

$$I_{\mathrm{H}}^{*} - I_{\mathrm{H}}^{\mathrm{W}} = \frac{v}{Aa_{\mathrm{H}}^{2}}\left\{\frac{(a_{\mathrm{H}} + a_{\mathrm{L}})B - Aa_{\mathrm{L}}}{Aa_{\mathrm{L}}(A - 2B)} + \frac{4}{3}\right\}, \tag{15c}$$

$$I_{\mathrm{L}}^{*} - I_{\mathrm{L}}^{\mathrm{W}} = \frac{v}{Aa_{\mathrm{L}}^{2}}\left\{\frac{(a_{\mathrm{H}} + a_{\mathrm{L}})B - Aa_{\mathrm{H}}}{Aa_{\mathrm{H}}(A - 2B)} + \frac{4}{3}\right\}. \tag{15d}$$

The signs of the first terms in braces of (15c) and (15d) are ambiguous, but there is clearly a possibility of over-investing. The sufficient (but not necessary) condition for the high-quality firm to over-invest is that $(a_{\mathrm{H}} + a_{\mathrm{L}})B < Aa_{\mathrm{L}}$, and for the low-quality firm that $(a_{\mathrm{H}} + a_{\mathrm{L}})B < Aa_{\mathrm{H}}$.

We summarize our findings in

Proposition 3. *The relationships between the socially optimal and profit-maximizing investment levels are the following:*

(a) *The socially optimal investment levels depend on the efficiency of investments, i.e.,* $I_H^W - I_L^W \geqslant (<) 0$, *if* $a_H \leqslant (>) a_L$.
(b) *Profit-maximizing firms that are equally efficient invest too much, i.e.,* $I_H^* - I_H^W > 0$ *and* $I_L^* - I_L^W > 0$ *for* $x = 0$.
(c) *Profit-maximizing firms under unequal efficiency will likely invest too much, i.e.,* $I_H^* - I_H^W > 0$ *and* $I_L^* - I_L^W > 0$ *is likely for* $x \neq 0$.

The intuition for Proposition 3 mirrors our earlier discussions regarding the relative difference in fixed costs of investment between firms. In the duopoly outcome, firms invest to decrease costs and relax price competition. The social welfare maximizer perceives no rents from relaxing price competition due to marginal cost pricing. Thus, we would generally expect the social welfare maximizer to invest less. Finally, we can show allowing for endogenous marginal willingness to pay increases investment for the high-quality firm relative to the profit maximizing outcome, but its effect on the low-quality firm remains ambiguous (see Appendix B for further details).

According to Proposition 3, there is a tendency for excessive investment by profit maximizing firms. The government can decrease this excess and increase the provision of environmental quality by introducing eco-labeling only under certain conditions. Suppose that obtaining an eco-label is linked to an auditing or application procedure, made by an independent government agency, which ensures that the firm has really installed a green technology. Then, charging firms a reasonable fee for auditing will always decrease excessive investments. Eco-labeling increases environmental quality if it tightens quality competition between the firms. Given our previous analysis, we know this can happen when the duoply shifts from more relaxed to tighter quality competition, i.e., when the cost difference shifts from $x<0$ to either $x=0$ or $x>0$ with investment. We can characterize this finding as

Proposition 4. *If eco-labeling involves costs that raise firms' effective unit cost of investment, then it decreases excessive investments in green technologies and increases environmental quality when the market under-provides it. If the marginal willingness to pay for high-quality goods is affected by the eco-label (investment), then the possibility of only the high-quality firm obtaining an eco-label is higher.*

Interestingly, this result holds even though we have not assumed the presence of environmental externalities from inefficient average quality provided by the market.

4. Externalities and social welfare

Thus far we have established that the profit-maximizing cases can involve different levels of quality. We now consider the presence of an explicit externality related to average environmental quality provided in the market.[14] The interpretation is that consumer utility is increasing in

[14] Cremer and Thisse [13] provide an analysis of the potential role of externalities in a different product differentiation context.

average quality. Like Cremer and Thisse we assume that the utility of each consumer is now written as

$$u^e = \theta s_k - p_k + \gamma \theta s_a \tag{16}$$

where the superscript "e" denotes the presence of a positive externality, and "s_a" denotes average market product quality. Average quality is defined by weighting the quality of both firms by their demands; given that market shares are equal, this weighted average quality is $s_a = (s_H + s_L)/2$. Notice that, because a single consumer cannot affect average quality by their actions, the externality term will be a constant in any consumer's optimization problem. It will not be a constant, though, when the social optimum is determined.[15]

Using (16), the social welfare function in (12) is now modified as follows:

$$SW^e = SW + \gamma(\bar{\theta} - \underline{\theta})\left[\frac{\bar{\theta} + \underline{\theta}}{2} s_a\right] = SW + \frac{\gamma(1+\gamma)}{2b}[(\bar{\theta} - \underline{\theta})(\bar{\theta} + \underline{\theta})^2]. \tag{17}$$

4.1. Environmental quality choice

Turning first to the determination of qualities, we differentiate (17) with respect to high- and low-quality levels, for a given investment, and find,

$$SW^e_{s_H} = \frac{3\bar{\theta} + \underline{\theta}}{4} - bs_H + \gamma\frac{(\bar{\theta} + \underline{\theta})}{2} = 0, \tag{18a}$$

$$SW^e_{s_L} = \frac{3\underline{\theta} + \bar{\theta}}{4} - bs_L + \gamma\frac{(\bar{\theta} + \underline{\theta})}{2} = 0. \tag{18b}$$

Solving these for the socially optimal qualities in the presence of the externality, we find

$$s_H^{we} = \frac{3\bar{\theta} + \underline{\theta} + 2\gamma(\bar{\theta} + \underline{\theta})}{4b}, \quad s_L^{we} = \frac{3\underline{\theta} + \bar{\theta} + 2\gamma(\bar{\theta} + \underline{\theta})}{4b}, \quad s_H^{we} - s_L^{we} = \frac{3\bar{\theta} + \underline{\theta}}{4b}. \tag{19}$$

Hence, again, qualities are not affected by the investments. Accounting for the externality increases both firms' qualities by an equal amount because the externality is symmetric.

More intuition can be gained by comparing optimal qualities across the various social welfare and profit-maximizing equilibria. We are specifically interested in finding cases where there is *underprovision* of quality in the profit-maximizing equilibrium relative to the social welfare optimum. Using (13a), (13b) and (19) we can see that the socially optimal qualities are always higher in the presence of an externality:

$$s_H^{we} - s_H^{w} = s_L^{we} - s_L^{w} = \frac{\gamma(\bar{\theta} + \underline{\theta})}{2b} \equiv \Omega > 0. \tag{20a}$$

[15] We could alternatively express consumer utility (Eq. (16)) as follows: $u^e = \theta_k s_k - p_k - \gamma E$, where $E = (\bar{s} - s_H)d_H + (\bar{s} - s_L)d_L$ and $\bar{s}$ denotes unabated emissions. Given that $d_H + d_L = (1/2) + (1/2) = 1$ we can express the utility function as $u^e = \theta_k s_k - p_k - \gamma(\bar{s} - s_a)$. Hence this formulation will lead to the same results.

While this is intuitive, it is more interesting to ask how the market solution relates to socially optimal provision of environmental quality. When both firms are equally efficient in investment, Eqs. (5′) and (19) imply that

$$(s_{\mathrm{H}}^{\mathrm{we}} - s_{\mathrm{H}}^{*})|_{a_{\mathrm{H}}=a_{\mathrm{L}}} = -\frac{\bar{\theta} - \underline{\theta}}{2b} + \frac{\gamma(\bar{\theta} + \underline{\theta})}{2b} \equiv \Phi,$$

$$(s_{\mathrm{L}}^{\mathrm{we}} - s_{\mathrm{L}}^{*})|_{a_{\mathrm{H}}=a_{\mathrm{L}}} = \frac{\bar{\theta} - \underline{\theta}}{2b} + \frac{\gamma(\bar{\theta} + \underline{\theta})}{2b} \equiv \Psi > 0. \tag{20b}$$

Hence, the quality difference depends on the taste spread term and on the size of the externality. The low-quality firm underprovides quality, while the high-quality firm may or may not underprovide. If the relative efficiency in investment differs we can show that this is important to the spread in quality,

$$(s_{\mathrm{H}}^{\mathrm{we}} - s_{\mathrm{H}}^{*})|_{a_{\mathrm{H}}\neq a_{\mathrm{L}}} = \Phi - \frac{2v(a_{\mathrm{H}} - a_{\mathrm{L}})}{3(\bar{\theta} - \underline{\theta})(A - 2B)a_{\mathrm{H}}a_{\mathrm{L}}},$$

$$(s_{\mathrm{L}}^{\mathrm{we}} - s_{\mathrm{L}}^{*})|_{a_{\mathrm{H}}=a_{\mathrm{L}}} = \Psi - \frac{2v(a_{\mathrm{H}} - a_{\mathrm{L}})}{3(\bar{\theta} - \underline{\theta})(A - 2B)a_{\mathrm{H}}a_{\mathrm{L}}}. \tag{20c}$$

Now, understandably, the quality difference depends on both the size of the externality and the relative efficiency of firms' investments in the green technology. When $(a_{\mathrm{H}} - a_{\mathrm{L}}) < 0$ so that high-quality firm is more efficient in investment and quality competition is less severe, both firms underprovide environmental quality according to (20c). However in the reverse case, when $(a_{\mathrm{H}} - a_{\mathrm{L}}) > 0$, which leads to more intensive quality competition, overprovision of quality may be possible. We present our main findings as

Corollary 1. *In the presence of environmental externalities, the possibility of overprovision of environmental quality by a duopoly decreases and that of underprovision increases.*

The possibility of underprovision of environmental quality by a high-quality firm is an important departure from the two-stage quality competition literature. When firms are identical, the possibility of under-provision in our model depends on the size of the externality. If, in addition, the high-quality firm is more efficient in investment, then this serves to relax quality competition and thus reinforces the effect of any externality. Thus, given duopolists' investment behavior, the needs of price competition determine how much the high-quality firm underprovides.

4.2. Technology investment choice

We finally comment on socially optimal technology investments in the presence of a positive average quality externality. The necessary conditions for investment under an externality are identical to the case of no externality (see Eqs. (14a) and (14b)). Thus, investments for quality provision are based on differences across firms in the efficiency of investment. Intuitively, quality choices are used to mitigate externalities, and investments are the means for a firm to achieve efficient quality cost reductions. If marginal willingness to pay depends on investment, then the presence of an externality increases investment for the high-quality firm, but investment for the low-quality firm remains ambiguous.

Thus our previous comparison of private and socially optimal investment levels continues to hold. How does the externality affect the role of eco-labeling in environmental policy? As Eqs. (20b) and (20c) demonstrate, the number of cases where the market underprovides quality is increased by the presence of an externality. Thus, again, eco-labeling might be used to achieve the dual goals of decreasing excessive investment and increasing environmental quality in the market (thereby mitigating externalities).

5. Conclusions

In most market settings, considerable emphasis has been placed on understanding investments firms make in "clean" technologies that can reduce pollution associated with production. At the same time there have been numerous empirical examples of firms obtaining the right to eco-label their goods. In its purest sense, eco-labeling provides firms with a means of differentiating products. Eco-labeling often induces or effectively requires firms to adopt green technologies, because these reduce the costs of providing high levels of environmental quality required to obtain or continue holding eco-labeling rights. Thus, to understand incentives for firms to achieve high environmental quality of their production processes and receive eco-labels, and to identify the socially optimal level of environmental quality in markets where goods are not homogenous in quality, we must first understand how firms make investment choices when products can be differentiated.

We therefore extend vertical product differentiation literature to incorporate a firm's choice of technology investment, which is made prior to environmental quality and price competition decisions. A Nash game is solved over three stages: in the first stage, each firm chooses the level of their technology investment, in the second stage firms choose their environmental quality levels conditional on this investment, and the third stage has the firms competing in product markets. The resulting duopoly solution is then compared to the socially first best solution, in cases where externalities are absent and present, and where consumer willingness to pay does and does not depend on firms' investments in technologies.

We showed firms' incentives to invest in technologies that can be used to provide increased environmental quality at lower costs, and can allow the firm to obtain and retain eco-labels, depend critically on differences in their cost structures. If firms are identical with respect to fixed costs of investment, then both firms will always invest in green technologies. If the high-quality firm is more efficient at investing, then the low-quality firm may or may not invest depending on the unit cost of investment. The opposite holds if the low-quality firm is more efficient in investing. We have also shown that market-driven incentives for high-quality firms to invest are reinforced if technology investment increases consumer marginal willingness to pay for environmental quality; but this is not true for a low-quality firm.

One of our most important findings is that the incentives to invest and the possibilities of corner solutions in investment differ depending on the efficiency of investment across firms. The differences in incentives to invest lead to different outcomes for the provision of environmental quality in a market. Surprisingly, quality competition will be highest among firms when the low-quality firm is more efficient in investing. Quality competition is lowest when the high-quality firm is more efficient in investing. Between these cases, i.e., when the firms are equally efficient in investment, we arrive at the basic results of quality competition characterized by Crampes and Hollander [12]. Our model therefore generalizes the vertical differentiation literature.

We have also compared the duopoly outcome with the socially optimal one. According to our results the possibility of a corner solution in investment never occurs at the social optimum. In the absence of environmental externalities, the quality dispersion and average quality chosen under the profit-maximizing firms again depends on their relative efficiency in investment. If the high-quality firm in more efficient in investing, then both firms underprovide quality. Interestingly, the presence of environmental externalities increases the possibility of underprovision in cases where the firms have identical cost structures, or where the low-quality firm is more efficient in investing.

Our analysis has potentially important implications for a government targeting the eco-labeling behavior of firms. Eco-labeling, by changing the firms' effective unit cost of investment, might be used to achieve the dual goal of reducing excessive investments and increasing average environmental quality, both in the absence and presence of externalities related to low average environmental quality. For example, a government interested in achieving a desired environmental quality target in markets where eco-labeling exists can use eco-labeling criteria and auditing/license charges as a means of decreasing the types of excessive investments we show could exist. However, the ability of the government to do this depends on the difference in cost structure between the firms, i.e., which firm is more efficient in terms of the effects of investment on its costs.

It is worth noting that we have conducted our analysis assuming consumers can observe the investment levels of firms, and firms can observe the preferences of consumers. An interesting area for further research would be to allow for asymmetric information between firms and consumers regarding environmental qualities of firms' products. This would require modeling of eco-labeling as a signal for consumers.

Acknowledgments

We would like to thank an anonymous referee for helpful comments and the Ph.D. student Chiara Lombardini-Riipinen for computational help. Amacher would like to thank Virginia Commonwealth University's Department of Economics for hospitality during his sabbatical, when this research was completed. Koskela thanks Department of Forestry of Virginia Tech and Research Department of the Bank of Finland for their hospitality. This paper is a part of the project "Studies in Environmental and Resource Economics" financed by the Academy of Finland. The funding is gratefully acknowledged.

Appendix A. Proof of a stable Nash equilibrium, and of investments as strategic substitutes, when willingness to pay is independent of investment

Re-express the first-order conditions for the investment game as follows:

$$V_{I_{\mathrm{H}}} = (y - z)\alpha'_{\mathrm{H}} - v = 0, \tag{A.1}$$

$$V_{I_L} = (y+z)\alpha'_L - v = 0, \tag{A.2}$$

with $y = \frac{2}{3}(\bar{\theta} - \underline{\theta}) > 0$ and $z = \frac{16bx}{27(\bar{\theta}-\underline{\theta})} < 0$.

The second-order conditions and the cross-derivatives are

$$V_{I_H I_H} = (y-z)\alpha''_H - \alpha'^2_H \frac{z}{x} < 0, \tag{A.3}$$

$$V_{I_L I_L} = (y+z)\alpha''_L - \alpha'^2_L \frac{z}{x} < 0, \tag{A.4}$$

$$V_{I_H I_L} = V_{I_L I_H} = \alpha'_H \alpha'_L \frac{z}{x} < 0. \tag{A.5}$$

Eq. (A.5) indicates that the firm's investments in technology are strategic substitutes.

Stability and uniqueness of the investment game requires that the condition $\Delta = \tilde{V}^H_{I_H I_H} \tilde{V}^L_{I_L I_L} - \tilde{V}^H_{I_H I_L} \tilde{V}^L_{I_L I_H} > 0$ holds for the optimum. Applying Eqs. (A.3)–(A.5) to this condition yields after some manipulation:

$$\Delta = (y+z)(y-z)\alpha''_H \alpha''_L - \frac{z}{x}[(y-z)\alpha''_L \alpha'^2_H + (y+z)\alpha''_H \alpha'^2_L]. \tag{A.6}$$

By rearranging, we get

$$\Delta = (y+z)(y-z)\alpha''_H \alpha''_L \left\{1 - \frac{z}{x}\left[\frac{\alpha'^2_H}{\alpha''_H} + \frac{\alpha'^2_L}{\alpha''_L}\right]\right\} > 0.$$

This gives as the sufficient condition for a unique and stable equilibrium, i.e., for $\Delta > 0$:

$$\frac{z}{x}\left[\frac{\alpha'^2_H}{\alpha''_H (y+z)} + \frac{\alpha'^2_L}{\alpha''_L (y-z)}\right] < 1 \tag{A.7}$$

due to the first-order conditions (A.1) and (A.2).

Appendix B. Analysis of the case when willingness to pay depends on investment

Denote the marginal willingness to pay for consumers of high- and low-quality goods as $\bar{\theta}(I_H, I_L)$ and $\underline{\theta}(I_H, I_L)$, respectively.

Assumption. Marginal willingness to pay is affected by investment according to the following derivatives: $\bar{\theta}_H > 0,\ \bar{\theta}_L = 0,\ \underline{\theta}_L > 0,\ \underline{\theta}_H = 0;\ \bar{\theta}_H = \underline{\theta}_L$.

Thus, investment by the high-quality firm affects the upper limit of marginal willingness to pay, while investment by the low-quality firm affects the lower limit. This assumption relies on the fact that the demand for each firm's product is satisfied by a separate group of consumers, following Anderson et al. [3, pp. 69–70].

B.1. Duopoly solution

The first-order conditions for optimal investments corresponding to (9a) and (9b) are

$$V^{\mathrm{H}}_{I_{\mathrm{H}}} = \underbrace{V^{\mathrm{H}}_{I_{\mathrm{H}}}\big|_{\bar{\theta}_{\mathrm{H}}=0}}_{-} + \underbrace{\tilde{\pi}^{\mathrm{H}}_{\bar{\theta}}\bar{\theta}_{\mathrm{H}}}_{+} = 0, \tag{B.1}$$

$$V^{\mathrm{L}}_{I_{\mathrm{L}}} = \underbrace{V^{\mathrm{L}}_{I_{\mathrm{L}}}\big|_{\underline{\theta}_{\mathrm{L}}=0}}_{+} + \underbrace{\tilde{\pi}^{\mathrm{L}}_{\underline{\theta}}\underline{\theta}_{\mathrm{L}}}_{-} = 0, \tag{B.2}$$

where $\frac{\partial \tilde{\pi}^{\mathrm{H}}}{\partial \bar{\theta}} = \frac{27(\bar{\theta}-\underline{\theta})^2}{24b} - \frac{2x}{3}[1 + \frac{4xb}{9(\bar{\theta}-\underline{\theta})^2}] > 0$ as $9(\bar{\theta}-\underline{\theta})^2 + 4xb \geqslant 0$—which is consistent with the condition for the incentive of low-quality firm to invest $9(\bar{\theta}-\underline{\theta})^2 + 8xb < 0$—and $\frac{\partial \tilde{\pi}^{\mathrm{L}}}{\partial \underline{\theta}} = -[\frac{27(\bar{\theta}-\underline{\theta})^2}{24b} + 2x + \frac{8xb}{9(\bar{\theta}-\underline{\theta})^2}] < 0$ when $\tilde{\pi}^{\mathrm{L}} > 0$. The first terms on the RHS of (B.1) and (B.2) correspond to the RHS terms of (9a) and (9b). The second RHS terms in (B.1) and (B.2) indicate how changes in willingness to pay from investment affect the profits of firms. They imply that incentives for the high-quality firm to invest increase, but incentives for the low-quality firm decrease when the willingness to pay depends on investments.

The second-order conditions and the cross-derivatives are

$$\frac{\partial^2 V^{\mathrm{H}}}{\partial I^2_{\mathrm{H}}} = \frac{\partial^2 V^{\mathrm{H}}}{\partial I^2_{\mathrm{H}}}\bigg|_{\bar{\theta}_{\mathrm{H}}=0} + \left[\frac{\partial^2 \tilde{\pi}^{\mathrm{H}}}{\partial \bar{\theta}^2}\bar{\theta}_{\mathrm{H}} + \frac{\partial \tilde{\pi}^{\mathrm{H}}}{\partial \bar{\theta}}\bar{\theta}_{\mathrm{HH}}\right] < 0, \tag{B.3}$$

$$\frac{\partial^2 V^{\mathrm{L}}}{\partial I^2_{\mathrm{L}}} = \frac{\partial^2 V^{\mathrm{L}}}{\partial I^2_{\mathrm{L}}}\bigg|_{\underline{\theta}_{\mathrm{L}}=0} + \left[\frac{\partial^2 \tilde{\pi}^{\mathrm{L}}}{\partial \underline{\theta}^2}\underline{\theta}_{\mathrm{L}} + \frac{\partial \tilde{\pi}^{\mathrm{L}}}{\partial \underline{\theta}}\underline{\theta}_{\mathrm{LL}}\right] < 0, \tag{B.4}$$

$$\frac{\partial^2 V^{\mathrm{H}}}{\partial I_{\mathrm{H}}\partial I_{\mathrm{L}}} = \frac{\partial^2 V^{\mathrm{L}}}{\partial I_{\mathrm{L}}\partial I_{\mathrm{H}}} = \frac{\partial^2 V^{H}}{\partial I_{\mathrm{H}}\partial L_{\mathrm{L}}}\bigg|_{\bar{\theta}_{\mathrm{H}}=0} + \underline{\theta}_{\mathrm{L}}\tilde{\pi}^{\mathrm{L}}_{\underline{\theta}\bar{\theta}}\bar{\theta}_{\mathrm{H}} < 0, \tag{B.5}$$

where $\tilde{\pi}^{\mathrm{L}}_{\underline{\theta}\bar{\theta}} = \tilde{\pi}^{\mathrm{H}}_{\bar{\theta}\underline{\theta}} < 0$. The negativity of (B.5) results directly from our assumptions and it shows that the dependence of the marginal willingness to pay reinforces the strategic substitutability of investments. The stability condition is given by

$$\Delta = \frac{\partial^2 V}{\partial I^2_{\mathrm{H}}}\frac{\partial^2 V}{\partial I^2_{\mathrm{L}}} - \left(\frac{\partial^2 V}{\partial I_{\mathrm{H}}\partial I_{\mathrm{L}}}\right)^2 > 0. \tag{B.6}$$

The proof of this follows the previous procedure presented in Appendix A.

304 G.S. Amacher et al. / Journal of Environmental Economics and Management 47 (2004) 284–306

B.2. Social optimum without externalities

The social welfare maximization problem under endogenous marginal willingness to pay yields

$$SW^{\theta}_{I_H} = \bar{\theta}_H\left[\left(\frac{SW}{\bar{\theta}-\underline{\theta}} + v(I_H + I_L)\right) + \frac{\bar{\theta}-\underline{\theta}}{2}\left(\frac{3}{4}s_H + \frac{1}{4}s_L\right)\right] - \frac{(\bar{\theta}-\underline{\theta})}{2}\alpha'_H - v = 0, \quad \text{(B.7)}$$

$$SW^{\theta}_{I_L} = \underline{\theta}_L\left[\left(-\frac{SW}{\bar{\theta}-\underline{\theta}} + v(I_H + I_L)\right) + \frac{\bar{\theta}-\underline{\theta}}{2}\left(\frac{1}{4}s_H + \frac{3}{4}s_L\right)\right] - \frac{(\bar{\theta}-\underline{\theta})}{2}\alpha'_L - v = 0, \quad \text{(B.8)}$$

where SW is defined in (12), and we employ an additional (common) assumption that $\underline{\theta} > bs_H$.[16] The socially optimal level of the high-quality firm's investment will increase (compare (B.3) and (14a), and note that $SW^{\theta}_{I_H} > SW_{I_H}$). For the low-quality firm, the effect is ambiguous (the first term in brackets of (B.4) is negative and the second term is positive). In fact, a corner solution may again exist where it is not socially optimal for the low-quality firm to invest at all. However, comparing (B.3) with (B.4) suggests that the high-quality firm always invests more than the low-quality firm at the social optimum.

B.3. Social optimum with externalities

The necessary conditions for socially optimal investments are

$$SW^{e\theta}_{I_H} = SW^{\theta}_{I_H} + \frac{\gamma(1+\gamma)}{2b}\bar{\theta}_H(\bar{\theta}+\underline{\theta})(3\bar{\theta}-\underline{\theta}) = 0, \quad \text{(B.9)}$$

$$SW^{e\theta}_{I_L} = SW^{\theta}_{I_L} + \frac{\gamma(1+\gamma)}{2b}\underline{\theta}_L\{(\bar{\theta}+\underline{\theta})(\bar{\theta}-3\underline{\theta})\} = 0, \quad \text{(B.10)}$$

where the first RHS terms are given in (B.3) with (B.4). The second RHS term in (B.9) is positive, while the second term in (B.10) is ambiguous. Hence, the presence of an externality increases investment for the high-quality firm, but investment for the low-quality firm remains ambiguous.

References

[1] G. Amacher, A. Malik, Bargaining in environmental regulation and the ideal regulator, J. Environ. Econ. Manage. 30 (1996) 233–253.

[2] G. Amacher, A. Malik, Pollution taxes when firms choose technologies, Southern Econ. J. 68 (2002) 891–917.

[3] S. Anderson, A. de Palma, J. Thisse, Discrete Choice Theory of Product Differentiation, MIT Press, Cambridge, MA, 1992.

[4] S. Arora, T. Carson, An experiment in voluntary environmental regulation: participation in EPA's 33/50 program, J. Environ. Econ. Manage. 28 (1995) 271–286.

[5] S. Arora, S. Gangopadhyay, Toward a theoretical model of voluntary overcompliance, J. Econom. Behav. Organ. 28 (1995) 289–309.

[16] This means the marginal willingness to pay of a consumer with the least intensive preference for quality exceeds the increment in marginal cost of achieving the highest quality level. See, for example, [13, Assumption 4, p. 580].

[6] N. Ashford, The influence of information-based incentives and negotiated environmental agreements on technological change. Abstracts of Working Papers in Economics (AWPE), Cambridge University Press and Fondazione Eni Enrico Mattei Note di Lavoro # 16/97, 1997.
[7] G. Biglaiser, J. Horowitz, Pollution regulation and incentives for pollution control research, J. Econ. Manage. Strat. 3 (1995) 663–684.
[8] P. Bontems, V. Requillart, Remarks on good vs. bad characteristics in vertical differentiation, Econ. Letters 70 (2001) 427–429.
[9] A.L. Bovenberg, F. van der Ploeg, Optimal taxation, public goods and environmental policy with involuntary unemployment, J. Public Econ. 62 (1996) 59–83.
[10] J. Bulow, J. Geanakoplos, P. Klemperer, Multimarket oligopoly: strategic substitutes and complements, J. Polit. Economy 93 (1985) 488–511.
[11] J. Campbell, Inspecting the mechanism. An analytical approach to the stochastic growth model, J. Monet. Econ. 33 (1984) 463–506.
[12] C. Crampes, A. Hollander, Duopoly and quality standards, Europ. Econ. Rev. 39 (1995) 71–82.
[13] H. Cremer, J. Thisse, On the taxation of polluting products in a differentiated industry, Europ. Econ. Rev. 43 (1999) 575–594.
[14] A. Dixit, The role of investment in entry deterrence, Econ. J. 90 (1980) 95–101.
[15] A. Dixit, V. Norman, Advertising and welfare, Bell J. Econ. 9 (1978) 1–17.
[16] G. Ecchia, L. Lambertini, C. Scarpa, On the regulation of vertically differentiated markets through minimum quality standards, Rivista Politica Econ. 912 (2002) 163–202.
[17] R. Goodland, Ecolabeling: Opportunities for Progress Toward Sustainability, Consumer's Choice Council, Washington DC, 2002, 52pp.
[18] C. Kiker, F. Putz, Ecological certification of forest products: economic challenges, Ecolog. Econ. 20 (1997) 37–51.
[19] R. Lahmandi-Ayed, Natural oligopolies: a vertical differentiation model, Int. Econ. Rev. 41 (2000) 971–987.
[20] U. Lehmann-Grube, Strategic choice of quality when quality is costly: the persistence of high-quality advantage, RAND J. Econ. 28 (2) (1997) 372–384.
[21] A. Matto, H. Singh, Eco-labelling: policy considerations, Kyklos 47 (1994) 53–65.
[22] R. McKitrik, A derivation of the marginal abatement cost curve, J. Environ. Econ. Manage. 37 (1999) 306–314.
[23] S. Milliman, R. Prince, Firm incentives to promote technological change pollution control, J. Environ. Econ. Manage. 17 (1989) 247–265.
[24] T. Moore, Repowering as a competitive strategy, Elect. Power Res. Inst. J. 20 (1995) 6–11.
[25] W. Nimon, J. Beghin, Are eco-labels valuable? evidence from the apparel industry, Amer. J. Agr. Econ. 81 (1999) 801–811,.
[26] P. Nunes, Y. Riyanto, Policy instruments for creating markets for biodiversity: certification and ecolabeling, FEEM Working Paper 72, Rome, Italy, 2001.
[27] OECD, Eco-labeling: actual effect of selected programs, Organization for Economic Co-Operation and Development, GD 97, 1997.
[28] L. Ozanne, P. Smith, Segmenting the market for environmentally certified wood products, For. Sci. 44 (1998) 379–389.
[29] E. Ravenswaay, J. Blend, Using ecolabeling to encourage adoption of innovative environmental technologies, Proceedings, Symposium on Flexible Incentives to Promote Adoption of Environmental Technologies in Agriculture, June 8–10, Gainesville, FL, USA, 1997.
[30] U. Ronnen, Minimum quality standards, fixed costs and competition, RAND J. Econ. 22 (1991) 490–504.
[31] A. Shaked, J. Sutton, Relaxing price competition through product differentiation, Rev. Econ. Stud. 49 (1982) 3–13.
[32] M. Spence, Entry, capacity, investment and oligopolistic pricing, Bell J. Econ. 10 (1977) 1–19.
[33] B. Spencer, J. Brander, International R&D rivalry and industrial strategy, Rev. Econ. Stud. 50 (1983) 702–722.
[34] T. Sterner, The Selection and Design of Policy Instruments in Environmental and Natural Resource Management, RFF Press, Washington, 2002.

[35] S. Swallow, R. Sedjo, Eco-labeling consequences in general equilibrium: a graphical assessment, Land Econ. 76 (2000) 15–27.
[36] S. Tarranto, S. Humphries, FSC Forest Certification Handbook, The Forest Management Trust, Gainesville, FL, USA, 2002, 55pp.
[37] D. Zhou, B. Spencer, I. Vertinsky, Strategic trade policy with endogenous choice of quality and asymmetric costs, J. Int. Econ. 56 (2002) 205–232.

[9]

Overcompliance, labeling, and lobbying: The case of credence goods

Stefanie Engel

Center for Development Research, University of Bonn, Bonn, Germany
E-mail: st.kirchhoff@uni-bonn.de

This paper presents a model of quality choice in the case of credence goods, i.e., when consumers cannot observe quality even after purchase. It shows that firms may voluntarily overcomply, i.e., produce high quality, even when doing so implies giving up short-run profits. This generalizes results on reputation effects derived in the IO literature for the case of experience goods. The crucial assumptions of the model are that there is a positive degree of monitoring of firms' claims and a positive probability that the firm is of an "honest type," i.e., always prefers to produce high quality. The result also helps explain why we see phenomena such as firms voluntarily overcomplying with environmental standards, food safety laws, etc. It is shown that overcompliance is more likely when consumers learn about all (positive and negative) monitoring results than when consumers only find out about firms that have been found cheating, as is often the case in practice. I further show that even firms that pretend to be producing high quality while really producing low quality may have an incentive to lobby for stricter monitoring. This helps explain, for example, why firms in Europe and the United States lobby for the implementation of voluntary environmental audits, third-party labeling agencies or other disclosure strategies.

Keywords: labeling, overcompliance, credence goods, quality choice

1. Introduction

The objective of this paper is to give an explanation of two empirical puzzles. The first puzzle is why some firms appear to voluntarily incur higher costs to produce a "high quality" good even when quality is a credence good, i.e., quality cannot be observed by consumers even after purchase. Examples include the increasingly common phenomenon of "green business," i.e., firms that voluntarily reduce their negative environmental impacts.[1] Similarly, we observe companies paying "fair wages" or reducing the health or safety risks associated with their products.[2] The second puzzle is that firms are actually observed lobbying for increased monitoring of their actions. For example, firms in Europe and the United States are lobbying for voluntary environmental audits or independent labeling agencies.[3]

The environmental, social, health or safety impacts of a product (denoted hereafter as product "quality") are often valued by consumers. For example, 83% of consumers in a 1993 poll stated that they were willing to pay more for environmentally sound products [4]. However, these quality aspects are credence goods because they usually cannot be observed directly by consumers even after purchase. Instead, consumers frequently rely on some type of advertising, product labeling or public disclosure of pollution information to form expectations of these product characteristics. Where firms simply advertise or issue their own labels, environmental and consumer-rights organizations have taken the role of monitoring firms' claims to some extent. For example, Greenpeace issues reports and poster campaigns of companies that have been found lying about their environmental impacts. Moreover, as Tietenberg [5] points out, there is an increasing trend for disclosure strategies as what he calls "the third wave in pollution control policy." For example, many countries have established independent labeling agencies. Under these systems, firms apply to the labeling authority for a license to use a particular label stating that their product is, for example, environmentally friendly, socially responsible or safe.[4] The labeling authority provides random monitoring of the firms' claims. If a firm is found cheating, its license to use the label is withdrawn. Other examples of disclosure strategies include the U.S. Toxic Release Inventory (TRI), which provides information to the public regarding releases of toxic substances into the environment. A similar program provides a public ranking of firms' environmental performance in Indonesia [5].

In some cases, consumers learn about "good and bad" monitoring results, i.e., about those firms that have been found cheating as well as those that have been found telling the truth. This is the case for public rankings such

[1] For a description of this phenomenon, see, for example, Tomer [1] and Smart [2].

[2] Several nongovernmental organizations market so-called Fair Trade coffee, which is produced by small farmers who are paid above-average prices. The increased offer of organic products is an example of reduced health risks.

[3] See Smart [2] and Spiegel [3] for examples.

[4] In the area of environmental claims, Germany's Blue Angel system is the oldest independent labeling program. Currently, at least 22 countries have government-sponsored, third-party eco-labeling systems. The United States has two private-sector systems (Green Seal and Scientific Certification Systems), and the European Union is currently launching a regional system.

as the TRI. Also, some consumer organizations issue rankings of products' safety or health impacts, informing consumers about the performance of all companies that have been monitored. However, frequently, consumers only find out about "bad news," i.e., firms that have been found cheating. For example, Greenpeace's poster campaigns only point out firms found lying about their environmental performance. Similarly, labeling agencies withdraw the labeling license of firms that have been found violating the labeling standards. However, consumers cannot distinguish between firms that have not been monitored and those firms that have been monitored and found to be complying with the licensing agreement; in both cases, consumers simply continue to see the label on the products.

Several models of firm's quality choice under asymmetric information have been presented in the IO literature, including Allen [6], Klein and Leffler [7], and Shapiro [8]. Tirole [9] presents a model based on the articles of Kreps and Wilson [10] and Milgrom and Roberts [11]. All of these papers are based on the idea that a firm may voluntarily incur higher costs to build up a high-quality reputation that yields an increase in consumers' willingness to pay for its products in the future. With the exception of Shapiro [8], all of these models focus on the case of an experience good, where quality can be observed after purchase. While Shapiro does allow for credence goods by using a very general reputation adjustment function, he does not analyze under what conditions the quality choice will be above minimum quality. The author explicitly states that any more detailed analysis requires the specification of a particular reputation adjustment process.

The case of credence goods has also been addressed in a few recent papers that focus on the provision of expert services, such as repairs and medical services. Wolinsky [12] examines the role of consumers' search for multiple opinions in disciplining experts. Emons [13] shows that a competitive market equilibrium with nonfraudulent behavior exists when firms are capacity constrained. Both explanations seem less applicable to the case of environmental, social or health impacts. This paper presents an alternative explanation of voluntary overcompliance in the case of credence goods based on the idea of reputation effects.

This paper builds on Tirole's [9] model to explain voluntary overcompliance in the case of a credence good. Tirole's model is chosen as a basis for mainly two reasons. First, by assuming two potential types of firms, one that will always overcomply and one that finds it cheaper to produce low quality, Tirole allows for the possibility that consumers' expectations about the firm's quality choice are incorrect *ex post*. This is important because we do observe firms that are selling their products at higher prices based on their promises of a higher quality, and are then found to not be complying with their promises. Second, contrary to the other models, Tirole does not rely on an infinite planning horizon. Since incentives for overcompliance can be shown to increase as the planning horizon (and thereby potential reputation gains), increases, a finite-horizon model is more general than an infinite-horizon model. The paper presented here differs from Tirole's model in that it considers quality to be unobservable by consumers even after purchase and allows for imperfect monitoring by an independent party. Moreover, Tirole's analysis only focuses on the conditions for the firm to choose a high-quality level. The analysis below is taken several steps further to discuss the effects of different information structures, as well as firms' lobbying incentives.

The structure of the paper is as follows. Section 2 describes the basic assumptions and time line of the model. Section 3 analyzes the sequential equilibria of the model. In particular, the conditions for an equilibrium in which a firm voluntarily incurs higher costs to overcomply with a positive probability are analyzed. This is done in section 3.1 for the case where consumers only find out about "bad monitoring results," and in section 3.2 for the case of "good and bad news."[5] It is shown that overcompliance is more likely in the latter case and that the establishment of a labeling agency may increase overcompliance in both cases. Section 4 analyzes firms' incentives to lobby for or against increased monitoring. It is shown that even a firm that finds it cheaper to produce low quality and cheats with a positive probability may lobby for increased monitoring. Section 5 concludes.

2. Assumptions

There are two periods, denoted by t ($t = 1,2$). In each period, there is a single firm selling a product of variable quality but fixed quantity.[6] For simplicity, I assume that only two quality levels exist: a high level (e.g., "biodegradable," "made of nontoxic materials only," "produced at fair wages") and a low level, representing the legal minimum quality or the least-cost technology. Firms use labeling to indicate the level of quality to consumers. There are two possible types of firm: an "honest type" and a "dishonest type." Let ρ denote the *ex ante* probability (at the beginning of period 1) that the firm is honest.

[5] This is not to say that labeling programs are always effective. For example, evidence from Colombia suggests that the public disclosure program was less effective in inducing firms to invest in environmental programs than command-and-control policies (see "El efecto del regulador y de la comunidad sobre el desempeño de la industria en Bogotá, Colombia," Documento 2002-05, Universidad de los Andes, Bogotá). I thank an anonymous reviewer for this point. In fact, the model presented here would suggest that the effectiveness of labeling programs to induce overcompliance may be limited in countries where environmental premia are small.

[6] The focus on a single firm is made for simplicity, following Tirole [9]. While the presence of quality premia often leads to some degree of market power, it is acknowledged that the case of a single firm ignores such issues as competitive advantage and the simultaneous presence of different quality levels in the market.

The honest type always prefers to produce the high quality. This may be either because he or she derives "moral" satisfaction from producing high quality, or because he or she actually finds it cheaper to produce high quality than to produce low quality. For example, Tomer [1] talks about the issue that firms might reduce their negative environmental impacts for moral or cost-savings reasons. Throughout the paper, the parameters of the honest type's problem are denoted by capital letters, while those of the dishonest type's problem are denoted by small letters. The present value of the honest type's expected profits, denoted by Π, is

$$\Pi = P_1 + \delta E(P_2) + (1 + \delta)V, \quad (1)$$

where P_t is the market price for the good in period t, δ is a discount factor, and V is the valuation of the moral satisfaction of high-quality production net of production costs. The expectation operator is used because the price in period 2 may depend on whether the firm is monitored or not.

The dishonest type maximizes the present value of his profits. He does not receive moral satisfaction from producing high quality and always finds it cheaper to produce low quality. Let the unit cost of production of the product for the dishonest type be c if he produces high quality and 0 if he produces low quality. The latter assumption is made for simplicity of notation and is not crucial to the results. c can be interpreted as the cost increase from improving the dishonest type's product quality beyond the legal minimum. In general, the firm chooses its probability of overcompliance, denoted by α ($0 \leq \alpha \leq 1$). It either overcomplies with probability one ($\alpha = 1$), produces low quality with probability one ($\alpha = 0$), or it may randomize between the two quality levels ($0 < \alpha < 1$). We are interested in finding conditions under which the dishonest type finds it profitable to produce high quality (i.e., overcomply) with a positive probability, despite the cost increase incurred.

All consumers are identical and possess demands for the product in each period. Consumers' willingness to pay in any given period is θ for a high-quality product, and 0 for a low-quality product. Without loss of generality, θ can be interpreted as the quality premium. Consumers cannot directly observe the quality level in any period and cannot acquire this information at reasonable cost. (For example, no technology exists for testing whether a paper product contains recycled fiber. The only way to find out is through on-site inspection. It is prohibitively costly for a consumer to perform on-site tests, or hire someone else to do so, every time the consumer purchases a product labeled to be made from recycled fibers.) However, consumers form expectations about a dishonest type's probability of overcompliance, denoted by $\hat{\alpha}$, ($0 \leq \hat{\alpha} \leq 1$). Throughout the analysis it is assumed that firms and consumers are risk neutral.

Note that the firm will always label its products as being of high quality, thereby trying to receive a positive price. At worst, consumers do not believe it and pay zero, which is the same the firm would get if it admitted to producing unsoundly.[7] Consumer or environmental organizations or an independent labeling authority monitor the firm's claims by random checking. Let the probability of being monitored with respect to the voluntary activity in any given period be γ, where $0 \leq \gamma \leq 1$. Monitoring is infallible. If a firm cheats and is monitored, its cheating will be detected. However, in the real world, the transaction costs of monitoring all firms in such an infallible manner may be arbitrarily high. Therefore, the monitoring authority may rationally choose not to monitor all firms in any given time. Alternatively, one could presume that all, or most, firms are monitored, but that monitoring is not infallible. (For example, monitoring might be done using standard probabilistic quality-control techniques that allow for some error.) γ would then correspond to the probability of being detected. In section 3, I analyze the sequential equilibria of the model taking the monitoring intensity (γ) as exogenously given. In section 4, I then consider the case where firms can affect the level of γ through lobbying and analyze firms' incentives to lobby for or against increased monitoring.

Below I analyze two possible information mechanisms. The first mechanism, denoted as "spreading only bad news," informs consumers only of firms that have been monitored and found cheating, i.e., falsely claiming to produce high quality. However, if the firm is not monitored or is found to be telling the truth, consumers do not obtain new information and revise their expectation for the next period of the probability of the firm being honest. As argued earlier, this tends to be the case with most third-party labeling systems as well as reports such as Greenpeace's book of Greenwash, i.e., firms found lying about their environmental impacts. The second information mechanism, denoted as "spreading both good and bad news," informs consumers of all monitoring results. Thus, consumers not only learn about firms found cheating, but also about firms found overcomplying (really producing high quality). This is the case, for example, with public disclosure programs or consumer reports that publish quality rankings.

The time line of the model is illustrated in figure 1. First, consumers form expectations about the firm's quality choice in period 1. These expectations determine consumers' maximum willingness to pay for the product and thereby the price (P_1), which can be charged by the firm in period 1. The monopolist then decides on a level of quality (high or low), in period 1, and consumers buy the good at the price P_1. At the end of period 1, the quality of the good

[7] This result no longer holds if there was a fine for cheating or a labeling fee. With a fine for cheating, monopolists by falsely labeling their product as high quality would risk paying a fine, possibly reducing their profits below zero. Similarly, with a labeling fee, the firm has to trade off the potential gain from labeling its products as high quality and getting away with it against the certain cost of the labeling fee.

t=1: Consumers form expectations about quality in t=1.
↓
Determines P_1.
↓
Firm chooses quality in t=1..
↓
Consumers purchase the product at price P_1.
↓
Random monitoring takes place.
↓
t=2: Consumers receive information about monitoring results.
↓
Consumers revise their expectations about firms type.
↓
Consumers form expectations about quality in t=2.
↓
Determines P_2.
↓
Firm chooses quality in t=2.
↓
Consumers purchase the product at price P_2.

Figure 1. Timeline of the model.

is monitored randomly. If the firm is monitored, consumers are informed about monitoring results at the beginning of period 2. Depending on the information mechanism, they receive only bad news or both good and bad news. Based on the information and their prior expectations of the behavior of dishonest types, consumers revise their expectations that the monopolist is honest, thereby determining the price (P_2) that the monopolist can charge in period 2. The monopolist then chooses the level of quality for period 2 and sells the good to consumers at the price P_2.

3. Equilibrium analysis

There are two possible explanations of our empirical puzzle of a firm's voluntary overcompliance in our model. One obvious explanation is that the firm is of the honest type. The second, more interesting explanation is that a sequential equilibrium exists in which the dishonest type produces the high quality with some positive probability in any period. The objective of the following two sections is to analyze the possible sequential equilibria of the game defined above and examine the conditions for an equilibrium in which the dishonest type overcomplies. This is done for each of the two information structures. A sequential equilibrium satisfies two conditions. First, the firm's quality decision is rational in each period. That is, given prices and consumers' expectations the firm chooses α to maximize its expected profits. Second, the firm's quality choice is consistent with consumers' beliefs about it (i.e., $\alpha = \hat{\alpha}$).

Before analyzing the equilibrium in detail, we can make two general observations. First, the dishonest type will always produce low quality in period 2. Since period 2 is the last period of the game, there are no reputation gains to be made by overcomplying in that period; the model becomes just like a one-period model. Second, consumers' maximum willingness to pay for the product is given by their expectation that the firm will produce high quality multiplied by their valuation of a high-quality product (θ). In a sequential equilibrium, consumers know that the dishonest type will always produce low quality in period 2. Therefore, their willingness to pay for the product in period 2 is given by the revised probability that the firm is honest (given monitoring information) multiplied by θ. Intuitively, we expect an overcompliance equilibrium to emerge when the increase in production costs associated with producing high quality is more than offset by the firm's reputation gains from fooling consumers into believing it may be honest, thereby receiving a positive price in the next period.

3.1. Spreading only bad news

Figure 2 illustrates the game for the case when consumers only learn about firms that have been found cheating. At the beginning of period 1, consumers form beliefs about the firm's type and behavior in period 1. They know that the firm is honest with probability ρ, in which case it always produces high quality, and dishonest with probability $1 - \rho$, in which case it could produce low or high quality in period 1. They also form beliefs of the probability with which the dishonest type overcomplies ($\hat{\alpha}$). Consumers' beliefs determine their maximum willingness to pay for the good and thereby the price in each period. In period 1, consumers' beliefs imply that their expectation of the probability of a high-quality product is $\rho + (1-\rho)\hat{\alpha}$. Therefore, the price in period 1 is

$$P_1(\hat{\alpha}) = [\rho + (1-\rho)\hat{\alpha}]\theta. \tag{2}$$

Consumers' willingness to pay in period 2 is given by their *ex post* expectation that the firm is honest, given their information on monitoring results, times θ. Therefore, the price in period 2 depends not only on consumer beliefs, but also on monitoring results (denoted by m). There are two possibilities here. If consumers receive bad news, i.e., find out that the firm has been cheating, they realize that the firm is dishonest because only the dishonest type ever produces low quality. In that case consumers pay zero in period 2 because they know that the dishonest type always produces low quality in $t = 2$. If consumers receive no news, they know that the firm could be honest (with a marginal probability of ρ), dishonest and producing high quality [with a marginal probability of $(1-\rho)\hat{\alpha}$], or dishonest, cheating, but not have been monitored [with a marginal probability of $(1-\rho)(1-\gamma)\hat{\alpha}$]. Using Bayes' rule, consumers' updated expectation that the firm is honest given they have not received any monitoring news (denoted by β), is given by the conditional probability

$$\beta = \frac{\rho}{\rho + \hat{\alpha}(1-\rho) + (1-\hat{\alpha})(1-\rho)(1-\gamma)} = \frac{\rho}{1-\gamma(1-\hat{\alpha})(1-\rho)} \geq \rho. \tag{3}$$

Since consumers know that only the honest type produces high quality in period 2, their willingness to pay for the good in that period if they have not received any monitoring news (denoted by $\bar{P}$), is equal to

$$\bar{P}(\hat{\alpha}) = \beta\theta. \tag{4}$$

Given consumers' beliefs and its implications for the prices that can be charged by the firm in each period, the firm chooses its probability of producing high quality (α). An honest firm always overcomplies, i.e., it chooses $\alpha = 1$. It receives $\bar{P}_1(\hat{\alpha}) + V$ in period 1, and $\bar{P}(\hat{\alpha}) + V$ in period 2 (figure 2).

A dishonest type chooses a level of α between 0 and 1 to maximize its expected profits. In calculating its expected profits the firm considers the effect of its action on its production costs as well as on the probability of being detected cheating, in which case he would receive a price of zero. When he overcomplies he receives $P_1(\hat{\alpha}) - c$ in period 1. There is no risk of being discovered cheating. Thus, he always receives $\bar{P}(\hat{\alpha})$ in period 2. When he produces the low quality in period 1, he saves production costs and just receives $P_1(\hat{\alpha})$ in period 1. However, he risks being monitored, with probability γ, in which case consumers find out his type and pay zero in period 2. If he is not monitored, he gets away with

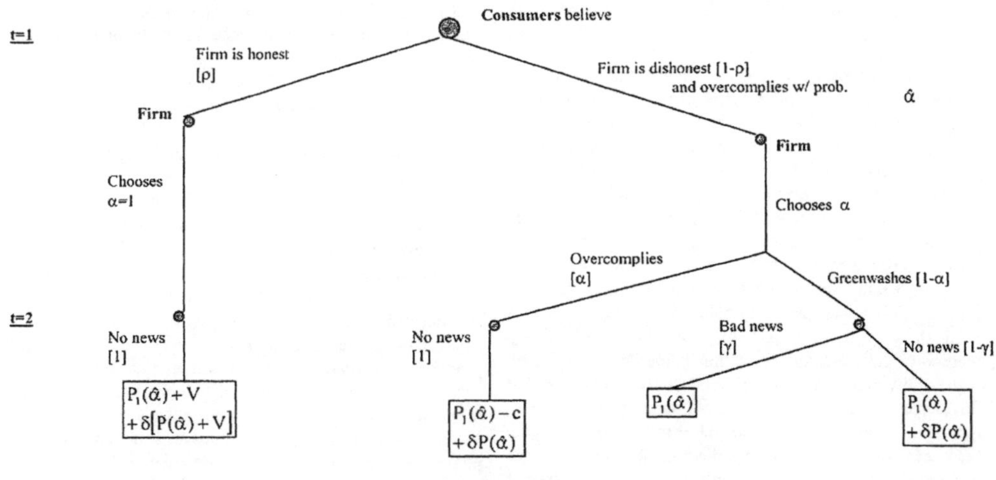

where $P_1(\hat{\alpha}) = \text{Prob(Firm overcomplies in } t = 1/\hat{\alpha}) \cdot \theta = [\rho + (1-\rho)\hat{\alpha}]\theta$

$P_2(\hat{\alpha}, m) = \text{Prob(Firm overcomplies in } t = 2/\hat{\alpha}, m) \cdot \theta$

$$= \text{Prob(Firm is honest }/\hat{\alpha}, m) \cdot \theta = \begin{cases} \bar{P}(\hat{\alpha}) = \dfrac{\rho}{1-\gamma(1-\hat{\alpha})(1-\rho)}\theta & \text{if "no news"} \\ 0 & \text{if "bad news"} \end{cases}$$

Figure 2. Spreading only "bad news."

Table 1
Equilibria when consumers receive only bad news.

Parameter values	Nongreen types behavior in period 1	Prices	Expected profits of the green type (Π) and the nongreen type (π)
$\gamma > \gamma^{**} \equiv \frac{c}{\delta\rho\theta}$	Overcomplies with probability 1 ($\alpha = 1$)	$P_1 = \theta$, $P_2 = \rho\theta$	$\Pi = \theta + \delta\rho\theta + (1+\delta)[V - C]$, $\pi = \theta - c + \delta\rho\theta$
$\gamma < \gamma^{*} \equiv \frac{c}{\delta\rho\theta + (1-\rho)c}$	Produce low quality with probability 1 ($\alpha = 0$)	$P_1 = \rho\theta$, $P_2 = \begin{cases} 0 & \text{if bad news,} \\ \bar{P}(0) = \frac{\rho}{1-\gamma(1-\rho)}\theta & \text{otherwise.} \end{cases}$	$\Pi = \rho\theta + \delta\frac{\rho}{1-\gamma(1-\rho)}\theta + (1+\delta)[V - C]$, $\pi = \rho\theta + \delta(1-\gamma)\frac{\rho}{1-\gamma(1-\rho)}\theta$.
$\gamma^{*} \leq \gamma \leq \gamma^{**}$	Randomizes, choosing high quality with probability $\alpha = \overline{\alpha} \equiv 1 - \frac{c - \rho\delta\gamma\theta}{(1-\rho)\gamma c}$	$P_1 = [\rho + (1-\rho)\alpha]\theta$, $P_2 = \begin{cases} 0 & \text{if bad news,} \\ \overline{\beta}\theta & \text{otherwise,} \end{cases}$ where $\overline{\beta} = \frac{\rho}{1-\gamma(1-\overline{\alpha})(1-\rho)} = \frac{c}{\delta\gamma\theta}$	$\Pi = [\rho + (1-\rho)\overline{\alpha}]\theta + \delta\overline{\beta}\theta + (1+\delta)[V - C]$, $\pi = [\rho + (1-\rho)\overline{\alpha}]\theta - \overline{\alpha}c + \delta(1 - (1-\overline{\alpha})\gamma)\overline{\beta}\theta$.

pretending to be honest and receives $\bar{P}(\hat{\alpha})$ in period 2. The dishonest type chooses its probability of overcompliance (α) to maximize its discounted sum of expected profits, denoted by $u(\alpha)$:

$$\max_{\alpha} \; u(\alpha) \quad \text{s.t.} \quad 0 \leq \alpha \leq 1,$$

$$\begin{aligned} \text{where } u(\alpha) &= \alpha[P_1(\hat{\alpha}) - c + \delta\bar{P}_2(\hat{\alpha})] \\ &\quad + (1-\alpha)[P_1(\hat{\alpha}) + \delta(1-\gamma)\bar{P}_2(\hat{\alpha})] \\ &= P_1(\hat{\alpha}) + \delta(1-\gamma)\bar{P}_2(\hat{\alpha}) + \alpha[\delta\gamma\bar{P}_2(\hat{\alpha}) - c]. \end{aligned} \tag{5}$$

The latter expression for $u(\alpha)$ helps understand the intuition of the game. Independent of the dishonest type's behavior, he receives $P_1(\hat{\alpha})$ in period 1. With probability $1 - \gamma$, the firm is never monitored and receives $\bar{P}(\hat{\alpha})$ in period 2, regardless of its actual behavior. The actual difference between producing low and high quality in period 1 lies in two factors. Producing high quality raises production costs in period 1 by c. In return it increases the price in period 2 from 0 to $\bar{P}(\hat{\alpha})$ in the case where the firm is monitored. In other words, by producing high quality the firm can fool consumers into believing it may be honest even if it is monitored. Thus, the term $\delta\gamma\bar{P}(\hat{\alpha}) - c$ represents the expected reputation gains from producing high quality minus the costs of doing so. It is this term that drives the firm's choice of α. If this term is positive, i.e., reputation gains outweigh costs, the firm's expected profits are increasing in α. In that case, the firm chooses α as large as possible and always overcomplies ($\alpha = 1$). Similarly, if the increase in production costs exceeds expected reputation gains, the firm always produces low quality ($\alpha = 0$). Only if reputation gains are exactly equal to the cost increase might the firm ever randomize between high and low quality. In fact in that case, the firm is indifferent with respect to its choice of α. Recall, however, that a sequential equilibrium requires that firms' actions are not only rational but also consistent with consumer beliefs. Therefore, in equilibrium, we must have $\alpha = \hat{\alpha}$.[8] In Appendix A, I formally prove that this requirement, together with the three possible solutions to the firm's problem, results in the three possible equilibria described in table 1. Table 1 shows the conditions on the parameter values required for each equilibrium as well as the prices and expected profits evolving in each equilibrium. The honest type's profits (Π) are given by equation (1), with $E(P_2) = \bar{P}\hat{\alpha}$. The expected profits of the dishonest type (denoted by π) are given by the maximized value in (5).

The conditions on the parameter values are phrased in terms of the monitoring intensity (γ) to facilitate the discussion in section 4 of what happens when firms can influence γ. As explained above, in a randomizing equilibrium, the firm is really indifferent between any level of overcompliance. However, the level $\overline{\alpha}$ defined in table 1 is the only one that satisfies the consistency requirement of a firm's behavior with consumer expectations.

The results are intuitive. The dishonest type's potential reputation gain from producing high quality in period 1 are due to the possibility that the firm is monitored, in which case producing high quality fools consumers into believing that the firm may be honest. This potential gain is larger the more likely the firm is to be monitored. Therefore, for small monitoring intensities ($\gamma < \gamma^*$), the firm always produces low quality, for medium monitoring intensities ($\gamma^* \leq \gamma \leq \gamma^{**}$), it randomizes, and for large monitoring intensities ($\gamma > \gamma^{**}$), the firm always produces high quality.[9] The prices are also intuitive. In a pure overcompliance equilibrium ($\alpha = 1$), consumers realize that both types produce high quality in period 1. Therefore, $P_1 = \theta$. Monitoring is not informative about the firm's type in this case. Therefore, the *ex post* probability that the firm is

[8] If $\alpha \neq \hat{\alpha}$, we cannot have an equilibrium. Consumers would adjust their expectations and/or firms would adjust their assumption about consumers' beliefs.

[9] As shown in Appendix D, it does not matter whether the borderline cases (γ^* and γ^{**}) are included in the randomization or pure equilibria.

honest is still ρ, and hence, the price in period 2 in the case of no new information is $\bar{P}(1) = \rho\theta$. In an equilibrium where the dishonest type always produces low quality, the probability of a high quality good in period 1 is just the probability of the firm being honest, and thus $P_1 = \rho\theta$. In that case, monitoring results are informative. If consumers receive no news, the firm could either be honest or dishonest and not have been monitored. Therefore, the price in period 2 in the case of no news is $\bar{P}(0) = \frac{\rho}{\rho+(1-\rho)(1-\gamma)}\theta = \frac{\rho}{1-\gamma(1-\rho)}\theta$.

The results in table 1 indicate that it is indeed possible to obtain a sequential equilibrium in which the dishonest type voluntarily overcomplies with legal quality standards. Combining the conditions for a pure overcompliance and a randomizing equilibrium and rearranging terms, we obtain

Result 1. When consumers only learn about firms found cheating, the dishonest type will overcomply with a positive probability (randomization or overcompliance with probability 1) whenever the condition $\gamma > \gamma^*$, or equivalently

$$\delta\gamma\frac{\rho}{1-\gamma(1-\rho)}\theta > c \qquad \text{(R1)}$$

is satisfied.

The left-hand side represents the maximum expected gains from overcompliance: the discounted value of consumers willingness to pay for the good in period 1 if the firm is monitored and overcomplying and consumers believe that only honest firms overcomply $\left(\delta\frac{\rho}{\rho+(1-\rho)(1-\gamma)}\theta\right)$ multiplied by the probability of being monitored. The right-hand side of (R1) shows the cost increase from overcompliance.

Result 1 implies that we expect to see overcompliance where the quality premium (θ) is relatively large, where the cost increase associated with high-quality production (c) is small, where the probability of being monitored (γ) is high, where the discount rate is low (δ is high), and where consumers believe that the probability that the firm is honest (ρ) is high.[10] All of these results are intuitive. The probability of being monitored (γ) will generally be higher under a third-party labeling system than when companies issue their own labels. Therefore, the existence of third-party labeling increases the likelihood of overcompliance.

The effect of ρ is particularly interesting because it implies that consumer expectations are, to some extent, self-fulfilling: If consumers believe that the probability that the firm is honest (ρ) is high, the dishonest type's potential gain from investing in reputation increases and overcompliance becomes more likely. And to the contrary, when consumers believe that the firm is unlikely to be honest, there is little gain from investing in overcompliance. Consequently, if third-party labeling increases consumer confidence in the firm being honest it is likely to increase the probability of overcompliance.

Examination of the value of α in a randomizing equilibrium (i.e., $\alpha = \overline{\alpha}$) in row 3 of table 1 implies that the probability with which the dishonest type overcomplies also is nondecreasing in the quality premium associated with environmental soundness (θ), the probability of being monitored (γ), the discount factor (δ), and the probability that the firm is honest (ρ), and is nonincreasing in the cost increment from overcompliance (c).

Note that this model allows for the dishonest type to pretend to be producing high quality while really producing low quality and receiving a positive price, as we sometimes observe in practice. This happens in period 1 whenever the monopolist is dishonest and produces low quality. It also happens in period 2 whenever the monopolist is of the dishonest type and has not been found cheating. Note that consumers' expectations are incorrect *ex post*, but correct *ex ante*, given that there exists a positive probability that the firm is of the honest type.

3.2. Spreading good news and bad news

Figure 3 presents the game graphically for the case where consumers are informed of all monitoring results. (I use primes to distinguish the variables in this case from the ones in the previous case.) The only difference to the case of spreading only bad news is that consumers can now distinguish between the case where the firm has not been monitored (no news) and the case where the firm has been monitored and found overcomplying (good news). Before, consumers received no news in both of these cases. Therefore, instead of one price ($\bar{P}$) in the case of no news we now have two prices: one if the firm was not monitored (denoted by $\bar{P}^n$ and another if the firm was monitored and found overcomplying (denoted by $\bar{P}^g$). If consumers receive no news they know that the firm was not monitored. In that case, the *ex post* probability that the firm is honest is still ρ, and hence

$$\bar{P}^n = \rho\theta. \qquad (6)$$

If consumers receive good news, i.e., the firm has been found overcomplying, the firm could either be honest or dishonest and overcomplying. Therefore, the ex post probability that the firm is honest is $\frac{\rho}{\rho+(1-\rho)\hat{\alpha}'}$, and

$$\bar{P}^g(\hat{\alpha}') = \frac{\rho}{\rho+(1-\rho)\hat{\alpha}'}\theta. \qquad (7)$$

Note that $\bar{P}^n \leq \bar{P}(\hat{\alpha}) \leq \bar{P}^g(\hat{\alpha}')\ \forall\hat{\alpha}' = \hat{\alpha}$. Intuitively, consumers are better informed if they receive both news than if they receive only bad news. The improved information allows them to pay a lower price to firms that have not been monitored than to firms that have actually been found overcomplying. When consumers only learn

[10] The latter is easily seen by rewriting $\frac{\rho}{1-\gamma(1-\rho)} = \frac{\rho}{\rho+(1-\rho)(1-\gamma)} = \frac{1}{1+(\frac{1}{\rho}-1)(1-\gamma)}$.

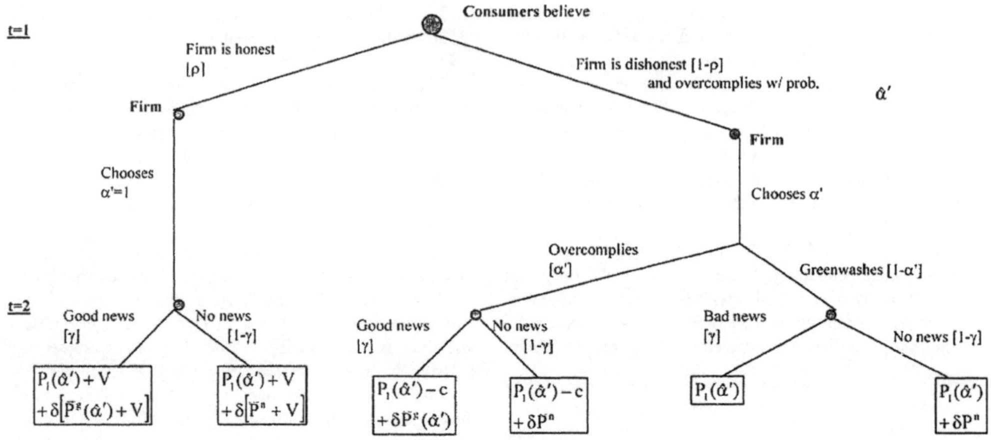

where $P_1(\hat{\alpha}') = \text{Prob}(\text{Firm overcomplies in } t=1/\hat{\alpha}')\cdot\theta = [\rho+(1-\rho)\hat{\alpha}']\theta$

$$P_2(\hat{\alpha}', m) = \text{Prob}(\text{Firm is honest}/\hat{\alpha}', m)\cdot\theta = \begin{cases} \bar{P}^g(\hat{\alpha}') = \dfrac{\rho}{\rho+(1-\rho)\hat{\alpha}'}\theta & \text{if "good news"} \\ P^n = \rho\theta & \text{if "no news"} \\ 0 & \text{if "bad news"} \end{cases}$$

Figure 3. Spreading the "good and bad news."

about bad news they pay an intermediate price in both cases. Since the price for firms found overcomplying is now higher than before, we would expect the improved information structure to increase the incidence of overcompliance. This is in fact the case and will be shown next. The price in period 1 is exactly the same as in the case of spreading only bad news, with α replaced by α'.

The dishonest firm again chooses its probability of overcompliance, α', to maximize its expected profits, given consumer's expectations, $\hat{\alpha}'$:

$$\max_{\alpha} \; w(\alpha') \quad \text{s.t.} \quad 0 \le \alpha' \le 1,$$

$$\begin{aligned} \text{where } w(\alpha') &= \alpha'[P_1'(\hat{\alpha}') - c + \delta(\gamma\bar{P}^g(\hat{\alpha}') + (1-\gamma)\bar{P}^n)] \\ &\quad + (1-\alpha')[P_1(\hat{\alpha}') + \delta(1-\gamma)\bar{P}^n] \\ &= P_1'(\hat{\alpha}') + \delta(1-\gamma)\bar{P}_2^n + \alpha'\left[\delta\gamma\bar{P}_2^g(\hat{\alpha}') - c\right] \end{aligned}$$

Again, the latter expression for expected profits helps understand the intuition behind this game. In period 1, the dishonest type receives $P_1(\hat{\alpha}')$ independent of its quality choice. The price it will receive in period 2 if not monitored, $\bar{P}_2^n$, is also independent of the firm's actions. The only effect of overcompliance as compared to producing low quality is that costs increase by c in period 1 and that the firm receives a positive price in period 2 if monitored $[\bar{P}^g(\hat{\alpha}')]$. The term $\delta\gamma\bar{P}^g(\hat{\alpha}') - c$, therefore, reflects the expected reputation gains from overcompliance net of the increase in production costs. The only difference to the case of spreading only bad news is that the potential gain from overcompliance is now higher since $\bar{P}^g \ge \bar{P}$. Again, there are three possible equilibria, depending on whether reputation gains are greater than, less than, or equal to the increase in production costs. Incorporating the requirement that consumer beliefs ($\hat{\alpha}'$) have to be equal to the firm's action (α') in equilibrium allows us to find the conditions for each type of equilibrium to emerge and the probability of overcompliance in a randomizing equilibrium. This is done in appendix B, and the results are presented in table 2. Table 2 also shows the prices and expected profits for each equilibrium. In this case, the honest type's in period 2 depends on whether he is monitored or not. His expected price is $E(P_2^t) = \gamma\bar{P}^g(\hat{\alpha}') + (1-\gamma)\bar{P}^n$.

Note that the condition for a pure overcompliance equilibrium is the same as in the case of spreading only bad news. This happens because in such an equilibrium both types of firms overcomply. Hence, consumers realize that monitoring yields no new information about the firms' type, no matter how monitoring results are passed on to them. However, as expected, the monitoring intensity that induces a switch from a low-quality equilibrium to a randomizing equilibrium is lower when consumers receive both good and bad news ($\gamma' < \gamma^*$). Combining the

Table 2
Equilibria when consumers receive both good and bad news.

Parameter values	Nongreen types behavior in period 1	Prices	Expected profits of the green type (Π') and the nongreen type (π')
$\gamma > \gamma^{**} \equiv \frac{c}{\delta\rho\theta}$	Overcomplies with probability 1 ($\alpha' = 1$)	$P_1' = \theta$, $P_2' = \rho\theta$	$\Pi' = \theta + \delta\rho\theta + (1+\delta)[V - C]$, $\pi' = \theta - c + \delta\rho\theta$
$\gamma < \gamma' \equiv \frac{c}{\delta\theta}$	Produce low quality with probability 1 ($\alpha' = 0$)	$P_1' = \rho\theta$, $P_2' = \begin{cases} 0 & \text{if bad news,} \\ \rho\theta & \text{if no news,} \\ \overline{P}^g(0) = \theta & \text{if good news.} \end{cases}$	$\Pi' = \rho\theta + \delta[\gamma\theta + (1-\gamma)\rho\theta + (1+\delta)[V - C]$, $\pi' = \rho\theta + \delta(1-\gamma)\rho\theta$
$\gamma' \leq \gamma \leq \gamma^{**}$	Randomizes, choosing high quality with probability $\alpha' = \overline{\alpha}' \equiv \frac{\rho(\delta\gamma\theta - c)}{(1-\rho)c}$	$P_1' = [\rho + (1-\rho)\overline{\alpha}']\theta$, $P_2' = \begin{cases} 0 & \text{if bad news,} \\ \rho\theta & \text{if no news,} \\ \bar{\beta}'\theta & \text{if good news,} \end{cases}$ where $\overline{\beta}' = \frac{\rho}{\rho + (1-\rho)\overline{\alpha}'} = \frac{c}{\delta\gamma\theta}$	$\Pi' = [\rho + (1-\rho)\bar{\alpha}']\theta + \delta[\gamma\bar{\beta}'\theta + (1+\gamma)\rho\theta] + (1+\delta)[V - C]$, $\pi' = [\rho + (1-\rho)\bar{\alpha}']\theta - \bar{\alpha}'c + \delta\left[\bar{\alpha}'\gamma\overline{\beta}'\theta + (1-\gamma)\rho\theta\right]$.

conditions for a pure overcompliance equilibrium and a randomizing equilibrium, we have

Result 2. When consumers are informed about all monitoring results, we expect to see overcompliance with a positive probability ($\alpha' > 0$) whenever $\gamma > \gamma'$, or equivalently

$$\delta\gamma\theta > c. \tag{R2}$$

The left-hand side of (R2) represents the maximum expected gains from overcomplying: consumers' discounted willingness to pay for the good if the firm is found overcomplying and consumers believe that only honest firms overcomply ($\delta\theta$) multiplied by the probability of being monitored. The right-hand side of (R2) shows the increased costs from overcompliance. This condition differs from the one derived in the case of spreading only bad news in that there consumers did not know whether the firm had actually been monitored and found overcomplying or just not been monitored. Therefore, consumers were willing to pay less in that case.

As in the case of only bad news, a positive degree of overcompliance is more likely the higher the discount factor, monitoring intensity, and quality premium, and the lower the cost increase from producing high quality. Note, however, that condition (R2) is independent of the *ex ante* probability that the firm is honest (ρ). However, the probability of overcompliance in a randomizing equilibrium ($\overline{\alpha}'$) is increasing in ρ as well and pure overcompliance becomes more likely as ρ increases (since γ^{**} is decreasing in ρ).

As shown formally in Appendix C, comparing results 1 and 2 as well as the probabilities of overcompliance in the randomizing equilibria for the two information structures ($\overline{\alpha}$ and $\overline{\alpha}'$) yields

Result 3. For any given values of the parameters, the dishonest type is at least as likely to overcomply when consumers are informed of all monitoring results than when they only find out if the firm is found cheating. Moreover, ceteris paribus, the probability of overcompliance in the case of a randomizing equilibrium is at least as large when consumers receive all monitoring results than when they receive only negative results ($\overline{\alpha}' \geq \overline{\alpha}$).

Result 3 implies that the incidence of overcompliance could potentially be increased if the monitoring organization publicized information about firms that have been found overcomplying in addition to the information about firms that have been found cheating. Result 3 also implies that, *ceteris paribus*, a given level of overcompliance can be reached at a lower monitoring intensity when spreading both news than when spreading only bad news. This is interesting because monitoring is likely to be costly for society. Therefore, if the two information structures have approximately equal costs, society would be at least as well off under systems such as public disclosure programs or product ranking reports, which inform consumers of all monitoring results, as under systems such as third-party labeling or poster campaigns on firms found cheating, which inform consumers only of bad news.

4. Lobbying incentives

The objective of this section is to shed some light on the second empirical puzzle by analyzing under what conditions firms may want to lobby for increased monitoring. Intuitively, we may expect honest types to lobby for increased monitoring in order to distinguish themselves from dishonest types. We will show that this is at least weakly the case. More surprisingly, however, we will see that even dishonest types may want to lobby for more monitoring.

In order to examine firms' incentives to lobby for or against stricter monitoring, we need to analyze how the firms' expected profits vary with the monitoring intensity. Note that, as indicated in tables 1 and 2, there are two critical levels of γ for each information structure. The first is the level at which the equilibrium switches from "greenwashing" to randomization, γ^* in the case of spreading only bad news and γ' in the case of spreading good and bad news. The second critical level of γ is the level at which the equilibrium switches from randomization to overcompliance, namely, γ^{**} for both information structures.

I first consider the behavior of the honest type's profits with respect to changes in the monitoring intensity. In Appendix D, I prove

Result 4. Under both information structures, the honest type's profits are strictly increasing in γ for $0 < \gamma < \gamma^{**}$ and independent of γ for $\gamma^{**} < \gamma < 1$.

Result 4 implies that the honest type has an incentive to lobby for increases in γ up to γ^{**} and is indifferent with respect to increases in γ for values of $\gamma > \gamma^{**}$. This result is intuitive. When $\gamma > \gamma^{**}$, both types overcomply with probability one. Therefore, consumers realize that monitoring yields no new information on the firm's type; the price in period 2 is independent of monitoring results and thus the honest type is unaffected by increased monitoring. However, when $\gamma < \gamma^{**}$ the parameter values are such that dishonest types will produce low quality (cheat) with a positive probability. In that case, honest types gain from increased monitoring because it helps them distinguish themselves from dishonest types. Therefore, in real-world terms, if lobbying costs and reputation gains to dishonest types are relatively low (i.e., dishonest types cheat with a positive probability), we expect to see those companies lobby for third-party labeling systems or voluntary audits that for moral or economic reasons always overcomply.

In the Appendix, I also show

Result 5. If consumers are informed of only bad news, the dishonest type's profits are decreasing in γ for $0 \leq \gamma < \gamma^*$, increasing in γ for $\gamma^* < \gamma < \gamma^{**}$, and independent of γ for $\gamma^{**} < \gamma \leq 1$. If consumers are informed of good and bad news, the dishonest type's profits are decreasing in γ for $0 \leq \gamma < \gamma'$, increasing in γ for $\gamma' < \gamma < \gamma^{**}$, and independent of γ for $\gamma^{**} < \gamma \leq 1$. For both information structures, if $c < (1-\rho)\theta$, the dishonest type always prefers values of $\gamma \geq \gamma^{**}$ over values of $\gamma < \gamma^{**}$.

The intuition behind the first part of result 5 is as follows. For low levels of monitoring intensity ($0 \leq \gamma < \gamma^*$ in the case of spreading only bad news and $0 \leq \gamma < \gamma'$ in the case of spreading both news) the dishonest type always produces low quality. He prefers less monitoring because it raises the chance that he will not be discovered, allowing him to receive a positive price in period 2. In the case of spreading only bad news, less monitoring also has the additional effect of lowering the price consumers are willing to pay in period 2 if they have received no monitoring news. This happens because with less monitoring it is more likely that the firm has not been monitored and maybe dishonest. However, as shown in appendix E, this effect is dominated by the effect of a reduced chance to be discovered. Therefore, under both information structures, the dishonest type prefers reductions in monitoring intensity when he is always producing low quality.

For high levels of monitoring intensity ($\gamma^{**} < \gamma \leq 1$) the dishonest type always overcomplies. Since consumers realize that monitoring yields no new information, both types' profits are independent of γ and the dishonest type is indifferent with respect to monitoring.

For intermediate values of γ, the dishonest type randomxizes. In this case, increased monitoring has several effects. As shown in Appendix E [equation (A13)], In the case of spreading only bad news, the change in profits can be written as

$$\left.\frac{\partial\pi}{\partial\gamma}\right|_{\gamma^*<\gamma<\gamma^{**}} = \frac{\partial\bar{\alpha}}{\partial\gamma}\left[(1-\rho)\theta - c + \delta\gamma\bar{\beta}\theta\right] + \frac{\partial\bar{\beta}}{\partial\gamma}\left[1-(1-\bar{\alpha})\gamma\right]\theta - \delta(1-\bar{\alpha})\bar{\beta}\theta. \quad (8)$$

The first term is positive and reflects the effect that increased monitoring has through the firm's probability of overcompliance. As γ increases, the firm overcomplies with a higher probability ($\frac{\partial\bar{\alpha}}{\partial\gamma} > 0$). This increases the first period price by $(1-\rho)\theta\frac{\partial\bar{\alpha}}{\partial\gamma}$, increases production costs by $c\frac{\partial\bar{\alpha}}{\partial\gamma}$, and increases the probability of obtaining a higher price in period 2 due to increased overcompliance by $\frac{\partial\bar{\alpha}}{\partial\gamma}\delta\gamma\bar{\beta}\theta$. The two latter effects cancel each other out in equilibrium, because in a randomization equilibrium reputation gains $(\delta\gamma\bar{\beta}\theta)$ are just equal to the cost increase from overcompliance (c). The second term in equation (8) represents the effect of increased monitoring through affecting consumer's *ex post* beliefs about the firm being honest if not found cheating. This term is negative because consumers realize that dishonest types overcomply with a higher probability. Therefore, they are willing to pay less in period 2. The last term in (8) is negative as well and reflects the fact that with increased monitoring the firm is more likely to be discovered when cheating. As shown in appendix E, the positive effect dominates the two negative effects, so that the total effect of increased monitoring is to increase the dishonest type's profits.

Similarly, in the case of spreading both good and bad news, the effect of more monitoring on the dishonest type's profits is

$$\left.\frac{\partial\pi'}{\partial\gamma}\right|_{\gamma'<\gamma<\gamma^{**}} = \frac{\partial\bar{\alpha}'}{\partial\gamma}\left[(1-\rho)\theta - c + \delta\gamma\bar{\beta}'\theta\right] + \frac{\partial\bar{\beta}'}{\partial\gamma}\delta\bar{\alpha}'\gamma\theta + \delta\left[\bar{\alpha}'\bar{\beta}'\theta - \rho\theta\right]. \quad (9)$$

The intuition behind the first term is exactly as in the case of spreading only bad news. The second term again is

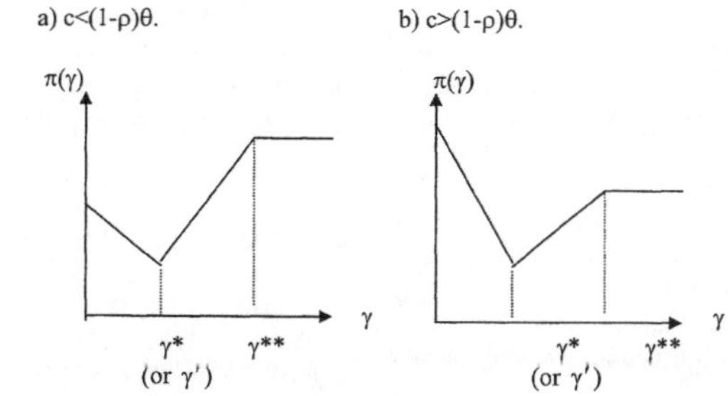

Figure 4. Dishonest type's profits as a function of monitoring intensity.

negative and reflects the effect of increased monitoring in reducing the *ex post* probability of being honest if monitored and found overcomplying, thereby reducing the price consumers are willing to pay in the case of good news. The third term reflects the fact that increased monitoring increases the probability of receiving a higher price if overcomplying and a lower price if cheating. Again, Appendix E shows that the positive effects dominate so that the dishonest type's profits are increasing in γ.

Result 5 is interesting because it implies that even those firms that find it cheaper to produce unsoundly and whose objectives are purely profit-maximizing (dishonest types) may have an incentive to lobby for increased monitoring. Because the behavior of the dishonest type's profits with respect to changes in monitoring intensity are not monotonic, his lobbying incentives generally depend on the initial value of γ and on the increase in γ that is expected to result from the introduction of third-party labeling or any other policy change under consideration. The dishonest type's profits as a function of γ are shown in figure 3a for the case where $c < (1 - \rho)\theta$ and in figure 3b for the case where $c > (1 - \rho)\theta$. Some interesting cases where the dishonest type has unambiguous lobbying incentives are described below (figure 4).

When reputation gains are such that the dishonest type voluntarily overcomplies with some positive but imperfect probability ($\gamma^* \leq \gamma \leq \gamma^{**}$ or $\gamma' \leq \gamma \leq \gamma^{**}$), the dishonest type always strictly prefers increases in monitoring intensity over remaining at the initial monitoring level. If reputation gains are so high that the dishonest type overcomplies with probability 1 ($\gamma \geq \gamma^{**}$), he is indifferent with respect to increases in γ. On the other hand, if reputation gains are low so that the dishonest type prefers to always produce low quality he looses from small increases in γ and may therefore be expected to lobby against third-party labeling or other policies that increase monitoring.

Finally, if the cost increase from high-quality production is relatively low, the quality premium is high, but the probability of being honest is low, such that the condition $c < (1 - \rho)\theta$ holds, then the dishonest type always prefers an increase in monitoring from values below γ^{**} to values above γ^{**}. In that case, the dishonest type always prefers an overcompliance equilibrium over a low-quality equilibrium.

In summary, we found that not only firms that always overcomply for moral or economic reasons, but also those firms that cheat with a positive probability may have an incentive to lobby for stricter monitoring. This may explain, for example, why we see firms lobbying for environmental audits or the introduction of third-party eco-labeling. Intuitively, honest firms lobby for stricter monitoring because doing so helps them distinguish themselves from dishonest types. Dishonest types lobby for more monitoring because the increased risk of being detected cheating as well as the reduction in consumers' willingness to pay if not found cheating may be outweighed by an increase in consumers' willingness to pay in period 1 because the dishonest type will overcomply with a higher probability.

5. Conclusions

I have presented a model based on the assumptions of a positive degree of monitoring and a positive probability that the firm always overcomplies for moral or economic reasons. The model shows that firms that find it cheaper to produce low quality may voluntarily overcomply with legal quality standards even when quality is a credence good. This helps explain why we see firms voluntarily reducing their products' negative environmental, social or health impacts even when these impacts cannot be directly observed by consumers after purchase. The results indicate that this happens when the quality premium associated with the voluntary activity is large, the associated cost increase is small, the monitoring intensity is high, the discount rate is low, and when consumers believe that the probability that the firm is honest is high. When these conditions are not satisfied, we can expect cheating, i.e., firms label their products as high quality when they are not. This is observed in reality, for example, with firms discovered lying about their environmental impacts (e.g., Greenpeace's Greenwash campaigns).

The results imply that the introduction of an independent labeling agency or other forms of increased monitoring of firms' claims are likely to increase the incidence of voluntary overcompliance. Moreover, I show that, *ceteris paribus*, overcompliance is more likely when consumers are informed of all monitoring results (e.g., through quality rankings or disclosure programs such as the TRI) than when consumers only learn about firms found cheating (e.g., Greenpeace's publications on Greenwash and most eco-labeling systems). Since monitoring tends to be costly while spreading the additional monitoring information may be relatively inexpensive, this implies that it might be generally preferable to inform consumers of all monitoring results.

I have also analyzed firms' incentives to lobby for or against increased monitoring. The results indicate that not only firms that always overcomply for moral or economic reasons, but also those firms that cheat with a positive probability may have an incentive to lobby for stricter monitoring. This may explain, for example, why we see firms lobbying for environmental audits or the introduction of third-party eco-labeling. Interestingly, the results indicate that not all firms who lobby for such systems are necessarily always overcomplying. Even firms that cheat with a positive probability may benefit from increased monitoring when the gains in period 1, because consumers realize that the firm is more likely to be overcomplying, outweigh the cost of being more likely to be discovered when cheating and receiving a lower price in period 2.

Several aspects were consciously left out of the present analysis and provide potential for future extensions. First, the model could be extended to more than two periods. We would expect overcompliance to become more likely as the time horizon increases since reputation gains from overcompliance would be obtained in more than one period. Second, an obvious extension would be to introduce competition into the model. While the presence of quality premia often leads to some degree of market power, the case of a single firm ignores such issues as competitive advantage and the simultaneous presence of different quality levels in the market. Finally, one might consider the introduction of a labeling fee into the model. We would expect this to lead to an additional type of equilibrium: the firm admitting to producing low quality. This is expected to happen when the certain cost of the labeling fee outweighs the expected gains from cheating. The model presented in this paper can serve as the basis for such future extensions.

Acknowledgements

I thank Bob Chambers and Ana Maria Ibañez for many helpful comments. Financing for work on this paper was provided by Fedesarrollo, Colombia.

Appendix A: Proof of results in table 1

Since the dishonest firm's objective function, $u(\alpha)$, is linear in α, three possible solutions to the firm's choice of α exist.

First, if $\delta\gamma\bar{P}(\hat{\alpha}) - c > 0$, $u(\alpha)$ is increasing in α. Thus, we have a corner solution at $\alpha = 1$. For this to be an equilibrium, we need $\alpha = \hat{\alpha}$. Therefore, the condition for $\alpha = 1$ to be a sequential equilibrium is

$$\delta\gamma\bar{P}(1) - c > 0, \quad \text{(A1)}$$

which, using the definition of $\bar{P}(\hat{\alpha})$, simplifies to the condition for $\alpha = 1$ in the left column of table 1 ($\gamma > \gamma^{**}$).

Second, if $\delta\gamma\bar{P}(\hat{\alpha}) - c < 0$, $u(\alpha)$ is decreasing in α. Thus, we have a corner solution at $\alpha = 0$. For this to be an equilibrium, we need $\alpha = \hat{\alpha}$, which requires

$$\delta\gamma\bar{P}(0) - c < 0, \quad \text{(A2)}$$

which, using the definition of $\bar{P}(\hat{\alpha})$ simplifies to the condition for $\alpha = 0$ in table 1 ($\gamma < \gamma^*$).

If neither inequality (A1) nor (A2) are satisfied (i.e., $\gamma^* \leq \gamma \leq \gamma^{**}$), we cannot have an equilibrium in which the dishonest type always overcomplies or always produces low quality. In that case, we must have $\delta\gamma\bar{P}(\hat{\alpha}) - c = 0$, i.e., the firm's objective function is independent of α. Thus, the firm is indifferent between all levels of α. Again, for an equilibrium, we need $\alpha = \hat{\alpha}$. Combining the two conditions, we have

$$\delta\gamma\bar{P}(\alpha) - c = 0,$$

which, using the definition of $\bar{P}(\hat{\alpha})$ solves for $\overline{\alpha}$ as defined in the last row of table 1. In this case we have a randomizing equilibrium. Q.E.D.

It can be shown that, given $\gamma^* \leq \gamma \leq \gamma^{**}$, we have $0 \leq \overline{\alpha} \leq 1$. Moreover, as shown in Appendix D, $\bar{\alpha}|_{\gamma=\gamma^*} = 0$ and $\bar{\alpha}|_{\gamma=\gamma^{**}} = 1$, so that it is irrelevant whether the borderline cases are included into the randomization or pure equilibria.

Appendix B: Proof of results in table 2

Again, the dishonest firm's objective function, $w(\alpha')$, is linear in α', and thus, three possible solutions to the firm's choice of α' exist.

First, if $\delta\gamma\bar{P}^g(\hat{\alpha}') - c > 0$, $w(\alpha')$ is increasing in α. Thus, we have a corner solution at $\alpha' = 1$. For this to be an equilibrium, we need $\alpha' = \hat{\alpha}'$. Therefore, the condition for $\alpha' = 1$ to be a sequential equilibrium is

$$\delta\gamma\bar{P}^g(1) - c > 0, \quad \text{(A3)}$$

which, using the definition of $\bar{P}^g(\hat{\alpha}')$ [equation (7)], simplifies to the condition for $\alpha' = 1$ in table 2.

Second, if $\delta\gamma\bar{P}^g(\hat{\alpha}') - c < 0$, $w(\alpha')$ is decreasing in α'. Thus, we have a corner solution at $\alpha' = 0$. For this to be an equilibrium, we need $\alpha' = \hat{\alpha}'$, which requires

$$\delta\gamma\bar{P}^g(0) - c < 0, \quad \text{(A4)}$$

which, using the definition of $\bar{P}^g(\hat{\alpha}')$ simplifies to the condition for $\alpha' = 0$ in table 2.

If neither inequality (A3) nor (A4) are satisfied, we cannot have an equilibrium in which the dishonest type always overcomplies or always produces low quality. In that case, we must have $\delta\gamma\bar{P}^g(\hat{\alpha}') - c = 0$, i.e., the firm's objective function is independent of α'. Thus, the firm is indifferent between all levels of α'. Again, for an equilibrium, we need $\alpha' = \hat{\alpha}'$. Combining the two conditions, we have

$$\delta\gamma\bar{P}^g(\alpha') - c = 0,$$

which, using the definition of $\bar{P}^g\hat{\alpha}'$ solves for $\bar{\alpha}'$ as defined in table 2. In this case we have a randomizing equilibrium.

Appendix C

Since the term $\frac{\rho}{1-\gamma(1-\rho)} = \frac{\rho}{\rho+(1-\rho)(1-\gamma)}$ in condition (R1) is less then one, it follows that (R1) implies (R2). Therefore, any parameter values that induce overcompliance when consumers receive only bad news also do so when consumers receive both news. However, some parameter values satisfy (R2) but not (R1). This implies the first part of result 3.

To show that, *ceteris paribus*, $\bar{\alpha}' \geq \bar{\alpha}$, note first that the comparison is only valid when the conditions for a randomizing inequality hold in both cases, i.e., inequalities $\gamma^* \leq \gamma \leq \gamma^{**}$ and $\gamma' \leq \gamma \leq \gamma^{**}$ hold. Note that the former implies the latter, so that the relevant region is given by $\gamma^* \leq \gamma \leq \gamma^{**}$, which using the definitions of γ^* and γ^{**} can be rewritten as

$$\delta\gamma\rho\theta \leq c \leq \delta\gamma\frac{\rho}{1-(1-\rho)\gamma}\theta. \qquad (A5)$$

Using the definition of $\bar{\alpha}$ and $\bar{\alpha}'$ from tables 1 and 2, we obtain

$$\begin{aligned}\bar{\alpha}' - \bar{\alpha} &= \frac{\rho(\delta\gamma\theta - c)}{(1-\rho)c} - 1 + \frac{c - \rho\delta\gamma\theta}{(1-\rho)\gamma c}\\ &= \frac{1}{(1-\rho)c\gamma}\left[\rho\delta\gamma^2\theta - \gamma c + c - \rho\delta\gamma^2\theta\right]\\ &= \frac{1-\gamma}{(1-\rho)c\gamma}[c - \rho\delta\gamma\theta] \geq 0,\end{aligned}$$

where the inequality follows from the first part of (A5). Q.E.D.

Appendix D: Proof of result 4

1. *Spreading only bad news*

Let us first focus on the case of spreading only bad news. The honest type's profits are given by equation (1). The only portion of these profits that varies with γ and across the different equilibria is $P_1 + \delta E(P_2)$. We can therefore focus our attention on analyzing how $P_1 + \delta E(P_2)$ changes when γ changes. In the case of spreading only bad news $E(P_2) = \bar{P}$ for the honest type, since monitoring him never leads to bad news. Using the equilibrium prices as summarized in table 1, we obtain the following results:

(a) If $\gamma > \gamma^{**}$ (dishonest type overcomplies with probability 1), then

$$P_1 + \delta\bar{P}|_{\gamma>\gamma^{**}} = \theta + \delta\rho\theta,$$

which is independent of γ. Therefore, the honest type's profits are independent of γ for values of $\gamma > \gamma^{**}$.

(b) If $\gamma < \gamma^*$ (dishonest type produces low quality with probability 1), then

$$P_1 + \delta\bar{P}|_{\gamma<\gamma^*} = \rho\theta + \delta\frac{\rho}{1-\gamma(1-\rho)}\theta.$$

Inspection shows that this term increases as γ increases. Therefore, the honest type's profits are increasing in γ for values of $\gamma < \gamma^*$.

(c) If $\gamma^* \leq \gamma \leq \gamma^{**}$ (dishonest type randomizes), then

$$P_1 + \delta\bar{P}|_{\gamma^*\leq\gamma\leq\gamma^{**}} = [\rho + (1-\rho)\bar{\alpha}]\theta + \delta\bar{\beta}\theta,$$

and taking the partial derivative with respect to γ (excluding the borderline points at which profits may not be differentiable), we have

$$\left.\frac{\partial(P_1 + \delta\bar{P})}{\partial\gamma}\right|_{\gamma^*<\gamma<\gamma^{**}} = (1-\rho)\theta\frac{\partial\bar{\alpha}}{\partial\gamma} + \delta\theta\frac{\partial\bar{\beta}}{\partial\gamma}. \qquad (A6)$$

Differentiating the expressions for $\bar{\alpha}$ and $\bar{\beta}$ in table 1 with respect to γ and simplifying, we get

$$\frac{\partial\bar{\alpha}}{\partial\gamma} = \frac{1}{\gamma^2(1-\rho)} > 0, \quad \text{and} \quad \frac{\partial\bar{\beta}}{\partial\gamma} = -\frac{c}{\delta\gamma^2\theta} < 0. \qquad (A7)$$

Thus, increase in monitoring has two effects on the honest type's profits. First, it increases the probability of overcompliance by the dishonest type, thereby increasing the price in period 1. Second, since consumers realize that dishonest types are more likely to overcomply, it reduces the price in period 2 in the case of no news.

Substituting (A7) into (A6), we obtain

$$\begin{aligned}\left.\frac{\partial(P_1 + \delta\bar{P})}{\partial\gamma}\right|_{\gamma^*<\gamma<\gamma^{**}} &= (1-\rho)\theta\frac{1}{\gamma^2(1-\rho)} - \delta\theta\frac{c}{\delta\gamma^2\theta}\\ &= \frac{\theta - c}{\gamma^2} > 0,\end{aligned}$$

where the inequality follows from the fact that in the randomization case

$$\gamma > \gamma^* \Rightarrow c < \delta\gamma \frac{\rho}{1-\gamma(1-\rho)}\theta$$
$$= \delta\gamma \frac{\rho}{\rho+(1-\gamma)(1-\rho)}\theta < \theta. \quad \text{(A8)}$$

Thus, the first effect dominates the second and the honest type's profits are increasing in γ when $\gamma^* < \gamma < \gamma^{**}$.

(d) So far we have analyzed the behavior of the honest type's profits with respect to changes in γ within each equilibrium. It remains to examine whether the profit function is continuous in γ at the points where a switch occurs from one equilibrium to another, i.e., at γ^* and γ^{**}.

- At γ^* we switch from the equilibrium where the honest type produces low quality with probability 1 to that where he randomizes. Note that

$$\bar{\alpha}|_{\gamma=\gamma^*} = 1 - \frac{1}{1-\rho}\left[\frac{1}{\gamma^*} - \frac{\delta\rho\theta}{c}\right] = 0. \quad \text{(A9)}$$

Also substituting $\bar{\alpha} = 0$ in the definition of $\bar{\beta}$ yields $\bar{\beta}|_{\gamma=\gamma^*} = \frac{\rho}{1-\gamma^*(1-\rho)}$. Therefore, at γ^*, the variable portion of the honest type's profit when the dishonest type randomizes is

$$P_1 + \delta\bar{P}|_{\gamma=\gamma^*} = \rho\theta + \delta\frac{\rho}{1-\gamma^*(1-\rho)}\theta,$$

which is the same as the honest type's profits when the dishonest type produces low quality with probability 1, evaluated at γ^*. Therefore, the honest type's profit function is continuous at γ^*.

- At γ^{**}, we switch from the randomizing equilibrium to one where the dishonest type overcomplies with probability 1. Rewriting $\bar{\alpha}$ as in equation (A9), but replacing γ^* by γ^{**}, it is easily seen that $\bar{\alpha}|_{\gamma=\gamma^{**}} = 1$. Consequently, $\bar{\beta}|_{\gamma=\gamma^{**}} = \rho$. Therefore, at γ^{**}, the variable portion of the honest type's profits when randomizing is

$$P_1 + \delta\bar{P}|_{\gamma=\gamma^{**}} = [\rho + 1 - \rho]\theta + \delta\rho\theta = \theta + \delta\rho\theta,$$

which is the same as when the dishonest type overcomplies with probability 1. Therefore, the honest type's profit function is continuous at point γ^{**}.

Parts (a) through (d) jointly imply that the honest type's profits are strictly increasing in γ up to γ^{**} and independent of γ for $\gamma > \gamma^{**}$. Q.E.D.

2. Spreading both good and bad news

A very similar argument holds in the case of spreading both good and bad news. In that case, $P_1' + E(P_2') = P_1' + \delta[\gamma P^g + (1-\gamma)P^n]$. In a pure overcompliance equilibrium ($\gamma > \gamma^{**}$), we have

$$P_1' + E(P_2')|_{\gamma>\gamma^{**}} = \theta + \delta[\gamma\rho\theta + (1-\gamma)\rho\theta] = \theta + \delta\rho\theta,$$

which is independent of γ. For $\gamma < \gamma'$, we have

$$P_1' + E(P_2')|_{\gamma<\gamma'} = \rho\theta + \delta[\gamma\theta + (1-\gamma)\rho\theta]$$
$$= \rho\theta(1+\delta) + \delta\gamma(1-\rho)\theta,$$

which is obviously increasing in γ. For $\gamma' < \gamma < \gamma^{**}$, the variable portion of profits are

$$P_1' + E(P_2')|_{\gamma\le\gamma\le\gamma^{**}} = [\rho + (1-\rho)\bar{\alpha}']\theta + \delta[\gamma\bar{\beta}'\theta + (1-\gamma)\rho\theta].$$

Taking the derivative with respect to γ, we obtain

$$\left.\frac{\partial(P_1' + E(P_2'))}{\partial\gamma}\right|_{\gamma'<\gamma<\gamma^{**}} = (1-\rho)\theta\frac{\partial\bar{\alpha}'}{\partial\gamma} + \delta\gamma\theta\frac{\partial\bar{\beta}'}{\partial\gamma} + \delta\theta(\bar{\beta}' - \rho). \quad \text{(A10)}$$

Using the definitions of $\bar{\alpha}'$ and $\bar{\beta}'$ in table 2, we have

$$\frac{\partial\bar{\alpha}'}{\partial\gamma} = \frac{\delta\rho\theta}{(1-\rho)c} > 0, \quad \text{and} \quad \frac{\partial\bar{\beta}'}{\partial\gamma} = -\frac{c}{\delta\gamma^2\theta} < 0. \quad \text{(A11)}$$

Thus, the first term in (A10) is positive, reflecting the fact that stricter monitoring leads the dishonest type to overcomply with a higher probability, thereby raising the price in period 1. The second term is negative, representing the effect on consumers' *ex post* expectations that the firm is honest. Consumers realize that the dishonest type overcomplies with a higher probability, and thus, the probability that a firm that was found overcomplying is indeed honest is lower, which lowers the price in period 2. The third term in (A10) is positive (since $\bar{\beta}' > \rho$) and reflects the fact that with more monitoring the honest firm is more likely to be found overcomplying and receive the higher price (P^g instead of P^n). Note, however, that, substituting (A11) into (A10), the total effect of stricter monitoring simplifies to

$$\left.\frac{\partial(P_1' + E(P_2'))}{\partial\gamma}\right|_{\gamma'<\gamma<\gamma^{**}} = \frac{\delta\rho\theta^2}{c} - \frac{c}{\gamma} + \delta\theta\left(\frac{c}{\delta\gamma\theta} - \rho\right)$$
$$= \delta\rho\theta\left(\frac{\theta}{c} - 1\right) > 0,$$

where the inequality follows from the fact that, in a randomizing equilibrium,

$$\gamma > \frac{c}{\delta\theta}(=\gamma') \Rightarrow c < \gamma\delta\theta < \theta. \quad \text{(A12)}$$

Thus, we have shown that the total effect of increased monitoring on the honest type's profit in a randomizing equilibrium is positive. To show that the honest type's profits are continuous at γ' note that $\bar{\alpha}'|_{\gamma=\gamma'} = 0$ and

$\bar{\beta}'|_{\gamma=\gamma'}=1$. Therefore, at γ', the variable portion of the honest type's profits when randomizing is

$$P_1' + \delta E(P_2')|_{\gamma=\gamma'} = \rho\theta + \delta\rho\theta + \delta\theta(1-\rho)\frac{c}{\delta\theta} = (1+\delta)\rho\theta + c(1-\rho),$$

which is the same as the profits from always producing low quality, evaluated at γ'. At γ^{**}, the results carry through exactly from the case of spreading only bad news. Q.E.D.

Appendix E: Proof of Result 5

1. Spreading only bad news

The dishonest type's profits for each equilibrium are given in table 1.

(a) If $\gamma > \gamma^{**}$ (dishonest type overcomplies with probability 1), then the dishonest type's profits are

$$\pi|_{\gamma>\gamma^{**}} = \theta - c + \delta\rho\theta,$$

which is independent of γ. Therefore, the dishonest type's profits are independent of γ for values of $\gamma > \gamma^{**}$.

(b) If $\gamma < \gamma^*$ (dishonest type produces low quality with probability 1), then the dishonest type's profits are

$$\pi|_{\gamma<\gamma^*} = \rho\theta + \delta(1-\gamma)\frac{\rho}{1-\gamma(1-\rho)}\theta = \rho\theta + \delta\frac{\rho}{\frac{\rho}{(1-\gamma)}+1-\rho}\theta,$$

which is obviously decreasing in γ. Therefore, the dishonest type's profits are decreasing in γ for values of $\gamma < \gamma^*$.

(c) If $\gamma^* \leq \gamma \leq \gamma^{**}$ (dishonest type randomizes), then the dishonest type's profits are

$$\pi|_{\gamma^*\leq\gamma\leq\gamma^{**}} = [\rho + (1-\rho)\bar{\alpha}]\theta - \bar{\alpha}c + \delta[1-(1-\bar{\alpha})\gamma]\bar{\beta}\theta.$$

Taking the partial derivative with respect to γ yields

$$\left.\frac{\partial\pi}{\partial\gamma}\right|_{\gamma^*<\gamma<\gamma^{**}} = \frac{\partial\bar{\alpha}}{\partial\gamma}[(1-\rho)\theta - c + \delta\gamma\bar{\beta}\theta] + \frac{\partial\bar{\beta}}{\partial\gamma}[1-(1-\bar{\alpha})\gamma]\theta - \delta(1-\bar{\alpha})\bar{\beta}\theta. \quad \text{(A13)}$$

Substituting from equation (A7) and the definitions of $\bar{\alpha}$ and $\bar{\beta}$, we can simplify (A13) to

$$\left.\frac{\partial\pi}{\partial\gamma}\right|_{\gamma^*<\gamma<\gamma^{**}} = \frac{1}{\gamma^2}(\theta - c) > 0,$$

where the inequality follows from (A8). Therefore, the dishonest type's profits are decreasing in γ for $\gamma^* < \gamma < \gamma^{**}$.

d) To show that the dishonest type's profits are continuous at γ^*, recall that at this point, $\alpha = 0$ and $\beta = \frac{\rho}{1-\gamma^*(1-\rho)}$. Substituting these values into the expression for the dishonest type's profits in a randomization equilibrium (from table 1), we get

$$\pi|_{\gamma=\gamma^*} = \rho\theta + \delta(1-\gamma^*)\frac{\rho}{1-\gamma^*(1-\rho)}\theta,$$

which is the same as the profits in a pure low-quality equilibrium. Similarly, at γ^{**}, $\bar{\alpha}=1$ and $\bar{\beta}=\rho$. Substituting into the expression for profits in a randomization equilibrium, we obtain

$$\pi|_{\gamma=\gamma^{**}} = \theta - c + \delta\rho\theta,$$

which is the same as the profits in a pure overcompliance equilibrium. Therefore, the dishonest type's profits are continuous at $\gamma = \gamma^*$ and $\gamma = \gamma^{**}$.

(e) It remains to be shown that when $c < (1-\rho)\theta$, then the dishonest type's profits are higher for any $\gamma > \gamma^{**}$ than for any $\gamma < \gamma^{**}$. We know that π increases between γ^* and γ^{**} and remains constant for $\gamma > \gamma^{**}$, so that any value of $\gamma > \gamma^{**}$ yields a higher π than any value of γ such that $\gamma^* < \gamma < \gamma^{**}$. However, π decreases between 0 and γ^*. Therefore, we need to show that the maximum value of π in this region, namely the value at $\gamma = 0$, is less than the value of π at $\gamma = \gamma^{**}$. Evaluating π at these values and comparing yields

$$\pi|_{\gamma=0} = \rho\theta + \delta\rho\theta < \pi|_{\gamma=\gamma^{**}} = \theta - c + \delta\rho\theta,$$

if and only if $c < (1-\rho)\theta$. Q.E.D.

2. Spreading both good and bad news

In this case, the dishonest type's profits for each equilibrium are given in table 2. For $\gamma > \gamma^{**}$, the dishonest type's profits are exactly as in the case of spreading only bad news.

For $\gamma < \gamma'$, the dishonest type's profits are

$$\pi'|_{\gamma<\gamma'} = \rho\theta + \delta(1-\gamma)\rho\theta,$$

which is obviously decreasing in γ. For $\gamma' \leq \gamma \leq \gamma^{**}$ (dishonest type randomizes), the dishonest type's profits are

$$\pi'|_{\gamma'\leq\gamma\leq\gamma^{**}} = [\rho + (1-\rho)\bar{\alpha}']\theta - \bar{\alpha}'c + \delta[\bar{\alpha}'\gamma\bar{\beta}'\theta + (1-\gamma)\rho\theta].$$

Taking the partial derivative with respect to γ yields

$$\left.\frac{\partial\pi'}{\partial\gamma}\right|_{\gamma'<\gamma<\gamma^{**}} = \frac{\partial\bar{\alpha}'}{\partial\gamma}[(1-\rho)\theta - c + \delta\gamma\bar{\beta}'\theta] + \frac{\partial\bar{\beta}'}{\partial\gamma}\delta\bar{\alpha}'\gamma\theta + \delta[\bar{\alpha}'\bar{\beta}'\theta - \rho\theta]. \quad \text{(A14)}$$

Using (A11) and the definitions of $\bar{\alpha}'$ and $\bar{\beta}'$, we can simplify (A14) to

$$\left.\frac{\partial\pi'}{\partial\gamma}\right|_{\gamma' < \gamma < \gamma^{**}} = \rho\delta\theta\left(\frac{\theta}{c} - 1\right) > 0,$$

where the inequality follows from (A12). Therefore, the dishonest type's profits are decreasing in γ for $\gamma' < \gamma < \gamma^{**}$.

To show that the dishonest type's profits are continuous at γ', recall that at this point, $\bar{\alpha}' = 0$ and $\bar{\beta}' = 1$. Substituting these values into the expression for the dishonest type's profits in a randomization equilibrium, we get

$$\pi'|_{\gamma=\gamma'} = \rho\theta + \delta(1 - \gamma^*)\rho\theta,$$

which is the same as the profits in a pure low-quality equilibrium. At γ^{**}, the argument carries through exactly from the case of spreading only good news.

The proof that the dishonest type's profits are higher for any $\gamma > \gamma^{**}$ than for any $\gamma < \gamma^{**}$ when $c < (1 - \rho)\theta$, carries through exactly from the case of spreading only bad news, with γ^* replaced by γ'. Q.E.D.

References

[1] J.F. Tomer, The human firm in the natural environment: a socio-economic analysis of its behavior, Ecol. Econ. 6 (1992) 119–138.

[2] B. Smart, *Beyond Compliance. A New Industry View of the Environment* (World Resources Institute, Washington, DC, 1992).

[3] Spiegel Special, *Wege aus der Weltkrise. Oko-Bilanz '95* (Spiegel-Verlag, Hamburg, Germany, February 1995).

[4] Greenberg/Lake, *New Priorities for Energy Policy*, Polling Report (Washington, DC, February 8, 1993).

[5] T. Tietenberg, Disclosure strategies for pollution control, Environ. Resour. Econ. 11(3–4) (1998) 587–602.

[6] F. Allen, Reputation and product quality, RAND J. Econ. 15 (1984) 311–327.

[7] B. Klein and K. Leffler, The role of market forces in assuring contractual performance, J. Polit. Econ. 81 (1981) 615–641.

[8] C. Shapiro, Premiums for high quality products as rents to reputation, Q. J. Econ. 98 (1983) 659–680.

[9] J. Tirole, *The Theory of Industrial Organization* (MIT Press, Cambridge, MA, 1988) pp. 125–126.

[10] D. Kreps and R. Wilson, Reputation and imperfect information, J. Econ. Theory 27 (1982) 253–279.

[11] P. Milgrom and J. Roberts, Predation, reputation, and entry deterrence, J. Econ. Theory 27 (1982) 280–312.

[12] A. Wolinsky, Competition in markets for credence goods, J. Inst. Theor. Econ. 151(1) (1995) 117–131.

[13] W. Emons, Credence goods and fraudulant experts, RAND J. Econ. 28(1) (1997) 107–119.

Part II
Empirical Indications of Effectiveness

[10]

ARE ECO-LABELS VALUABLE? EVIDENCE FROM THE APPAREL INDUSTRY

WESLEY NIMON AND JOHN BEGHIN

Using apparel catalog data from the United States, we estimate hedonic price functions to identify market valuation of environmental attributes of apparel goods. We identify a significant and robust premium for the organic fibers embodied in the apparel goods. We also find a discount for the "no-dye" label. We do not, however, find any evidence of a premium for environment-friendly dyes. We further investigate the pricing behavior of apparel suppliers for potential heterogeneous pricing of the organic-fiber attribute and find no evidence of different premia across firms.

Key words: dyes, eco-labels, hedonic price, organic-cotton apparel.

Environmentally friendly labels, or eco-labels, have been appearing on a number of products in the United States and Europe (Wasik, OECD). These labels are the manifestation of industries' efforts to become or to be perceived as environment-friendly. They also reflect environmental regulatory changes, especially in regulation-minded European countries. In some instances, such as the dolphin-safe tuna case, these labeling initiatives have preceded regulation mandating the touted environmentally friendly attribute (Teisl, Roe, and Hicks). A similar recent occurrence concerns eco-labels in the textile/apparel industry. For example, in the United States, agencies, such as the California Certified Organic Farmers and the Texas Organic Cotton Cooperative, certify organic cotton farmers and allow firms selling apparel made from their cotton to market them as certified organic cotton apparel.

Despite some qualitative analysis, there has been minimal quantitative economic work on these labels (Teisl, Roe, and Hicks is a notable exception). We investigate the extent of market valuation, if any, of the environmental claims made by apparel manufacturers. Particularly, we estimate the price premium for the "organic-cotton apparel," "environmentally friendly dyes," and "no-dyes" appellations, as well as the existence of a health component in addition to the non-use valuation of environmental characteristics of apparel goods.

Second, our disaggregated data set provides an opportunity to investigate pricing behavior by retail apparel suppliers. The green apparel market predominantly consists of a presumably competitive fringe of small mail-order firms, and of one large firm, Patagonia, which is really the only candidate that might wield market power in the organic cotton apparel industry. Patagonia's nearly $15 million (in 1995) sportswear division exclusively uses organic cotton (Imhoff). This firm's potential market power could raise the marginal price premium above its marginal cost. Price setting of attributes by this single firm would complicate the interpretation of the results. The eventual premium would have two components, the marginal cost of supplying the environmental attribute and the price markup resulting from the market power.

Our analysis is based on the hedonic price function (Freeman, Palmquist). Recent notable contributions include Osborne; Michaels and Smith; as well as Palmquist, Roka, and Vukina; and others. The major contribution of our article, in reference to the existing literature, is to address the valuation associated

---

Wesley Nimon is a PhD candidate and an instructor at North Carolina State University and John Beghin is associate professor of economics and head of the trade and agricultural policy division of the Center for Agricultural and Rural Development at Iowa State University.

The authors thank, without implicating, participants at the AAEA meetings, the 1997 Camp Resources Conference, the 1997 Southern Economics Association Meetings, and workshop participants at Iowa State University, North Carolina State University, and the U.S. Department of Agriculture, and Mary Bohman, Alastair Hall, Tom Johnson, Jeff Kraus, Barry Krissoff, Ray Palmquist, Brian Roe, Jim Rucker, Kevin Smith, Kevin Sweeney, Mario Tiesl, Walter Thurman, and four anonymous referees for discussion and comments. The authors are indebted to Kevin Smith for his suggestion to use catalog data. Financial support from USDA-NRI project No. 9603068 is acknowledged.

with ecological claims embodied in the labels. In addition, we attempt to investigate the pricing behavior of organic apparel suppliers within the hedonic approach. The treatment of the latter issue has been difficult and limited (see Osborne and Smith, and Feenstra and Levinsohn).

We find a significant and robust price premium for organic cotton, but little evidence of a premium for environmentally friendly or low-impact dyes. The average "organic" markup is about 33.8% of the apparel price for a large selection of apparel items. Further, apparel items sold without dyes at all sell at a substantial discount of about 15%. This discount is consistent with actual cost savings implied by the absence of the dying process. Further, our results suggest that the organic-fiber attribute appears to be priced similarly across the organic apparel retailers included in our data set.

Background on Eco-Labeling and the Organic Cotton Market

Eco-labels present some difficulties in the international context, because requirements to establish the label may serve as de facto trade barriers (Zarrilli, Jha, and Vossenaar). Despite this shortcoming, voluntary labels bring an unusual consensus from many fronts, appealing to economists, ecological groups, and industry groups. In markets where consumers show preferences for "green" products, firms support eco-labels as they offer a way to differentiate their product from others in the same class. Ecological groups generally support eco-labeling initiatives as a way to encourage firms to reduce their products' adverse environmental impacts (Wasik).

Economists like the labels because they are voluntary and let the market reveal the potential valuation of environmental attributes, and because they move the market closer to full information by transforming a credence attribute into a search one, assuming that the labels contain accurate information (Nelson). Labels provide a market-oriented instrument to achieve environmental goals, and thereby avoid the inefficiencies associated with mandatory standards, or bans.

From a conservationist perspective, eco-labels are educational tools to inform consumers about the environmental impacts of the labeled product, and thereby induce a change in purchasing behavior that mitigates the product's environmental damage. This aspect appeals to granola types and environmental interest groups because firms fund the educational program as part of their promotional efforts.

The U.S. organic cotton market emerged a few years ago, and it has been growing rapidly [from 1991 with 800 acres of organic cotton to 1994 with 36,000 acres (Imhoff)]. Organic cotton is grown without the use of any synthetic insecticides, fungicides, herbicides, or fertilizers. Conventional cotton farming methods are much more chemically dependent than most crops, and it is this fact that has spurred an interest in organic cotton textiles (Bowen; Imhoff, Grose, and Carra). There is evidence of groundwater contamination caused by agricultural chemicals, and, since cotton farming is exceptionally dependent on these chemicals, it seems a reasonable crop on which to concentrate these concerns. In a recent document the U.S. Environmental Protection Agency (U.S. EPA) made the argument that over the last twenty years there has been significant investment in point source pollution reduction technologies, but relatively little in nonpoint source pollution control. Apodaca reports that consumers who prefer clothing made from organic cotton do so primarily for environmental concerns. The price premium for the environmental attributes largely reflects the consumer's valuation of the decreased environmental impact of organic cotton apparel.

Health concerns, especially for young children, represent another potential source of benefits and demand for organic apparel. There is little evidence that conventional cotton apparel contains levels of pesticide residues that cause any adverse health effects. Clothing catalogs included in our data base, such as "The Natural Baby Catalog," nonetheless claim that organic cotton clothes are "healthy for babies because there are no pesticide or chemical residues to be absorbed through the skin." They go so far as to imply potential toxicity of conventional cotton and raise fears of contamination by dioxins resulting from the bleaching and dying processes. The Texas Organic Cotton Marketing Cooperative in pamphlets it distributes has insinuated similar fears. Industry groups, such as the National Cotton Council, have challenged this alarmist view. Since defoliates are the only chemicals applied after the cotton boll opens, the chemical residue on the fiber is negligible. In fact, major organic-cotton growers, such as Sally Fox of Natural Cotton Colors, Inc., a main producer of naturally colored and organic cotton in the United States, concurs that health benefits from organic cotton apparel may be minimal despite contradictory claims (Burnett, Wakelyn, and Gordon).

Since these health claims have not been thoroughly debunked, consumers may believe in the existence of valuable health benefits, especially for youngsters, from wearing organic-cotton apparel. If there were perceived health benefits then one would expect the price premium for organic apparel to contain an additional health component, beyond the environmental benefit associated with baby organic items and with the absence of dyes. We try to shed some light on that question by investigating the existence of premia associated with organic apparel items for young children. This roundabout way to investigate the health claims is dictated by the lack of direct categorical data available on environmental benefits claims versus health benefits claims in the labels. Our conjecture is that if there were a health component in the organic premium, it would manifest itself most strongly in items for children, who are perceived to be more susceptible to toxins.

The Hedonic Framework

In this section we introduce the hedonic framework leading to the estimated equations, and then explain how we intend to test pricing behavior. We refer readers to Freeman or Rosen for a more detailed presentation of the hedonic framework. We are interested in identifying price premia/discounts associated with the environmental characteristics of the apparel advertised by the eco-label. A first-stage hedonic price function is an appropriate approach to our problem. It allows us to estimate the implicit prices of the attributes of a good, including its environmental characteristics. Rosen developed the seminal model of a competitive market for a differentiated good in which consumers' and producers' interactions generate an equilibrium. Hedonic prices are the implicit prices of attributes of a differentiated product.

The producer is assumed to maximize profits. The offer function, $\phi(\mathbf{z}; \pi, B)$, is the minimum price the firm is willing to offer a bundle of attributes $\mathbf{z}$, given some level of profits, π, and some cost structure parameterized by B. This $\mathbf{z}$ accounts for the possibility that some firms are better suited to produce a particular bundle of attributes than are others. All firms maximize profit, $\pi = MP(\mathbf{z}) - C(M, \mathbf{z}; B)$, with respect to the arguments of the cost function, M and $\mathbf{z}$, where M is the quantity produced of the good and $\mathbf{z}$ is the vector of attributes of the good. The assumption here is that each firm only produces a single bundle of the good. It produces the bundle for which it is best suited, given its underlying cost function summarized by the parameter B. In this way the model only treats joint production in the limited sense that a firm may own two plants producing different bundles, but there are no cost dependencies necessary for meaningful joint production. Differentiation of profits with respect to $\mathbf{z}$ gives $\partial\pi/\partial\mathbf{z} = M(\partial P/\partial\mathbf{z}) - \partial C/\partial\mathbf{z} = 0$, or

$$(1) \quad P_{z_i} = \frac{C_{z_i}}{M}.$$

Hence, the offer price of the ith attribute must equal the per unit marginal cost of that characteristic. In a competitive market firms cannot influence prices. Thus, the firm maximizes profit when it charges the market price (i.e., the highest price it can) for each of the attributes of the good it produces. The maximum profit then occurs when $P(\mathbf{z}) = C_M(M, \mathbf{z}, B)$. The firm's offer function, $\phi = \phi(\mathbf{z}; \pi, B)$, is the price at which the firm would be willing to make available a particular product. It is defined implicitly as $\pi = M\phi - C(M, \mathbf{z}; B)$, and must equal the marginal cost of production, $\phi = C_M(M, \mathbf{z}, B)$. A set of production "indifference" curves can be found from the necessary condition for profit maximization with respect to the output quantity M, yielding

$$(2) \quad \phi = C_M(M, \mathbf{z}, B).$$

The consumer problem mirrors the producer problem. Consumers maximize utility subject to a budget constraint $[\max U(x, z_1, \ldots, z_n)$ s.t. $y = x + P(z_1, \ldots, z_n)]$, where x is a *numéraire* whose price is unity, the $\mathbf{z}$'s are attributes of the differentiated good being purchased, and y is consumer's income. By assumption each consumer purchases only one unit of the good, which contains the utility-maximizing bundle of characteristics. The consumer's bid function $\theta(z_1, \ldots, z_n; u, y)$ mirrors the offer function and is the amount the consumer is willing to pay for a particular bundle of characteristics, given his or her income and some fixed level of utility. Noting that $x = y - \theta$, this bid function is defined such that

$$(3) \quad U(y - \theta, z_1, \ldots, z_n) = u.$$

Differentiating equation (3) with respect to z_i yields $(\partial U/\partial x)(\partial x/\theta)\theta z_i + \partial U/\partial z_i$ and since $\partial x/\partial\theta = -1$,

$$(4) \quad \theta_{z_i} = \frac{U_{z_i}}{U_x}.$$

Thus, θz_i is the marginal rate of substitution between the attribute z_i and "money," that is, x. In other words θz_i is the marginal implicit valuation the consumer places on the attribute z_i at a given level of utility and income. When considering all attributes of the good, $\theta(z_i, \ldots, z_n; u, y)$ is the maximum consumers are willing to pay for the product while $P(\mathbf{z})$ is the minimum they must pay. In this case utility is maximized when $\theta(\mathbf{z}^*; u^*, y) = P(\mathbf{z}^*)$ and when $\theta z_i = P_i$. Utility is maximized when consumers choose a bundle such that the price they are willing to pay for each attribute exactly equals what they have to pay. Since consumers are paying $P_i(\mathbf{z})$ for each attribute and firms are charging $P_i(\mathbf{z})$, then the equilibrium condition is $P(\mathbf{z}) = \phi z_i(\mathbf{z}^*; \pi^*, B) = \theta z_i(\mathbf{z}^*; u, y)$.

If there are many firms and many consumers then the tangency points of consumers' and producers' indifference surfaces trace out the hedonic price function representing the price of the good holding all the attributes constant except one. Further, Rosen establishes that in the long run the supply side of the market completely characterizes the hedonic price function, which represents the minimum price at which characteristics can be bought and produced. This long-run equilibrium implies that the marginal price of a particular attribute, evaluated at a fixed level of the attribute, should be equal across firms, holding other things constant, and should correspond to the lowest marginal cost of producing that characteristic's particular level.

This equilibrium characterization with entry allows us to derive implications on marginal cost pricing based solely on information on pricing behavior of organic apparel suppliers. The test checks for equality of the marginal prices of environmental attributes across firms holding other things equal. This simple test is consistent with the more elaborate test of price-setting behavior proposed by Osborne and Smith based on cost of production. Using Feenstra's framework, the latter authors look for evidence of price markup above marginal cost of characteristics. The lack of data on cost precludes us from adopting this preferable and sophisticated methodology. Our approach presents the drawback of not spelling out the economic behavior or structure leading to a well-specified alternative hypothesis. However, our test presents the virtue of simplicity and abstracts from the strong structure of the Feenstra and Osborne and Smith approach necessary to derive well-defined markup behavior.

Data

Our data are unconventional, but reliable. We collected price and characteristic data for apparel from retail order catalogs (mail catalogs and others). The data were collected between May and October of 1996. This was done first for organic items. The search for the organic apparel firms combined internet searches and was extensive if not exhaustive. We believe we nearly collected the whole population of the retail organic cotton apparel industry providing catalogs.[1] Besides the internet, sources included various apparel wholesalers and retailers, and organic cotton producer associations. The incentive structure faced by firms reinforces our belief in the extent of our search. Firms want to be noticed and catalogs are powerful signals. We only found four catalogs jointly offering both conventional and organic cotton apparel, which is not so surprising, given the organic-conventional antagonism used in organic marketing.

The data search for conventional cotton apparels proceeded in a similar fashion, but since the conventional market is so vast, an attempt was made to get a sample of retailers across the whole spectrum of quality and styles. For each firm and for both apparel types, we have a large cross section of items (pants, T-shirts, long-sleeve shirts, etc.). In total we have twenty-seven apparel firms in our data set, fifteen of which are specialized in organic apparel. The data set, including the list of firms and items, is available from the authors.

[1] There are organic apparel retail shops, which do not advertise by catalog, and we do not cover those outlets.

The data set has 794 observations in total with 364 observations for conventional-apparel observations and 430 observations for the organic-apparel ones; 750 of the observations are usable. We have 117 observations containing synthetic/cotton fiber blends, both in organic and conventional observations. The number of observations varies by item and by catalog. Most catalogs have a large number of observations, except for four catalogs with less than ten observations each. By contrast, several item categories have a limited number of observations.

In our empirical investigation we pondered the issue of a "reasonable" number of item categories. We initially used two criteria (stepwise regression and power analysis) to disqualify some item categories, but found that removing some item categories would obscure the identification of potential outliers. Therefore, we chose to keep the initial large set of items—in excess of forty items. In the empirical estimation we investigate at length the potential presence of outliers on an individual observation basis, by catalog groupings and by item groupings.

The data then consist of prices and categorical dummies for items, catalogs, gender, age, type of textile fiber used (organic cotton, conventional cotton), type of dyes, and an estimate of quantities sold by item. The categories are useful to capture fixed effects specific to catalogs and items because we do not include all the attributes consumers may value and cost characteristics of suppliers (variable B in the hedonic function). One central concern is that the organic premium may reflect style or craftsmanship specific to a firm. The catalog-fixed effects correct for that possibility.[2] From the National Cotton Council's *Cotton Counts Its Customers* we constructed quantity estimates of each item produced. These quantity estimates are not specific to firms and they represent the best available palliative solution to the lack of actual sales data. The quantity estimates are used as weights in our econometric estimation. The weights are important because they reflect the relative contribution of item types to the average organic premium.

The National Cotton Council also provided the cotton-weight of items, which allows us to do two things. First, we can further quantify the organic/conventional cotton attribute by computing an approximation of the organic cotton content by item. This approximation provides a scalar quantitative measure of the organic attribute as in Rosen, rather than a dichotomous/categorical definition of the environmental attribute (organic or conventional). Similarly, we treat the conventional cotton fiber content of regular apparel and the other fiber content for blend apparel items as two other continuous characteristics. Hence, with this quantitative definition, apparel items are defined on three dimensions representing their content of the three fiber types.

There are two ways to define the environmental attribute associated with organic cotton. The categorical approach avoids the potential pitfall of the quantitative definition of the organic attribute (a 100% organic shirt may not be twice as green as a 50/50 blend) but it does not allow us to test thoroughly the pricing behavior of organic-apparel firms. Indeed, the firm dummy variable is identical to the interaction terms (organic dummy*firm dummy) for organic-apparel firms, except for four firms selling both conventional and organic apparel goods. The quantitative definition of the organic cotton content, the percentage of the fiber content that is made from organic cotton, partially overcomes this problem because we have much variation in the fiber content and colinearity of the former type does not arise between organic fiber content and the firm categories.

To obtain an output price by unit of fiber content, we normalize the price by the amount of cotton included in the item. This normalization allows us to reduce the prices to a common unit. Hence, the attribute valuation included into the output price is also normalized. The normalization allows us to think in terms of an organic cotton premium in reference to a price per unit of fiber weight. The mean price per pound of cotton is $11.18 with a standard deviation of 5.29. Other normalizations are possible. One could normalize prices by square meters of fabric included in apparel goods rather than by fiber weight.[3]

[2] We also estimated our hedonic equation using the data from the four catalogs offering both conventional and organic items. We also include three catalog-fixed effects. This approach is closer to the spirit of experimental design and should eliminate most style effect captured by the organic premium. We found a significant premium of 25.4% which is a bit lower than with our full data set, partly because many items are not represented in this smaller data set of 177 observations.

[3] The normalization by square meters of fabric results in fewer usable observations due to lack of information on fabric content for some items. Estimation results based on the fabric normalization strongly suggest similar qualitative results as the ones obtained with the fiber content normalization.

Table 1. Estimation of Hedonic Price Equations

Variables of Interest	Specification 1: Core 1 Only		Specification 2: Core 2 Only		Specification 3: Core 1 and Health Variables		Specification 4: Core 2 and Health Variables	
	Estimated Premium	P-value	Estimated Premium	P-value	Estimated Premium	P-value	Estimated Premium	P-value
LI	4.8%	0.3810	4.9%	0.4059	4.8%	0.3821	5.1%	0.3929
ND	−14.8%	0.0101	−14.8%	0.0145	−17.6%	0.0024	−17.6%	0.0038
Baby	10.7%	0.2899	14.2%	0.1994	−8.2%	0.5687	−4.6%	0.7938
Organic	33.8%	0.0001			31.7%	0.0001		
Porganic			33.2%	0.0001			31.2%	0.0001
Org*Baby					−18.7%	0.8797		
Porg*Baby							−14.9%	0.8448
ND*Baby					−10.7%	0.7923	−21.2%	0.8308
LI*Baby					−22.0%	0.9284	−26.5%	0.9871

Note: Other coefficients not reported include ones for items, catalogs, age, and gender. LI refers to low impact dyes, ND refers to the use of no dyes, Porganic refers to the percentage of organic fibers in each item, and Porg*Baby is the interaction between Porganic and the Baby dummy.

A potential drawback of our data is that prices are not realized prices since we do not consider actual transactions. The prices we use represent rational expectations of prices by apparel firms. To address this problem, we checked the stability of prices over time by revisiting our data sources at a later date for possible sales at discounted prices, although it is possible that some catalogs may discount and move their remainders onto other merchandisers in order not to spoil their reputation. We did not find evidence of poor expectations corrected by discounted sales decisions within catalogs. Another argument supporting the use of supply prices is found in Osborne and Smith. They find that posted and realized prices were very close in their investigation of the North Carolina beach rental markets. Other studies have used this "catalog-price" approach, ignoring realized data (e.g., Raff and Trajtenberg).[4]

Empirical Estimation and Results

The hedonic price equation is estimated for the apparel data using a semi-logarithmic functional form. Although there is no general consensus on a preferred "best" functional form, the semi-log form was chosen because it is widely used and an accepted practice in the hedonic literature (Palmquist; Graves et al.; Cropper, Deck, and McConnell).

The first goal of our empirical exercise is to identify the implicit price or premium associated with the organic cotton component of apparel. We present two sets of results with both the categorical and quantitative definitions of the organic attribute and using normalized prices. Using weighted least squares, the logarithm of the price of each piece of apparel, normalized by the pounds of cotton it contains, is regressed onto "core" attributes of the observations using the two following specifications:

$$\ln(\textit{price/fiber content}) = \beta + \boldsymbol{\alpha}^{T}\,\textit{items} + \boldsymbol{\gamma}^{T}\,\textit{catalogs} + \boldsymbol{\mu}^{T}\,\textit{gender} + \boldsymbol{\lambda}^{T}(\textit{age categories}) + \boldsymbol{\varphi}^{T}\,(\textit{dye types}) + \eta\,\textit{organic}$$

(specification 1 in table 1)

and

$$\ln(\textit{price/fiber content}) = \beta + \boldsymbol{\alpha}^{T}\,\textit{items} + \boldsymbol{\gamma}^{T}\,\textit{catalogs} + \boldsymbol{\mu}^{T}\,\textit{gender} + \boldsymbol{\lambda}^{T}(\textit{age categories}) + \boldsymbol{\varphi}^{T}\,(\textit{dye types}) + \eta_1\,\textit{percent organic} + \eta_2\,\textit{percent synthetic}$$

(specification 2 in table 1)

where the various categories are expressed as dummy variables for items (socks, T-shirts, etc.), catalogs, gender (male, female, unisex), age (baby, youth, adult), dye type (low impact or no dyes), organic cotton categories (conventional cotton in the intercept), and shares of organic cotton and synthetic fibers in total fiber content.

We use weighted least squares (WLS) for

[4] Catherine Kling reminded the authors that in pre-Rosen days, this approach was actually popular and called the "Sears Catalog Method."

two reasons. First, we want to obtain a better estimate of the aggregate premium by giving greater weight to items produced in larger quantities. Second, we believe that the variance in the price of items sold in small quantities may be greater than that of those sold in large quantities. Firms tend to use "small-quantity" items as "loss leaders" to attract customers. Doing the same with large-quantity items would be prohibitively expensive.[5]

The series of fixed effects by items, catalogs, and demographics captures the omitted attributes and cost structure, *B*, specific to each supplier and item type.[6] This approach is well accepted in hedonic regressions (for example, see Feenstra and Levinsohn). Table 1 shows that the estimated organic premium is robust to specification changes.

Table 1 reports results on the variables of interest for our investigation. The full results are available from the authors. As shown in column 1 of table 1, the sign on organic is positive as expected and significant at the $\alpha = 0.0001$ level. The parameter estimate is 0.292317 and the Halvorsen-Palmquist-Kennedy formula (Halvorsen and Palmquist, Kennedy) implies an average premium of 33.8%. The second column reveals similar results using the quantitative definition of the organic attribute. The estimated premium is 33.2% using the latter definition. The premium reflects cost differences between conventional and organic apparel. First, organic cotton costs more because it is more expensive to produce than conventional cotton (a 45% premium in 1995). Based on the cotton content of items, this difference in the cost of the raw fiber alone would result on average in an organic apparel premium of 12.5% (a weighted average over all items using the 1995 cotton price premium). Additional substantial costs originate in ginning and spinning as the gins and mills must be shut down and cleaned before organic cotton runs, usually small, are spun. The weaving/knitting process is the same for conventional and organic textile, but there are usually small-run penalties here as well. When the fabric is cut, a higher value of the scraps goes to waste. And, again, while the sewing process is the same, there are often small-run penalties. A consensus estimate of the increased cost of production prior to marketing is between 20% to 40%.[7] Finally, the premium has to be interpreted cautiously because it reflects the additional willingness to pay for the organic attribute for consumers willing to buy such items. Many consumers do not buy organic apparel, and our estimated premium could not be readily used to infer welfare implications.

Many of the organic cotton clothes also employed environmentally friendly dyes or no dyes. The two most common claims made were that the dyes used in the manufacturing process were low impact (LI) or that they used no dyes at all (ND). Our specification allows one to test if these claims add to the purchase price. Table 1 indicates that an LI claim does not add any significant premium. On the other hand if the manufacturer used no dyes at all, then one could expect a 14.8% discount in the purchase price. These two results appear to be driven by cost implications and are consistent with consensus estimates of the cost of dying textiles. Obviously, it is less expensive not to dye an item than the opposite. A median consensus estimate of the additional cost of dying an apparel item is about 15%. Dying cost is fairly homogenous across dye types although some colors are more expensive. The lack of significance of the "environmentally friendly dye" dummy probably reflects this lack of cost difference for different dyes.[8]

In the third and fourth columns of table 1 we report results for two specifications which allow for the interaction of ecological attributes with the baby category. Our hypothesis is that if there were a health component to the organic premium it would show up most strongly in baby clothes, as children are commonly assumed to be more susceptible to toxins. If present, the organic-baby interaction term in specifications three and four should capture this. Interestingly, there appears to be no significant additional premium for organic baby apparel in the weighted-least-squares estimation. Inspection of the quantity weights used reveals that the weights for nonbaby items are on average five times that of baby items. The observations in which such a baby premium may be present are heavily discount-

[5] We thank an anonymous referee for this suggestion.

[6] Demographics affect cost because the sophistication of given items and their cost tend to change with gender and age. For example, baby items use simpler cuts and stitching patterns than their adult counterparts.

[7] The figure of 20% was provided by K. Sweeney, Corporate Communication Office at Patagonia, and Trisha Feenly at Seventh Generation (now renamed Harmony) indicated that the cost increase is between 20% to 40% depending on items.

[8] These consensus figures and information on cost of production come from Jeff Kraus and J. Rucker of the North Carolina State University School of Textiles.

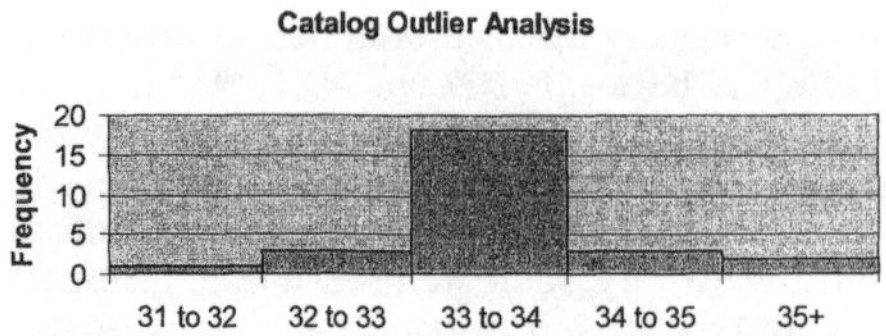

Figure 1. Price premium with catalog deletion

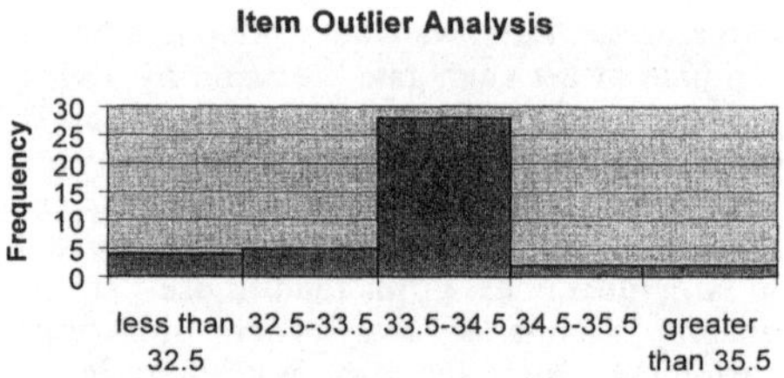

Figure 2. Organic price premium with item deletion

ed. While quantity weighting is judicious to obtain economically meaningful estimates of the aggregate premiums of the attributes, it becomes much harder to detect an additional organic-baby premium if one existed.[9]

We now report on the robustness of the organic premium result. We analyze three potential sources of influence: catalogs, items, and single observations. The item and catalog sources represent obvious candidates for group-outlier analysis, and these groups are less detectable running the usual outlier analysis which deletes one observation at the time (Belsley, Kuh, and Welsch).

The number of catalogs, especially conventional apparel catalogs, included in the data set is not exhaustive. A legitimate question is whether the inclusion of any one particular catalog is driving the results. If the results were sensitive to the inclusion/deletion of specific catalogs, then our results would lack generality and hence relevance. Group-outlier analysis focusing on the organic coefficient shows that dropping one catalog's observations at a time can generate an estimated premium ranging from 31.8% to 37.4% while the model predicts 33.8% with the full data set. This tight range suggests that no single firm is influential. Figure 1 plots the estimated organic premia after having deleted one catalog's observations at a time. A lower premium corresponds to a deleted organic catalog commanding a higher premium or to a deleted conventional catalog associated with low prices.

The histogram in figure 1 appears normal, suggesting that the sample chosen is a representative one. Since Patagonia is the largest player in the organic cotton apparel market, a priori, if any firm exerted market power with respect to this characteristic it would be Patagonia. The deletion of Patagonia causes a moderate fall in the estimated premium. There are several other catalogs whose deletion results in a larger drop.

To analyze this issue further, we calculate the DFBETAS (Belsley, Kuh, and Welsch) for the two catalogs whose deletion results in the largest decrease in the estimated premium. The Earth Care catalog results in the largest decrease, but only one of the six observations has a DFBETA statistic larger than 0.073 (the critical value). Similarly, the Back to Basics catalog, whose deletion results in the second largest reduction in the estimated premium, has only two of its seventeen observations providing significant DFBETA statistics. In summary, we cannot find a systematic outlying pattern associated with these two catalogs. If they commanded a brand-related premium it should be reflected in a majority of the observations, which is not the case. Furthermore, there are six firms whose deletion results in lower estimated premiums than that of Patagonia. The stylized facts suggest that if Patagonia doesn't wield any market power with respect to this characteristic, then it is unlikely that any of the others do either.

Next, we explore the sensitivity to the deletion of apparel items. A similar pattern emerges as in the catalog deletion exercise. Figure 2 provides a histogram of the organic price premia obtained in forty-one separate regressions, each deleting all the observations of one of the forty-one items. The majority of the estimated values fall around the central value (about 33.8%) and the chart suggests a bell-shaped distribution. Although the changes in the estimated premium fluctuate within a fairly tight band, one item group, sweatshirts, exhibits the strongest influence. Its deletion induces a jump in the premium to 40%.

[9] Under OLS, a substantial and significant premium emerges for organic baby clothes. Less confidence is attached to the estimated premiums in these regressions because of lack of weights. They are suggestive that such an additional premium may exist. Conversely, the OLS runs do not show any significant interaction between the dyes and baby dummies. Investigation based on actual sales data will be necessary to settle the question of potential health motives captured by an additional "baby organic" premium.

At the opposite tail, deletion of turtlenecks induces the largest drop in premium, to about 29%. Although the variation is not extreme, it indicates the importance of accounting for item-specific premia when analyzing pricing behavior. We incorporate this into our investigation of potential heterogeneous pricing by some suppliers.

Testing Pricing Behavior

As mentioned in the data section, we have two options to test the pricing behavior of organic apparel suppliers, which are based on the two alternative measures of the organic attribute. We present results based on both definitions. The test using the categorical definition is limited because we can only test the pricing behavior of four individual suppliers selling both conventional and organic apparel, with respect to the pricing behavior of all other organic apparel suppliers as a single pricing entity. As explained before, if a firm sells only organic apparel items, then the organic dummy everywhere equals 1 and the interaction term (organic-catalog) is perfectly collinear with the catalog dummy. The test is interesting, however, because two of the six catalogs, whose deletion creates the largest drop in the estimated premium in the outlier analysis, are included in the test. They are the Tomorrow's World and Natural Baby catalogs.

The second definition of the organic attribute allows us to test ten organic apparel suppliers who all use fiber blends against the other organic catalogs offering only 100% organic items. In this second approach we can also test Patagonia's pricing strategy. A set of linear restrictions imposes equality of the organic premium among all catalogs—each interaction term between the organic variable and the catalog dummy is restricted to be equal to zero for all tested firms (four in the first approach and ten in the second approach). If all the organic-catalog interaction terms are jointly equal to zero, then this is evidence that the organic markup is constant across the tested firms. Given the potential interaction between items and the organic attribute noted in the outlier analysis, we include an item by organic interaction term. Based on the first definition of the organic attribute, the joint F-test for the restriction that the four interaction terms are jointly equal zero indicated that the restriction cannot be rejected (p-value equal to 0.65), giving support for the equal-pricing hypothesis. The second approach provides a more comprehensive test of the "organic" pricing by ten separate firms. The F-test again indicates that this restriction cannot be rejected, offering further support for the similar pricing hypothesis. Given the evidence of the ease of entry in the organic apparel market, we see our evidence as consistent with competitive pricing as opposed to a similar markup above marginal cost for all the organic firms.

Conclusions

Our results appear promising. We identify a robust organic premium, which on average is about 33.8% of the apparel price expressed in dollars per unit of fiber content. We also found a discount for the "no-dye" attribute, which mostly reflects the cost savings resulting from this simplified production. However, we could not find any evidence of a premium associated with environmentally friendly dyes, and we did not find conclusive evidence of additional premia for baby-organic items. Hence, the health motive, if any, does not seem to be revealed in the pricing of baby organic items.

Influential data points were found, but the results relative to the organic premium appear robust, with most estimates of the premium locating around a central value of 33.8%. Our findings on eco-labels are specific to the apparel industry, and they await corroboration or refutation from ongoing investigation of eco-labels in other markets.

Finally, we find no evidence of heterogeneous pricing of the organic attribute by the organic-apparel firms included in our data set. With the provision of no barriers to entry, the premium for the organic attribute should tend toward a competitive level and should reflect the differences in cost of producing organic and conventional apparel. This reminds us of Perloff's "quasi-theorem à la Chicago" based on the observation that when entry is not restricted, competitive pricing is often rapidly achieved when the number of firms in an industry increases beyond "a few."

The circumstantial evidence gathered from industry sources on the cost implications of using organic fibers and dyeing processes in textiles supports the conjecture that our estimated organic premium and no-dye discount reflect these cost differences. Since our findings suggest that premia and discounts tend to reflect cost differentials, consumers appear pivotal in the determination of the equilibrium levels and types of environmental attributes embodied in apparel items. This implication rationalizes the presence of attributes that may not yield any positive premium but may be perceived as "required" by consumers and in turn may be deemed necessary by manufacturers.

We reflected on cost differences between conventional and organic apparel. Organic cotton is more expensive to produce than conventional cotton (a 45% premium in 1995). Based on the cotton content of items, this difference in the cost of the raw fiber alone would result, on average, in an organic apparel premium of 12.5% (a weighted average over all items using the 1995 cotton price premium), and would explain about 37% of the premium for the organic attribute. Our results provide some limited insight into the interaction between organic cotton and its use in apparel but do not tell us anything about how large the demand for organic cotton is or will be in the future. An investigation of the determinants of organic cotton supply and a better sense of the expansion of organic fiber use are necessary to assess the viability of the organic cotton market.

[Received March 1998; accepted February 1999.]

References

Apodaca, J. "Market Potential of Organically Grown Cotton as a Niche Crop." Natural Fibers Research and Information Center, Bureau of Business Research, University of Texas at Austin. Paper presented at the Beltwide Cotton Conference in Nashville TN, January 1992.

Belsley, D.A., E. Kuh, and R.E. Welsch. *Regression Diagnostics.* New York: John Wiley & Sons, 1980.

Bowen, N. *World Agrochemical Markets.* London: PJB Publications, March 1991.

Burnett, P.C., P.J. Wakelyn, and M.B. Gordon. "Availability and Supply of Natural Colored Cottons and Organic Cotton and Other Considerations." Paper presented at the ITMF Joint Cotton Committee Meeting, San Paulo, Brazil, 18 September 1994.

Cropper, M.L., L.B. Deck, and K.E. McConnell. "On the Choice of Functional Forms for Hedonic Price Functions." *Rev. Econ. and Statist.* 70(November 1988):668–75.

Feenstra, R.C. "Exact Hedonic Price Indexes." *Rev. Econ. and Statist.* 77(November 1995): 634–53.

Feenstra, R.C., and J.A. Levinsohn. "Estimating Markups and Market Conduct with Multidimensional Product Attributes." *Rev. Econ. Stud.* 62(January 1995):19–52.

Freeman, A.M., III. *The Measurement of Environmental and Resource Values. Theory and Methods.* Washington DC: Resource for the Future, 1993.

Graves, P., J.C. Murdoch, M.A. Thayer, and D. Waldman. "The Robustness of Hedonic Price Estimation: Urban Air Quality." *Land Econ.* 64(August 1988):220–33.

Halvorsen, R., and R. Palmquist. "The Interpretation of Dummy Variables in Semilogarithmic Equations." *Amer. Econ. Rev.* 70(June 1980):474–75.

Imhoff, D. "Growing Pains: Organic Cotton Tests the Fiber of Growers and Manufacturers Alike." *Farmer to Farmer,* December 1995. Reprinted online, available HTTP: http://www.simplelife.com (March 1998).

Imhoff, D., L. Grose, and R. Carra. "Organic Cotton Exhibit." *Simple Life,* mimeo distributed by the Texas Organic Cotton Marketing Cooperative, O'Donnell TX, 1996.

Kennedy, P.E. "Estimation with Correctly Interpreted Dummy Variables in Semilogarithmic Equations." *Amer. Econ. Rev.* 71(September 1981):801.

Michaels, R.G., and V.K. Smith. "Market Segmentation And Valuing Amenities With Hedonic Models: The Case of Hazardous Waste Sites." *J. Urban Econ.* 28(September 1990): 223–42.

Nelson, P. "Information and Consumer Behavior." *J. Polit. Econ.* 78(March–April 1970): 311–29.

Organization for Economic Cooperation and Development (OECD). "Eco-labelling: Actual Effects of Selected Programmes." OCDE/GD (97) 105, Paris, 1997. Online, available HTTP: http://www.oecd.org/env/eco/books.htm#trademono.

Osborne, L. "Market Structure, Hedonic Models, and the Valuation of Environmental Amenities." PhD dissertation, North Carolina State University, 1995.

Osborne, L., and V.K. Smith. "Environmental Amenities, Product Differentiation, and Market Power." Mimeo, 1997.

Palmquist, R.B. "Hedonic Methods." *Measuring the Demand for Environmental Improvement.* J.B Braden and C.D. Kolstad, eds. Amsterdam: Elsevier, 1991.

Palmquist, R.B., F.M. Roka, and T. Vukina. "Hog Operations, Environmental Effects, and Residential Property Values." *Land Econ.* 73(February 1997):114–24.

Perloff, J. "Industrial Organization Lecture Notes." Mimeo, University of California-Berkeley 1985.

Raff, D.M.G., and M. Trajtenberg. "Quality-Adjusted Prices for the American Automobile Industry: 1906–1940." NBER Working Paper Series, Working Paper No. 5035, February 1995.

Rosen, S. "Hedonic Prices and Implicit Markets: Product Differentiation in Pure Competition." *J. Polit. Econ.* 82(January–February 1974):34–55.

Teisl, M.F., B. Roe, and R.L. Hicks. "Can Eco-labels Tune a Market? Evidence from Dolphin-Safe Labeling." Paper presented at AAEA annual meeting, Toronto ONT, 27–30 July 1997.

U.S. Environmental Protection Agency (U.S. EPA). *National Water Quality Fact Inventory: 1990 Report to Congress.* EPA 503-9-92-006, April 1992.

Wasik, J.F. *Green Marketing and Management: A Global Perspective.* Cambridge MA: Blackwell Business, 1996.

Zarrilli, S., V. Jha, and R. Vossenaar, eds. *Eco-labelling and International Trade.* New York: St. Martin Press, 1997.

[11]

Can Eco-Labels Tune a Market? Evidence from Dolphin-Safe Labeling[1]

Mario F. Teisl[2]

Department of Resource Economics and Policy, University of Maine

Brian Roe

Department of Agricultural, Environmental and Development Economics, Ohio State University

and

Robert L. Hicks

Department of Coastal and Ocean Policy, The College of William and Mary

Received July 29, 1997; revised August 16, 2000; published online August 21, 2001

In this paper we test whether the dolphin-safe labels altered consumer purchases of tuna. We also provide a partial measure of the total welfare effects of the dolphin-safe labeling policy. The results confirm our hypothesis that the dolphin–tuna controversy and the subsequent implementation of dolphin-safe labeling affected consumer behavior. Further, the paper provides market-based evidence that consumers can respond to eco-labels; the dolphin-safe label increased the market share of canned tuna. The welfare analysis provides a partial measure of society's willingness to pay to avoid personally contributing to dolphin mortality as a result of tuna fishing.

INTRODUCTION

The environmental characteristics of products have become increasingly important to consumers [40, 52]. Firms have responded by placing eco-labels on products that highlight the item's environmental attributes and by introducing new, or redesigned, "green" products [50]. Governments and nongovernmental organizations have also responded by organizing, implementing, and verifying eco-labeling programs that cover thousands of products in more than 20 countries [51], while international efforts to standardize environmental labeling schemes have also emerged [12, 22]. From a policy perspective, one aim of eco-labels is to educate consumers about the environmental impacts of the product's manufacture, use, and disposal, thereby leading to a change in purchasing behavior and ultimately, to a reduction in negative impacts. Further, eco-labeling policies may promote environmental objectives without production site command and control methods and are

[1] Maine Agricultural and Forest Experiment Station Number 2449.

[2] Address correspondence to: Mario Teisl, Department of Resource Economics and Policy, 5782 Winslow Hall, University of Maine, Orono, ME 04469-5782.

seen as a way of meeting global environmental objectives while complying with international trade agreements.[3]

In order for eco-labels to achieve policy objectives, consumers must hold preferences for certain environmental amenities and respond to the information presented on eco-labels by altering purchases toward eco-labeled goods. Their widespread use suggests that eco-labeling is perceived as an effective method of altering consumer behavior. However, few studies have attempted to identify the behavioral effectiveness of eco-labeling programs. Evaluating the policy effectiveness of eco-labeling programs requires understanding how information affects market behavior. Evaluating the economic efficiency of labeling programs requires measuring the benefits and costs of such information.

In this paper we measure the policy effectiveness of dolphin-safe labeling of canned tuna by estimating a demand system for the canned protein market by using retail level data and testing whether the dolphin-safe labels altered consumer purchases of tuna. To gain insight into the efficiency effects of information we use the estimated demand system to provide a partial measure of the total welfare effects of the dolphin-safe labeling policy using the assumption that any shifts in demand resulting from the dolphin-safe labels would reflect a type of use value.[4] Our welfare measure is partial in the sense that it only captures the values of individuals who are canned tuna purchasers. Other individuals may also obtain a welfare gain due to the increased protection of dolphin brought about by the dolphin-safe labeling of tuna.

I. BACKGROUND AND LITERATURE REVIEW

A. Labeling and Consumer Choice

Although implementation of eco-labels is widespread, research concerning its impact and effectiveness is limited and aggregate quantitative results are rare. Much of the research has measured effectiveness either by identifying changes in consumer awareness after exposure to label information [25, 26, 55] or by asking consumers whether eco-labeling programs would affect their purchase behavior [13]. However, a change in awareness does not necessarily translate into a change in behavior [6, 49] and consumers do not necessarily follow their own purchasing assertions [2, 23].

Market-based research investigating other types of labels (e.g., nutrition labels, quality seals) has demonstrated that a change in labeling can change market behavior [38, 44]. However, the apparent effectiveness of these (non-eco) labeling programs may not be applicable to eco-labeling because these non-eco-labels

[3] The dolphin-safe label used on tuna is an example of how an eco-label can bypass trade barriers. Under the Marine Mammal Protection Act, which bans the import of tuna when incidental dolphin kills exceed certain limits, the U.S. placed an embargo on Mexican tuna in February 1991, followed by embargoes on tuna from 11 other countries. In September 1991, a panel of the General Agreement on Tariffs and Trade (GATT) agreed with Mexico that the U.S. embargo violated the GATT agreement. The Dolphin Protection Consumer Information Act, passed in 1990, enforces a dolphin-safe standard in labeling and circumvents the GATT ruling.

[4] Bockstael and Stand [5] hypothesize this value reflects a change in the moral quality of the tuna. That is, by purchasing dolphin-safe tuna, the consumer can avoid personally contributing to dolphin mortality.

provide information about the characteristics of the product. Eco-labels, like the dolphin-safe labels studied here, differentiate products with respect to characteristics of the product's production.

B. Information and Environmental Valuation

It has been well established in the stated-preference literature that the provision of information affects the value that a survey respondent places upon an environmental commodity [1, 3, 8]. The issue for researchers in this area is providing the appropriate amount of information such that respondents have a clear definition of the public good they are supposed to value without affecting the underlying preferences or causing certain information biases [14]. That is, the information set used by the respondent to value the commodity of interest should reflect the information set that would be invoked during the choice process in a true market, if such a market existed for the commodity in question. The availability of eco-labeling turns this concern about appropriate information disclosure on its side. Instead of treating environmental information as something that must be controlled within a survey instrument so as not to bias resulting value estimates, eco-labeling makes information disclosure a policy variable which may be used to achieve certain social objectives. Effectively, labeling decreases the search cost for the information and may signal the importance of the environmental information. Hence, labeling may affect behavior by influencing the number of attributes that a consumer considers during a choice occasion. Furthermore, labeling may affect the implicit weights that consumers assign to each attribute. Rather than trying to minimize valuation biases due to information, regulators may find it socially optimal to supply environmental information via labeling such that market behavior is affected.

C. The Tuna–Dolphin Issue

Canned tuna is the most important seafood product in the United States in terms of quantity consumed [48]. Much of this tuna comes from the Eastern Tropical Pacific (ETP hereafter), which accounts for almost one-fourth of the world's tuna catch [41]. It is well known that dolphins often associate with large yellowfin tuna in the ETP. With adoption of purse seine technology[5] approximately 40 years ago, dolphins have been targeted by fishermen in the ETP as a way of finding and netting tuna. Once located, fishermen capture tuna schools by encircling the dolphins with the seine, tightening the seine, and then hauling the tuna onboard. Dolphins can be killed during the process, and during the early years of purse seining no industry-wide effort was made to mitigate dolphin deaths.

From 1960 to 1972 the National Research Council estimates that on average more than 100,000 dolphins were killed each year by the U.S. tuna fleet [41] and, in the early 1970s, biologists warned that commercial tuna operations in the ETP threatened dolphin populations. However, several governmental initiatives helped decrease dolphin mortality. The Marine Mammal Protection Act was passed in

[5] The seines used in the ETP are very large nets, approximately 1 mile long and 600 feet deep. The purse seine technology involves encircling the tuna with this net, closing the bottom of the net with a purse line, and hauling the catch onboard [41].

1972 and by 1975, in response to an amendment to the Marine Mammal Protection Act, the U.S. tuna fleet began preventive measures aimed at reducing dolphin kills. The 1988 reauthorization of the Marine Mammals Protection Act insured that there was full observer coverage on U.S. flag vessels. The observers documented if dolphins were encircled or killed as a result of fishing operations and provided the means to eventually institute a dolphin-safe labeling and verification program. Many of the member nations of the Inter-American Tropical Tuna Commission also instituted or expanded their own observer programs in subsequent years. Consequently, the National Marine Fisheries Service and the Inter-American Tropical Tuna Commission were able to verify dolphin related activities for each vessel including logbook data each time a vessel set its net in the water (commonly referred to as "a set"). The cumulative effect of these changes in terms of dolphin mortality has been dramatic. Dolphin deaths due to tuna fishing dropped to 25,000 in 1991 [41], with current mortality levels of less than 5000 per year [43].

Also during the late 1980s consumers became aware that the harvest of yellowfin tuna in the ETP caused the incidental mortality of dolphins. Media attention to the issue spawned significant controversy and, eventually, calls for consumer boycotts of canned tuna. The high-water mark of media attention and public concern coincided with the national televising of the Sam La Budde (Earth Island Institute) video which showed dolphins dying as a result of fishing operations in the ETP. The three largest tuna canners in the U.S. market announced a dolphin-safe labeling policy in April 1990 [45]. Other canners quickly followed. The U.S. government responded by passing the Dolphin Protection Consumer Information Act of 1990 which mandates that tuna products cannot be labeled as "dolphin safe" unless dolphins were not used to capture tuna for the entire fishing trip, as verified by a sanctioned observer aboard the boat. The combination of U.S. laws and embargoes in effect closed the U.S. market for tuna caught using dolphins [31]. Internationally, a nonprofit wildlife conservation group began "The Flipper Seal of Approval" [52], a dolphin-safe labeling program.

The International Dolphin Conservation Program Act [43] was passed as a domestic endorsement of the La Jolla Agreement and the Panama Declaration signed by participants in the ETP tuna fishery. These agreements set quotas on dolphin kills and reduce quotas through time. Provided that producers harvest tuna in a manner consistent with these agreements, the Act also allows for the tuna from foreign producers to be imported into the United States. Under the Act, the definition of dolphin safe is now much less restrictive, allowing fishermen to encircle dolphins while capturing tuna so long as no dolphins are killed.[6] Additionally, the dolphin-safe definition applies on a set by set basis rather than for the entire trip.[7] Under the old dolphin-safe language, one set of targeting dolphins would cause all tuna caught from the trip to fail to meet the dolphin-safe definition.

Also during the late 1980s there was a significant change in the quality of canned tuna associated with the change in tuna harvesting techniques. Tuna caught in association with dolphins tends to be large yellowfin tuna that is ideally suited for

[6] If the Secretary of Commerce finds that fishing activities are "having a significant adverse impact on any depleted stock," producers are required to not encircle and not kill dolphins while fishing in order to meet a more restrictive dolphin-safe definition.

[7] Vessels are required to maintain a separate storage well for tuna that is dolphin safe.

canning and is of high quality [53]. As fishermen decreased fishing effort directed at dolphins, the composition of their catch changed from large yellowfin to smaller yellowfin and skipjack tunas of relatively lower quality for canning.

Evidence that consumers reacted to the dolphin-safe label is difficult to find. A study by Wallstrom and Wessells [54], based on data from scanned grocery store transactions in metropolitan areas, documents an initial negative market reaction to public exposure of tuna harvesting practices in 1988 but found no significant demand effect from the labeling program. Bockstael and Strand [5] also found no significant label effect on U.S. demand for Latin American tuna imports. In addition to specific problems,[8] both studies use data with relatively short postlabel time series (30 months). However, demand shifts due to a change in labeling may occur over a relatively long time period as the label is noticed by and the information diffuses through the population [38]. In addition, consumers who notice the label may initially doubt the veracity of the label information. Anecdotal evidence suggests that it may have taken up to four years for consumers to gain full confidence in the dolphin-safe label. The data set used in this study has 66 months of postlabel time series, which allows testing for time dependent label effects.

II. THEORETICAL FRAMEWORK

To provide a modeling framework to measure changes in consumer behavior and welfare due to changes in environmental information one needs to know how environmental awareness enters an individual's utility function (here defined in terms of a purchase occasion or decision). The utility evaluation can be represented by the indirect utility function

$$V^S = V(\mathbf{A}^S, \mathbf{q}, Y, \mathbf{p}), \tag{1}$$

where $\mathbf{A}^S$ denotes a vector of environmentally related assessments for m products given information set S (i.e., $\mathbf{A}^S = [A_1^S, \dots, A_m^S]$), $\mathbf{q}$ denotes a vector of other quality characteristics (e.g., taste or texture), $\mathbf{p}$ is a corresponding vector of prices, and Y denotes income. V^S is increasing in $\mathbf{q}$ and Y, and decreasing in $\mathbf{p}$.

The technology that extracts and translates environmental information into an assessment of a product's environmental impact can be viewed as a "household production" process by which an individual combines his or her prior environmental knowledge, cognitive abilities, time, and the environmental information presented during the purchase decision. Thus, we could model the assessment process during the purchase decision as

$$A_j^S = f(S_j, G, t_j), \tag{2}$$

where A_j^S denotes the (subjectively) assessed environmental impact of purchasing good j given information set S, S_j is the environmental information displayed about product j at the point of purchase (e.g., a dolphin-safe label on a can of

[8] A potential problem with the Wallstrom and Wessells study [54] is their approach in estimating demand (use of least squares on single-equation models without apparent testing/correcting for autocorrelation), leading to a probable omitted variable bias and misidentification of label effects. The Bockstael and Strand study [5] may be limited by trying to estimate changes in consumer demand with import/export data.

tuna), G denotes the consumer's prior stock of environmental information which may include information from news accounts, firm-provided advertising, and public education campaigns, and t_j denotes the time that the individual devotes to processing S_j.

The objective level of the environmental impact characteristics represented by the information variable S is denoted by θ. For example, if S represents a dolphin-safe claim on a canned tuna label, then θ denotes that the production of the tuna led to no actual dolphin deaths. θ is separate from the assessment function because the individual does not observe it at the time of purchase except through the variable S. Although θ may be unobservable to the consumer at the time of the purchase decision, we include it within the discussion to distinguish between the factor that affects consumer decisions, S, and the one that ultimately determines the environmental impact of production, θ.

The subject of interest is to measure the welfare effect of a change in information that accurately reflects a change in the true environmental impact of the good (e.g., the provision of dolphin-safe labeling of tuna that corresponds to a change in commercial tuna fishing practices). We begin by presenting expenditure functions under two alternative states of information. Denote

$$e(\mathbf{A}^0(\cdot), \mathbf{q}, U, \mathbf{p}) \tag{3}$$

as the expenditure function when the environmental impact of a product is at level 0 and the information about the environmental impact accurately reflects that level ($S = 0$ and reflects the state of θ). Likewise,

$$e(\mathbf{A}^1(\cdot), \mathbf{q}, U, \mathbf{p}) \tag{4}$$

is the expenditure function when the environmental impact of a product is at level 1 and the information about the environmental impact reflects the new level ($S = 1$ and reflects the state of θ).

Although not commonly framed as such, compensating variation (CV) measures the change in individual welfare when the qualities of a good changes (denoted as a move from $\theta = \theta^0$ to $\theta = \theta^1$) along with a corresponding change in information about the quality change ($S = 0$ to $S = 1$). During a purchase occasion, CV can be expressed as

$$\mathrm{CV} = e(\mathbf{A}^1(\cdot), U^0, \mathbf{p}^0) - e(\mathbf{A}^0(\cdot), U^0, \mathbf{p}^0), \tag{5}$$

where U^0 denotes the initial utility level (q is dropped for simplicity).

If the change in θ is positive then $\mathrm{CV} \geq 0$. Here CV measures the maximum amount of money an individual is willing to pay to gain the better quality, given that he or she knows about the quality change (i.e., he or she has the correct information about the quality change so that he or she is free to alter consumption). Alternatively, if the change in θ is negative then $\mathrm{CV} \leq 0$, which denotes the maximum amount of money an individual must be given to accept the (known) poorer quality.

III. THE MODEL

Typically, economists do not have complete price and quantity information for all goods and must be satisfied estimating demand systems for a subset of the universe of goods. Our analysis of the impact of the dolphin-safe label focuses on canned tuna and its substitutes. Although the data here are complete with respect to expenditures on canned tuna and its hypothesized substitutes, they do not include market-related information on all other goods. A common approach to demand estimation under this lack of information is to assume a weakly separable utility function and estimate a set of conditional demand functions which are a function of the prices of the goods of interest and of expenditures on the subset of goods.[9] A problem with conditional demand functions is that the conditional demand systems can produce welfare measures that are biased [36, 37]. However, the conditional compensating variation measure is conservative in that it always understates the true compensating variation [37].[10]

For the problem at hand, it is essential that a utility-theoretic demand model be estimated so that the relevant parameters of the underlying expenditure function can be recovered. The demand system was estimated using Deaton and Muelbauer's almost ideal demand system (AIDS) expanded to include information effects [16, 42]. The AIDS model is chosen because it satisfies the axioms of choice, allows imposition of adding up, homogeneity, and Slutsky symmetry restrictions, allows some forms of aggregation, and is integrable [15]. In addition, AIDS models have been shown to be equivalent or superior to other common demand specifications, e.g., translog [39]; log–translog, CES-transformed AIDS [10]; Rotterdam, CBS [18]; LES, indirect addilog, general dynamic AIDS, indirect translog [20]; and linearized AIDS [7].

The demand system includes canned tuna and three substitute meat products: shelf-stable luncheon meats (both sliced and deli-pouch), other canned seafood products (salmon, shrimp, and sardines), and canned red meat (beef, barbecued beef, corned beef, corned beef hash, deviled ham, potted meat spread, other meat spreads, spiced meats).[11] We do not separate canned tuna into dolphin-safe (labeled) and non-dolphin safe (unlabeled) segments because virtually all tuna marketed after 1990 was dolphin safe (labeled). That is, at no point in time is there any discernible cross-sectional variation with respect to dolphin-safe labeling.

The expanded AIDS model begins with the household's conditional expenditure (hereafter, expenditure) function

$$\log\{e_h(p, A(\cdot), \Psi, U)/K_h\} = a(p, A(\cdot), \Psi) + b(p)U, \qquad (6)$$

[9] Weak separability is a necessary and sufficient condition for the existence of conditional demand functions [35].

[10] An alternative approach is to specify an incomplete system of demand equations that include equations representing the goods of interest with an additional equation representing spending on all other goods (through the construction of a composite commodity). When the demand system is integrable then this approach can be used to calculate exact welfare measures from the parameters of the incomplete demand system [33, 34]. We do not follow this approach because we lack sufficient data to construct this composite commodity.

[11] Either of two conditions allows aggregation across goods: (1) when the prices of the group of goods to be aggregated move in parallel or (2) when the utility function is weakly separable across the different groups of aggregated goods [15].

where p denotes prices, $A(\cdot)$ denotes the awareness function which may be influenced by label and media-provided information, Ψ denotes other demand influences (such as product seasonality), U denotes utility, and K_h is a general measure of household size to deflate the budget of the household to a "needs-corrected" per capita basis [15]. Here, during time t,

$$a(p, A(\cdot), \Psi) = \alpha_0 + \sum \alpha_i \log P_{it} + (1/2) \sum \sum \gamma_{ij} \log P_{it} \log P_{jt},$$

$$\alpha_i = \left\{ \xi_i + \tau_i T_t + \rho_{1i} M_t + \rho_{2i} V_t + \phi_i L_t + \sum \omega_j A_{jt} + \theta_{1i} iRt + \theta_{2i} R_{2t} \right\},$$

and

$$b(p) = \beta_0 \Pi P_i^{\beta_j},$$

where T_t is a time trend, M_{it} denotes the number of newspaper and magazine articles discussing the tuna/dolphin issue and is included to capture any changes in demand due to news media coverage of the tuna-dolphin controversy, and V_t denotes a three-month time period immediately following the news release of the Sam La Budde video. L_t denotes the diffusion of the label information across the population. The functional form of the label information diffusion process is modeled as a logit CDF occurring over an 18-month period. Specifically, L_t is zero pre-April 1990, follows a logit probability curve until November 1991 and is equal to one afterward.[12] This S-shaped functional form is similar to others modeling information diffusion (e.g., see [9, 28]). A_{jt} denotes the level of retail-sponsored advertising used to promote the jth good; R_{1t} and R_{2t} represent "religious–seasonal" indicator variables. Specifically, R_1 is equal to one during the months of February and March (Lent), zero otherwise; R_2 is equal to one during the month of December (winter holidays), zero otherwise; and P_{jt} is the price of (composite) good j.

Taking the derivative of (6) with respect to log P_i and appropriate substitution of U provides the expenditure share equation for the ith good during time t,

$$W_{iht} = \xi_i + \tau_i T_t + \rho_{1i} M_t + \rho_{2i} V_t + \phi_i L_t + \sum_j \omega_{ij} A_{jt}$$
$$+ \theta_{1i} R_{1t} + \theta_{2i} R_{2t} + \sum_j \gamma_{ij} \log P_{jt} + \beta_i \log(X_h / K_h P^*)_t, \qquad (7)$$

where W_{iht} is the share of household expenditure on good i, X_h is household expenditures on the goods in the system, and $\log P^* = a(p, A(\cdot), \Psi)$.[13]

[12] We considered two other functional forms to represent the label effect on budget shares. One specification included a dummy variable to denote the change in the information policy along with an interaction term between the time variable and the label policy dummy. With this specification both the label-dummy variable and the label–time interaction term were significant and positive, indicating a label effect. However, the parameter on the label dummy was relatively small, indicating that the effect of the label was small immediately after the labels entered the market but that the effect continued to grow over time. In the second specification, information diffusion was modeled as a log of the time trend variable. In this specification the label parameters were not significant. The presented specification allowed for the best fit of the model.

[13] Most empirical studies using AIDS models have used the Linearized AIDS model which substitutes P^* with Stone's price index. Although simpler to estimate, the Linearized AIDS provides inconsistent parameter estimates [7].

A limitation often encountered by researchers is that data on market purchases are often aggregated at some level. A benefit of the AIDS framework is that, as a member of the price independent generalized log (PIGLOG) class of expenditure functions, it fulfills the conditions required for exact nonlinear aggregation. That is, the share equations and the expenditure function derived from the AIDS model can be seen as coming from a single representative household. Thus the parameters of a household's expenditure function can be (approximately) recovered even though the share equations are estimated using aggregate data.

Following Deaton and Muelbauer, the individual household share equations can be aggregated across households by multiplying W_{iht} by individual household expenditure, summing over all households, and dividing by the aggregate expenditure,

$$W_{it} = \alpha_i^* + \sum_j \gamma_{ij} \log P_{jt} + \beta_i \log(X_t/P_t^*), \tag{8}$$

where W_{it} is the share of aggregate expenditure spent on good i in the aggregate expenditure of all households, $\alpha_i^* = \Sigma_h\{X_{ht}\alpha_i/\Sigma_h X_{ht}\}$, and $\log X_t = \Sigma_h\{X_{ht}\log(X_{ht}/k_h)/\Sigma_h X_{ht}\}$.

Deaton and Muelbauer note that one can find an aggregate index, K, such that $\log(\mathbf{X}/K) \equiv \Sigma_h\{X_{ht}\log(X_{ht}/k_h)/\Sigma_h X_{ht}\}$, where $\mathbf{X}$ is average household expenditure and $\mathbf{X}/K$ is the representative expenditure level. If cross-sectional data on X_{ht} and k_h are available then one could calculate a value for K. Typically, this type of data is unavailable so economists have generally approximated $\mathbf{X}/K$ with per capita "conditional" expenditures (i.e., total expenditures on the goods within the system/population). Given the above, the AIDS model provides a method of approximating the change in welfare of a "representative household" while using aggregate data.

We estimate the above nonlinear system of equations using iterated seemingly unrelated regression. We correct for potential autocorrelation by following the procedures outlined by Berndt and Savin [4, 42]. Denote the system of four share equations as

$$\mathbf{w}_t = \Pi \mathbf{Z}_t + \mathbf{v}_t, \tag{9}$$

where $\mathbf{w}_t$ is the 4×1 vector of dependent variables, Π is the $4 \times k$ matrix of unknown parameters, $\mathbf{Z}_t$ denotes the $k \times 1$ vector of variables denoted by the right-hand side of (8), and $\mathbf{v}_t$ is the 4×1 vector of errors. Assume the vector of errors exhibits first-order autocorrelation; i.e.,

$$\mathbf{v}_t = R\mathbf{v}_{t-1} + \mathbf{e}_t, \tag{10}$$

where $\mathbf{R}$ is a (4×4) matrix of unknown parameters, r_{ij} $(i, j = 1, \ldots, 4)$, and the $\mathbf{e}_t$ are independent, normally distributed random vectors with mean vector zero and covariance matrix Ω. Because of the adding up condition, $(\zeta'\mathbf{w}_t = 1$, where ζ is a 4×1 vector of ones) one of the share equations is dropped during estimation. The adding up condition implies $\zeta'\mathbf{v}_t = 0$, and since $\mathbf{v}_{t-1}$ and $\mathbf{e}_t$ are independent, it also follows that $\zeta'\mathbf{R} = k$, where k is an unknown constant, and $\zeta'\mathbf{e}_t = 0$. Deleting the last equation from (9) and (10) gives

$$\mathbf{w}_t^* = \Pi^* \mathbf{Z}_t + \mathbf{v}_t^* \tag{11}$$

and

$$\mathbf{v}_t^* = \mathbf{R}^* \mathbf{v}_{t-1} + \mathbf{e}_t^*, \tag{12}$$

where $\mathbf{w}_t^*$, $\mathbf{v}_t^*$, and $\mathbf{e}_t^*$ are equal to $\mathbf{w}_t$, $\mathbf{v}_t$, and $\mathbf{e}_t$ with the last element removed and Π^* and $\mathbf{R}^*$ are equal to Π and $\mathbf{R}$ with the last row removed. The above system of equations cannot be estimated because the $\mathbf{R}^*$ matrix is not square (3×4). However because $\zeta' \mathbf{v}_t = 0$, we can rewrite $\mathbf{R}^*$ as $\underline{\mathbf{R}}^*$ where $\underline{\mathbf{R}}^*$ is a square (3×3) matrix with elements $\underline{r}_{ij} = r_{ij} - r_{i4}$ $(i, j = 1, \dots, 3)$. Allowing for the most general specification of the autocorrelation parameters only requires that each column of the $\underline{\mathbf{R}}^*$ matrix sums to zero.

The models were estimated both with and without the adding-up ($\Sigma \xi_i = 1$; $\Sigma \tau_i = 0$; $\Sigma \phi_i = 0$; $\Sigma \rho_{1i} = 0$; $\Sigma \rho_{2i} = 0$; $\Sigma \omega_{ij} = 0$; $\Sigma \theta_{1i} = 0$; $\Sigma \theta_{2i} = 0$; $\Sigma \gamma_{ij} = 0$), homogeneity ($\Sigma \beta_i = 0$), and symmetry ($\gamma_{ij} = \gamma_{ji}$) conditions imposed on the system. Using the joint test procedure of Gallant [17], we found no significant difference between the restricted and unrestricted models ($\chi^2_{(0.05, 16)} = 24.24$). Reported results are based on the restricted (theoretically consistent) models.

There is some question on whether to restrict α_0; Deaton and Muelbauer suggest that α_0 be restricted to some small positive number because this parameter can be interpreted as the natural log of subsistence monthly outlay on goods in the system when prices are unity (i.e., $\log e(\cdot) = \alpha_0$ when $U = 0$ and all $P_{jt} = 1$). In addition, many published articles using AIDS models do not report whether they impose a restriction on α_0, nor do they commonly report an estimate for α_0. In response, we estimated the model without restricting α_0; the unrestricted α_0 was estimated as equal to -4.6, significant at the 5% level. We also estimated the model by restricting α_0 to equal zero. A Chow test indicates that the vector of parameters across the restricted and unrestricted models are not significantly different ($\chi^2_{(0.05, 1)} = 0.75$). However, similarity in parameter vectors does not necessarily imply similarity in welfare estimates.[14] Given the above we report both sets of estimation results.

Identification of structural shifts in a time series due to a change in information is often problematic. Many authors (e.g., [29, 30]) simply assume that changes in parameter vectors are due to a change in information availability. However, estimated parameters may be spuriously correlated with nonlinear trends in the data series. We use a "switching regressions" approach [32] to test whether our hypothesized label variable correctly identifies the timing of the structural break in the data (announcement of the dolphin-safe tuna-labeling program in April 1990). In this approach, the demand system is estimated repeatedly with different possible structural breakpoints. The estimation providing the largest likelihood value is then compared with a model estimated with no breakpoint (i.e., a demand system estimated with no label variable). The likelihood values indicate that our hypothesized breakpoint does in fact provide the best fit of the data (Table I). We also find that this model is significantly different than a demand system estimated without a label variable.[15]

[14] We thank an anonymous reviewer for this insight.

[15] A CUSUM [24] test was also performed on the tuna equation; the analysis confirmed a structural change in the data beginning in the 25th time period (the month during which the dolphin-safe labeling policy was announced). Results of this analysis are available from the first author.

TABLE I
Selected Likelihood Values Testing the Timing of the Label Variable[a]

Time period	Likelihood value
20	−49.11
23	−47.94
24	−46.74
25	-45.88^{A}
26	-45.98^{A}
27	-47.06^{B}
30	-47.39^{B}
No label	-47.21^{B}

[a] Likelihood values sharing a superscript are not significantly different from each other.

The AIDS model provides estimated share equations that can be used to calculate the market effects of the labeling program and to provide estimates of the relevant parameters of the expenditure functions. The expenditure functions thus derived can be used to calculate the CV measure for a representative consumer.[16, 17] By substituting the parameters of the demand system into expression (5) we obtain an estimate for the value of dolphin-safe labeling for a representative canned tuna user:

$$CV = X\left[\exp\left[\sum_i \{\{\phi_i L_t\}^* \log P_{it}\}\right] - 1\right]. \tag{13}$$

This CV is the average amount of money households would need to be compensated per month if the labeling program had not existed. Given that prices and the labeling effect (movement along the logistic information diffusion curve) are not stationary through time, CV will also change through time as prices change and the information about the label is processed by more consumers. This measure of CV is not equivalent to a hypothetical valuation approach that asked the question, "How much would you pay to prevent all dolphin mortality in the ETP from commercial fishing operations?" The dolphin-safe label does not ensure consumers that no dolphins are being killed in the ETP as a result of fishing operations; it ensures the consumer that no dolphins were killed as a result of fishing operations corresponding to a purchase of canned tuna.

A confidence interval (CI) around the mean CV estimate is constructed using a Taylor-series expansion [19]. Specifically, $CI = CV \pm t_{(1-\alpha)/2}\sqrt{\mathbf{ZCZ'}}$, where $\mathbf{Z}$ de-

[16] To confirm that our underlying utility function was well behaved we tested the concavity of the expenditure function using the procedure outlined in [11]. Three of the four eigenvalues were consistent with negative semidefiniteness (−0.1163, −0.0737, −0.0037). Although the fourth eigenvalue was positive it was small (0.0007).

[17] LaFrance and Hanemann [33] show that estimation of the welfare effects due to a change in a nonprice parameter is only valid if one assumes that the constant of integration is not affected by changes in the nonprice parameter. Unfortunately, this assumption has no behavioral consequences and is thus untestable. As a result, "it is generally impossible to measure unequivocally welfare changes from non-market effects using incomplete systems of market demand functions" (p. 272).

notes a row vector composed of the partial derivatives of the CV expression with respect to the estimated parameters (here, $\partial CV/\partial \phi_i$) and **C** denotes the covariance matrix of the parameters.

IV. DATA

Market-related information (market share, retail-sponsored advertising indices, and share prices) was obtained from SCANTRACK® scanner data supplied by the A.C. Nielsen Marketing Research Company. The scanner data include monthly sales, price, and retail-support information for all scannable food items in 3000 supermarkets in the United States with at least $2 million in annual sales (estimated to cover 84% of all supermarket sales). Nielsen then projects the data to provide national estimates for all supermarkets with annual sales of over $2 million. For the analysis, package sizes and prices are converted into standardized 16-oz. equivalent units. The retail-sponsored advertising indices denote the percent of stores in the SCANTRACK® data that used a special display, coupon, or advertisement to promote the specific product during the month. This study uses a monthly data series beginning in April 1988 and ending in December 1995.

Information used to construct the media index was obtained using the DIALOG Information Retrieval Service. The newspaper and magazine databases accessed through DIALOG include the full text of all news stories and feature articles in over 60 newspapers and 400 different magazines. We searched the databases for the following keyword combinations: tuna/dolphin, dolphin-safe/tuna, dolphin-safe/label. Results of the searches were scanned to eliminate any articles that were deemed inappropriate. The index was constructed as a sum of all articles discussing the tuna–dolphin issues during the relevant time period. Articles were not coded based upon the articles content nor weighted by the prominence of each article within the newspaper/magazine because previous research indicates that these "corrections" do not add additional information. Essentially, the number of negative and positive articles tends to be highly correlated [46] and article prominence is correlated with reporting frequency [53].

V. RESULTS AND DISCUSSION

The significant, negative coefficient on the time-trend variable in the tuna and other seafood equations indicates that, *ceterus paribus*, there exists a downward trend for canned tuna and other seafood market share (Tables II and III). Conversely, the significant, positive coefficient on the time-trend variable in the luncheon meat equation indicates that there exists an upward trend for this product. Note that both these trends exist even though we hold constant for the information effects surrounding the tuna–dolphin controversy via the video, media, and labeling coefficients. These trend coefficients may reflect the U.S. population's changing preferences for these different products. This interpretation may also be indicated by the positive, significant coefficient on the expenditure variable in the luncheon meat equation, indicating that expenditures on luncheon meat expand relatively more rapidly as total expenditures increase. The negative time trend for canned tuna could also be a result of the changing quality of canned tuna. Due to

TABLE II
Estimated Share Equations, α_0 Restricted to Zero[a]

	Tuna	Luncheon meat	Seafood	Red meat
Intercept (ξ)	0.31904*	0.47335*	0.11055*	0.09706
	(0.01164)	(0.0123)	(0.00541)	
Expenditure (β)	−0.00357	0.01785*	−0.00362	−0.01066
	(0.00514)	(0.0073)	(0.00414)	
Time trend (τ)	−0.0003*	0.00055*	−0.00024*	−0.00001
	(0.00005)	(0.00007)	(0.00004)	
Label (ϕ)	0.00726**	−0.01661*	0.00861*	0.00074
	(0.00385)	(0.00597)	(0.00325)	
Lent (θ_1)	0.01238*	−0.0218*	0.00644*	0.00298
	(0.00211)	(0.00304)	(0.0017)	
December (θ_3)	−0.00519*	0.00294	−0.00113	0.00338
	(0.00226)	(0.00319)	(0.00173)	
Video (ρ_2)	−0.00712*	−0.00034	0.0059*	0.00156
	(0.00339)	(0.00537)	(0.00294)	
Media (ρ_1)	0.00024	−0.00057	0.00059*	−0.00026
	(0.00025)	(0.00038)	(0.00021)	
Advertising (ω)				
Tuna	0.00035*	−0.00032*	0.00003	−0.00006
	(0.00004)	(0.00005)	(0.00002)	
Luncheon meat	−0.00032*	0.00047*	−0.00022*	0.00007
		(0.00008)	(0.00003)	
Seafood	0.00003	−0.00022*	0.00024*	−0.00005
			(0.00002)	
Red meat	−0.00006	0.00007	−0.00005	0.00004
Log price (γ)				
Tuna	0.1402*	−0.14471*	0.0152*	−0.01069
	(0.01421)	(0.01444)	(0.00459)	
Luncheon meat	−0.14471	0.17317*	−0.01982*	−0.00864
		(0.02531)	(0.00702)	
Seafood	0.0152	−0.01982	0.00139	0.00323
			(0.00399)	
Red meat	−0.01069	−0.00864	0.00323	0.01610
Adj. R-square	0.9443	0.9468	0.8997	0.7932

[a] Standard errors are reported in parenthesis.
* denotes significance at the 0.05 level.
** denotes significance at the 0.10 level. Note: standard errors and significance levels are not reported for restricted parameters.

the dolphin-safe criteria, fishermen began landing smaller yellowfin and skipjack tuna less suitable as a canned product than large yellowfin tuna.

The significant coefficients on the video variables indicate that the negative media activity surrounding the tuna–dolphin issue seems to have depressed sales of canned tuna (Fig. 1) and increased market share of other canned seafood. The labeling coefficient in the tuna-share equation is significant and positive, indicating that the presence of dolphin-safe labels increased tuna market share and, given the functional form of this variable, that this increase in market share increased over a period of several months. The label coefficient is negative in the luncheon meat equation. Thus the presence of dolphin-safe labels seems to have caused declines in the luncheon meat markets. Although predicted market share of tuna is declining through time, the labeling program slowed this decrease so that by

TABLE III

Estimated Share Equations, α_0 not restricted[a, b]

	Tuna	Luncheon meat	Seafood	Red meat
Intercept (ξ)	0.34570*	0.43570*	0.10638*	0.11222
	(0.03055)	(0.03824)	(0.02029)	
Expenditure (β)	−0.00363	0.01293**	−0.00064	−0.00865
	(0.00501)	(0.00705)	(0.00377)	
Time trend (τ)	−0.00031*	0.00048*	−0.00022*	0.00004
	(0.00005)	(0.00007)	(0.00004)	
Label (ϕ)	0.00904*	−0.01415*	0.00774*	−0.00263
	(0.00385)	(0.00594)	(0.00322)	
Lent (θ_1)	0.01205*	−0.02087*	0.00594*	0.00289
	(0.00209)	(0.00306)	(0.00171)	
December (θ_3)	−0.00454*	0.00460	−0.00195	0.00189
	(0.00223)	(0.00310)	(0.00160)	
Video (ρ_2)	−0.00712*	0.00193	0.00534**	−0.00015
	(0.00344)	(0.00530)	(0.00290)	
Media (ρ_1)	0.00025	−0.00066**	0.00065*	−0.00024
	(0.00024)	(0.00037)	(0.00021)	
Advertising (ω)				
Tuna	0.00034*	−0.00032*	0.00003	−0.00005
	(0.00004)	(0.00005)	(0.00003)	
Luncheon meat	−0.00032*	0.00050*	−0.00023*	0.00005
		(0.00008)	(0.00004)	
Seafood	0.00003	−0.00023*	0.00023*	−0.00003
			(0.00002)	
Red meat	−0.00005	0.00005	−0.00003	0.00003
Log price (γ)				
Tuna	0.14146*	−0.12529*	0.01417*	−0.03034
		(0.01396)	(0.01462)	(0.00471)
Luncheon meat	−0.12529*	0.16854*	−0.02804*	−0.01521
		(0.02407)	(0.00722)	
Seafood	0.01417*	−0.02804*	0.00421	
			(0.00400)	
Red meat	−0.03034	−0.01521	0.00967	0.03588
Adj. R-square	0.94340	0.94880	0.90320	0.81880

[a] Standard errors are reported in parenthesis.

[b] The estimated $\alpha_0 = -4.6885$ with a standard error of 1.8702, which is significant at the 5% level.

* denotes significance at the 0.05 level.

** denotes significance at the 0.10 level. Note: standard errors and significance levels are not reported for restricted parameters.

December 1995 tuna market share with the labeling program is roughly 1% higher than the estimated market share in absence of the label (Fig. 1).

The results demonstrate that the impact of a labeling program may not be felt immediately as the best fitting functional form for the label variable takes on an S-shape similar to previous models of information diffusion. We suggest two possible reasons for this temporal effect. The first is the presence of a pure information diffusion process. A few interested or knowledgeable consumers immediately notice the label change while others only notice and react to the label after several in-store exposures or after discussions with friends and neighbors who

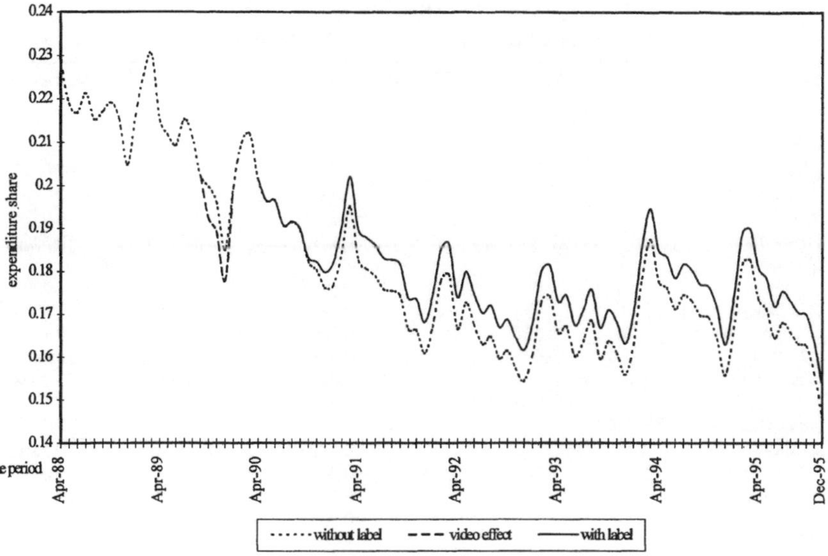

FIG. 1. Market effects of information: Expenditure share of canned tuna.

have already noticed and reacted to the label. The second effect may derive from a verification effect. That is, some consumers may have noticed the new labels but did not believe the label information. As information that verifies the label claim is presented to these consumers (from media, lawmakers, etc.), they may have increased their canned tuna consumption to reflect their level of trust. Diffusion of awareness about the dolphin-safe label and increases in perceived believability may explain the temporal increase in tuna's market share.

The coefficients on the "own" advertising variables are positive in all equations and significant in the tuna, luncheon meat, and other seafood equations, indicating that increases in product advertising seem to increase market share. Conversely, luncheon meat advertising seems to decrease tuna and other seafood market share. Similarly, tuna and other seafood advertising decreases luncheon meat market share. The coefficients on the "religious–seasonal" variables (θ_1 and θ_2) indicate that the market share of tuna and seafood increases during Lent while the market share of luncheon meats decreases. In addition, during the winter holiday season the market share of tuna decreases.

All of the own-price elasticities (both Marshallian and Hicksian) are negative, indicating downward sloping demand (Table IV).[18] The positive signs on most of the Hicksian cross-price elasticities indicate that most of the goods are substitutes to one another. The price elasticities all fall between the ranges of -1 and 1. Though this is partly a result of the restrictions placed on the estimated parameters, it does provide a realistic range of price responses for food items. The

[18] The elasticity formulas for the AIDS model are reported in numerous places [7, 22]. The introduction of labeling, advertising, and other variables in our AIDS model does not significantly change these formulas so we will not report them here.

TABLE IV

Estimated Elasticities for Canned Meat Products

	Tuna	Luncheon meat	Seafood	Red meat
	α_0 restricted to zero			
Marshallian				
Tuna	−0.2664	−0.7421	0.0806	−0.0535
Luncheon meat	−0.2340	−0.7438	−0.0334	−0.0169
Seafood	0.2353	−0.2604	−0.9753	0.0538
Red meat	−0.0803	−0.0215	0.0384	−0.8358
Hicksian				
Tuna	−0.0778	−0.1195	0.1471	0.0503
Luncheon meat	−0.0364	−0.0916	0.0363	0.0918
Seafood	0.4171	0.3401	−0.9111	0.1539
Red meat	0.0925	0.5489	0.0993	−0.7407
Expenditure	0.9814	1.0281	0.9465	0.8992
	α_0 not restricted			
Marshallian				
Tuna	−0.2595	−0.6419	0.0752	−0.1550
Luncheon meat	−0.2021	−0.7453	−0.0458	−0.0272
Seafood	0.2111	−0.4085	−0.9372	0.1440
Red meat	−0.2686	−0.0996	0.0978	−0.6477
Hicksian				
Tuna	−0.0710	−0.1956	0.1418	0.0512
Luncheon meat	−0.0060	−0.0980	0.0234	0.0807
Seafood	0.4014	0.2199	−0.8700	0.2487
Red meat	−0.0922	0.4828	0.1601	−0.5506
Expenditure	0.9811	1.0204	0.9906	0.9181

magnitudes of these results are similar to other meat demand studies [16, 27]. Conforming to expectations, the positive signs of the expenditure elasticities indicate that the share of expenditures spent on an item increases as expenditures on proteins increase.

The confidence intervals indicate that the mean monthly CV estimates for a representative household are different based on whether α_0 is restricted (Table V). In general, the CV estimates from the unrestricted equation are approximately three times greater than the CV estimates from the restricted equation. In addition, the confidence intervals indicate that the mean monthly CV estimates from the restricted equation are not consistently significantly different from zero until 10 months after the announced dolphin-safe labeling policy, whereas the CV estimates from the unrestricted equation are significantly different from zero immediately after the announcement. Similar to the market results, to accurately measure all of the benefits from the labeling program one must look beyond instantaneous effects. Notably, once the label effect is fully realized, the magnitude of the CV estimates remains relatively stable across the remaining time period. Variation in reported CV after 1991 is due solely to changes in relative prices and incomes; CV is relatively lower (higher) during months in which the price of tuna was relatively lower (higher). The U.S. aggregate annual CV from the restricted equation is approximately \$6 million (monthly per household CV $*$ 12 months/year $*$ 92.62 million households), whereas the comparable estimate from the unresricted equation is about \$15 million (Table VI).

TABLE V
Estimated Representative Household Compensating Variations for Selected Months[a]

	α_0 restricted to zero			α_0 not restricted		
Time period	Lower 95% confidence interval	Mean	Upper 95% confidence interval	Lower 95% confidence interval	Mean	Upper 95% confidence interval
Pre-April 1990	n/a	n/a	n/a	n/a	n/a	n/a
May	−$0.000041	−$0.000040	−$0.000039	$0.000069	$0.000070	$0.000071
June	−$0.000077	−$0.000072	−$0.000067	$0.000015	$0.000019	$0.000024
July	−$0.000021	−$0.000010	$0.000001	$0.000035	$0.000046	$0.000057
August	−$0.000031	−$0.000002	$0.000027	$0.000128	$0.000156	$0.000184
September	−$0.000065	$0.000029	$0.000123	$0.000363	$0.000455	$0.000547
October	$0.000165	$0.000344	$0.000523	$0.001247	$0.001424	$0.001601
November	$0.000368	$0.000748	$0.001128	$0.002726	$0.003096	$0.003466
December	−$0.000615	$0.000205	$0.001025	$0.003945	$0.004755	$0.005565
January 1991	−$0.000559	$0.000381	$0.001321	$0.006290	$0.007210	$0.008130
February	$0.003279	$0.004469	$0.005659	$0.011168	$0.012348	$0.013528
March	$0.005059	$0.006709	$0.008359	$0.013594	$0.015224	$0.016854
April	$0.002867	$0.004087	$0.005307	$0.011327	$0.012537	$0.013747
May	$0.002426	$0.003676	$0.004926	$0.012032	$0.013262	$0.014492
June	$0.000717	$0.002397	$0.004077	$0.009387	$0.011017	$0.012647
July	$0.003184	$0.004604	$0.006024	$0.010448	$0.011848	$0.013248
August	$0.004621	$0.006061	$0.007501	$0.013061	$0.014481	$0.015901
September	$0.003108	$0.004868	$0.006628	$0.011772	$0.013502	$0.015232
September 1992	$0.006901	$0.008641	$0.010381	$0.015047	$0.016767	$0.018487
September 1993	$0.002485	$0.004045	$0.005605	$0.010514	$0.012054	$0.013594
September 1994	$0.003200	$0.004640	$0.006080	$0.010555	$0.011975	$0.013395
September 1995	$0.003946	$0.005416	$0.006886	$0.011624	$0.013074	$0.014524

[a] All values are in 1990 dollars.

VI. CONCLUSIONS AND LIMITATIONS

The results support our hypothesis that the dolphin–tuna controversy and the subsequent implementation of dolphin-safe labeling affected consumer behavior. Further, the paper provides market-based evidence that consumers can respond to eco-labels. The statistical significance of the label coefficient in the tuna share equation indicates the dolphin-safe label did increase the market share of canned tuna and that this increase continued over time. When measuring market or welfare effects of labeling programs it is important to recognize that consumers may not respond instantaneously to these programs. After the introduction of the label, shares of other products decreased as people substituted back to tuna, providing an interesting look at how people substitute between products based on moral or ideological grounds. Finally, the welfare analysis may provide a partial measure of society's willingness to pay to avoid personally contributing to dolphin mortality as a result of tuna fishing.

In addition to changes in consumer behavior the presence of eco-labeling may alter manufacturer behavior. That is, if a significant portion of the consumer population demands environmentally friendly products, the presence of an eco-labeling program may provide firms an incentive to differentiate and market their

TABLE VI

Estimated U.S. Aggregate Compensating Variations for Selected Months[a]

Time period	α_0 restricted to zero	α_0 not restricted
Pre-April 1990	n/a	n/a
May	*−$44,313*[b]	$77,630
June	*−$79,930*	$21,668
July	*−$11,602*	$50,586
August	*−$2,655*	$173,358
September	*$32,699*	$505,764
October	$382,005	$1,582,594
November	$830,903	$3,441,224
December	*$228,347*	$5,284,611
January 1991	*$423,190*	$8,013,818
February	$4,967,483	$13,723,742
March	$7,456,189	$16,920,634
April	$4,541,998	$13,933,747
May	$4,085,882	$14,740,139
June	$2,664,550	$12,244,243
July	$5,116,661	$13,168,659
August	$6,736,335	$16,094,725
September 1991	$5,410,688	$15,007,204
September 1992	$9,603,627	$18,635,730
September 1993	$4,495,675	$13,397,597
September 1994	$5,156,843	$13,309,844
September 1995	$6,019,270	$14,531,233

[a] All values are in 1990 dollars.

[b] Italicized aggregate estimates are based on household estimates not significantly different from zero.

products along an environmental characteristic(s). An increase in supply of these environmentally friendly products may increase consumer purchases simply through greater availability without changes in individual awareness.

Evidence that providing information to consumers through labeling can alter behavior is likely to be welcomed by policymakers because this sort of policy tool may be easier to implement and enforce than explicit production site regulations. Furthermore, recent trade agreements restrict the use of economic sanctions as a way of influencing trading partners. However, labeling allows market behavior to influence trading partners while complying with trade agreements.

This study is limited in that it identifies the effects of the labeling program only through changes in aggregate consumption before and after the introduction of labeling due to simultaneous implementation by the major tuna industry suppliers; no observable cross-sectional variation in labeling occurred in our data. Further research involving cross-sectional variation in eco-labeling from other markets will broaden our understanding of the efficacy of such programs in terms of altering consumption.

Another critique of this research involves identifying the true marginal effect of the on-package labeling. That is, would consumption patterns have changed in a different fashion if all the same events took place (media stories, release of the video, Congressional action, supplier announcements concerning changes in production practices) except that no label was ever placed upon a package? What role does the label play in educating the public about the change in supplier policy? Answers to these questions would help to identify the effect of various informational policies and determine their substitutability and complementarity. As in the study of food labels, a better understanding of such possible effects and improved calibration of models may require both experimental and quasi-experimental approaches.[19]

The 105th Congress has redefined the meaning of dolphin safe. Under The International Dolphin Conservation Program Act [43], a tuna product could be labeled dolphin safe if no dolphins were killed while capturing tuna. This would allow fishermen to chase and encircle dolphin schools and still meet the dolphin-safe requirement so long as no dolphins are killed. Given the paucity of quantitative research on the impact of the original dolphin-safe labels, it is not clear how a new definition would impact consumer behavior. Our results indicate it might take consumers some time to fully adjust to the new definition.

REFERENCES

1. I. Ajzen, T. C. Brown, and L. H. Rosenthal, Information bias in contingent valuation: Effects of personal relevance, quality of information, and motivational orientation, *J. Environ. Econom. Management* **30**, 43–57 (1996).
2. W. Bailey and M. A. Eastlick, Study shows many claiming to be green don't buy green, *Northwest Arkansas Times*, Nov. 12, 1993.
3. J. C. Bergstron, J. R. Stoll, and A. Randall, The impact of information on environmental commodity valuation decisions, *Amer. J. Agric. Econom.* **72**, No. 3, 614–621 (1990).
4. E. R. Berndt and N. E. Savin, Estimation and hypothesis testing in singular equation systems with autoregressive disturbances, *Econometrica* **43**, 937–957 (1975).
5. N. E. Bockstael and I. E. Strand, "Economic Sanctions and Environmental Preferences: Protecting Dolphins and Tuna Markets," Mimeo, Dept. of Agricultural and Resource Econ., University of Maryland (1995).
6. Bonneville Power Administration, "Report on Market Research and Program Recommendations: Bonneville Power Administration Regionwide Promotion of Energy-Efficient Appliances," B. G. William Lesh, Inc. (1985).
7. A. Buse, Evaluating the linearized almost ideal demand system, *Amer. J. Agr. Econom.* **76**, 781–793 (1994).
8. T. A. Cameron and J. Englin, Respondent experience and contingent valuation of environmental goods, *J. Environ. Econom. Management* **33**, 296–313 (1997).
9. O. Capps, Jr. and J. D. Schmitz, A recognition of health and nutrition factors in food demand analysis, *Western J. Agr. Econom.* **16**, No. 1, 21–35 (1991).
10. R. M. T. Castagnini, T. L. Cox, and C. F. Perali, On the choice of functional forms in demand analysis: An application to beverage consumption in Italy 1960–1990, *presented at* the Annual Meeting of the American Agricultural Economics Association, 1996.
11. J. A. Chalfont and K. J. White, "Estimation and Testing in Demand Systems with Concavity Constraints," Department of Agricultural and Resource Economics Working Paper 454, University of California (1987).
12. S. W. Chang, GATTing a green trade barrier: Eco-labelling and the WTO agreement on technical barriers to trade, *J. World Trade* **31**, No. 1, 137–159 (1997).
13. D. Chase and T. K. Smith, Consumers keen on green but marketers don't deliver, *Advertising Age* June 29, 1992.

[19] For example, see Teisl and Levy [47] regarding a market-based experimental approach to determining the effects of nutrition labels.

14. R. G. Cummings, D. S. Brookshire, and W. D. Schulze, "Valuing Environmental Goods: An Assessment of the Contingent Valuation Method," Rowan and Allenheld, Totowa, NJ (1986).
15. A. S. Deaton and J. Muelbauer, An almost ideal demand system, *Amer. Econom. Rev.* **70**, 312–326 (1980).
16. J. Dyck, "Demand for Meats in Japan: A Review and Update of Elasticity Estimates," AGES88-525, U.S. Department of Agriculture, Economic Research Service, Agriculture and Trade Analysis Division.
17. A. R. Gallant, "Nonlinear Statistical Models," Wiley, New York (1987).
18. X. M. Gao, E. J. Wailes, and G. L. Cramer, A synthetic demand system: An application to U.S. consumer demand for rice and selected rice substitutes, *Rev. Agr. Econom.* **16**, 27–38 (1994).
19. W. H. Greene, "Econometric Analysis," 4th ed., Prentice Hall, Upper Saddle River, NJ (2000).
20. R. D. Green, Z. A. Hassan, and S. R. Johnson, Selection of nonnested demand models, *Canad. J. Agr. Econom.* **43**, 485–499 (1995).
21. R. D. Green and J. Alston, Elasticities in AIDS models: A clarification and extension, *Amer. J. Agr. Econom.* **73**, 874–875 (1991).
22. E. F. Greenberg, ISO 14000 to change business environment, *Packaging Dig.* **33**, No. 4, 20–22 (1996).
23. R. Gutfield, Shades of green: Eight of 10 Americans are environmentalists, or at least say so, *Wall Street J.*, August 2, 1991.
24. A. K. Han and D. Park, Testing for structural change in panel data: Application to a study of foreign trade in manufacturing goods, *Rev. Econ. Statist.* 135–142 (1989).
25. R. V. Hartwell and L. Bergkamp, Eco-labeling in Europe: New market-related environmental risks? *BNA Internat. Environ. Daily*, October 20, 1992.
26. S. Hashizume, "Environmental Labeling in Japan: The Eco-Mark," Japan Environ. Assoc. (1992).
27. J. Hayes, T. Wahl, and G. Williams, Testing restrictions on a model of Japanese meat demand, *Amer. J. Agr. Econom.* **72**, 556–566 (1990).
28. D. Horsky, A diffusion model incorporating product benefits, price, income and information, *Market. Sci.* **9**, No. 4, 342–365 (1990).
29. P. M. Ippolito and A. D. Mathios, Information, advertising and health choices: A study of the cereal market, *RAND J. Econom.* **21**, 459–480 (1990).
30. P. M. Ippolito and A. D. Mathios, "Information and Advertising Policy: A Study of Fat and Cholesterol Consumption in the United States, 1977–1990," Bureau of Economics Staff Report, Federal Trade Commission (1996).
31. J. Joseph, The tuna–dolphin controversy in the Eastern Pacific Ocean: Biological, economic, and political impacts, *Ocean Develop. Internat. Law* **25**, 1 (1994).
32. G. G. Judge, R. C. Hill, W. E. Griffiths, H. Lutkepohl, and T. C. Lee, "Introduction to the Theory and Practice of Econometrics," Wiley, New York (1988).
33. J. T. LaFrance and M. Hanemann, The dual structure of incomplete demand systems, *Amer. J. Agr. Econom.* **71**, 262–274 (1989).
34. J. T. LaFrance, "Weak Separability and Two-Stage Budgeting: Estimation and Welfare Measurement," Agricultural Economics and Economics Department Staff Paper 89-3, Montana State University (1989).
35. J. T. LaFrance, When is expenditure "exogenous" in separable demand models? *Western J. Agr. Econom.* **16**, 49–62 (1991).
36. J. T. LaFrance, "Incomplete Demand Systems, Weak Separability and Weak Complimentarity in Applied Welfare Analysis: Expanded Version," Department of Agricultural Economics Working Paper 77, University of Arizona (1992).
37. J. T. LaFrance, Weak separability in applied welfare analysis, *Amer. J. Agr. Econom.* **75**, 770–775 (1993).
38. A. Levy, O. Mathews, M. Stephenson, J. Tenney, and R. Schucker, The impact of a nutrition information program on food purchases, *J. Public Policy Market.* **4**, 1–13 (1985).
39. A. Lewbel, Nesting the aids and translog demand systems, *Internat. Econom. Rev.* **30**, 349–356 (1989).
40. L. Loro, Green marketing comes to computers, *Advertising Age*, September 29, 1993.
41. National Research Council, "Dolphins and the Tuna Industry," National Academy Press (1992).
42. N. E. Piggott, J. A. Chalfont, J. M. Alston, and G. R. Griffith, Demand response to advertising in the Australian meat industry, *Amer. J. Agr. Econom.* **78**, No. 2, 268–279 (1996).

43. PL 42, 105th Congress, The International Dolphin Conservation Program Act.
44. SAMI/Burke, "The SAMI/Burke Good Housekeeping Seal Study," Good Housekeeping Institute (1988).
45. P. Shabecoff, Three companies to stop selling tuna caught with dolphins, *New York Times*, April 13, 1990.
46. M. E. Smith, E. O. van Ravenswaay, and S. R. Thompson, Sales loss determination in food contamination incidents: an application to milk bans in Hawaii, *Amer. J. Agr. Econom.* **70**, 513–520 (1988).
47. M. F. Teisl and A. S. Levy, Does Nutrition Labeling lead to Healthier Eating? *J. Food Distribution Res.* **28**, No. 3, 18–27 (1997).
48. U.S. Dept. of Agriculture, "Continuing Survey of Food Intakes by Individuals" (1991).
49. U.S. Environmental Protection Agency, "Environmental Labeling in the United States: Background Research, Issues and Recommendations," Applied Decision Analysis (1989).
50. U.S. Environmental Protection Agency, "Assessing the Environmental Consumer Market" (1991).
51. U.S. Environmental Protection Agency, "Status Report on the Use of Environmental Labels Worldwide" (1993).
52. U.S. Environmental Protection Agency, "Determinants of Effectiveness for Environmental Certification and Labeling Programs" (1994).
53. E. O. van Ravenswaay and J. P. Hoehn, The impact of health risk and food demand: A case study of Alar and apples, *in* "Economics of Food Safety" (J. Caswell, Ed.), Elsevier Science, New York (1991).
54. P. Wallstrom and C. R. Wessells, "Analysis of Consumer Demand for Canned Tuna: Impact of Dolphin-Safe Controversy," *presented at* the VII Conference of the International Institute of Fisheries Economics and Trade, July 1994.
55. S. Yamaki, "Critical Analysis of the Vermont Household Hazardous Product Shelf-Labeling Program on Consumer Awareness: A Survey Evaluation" (1992).

[12]

Environmental labeling and consumers' choice—an empirical analysis of the effect of the Nordic Swan

Thomas Bue Bjørner,[a,*] Lars Gårn Hansen,[a] and Clifford S. Russell[b,c]

[a] *AKF, Institute of Local Government Studies, Nyropsgade 37, DK-1602 Copenhagen V, Denmark*
[b] *Bowdoin College, Brunswick, ME 04011, USA*
[c] *Vanderbilt University, Nashville, TN 37212, USA*

Received 14 May 2002; accepted in revised form 3 June 2003

Abstract

Provision of information on the environmental effects of consumption is often put forward as an appealing alternative to traditional means of environmental regulation such as permits and environmental taxes. When consumers in opinion polls are asked if their purchasing decisions would be influenced by information on environmental or ethical aspects of products, the majority seem very ready to say yes. However, evidence for actual behavior along these lines is still limited. In this paper a mixed logit model is used to quantify the effect of a certified environmental label (the Nordic Swan) on consumers' brand choices of toilet paper, paper towels and detergents, using a large Danish consumer panel with detailed information on actual purchases from the beginning of 1997 to January 2001. It is found that the Nordic Swan label has had a significant effect on consumers' brand choices for toilet paper, corresponding to a marginal willingness to pay for the certified environmental label of 13–18% of the price and it also appears that information on environmental performance has had an effect on consumers' choice of detergent.

JEL classification: C25; D12; D64; Q28

Keywords: Environmental labeling; Information provision; Brand choice; Panel mixed logit

1. Introduction

When political or administrative policy makers set out to directly influence private sector environmental behaviors via prohibitions, permits, or even "prices", tensions and conflicts are usually explicit and long lasting. The provision of information appears less confrontational than

*Corresponding author. Fax: +45 33 15 28 75.
E-mail addresses: tbb@akf.dk (T.B. Bjørner), lgh@akf.dk (L.G. Hansen), russells@tidewater.net (C.S. Russell).

traditional regulatory rule or price making, and it seems that public programs that disseminate information about the environmental attributes of products or companies are becoming an increasingly popular tool of government agencies and organizations, with the intent to improve the environmental performance of environmental actors. Sceptics have, on the other hand, been concerned that such programs may not significantly influence the behavior of producers or consumers. For instance, although consumers in opinion surveys often maintain that they will reflect information on environmental or ethical aspects of goods or companies in their purchasing decisions ("green consumption"), there has been little empirical evidence of actual behavior along these lines.

A label indicating that the labeled brand is more environmentally friendly than unlabeled brands of the given consumer product is often used as a simple way to present complex environmental information to consumers. Examples include the European Union's Ecoflower, the German Blue Angel, and the Nordic Swan. The general idea of these programs is that credible environmental information will affect consumer brand choice and increase the market shares of companies with relatively clean technologies or environmentally friendly products.

In this paper we present an analysis of the effect of the Scandinavian environmental label (Nordic Swan) on Danish consumers' choices among different brands of toilet paper, paper towels[1] and detergents. The analysis is based on an extensive panel data set with weekly purchase diary data for 1,596 Danish households (on average) from 1997 to the beginning of 2001. During this period a number of brands of the consumer products being analyzed were awarded the Swan label, so the data include information on purchases both before and after the label was awarded. This allows us to control for potential confounding effects of unobserved (or unmeasurable) brand quality differences, when estimating the effect of the label. In the consumer panel data there are repeated choice observations for each household. In order to take into account that decisions made by each of the households may be correlated over time due to unobserved differences in brand preferences we use the random parameter (mixed) logit model for repeated choice described by Revelt and Train [28] to estimate a random effects model for brand choice. More precisely, we allow the (alternative specific) brand dummies to be random parameters instead of constants. An additional advantage of the mixed logit framework is that it does not exhibit the IIA property associated with standard logit models (see, e.g., [22,36]).

In addition to the presence or absence of the Swan, other explanatory variables include the prices of the different brands, the level of advertising the brands, whether the brand was purchased on sale (as a special offer), and finally indicators of results from a widely published consumer test report that has ranked different brands of detergents.

It is found that the Swan label has had a significant effect on Danish consumers' choices of toilet paper and it also appears that information on environmental performance (deriving either from the label or from consumer test reports) has had an impact on consumers' choice of detergent. Marginal willingness to pay for Swan-labeled toilet paper ranges between 13% and 18% of the price. The effect of the Swan label on the choices of paper towels was lower than for toilet paper (and often not significantly different from zero). Paper towels are not purchased frequently by all Danish households, as it is still common in Denmark to use dishcloths or other similar products, which can be washed out and reused. Hence, it seems likely that "green"

[1] For clarification: "Paper towels" (in the US) are known as "kitchen rolls" in the UK (or "køkkenruller" in Danish).

consumers prefer to rely mainly on dishcloths instead of purchasing paper towels. In addition, it has also been suggested that environmental labeling may be more effective on products purchased more frequently, as consumers may feel that it would make a greater environmental impact.

In the following section we briefly describe the Nordic Swan label and review the literature pertaining to information provision and labeling. The model is presented in Section 3 and the data are described in Section 4. Results are presented in Section 5. Concluding remarks are offered in the last section.

2. Background

One important distinction between different types of information provision programs concerns the motivation of the end user of the information. Some labels give information that is directly applicable to the end-user. For example, hazard warning labels, such as those on cigarettes and domestically used pesticides, are intended to protect the user or those around him. Other labels give information that is applicable only to the extent that the user has some degree of concern for wider, more diffuse environmental effects on which the consumer's individual action can hardly make even a tiny impact.[2] In addition to environmental labels like the Nordic Swan this also includes "ethical" labels like the ones that are related to animal welfare (e.g. the US dolphin-safe tuna label) or "fair trade" with developing countries (e.g. the Max Havelaar and the TransFair labels).

Because of the well-known free riding problem of public goods, one may be skeptical of the whole idea that labels on attributes that do not yield the consumer significant direct benefits should have any effect on purchase decisions. However, early economic contributions, for example Sen [31], attributed such behavior to what Sen called "sympathy" or "commitment" (and others refer to as "altruism") or to the fear of acting socially irresponsibly (non-instrumental or symbolic behavior).[3]

More recently it has been shown by Kennedy et al. [16] that there can be a market failure in the (private) provision of information on polluting products, which makes a case for government action as provider of information. In addition, provision of information may potentially have a positive effect on the environmental commitment of the individuals, whereas fiscal incentives run the risk of crowding out proenvironmental motivation as suggested by Frey and Oberholzer-Gee [11].[4]

[2] A number of labels capture characteristics with mixed private and public good dimension, for example labels pointing out cost savings that flow from environmentally friendly actions, such as energy usage programs or organic food labels, which are produced in more environmentally friendly ways, but also often perceived as being healthier for the user.

[3] A number of recent contributions have discussed the nature and implications of altruistic preferences in different areas of economic theory including environmental economics (see e.g. [15,21,25]) and game theory (see e.g. [2,3]).

[4] Note, however, that in the theoretical model by Kennedy et al. [16] it is found that the provision of information is especially important when corrective taxes are not available for attacking the externality. Another view is offered by Stern [32], who—based on a survey of empirical studies—suggests that information and fiscal incentive have synergistic effects on behavior.

A number of recent theoretical papers have analyzed the impact of information where this takes the form of an environmental label. This includes the effect of a label on the technology investments in companies [9], market feedback mechanisms in a general equilibrium framework ([20,33]) and the effect of labeling on the asymmetric information problem related to the credence nature of environmental quality of consumer products ([8,17,19]). It appears from these contributions that environmental labels in some situations could have adverse effects on the environment and/or welfare. Nevertheless, it still appears to be the general view that the market pressure created by "green" consumers and investors provides reasons for optimism because it is a legitimate way for people to express their preferences and thus encourage socially beneficial outcomes (see e.g. [18,24]). From an empirical point of view the main question seems to be how strong these effects are.

The proliferation of "green products" and "green advertising" suggests that at least some firms believe that consumers are willing to pay a price premium for environmentally friendly products. There are also examples of companies whose sales dropped precipitously as a result of being labeled environmentally unfriendly [7]. Finally, there is a great deal of survey evidence that consumers will readily express a willingness to incorporate environmental information into consumption decisions (see Russell and Clark [30]). Nevertheless, there are only few empirical analyses of whether consumers will actually do so. Indeed, according to EPA [10] and OECD [27] most evidence is anecdotal.

Altogether we have identified five empirical studies published in academic journals or editions. These are summarized in Table 1.[5] In three cases it appears that the presence of the label or the information about environmental attributes has increased willingness to pay for the good or increased sales. In the remaining two cases results are mixed.

Methodologically, the work of Bennett et al. [4]—see also Blamey and Bennett [6]—is the closest to the study presented in this article as they also use real market data in a discrete choice model. However, there are two important differences. First, Bennett et al. [4] investigated the effect of two environmental attributes, while we seek to identify the effect of a hybrid environmental label certified by a third party. Second, our study includes information on a number of brands both before and after they began to display the label. This helps to avoid confounding effects with other unobserved brand characteristics. In a study also based on real market data Teisl et al. [35] used aggregate time-series data to investigate the effect of the dolphin-safe label on the overall market share of canned tuna. The effect of the dolphin-safe label is identified only through changes in aggregate consumption of tuna before and after the label (all brands of tuna labeled). As noted by the authors the lack of cross sectional variation in labeling is a limitation in the data. Thus, it is possible that the significant effect of the dolphin-safe label could be caused by unaccounted-for market trends. In our study the different brands of toilet paper, paper towels and detergents were awarded the label at different times, so the risk that the estimated coefficient on the presence of the label just picks up a market trend is much reduced.

[5] As it appears from Table 1 seven different studies have actually been found, but in two cases there are overlaps with data applied in another study.

Table 1
Empirical studies on the effect of environmental labels and product characteristics

Reference	Method/data	Market	Type of label or environmental characteristic	Did label have an effect?
Henion [13]	Real market experiment in four stores	Detergents	Content of phosphate	Yes
Teisl et al. [35]	Real market behavior using aggregate monthly time series data (using an 'almost ideal demand system')	Canned seafood and substitute meat products	Dolphin-safe label	Yes
Blamey and Bennett [6] Bennett et al. [4]	Real market behavior in discrete choice models (also combined with stated preference data)	Toilet paper	Unbleached Recycled	No Yes
Nimon and Beghin [24]	Hedonic regression using catalog prices	Apparels	Environmentally friendly dyes Organic cotton	No Yes
Roe et al. [29] Teisl et al. [34]	Hypothetical market (validated with hedonic regression based on electricity prices)	Electricity	Certified green electricity	Yes

3. Model for brand choice

In this paper we estimate a repeated mixed logit model of brand choice with the aim of deriving household willingness to pay for one specific good attribute: The Swan label. For each choice situation the repeated logit model requires a vector describing the price and other attributes of each available brand. The estimated parameters can then be interpreted as the utility value generated by consumption of the corresponding good attributes which in turn allows calculation of willingness to pay for each attribute.

A key assumption typically employed in the repeated logit model is that the estimated coefficients (i.e. the utility values generated by the consumption of the attributes) do not vary across choice situations. This implies that the utility value generated by the consumption of any given brand does not vary systematically across choice situations. The assumption seems reasonable for lumpy goods (such as recreational sites, etc.) to which the repeated logit model is typically applied.[6]

The goods we analyze (toilet paper, paper towels and compact detergent) are not lumpy, so the consumed quantity may vary between choice situations. If the quantity consumed does change (e.g. according to the current purchasing price), so may the utility generated by consumption of a given brand.[7] Thus, there may be different levels of utility "to share" between the attributes if consumed quantity changes. This indicates that the use of a linear logit model may be problematic, because the parameters we wish to estimate may vary systematically with the price attribute of a given brand.

One way to address this would be to integrate a model of the consumers' quantity decision. However, the goods in question here are also easily stored, so the quantity *purchased* in a given period of time (which we observe in the data) is not necessarily equal to the quantity *consumed* in that period (which we do not observe). In fact, we suspect that the variation in purchase quantity and frequency we see in the data primarily reflects consumers' stocking up when coming across a good offer. This approach would then also require an inventory model and modeling of the price expected to apply when stocks are replenished making estimation and interpretation even more challenging.

Instead of modeling price expectations and inventory decisions we make an assumption that ensures that the consumed quantity (of good utility units) per period of time is constant over the data period. The assumption made is that the price expected to apply when stocks are replenished is constant over the analyzed period (i.e. that the expected cost of using the goods does not change over the data period). Within the utility foundation developed in Appendix A this is sufficient to ensure that the utility value of attributes is constant across purchase situations, and that this evaluation is sufficient for choosing which brand to buy in any given purchasing situation independently of the quantity decision. This ensures that the repeated logit model of brand choice is valid in our context and that estimated parameters can be interpreted as the marginal utility value of the corresponding attributes. Thus, the assumption allows marginal willingness to pay for attributes to be derived with the relatively simple multinomial setup without modeling expectations or inventory behavior.

Though the above assumption may seem restrictive we cannot detect any strong trends in the mean purchase prices of different brands over the data period, which suggests that the assumption of constant expected cost over the data period is reasonable.

The households' choice of brand of toilet paper, paper towels and detergents is estimated using a mixed logit model, which does not impose the same restrictions on the substitution pattern between brands as the standard logit model [23,36].

With consumer panel data there are repeated brand choice observations for each household. It is likely that decisions made by each of the households are correlated over time due to unobserved differences in brand preferences between the households. Different approaches have been adopted

[6] Analytically, the structure of the problems is similar to repeated discrete-choice models of recreational demand, see e.g. [23] for (repeated) nested logit models and [14] for a comparison of the repeated nested logit and the mixed logit model.

[7] Under the standard assumption of decreasing marginal utility, the utility derived from consuming any given attribute of the marginal unit will vary with the quantity consumed.

by marketing researchers to account for the correlation in brand choices, see e.g. [1,12]. The model we apply is adopted from Revelt and Train [28], who present a mixed logit model allowing for repeated choices by each household.

Household i's utility from choosing alternative j during week t is given by

$$U_{ijt} = \beta' X_{jt} + \alpha_{ij} D_j + \varepsilon_{ijt} \tag{1}$$

Here X_{jt} is a vector of observable brand attributes and β is the associated vector of coefficients, while ε_{ijt} are unobserved random terms following independent identical extreme value distributions. In this study the observed brand attributes include the Swan variable (dummy), price, on sale (as a special offer) and advertising. In addition, we also include (alternative specific) brand dummy variables denoted D_j (for identification one of these is restricted to be zero). The parameters on these variables (α_{ij}) capture the effect of unobserved brand-specific attributes on the utility of each brand. These parameters, which are allowed to vary over households, capture heterogeneity in households' tastes for different brands. The variation in (α_{ij}) over households implies that the unobserved utility of each alternative is necessarily correlated over time for each household. Hence, it induces correlation in utility over choice occasions.

It is assumed that α_{ij} are random effects that for each brand follow an independent normal distribution: $\alpha_{ij} \sim \text{iidN}(\alpha_j, \sigma_j^2)$. Thus, α_j may be interpreted as the utility effect of (unobserved) differences in quality, brand reputation, appeal of packaging etc. of the mean household, while σ_j is a measure of how tightly individual households' utility effects are distributed around α_j (i.e. if σ_j is small then households have similar utility of the brand in question).[8]

Since marginal WTP by definition is the income reduction needed to ensure constant utility when an attribute is added to the marginal good, the marginal WTP for attribute l is given by (see Appendix A):

$$WTP_l = -\frac{\beta_l}{\beta_p} \tag{2}$$

where β_p is the coefficient on price from Eq. (1).

4. Data

4.1. Data sources

Data on weekly purchase of toilet paper, paper towels and detergents were obtained from GfK Denmark for the period January 1, 1997, to January 31, 2001. These data are taken from shopping diaries kept by a wide sample of Danish households (an average of 1596 diaries are returned weekly).[9]

[8] The coefficients on the other brand attributes (X_{jt}) could also be random parameters. However, the estimation of the model with random parameters on the brand dummies alone turned out to be rather involved (with up to 32 brands a large number of normal distribution parameters were to be estimated), so the simplifying assumptions (i.e. fixed parameters on the other attributes) were required.

[9] The GfK data contain information on households' purchases of food, personal products (shampoo, deodorant), cleansing agents, etc. According to GfK about 80% of each household's budget for "everyday necessities" are covered

In order to simplify the modeling task we focus on the market for typical package sizes (in Denmark), which are 6–10 rolls of toilet paper, and 3–4 rolls of paper towels. These package sizes cover about 85–90% of all observations and 70–80% of all rolls sold.[10] For detergents, we focus on the highly concentrated powdered detergents, which are known as "compact" detergents on the Danish market. The compact detergents account for more than half of all the packages sold. Liquid detergents and powder with low concentration are omitted, but it is only compact brands of detergents that have obtained the Swan label. Altogether this leaves 45,796 purchases of toilet paper, 28,198 purchases of paper towels and 16,738 purchases of detergents in the final data set covering a period of 213 weeks.

Brands are recorded at a very detailed level by GfK (at bar code level). To reduce the number of choices we have "aggregated" into a smaller, but still fairly detailed number of brands. Altogether, we obtain 32 different brands of toilet paper, 26 different brands of paper towels and 15 different brands of (compact) detergents (the last brand numbers being "other brands"). Note that some of these brands were not available over the whole period of time as new brands have emerged while others have been removed from the market. Some brands are sold in a variety of stores, while others are only available in one chain of supermarkets ("house brands"). Likewise, the different stores and supermarket chains in the data set have been aggregated into 12 different chains or shop categories which have similar pricing policies and the same selection of brands. A more detailed description of the data, including the definition of brands and shops, can be found in Bjørner et al. [5].

An index for the weekly marketing effort on Danish television, in newspapers and in a number of weekly magazines of each brand was kindly made available by the Danish Gallup Adfacts. Basically, this index is based on the "standard price" of placing commercials in a certain medium (e.g. buying a time slot in prime time). No effort has been made to judge among the "qualities" of the placed advertisements (or the cost of the advertising agency to produce the commercial). In addition, it should also be noted that the definition of brands by the Danish Gallup Adfacts did not correspond unambiguously to the brand definition based on the GfK codes, so some judgement was used when determining the marketing effort for each brand.

Information about which brands have qualified to use the Swan label (and when they qualified) was obtained from Eco-labeling Denmark (i.e. the Danish labeling authority) as well as from producers and/or sales organizations. As it turned out there were some differences between the time the label was officially awarded and the date a brand was available to consumers with the label (for example because old stocks without the label were being cleared).

Altogether 13 of the 32 different brands of toilet paper and 9 of the 26 brands of paper towels obtained the label in the period analyzed. For detergents only 3 of the defined brands have been

(*footnote continued*)
by the consumer diary. All data are self-reported by the households. GfK recommends that the diaries are filled in directly after each shopping trip to avoid problems with forgotten purchases. For each recruited household a "diary keeper" is chosen. Information recorded includes (for each item purchased) the following information: Bar code and/or brand, volume (e.g. eight rolls), units (e.g. three packages), price, special offer or not, store and other information depending on the type of commodity (e.g. organic or conventional for dairy products).

[10] Over time the market share of these package sizes has been fairly stable. In addition there are also Swan labels available on larger packages of toilet paper (e.g. 12–16 rolls) and paper towels. Therefore, it seems unlikely that the omission of non-typical package sizes will bias the estimated effects of the Swan label.

Table 2
Share of observations with label

	1997	1998	1999	2000	Jan 2001	All
Toilet paper						
– *N* observations	10,756	11,330	11,593	11,045	1.072	45,796
– % with label	**4%**	**13%**	**23%**	**29%**	**35%**	18%
Paper towels						
– *N* observations	6,854	7,141	6,921	6,620	662	28,198
– % with label	**15%**	**21%**	**25%**	**26%**	**28%**	22%
Detergents						
– *N* observations	3,918	4,508	4,050	3,802	460	16738
– % with label	**0%**	**4%**	**9%**	**24%**	**25%**	9%

labeled. In most cases the labeled brands have also been available without the label, but in four cases (two for toilet paper and two for paper towels) the brand was introduced to the market with the Swan label already on. In these cases the coefficients on the brand dummies are confounded with the label dummies, so these brands do not add to the identification of the effect of the Swan. All 3 labeled brands of detergents were available on the market also before they were labeled, but for two of the brands there are few observations with the label, either because the brand has a very low overall share of the market (Tusindfryd) or because the brand was labeled very late in the period analyzed (Neutral). Thus, these two brands will only give a limited contribution to the empirical identification of an effect of the label on the choice of detergents. The effect of the Swan label on detergents will therefore be tied closely to only one brand (Bluecare), which was labeled first in 1998 and has been sold as a house brand only in shops 1–3.

The number of observations and the share of observations (i.e. purchase of some quantity of a particular brand) with the Swan label are given in Table 2. Evidently, there has been a large increase in the share of labeled observations, which largely reflects an increasing number of the available brands having obtained the Swan label. The point of the model described in Section 3 is precisely to disentangle the supply side effects of more brands being labeled from the possible shifts in consumers' brand choice generated by labeling.

4.2. Choices and variables in the model

Before estimating the model it is necessary to consider the relevant number of alternative brands in the choice situation of the individual household. One option is to restrict the choice set of each choice occasion to the number of brands available in the shop in which purchases were actually made. However, some households may change shops when a brand sold in another shop changes its attribute vector. In particular some (committed) households may choose to take their business to another shop if this allows them to obtain a Swan labeled brand. A choice model with all brands being made available to all households may be able to capture such behavior. This model, on the other hand, implies the somewhat unrealistic assumption that each household is informed about the price and special offers in shops it has not visited in a given week. In addition, some brands are not available to all households because of geographical restrictions in supply.

Two separate sets of choice models will be estimated, where the households are allowed to choose (1) only among the brands available in the shop in which the purchase was actually made and (2) between all brands available in the market (at that week).

For any choice situation the only price (P) recorded in the data is that of the purchased brand. To obtain a measure of the price for each of the other brands in the choice set we use the average price for all purchases of each brand within the same week (prices are measured as price per roll or per standard wash for each brand).[11] In the consumer diaries each consumer indicates whether a chosen brand was purchased on sale as a special offer. At the individual level this is a dummy variable, but in order to obtain information on the sale offers of other brands an ONS variable was calculated in the same way as the price variable. In most cases this variable has the value 0 or 1, but it may also be a ratio between 0 and 1 (households may give contradicting response to the question on whether the good was purchased on sale). The marketing variable is denoted ADV,[12] while the label (dummy) variable is called SWAN. Summary statistics for the main explanatory variables is provided in Table 3.

In 1999, a consumer test carried out on compact detergents was reported in the Danish Consumer magazine (*Råd & Resultater no. 10. 1999*). Results of what appeared to be a thorough washing test as well as the environmental performance of the different compact brands were described in the test. As expected the two brands that carried the Swan in 1999 (Bluecare and Tusindfryd) got a top score in environmental performance. Two of the largest brands on the market (Ariel and Omo) were pointed out as having a bad environmental performance. A variable denoted CTESTENV takes the value 1 and −1 for the brands that scored high and low on environmental performance. More surprisingly, the washing tests also gave a top score to the two Swan labeled brands (Bluecare and Tusindfryd), while one brand (Neutral) obtained a low score. The variable CTEST (for washing test) takes the value 1 and −1 for these three brands.

The consumer test report for compact detergents obtained a lot of press. Bluecare and Tusindfryd were pointed to as "winners". Besides emphasizing that the most environmentally friendly brands on the market were the best at cleaning dirty clothes it was also noted that the price of these brands was relatively low. Ariel and Omo were described as the losers of the test. In addition to having the worst environmental performance it was emphasized that they were expensive without washing better than the most environmentally friendly brands (despite the fact that they actually were in the runner up group in the laundry test). As a way to control for the

[11] In the choice model conditioning on the supply of brands in each shop we use the average price based on observations in the respective shops. In the choice model including all brands it is all observations in the week for the particular brand (across all shops). For some of the less frequently sold brands there are weeks with zero purchases. In these cases we rely on the monthly mean of the "standard" price (i.e. not including on-sale prices). Inspections of average weekly prices in the same shop suggest that these are fairly stable on a week to week basis (when the brand is not on sale).

[12] In the models presented we use marketing effort for the week in which the purchase was actually taking place. We have experimented with other formulations of this variable, which allow for a gradual decrease over time in the effect of the marketing effort (a stock approach to marketing effort), but this did not (qualitatively) change the results. However, note that a good brand reputation earned from past marketing effort will be captured by the parameters on the brand dummies.

Table 3
Summary statistics for the explanatory variables

	Mean	Std	min	max
Toilet paper ($N=45{,}796$)				
P	2.100	0.739	0.833	5.269
ONS	0.314	0.330	0	1
SWAN	0.176	0.381	0	1
ADV	0.003	0.022	0	0.316
Paper towels ($N=28{,}198$)				
P	3.750	1.054	1.738	7.983
ONS	0.319	0.328	0	1
SWAN	0.223	0.416	0	1
ADV	0.007	0.040	0	0.610
Detergents ($N=16{,}738$)				
P	1.522	0.446	0.423	3.342
ONS	0.498	0.304	0	1
SWAN	0.094	0.292	0	1
ADV	0.104	0.184	0	1.064

Note: Price (P) is measured in DKK per roll or per (standard) wash. One $ was equal to 5.94 DKK on 1.1.1997 and 8.02 DKK on 31.1.2001.

publicity effect of the consumer report a dummy variable for good publicity (CGODPUB) takes the value 1 for Bluecare and Tusindfryd, while another dummy variable (CBADPUB) takes the value 1 for Ariel and Omo.

In order to identify and control for the effects of the consumer report three alternative regressions were carried out for detergents. First, without any indicators of the consumer test, next with CTESTENV and CTEST (to control for the *results* of the consumer test) and finally instead with CGODPUB and CBADPUB (focusing on the *publicity* effect of the consumer report).[13]

Basically, we would like to control for all factors that over time influence the sale of different brands of the products analyzed. Most supermarket chains distribute weekly sales flyers stressing the special offers in the week. It is a widely held view of actors on the Danish market that these sales flyers have a large influence on (at least some) consumers' behavior. Brands on sale are also often given a better location in the store during the week of promotion. Systematic information on the exposure of the different brands in these sales flyers (or other promotional activities in the shops) was not available, but the exposure of brands in the sales flyers is very likely to be correlated with the on-sale variable.[14] The ONS variable is included in the model in order to

[13]A referee suggested that the effect of the consumer reports might diminish over time. A number of modelling exercises indicated that there was a diminishing effect over time, though it appeared only to be modest (perhaps because the time period after the test covered by data is fairly short). Further, the effect on the SWAN parameter of different specifications was negligible and therefore we have kept the simple formulation with a constant effect of the consumer report.

[14]Part of the marketing literature has devoted attention to the empirical problems associated with assessing the effect of coupons. It is therefore worth noting that coupons are almost never used in the Danish market.

account for the promotional effects of a sale, beyond the price changes associated with a special offer (the latter effect should in principle be captured by the price variable).

When conditioning on shops we also control for 'overall' changes in supply strategy of each shop, as we can observe when a certain shop no longer sells a particular brand (to be more precise, we observe if none of the respondents purchases a brand from the shop). However, a number of other supply decisions of the shop managers—such as the choice of more or less favorable locations in the stores—are not controlled for in the model.

5. Results

5.1. Estimation results

We start by presenting results when the choice set is determined by the brands available in the shop in which the consumers purchased the good in question.[15] In all presented estimations, normally distributed random effects are included (as parameters on the brand dummies) in order to control for differences in unobserved brand quality (and heterogeneity across households in their perception of this quality). Hence for each brand (except for one base brand) we estimate the mean and the standard deviation (α_j and σ_j) of the household specific brand dummy coefficient. For presentational ease the estimated coefficients to these variables are not included in the tables.[16]

Consider first the estimates of the coefficients in model 1 for toilet paper, paper towels and detergents as shown in Table 4 (choice set conditioned on brands in shop). As expected, the estimated coefficient capturing the effect of the price (P) is negative, while the corresponding coefficient for the on-sale variable (ONS) is positive. The estimated coefficient on media marketing (ADV) is significant at the 5% level and positive only for some of the estimated models for detergents.

The estimate of the SWAN coefficient is positive for all goods. The coefficient is significant at a 1% level for toilet paper and at a 10% level for paper towels. For detergents the size and significance of the coefficient on the SWAN depend on the way that the effect of the consumer report is controlled for in the model. When the consumer report is not controlled for the coefficient is highly significant, but when we control for the results of the consumer test the coefficient on the SWAN is reduced and only significant at a 5% level. Finally, when we instead control for the "publicity effect" of the consumer report the coefficient is insignificant.

In the last presented model for detergents, where the publicity effect of the consumer test is controlled for, the increase in the sale of Bluecare and Tusindfryd after the consumer report was released is effectively "dummied out" of the model, in the sense that nearly all label observations after the release of the consumer test at the end of 1999 are confounded with the publicity variable CGODPUB. It is difficult to know from this regression whether the increases in the purchases of

[15] Estimation was carried out with Gauss 3.5 using the simulated maximum likelihood method as described in Revelt and Train [28]. The program used was an adaption of Kenneth Train's program code, which he kindly has made available at http://elsa.berkeley.edu/~train/software.html. 500 Halton draws were used for estimation.

[16] Most of the σ_j are (highly) significant, which suggests that it is important to take into account heterogeneity in household brand preferences and the correlation among decisions over time.

Table 4
Mixed repeated logit model for toilet paper, paper towels, and detergents (brands in choice set conditioned on shop[a])

	Toilet paper[b]				Paper towels			
	Model 1		Model 2		Model 1		Model 2	
	Coeff.	*t*-ratio	Coeff.	*t*-ratio	Coeff.	*t*-ratio	Coeff.	*t*-ratio
P	−0.518***	−9.17	−1.267***	−16.29	−0.194***	−6.73	−0.688***	−22.62
SWAN	0.327***	3.90	0.462**	2.30	0.145*	1.71	0.098	1.16
ONS	1.322***	20.21			1.466***	28.06		
ADV	−0.827*	−1.71	−0.738	−1.46	0.292	1.32	0.153	0.68
N-obs.		21,701[b]		21,701[b]		25,238		25,238
N-househ.		1,466[b]		1,466[b]		2,483		2,483
log *L*		−23,599.76		−24,065.81		−27,010.4		−27,872.35

	Detergents (without results of consumer report)				Detergents (with results of consumer reports)				Detergents (with publicity effect of consumer report)			
	Model 1		Model 2		Model 1		Model 2		Model 1		Model 2	
	Coeff.	*t*-ratio	Coeff.	*t*-ratio	Coeff.	*t*-ratio	Coeff.	*t*-ratio	Coeff.	*t*-ratio	Coeff.	*t*-ratio
P	−0.596***	−7.03	−1.953***	−23.97	−0.726***	−8.12	2.018***	−25.01	−0.716***	−8.66	−2.036***	−25.48
SWAN	1.220***	4.88	1.267***	5.40	0.581*	2.15	0.545*	1.83	0.124	0.41	−0.120	−0.39
ONS	1.642***	21.54			1.572***	21.01			1.583***	21.44		
ADV	0.231**	2.31	0.282**	2.76	0.219*	2.20	0.196*	1.89	0.184*	1.81	0.175*	1.70
CTEST					1.771***	7.48	1.724***	7.60				
CTESTENV					0.433***	5.57	0.589***	7.95				
CGODPUB									3.177***	8.16	3.335***	10.03
CBADPUB									−0.615***	−5.78	−0.815***	−7.77
N-obs.		16,218		16,218		16,218		16,218		16,218		16,218
N-househ.		2,028		2,028		2,028		2,028		2,028		2,028
log *L*		−16,758.40		−17,454.75		−16,604.30		−17,143.75		−16,451.05		−17,035.83

Note: Coefficients to the brand dummies (mean and s.d. of the normal distribution) estimated, but not shown here. The *t*-ratio is calculated using robust standard errors. One asterisk (*), two asterisks and three asterisks indicate that the coefficient is significant at a 10% level, 5% level and 1% levels, respectively. 500 Halton draws were used for estimation of the parameters.

[a]Purchases from the Aldi supermarket chain not included, because this chain only sells "housebrands" (since there is no overlap with brands sold in other supermarkets, it is not possible to identify the coefficients to the brand dummies of the Aldi brands in the model with choice conditioned on shop).

[b]For toilet paper we use only a random sample of half the households (due to the size of the whole data set for toilet paper it was not feasible to estimate the model for all the households).

Bluecare and Tusindfryd after the release of the consumer report (i.e. the positive coefficient on CGODPUB) derived from the good results in the washing test, the emphasis of the relative low price of the two brands, or the emphasis on the two brands being Swan labeled (and testing as environmentally friendly). However, the results from the model with the results of the consumer report indicate that both the (private) washing quality of detergents (captured by CTEST) as well as the environmental performance (captured by CTESTENV) are significant with the expected sign.[17] This suggests that information on environmental performance, deriving either from an environmental label or from widely known consumer test reports, has had a significant influence on respondents' brand choice.

Changes in P are highly correlated with the ONS variable. In model 2 (for toilet paper, paper towels and detergents, respectively) ONS has been excluded as an explanatory variable. This has a large impact on the estimated price coefficient, which increases in absolute value by a factor of approximately 3. Thus, it appears that ONS also captures some of the effect of the price changes. As is normal with highly correlated variables, this is open for interpretation.

For toilet paper and detergents the effect of the Swan label is fairly similar in model 1 and model 2. E.g. the coefficient on SWAN increases from 0.327 in model 1 to 0.462 in model 2 for toilet paper, while it decreases from 0.145 in model 1 to 0.095 in model 2 for paper towels (and is no longer significant at the 10% level). Hence, the inclusion/omission of ONS does not seem to be important for an evaluation of the effect of the Swan label as its effect on utility is fairly stable. However, when we later use the equations to estimate marginal WTP for the Swan attribute the change in the coefficient on P by a factor of 3 yields a similar change in marginal WTP. So the inclusion/omission of ONS is by no means trivial.

Results of choice models where all brands are included in the choice set are given in Appendix B. By and large the estimation results from this specification of the choice set are similar to the results given in Table 4. Hence, coefficients on P and ONS have the expected signs and are generally significant, while the effect of advertisement (ADV) is positive for detergents and paper towels. As in Table 4, the positive sign on ADV is only significant in a number of the models for detergents. However, a significant negative sign on the coefficient to ADV is found for toilet paper, when all brands are included in the choice set.

The significant negative sign of advertisement on choice of toilet paper is not plausible, but as described in Section 4.1 the definition of brands from the data source of advertisement did not correspond unambiguously to the brand definitions from the consumer panel. However, omission of ADV for toilet paper does not have any impact on the sign and size of the coefficient on SWAN (which is our main concern here). It is also worth noting that advertising expenditures are considerably higher for detergents than the paper products, i.e. the direct cost of advertising measured by the index is five times higher for detergents as compared with the sums used on both toilet paper and paper towels, even though household expenditures on detergents are lower than the expenditures on toilet paper and paper towels (taken together). Hence, it seems that market actors agree that the return on advertisement spending is higher for detergents, which corresponds well with the significant positive coefficient on ADV in most of the estimated models for detergents.

[17] Note, however, that the log-likelihood value is higher in the model with the publicity effect of the consumer reports as compared with the models, which includes only the test results. Therefore, the more interpretable results of the latter model may be subject to omitted variable bias.

With respect to the effect of the label the coefficient on SWAN is significant and in the same range as in Table 4 for toilet paper. For paper towels the coefficient on SWAN is slightly lower and not even significant at the 10% level. For detergents the effect of the SWAN depends on the inclusion of indicators for the consumer test (or the publicity effect of this) in the same way as before.

Summarizing, it appears that the Swan label had a significant influence on the choice of toilet paper, while the impact on paper towels and detergents is less clear. For paper towels the effect of the Swan label appears to be lower, and only significant at a 10% level in one of the four presented models. For detergents the effect of the environmental label is to a large extent confounded with the results of a consumer test report that had a large impact on the market in the period analyzed. Nevertheless, results suggest that information on environmental performance, deriving either from an environmental label or from the widely known consumer test reports, has had a significant influence on respondents' brand choice.

5.2. Marginal WTP

Based on the estimates of the coefficient on P own- and cross-price elasticities of (probability of) brand choice can be calculated for each alternative in the choice set. When ONS is included (model 1) choice of brand is, in general, price inelastic, as most of the own price elasticities are below unity. When ONS is omitted (model 2) choice is price elastic as most elasticities are above unity. It seems reasonable that different brands of toilet paper, paper towels, or detergents (located next to each other in the same shop) should be close substitutes, and that the price elasticities of choice should be larger than minus one. Hence, model 2 seems to us to imply the most plausible results with respect to the price effects.

In the linear utility model applied here an "implicit price" or marginal WTP for a certified environmentally friendly brand may be obtained by dividing the estimated coefficient on SWAN by the coefficient on P (Eq. (2) above). The marginal WTP estimates for the Swan label are presented in Table 5.

Looking first at the marginal WTP for Swan-labeled toilet paper, it is 2–3 times higher in model 1 than in model 2. This difference is largely related to the difference in the estimated coefficient on P between the two models. As price elasticities appear more reasonable in model 2 focus will be on marginal WTP from this model. In the model where the choice set is conditioned by shop the marginal WTP is 13% of the price of labeled toilet paper (significant at the 5% level), while it is 18% of price in the model where all brands are included in the choice set (significant at the 1% level).

The marginal WTP for labeled paper towels is considerably lower than for toilet paper. The WTP is only significant at the 10% level in the model with the on-sale variable included when conditioning on the brands in shop. Looking, finally, at the WTP for detergents the correlation between the effect of the Swan label and the effect of consumer test report makes it difficult to draw a strong conclusion. However, in the model where we have controlled for the results of the consumer report we generally obtain a significant WTP. In the models without the on-sale variable the marginal WTP ranges from 17% (significant at 10%) to 29% (not significant). This is slightly higher, but in the same range as for toilet paper.

It is worth noting that the results for detergents derive from only a few brands becoming labeled (see Section 4.1), while the more robust results for toilet paper are obtained for a larger number of

Table 5
Marginal willingness to pay for environmentally labeled goods

	Conditioning on brands in shop				All brands in choice set			
	Model 1 (incl. of on-sale)		Model 2 (excl. of on-sale)		Model 1 (incl. of on-sale)		Model 2 (excl. of on-sale)	
	WTP (DKK per roll/wash)	% of price	WTP (DKK per roll/wash)	% of price	WTP (DKK per roll/wash)	% of price	WTP (DKK per roll/wash)	% of price
Toilet paper	0.63***	22.9%***	0.37**	13.3%**	1.79**	65.2%**	0.50***	18.1%***
Paper towels	0.75*	15.1%*	0.14	2.9%	0.44	8.9%	0.10	1.9%
Detergents								
– with result of consumer report	0.80**	50.3%**	0.27*	17.0%*	0.91***	57.1%***	0.46	29.2%
– with publicity effect of consumer report	0.17	10.9%	−0.06	−3.7%	0.39	24.7%	0.18	11.3%

Based on parameter estimates from Table 4 and Appendix B.
One asterisk (*), two asterisks and three asterisks indicate that the marginal WTP is significant at a 10% level, 5% level and 1% levels, respectively. Significance levels for WTP were calculated using the Delta Method.

brands becoming labeled at different points in time. For paper towels several brands became labeled, though not as many as for toilet paper.

Initially, the differences in the results for toilet paper and paper towels may appear surprising as these two paper products in many aspects are very similar (except for the height of the roll). But, closer examination reveals that toilet paper and detergents are purchased by practically all households, whereas paper towels are purchased frequently by a smaller fraction of households. The smaller number of observations for paper towels, combined with the lower number of labeled brands, does also (other things being equal) suggest that it will be more difficult to identify an effect of the Swan on choice of paper towels. Beyond that, O'Brien and Teisl [26] note that focus group participants stated that they were more inclined to purchase environmentally labeled brands if the product was a frequently purchased item, because this would allow them to make a greater environmental impact. In addition, it also seems likely that households frequently using paper towels are likely to be on average less environmentally concerned than households not using paper towels. That is, there are reusable substitutes for them in most applications such as the dishcloth still used widely in Denmark. The need to wipe up spills is probably constant across households, but the choice set for method is broader than just brands of paper towels. The "green" consumers may well be choosing to use a reusable alternative rather than *any* brand of paper towels.[18]

5.3. Discussion of alternative model specifications

A number of alternative specifications of the model have been estimated, some of which will briefly be described here.[19] In the models presented above, the effect of the Swan label is assumed constant over the whole time period. General consumer surveys carried out between 1997 and 2000 suggest that the proportion of Danish consumers recognizing/understanding the Swan label has increased since Denmark joined the Swan label scheme and one would therefore expect the effect of the Swan label to be lower in the beginning of the time period analyzed. In order to account for such an effect alternative models have been estimated where the SWAN dummy has been multiplied by the share of the general population understanding the label at a certain point in time (based on linear interpolation of point estimates from three similar surveys on the recognition/understanding of the Swan label). However, the qualitative results with this specification of the effect of the label are similar to the ones presented in the paper and the pseudo R^2's of the respective models are almost identical. In addition, it also appears that survey results for recognition/understanding of the Swan label are very sensitive to the formulation of the question as two other surveys using a different formulation of the questions indicate a considerably higher level of understanding.

We have also explored whether the labeling has a stronger effect on households with certain socioeconomic characteristics (by interacting these with the SWAN variable). It is interesting to note that little consistency could be found in the effect of the Swan for respondents with different

[18] One referee also suggested that the result may derive from health concerns of the paper towel users. A dishcloth may harbor and spread food pathogens, so paper towel users may simply be more concerned about food safety (as compared with environment).

[19] For a more detailed description of most of the alternative specifications see [5].

levels of education or income. That is in some cases the effect of the label varied significantly with level of education and income, but the directions were not consistent across different products groups.[20] With respect to income a similar result was obtained in [34]. Results were most consistent across product groups for gender and children in the household. Males shopping (on their own) as well as households with children were generally *less* likely to be influenced by the Swan.

Finally, in the models presented above heterogeneity in consumer preferences was reflected in the random parameters on the brand dummies. As an alternative we have also tried to model the heterogeneity by allowing the brand preferences to be functions of socioeconomic characteristics like income level, age, etc. This was carried out by including interaction terms between the brand dummies and income level and age (with fixed coefficients estimated on these interaction terms). However, the log-likelihood value is considerably higher in the model where random parameters are included as opposed to socioeconomic interaction terms.[21] This indicates that heterogeneity in brand preferences is not captured by the observed characteristics of the households. A similar result is noted by [28].

6. Summary and conclusion

In this study we have estimated models for consumers' choices among different brands of toilet paper, paper towels and detergents in order to trace effects of the Nordic certified environmental label on consumer preferences. In the period covered by the data a number brands obtained the label at different points in time and the data include information on purchases before and after the brands were labeled, which allows us to control for (unobserved) differences in the qualities of the different brands. In addition, this same fact (that brands have obtained labels at different points in time) makes it unlikely that the identified effect of the label is confounded with unaccounted for trends over time in the respective markets. Models with different numbers of brands in the choice set have been estimated as a rough way to describe alternative decision structures of the individual.

It appears from this analysis that the environmental label has had a significant effect on the choice of toilet paper, where marginal willingness to pay for certified environmentally friendly brands ranges from 13% to 18% of the price. For paper towels the effect of the Swan label is lower (and often insignificant), perhaps because green consumers in Denmark choose to avoid using paper towels and instead rely more on a dishcloth to wipe up spills etc.

The effect of the label on choice of detergents was more difficult to identify, because fewer brands have obtained the label and because sales of these brands were also influenced by results of a consumer report, which had a large impact on the detergents market in Denmark. Results

[20] As an example for toilet paper high income was associated with a significant increase in the effect of the SWAN, but a significantly lower effect of the SWAN was found for respondents with high income when purchasing detergents (for paper towels there was no significant differences in the effect of the Swan between income groups).

[21] As an example the pseudo R^2 of the paper towels equation with all brands in the choice set (and with ONS) is 0.38 in the model with random parameters on the brand dummies, while it is only 0.16 with fixed parameters on the interaction terms between the brand dummies and income and age (even though there are more estimated parameters in the latter model).

indicate, however, that the marginal willingness to pay (as a percentage of price) for certified environmentally labeled brands of detergents may be in the same range as for toilet paper. In any case, results for detergents also suggest that a good (bad) environmental reputation, obtained in this case from the consumer test report, can result in a significant increase (decrease) in sales.

The fact that (Danish) consumers are willing to act on an environmental label (or environmental product information released by a credible consumer test report), even though this does not yield any direct benefit to them, indicates the presence of altruistic motives of some kind. It seems that a number of conditions conducive to the success of environmental labels are to be found in Denmark. Most prominently, there is a great confidence in the government (which certifies the label), environmental issues receive substantial attention in the media, and there appears to be wide acceptance of a policy of pursuing relatively ambitious environmental goals. So it is an important, but unanswered question whether the positive effect of the environmental Swan label found in the Danish market will also be found in other countries, especially those outside the Nordic group.

Acknowledgments

The research was supported by the Danish Social Science Foundation. The authors are grateful for invaluable research assistance from Laura Mørch Andersen and Tore Olsen, as well as for comments on previous versions of the paper from Nick Hanley, Martin Browning, Christopher Clark, Christian Hjorth-Andersen and two anonymous referees. Finally, we are grateful for the editor's helpful suggestions with respect to the specification of the model.

Appendix A. Utility foundation for brand choice of storable goods

Let Ω denote household utility per unit of time, which is assumed to depend on sub-utility derived from consuming the good in question (denoted V) and sub-utility of consuming other goods, which for simplicity is set equal to the part of income allocated to other goods, i.e.

$$\Omega_j = \Omega(V, Y - q_j p_j) \tag{1}$$

where q_j is the quantity consumed of the good in question, p_j is the per unit price (the index j indicates that brand j of the good is consumed) and Y is total household income, so that income allocated to other goods becomes $Y - p_j q_j$. For notational simplicity, the household and time period indexes (i, t) have been suppressed. We further assume that good sub-utility is linear in the consumed good attributes, i.e.,

$$V(q_j, X_j, a_j) = q_j(bX_j + a_j),$$

where X_j is the vector of per unit observed attributes of the consumed brand, b is the vector of parameters of the linear sub-utility function for the observable good attributes and a_j captures the sub-utility effect of unobserved brand specific attributes.

Since V is linear in q_j we may normalize with respect to V-sub-utility which is an advantage in the following. Defining c_j as the per sub-utility unit cost of consuming brand j (so that

430 T.B. Bjørner et al. / Journal of Environmental Economics and Management 47 (2004) 411–434

$c_j = p_j/(bX_j + a_j)$) Eq. (1) may be written as:

$$\Omega_j = \Omega(V, Y - Vc_j). \tag{1a}$$

The goods in question are easily stored, so that consumers typically will buy goods for stock, taking advantage of sales and special offers while consuming from stock based on their expectation of the cost per sub-utility unit faced when stocks are replenished. Thus, the household is faced with two distinct (although linked) problems in the short run: the problem of *consuming* out of stocks and the problem of *purchasing* to replenish stocks. The fact that consumption and purchases may deviate substantially in the short run and that it is rational to base consumption decisions on the expected cost of replenishing stocks would make consumption and inventory modeling a challenge. Instead, we assume that (unobserved) expected cost of replenishing stocks is constant over the data period. This makes developing an explicit model of consumption and stocks unnecessary for our purposes—it turns out that modeling household brand choice in the purchasing situation is sufficient.

Consider the consumption decision implied by the normalized utility equation (1a) where we let c^e denote the cost per sub-utility unit V that the consumer expects to encur when stocks are replenished. The optimal quantity of sub-utility to consume (denoted V^*) is set according to the first-order condition derived from (1a), i.e. so that:

$$\Omega_v - \Omega_Y c^e = 0.$$

Thus, when optimal consumption is implemented we have that marginal utilities $\Omega_V^* = \Omega_V(V^*(c^e), Y - c^e V^*(c^e))$ and $\Omega_Y^* = \Omega_Y(V^*(c^e), Y - c^e V^*(c^e))$ only depend on the expected replenishing cost (trivially if brand j is consumed the optimal quantity to consume is $q_j^* = (V^*(c^e)/(bX_j + a_j))$.

The key assumption we make here is that the *expected* replenishing cost per sub-utility unit (c^e) is constant over the estimation period. The assumption implies that although the purchased brand and price may vary between purchasing situations the mean price per sub-utility unit that the consumer expects to pay in the future does not change over the data period. Subject to this assumption Ω_V^* and Ω_Y^* will be constant across different purchasing situations, which, as we shall see below, allows us to discover willingness to pay for attributes through a multinomial model of brand choice. Though restrictive, the assumption does not seem unreasonable as the mean prices of brands overall are very stable over the data period.

When the household is faced with a purchasing opportunity, it must choose which brand to purchase, if any, as well as the quantity (Q) to purchase. In the purchasing situation good attributes should be evaluated at their marginal utility value in consumption, as should costs. The marginal effect on utility of *purchasing* a unit of brand j is the marginal utility value in consumption of the purchased attributes of j net of the marginal utility of the reduction in consumption of other goods needed to finance the purchase. Using the original formulation of the utility equation (1) and letting U_j denote marginal utility of purchasing a unit of brand j we have:

$$U_j = \Omega_V^*(bX_j + a_j) - \Omega_Y^* p_j, \tag{2}$$

where Ω_V^* and Ω_Y^* will be constant across purchasing situations given the assumed constancy of expected replenishing costs. The optimal quantity to purchase in the given purchasing situation is restricted by the utility costs of transporting goods home, storing them, etc. Letting TU_j denote the total utility effect of purchasing Q_j units of brand j we have:

$$TU_j = U_j Q_j - C(Q_j),$$

where $C(\cdot)$ denotes the purchase specific utility costs which are assumed to be increasing and convex. Let Q_j^* denote the optimal purchase volume given that brand j is purchased. Q_j^* is set to satisfy the corresponding first-order condition:

$$\frac{dTU_j}{dQ_j} = U_j - C_Q(Q_j) = 0.$$

Although the $C(\cdot)$ function typically varies from purchasing situation to purchasing situation (the purchase specific utility costs would typically depend on distance to the shop and on whether the buyer is transporting purchases on foot, on a bike, on a bus, or in a car) the utility cost function can reasonably be assumed identical across brands for the given purchasing situation.

Let us now consider the choice of brand. In a given purchasing situation the household will prefer brand h over brand j when

$$U_h Q_h^* - C(Q_h^*) > U_j Q_j^* - C(Q_j^*),$$

i.e., when utility of buying the optimal quantity of brand h (given that brand h is chosen) is greater than utility of buying the optimal quantity of brand j (given that brand j is chosen). By the first-order condition this is satisfied if, and only if, marginal utility of purchasing brand h is greater than marginal utility of purchasing brand j, i.e.,

$$U_h > U_j$$

$$\Leftrightarrow$$

$$[\Omega_V^* b]X_h + [\Omega_V^* a_h] - [\Omega_Y^*]p_h > [\Omega_V^* b]X_j + [\Omega_V^* a_j] - [\Omega_Y^*]p_j.$$

This expression is independent of purchase specific costs $C(\cdot)$ and given the assumption of constant expected replenishing price (ensuring that Ω_V^* and Ω_Y^* are constant) the parameters to be estimated (indicated by []) do not vary between purchase situations. Adding random components e_h and e_j to capture the effect of unobservable variables the probability of choosing brand h over brand j in any given purchase situation is

$$P_h = Prob(U_h - U_j > e_j - e_h) \quad \text{for all } j \neq h.$$

This relationship is straightforward to estimate within the logit setup (see Section 3).

The model's main advantage is that even though the complex decision problems of household consumption and purchased quantity are not modeled, the parameters that are identified suffice for calculating willingness to pay (WTP) for the Swan label attribute at the consumption margin—which is the focus of attention here. Since marginal WTP by definition is the income reduction needed to ensure constant utility, when an attribute is added to the marginal good, the WTP for attribute l is given by

$$[\Omega_V^* b_l] - [\Omega_Y^*]WTP_l = 0$$

so that $WTP_l = [\Omega_V^* b_l]/[\Omega_Y^*]$. Using the specification in Section 3 this is estimated by $-\beta_l/\beta_p$.

Table 6
Mixed repeated logit model for toilet paper, paper towels and detergents (all brands in choice set)

	Toilet paper[a]				Paper towels			
	Model 1		Model 2		Model 1		Model 2	
	Coeff.	t-ratio	Coeff.	t-ratio	Coeff.	t-ratio	Coeff.	t-ratio
P	−0.202	−1.47	−0.932***	−17.39	−0.242***	−8.79	−0.623***	−24.0
SWAN	0.362***	2.78	0.464***	6.87	0.106	1.28	0.059	0.75
ONS	1.145***	16.67			1.006***	21.82		
ADV	−1.058**	−2.51	−0.878**	−2.45	0.367*	1.86	0.135	0.68
N-obs.		24,356[a]		24,356[a]		28,198		28,198
N-househ.		1,492[a]		1,492[a]		2,538		2,538
Log L		−51328.26		−51481.09		−56756.83		−57230.00

	Detergents (without results of consumer report)				Detergents (with results of consumer reports)				Detergents (with publicity effect of consumer report)			
	Model 1		Model 2		Model 1		Model 2		Model 1		Model 2	
	Coeff.	t-ratio	Coeff.	t-ratio	Coeff.	t-ratio	Coeff.	t-ratio	Coeff.	t-ratio	Coeff.	t-ratio
P	−0.775***	−7.75	−1.574***	−20.00	−0.922***	−8.82	−1.605*	−6.93	−0.866***	−8.32	−1.575***	−19.95
SWAN	1.386***	8.23	1.461***	8.64	0.837***	3.39	0.744	1.24	0.340	1.49	0.282	1.29
ONS	0.939***	11.29			0.827***	9.62			0.839***	9.77		
ADV	0.253***	2.98	0.284***	3.35	0.237***	2.74	0.245	1.13	0.160*	1.88	0.210**	2.44
CTEST					1.830***	9.14	1.808	1.16				
CTESTENV					0.405***	5.75	0.489***	6.36				
CGODPUB									3.076***	12.93	3.155***	12.88
CBADPUB									−0.562***	−6.70	−0.628***	−7.04
N-obs.		16,738		16,738		16,738		16,738		16,738		16,738
N-househ.		2,055		2,055		2,055		2,055		2,055		2,055
log L		−27039.90		−27141.82		−26605.66		−26616.06		−26478.85		−26526.98

Note: Coefficients to the brand dummies (mean and s.d. of the normal distribution) estimated, but not shown here. The t-ratio is calculated using robust standard errors. One asterisk (*), two asterisks and three asterisks indicate that the coefficient is significant at a 10% level, 5% level and 1% levels, respectively. 500 Halton draws were used for estimation of the parameters.

[a] For toilet paper we use only a random sample of half the households (due to the size of the whole data set for toilet paper it was not feasible to estimate the model for all the households).

Appendix B

Mixed repeated logit model with all brands in choice set is given in Table 6.

References

[1] G.M. Allenby, P.J. Lenk, Reassessing brand loyalty, price sensitivity, and merchandising effects on consumer brand choice, J. Business Econom. Statist. 13 (3) (1995) 281–289.

[2] J. Andreoni, Privately provided public goods in a large economy: the limits of altruism, J. Public Econom. 35 (1988) 57–73.

[3] J. Andreoni, Cooperation in public goods experiments: kindness or confusion?, Amer. Econom. Rev. 85 (1995) 891–904.

[4] J. Bennett, R. Blamey, J.J. Louviere, M. Morrison, Green product choice, in: J. Bennett, R. Blamey (Eds.), The Choice Modeling Approach to Environmental Valuation, Edward Elgar, Cheltenham, UK, 2001.

[5] T.B. Bjørner, L.G. Hansen, C.S. Russell, T. Olsen, The Effect of the Nordic Swan Label on Consumers' Choice, AKF Forlaget, Copenhagen, 2002. (http://www.akf.dk/eng2002/swan.htm).

[6] R. Blamey, J. Bennett, Yea-saying and validation of a choice model of green product choice, in: J. Bennett, R. Blamey (Eds.), The Choice Modeling Approach to Environmental Valuation, Edward Elgar, Cheltenham, UK, 2001.

[7] F. Cairncross, Costing the Earth: The Challenge for Governments, the Opportunities for Business, Harvard Business School Press, Boston, MA, 1992.

[8] T.N. Cason, L. Gangadharan, Environmental labeling and incomplete consumer information in laboratory markets, J. Environ. Econom. Manage. 43 (2002) 113–134 doi:10.1006/jeem.2000.1170.

[9] C. Dosi, M. Moretto, Is eco-labelling a reliable environmental policy measure?, Environ. Res. Econom. 18 (1) (2001) 113–127.

[10] EPA, Determinants of Effectiveness for Environmental Certification and Labeling Programs, EPA 742-R-94-001, United States Environmental Protection Agency, Office of Pollution Prevention and Toxics, Washington, D.C., April 1994.

[11] B. Frey, F. Oberholzer-Gee, The cost of price incentives: an empirical analysis of motivation crowding-out, Amer. Econom. Rev. 87 (1996) 746–755.

[12] P.M. Guadagni, J.D.C. Little, A logit model of brand choice calibrated on scanner data, Market. Sci. 2 (3) (1983) 203–238.

[13] K.E. Henion, The effect of ecologically relevant information of detergent sales, J. Market. Res. 9 (1) (1972) 10–14.

[14] J.A. Herriges, D.J. Phaneuf, Controlling for correlation across choice occations and sites in a repeated mixed logit model of recreation demand, Amer. J. Agric. Econom. 84 (4) (2002) 1076–1090.

[15] O. Johansson, Optimal Pigovian taxes under altruism, Land Econom. 73 (1997) 297–308.

[16] P.W. Kennedy, B. Laplante, J. Maxwell, Pollution policy: the role for publicly provided information, J. Environm. Econom. Manage. 26 (1) (1994) 31–43.

[17] S. Kirchhoff, Green business and Blue Angels: a model of voluntary overcompliance with asymmetric information, Environm. Res. Econom. 15 (4) (2000) 403–420.

[18] T.P. Lyon, J.W. Maxwell, "Voluntary" approaches to environmental regulation: a survey, in: M. Franzini, A. Nicita (Eds.), Economic Institutions and Environmental Policy Aldershot, Ashgate Publishing Ltd., Hampshire, 2002.

[19] C.F. Mason, On the economics of eco-labeling, Paper presented at the EAERE 2001 Conference in Southampton, 2001.

[20] A. Mattoo, H.V. Singh, Eco-labeling: policy considerations, Kyklos 47 (1) (1994) 53–65.

[21] K.E. McConnell, Does altruism undermine existence value?, J. Environ. Econom. Manage. 32 (1997) 22–37.

[22] D. McFadden, K. Train, Mixed MNL models for discrete response, J. Appl. Econom. 15 (2000) 447–470.

[23] E.R. Morey, R.D. Rowe, M. Watson, A repeated nested-logit model of Atlantic salmon fishing, Amer. J. Agric. Econom. 75 (3) (1993) 578–592.

[24] W. Nimon, J. Beghin, Are eco-labels valuable?, evidence from the apparel industry, Amer. J. Agric. Econom. 81 (4) (1999) 801–811.

[25] K. Nyborg, Homo Economicus and Homo Politicus: interpretation and aggregation of environmental values, J. Econom. Behav. Org. 42 (3) (2000) 305–322.

[26] K.A. O'Brien, M.F. Teisl, Factors affecting consumer valuation of environmentally labeled forest products, Paper presented at the 2001 meeting of the Northeastern Association of Resource and Environmental Economists, 2001.

[27] OECD, Eco-Labelling: Actual Effects of Selected Programmes, OECD/GD(97)105. OECD, Paris, 1997.

[28] D. Revelt, K. Train, Mixed logit with repeated choices: households' choices of appliance efficiency level, Rev. Econom. Statist. 80 (4) (1998) 647–657.

[29] B. Roe, M.F. Teisl, A.S. Levy, M. Russell, US consumers' willingness to pay for green electricity, Energy Policy 29 (11) (2001) 917–925.

[30] C.S. Russell, C.D. Clark, The potential effectiveness of the provision of consumer information on product environmental characteristics as a regulatory tool, Paper presented at the SØM-conference November 1999 in Copenhagen, Denmark, 1999 (www.akf.dk/som/index.html).

[31] A. Sen, Behavior and the concept of preferences, Economica 40 (1973) 241–259.

[32] P.C. Stern, Information, incentives, and proenvironmental consumer behavior, J. Consumer Policy 22 (4) (1999) 461–478.

[33] S.K. Swallow, R.A. Sedjo, Eco-labelling consequences in general equilibrium: a graphical assessment, Land Econom. 76 (1) (2000) 28–36.

[34] M.F. Teisl, B. Roe, A.S. Levy, Ecocertification: why it may not be a "field of dreams", Amer. J. Agric. Econom. 81 (5) (1999) 1066–1071.

[35] M.F. Teisl, B. Roe, R.L. Hicks, Can eco-labels tune a market? evidence from dolphin-safe labeling, J. Environ. Econom. Manage. 43 (2002) 339–359 doi: 10.1006/jeem.2000.1186.

[36] K. Train, Discrete Choice Methods with Simulation, Cambridge University Press, Cambridge, 2003.

[13]

Eco-labels for credence attributes: the case of shade-grown coffee

BRUCE A. LARSON*

Department of Agricultural and Resource Economics, University of Connecticut, USA. E-mail: blarson@canr.uconn.edu

ABSTRACT. Using the case of shade-grown coffee, this paper examines the market impacts of 'eco-labels' for credence attributes. First, the Mattoo and Singh (1994) test is conducted for the case of shade-grown coffee to investigate the market impacts of a shade label. This analysis in section 2 shows that a shade label could 'pass' the test, but the market impacts are likely to be minor. Section 2 also shows how to use estimates of supply, potential demand, and price elasticities of demand and supply to predict eco-label premiums in the post-label equilibrium. And second, given the importance of consumer demands for eco-label impacts, and since the theoretical foundations of demand for eco-labeled items are not well developed in the literature, Section 3 takes a closer look at the microeconomics of labels and consumer demand. A nested constant-elasticity-of-substitution preference structure is used to derive theoretically consistent Marshallian demands for shade and non-shade coffee. A numerical simulation shows how relative prices and consumer preferences for the credence attribute and variety are both important factors in demand creation of labeled items.

I. Introduction

Eco-labels provide information to consumers on credence attributes of products. Some labels are voluntary programs and some are mandatory regulations, and some programs focus on consumption attributes (for example, food safety issues) and some focus on production process attributes (for example, sustainably harvested tropical timber). The focus of this paper is on voluntary eco-label programs designed to provide consumers with additional information on production process attributes.[1] By

* This work was supported by the Food Marketing Policy Center at the University of Connecticut. The author thanks, without implicating, Shinano Hayashi for research assistance during the completion of this analysis, and Ron Cotterill, Jeremy Foltz, Kathy Sergerson, Sara Scatasta, the Editor, and three reviewers for their substantive comments and suggestions during the completion of this analysis.

[1] For reference, eco-label programs were operating in at least 17 countries as of 1995. Such programs covered over 200 categories of products including fresh and processed foods, textiles and apparel, natural resources, and manufactured products (Vossenaar, 1997). Teisl and Roe (1998) provide a concise overview of various types of labeling schemes. Also see Karl and Orwat (1999), Markandya (1997), and Golan, Kuchler, and Mitchell (2000).

providing such information to consumers, it is hoped that an eco-label will stimulate additional demand for the labeled item and create a price premium for the labeled item in the market. As a result, the label program would help to make production of the eco-labeled product more competitive *vis-à-vis* conventionally produced items, increase supply of the labeled item, reduce supply of non-labeled items, and reduce negative externalities (or increase positive externalities) (Mattoo and Singh, 1994).

Given the interest in eco-labels, and the rapidly growing descriptive literature on eco-label programs (see, for example, the various chapters in Zarrilli, Jha, and Vossenaar, 1997), there has been limited theoretical analysis of the consumer response to labeled products, the resulting market impacts, and the eventual impact on reducing environmental externalities. Mattoo and Singh (1994) and Swallow and Sedjo (2000) begin to analyze the market impacts from introducing an eco-label program. Mattoo and Singh (1994) is a partial equilibrium analysis focused on voluntary labeling of a credence attribute for an otherwise homogeneous product. In the post-label equilibrium, 'eco-consumers' only purchase the labeled item, while 'non-eco-consumers' only purchase the item with the lowest price.[2]

Mattoo and Singh (1994) suggest an empirical test based on excess potential demand in the pre-label market to determine if an eco-label scheme would lead to a post-label equilibrium where: (1) larger quantities of the labeled item are sold as compared to the pre-label equilibrium; and (2) the labeled item is sold for a higher price than the conventional item due to higher marginal production costs. Besides the simple examples discussed in Mattoo and Singh (1994), this empirical test has not been directly used to evaluate potential eco-label schemes.

While the preference structure is not developed in detail in Mattoo and Singh (1994), more recent analyses by Bureau, Marette, and Schiavina (2000) and Nimon and Beghin (1999a, b) have adapted a simple preference structure for quality differentiation from Mussa and Rosen (1978). These analyses allow for products to be differentiated on two attributes (that is, an experience attribute such as taste as well as a credence attribute such as a production externality). Given a underlying linear utility structure for the labeled and non-labeled item and a fixed budget share allocated to the product category, the ratio of constant marginal utilities and relative prices determines which item is purchased (labeled or non-labeled).[3] As a result, some consumers only buy the labeled item, and some consumers only buy the non-labeled item. Although adequate for their purposes, such extreme assumptions on preferences (additively separable and linear) are unlikely

[2] Swallow and Sedjo (2000) focus on secondary market impacts ('general equilibrium' impacts) of an eco-label that is a mandatory product standard (only certified products are allowed on the market). As in the Mattoo and Singh (1994) analysis, 'eco-consumers' preferences are such that their marginal utility from consuming the labeled product is always higher than their marginal utility from consuming the non-labeled product.

[3] Recall that linear preferences yield linear indifference curves, so that corner solutions where only one item is purchased is common.

to be reasonable for a wide range of items that consumers regularly purchase.

Using the case of shade-grown coffee, the purpose of this paper is to provide two contributions to the existing literature. First, the empirical test outlined in Mattoo and Singh (1994) is implemented. While it is likely that a shade label could pass the test, it is also likely that the market impacts of the label would be minor in terms of additional production induced by the label and the creation of any price premium for shade coffee. This analysis also shows how to combine estimates of 'potential' demand (that is demand for the labeled item with no price premium) and price elasticities of demand and supply to predict market price premiums for the labeled item. Not surprisingly, large potential demand that is also price inelastic will support higher price premiums in the market, but even some minor positive supply response will substantially diminish the resulting market price premium for the eco-labeled item.

And, second, the structure of consumer demand for labeled and non-labeled products are investigated using fairly general assumptions on consumer preferences. This analysis shows how preferences for credence attributes, preferences for other attributes of the product (for example, taste), and basic preferences for variety combine to generate demands for eco-labeled items. While strong preferences for the credence attribute generate large potential demand, preferences for variety also make demands relatively more inelastic with respect to price. In other words, simply selling the 'difference' of the labeled item can be just as important for demand creation as selling the environmental 'goodness' of the product.

The paper is outlined as follows. Section 2 introduces the topic of 'shade-grown' coffee, where the style of coffee production affects habitat services provided by a coffee plantation. The Mattoo and Singh (1994) test is then implemented, and the resulting market impacts in terms of quantities supplied and price premiums are discussed. Section 3 then develops a general consumer utility framework to derive theoretically consistent demands for eco-labeled items. Coffee, again the reference item, is typical of many processed food products widely purchased by consumers in developed economies: it is purchased regularly, it is purchased in varying quantities and qualities, there are competing suppliers, and each supplier may have several competing brands. Section 4 concludes.

2. Shade coffee and the Mattoo and Singh 'Test'

Background on shade coffee

While a few thousand species of the *Coffea* genus exist in the world, the species commonly called *Arabica* accounts for about 75 per cent of world production and *Robusta* accounts for most of the rest (Studivant, 1998). *Arabica* coffees grow at higher altitudes, are considered of higher quality, and, therefore, command higher prices per pound than *Robusta* coffees. *Arabica* coffees are essentially the only type of fresh brewed coffee consumed in the United States. Within the category *'Arabica'*, there are numerous varieties grown in different countries around the world.[4]

[4] Unless otherwise noted, the remainder of this paper focuses only on *Arabica* coffees.

Table 1. *US coffee imports for 1998 (green bean)*

Country	*Quantity (60 kg. bags)*	*Share of total imports*	*Values*	*Price per kilogram*	*Price per pound*	*% Average retail price*
El Salvador	500,700	0.03	88,113,346	2.93	1.33	0.35
Honduras	633,871	0.03	123,602,646	3.25	1.48	0.39
Peru	770,614	0.04	122,518,253	2.65	1.20	0.32
Costa Rica	770,983	0.04	154,118,238	3.33	1.51	0.40
Indonesia	1,272,991	0.07	150,957,815	1.98	0.90	0.24
Vietnam	1,508,202	0.08	142,228,022	1.57	0.71	0.19
Guatemala	1,563,363	0.08	299,838,252	3.20	1.45	0.39
Mexico	2,470,628	0.13	466,643,933	3.15	1.43	0.38
Brazil	2,687,770	0.14	371,546,261	2.30	1.05	0.28
Columbia	3,410,206	0.18	655,293,072	3.20	1.46	0.39
Total	18,997,935	0.82	3,062,636,560	2.69	1.22	0.32

Sources: All data from FAO (2001) except the average retail price of $3.756 per pound in 1998 is from the (USDA FAS 1999).

Table 1a. *Coffee consumption per capita (kg.) in the US and Europe*

	1992	*1998*
United States	4.33	4.09
Europe	5.78	5.4

Source: USDA FAS 1999.

The United States imports about 2.5 billion pounds of green coffee each year, with over 80 per cent coming from ten producing countries (USDA FAS, 1999).[5] Table 1 provides summary information on US coffee prices and quantities of US coffee imports by country of origin for 1998. For 1998, the US imported over $3 billion of green coffee (with another $175 million of roasted coffee). The largest import quantities came from Columbia (18 per cent market share), Brazil (14 per cent), Mexico (13 per cent), Guatemala (8 per cent), and Vietnam (8 per cent). In terms of import values, the largest import values came from Columbia, Mexico, Brazil, Guatemala, and Indonesia, although imports from Vietnam have grown quickly in recent years.[6]

While coffee bushes were traditionally planted under a canopy of shade trees because they could not tolerate direct sun in tropical growing conditions, new 'sun-tolerant' varieties were developed this century that can be grown in direct sun. Thus, as a rough approximation, coffee production

[5] For reference, 1 pound roasted coffee = 1.19 pounds green bean coffee, which is used to convert roasted coffee into green bean equivalent (GBE units).

[6] The coffee import market can also be separated into standard qualities of *Arabica* coffees and higher-valued 'speciality' coffees that are also *Arabicas*. Of total production of *Arabicas*, perhaps 10 per cent grown is of high enough quality to be considered specialty, although speciality coffee imports into the U.S. have grown to about 30 per cent of total coffee imports (Sturdivant, 1998).

can be considered along a continuum from traditional-shade methods to 'modern' sun-tolerant methods.[7] Shade methods commonly involve relatively small farms, relatively smaller use of chemical inputs, lower densities of coffee bushes per hectare, and a forest canopy for shade. Such forest canopies can include commercially useful trees (fruit trees, fuel, building materials). Sun methods typically involve higher density plantings per hectare, more chemical inputs, and more capital.

In general, yields are higher using sun methods as compared to shade methods (Gobbi, 2000; Rice and Ward, 1996). As one indication of yield differences between sun and shade methods, average yields in El Salvador are around 0.90 MT per hectare and average yields in Costa Rica are around 1.5 MT per hectare, with El Salvador generally using more shade methods and Costa Rica using sun methods (see, for example, Harner, 1997).[8] Yield differences are not profit differences, however, and it is likely that, for any given farm, shade methods could provide lower marginal costs at lower output levels and sun coffee methods could provide lower marginal costs at higher output levels. Given the substantial switch to sun methods during the nineteenth century, it seems logical to conclude that sun methods have been more profitable in such locations.

It has become recognized that shade coffee plantations can provide on-site environmental benefits in terms of biodiversity conservation (see, for example, Gobbi, 2000). For example, the CEC (1999) reports that shade coffee plots are the only 'forested' areas in many areas of Mexico. The Smithsonian Migratory Bird Center has recently promoted the benefits of shade-grown coffee areas as habitats for many bird species (see Smithsonian, 2001).[9]

Shade labels have begun to emerge on coffees sold in the United States to market the positive production externalities generated by shade coffee plantations (see, for example, Rainforest Alliance, 1997). For example, 'Bird Friendly' is a shade-grown seal issued by the Smithsonian Migratory Bird Center. 'Café Audubon' is a label for coffees sponsored by the National Audubon Society, with the Smithsonian's Bird Friendly seal, and is distributed by ECO Organic Coffee Company. 'Song Bird Coffee' is sponsored by the American Birding Association and distributed by Thanksgiving Coffee Company. While such labels exist, quantities sold under such labels remain a very minor share of the US market.

[7] *Arabica* coffees are grown using both sun and shade methods. Rice and Ward (1996), CEC(1999), and Gobbi (2000) provide more complete background on shade coffee production.

[8] These data are for green bean equivalent in 1997 and are from the FAOSTAT data set of the Food and Agricultural Organization (see, FAO, 2001). For different production methods within El Salvador, Gobbi (2000) reports that sun yields are 180 per cent higher than some shade methods (for example, commercialized polyculture) but only 20 per cent higher than other shade methods (for example, 'technified' shade).

[9] The CEC (1999) also suggests that shade coffee is somehow more 'healthy' to consumers than sun coffee, although the logic is tenuous at best. Coffee is not consumed for its nutrient value, and the roasting process (burning at high temperatures) eliminates pesticide residues and microbes.

The Mattoo and Singh test

The basic logic behind the Mattoo and Singh (1994) framework and empirical test is outlined in figures 1 and 2. In the pre-label market outlined in figure 1, consumers cannot determine whether an otherwise homogeneous good is produced using two different production methods, shade or sun. Since the consumer cannot identify this credence attribute (which production process is used), there is one aggregate demand labeled D^b in the

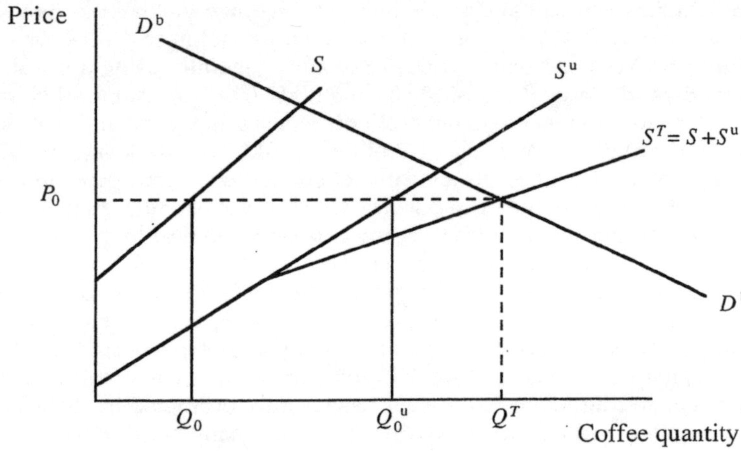

Figure 1. *The 'pre-label' market*

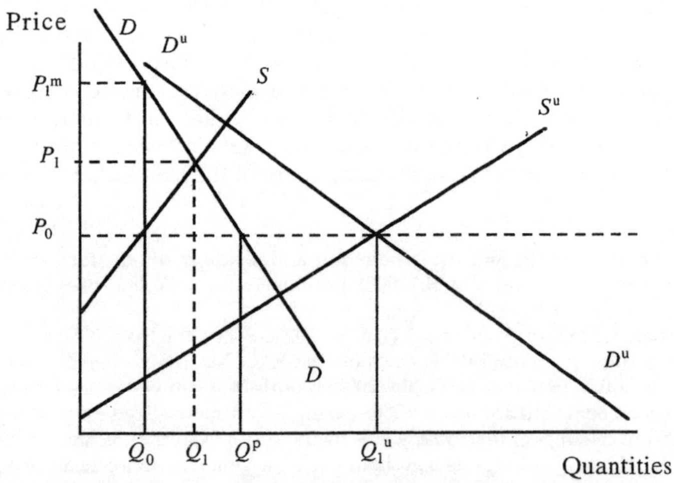

Figure 2. *The post-label market with premium* $(P_1 - P_0)/P_0$

pre-label market. There is supply of shade-labeled coffee (denoted as S) and supply of unlabeled sun coffee (denoted as S^u) in the market, and aggregate supply is $S^T = S + S^u$. Given this initial situation, the market clears at total quantity Q^T and price P_0. With this market price P_0, existing quantities are Q_0 for shade and $Q_0{}^u$ for sun coffee.

After a label exists, as shown in figure 2, consumers can identify the credence attribute and, as a result, the market is now differentiated into shade-grown and sun-grown products. As drawn in figure 2, the possibility exists that separate markets can clear with the point (Q_1, P_1) for shade-labeled coffee and the point ($Q_1{}^u$, P_0) sun coffee.[10] The outcome drawn in figure 2, where a label creates a price 'premium' for the labeled item ($P_1 > P_0$) and larger quantities sold ($Q_1 > Q_0$), is the hoped-for outcome of labeling schemes.

For the outcome depicted in figure 2 to hold, Mattoo and Singh (1994) show that potential demand for shade coffee in the pre-label situation, denoted as Q^p in figure 2, must be greater than actual supply of shade coffee in the pre-label situation (denoted as Q_0 in figures 1 and 2). This potential excess demand, $Q^p - Q_0$, provides the logic of the Mattoo and Singh (1994) 'empirical test'. If data can be developed to estimate supply and demand at pre-label prices, excess potential demand can be computed directly. With excess potential demand, the post-label market clears at some point (Q_1, P_1). The change in land areas needed to provide the additional shade production from Q_0 to Q_1 could also then be used to evaluate the additional positive environmental externalities created by the label.

A shade-coffee label could pass the test

Given that the production systems of major coffee exporting countries are fairly well known, it is possible to estimate existing quantities of non-labeled shade-grown coffee that is sold in the US. Recent information on consumer preferences for shade coffee also can be used to discuss potential market demands in the near future for this product.

Table 2, part A provides basic information to conduct the 'test' discussed in Mattoo and Singh (1994). Information on current demand in the US and consumer survey data (CEC 1999) can be used to hazard a guess at existing potential demand, Q^p, for shade coffee in the United States. As reported in table 1, the US imports about 2.5 billion pounds of coffee annually. If 30 per cent of imported quantities are 'specialty' coffees (Sturdivant, 1998), the existing market for specialty coffees (for example, high-quality dark roasts, blends, coffees of specific origins) would be about 750 million pounds annually. The CEC (1999) reports that around 50 per cent of US consumers would buy shade coffee with no price premium assuming that taste is the same as a high-quality coffee. If this were the case, and shade-labeled coffee were to compete in the specialty coffee market (CEC, 1999), the potential demand for shade coffee with no price premium could be

[10] For simplicity, figure 2 assumes that the post-label price for the unfriendly items remains the same as the pre-label market price P_0. This outcome is likely if the market share of the labeled item is a small portion of the overall market.

Table 2. *The Mattoo and Singh test*

Part A: Does excess demand exist?	*Notation*	*Numbers*				
Speciality coffee imports (pounds per year)		750,000,000				
% Consumer willing to buy with no price premium relative to speciality coffee		0.5				
Potential demand for shade coffee	Q^P	375,000,000				
Existing supply of shade coffee	Q_0	325,000,000				
Pre-label excess demand	$Q^P - Q_0$	50,000,000				
Part B: What about price premiums for shade coffee?						
Potential demand for shade coffee	Q^P	375,000,000				
Existing supply of shade	Q_0	325,000,000				
Excess demand	$Q^P - Q_0$	−50,000,000				
			Case 1	Case 2	Case 3	Case 4
Elasticity of demand	E_{Dp}		−4.4	−0.5	−0.5	−0.5
Elasticity of supply	E_{Sp}		0	0	0.5	2
Price premium	$(P_1 - P_0)/ P_0$		0.03	0.27	0.14	0.06

where $(P_1 - P_0)/ P_0 = [Q_0 - Q^P] / [Q^P E_{Dp} - Q_0 E_{Sp}]$

estimated at about 375 million pounds per year.[11] As a starting point, at least, 375 million pounds annually is used as an initial estimate of potential shade coffee demand (Q^p in figure 2).

Regarding quantities sold in the pre-label market, current information suggests that most of the coffee imported into the US from El Salvador and Mexico is currently grown under some form of shade methods (CEC, 1999; Harner, 1997; Rice and Ward, 1996). As a starting then, based on the quantities reported from table 1, shade-grown coffee currently accounts for at least 13 per cent of the total market for all coffee in the United States, which is equivalent to 325 million pounds of shade coffee per year.[12] Thus, 325 million pounds is used as an initial estimate of Q_0 in figure 2.

With an initial estimate of 325 million pounds of shade coffee sold and potential demand if labeled at 375 million pounds annually, excess demand could be 50 million pounds annually in the United States for shade coffee. In short, according to the Mattoo and Singh (1994) test, there could be a positive outcome – more shade coffee with a higher price.

What about a price premium for shade-labeled coffee?

While a shade coffee label might pass the Mattoo and Singh (1994) test, excess demand in the pre-label situation says little about the magnitude of the price premium in the post-label equilibrium, where $(P_1 - P_0)/P_0$ is used to define the premium as a percentage above the price of the non-labeled item. As shown in figure 2, if the quantity of shade coffee supplied was fixed at Q_0 in the post-label market, then the maximum level of the price premium could be $(P_1{}^m - P_0)/P_0$. If shade coffee supply responds to higher prices, however, then $(P_1 - P_0)/P_0$ would be the market premium associated with the higher production level Q_1. Using existing information regarding price elasticities of demand and supply for coffee, these premiums are calculated and discussed below.

To approximate the maximum premium, first note that $[Q_0 - Q^p]/Q^p = -0.133$. As a result, the maximum price premium $(P_1{}^m - P_0)/P_0$ would need to be high enough to reduce demand by 13.3 per cent. Letting $D = D(p)$ represent a constant elasticity demand function for labeled coffee D in figure 2, the standard definition of a demand elasticity with respect to price, $E_{Dp} = (\Delta D/\Delta p)^*(p/D)$, can be rearranged to yield $\Delta p/p = (\Delta D/D)/E_{Dp}$. Evaluating this expression where $\Delta D/D = [Q_0 - Q^p]/Q^p = -0.133$, the maximum premium is $(P_1{}^m - P_0)/P_0 = -0.133/E_{Dp}$. An

[11] This reported percentage willing to buy with no price premium is consistent with other related empirical literature. For example, the data reported in Govindasamy and Italia (1998) suggest that about 80 per cent of individuals sampled would be willing to purchase fresh produce labeled as grown using integrated pest-management methods. This potential demand figure of 375 million pounds annually may be too high given that the CEC (1999) information is more likely to be described as 'willing to try', rather than will 'only buy' shade for the same price. At the same time, the calculations above assume that shade-labeled coffees would be competing primarily in the market for high-quality coffees.

[12] Since shade methods are used in other countries as well, it is likely that this 13 per cent figure is too low.

estimate of a demand elasticity E_{Dp} is needed to estimate this maximum price premium.

While existing information on actual price elasticities of demand for shade coffee does not exist, the existing literature does suggest that total coffee demand is fairly inelastic with respect to price, but that brand elasticities are substantial. For example, Okunade (1992) estimates a price elasticity of demand in the US equal to −0.34, Akiyama and Variangis (1990) report an estimate of −0.46 for the US and −0.17 for Germany, and Feuerstine (2002) reports an estimate of −0.18 for Germany. While demand for total coffee purchased by a consumer may be fairly inelastic, Krishnamuthi (1991) and Bell, Chiang, and Padmanabhan (1999) provide evidence that price elasticities for brands of coffee are very elastic (for example, −3, −4, etc.). In other words, prices do not change the total quantity of coffee purchased (for example 1 kilogram), but pricing does have a large impact on the brands purchased.

While there has been little analysis of consumer demand for shade coffee, the CEC (1999) also provides some information that can be used as a starting point. The CEC (1999, figure 3, p. 14) suggests that 58 per cent of US adults might buy shade coffee if the price was the same as other non-shade alternatives assuming the same taste (a \$0 premium). Demand would fall to 22 per cent for a \$1 premium. If \$7 per pound was an average retail price for specialty coffee, the implied elasticity is $E_{Dp} = -4.4$ (that is, there is a 62 per cent fall in demand for a shade for a 14 per cent increase in its price from the \$7 per pound base for both brands). This implied elasticity of coffee labeled as shade with respect to price is comparable to the estimates of coffee brand elasticities found in the literature.

As reported under case 1 in table 2 (part B), if demand for shade coffee is as elastic as reported in the CEC (1999) with $E_{Dp} = -4.4$, then $(P_1^m - P_0)/P_0 = -0.133/E_{Dp} = -0.133/(-4.4) = 0.03$. In other words, holding supply constant at Q_0, a 3 per cent price premium would eliminate excess demand if consumer demand for shade coffee is as elastic as implied by the CEC (1999). If, on the other hand, as reported under case 2 in table 2 (part B), demand for shade coffee is inelastic with $E_{Dp} = -0.5$, then $(P_1^m - P_0)/P_0 = -0.133/E_{Dp} = -0.133/(-0.5) = 0.27$. In this case, holding supply constant at Q_0, a 27 per cent price premium would eliminate excess demand.

Of course the actual premium $(P_1 - P_0)/P_0$ based on new production levels Q_1 in figure 2 will depend on demand and supply conditions for shade coffee. To compute the actual premium, let $S = S(p)$ represent a constant-elasticity supply schedule for labeled coffee S in figure 2, and let $E_{sp} = (\Delta S/\Delta p)^*(p/S)$ represent the supply elasticity with respect to price. Using these definitions of demand and supply elasticities and the fact that excess demand in figure 2 must be eliminated by some combination of a decrease in demand or an increase in supply (that is, $Q^p - Q^0 = (Q^p - Q_1) + (Q_1 - Q_0)$), it is possible to solve for the equilibrium premium as $(P_1 - P_0)/P_0 = [Q_0 - Q^p] / [Q_p E_{Dp} - Q_0 E_{Sp}]$.

In sum, four fairly simple pieces of information are needed to develop an estimate of the market premium for the labeled item (shade coffee in this case). Estimates of Q^p, Q_0, and E_{Dp} have been discussed above, so only an estimate of a supply elasticity E_{sp} is needed to compute the market premium. There is little recent evidence in the literature on coffee supply response to higher prices, in large part due to the perennial crop nature of coffee production. Existing literature suggests that coffee supply is fairly inelastic in the short run (at least increasing supply) because producing area is essentially fixed and supply therefore is essentially a harvesting decision (see, Wickens and Greenfield, 1973; Parikh, 1979; and Arak, 1969). In the longer run, supply could be fairly responsive to sustained price increases.

To show a range of possibilities, case 3 and case 4 reported in table 2 (Part B) use the inelastic demand assumption from case 2 along with two supply elasticities (an inelastic case with $E_{sp} = 0.5$ and an elastic case with $E_{sp} = 2$) to show how some supply response affects the final price premium. As shown in case 3, a fairly low supply elasticity (0.5) reduces the price premium from 27 per cent for case 2 to about 14 per cent for case 3. If supply is substantially more responsive to price, as considered in case 4, the premium falls to 6 per cent.

The numbers and calculations provided above are intended to provide a first level analysis of the US market implications of an eco-label for shade coffee production. To date, such analysis has been consistently missing in the eco-label literature. Future research can attempt to provide more precise information on the key needed pieces of information. It is also possible that, because of the presence of a shade label, consumers would become more informed over time about this topic and, as a result, preferences for the credence attribute would adjust over time. In terms of figure 2, this would imply that demand for the labeled product D would shift out over time, which could act to sustain and/or increase premiums in the future. Thus, the analysis in figure 2 and the calculations above are probably best considered the short run implications of an eco-label.[13]

What about the additional environmental benefits of the label?

While the above analysis suggests that a shade coffee label might pass the Mattoo and Singh test, passing the test says little about the ability of an eco-label to generate additional environmental benefits. For example, if existing average yields of shade coffee are about 1980 pounds per hectare (using an average of 0.9 MT/H for El Salvador), an additional 25,000 hectares of shade coffee would be needed to supply the excess demand of 50 million pounds per year holding prices constant (the no-premium case). Thus, even if supply expanded to meet all the excess demand at current prices, the environmental benefits from 25,000 hectares allocated to shade production would be the maximum initial impact possible. While perhaps important for a local growing area, 25,000 hectares represents about 3 per cent of existing coffee hectares just in Mexico (and less than 1 per cent of coffee hectares in Mexico and Central America). Of course with constant prices, there is no reason why producers would increase shade production by 25,000 hectares, and the final increase in shade area would be associated with the production level Q_1 from figure 2.

Given the minor adjustments in aggregate production of shade coffee implied by these calculations, it is likely at least in the shorter run that

[13] Of course this analysis assumes competitive markets. Given the potential for non-competitive retail markets for processed foods, future research needs to consider if and how much of any retail premium would actually be passed down to producers through the marketing chain.

there would be relatively few *additional* environmental benefits generated by a shade label. Looking at this issue from another perspective, however, a main benefit of a label might be to slow conversion of coffee plantations in the future from shade to sun methods. Slowing this conversion could clearly be important, and future research needs to look at farmer decisions regarding production methods and the relative returns to different coffee production systems in different climates and elevations.

3. A closer look at preferences and demands for credence attributes

The Mattoo and Singh (1994) model and the discussion provided in section 2 emphasize the importance of consumer demand for the impacts of eco-labels in competitive markets. Clearly, an eco-label will create higher price premiums when there is substantial excess demand ($Q^p - Q_0$) in the pre-label market and if consumer demand for the labeled item are very inelastic. Since the existing eco-label literature does not provide explicit utility-maximizing frameworks, it is not clear what preference structures would generate such demands. Strong 'preferences' for the credence attribute are usually discussed, which might explain strong demands without a premium, but such preferences do not necessarily imply inelastic demands.

To begin to fill this gap in the literature, this section develops an explicit utility-maximizing framework to investigate the structure of demand for eco-labeled items. To develop this analysis, a nested constant-elasticity-of-substitution (CES) utility structure is used to model demand for labeled and non-labeled products. As is well known, a CES utility structure allows for a range of utility structures as special cases (linear, Leontief, Cobb–Douglas, etc.) based on the magnitude of the elasticity of the substitution parameter. For example, the utility framework used in the previous quality differentiation models (Nimon and Beghin, 1999a,b; and Bureau, Marette and Schiavina, 1998) is a special case of a CES utility framework with an infinite elasticity of substitution between labeled and non-labeled goods.

The 'pre-label' problem is considered first, followed by the 'post-label' situation. Example numerical simulations of post-label consumer demand for the labeled and non-labeled item shows clearly how preference structures interact to affect final consumer demand.

The 'pre-label' consumer problem

For notation in the pre-label situation, let C represent coffee purchased at price p_c let D represent all other beverages purchased at price p_d, and let I represent income allocated to beverages. Let $U(C,D)$ represent a constant-elasticity-of-substitution (CES) function for beverages, where:

$$U(C, D) = (C^{-\alpha} + D^{-\alpha})^{-\frac{1}{\alpha}} \tag{1}$$

In the utility function (1), the parameter α determines the elasticity of substitution σ_α, where $\sigma_\alpha = 1/(1+\alpha)$ (see Varian, 1984, p. 129 and Intrilligator, 1971, p. 187). The parameter α can be between -1 to $+\infty$ for (1) to be a utility function. The utility function in (1) is linear as α approaches -1, is Cobb–Douglas as α approaches 0, and is Leontief as α

approaches $+\infty$. Thus, by varying the parameter α, a range of preferences can be explored.

Given the above utility function, the dual expenditure function for the CES utility function in (1) can be written as

$$e(p_c, p_d, U) = [p_c^{\kappa} + p_d^{\kappa}]^{1/\kappa}\, U \tag{2}$$

where $\kappa = \alpha/(1 + \alpha)$ (for example, see Varian, 1984, p. 130 and page 31–32).[14]

Equating the expenditure function to total income I and inverting provides the indirect utility function

$$V(p_c, p_d, I) = [p_c^{\kappa} + p_d^{\kappa}]^{-1/\kappa} I \tag{3}$$

Using Roy's identity, the Marshallian demand for coffee and total coffee expenditures can be written as:

$$C(p_c, p_d, I) = I\,[p_c^{\kappa} + p_d^{\kappa}]^{-1}\, p_c^{\kappa-1}$$

$$\text{and } M = p_c C = I\,[P_c^{\kappa} + p_d^{\kappa}]^{-1}\, p_c^{\kappa} \tag{4}$$

This pre-label demand situation in (4) is equivalent to the pre-label demand D^b in figure 1.

The 'post-label' consumer problem

After a labeling program exists, consumers can now identify in advance the type of coffee they are purchasing. In effect, the coffee market is now separated into two commodities, and the household can allocate its income to three goods (labeled coffee, non-labeled coffee, and other drinks). For notation in the post-label situation, let G represent the quantity of labeled coffee (for example, 'environmentally-friendly', 'good') purchased at price p_G, and let B represent other coffee ('bad') purchased at price p_B. Let M continue to represent income allocated to all coffee purchases.

In the pre-label equilibrium, total coffee consumed C was a composite commodity of the form $C = G + B$ by default because consumers were only able to observe C and not G or B individually. As a result, utility was a simple function of total coffee purchases in (1). In the post-label situation, individuals can now express demand directly for G and B. While in the total post-label situation, individuals can express demand directly for G and B. While total coffee purchases in pounds or kilograms remain $G + B$, each item can provide utility directly to the consumer, so that the general utility structure is now $U = U(G,B,D)$.

As an example, consider the following nested CES structure, where a sub-utility function for coffee consumption is:

$$\chi(B, G) = [(\theta\gamma G)^{-\beta} + B^{-\beta}]^{-1/\beta} \tag{5}$$

where the parameter β determines the elasticity of substitution (σ_β) between labeled and non-labeled coffee, with $\sigma_\beta = 1/(1 + \beta)$. The

[14] The expression of the expenditure function looks slightly different than that in Varian (1984). I use the convention in Intrilligator (1971) that the parameter α is between -1 and positive infinity, while Varian uses the convention that $\rho = -\alpha$, with ρ between 1 and negative infinity.

parameter θ in (5) is related to product attributes such as taste, and the parameter γ is related to credence attributes identified by the label (for example, shade grown). The overall utility function is $U = U[\chi(B, G), D]$, with the function $U = U[\chi, D]$ defined in (1) above.

By labeling coffee, consumers can now distinguish production methods used for coffee growing, and they can also evaluate its taste relative to the non-labeled coffee. Intuitively, $\theta > 1$ in (5) implies that a consumer prefers the taste of labeled coffee type G as compared to coffee type B, and $\gamma > 1$ implies a consumer prefers the credence attributes of the labeled product relative to the non-labeled item. In combination, when $\theta\gamma > 1$, a unit of labeled coffee provides more utility than a unit of non-labeled coffee.[15]

Besides preferences for taste and credence attributes, consumers may also have preferences for variety, where preferences for variety can be related to the parameter β. Recall that as β becomes negative and close to -1, the sub-utility function $\chi = \chi(B, G)$ becomes linear in B and G. In this case, consumers have no preference for variety, and they consume at a corner solution of only one type of coffee. However, as β becomes positive and large, the sub-utility function $\chi = \chi(B, G)$ begins to approach a fixed proportions function (Leontief). Fixed proportions implies that a consumer strongly prefers variety; in other words, even as relative prices change, the consumer continues to buy both products. In sum, preferences related to taste, credence attributes, and variety will influence the post-label outcome.

To analyze this post-label situation, it is again convenient to begin with the expenditure minimization problem for the coffee sub-utility problem, where

$$e_c(p_g, p_b, \chi) = \min_{g, B} (p_g g + p_b B \; s.t. \; [g^{-\beta} + B^{-\beta}]^{-1/\beta} = \chi)$$
$$where \; g = \theta\gamma G \; and \; p_g = p_G/(\theta\gamma) \tag{6}$$

This coffee expenditure function just shows the minimum amount of money that is needed to produce 0 units of the composite coffee commodity when a label exists. For reference, $g = \theta\gamma G$ can be considered the 'effective' units of labeled coffee consumed, and $p_g = p_G/(\theta\gamma)$ is the effective price of labeled coffee, with p_G being the actual market price facing consumers.

As with the previous expenditure function for the pre-label case, the expenditure function in (6) can be written as:

$$e_c(p_g, p_b, \chi) = p_\chi(p_g, p_b, 1)\,\chi$$
$$\text{where } p_\chi(p_g, p_b, 1) = [p_g^\delta + p_g^\delta]^{\frac{1}{\delta}} \tag{7}$$
$$\text{and } \delta = \frac{\beta}{1 + \beta}$$

In this post-label situation, the 'composite' price $p_\chi = p_\chi(p_g, p_b, 1)$ in (7)

[15] Allowing for product and process attributes to be included directly in the utility function is important for coffee decisions and probably the case for a variety of food products. As reported by CEC (1999), while consumer surveys may document potential demand for labeled coffee (for example, shade), taste considerations could slow the growth in demand.

shows the cost to the consumer of 'creating' one unit of the composite commodity coffee given prices for the labeled and non-labeled items.

Using Shephard's lemma, the Hicksian demand for labeled coffee is

$$G(p_g, p_b, \chi) = \frac{g(p_g, p_b, \chi)}{\theta\gamma} = \frac{\partial\, p_\chi(p_g, p_b, 1)}{\partial p_g}\frac{\chi}{\theta\gamma} \tag{8}$$

and the Hicksian demand for non-labeled coffee is

$$B(p_G, p_b, \chi) = \frac{\partial\, p_c(p_g, p_b, 1)}{\partial p_b\, \chi} \tag{9}$$

In this post-label situation, the 'composite' price $p_\chi = p_\chi(p_g, p_b, 1)$ can be substituted into equation (4) to determine the total Marshallian demand for the coffee composite χ and total expenditures on coffee M as

$$\chi\,(p_\chi\,(p_g, p_b, 1), p_d, I) = I\,[p_\chi\,(p_g, p_b, 1)^k + p_d^{\,k}]^{-1}\, p_\chi(\,p_g, p_b, \delta)^{\,k-1}$$
$$\text{and } M = p_\chi\,\chi \tag{10}$$

As a last step, the Marshallian demand for the coffee composite (10) can be substituted back into the Hicksian demand for labeled coffee (8) and non-labeled demand in (9) to determine the Marshallian demands for labeled and non-labeled coffee.

Numerical examples of the nested CES consumer model

Rather than presenting somewhat tedious comparative static derivations of how demand changes as prices and other parameters of the model change, numerical simulations of the model outlined in (1)–(10) are used here to discuss how the combination of consumer preferences for credence attributes, variety, and relative prices combine to create consumer demand for the eco-labeled item. Four types of consumers are analyzed: type I consumers do not care about the credence attribute or variety; type II consumers care strongly about the credence attribute but not about variety; type III consumers do not care about the credence attribute but type III consumers have stronger preferences for variety; and type IV consumers have strong preferences for both the credence attribute and variety. The basic parameters for the simulation model based on equations (1)–(10) for these four types of consumers can be found in table 3 (under the appropriate column headings).

For these four types of consumers, figure 3 shows how demand for the labeled item adjusts to higher prices, where price is expressed as relative prices, p_G/p_b, representing a price premium for the labeled product. Consider first the results for the type I consumer. In short, while this consumer is indifferent between the labeled and non-labeled coffees if there is no price premium, $p_G/p_b = 1$, in which case even a type I consumer could buy a relatively high share of labeled coffee, any premium for the labeled item would lead to a large reduction in labeled purchases. In figure 3, for example, a 30 per cent price premium for labeled coffee would reduce demand for labeled coffee by 87 per cent, with an implied price elasticity of demand of −2.9.

If the US market contains a large number of type I consumers, eco-label programs will not be able to depend on retail price premiums to help make

Table 3. *CES initial parameters and information for type I–IV consumers*

General parameters	Notation	All consumers			
CES basic utility parameter	α	1.00			
Income allocated to all beverages	I	100			
Pre-label market prices					
Coffee price	p_c	1.00			
Other beverages price	p_d	1.00			
Post-label information					
Taste preference parameter	θ	1.00			
Labeled coffee price	p_G	1.00–1.90			
Non-labeled coffee	p_b	1.00			
Parameters that change across consumer types		*Type I*	*Type II*	*Type III*	*Type IV*
CES coffee sub-utility parameter	β	−0.90	−0.9	1.00	1.00
Credence preference parameter	γ	1.00	2.00	1.00	2.00

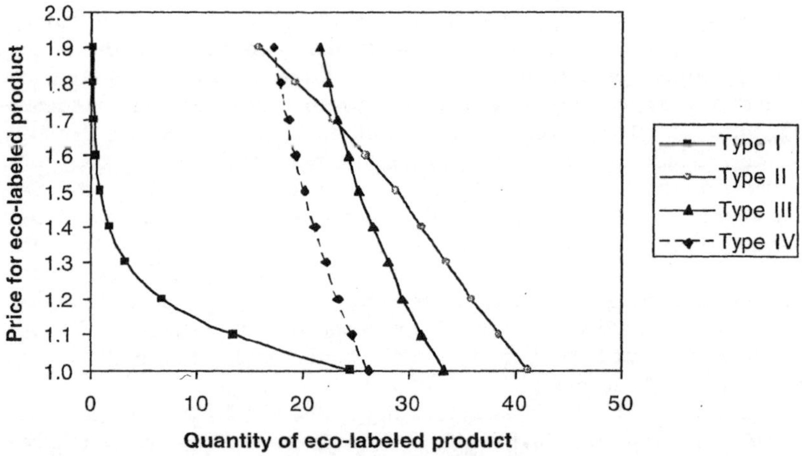

Figure 3. *The effect of preferences on consumer demands (Type I–IV consumers)*

shade coffee more profitable *vis-à-vis* sun methods. Groups developing eco-labels for shade-grown coffee would need to focus on reducing production costs so that shade can simply out-compete sun methods. At the same time, the logic of the 'fair-trade' movement could make sense, where new marketing channels are created, through reductions in profits elsewhere in marketing channels, so that farmers can receive higher farm-gate prices while retail prices can remain constant or fall.

Second, consider the results for the type II consumer presented in figure

3. For these consumers, with no price premium, they would consume essentially only the labeled item. As the premium increases, however, these consumers still respond in a normal fashion by reducing purchases of the labeled item and increasing purchases of the non-labeled item (even though they have 'strong' environmental preferences). In this case, a 30 per cent premium would reduce demand for the labeled item by about 19 per cent, with an average price elasticity of demand of −0.63.

If the US market contained a relatively large share of type II consumers, developing an eco-label for US sales could lead to large demands when relative prices $p_G/p_b = 1$. Demand could stay relatively strong with substantial premiums if supply remains constant in this case. As noted in case 3 of table 2 (part B), however, inelastic demand for the labeled item combined with some positive supply response creates fairly modest price premiums.

Next, consider the case of the type III consumer presented in figure 3. Type III consumers have no environmental preferences (same as type I consumers) but they like variety. As shown in figure 3, with no price premium, type III consumers purchase less of the labeled item as compared to type II consumers. As the premium increases, however, type III consumers also have fairly inelastic demand. For these type III consumers, a 30 per cent premium only reduces demand by about 16 per cent, with an implied price elasticity of demand equal to −0.53. Thus, while type III consumers tend to buy less of the labeled item as compared to type II consumers (strong environmental preferences), type III consumer demand is also fairly price inelastic.

And last, the results for the type IV consumer (likes the credence attribute and variety) are also presented in figure 3. Type IV consumers are simply a combination of type II and type III preferences. Because type IV consumers have strong environmental preferences, they buy more of the labeled item than type III consumers when there is no premium. Because they also like variety, however, they buy less of the labeled item as compared to the type II consumer. The demand functions for these type IV consumers remain less responsive to price than type II consumers, however. For example, a 30 per cent premium reduces demand by about 15 per cent, with an average elasticity of −0.5.

For each of these examples, as noted in table 3, it was assumed that other attributes of the commodity were identical (that is, no taste differences). Any taste differences between the labeled and non-labeled coffees (for example, if the parameter $\theta = 0.8$), would imply that the demand schedules for the labeled coffee would fall. As experience with the organic produce market shows, environmental and potential food safety attributes of products do not easily substitute for taste and appearance attributes.

4. Conclusion

While numerous types of eco-labels for credence attributes of products (that is, non-product-related production and process method attributes) exist in the US and elsewhere, there is surprisingly little analysis of their usefulness as a policy instrument to affect market prices, increase supplies of the labeled item, and manage production externalities. Using the topic of 'shade-grown' coffee, this paper discusses and analyzes the implications of a label for shade-grown coffee in the US. While shade-grown coffee might pass the Mattoo and Singh (1994) test, the resulting market implications are likely to be modest.

Given the importance of how consumers adjust to eco-labels, and given the lack of research on the theoretical foundations of demand for eco-labeled items, section 3 develops a representative consumer model based on a nested CES utility structure. This framework allows a relatively easy comparison of utility-maximizing outcomes across a rather wide range of preference structures. As shown in section 3, preferences for the credence attribute and variety are both important and supporting motives to purchase labeled items. Strong preferences for the credence attribute help to generate large demand in the absence of any price premium for the labeled item, while preferences for variety help to make such demand less responsive to higher prices. Future research should investigate the implications of variety for marketing the 'differentness' of eco-labeled products as opposed to (or in tandem with) the 'environmental-ness' of the items.

References

Arak, M. (1969), 'Estimation of asymmetric long-run supply functions: the case of coffee'. *Canadian Journal of Agricultural Economics* **17**: 15–22.

Bell, D.R., J. Chiang, and V. Padmanabhan (1999), 'The decomposition of promotional response: an empirical generalization', *Marketing Science* **18**: 504–526.

Bureau, J.-C., S. Marette, and A. Schiavina (1998), 'Non-tariff trade barriers and consumer's information: the case of the EU–US trade dispute over beef', *European Review of Agricultural Economics* **25**: 463–487.

Caswell, J. A. (1998), 'How labeling of safety and process attributes affects markets for food', *Agricultural and Resource Economics Review* **27**: 151–158.

Center for Environmental Cooperation (CEC) (1999), 'Measuring consumer interest in Mexican shade-grown coffee: an assessment of the Canadian, Mexican and US markets', October, available on-line at www.cec.org.

Coffee, Sugar, and Cocoa Exchange, Inc. (CSCE) (1997), 'Coffee economics', downloaded on 24 October at www.csce.com/econ/coffee.html.

Food and Agricultural Organization (2001), FAOSTAT Homepage at http:/apps.fao.org.

Feuerstein, S. (2002), 'Do coffee roasters benefit from high prices of green coffee?', *International Journal of Industrial Organization* **20**: 89–118.

Gobbi, J.A. (2000), 'Is biodiversity-friendly coffee financially viable? An analysis of five different coffee production systems in western El Salvador', *Ecological Economics* **33**: 267–281.

Golan, E., F. Kuchler, and L. Mitchell (2000), 'Economics of food labelling', Agricultural Economics Report No. 793, Economic Research Service, United States Department of Agriculture, December.

Govindasamy, R. and U. Italia (1998), 'A willingness-to-purchase comparison of integrated pest management and conventional produce', *Agribusiness* **14**: 403–414.

Harner, C. (1997), 'The coffee industry in El Salvador sustainability analysis', INCAE Working Paper – ENV 1.7, not for distribution, Alajuela, Costa Rica, Jul.

Intrilligator, M.D. (1971), *Mathematical Optimization and Economic Theory*, Englewood Cliffs, NJ: Prentice-Hall.

Karl, H. and C. Orwat (1999), 'Economic aspects of environmental labelling', in H. Folmer and T. Tietenberg (eds), *The International Yearbook of Environmental and Resource Economics 1999/2000*, Cheltenham, UK: Edward Elgar Publishing, Chapter 3, pp. 107–169.

Krishnamuthi, S.P.R. (1991), 'An empirical analysis of the relationship between brand loyalty and consumer price elasticity', *Marketing Science* **10**: 172–183.

Markandya, A. (1997) 'Eco-labeling: an introduction and review', in S. Zarrilli, V. Jha, and R. Vossenaary (eds), *Eco-Labeling and International Trade*, London: MacMillan Press (published for and on behalf of the United Nations), chapter 1, pp. 1–20.

Mattoo, A. and H.V. Singh (1994), 'Eco-labeling: policy considerations', *Kyklos* **47**: 53–65.

Mussa, M. and S. Rosen (1978), 'Monopoly and product quality', *Journal of Economic Theory* **18**: 301–307.

Nimon, W. and J. Beghin (1999a), 'Eco labels and international trade in textiles', *American Journal of Agricultural Economics* **81**: 801–811.

Nimon, W. and J. Beghin (1999b), 'Ecolabels and international trade in the textile and apparel market', *American Journal of Agricultural Economics* **81**: 1078–1089.

Okunade, A.A. (1992), 'Functional forms and habit effects in the US demand for coffee', *Applied Economics* **24**: 1203–1212.

Parikh, A. (1979), 'Estimation of supply functions for coffee', *Applied Economics*, **11**: 43–54.

Rainforest Alliance (1997), 'The eco-O.K. certification program: market incentives to reduce the environmental impacts of tropical agriculture', downloaded on 12 November at www.rainforest-alliance.org/okr.html#CoffeeCertification.

Rice, R.A. and J. R. Ward (1996), 'Coffee, conservation, and commerce in the western hemisphere: how individuals and institutions can promote ecologically sound farming and forest management in northern Latin America', Natural Resources Defense Council and the Smithsonian Migratory Bird Center, June, downloaded at http://www.nrdc.org/nrdcpro/cc/cptinx.html.

Rice, R., A. (1998), 'A rich brew from the shade', Americas, Organization of American States, April, pp. 52–59.

Smithsonian (2001), Migratory bird center, shade-grown coffee web-site at natzoo.si.edu/smbs/research/coffee.htm.

Sturdivant, S. (1998), 'Speciality coffee and foodservice', paper presented to the Specialty Coffee Association of America, 12th Annual Conference and Exhibition, San Francisco, California, available at www.stoneworks.com/scaa1_archive/sturdivant.html

Swallow, S.K. and R.A. Sedjo (2000), 'Eco-labeling consequences in general equilibrium: a graphical assessment', *Land Economics* **76**: 28–36.

Teisl, M.F. and B. Roe (1998), 'The economics of labeling: an overview of issues for health and environmental disclosure', *Agricultural and Resource Economics Review* **27**: 140–150.

US Department of Agriculture, Foreign Agricultural Service (USDA FAS) (1999), 'Tropical products: world markets and trade', Circular FRTROP 1–99, March, Online at http://www.fas.usda.)gov/htp2/tropical/1999/99-03/troptoc.htm.

Varian, H.R. (1984), *Microeconomic Analysis*, second edition, New York: W.W. Norton & Co.

Vossenaar, R. (1997) 'Eco-labeling and international trade: the main issues', in S. Zarrilli, V. Jha and R. Vossenaary (eds), *Eco-Labeling and International Trade*, London: Macmillan Press (published for and on behalf of the United Nations) chapter 2, pp. 21–36).

West, K. (1995), 'Eco-labels: the industrialization of environmental standards', *The Ecologist* **25**: 16–20.

Wickens, M.R. and J.N. Greenfield (1973), 'The econometrics of agricultural supply: an application to the world coffee market', *Review of Economics and Statistics* **55**: 433–440.

Zarrilli, S., V. Jha, and R. Vossenaary (1997), *Eco-Labeling and International Trade*, London: MacMillan Press (published for and on behalf of the United Nations).

[14]

CHINA'S ENVIRONMENTAL LABELING PROGRAM

Jimin Zhao
Stanford University

Qing Xia
Chinese Research Academy of Environmental Sciences

China's environmental labeling program was created because the government wanted to improve environmental management, and some Chinese enterprises expected that labeling would help eliminate non-tariff barriers for their exports and allow them to expand their domestic market shares. The labeling program, which is managed by the Chinese government, is a voluntary third-party certification system based on labeling procedures developed in Organization for Economic Co-operation and Development countries. Thus far, China's labeling program has increased consumer awareness of labeled products and encouraged some enterprises to adopt cleaner technologies in products that have close relationships with consumers' health. However, the effectiveness of the program is very limited because only a small number of product categories are included in the program, consumers' purchasing behavior in China has been influenced by environmental labels for only several types of products, and relatively few enterprises participate in this program. The following measures would improve the effectiveness of the labeling program: enhancing public awareness of environmental labeling and environmental protection, increasing the number of product categories involved in environmental labeling, displaying more information on labels, and integrating the labeling program into China's efforts to promote cleaner production and to encourage compliance with ISO 14000 standards.

Introduction

According to the Organization for Economic Co-operation and Development (OECD 1991) and the Secretariat of the United Nations Conference on Trade and Development (UNCTAD), "environmental labeling" consists

Address requests for reprints to: Jimin Zhao, c/o Duc Wong, Department of Civil and Environmental Engineering, Stanford University, Stanford, CA 94305-4020, USA.

of the voluntary granting of labels by a private or public body for display on a product that meets specific criteria; the label serves to inform consumers that the product is environmentally more friendly than other functionally similar products (OECD 1994). Using the OECD and UNCTAD concepts of labeling, a product can carry an environmental label if an analysis of the product's entire "lifecycle" (production, use, and disposal) demonstrates that the product is environmentally friendlier than similar, competing products. In 1993, the Secretariat of the General Agreement on Tariffs and Trade (GATT) expanded the definition of environmental labels by distinguishing among the following: eco-labeling, which refers to labels awarded based on a lifecycle analysis, single-issue labels, which describe one characteristic of a product; and negative-issue labels, which provide warnings of safety and health problems associated with a product.

In 1993, as part of an effort to improve environmental protection and expand exports, China created an environmental labeling program that is similar to the one used in Germany. The China Certification Committee for Environmental Labeling of Products (CCELP) was established in May 1994 to implement and manage this program. Currently China's environmental labeling program includes eco-labels and voluntary single-issue labels, but it does not include mandatory single-issue labels.[1] As of the end of 1997, China had issued environmental labels for 22 product categories. This paper examines the motivations for creating China's labeling program, the program's administrative structure, and technical requirements regarding product selection and the application of labels. The effectiveness of the prospects for China's environmental labeling program also are examined.

Development of the Program

Motivation for Labeling Program

Robust economic growth in the early 1990s placed extraordinary pressure on China's already insufficient natural resources and threatened many fragile environments. Environmental officials became increasingly convinced that traditional "command and control" approaches to environmental protection would not be enough to solve China's environmental problems. In 1992, Xie Zhenhua, the Administrator of the National Environmental Protection Agency (NEPA),[2] argued that raising the public's environmental awareness and encouraging people to become concerned with environmental protection on their own initiative were the keys to improving environmental protection in China. NEPA officials had become particularly inter-

---

[1] China has negative labels that indicate a product is flammable, poisonous, or dangerous to health, but these labels are not part of China's environmental labeling program.

[2] Prior to March 1998, China's central environmental protection agency was called the National Environmental Protection Agency. In that month, the agency was made a full ministry, given new responsibilities, and renamed the State Environmental Protection Administration (SEPA).

ested in approaches to enhancing consumer awareness using eco-labeling programs such as the programs adopted in Germany and Japan. In those countries, an environmental label appears on products whose environmental attributes satisfy certain labeling criteria. People in those countries can make their purchasing choices by considering not only price and the functional characteristics of products, but also the environmental attributes of products. Depending on the extent to which consumers choose labeled products, environmental labeling creates an incentive for manufacturers to account for the environmental impacts of these products throughout their life cycles. NEPA wanted to use environmental labeling as a tool for increasing citizens' awareness of environmental issues and strengthening manufacturers' environmental performance (CCELP 1995).

During the same period—the early 1990s—environmental labeling programs of some countries became non-tariff barriers for certain products made in China. For example, silk cloth from the Xiamen Silk Export Company was welcomed by the company's French partners, who intended to import the silk to France. However, because the silk cloth had no environmental label, the Xiamen Silk Export Company's French partners could not get an import quota from the French government, and the business transaction failed (CCELP 1995). Another example involves Chinese enterprises that exported household refrigerators to European countries. During the early 1990s, demand for household refrigerators that did not have environmental labels (and contained chloroflurocarbons) plummeted in some European countries. China's export of refrigerators dropped from 560,000 units in 1988 to 230,000 units in 1991, a decrease of 58%, and revenues from exported refrigerators decreased by 62% over the same period (Xia et al. 1995).

Several companies, including the Xiamen Silk Export Company and some household refrigerator manufacturers, asked NEPA to establish an environmental labeling program in China. They also tried to develop or adopt clean technologies like the ones used in countries with environmental labeling programs. Enterprises that encouraged the adoption of a labeling program also expected that the program would allow them to improve their reputation with Chinese consumers, thereby allowing them to expand their shares of domestic markets. Before the government initiated the labeling program, some household refrigerator enterprises had designed their own environmental labels and pasted them on their products, and others obtained environmental labels from other countries such as Germany (Zhu and Hu 1994).

Organizations in the government and in industry supported the establishment of an environmental labeling program in China. In 1992, Xie Zhenhua charged NEPA's Department of Science, Technology, and Standards with the task of creating a Chinese environmental labeling program. Officials from this department attended international environmental labeling confer-

FIGURE 1. China's environmental label.

ences to learn about experiences of countries that had labeling systems. They also asked the Chinese Research Academy of Environmental Sciences (CRAES) to study labeling programs in other countries and investigate the feasibility of establishing a program in China. Based on feasibility studies and other preparatory work by NEPA and CRAES, on March 31, 1993, NEPA issued a document entitled "Launch Environmental Labeling in China" to provinces, autonomous regions, and municipalities directly under the central government. This marked the beginning of the environmental labeling program in China.

China's labeling program aims to help citizens become environmentally responsible in their daily lives and increase their awareness of environmental issues. The program attempts to encourage enterprises to utilize resources and energy rationally, and to develop and produce environmentally friendly products. It also serves to enhance the competitiveness of Chinese enterprises in those international markets where the environmental attributes of products influence consumers' choices.

Development of the Program

In April 1993, NEPA published notices in *China Environment News, People's Daily,* and *Health News* inviting the public to submit designs for China's environmental label. In response, NEPA received 710 designs. Environmental specialists at CRAES reviewed the designs and made a selection, which was subsequently approved by NEPA. The final design for the label was published in *China Environmental News* on August 25, 1993. The central part of the label is composed of green hills, blue water, and the sun, and it is surrounded by ten rings (Figure 1). The hills, water, and the sun represent the natural environment, whereas the ten closely

TABLE 1. Development of China's Environmental Labeling Program

Date	Activities
March 31, 1993	NEPA issued "Launch Environmental Labeling in China" to mark the start of China's environmental labeling program
August 25, 1993	Design of China's environmental label was finalized
May 17, 1994	CCELP was established
May 30, 1994	NEPA issued technical criteria for the first seven product categories
July–November 1994	NEPA issued first environmental labels to two enterprises
December 1997	Twenty-two categories of products were included in the program

united rings represent the participation and efforts of the public to protect the environment. In Chinese, the character for the words "ring" and "environment" are identical. Viewed as a whole, the graphic part of the label means "human beings should unite to protect the environment they live in."

On May 17, 1994, NEPA established the China Certification Committee for Environmental Labeling of Products to manage China's environmental labeling program. A few weeks later, on May 30, NEPA approved and issued "Technical Requirements for the First Seven Categories of Products Eligible for Environmental Labels," which contained technical certification requirements for seven categories of products. Between July and November 1994, CCELP certified two enterprises to use environmental labels on three types of products. This marked the beginning of the operation of China's certification system. By the end of 1997, 22 categories of products had been included in China's labeling program. The development of the program is summarized in Table 1.

Administrative Structure

The principal participants in China's environmental labeling program include CCELP, SEPA, the State Bureau of Quality and Technology Supervision (SBQTS), the CCELP Secretariat, environmental protection bureaus (EPBs), and enterprises (Table 2). The program is dominated by government organizations, such as SEPA, SBQTS, and EPBs. Moreover, among the 24 members of CCELP, 15 are officials from agencies. The director and deputy director of CCELP are the administrator and vice administrator of SEPA, respectively.[3] Designers of the labeling program believe this type

[3] Although the government is a necessary participant in all national labeling programs, its active role can range from simply providing financing to active management. In the Japanese program, all technical research and administrative support is provided by the government. In the German, Austrian, and Canadian programs, these duties are shared by government officials and certification authorities. In the case of Austria, a consumer organization and a private certification authority are in charge of the labeling program. The Swedish, Finnish, and French programs rely on standardization institutes to oversee the process, from development of product categories and criteria, to signing of contracts with manufacturers for the use of labels (OECD 1991).

TABLE 2. Key Actors in China's Environmental Labeling Program

Actors	Functions
SEPA (NEPA before March 1998)	Higher level administrative agency of CCELP and leading agency for the labeling program. SEPA also approves and issues categories of products that are included in the program and criteria for labeling products.
SBQTS	Guiding agency for labeling program. SBQTS supervises the program to ensure its fairness and validity.
CCELP	Authorized agency for program inplementation. CCELP provides policy guidance to the CCELP Secretariat, reviews proposals about product categories and criteria, and approves the certification of labels.
CCELP Secretariat	Staff to perform day-to-day work for CCELP regarding product selection, criteria formulation, field investigation of enterprises that wish to use labels, and management of use of labels.
EPBs	Assistants to CCELP Secretariat in preliminary reviews of enterprises seeking certification.
Enterprises	Applicants and recipients of environmental labels.

CCELP = Certification Committee for Environmental Labeling of Products; EPB = Environmental Protection Bureau; NEPA = National Environmental Protection Agency; SEPA = State Environmental Protection Administration; SBQTS = State Bureau of Quality and Technology Supervision.

of administrative structure will maintain the program's fairness, credibility, and transparency and thereby increase the public's trust in the program (Xia et al. 1995).

CCELP and the CCELP Secretariat are in charge of implementing China's labeling program. CCELP, which includes officials and specialists from environmental, economic, academic, and product quality organizations, is the government agency authorized to manage the labeling program. CCELP reviews guidance for environmental labeling, proposes product categories for labeling certification, approves of permits issued to enterprises for the use of labels, and deals with appeals related to certification.

The CCELP Secretariat, which is a division within CRAES, is responsible for the day-to-day work under the guidance of CCELP. As a national environmental research agency, CRAES provides personnel and technical support to the CCELP Secretariat, the organization that proposes categories of products eligible to be considered for environmental labels. The Secretariat also drafts technical requirements that enterprises must satisfy to receive labels, conducts field investigations of enterprises that apply for labels, and submits preliminary comments on products that are being considered for labels.

Both SEPA and SBQTS provide leadership and guidance for the labeling program. As CCELP's supervising administrative agency, SEPA guides the work of CCELP by approving and issuing (1) categories of products that can receive environmental labels, and (2) criteria that must be satisfied by enterprises that wish to use labels. As the agency authorized to supervise Chinese certification systems, SBQTS is responsible for ensuring that the labeling program is administered properly based on national laws and regulations and the common practices of international certification systems. Local EPBs become involved in the labeling program by conducting preliminary reviews of applications submitted by enterprises that want to use labels.

Legal Requirements Governing the Use of Labels

CCELP conducts its certification work based on "Regulations of the China Certification Committee for Environmental Labeling of Products" and "Use and Management Guidelines on Certification Credentials and Labels for Environmental Labeling of Products." CCELP has registered the design of China's environmental label with the State Trademark Administration, and that design is the only one that can be used officially for an environmental label. China's environmental label is protected by the Trademark Law of People's Republic of China, the Product Quality Law of People's Republic of China, and the Anti-unlawful Competition Law.

When an enterprise becomes certified by CCELP to use environmental labels, it enters into a formal contract with CCELP. This contract clarifies the responsibilities of enterprises as users of labels and of the CCELP as the issuer of labels. The contract includes conditions on the use of labels, the contract period, the supervision to be conducted by CCELP, and penalties that can be imposed if the enterprise violates contract conditions.

Implementation: Voluntary Third-Party Certification System

China's environmental labeling program is a voluntary third-party certification system that is based on labeling procedures developed in OECD countries, with some changes that reflect China's current economic, social, and environmental situation. The overall procedure includes three main stages: (1) selecting product categories; (2) assessing environmental impacts of products throughout product lifecycles, and establishing criteria for qualifying products; and (3) applying for certification (i.e., the right of an enterprise to use labels on its products).

Product Category Selection

Any enterprise or individual in China can propose product categories for environmental labeling to the CCELP Secretariat. After conducting its own analysis, the Secretariat presents proposed product categories to the CCELP for review. In turn, CCELP presents recommended product categories to SEPA and SBQTS for approval. After approval by SEPA and SBQTS, product categories for environmental labeling are announced in *China Environmental News* and other newspapers.

TABLE 3. Categories of Products Receiving Environmental Labels (at the end of 1997)

Target Group	Product Category
Recyclable and reusable	Toilet paper (using recycled paper)
	Soft drinks (using reusable or recyclable containers)
Low water and energy saving	Energy-saving fluorescent lamps
	Energy-saving and low mercury content two-sided fluorescent light tubes
	Energy-saving electric ballast
	Energy-saving and low-noise air conditioners
Low toxicity and low pollution	Unleaded gasoline
	Water-based paints
	Rechargeable dry batteries free of mercury, cadmium, and lead
	Silk cloth
	Mothproof woolen fabric
	Dry batteries free of mercury
	Non-phosphorus laundry detergent
	Harmless glue
	Spray insecticides
	Low-noise washing machines
High biodegradability	Biodegradable plastic products
Low emission	Low emission gas ranges
	Low CFC household refrigerators
	Non-CFC aerosol products

In theory, all product categories can be considered for the environmental labeling program. In practice, CCELP focuses primarily on the following: (1) products whose impacts are closely linked with human health; (2) products made from substances related to global environmental problems; (3) products that cause significant environmental impacts during production, use, or disposal; and (4) export products that are included in environmental labeling programs in other countries. Based on these guidelines, CCELP has targeted products that are recyclable and reusable and products that can be manufactured in ways involving low toxicity, low energy and water use, high biodegradability, and low pollutant emissions. Table 3 lists the product categories included in China's environmental labeling program by the end of 1997, most of which are from the light industry sector.

Because an important objective of the labeling program is to promote Chinese exports, CCELP pays attention to products that have export potential. For example, China's earlier failure to enter the French silk market was linked to environmental labels; thus, CCELP selected silk cloth as one of the first seven types of products that could display labels. Another product category tied to exports is refrigerators manufactured without chloroflurocarbons (CFCs) or with low levels of CFCs. These products were also selected because of their importance in China's implementation of the *Montreal Protocol on Substances That Deplete the Ozone Layer* (Xia et al. 1995; Zhao and Ortolano 1999). CCELP also focuses on product categories included in the labeling programs of other countries. For example, toilet paper, water-based paints, batteries, and refrigerators are all commonly labeled in other countries with labeling programs.

Criteria Development

After product categories are identified, the CCELP Secretariat assesses the environmental impacts of the products and drafts criteria or technical requirements for the products. Once these criteria are reviewed and revised by CCELP in collaboration with environmental and technical experts, they are submitted to SEPA for approval and issuance.

Technical requirements or criteria for products are formulated based on the following four principles:

1. *Lifecycle assessment and key process assessment.* Effective environmental assessment requires analysis of effects over the lifecycle of a product, including production, distribution, use, and disposal. China's assessment criteria focus primarily on pollutants and lifecycle stages linked either to major impacts or to special requirements, such as requirements of the Montreal Protocol. The Protocol has influenced the environmental standards for household refrigerators, which include only conditions on a single input, CFCs. As another example, because of the impacts of lead on human health, gasoline standards concern only the content of lead. Silk provides a different example. Standards for silk involve the entire silk production process from use of pesticides for mulberry trees, to use of dyes, packaging, and recycling and reuse of silk. The broader standards were set to meet requirements of nations that receive exports from China's silk industry.
2. *Product quality assurance.* Products selected for China's environmental labeling have to meet quality and safety requirements based on national standards. Good quality is a key issue for consumers, and consumers' preference is important for the success of the labeling program.
3. *Consistency with criteria in other countries.* China's exports can benefit from having China's labeling requirements consistent with labeling requirements in other countries, and this plays a role in setting criteria. For example, CCELP referred to Canadian labeling requirements when it formulated criteria for water-based paints, and it considered German criteria when it developed criteria for silk cloth (Xia et al. 1995).
4. *Practicability and suitability.* The criteria must be suitable to China's technical conditions in product manufacturing and product testing.

The criteria must neither be so strict that no enterprise can satisfy them, nor so lax that most enterprises can meet them. CCELP tries to establish criteria so that they can be satisfied by about 20% of enterprises manufacturing the products (Xia et al. 1995). Criteria also are set in a manner consistent with available testing technologies.

Application for an Environmental Label

China uses a voluntary certification system in which a third party, CCELP, grants certificates that allow enterprises to label products that meet technical criteria and requirements. To receive a certificate, an enterprise applies to CCELP for the testing and evaluation of its products. Applications can only be made for products previously identified by CCELP as being among product categories having technical criteria and requirements for labeling.

An enterprise is charged a fee for a certificate that allows it to use environmental labels. The charge also includes payments for applications, investigation of enterprises, testing of products by designated inspection centers, label use, and annual fees. An additional fee is charged each time a certificate is renewed because the products must be reevaluated. The total fee for obtaining a certificate ranged form 15,000 to 50,000 yuan per product ($1,807 US to $6,024 US, at the March 1999 rate of exchange).

Any enterprise can apply to use environmental labels provided it has a business license from the appropriate industrial and commercial administrative agencies, and its products satisfy official quality control certification requirements.[4] The enterprise must meet national and local pollutant discharge standards, and it must not have been subjected to penalties from EPBs within 1 year of the date of its application for an environmental label. Before an enterprise can receive a certificate that allows it to display environmental labels on a particular product, it must carry out the application procedure shown in Figure 2, a process that takes about 2 to 3 months.

An enterprise applying for a certificate to use environmental labels must first obtain and complete four copies of the "Application Form for Environmental Labeling of Products." This form is available from provincial-level EPBs. Once the provincial-level EPB receives the completed application, it conducts a preliminary examination. The EPB prepares four copies of its comments on the environmental performance of the enterprise and the environmental attributes of the product: one copy for the enterprise, two for the CCELP Secretariat, and one for EPB files.

If a product has a satisfactory preliminary assessment, the CCELP Secretariat sends a team to conduct a field investigation of the enterprise. The investigation, which typically lasts for 1 to 2 days, includes an examination

[4] Quality control certification is required by the government for products to enter markets in China. The certification is approved and issued by a department within the SBQTS.

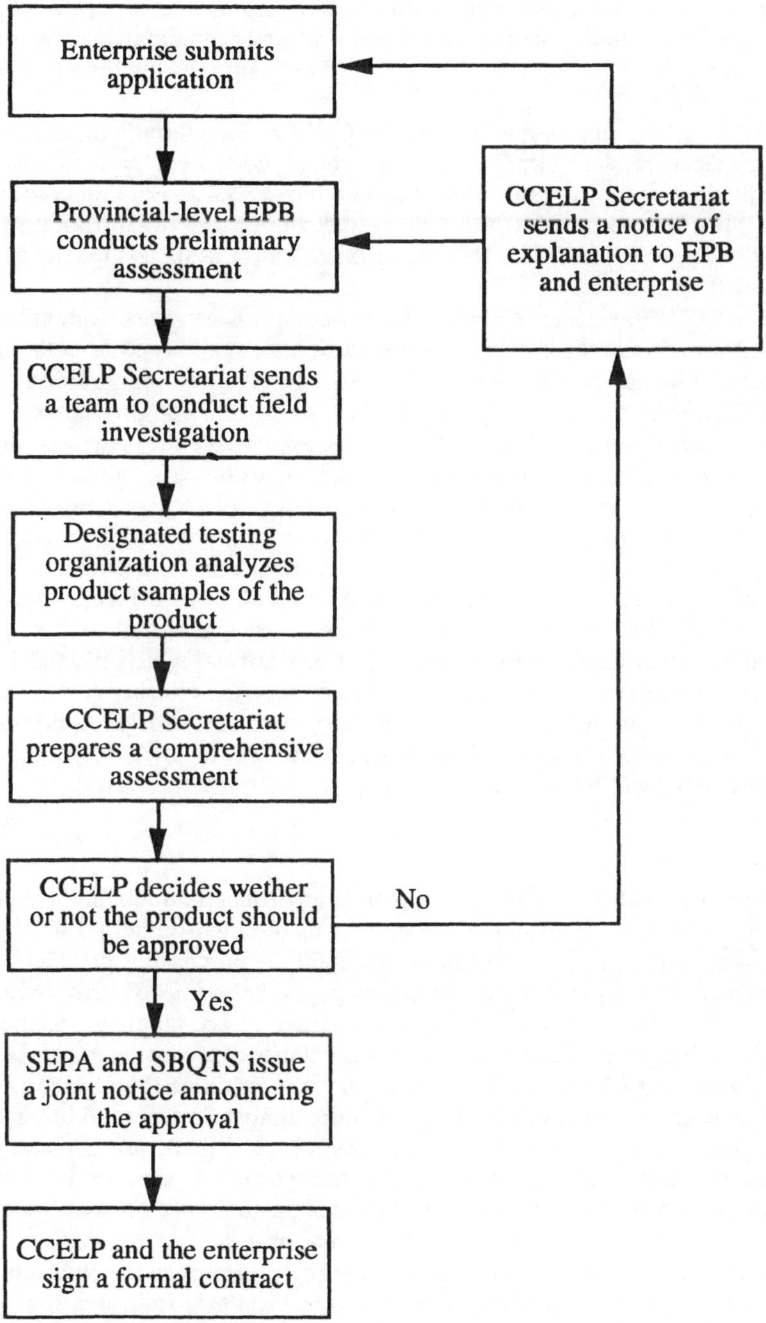

FIGURE 2. Procedure followed in applying for environmental labels.

of production conditions and pollution discharges, and a review of the enterprise's systems for quality control and environmental management. Following the investigation, the Secretariat's team informs the enterprise of the results.

After conducting its field study, the CCELP Secretariat's investigation team selects random samples of the product, and the enterprise sends the samples to a testing organization designated by CCELP. For large products that must be examined at the factory site, the testing organization sends staff to the enterprise. The testing organization sends its test results to the CCELP Secretariat.

The CCELP Secretariat next prepares a comprehensive assessment based on information in the enterprise's application form and reports of the field investigation and product testing. The Secretariat's evaluation then is sent to CCELP members, who decide whether the enterprise's product is qualified for certification. SEPA and SBQTS list approved enterprises and products that have received authority to display environmental labels in a joint public announcement. For a product that does not receive approval, the CCELP Secretariat sends a notice to the enterprise indicating reasons for the disapproval. A copy of this notice also is sent to the provincial EPB.

The final step in the procedure involves creation of a formal contract between CCELP and the enterprise that received the permit to display an environmental label on its product(s). The contract, which is valid for 3 years, clarifies the rights and responsibilities attached to using environmental labels. If (during the 3 years) the enterprise cannot satisfy requirements involving annual surveys and monitoring, the right to display the environmental label will be revoked.

Effectiveness

According to OECD (1991), an effective environmental labeling program is one that raises consumers' awareness of the environmental effects of products and changes consumers' behavior in purchasing products, improves manufacturers' designs of products, and helps improve environmental quality. The steady growth in the number of eco-labeling schemes in the world suggests that awareness and acceptance of environmental labeling are increasing (Appleton 1997). With the exception of the German Blue Angel scheme, which has been in existence almost 20 years, all other eco-labeling schemes are relatively new. As a result, there are a paucity of statistical data available to conduct quantitative analysis on effectiveness. In addition, it is difficult to distinguish between the influence of environmental labeling and effects of other environmental programs.

Currently, only preliminary, qualitative analyses exist on the environmental effectiveness of labeling schemes. OECD (1997) indicates that eco-labeling has only been moderately successful with individual consumers.

UNCTAD finds that, in Scandinavia, consumers show a strong preference for labeled goods in the paper and detergent sectors, but preference for labeled products is much weaker in other sectors (Zarrilli et al. 1997). OECD and UNCTAD reports indicate that the German Blue Angel program has raised consumer consciousness and resulted in market changes in certain sectors, that Singapore's Green Label system has shown mixed results, and that the Canadian Environmental Choice Programme and the Korean Ecomark system have resulted in few changes in consumer preferences (Appleton 1997).

Because of limitations on available data and assessment methodologies, our assessment of the effectiveness of China's environmental labeling program is preliminary and qualitative. The assessment considers the labeling program's influence on the following: consumers' awareness of environmental labeling and purchasing behavior, manufacturers' behavior, and benefits to the environment.

Response of Consumers to Labeling

Public awareness is the driving force behind the success of any environmental labeling program. During the past 4 years, China's environmental label has appeared on 3.5 million products. In other words, 3.5 million "green messengers" have entered households to spread the word about environmentally friendly products and thereby enhance people's sensitivity to environmental issues. Citizen awareness of China's environmental labeling system is expanding. According to recent surveys in Beijing, only 1% of the people knew about China's environmental label in 1995, but this number jumped to 30% in 1997.[5] However, these data were obtained from a small sample survey conducted in Beijing, a city where people have relatively high education levels and more access to information. Thus, even though the awareness of environmental labeling is increasing, awareness at the national level is lower than 30%. Moreover, even the 30% figure is low compared to levels reported in some OECD countries.[6]

Part of the increased public knowledge of environmental labeling can be explained by CCELP's efforts to disseminate information about the program. CCELP uses newspapers and TV programs to educate the public about products with environmental labels, technical requirements that must be satisfied to receive labels, and procedures enterprises must follow to apply for labels. This type of publicity has helped consumers to become knowledgeable about labeled products, and it has encouraged enterprises to apply for environmental labels (Pan 1998).

---

[5]The surveys were conducted by researchers at CRAES in four major stores in Beijing in April 1995 and June 1997, respectively. More than 500 people including Beijing citizens, students, soldiers, and tourists were interviewed.

[6]In countries such as Germany, Canada, and Japan, the level of consumer awareness to eco-labels seems to vary between 45% and 50% (OECD 1997).

In addition, CCELP and SEPA (formerly NEPA) use special days, such as World Environmental Day and Consumer Day, to disseminate information about environmental labeling. For example, in the context of World Environment Day of 1997, NEPA, the Beijing Commercial Association, CCELP, and the Deshili Plastics Group jointly organized a 1-month publicity activity in Beijing with the theme: "eliminate white pollution for life on earth." In this context, "white pollution" refers to pollution resulting from the disposal of *white* plastic materials. China Central Television (CCTV), *Beijing Market News*, and *China Environmental News* all reported on this 1-month of publicity activities. In 1998, CCELP and SEPA again used World Environmental Day to disseminate information about environmental labeling. The theme for 1998 was "promote green consumption for life on earth."

Another reason for the public's increased awareness of environmental labeling is extensive TV and newspaper advertising by manufacturers of household refrigerators. During 1995, several large refrigerator manufacturers conducted "advertising wars" on TV and in newspapers to sell their non-CFC refrigerators, prominently displaying environmental labels. Some even misled consumers by implying that use of CFC refrigerators could cause direct harm to human health. For a period, people were saturated with terms such as "green" and "non-CFC" (*wufu*) from these ad campaigns; as a result, many of them became familiar with environmental labels (Zhao and Ortolano 1999). According to the previously mentioned surveys conducted in Beijing in 1997, 72% of the people who knew about environmental labels learned about them from ads for refrigerators.

Even though awareness is increasing in large cities, most Chinese citizens still do not know about labels. In addition, awareness of environmental labels does not mean that consumers will buy labeled products. Chinese consumers' purchasing behavior are influenced more by price, product quality, after-sale service, and advertising campaigns than they are by environmental labels. In general, the price of labeled products is 10% to 20% higher than the same type of products without labels. For example, non-CFC refrigerators generally cost 200 yuan higher than CFC refrigerators.[7] In China, consumers only appear to show a preference for labeled products in one or more of the following circumstances: products are reformulated to eliminate direct harm to human health, products get consumers' attention through extensive ads programs, and products are cheaper using production technologies adopted to qualify for labels. For example, traditional paints contain organic and toxic substances that harm human health, and some enterprises (such as Suzhou New Type of Building Material Plant) have adopted Japanese technology and equipment to produce non-toxic and low-pollution paints. These new paints are 10% to 20% less expensive than

[7]These points about the prices of labeled products are from interviews with an official at Office of Household Electrical Appliance Industry and staff at the CCELP Secretariat in December 1998 and March 1999, respectively. Unfortunately, data on market shares held by firms using labels are not available.

traditional paints. Because of their reduced impacts on human health and lower price, water-based paints with environmental labels have captured an increasing fraction of the market share for indoor paints in China. More enterprises manufacturing paints have been trying to obtain environmental labels, which demonstrates the influence of consumer preference on behavior change of enterprises.[8]

The situation of non-phosphorous laundry detergents with environmental labels contrasts sharply with the example involving indoor paints. Non-phosphorous detergents can reduce water pollution, because phosphorous in detergents can cause excessive nutrition (and therefore algal blooms) in water bodies. However, sales of non-phosphorous detergents lag behind sales of other detergents. A reporter from *Beijing Youth Daily* visited many stores in Beijing and interviewed sales representatives and consumers to learn why. She found that the price of non-phosphorous detergents is not high compared to other detergents; thus, price is not the issue. The reports found that sales were relatively low because detergents with environmental labels did not have memorable ads on TV and were not well known by consumers (Fangfang 1998). The most popular detergents in China's markets are products containing phosphorous made by joint ventures between Chinese enterprises and companies from the United States, the United Kingdom, and Germany. These joint ventures have the resources to invest heavily in TV commercials. However, most companies that produce non-phosphorous detergents are small, new firms with low financial capabilities, and they cannot invest in extensive ad campaigns (Fangfang 1998). These companies have attempted to use environmental labels to attract consumers, but the labels have not yet had much effect.

Behavior Changes of Enterprises

Environmental labeling typically raises the direct and indirect costs for producers. Direct costs consist of application fees as well as annual fees for the use of labels; indirect costs refer to increased production costs involved in changing raw materials or productive processes to satisfy criteria for labeling certification. The direct and indirect costs may be shifted to consumers in the form of price increases. Producers must weigh the disadvantages of increasing prices against the advantage of increased sales that might accompany the use of labels.

The rising number of Chinese enterprises that have been qualified to display environmental labels on their products demonstrates the increased awareness of environmental protection by China's enterprises and signals a shift to environmental friendly performance. In 1994 only three products

[8] Interviews with staffs at the CCELP Secretariat in March 1999.

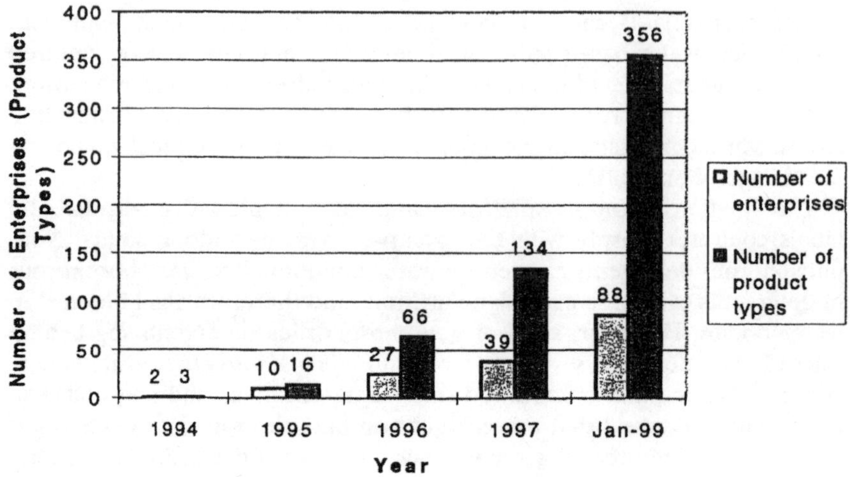

FIGURE 3. Number of products and enterprises that have received environmental labels. From *China Business Times*, June 5, 1988, and interviews with staff at the CCELP Secretariat, March 1999.

in two enterprises received environmental labels; 356 products in 88 enterprises received environmental labels as of January 31, 1999 (Figure 3).

According to the CCELP Secretariat, the 88 enterprises that had been certified to display environmental labels as of January 31, 1999 were distributed as follows: 26—low or non-CFC household refrigerators; 12—unleaded gasoline; 14—water-based paints; 13—non-phosphorous laundry detergent; and 10—biodegradable plastic products. Only one or two enterprises were in other sectors that had labeled products. About 57% enterprises that have labels are large-scale enterprises; small firms in certain sectors also are involved, e.g., non-phosphorous detergents and biodegradable plastics usually are manufactured by small enterprises. In terms of types of ownership, 34 enterprises were collectively owned, 25 were state-owned enterprises, and 25 were joint ventures. Only four enterprises were classified as foreign or individually owned.

Many enterprises that have environmental labels expected to enhance their image and to obtain economic benefits by displaying labels.[9] For example, a Chinese joint venture with German Benz Company applied for China's environmental label after they had already obtained Germany's label; Chinese media viewed this action as a demonstration of the company's high environmental awareness (Pan 1998). As another illustration, Heluxue Ice Cream Company developed a policy to only purchase freezers that

[9] Interviews with CCELP Secretariat officials in August 1998.

displayed environmental labels. Qingdao Aokema Electric Company, another domestic enterprise that produces freezers, went through the labeling certification process to obtain a major contract from Heluxue Ice Cream Company (Pan 1998).

The household refrigeration sector has been particularly active in China's environmental labeling program. In response to assistance from the Multilateral Fund (MLF) of the Montreal Protocol, and influences of the global market, multinational corporations, and intense (and occasionally misleading) advertising about non-CFC products, many household refrigerator manufacturers have developed technologies that use CFC substitutes. As of July 1998, 27 refrigerator manufacturing enterprises had received environmental labels for their refrigerators.

CCELP has tried to cooperate with enterprises and stores to promote the production and sale of environmentally friendly products. During the initial stages of the labeling program, CCELP initiated a "green consumption" campaign that encouraged consumers to purchase products with environmental labels. This campaign prompted some enterprises to apply for and display environmental labels. In 1997, CCELP extended its "green consumption" campaign to a "green marketing" campaign by adding stores into its program. In response to CCELP's green marketing campaign, 24 large stores in Beijing sent a letter to other stores throughout the entire country to request their participation in green marketing activities, such as advocating the use of biodegradable packaging material and selling products with environmental labels. The behavior of these stores has motivated more enterprises to apply for labels (Pan 1998).

The examples cited are, of course, not representative. There are millions of enterprises in China, but, as of January 31, 1999, only 88 had chosen to participate in the labeling program. The fear of losing market shares due to increased price of labeled products blocked many enterprises from changing technology to qualify for labels. The limited financial and technical capabilities of many enterprises also prevent them from changing their behavior to meet labeling requirements.

Environmental Benefits

It is difficult to estimate the environmental benefits resulting from China's environmental labeling program. Because the product categories are very limited, it is not yet possible to expect large environmental benefits from the program. Although some researchers believe that the environmental effects of labeling can be measured in terms of pollution avoidance (OECD 1991), it is difficult to distinguish the environmental gains achieved through labeling from those attained through environmental regulations. The authors' personal experience with environmental benefits from China's environmental labeling are limited to one area: CFC reduction. In this case, environmental labeling is one of several factors contributing to CFC reduction (Zhao and Ortolano 1999). The finding that environmental benefits of China's labeling program are limited is consistent with other research. For example, after assessing the European Union's eco-labeling program for the paper products sector, Erskine and Collins (1997) found little evidence that eco-labeling yielded significant environmental benefits.

China's labeling program is still under development. Consumers' awareness of environmental labeling has increased, and more enterprises are becoming aware of environmental labeling and attempting to use labels to maintain or increase market share for their products. However, notwithstanding that consumers' awareness of environmental labeling has increased, it is still very low, and it has influenced purchasing behavior in only a few sectors, such as household refrigeration and paints. At this point, it remains difficult to evaluate the environmental benefits resulting from China's labeling program.

Challenges and Prospects

Public Awareness on Environmental Labeling and Environmental Protection

Until citizen awareness of environmental labeling is raised, China's labeling program will do little to change consumers' behavior because consumers will not recognize the significance of environment labels. In China, awareness of the environmental labeling program is still relatively low. The need to increase awareness is a great challenge to organizations involved in the labeling program. How can this be changed? The relatively high level of consumer awareness in Sweden and Germany is attributed to the involvement of environmental non-governmental organizations, consumer organizations, and the media (OECD 1997). Cases involving China's laundry detergent industry and the household refrigeration sector reinforce the importance of the media's role in influencing consumers' awareness and purchasing behavior. In China, SEPA and other government agencies could finance programs to introduce environmentally labeled products on TV or in newspapers as a means to help people understand the difference between products with and without labels.

Mutual Recognition and Environmental Labels

One objective of China's labeling program is to improve exports by granting environmental labels to products produced for export. However, nations have not created a single, internationally recognized environmental label based on a common set of technical requirements for products. Consequently, even when enterprises become certified to display China's environmental label, their exported products often are not accepted as being environmentally friendly by other countries. For example, one enterprise received certification to display China's environmental label for its silk fabrics, and it packed its products for export to France and Germany in paper according to requirements set out by those countries. Despite these efforts, the silk still could not satisfy the technical criteria used by France and Germany for environmental labeling; thus, it was returned to China. This discouraged other enterprises from applying to use China's environmental labels for their exported products.

It would be ideal if a product manufactured in China and certified by SEPA to carry an environmental label were recognized for its positive environmental attributes by other countries. Although this has not yet happened, discussions of this subject took place between NEPA and the US Environmental Protection Agency in 1996. The US and China are particularly interested in recognizing each other's refrigerator standards with respect to energy consumption and non-use of CFCs, but they have not yet reached agreement on this (personal interview with SEPA officials, September 1998). According to CCELP officials, foreign companies may obtain China's environmental labels for their products through the same process as Chinese companies. If a foreign, well-recognized national certification program has already approved a particular product, CCELP can simplify the application process (personal interview with CCELP officials, March 1999).

Technical Criteria

Technical criteria for labeling present additional challenges. Should the label be restrictive or wide ranging? If labels are to enhance the environmental characteristics of products, technical criteria should be strict. But if too few products qualify for environmental labels, then few consumers will be exposed to the labels, and consumer awareness of environmental issues will not be raised significantly. An environmental labeling program must balance between having stringent technical criteria and a large number of labeled products. It is also difficult to decide which products and processes are least harmful to the environment. For example, is it more important to reduce the use of paper products by encouraging parents to use cloth diapers, or is it better to save the water and energy required to wash and heat cloth diapers by encouraging parents to use disposable diapers?

The existing environmental labeling program only requires manufacturers to display labels. On its own, the label does not provide specific information about the difference in environmental impacts of labeled products compared to products without labels. The label also cannot distinguish the environmental effects among all labeled products within a particular product category. This limitation has caused problems. For example, some household refrigerator manufacturers have complained that the environmental label is the same for both non-CFC and low-CFC refrigerators. This reduces incentives for refrigerator manufacturers to move forward toward a complete phaseout of CFCs in refrigerator manufacturing. An environmental label would be more effective in advising consumers if it somehow identified the main criteria that a labeled product had to satisfy or the degree of "environmental friendliness" of the product (e.g., using a "scoring" system from 1 to 10 with all labeled products given score based on their environmental impacts).

Combining Labeling with Cleaner Production and ISO 14000

In China, most environmental programs are not coordinated with each other—they are generally implemented separately, and the environmental labeling program is no exception. However, environmental labeling has close connections with China's efforts to promote cleaner production and to encourage enterprises to develop certified environmental management systems under International Standards Organization (ISO) 14000 guidelines. Cleaner production involves use of an integrated preventive environmental strategy for processes, products, and services to reduce environmental impacts throughout the lifecycle of a product. Thus, cleaner production provides foundation for an enterprise to obtain an environmental label for its products. Although China began emphasizing cleaner production in the early 1990s, no connection yet exists between the environmental labeling program and China's cleaner production activities. One way government offices promoting labeling and cleaner production could cooperate is by focusing on the same industrial sectors. In this instance, the labeling certification criteria could be based on particular clean technologies, and cleaner production policy advocates could promote those technologies to make it easier for enterprises to obtain labels. Moreover, the prospect of obtaining environmental labels might encourage enterprises to adopt cleaner production methods. In this circumstance, the efficiency of both sets of government activities would be improved.

In 1996, China began to introduce and implement ISO 14000 to strengthen environment management systems within Chinese enterprises. The creation of an effective environmental management system within enterprises is an important criteria for certifying an enterprise for environmental labeling of its products. Moreover, ISO 14024 specifically relates to environmental labeling and divides labeling scheme into three classes. Type I labels are lifecycle labels; type II labels are based on manufacturers' claims concerning individual environmental characteristics; and type III labels provide environmental information resulting from an independent scientific assessment of established criteria concerning a particular class of products. CCELP plans to incorporate ISO 14024 into its own environmental labeling program. Linkages between the ISO 14000 series of guidelines and China's environmental labeling program provide a unique opportunity to use ISO 14000 to advance the objectives sought by China's labeling program and its cleaner production policies.

We are grateful to the following individuals for their comments and help on a draft of this paper: Katherine Kao Cushing, Donald Kennedy, Liu Chang, Liu Zunwen, Leonard Ortolano, Read Vanderbilt, and two anonymous reviewers.

References

Appleton, A.E. 1997. *Environmental Labelling Programmes: International Trade Law Implications.* London: Kluwer Law International.

Chinese Certification Committee for Environmental Labeling Products. 1995. Study and Application of China's Environmental Labeling (Zhongguo Huanjing Biaozhi Yanjiu he Yingyong), Working Report. Unpublished.

Erskine, C., and Collins, L. 1997. Eco-labeling: Success or failure? *The Environmentalist* 17;125–133.

Fangfang. Environmental friendly detergent: Why prices are good but sales are not good? (Huanjing Xidiji: Jialian qiue Buhaoxiao). *Beijing Youth Daily* May 31, 1998.

Organization for Economic Co-operation and Development. 1991. *Environmental Labeling in OECD Countries.* Paris: OECD.

OECD. 1994. *Life-Cycle Management and Trade*, OECD. Paris: OECD.

OECD. 1997. *Eco-Labelling: Actual Effects of Selected Programmes.* Paris: OECD.

Pan, H. 1998. Environmental labeled products are occupying the life space of citizens (Huanjing Biaozhi Chanpin Zhengzai Zhanju Shimin Shenghuo Kongkian). *China Environment News* June 2.

Xia, Q., Liu, Z., and Wang, Y. 1995. *Environmental Labeling (Huanjing Biaozhi).* Beijing: China Environmental Science Press.

Zarrilli, S., Jha, V., and Vossenaar, R. 1997. *Eco-Labeling and International Trade.* New York: St. Martin's Press, Inc.

Zhao, J., and Ortolano, L. 1999. Implementing the Montreal protocol in China: Use of cleaner technology in two industrial sectors. *Environmental Impact Assessment Review* 19(5):000–000.

Zhu, X., and Hu, S. 1994. Suggestions on establishing China's environmental labeling program (Dui Zhongguo Kaizhan Huangjing Biaozhi Gongzuo de Jiangyi). In *SEPA, Collection of Environmental Labeling Literature in China and Abroad (Guoneiwai Huangjing Biaozhi Wenji).* Beijing: China Environmental Science Press.

Part III
Factors Impacting the Success and Design of Eco-labelling Programmes

[15]

Information as a regulatory instrument to price biodiversity benefits: certification and ecolabeling policy practices

PAULO A.L.D. NUNES [1,2,3,*] and YOHANES E. RIYANTO [4]

*[1]Department of Economics, Ca' Foscari University Venice, San Giobbe, I-30121 Venice, Italy; [2]Centre of Excellence for Sustainable Development, Ca' Foscari University Venice, Palazzo Giovanelli, I-30135 Venice, Italy; [3]Fondazione Eni Enrico Mattei, Palazzo Querini Stampalia, I-30122 Venice, Italy; [4]Department of Economics, Faculty of Arts and Social Sciences, National University of Singapore, AS2 1 Arts Link, Singapore 117570; *Author for correspondence (e-mail: pnunes@unive.it; fax: +39-041-2711-461)*

Received 17 May 2003; accepted in revised form 9 March 2004

Key words: Biodiversity benefits, Biodiversity oriented products, Biodiversity segmented market, Command and control, Information provision instrument, Market failure, Market regulation, Non-market values, Public good character

Abstract. In this paper we address the issue of market failure arising from the non-existence of (market) prices for biodiversity, and also present and discuss alternative policies to cope with it. Particular attention is given to certification and ecolabeling of policies. First, we critically survey the role of certification and ecolabeling as an information provision instrument. Second, we provide a comprehensive view on basic foundations and crucial issues that underpin the design of a certification and ecolabeling policy. Finally, we present some case studies to draw some lessons from current certification and ecolabeling policy practices.

Introduction

Biodiversity can be broadly defined as variety of life on earth. It provides a wide range of benefits to human activities. Most biodiversity benefits have a public good character – showing either non-rivalry, or non-excludability in the consumption. In concrete terms, this means that the one's consumption of a public good does not change the consumption opportunities to the remaining elements of the society. In addition, if one person gets access to or consumes biodiversity benefits, others cannot be excluded from enjoying them. A well-know example of a biodiversity benefits with a public good character is the non-use value of biodiversity, i.e. the benefit that one can derive simply by knowing the that diversity exists regardless of any human use. Consequently, traditional market price mechanisms will have difficulties to express the value of biodiversity benefits. One way to solve this is to ask people directly how much they would be willing to pay for biodiversity benefits to which they have access (see Nunes and Van den Bergh 2001). Unfortunately, individual consumers do not tend truthfully to reveal their willingness to pay. In such circumstances, the policy makers need to agree upon the design and

2010

implementation of an incentive compatible policy instrument, i.e. a policy instrument that induces individual consumers to truthfully reveal their own willingness to pay for biodiversity benefits and this way signaling the demand in the biodiversity segmented market.

In addition, biodiversity benefits also exhibit production and consumption externalities (see Nunes et al. 2001). No market prices exist to really reflect the value of these externalities. As a result, biodiversity externalities are not taken into account in the production and consumption decisions of economic agents. This exclusion hinders the proper functioning of the market mechanism, and therefore will lead to inefficiencies in the allocation of resources in the economy. Such inefficiencies bring about biodiversity quality-degradation, diminishing the ability of our ecosystems to keep up with the rapid progress of economic development. In the long run, such an imbalanced development pattern decreases our well-being (Nunes et al. 2003). In this context, value signaling through certification and ecolabeling can play a crucial role in assisting economic agents to price biodiversity benefits and to internalize them, greasing the wheel of the market mechanism. Hence, the market failure problem can potentially be mitigated. An efficient biodiversity-conservation, which will maintain the balance between our ecosystem's quality and economic development, can then be realized.

This paper presents a discussion and comprehensive survey on the role of certification and ecolabeling in guiding economic agents to value the biodiversity externalities. Certification and ecolabeling is thus seen here as an alternative policy instrument to mitigate the problem of market failure. The next section reviews some government policy instruments dealing with the problem of market failure. Then a more comprehensive review of certification and ecolabeling policy in the domain of biodiversity is given. General ideas will be made concrete with three examples of certification policy. The penultimate section discusses some caveats, after which conclusions are given.

Government policies dealing with the problem of market failure

Direct government intervention: taxes and standards

One possible way of addressing the problem of market failure is through direct government intervention. This involves, for instance, an introduction of taxes or standards in the markets. The best known tax instrument is the optimal or 'Pigouvian' tax, which at the optimal equilibrium is set equal to the marginal social costs caused by the economic activity responsible for the environmental externality. This taxation restores a resource allocation with biodiversity externalities to a social optimum. Such policy requires however, an important amount of accurate information concerning the general public's map of preferences with respect to biodiversity non-market benefits. Such information plays a crucial role in the design of an effective tax system. Thus government needs to conduct thorough research on the assessment of the economic value of biodiversity non-market benefits. This is a fairly costly activity.

The use of standards, referred to in this paper as 'command-and-control system', is especially attractive from the view point of its effectiveness. This is because the government directly dictates a clear quantity target (restriction) that has to be followed by market participants. In the area of biodiversity and nature conservation, an example of this type of policy is the limitation on the number of visitors to certain natural recreation areas (biodiversity sensitive areas). More recently, the settling of a fixed number of possible bioprospecting market contracts between the state and pharmaceutical industries is another example of this command-and-control policy. These contracts are signed to support the search of genetic codes contained in living organisms that can be used for the development of chemical compounds that have commercial value in agricultural, industrial, or pharmaceutical applications (Simpson et al. 1996). The most noted example of these agreements is the pioneering venture between Merck and Co., the world's largest pharmaceutical firm, and 'Instituto National de Biodiversidad' (INBio) in Costa Rica. At the time of contract signing, in 1991, Merck paid Costa Rica approximately one million dollars and agreed to pay royalties whenever a new commercial product is put into the market.

However, adopting a command-and-control policy also implies embracing high monitoring and enforcement costs. For instance, in the case of bioprospecting contracts, the government – owner of the resource – has to determine the acceptable threshold level of samples of compounds of plant, fungus and bacterial origin in the country that will be submitted to a pharmaceutical screen and which will be reflected in the market agreement. In addition, the government may have to set-up a regulatory body, which monitors the amount of bioprospecting samples that a given pharmaceutical firm has access to and then enforces penalties upon any violation.

In a world of asymmetric information, in which firms do not have an incentive to truthfully reveal their own information, the government may need to regulate and monitor intensely firms' conduct. Environmental auditing could be an example of such a monitoring activity. From an economic point of view, the discussion and evaluation of these two instruments of environmental policy is traditionally done on the basis of their efficiency features (Baumol and Oates 1988). Effectiveness and distribution effects, such as equity and fairness, often work as secondary policy evaluation criteria. The most common comparison is between uniform standards and uniform taxes. Taxes are attractive as they provide better incentives than standards to change individuals' behavior, for example, recreationists' behavior when visiting natural areas. Thus, taxes may lead to a more efficient outcome in term of the social welfare than standards.

Independently of the policy instrument chosen by the government, there are some factors that hinder the effectiveness of its involvement. First, governmental intervention involves high administrative costs because the government

2012

may have to establish a monitoring and enforcement agency. Second, governmental intervention may not be effective if the flow of information between firms and the government is characterized by a strong asymmetry. In fact, the capacity to design an effective government policy is hindered by the presence of information imperfection, in the majority of real-world situations. Finally, such policies may create bureaucratic inefficiency. Bureaucrats may pursue are prone to be influenced by market participants via lobbying activities (see Milgrom 1988). Even if the rent-seeking behavior is absent, direct government involvement may also create a disincentive for market participants to innovate or to employ the most efficient method of production. In this context, a policy instrument based on a market creation mechanism can be a valid alternative to direct government intervention. Policy instruments based on market creation mechanisms are discussed in the following section.

Market creation mechanism: information provision

In the context of biodiversity benefits, policy instruments based on market creation mechanism are characterized by the provision of information with respect to market and non-market biodiversity benefits. In other words, the certification mechanism informs the consumer about the product characteristics that are not directly related to its consumption or use, however provide environmental benefits, including the conservation of the biodiversity. In turn, the provision of such information works under one of two basic conceptual frameworks: ecolabeling and certification. Ecolabeling refers to the policy scheme that is characterized by the evaluation of a product, or product characteristics, against particular specifications. The idea here is to measure and confront specific characteristics attributed to the product's origins to detailed ecological, social and economic specifications. This information is, along with the product, provided to the consumer. Certification refers to the policy scheme that is characterized by an evaluation of the product's underlying management system against particular management specifications. The latter involves the identification and monitoring of the supply chain including the transport and processing of raw materials, secondary manufacturing and, finally, retail distribution. As before, such information is provided to the consumer.

Therefore, the overall objective of ecolabeling and certification policies is to link the consumers who wish to favor more biodiversity or socially responsible sounding products with the producers of these products and the raw materials and processes from which they are made, creating a separated market for these differentiated products. The respective market price will incorporate biodiversity benefits and therefore contribute to a better allocation of biodiversity. In either case, the government is not directly involved in the process. However, it has a crucial role in providing a favorable environment that helps to enhance the effectiveness of a certification and ecolabeling policy. For example, a credible scheme must evaluate the integrity of the producer's claims and the authenticity of product origin. Furthermore, the evaluation scheme needs to be seen as objective and impartial. Therefore, the success of certification and ecolabeling strategy *per se* may prove to be difficult to achieve. For this reason, this strategy often goes hand-in-hand with other micro-economic policies, giving rise to a 'mixed policy.' This government strategy will be analyzed in the following section.

Mixed policy

The core of this policy strategy is in the combination of direct government involvement (see above) and the market creation mechanism (see above) policies. The goal of this policy is to circumvent the weaknesses and inefficiencies that may occur when adopting either the command-and-control policy or the market mechanism approach.

Mixed policy is characterized by two features. They refer to the establishing of a quantity or quality standard and introducing the possibility of trade among economic agents. The first feature makes sure that an upper limit of biodiversity damages is set at regional or national level. This threshold level is reminiscent of the command and control system. The second feature assures flexibility and efficiency (cost-effectiveness) at the level of individual agents, and leads to equal costs of biodiversity damage reduction at the margin among all individuals and firms. This is because the certified credits can be sold to another participant who needs to satisfy the threshold level with respect to biodiversity damage. This type of trade possibility enhances agents' incentives to conform to the biodiversity damage standard.

This instrument has not yet led to applications in the area of land use, nature conservation and biodiversity protection. An exception is the Dutch bio-label energy system (see below for a more detailed discussion of this policy instrument).

A critical evaluation of certification and ecolabeling instruments

The framework

Certification is an integral part of a policy directed to induce the working of the market mechanism, without direct government involvement in the supply and demand forces. The goal of biodiversity certification policy is to inform consumers that a product has been processed in a production method that is biodiversity sound. Thus, in making their purchasing decisions, consumers will be exposed to a choice between buying biodiversity and non-biodiversity oriented products. If some consumers prefer to buy the former, then the policy maker can effectively segment the market. A new market-niche, i.e. the market for the biodiversity-oriented products, will then emerge. In order to sustain the

2014

biodiversity segmented market the policy maker should enhance the role of this new market-niche. An efficient working of the biodiversity segmented market crucially depends on the flow of information across demand and supply forces. In his seminal article, Akerlof (1970) showed that the presence of an informational problem could lead to a market failure.[1] The latter is, in turn, the cornerstone of any certification and ecolabeling policy instrument. This section critically analyzes the role of certification as an information provision instrument. Figure 1 presents an overview of the structure that will steer our comprehensive discussion and analysis.

The crucial questions to ask here are; to what extent would the certification and ecolabeling policies be successful for creating markets for biodiversity? Would an emerging segment-of-market where biodiversity benefits are internalized in the respective market price be, in fact, an effective tool to stimulate an efficient working of markets related to biodiversity oriented products, and thereby achieve a better allocation of such a scarce resource? Should other policy measures be complemented with this policy? The answers to these questions are discussed in detail in the following sections.

Market and non-market benefits of biodiversity

Our starting point in answering those questions is the identification of the type of the market and non-market biodiversity benefits. What we mean by the non-market biodiversity benefits are the net marginal social-benefits of consuming the goods that are accrue to society and which are not explicitly considered by producers in their strategic decision-making. Alternatively, market biodiversity benefits refer to the benefits that are indirectly related to the use or to the consumption of the good and are taken into account in market price determination. The participation of consumers in markets for these differentiated, biodiversity-oriented products usually permits the internalization of the market values immediately. If consumers internalize the non-market benefits of biodiversity, they are willing to pay for these benefits.[2] Then the question is: can consumers, who purchase the environmentally friendly good, internalize the non-market benefits of the good? If the answer to this question is yes, we can then categorize the good as a *private good.* Otherwise, the good is categorized as a *public good.*

[1] In the literature we distinguish two types of informational problem, i.e. the hidden information (adverse selection) and hidden action (moral hazard) problem. Hidden information refers to a case in which one party knows more about her true type than the other party before a contract (relationship) is initiated. Hidden action refers to a case in which one party knows more about her type (effort) after a contract (a relationship) is initiated.

[2] The Environmental Protection Agency (EPA 1993) reported that 'several surveys indicate that a majority of Americans consider themselves to be environmentalists and would prefer to buy products with a lessened environmental impact when the quality and costs are comparable.'

2015

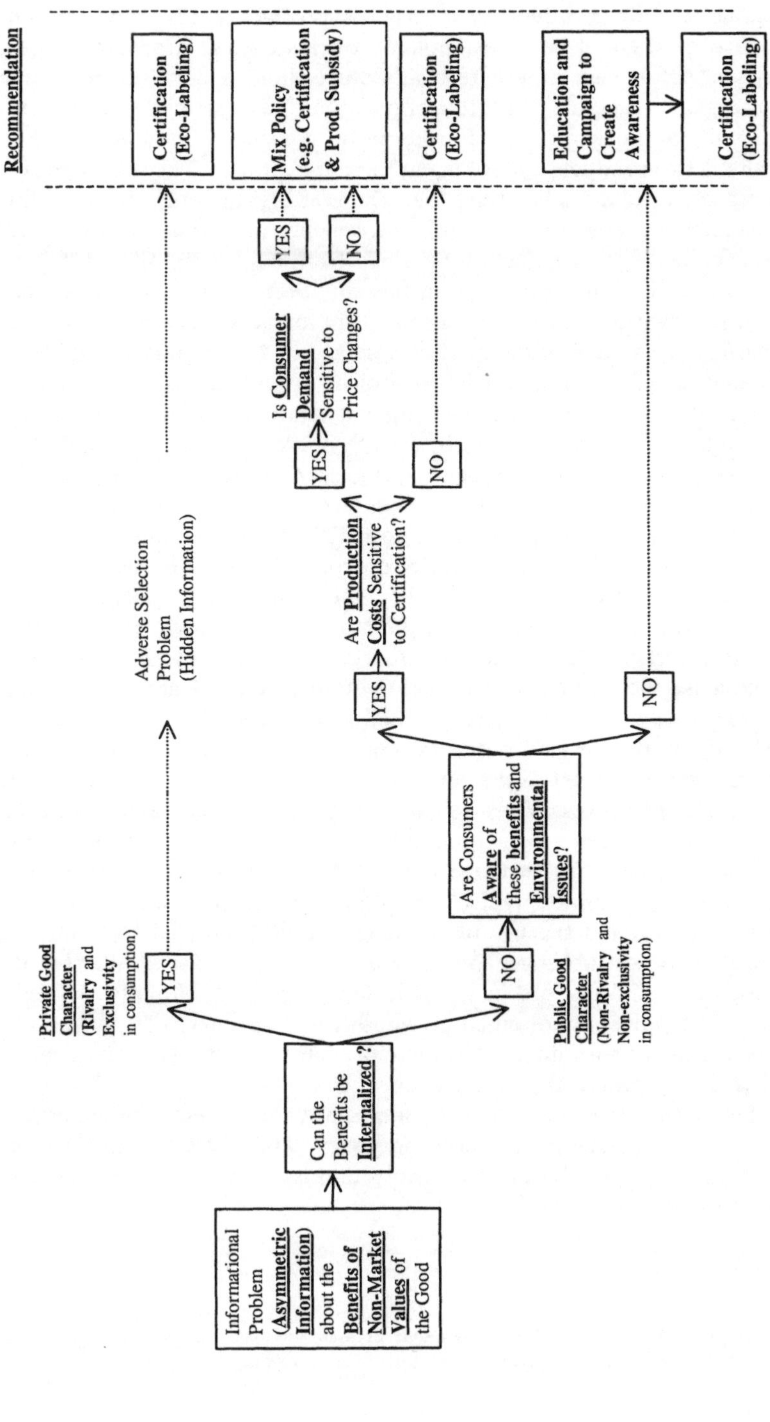

Figure 1. A comprehensive view on the certification and ecolabeling policy.

2016

An organic vegetable product, i.e. a vegetable product that is grown without chemical fertilizers, could be an example of a private good. Consumers may, subjectively or not, believe that there is a difference in taste between an organic and a non-organic vegetable. It is often argued that the organic vegetable tastes better than the non-organic one. Furthermore, consumers may also able to distinguish the two products by looking at their appearance. They also perceive that the organic product is healthier than the non-organic one. Hence in this case, consumers are able to internalize the benefits of consuming the good. They can experience the satisfaction obtained from consuming the good.[3]

Nonetheless, it does not mean that in this situation consumers will not face any further informational problem. The problem remains, since consumers are lacking perfect knowledge of the quality standard of the organic vegetable. A quality standard certification can be helpful in providing assurance to consumers. Here, the role of certification is simple. It acts as an instrument to resolve the standard hidden information problem. The market is also accommodating here, in the sense that if there are enough consumers who want to buy the organic vegetable, then some producers will have an incentive to enroll in this market segment. They have a proper incentive to do so. They also have interest in using certification to try to differentiate themselves from the other producers.

Things become more complicated when the biodiversity benefits have a public good character, in the sense that the biodiversity benefits of consuming the good are experienced by the society at large, which may include those who do not purchase the good. Two examples of such a good are biodiversity-oriented electricity and biodiversity-oriented forest products. Let us take the case of electricity. Recent environmental policies in the electricity sector seem to give unequivocal support to the belief that electricity that is produced using biodiversity friendly processes has a significant, positive social value. However, nobody can internalize the benefits and thereby exclude anybody else from enjoying the benefits. Furthermore, the nature of the biodiversity-oriented electricity products cannot really be physically distinguished from the remaining ones. Making the use of the language of Industrial Organization Theory, electricity is a homogeneous product. The same case applies for the biodiversity-oriented forest products, for instance plywood. Nobody can really distinguish the biodiversity-oriented plywood from the other type plywood by just inspecting the two products. It is therefore harder to design a certification policy in this case than in the previous one.

According to Figure 1, there are three important components that determine the success of the ecolabeling/certification policy. The first refers to the 'consumers' awareness' of the biodiversity protection issues and the social benefits of having a clean environment.

[3](van Ravenswaay 1995, 1996) indeed shows that organic products exhibit a high consumer awareness of the environment and private benefits of these products.

2017

Consumers' awareness

Consumers' awareness is a necessary condition to achieve an efficient certification policy when the biodiversity benefits have public good characteristics. However to be sufficient it should be coupled with consumers' willingness to take into account the social benefits in their consumption behavior. Of course there will be no interest for using biodiversity certification schemes when consumers are not at all aware of the respective biodiversity benefits.

If consumers have no awareness with respect to non-market biodiversity benefits, then there is a crucial and urgent action that has to be taken before the policy makers can launch the certification policy. Consumer awareness may take many years to develop (see van Ravenswaay and Blend 1997). Hence, the policy makers should launch extensive information campaigns, targeting the general public, as well as initiate formal education programs about the values and the benefits of having a clean, and biodiverse environment. Let's imagine a country or a society in which the public is not at all aware about the need to sustain biodiversity and to protect the environment. In such society, there is no use in implementing a certification policy, because it is doomed to fail (see Salim et al. 1997). Once there is sufficient consumer awareness about the biodiversity benefits, environmental friendly products and management processes then there maybe be a willingness to pay a price premium for ecolabeled or certified products. In other words, consumer is a necessary condition for policy design. However, in order to guarantee the success, the policy maker will need also to deal with the nature of marker supply and the nature of consumer demand.

Firms' incentive to endorse certification and ecolabeling policies

In addition to consumers' awareness, the degree of sensitivity of firms' production costs to the adoption of the certification schemes will also play an important role in determining the success of the certification policy in the case of a biodiversity benefits with a public good character. If production costs are not sensitive to certification, then producers may have sufficient incentive to accommodate the certification policy. However, in most cases the adoption of certification and ecolabeling policies will increase a firms' production costs because producers may have to install new production technologies, or may have to avail themselves of certain kind of inputs in order to satisfy the environmental standards that are stipulated in the product 'bio' label – see (van Ravenswaay and Blend 1997) for more details. Therefore, adopting certification implies higher production costs, which force them to increase the price. This, in turn, may damage the firm's market competitiveness. Furthermore, if quantity demanded by the consumer is sufficiently sensitive to changes in prices, then an increase in price means a reduction in the overall quantity demanded that, in turn, is reflected in a reduction in the firm's profits. Hence,

2018

hardly any producers will choose to engage in this new market segment and embrace the market for biodiversity oriented products. There are simply not enough incentives for the market mechanism to work.

In this setting policy makers need to complement (or combine) certification policy instruments with other micro-economic policies aimed at providing enough incentive for producers to adopt certification. Examples of such mix policies are, for instance; an input subsidy, technical assistance provision, and a R&D subsidy.

It is also worth noting that even if production costs are not sensitive to the certification, it does not mean that a certification policy is always advisable. Under marketing or price complementarities between environmental and conventional innovation production lines, a certification or ecolabeling requirement may also stimulate investment in the production technology of the conventional product (see Mattoo and Singh (1994) for more details). This in turn may lead to an increase in the output of the conventional product, thus making the certification policy miss its goal in stimulating biodiversity oriented management practices and respective niche-market. If this prevails, then certification should again be complemented with other policies. In this context, Dosi and Moretto (1998) propose a restriction (rationing), i.e. awarding certificates and ecolabels only to production lines that meet the environmental innovation criteria without increasing investments in the production technology of the conventional product.

Sensitivity of consumer demand to production costs

According to Figure 1, the last crucial component that determines the success of the certification policy in the case of a public biodiversity benefit is the nature of the consumer demand. As previously mentioned, if consumer demand is sufficiently elastic, an increase in price means a reduction in profits. This elasticity depends on the consumers' preference toward biodiversity oriented goods, which in turn depends on the consumers' awareness of issues related to biodiversity non-market benefits. The degree of consumer awareness with respect to biodiversity benefits varies across countries with different socio-economic status. For instance, one would tend to agree that the degree of consumer awareness is lower in the developing countries than in the developed countries.

Since the awareness of biodiversity benefits is generally lower in the developing countries, consumers in these countries tend not to have a high willingness to pay for the biodiversity-oriented products. As a result, while the introduction of an ecolabel or certification scheme raises the production costs, then producers' profits will inevitably decrease. Without any further effort, the certification scheme would probably fail. In order to avoid such failure other policy strategies should be considered. For instance, policy makers can launch a certification and an environmental campaign at the same time. NGOs could be encouraged to disseminate information that can increase consumer awareness. Alternatively, the government can interfere in the market functioning, directly influencing the supply and demand forces in the biodiversity segmented market.

A command-control policy within the certification and ecolabeling

The government can impose certain restrictions in the segmented market of biodiversity products. For instance the government can require producers to achieve a certain output of biodiversity products and in exchange producers obtain a certificate of compliance. The government can, at the same time, set biodiversity compliance thresholds to consumers. Thus, rather than facilitating the working of market mechanism to sustain the certification and ecolabeling policy, the government directly motivates producers and consumers to embrace the certification and ecolabeling policy. Figure 2 summarizes this idea.

For example, let us consider a scheme in which the number of certificates that can be issued by a seller depends on the quantity that he produces to the biodiversity segmented market. The number of 'biodiversity certificates' that can be issued by a seller depends on its production to the biodiversity segmented market. Therefore, the two markets co-exist, i.e. the market for the conventional product itself and the market for biodiversity-oriented products. If the demand exceeds the supply in the market for biodiversity-oriented products, the producers will be motivated to generate more biodiversity certificates to satisfy the demand. This implies that there will be more production on the biodiversity market segment. The revenue generated from trading bio-certificates compensates for the potential increase in production costs. Of course, one crucial aspect to take into account here is the determination of the 'settlement' price for the bio-certificates. It should be set lower than or equal to the reservation value of the consumers. The magnitude of this reservation value

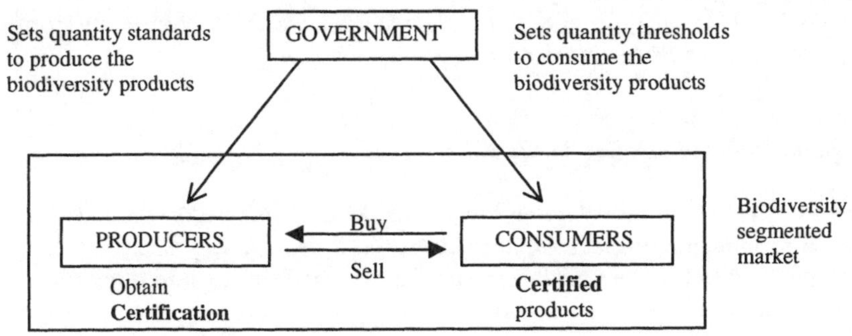

Figure 2. Command and control policy and the certification and ecolabeling.

2020

reflects the higher premium that consumers are willing to pay in order to have access to the products in the biodiversity segmented market.

Intermediate conclusions

It is clear that whenever (i) consumers are aware of the non-market biodiversity benefits, (ii) consumers are able to internalize biodiversity benefits, and (iii) the production costs are not too sensitive to the certification schemes, consumers are willing to pay a price premium and in this context certification policy can be sustained as an instrument for biodiversity conservation. Nevertheless, policy makers should still be concerned with the possibility that the product differentiation created by certification can, under certain conditions, lead to an increase in the demand of both biodiversity oriented market segment and remaining ones. The design of an efficient policy strategy becomes much more complex whenever the biodiversity benefits have a public good character. In this context, consumers are unable to fully internalize biodiversity benefits and thus ecolabeling and certification policy *alone* is likely to fail. Therefore, the design of an efficient policy strategy will involve an additional government intervention. For example, the government can launch extensive information campaigns and provide additional educational programs, aiming at the rising of consumer awareness of biodiversity benefits. It can also extend an input subsidy to induce producers to endorse certification, as well as provide output subsidy for 'bio' products. Finally, the policy maker is also able to make use of certification and ecolabeling policies together with command and control market interventions.

Certification and ecolabeling policy in practice

In order to move away from the general level of discussion, some illustrations are presented in which the aspects viewed in Section 'a critical evaluation of certification and ecolabeling instruments' come together. These are no means exhaustive. Our purpose here is to draw some lessons of experience from the current practices of certification policy.

Organic food and labeling in the Netherlands[4]

The European Common Agriculture Policy aims at, *inter alia*, the reduction of the environmental impact of agriculture, and reinforcing the path towards sustainable development practices, such as organic farming and integrated crop management. In the last decade all European countries have experienced

[4]See (van der Grijp and den Hond 1999) and van Bellegem et al. (1999) for more details.

2021

growth in the number of organic farms. There are, however, considerable variations among European countries. In general, European Countries can be distinguished in terms of their organic farm development into four groups, namely: (a) booming countries (Denmark, Finland and Italy); (b) stabilizing countries (Austria, Germany and Sweden); (c) high potential countries (Greece, Ireland, Norway, Portugal and Spain), and (d) lagging behind countries (Belgium, France, Luxembourg, the Netherlands and UK). These differences are associated with a wide range of factors, such as the diversity of national labor markets, the variation in consumer awareness with respect to ecological issues, the distinct direct government interventions to influence market supply and demand forces, and various labeling and certification strategies for the organic products. The latter implies the use of clear and accurate information on the organic status of products.[5] In the Netherlands, for example, it has been suggested that the diverse, fragmented and often unclear set of certification schemes for organic products created much confusion and skepticism among consumers and this can explain why the Netherlands is one of the countries lagging behind in organic food production.

As mentioned earlier, the capacity to distinguish organic products plays an important role in creating markets for biodiversity. Unfortunately, in the Netherlands this capacity is hindered by the fact that there are in general too many logos. This calls for a further standardization policy. Such a policy should also resolve the question of who should be given the right to standardize.

Another obstacle for the creation of markets for biodiversity is the fact that in the Netherlands most agricultural products are sold loose or are only packed late in the commercial chain. Take, for instance, fruit and meat products. For the creation of an efficient market, it is desirable that a reliable chain of products should be established. This can actually be achieved by individual, separate sales firms for organic products (organic butchers and health food stores) and by a clear, uniform and transparent labeling system under the principles and guidelines of an accredited, independent third party authority. This third party authority should be responsible for assessing the different food retailers' (e.g. supermarket chains) efforts to separate product streams so as to create a distinguishable product for the consumer. Furthermore, the prospect of achieving a successful organic market creation will increase if a clear, uniform and transparent labeling system goes hand-in-hand with fiscal and environmental micro-economic policies.

[5]For example, when a product has fulfilled a full standard requirement, i.e. a requirement stipulating that at least 95% of the ingredients of this product are certified organic in nature, the product may then be labeled as a certified-organic product. Alternatively, if less than 95% but not less than 70% of the ingredients of a product are of certified organic origin, the product will be called a *product made with organic ingredients*. The proportion of each organic ingredient will be clearly stated. Finally, if less than 70% of the ingredients are of certified organic origin, such a product cannot be called an organic product (IFOAM 2000).

2022

Energy market and certification: the Dutch bio-label system [6]

In the Netherlands, a system similar to the bio-certificate system was introduced starting in January 1998. The aim of the system was to increase the Dutch consumption of renewable energy to 10% of the total energy consumption by 2020, and also to induce producers to increase the production of the renewable energy products.[7] There are two phases of implementation. In the first phase, the sale of the bio-labels to consumers is based on voluntary purchase at a premium price. In the second phase, which started from the year 2001, the sale is determined by a minimum quota of renewable energy consumption.

In this system, renewable energy producers receive a bio-label for each 10,000 kW h of renewable electricity they produce, which is valid for 1 year. Thus, there is no possibility of inter-temporal fulfillment of the quota. The renewable electricity is sold to the same market as the non-renewable one. Thus, their price is the same. To compensate producers for keeping the selling price constant despite the higher production costs, the consumers are requested to attain a minimum quota of renewable energy consumption and producers are allowed to collect the revenues from selling the certificate that demonstrates that each household in complying the minimum quota with respect to the consumption of renewable energy. As the system was recently started, it is hard to evaluate its performance. Thus, it remains to be seen how successful the system will be in inducing production and investment in renewable energy products. Nevertheless, it should be noted that the design of the system is quite ingenious. It *limits* the direct role of the government, and allows the market mechanism to contribute in achieving the biodiversity goals. Furthermore, it will be less costly for the government, as the government can then reduce the government spending on input and R&D subsidies.

Nevertheless, there are two main obstacles for the proper working of the system, namely the credibility of the system, i.e. the settlement of the minimum quota and the determination of the price-ceiling for the certificates, i.e. the highest price consumers are willing to pay for the bio-label. The government should ensure that those who do not comply with the minimum level of consumption with respect to of renewable energy be sufficiently penalized. For instance, if consumers and distributors cannot fulfill the quota, then they are required to pay a non-compliance fee or to bio-labels at a higher price.

The price-ceiling is easier to be determined if the government knows the consumers' reservation value for the bio-labels. Unfortunately, this reservation value is private information of individual consumers. If the price-ceiling is set too low, then there will not be enough incentive for producers to invest in

[6]See Nielsen and Jeppesen (2000) for a detailed review.

[7]The production of renewable energy is costlier than the production of non-renewable energy, hence increasing the production of renewable energy implies increasing production costs for the producers.

2023

renewable energy products. However, if it is set too high, consumers will not find it individually rational to buy bio-labels, and if they are required to purchase the bio-labels anyway, then consumers will be worse off.[8]

Timber market and certification: the Indonesian case[9]

A timber certification is a process that results in a written statement attesting the origin of the woods and its status and qualifications. This certification typically includes two main components: (a) The forest management system certification, and (b) the timber product certification. The forest management system certification covers forest inventory, management planning, harvesting, road construction and other related activities, as well as the environmental, economic and social aspects of forest activities. The timber product certification can be used to validate any type of environmental claim made by a producer, and to provide objectively stated facts about the market system, the timber products and their forest of origin. These facts are normally not disclosed by the producer or manufacturer (Barron 1994; Baharuddin 2001).

The main focus of a timber certification policy is based on the standard of forest management as well as on broader environmental credentials. The later includes certification of environmental process-related issues such as plant pollution conditions, energy use, transport, disposal methods, etc. For instance, the content of the re-cycled or wastepaper used or the processes used to manufacture the product (e.g. whether the product is chlorine-free) are of interest. This last aspect is of crucial importance particularly for the pulp and paper industry, which represents a significant share of revenue within the timber industry activities. Furthermore, regulations are increasingly being introduced for packing and packaging, specifying the type of material that may be used and re-used, together with specific recycling materials as well as systems of recovery or return that must be followed (Bourke 2000a, b).

In an international context, two different schemes of timber certification emerge. On the one hand, there are the principles, guidelines and criteria for accreditation set by the Forest Stewardship Council (FSC) system (e.g. SGS Qualifor, SCS, Rainforest Alliance, Soil Association). On the other hand, there is the International Organization for Standardization (ISO) system and its 1400 series standards relating to environmental management tools and systems to measure a company's practices. There is a consensus that these schemes

[8]One can actually apply valuation methods in order to retrieve individuals' preferences and willingness to pay. There are two groups of valuation methods: revealed and stated preference methods (Braden and Kolstad 1991; Garrod and Willis 1999). Revealed preference methods use existing market data to retrieve individuals' preferences for biodiversity benefits whereas stated preference methods are based on collecting data by means of questionnaires (Carson et al. 1992; Nunes 2002). The latter are often supported by multidisciplinary research teams in order to describe and present accurately the environmental good under consideration.

[9]For a complete account of the study see Salim et al. (1997).

2024

complement each other. However, FCS is largely supported by NGOs whereas ISO accreditation is perceived to be heavily influenced by the industrialist lobby. In the Indonesian context, it is very important that a reliable certification scheme need to issued by a third party in order to avoid widespread documented corruption in the sector. The FSC has recently reported that about 17.7 million ha (FSC 2000). This represents about 0.5% of the world's forest area. Little of this relates to tropical areas. As a matter of fact, about 86% of the certified area is in temperate countries, largely Europe and North America. In addition, a new European certification process, the Pan-European Forest Certification Framework, has been launched with governing bodies established in countries like Austria, Belgium, Czech Republic, France, Finland, Ireland, Norway, Portugal, Spain, Sweden and Switzerland. In addition Southeast Asian countries[10] have established a national set of criteria for the auditing of forest management on logging concessions, as well as the ecolabeling of products of those concessions in light of the International Tropical Timber Organization guidelines.

For the sake of its extensive experience in the area of forest management and regulation, it may be useful to discuss in more detail the experience of the Indonesian timber industry and certification. Indonesia has benefited by the rapid growth of the forestry sector. This sector contributed US$ 2 billions in earnings during the 1980s and US$ 9 billion in 1994. Such a massive contribution puts the Indonesian forests under considerable heavy pressure. Recognizing these pressing problems, the Ministry of Forestry took several measures, including the initiative to create an Ecolabeling Working Group. This task force played a crucial role in the design of a set of criteria and indicators of sustainable forest management and in establishing the Indonesian Ecolabeling Institute. It is a consultative, objective and independent third party authority that is designed to allow producers to measure their management practices against standards and to demonstrate compliance with those standards. Today, after almost a decade of timber certification, the balance shows that the success of certified timber policies was particularly strong among specific timber products, the niche markets. For example, a timber commodity such as plywood, representing 70% of all Indonesia's forestry exports, has received very little eco-market attention. This fact has been justified by the concern of forest managers and forest stakeholders, who often assume that certification programs create obstacles to the trade since firms' additional costs will damage their position in international competition ranking. In addition, the demand for certified timber seems to be strongest in eco-sensitive countries such as Germany and the Netherlands and virtually non-existent in countries such as Japan and Korea. The latter constitute the main commercial partners of Indonesian forestry products, leaving only a small percentage of exports for eco-sensitive countries.

---

[10]Indonesia, Malaysia and China represent 75% of the forest products exports among the developing countries that, in turn, represent 15% of the world trade of forest products.

2025

Some caveats

This section briefly covers some caveats of the certification and ecolabeling policies that deserve further attention. First, it is important to find a clear, concise and trustfully criteria, which are underpinning the definition of certification and ecolabeling policy practices. In this respect, there should be a consensus on the common schemes to be adopted. Too many criteria create confusion for consumers, and in turn it may affect the credibility of the certification policy. Related to this, the questions of which institution should have the authority to set the common standard and at what level (i.e. regional, national, or international) this institution should be are crucial. Finally, and admitting that a common standard will be set at an international level, there should be a good coordination of certification policy between the institution which sets the standard and WTO. One of the commonly cited issues concerning the setting up of a common international scheme is the feasibility of what is commonly called *cradle to the grave* certification. This type of policy subjects products to a thorough evaluation including production and processing methods (FAO 1995). This type of certification policy is often seen as an intangible barrier to enter to the international market for developing countries' products. Hence, special attention is required so as to avoid using certification and ecolabeling policy as a barrier to the free-trade policy campaign of the WTO.

Concluding remarks

This paper provides a common ground for discussions on issues related to the use of information as a regulatory instrument for policy management. In particular, we focused attention on certification and ecolabeling policy practices dealing with the problem of market failure caused by the inexistence of prices for biodiversity non-market benefits. The complexity of the range of biodiversity benefits and the wide range of possible policy strategies constitute the basis for the proposed comprehensive view on the certification and ecolabeling policy. We conclude that the success of certification and ecolabeling as a policy instrument depends on the nature of some crucial factors, including the ability of the proposed policy instrument to internalize externalities caused by the inexistence of market prices for biodiversity benefits. The success of such an internalization attempt depends on, first, the public good nature of the non-market biodiversity benefits, second, on the application of appropriate economic valuation methodologies to assess the monetary magnitude of the non-market biodiversity benefits, and, finally, on the characteristics of the market supply and demand mechanism. The latter include, for example, the level of consumers' awareness with respect to biodiversity-sound products and producers' propensity to embrace the certification and ecolabeling scheme.

2026

In addition, we also stress that in some situations the certification and ecolabeling policy *alone* is not sufficient instrument to price biodiversity benefits. Indeed, mixed policies strategies such as, an example, the Dutch bioenergy certificate system, which involves a certification and direct government intervention in the supply or demand forces. Moreover, and observing closely the Indonesian's forest certification setting, one realizes that it is imperative to bring national initiatives closer together, to encourage an international common ground for certification and ecolabeling, and to avoid using such a policy instrument for the encouragement of unfair international trade practices.

Finally, it is also important to have a certification and ecolabeling policies that are sufficiently flexible to allow mutual recognition among countries, as well as meeting the demand of weak sensitive markets, remembering each country's unique environment and cultural characteristics.

Acknowledgments

The authors thank the participants of the OECD-World Bank International Workshop on Market Creation for Biodiversity Products and Services, OECD Headquarters Paris, 25–26 January 2001, for their valuable comments on an earlier version of the manuscript. P.N. acknowledges the *Fundação para a Ciência e a Tecnologia* for financial support. This research was initiated when Y.E.R. was still affiliated to the Department of Economics, University of Groningen, The Netherlands.

References

Akerlof G. 1970. The market for 'Lemons': qualitative uncertainty and the market mechanism. Quarterly Journal of Economics 89: 488–500.

Baharuddin H.G. 2001. Timber certification: an overview. available at http://www. fao.org.

Barron D.E. 1994. 'Sustainable forest certification', paper presented at the 75th Anniversary Annual Meeting of the Woodlands. Canadian Pulp and Paper Association, Alberta, Canada.

Baumol W.J. and Oates W.E. 1988. The Theory of Environmental Policy. Cambridge University Press, Cambridge, UK.

Bourke I.J. 2000a. 'Certification and labeling of paper and other forest products: where are we heading?'. FAO Committee on Paper and Wood Products, 41st Session, Rotorua, New Zealand.

Bourke I.J. 2000b. 'Trade and forestry: agreement on the application of sanitary and phytosanitary measures and agreement on technical barriers to trade'. FAO discussion paper, Rome, Italy.

Braden J.B. and Kolstad C.D. (eds), 1991. Measuring the Demand for Environment Quality. Elsevier Science Publishers, Amsterdam, The Netherlands.

Carson R.T., Mitchell R.C., Hanemann W.M., Kopp R.J., Presser S. and Ruud P.A. 1992. 'A contingent valuation study of lost Passive Use Values Resulting from the Exxon Valdez Oil Spill'. Report Prepared for the Attorney General of the State of Alaska, Washington, DC.

Dosi C.M. and Moretto 1998. 'Is Ecolabeling a Reliable Environmental Policy Measure?', mimeo. Department of Economics, University of Padova, Italy.

EPA 1993. Status Report on the Use of Environmental Labels Worldwide. Office of Pollution Prevention and Toxic, Environmental protection Agency, Washington, DC.

Forest Stewardship Council - FSC 2000. Forest Stewardship Council Web Site, http://www.fscoax.org/principal.htm.

Garrod G. and Willis K.G. 1999. Economic Valuation of the Environment: Methods and Case Studies. Edward Elgar, Cheltenham, UK.

FAO – Committee on Commodity Problems 1995. Developments in International Protection Legislation and ecolabeling. CCP: JU 95/7, Rome, Italy.

IFOAM 2000. Basic Standards for Organic Production and Processing, International Federation of Organic Agriculture Movements, General Assembly, Basel, Switzerland.

Mattoo A. and Singh V.S. 1994. Ecolabeling: policy considerations. Kyklos 47: 53–65.

Milgrom P. 1988. Employment contracts, influence activity, and efficient organization. Journal of Political Economy 96: 42–60.

Nielsen L. and Jeppesen T. 2000. Green Electricity Certificates – A Supplement to the Flexible Mechanisms of the Kyoto Protocol, mimeo. Department of Economics, University of Southern Denmark-Odense, Denmark.

Nunes P.A.L.D., van den Bergh J.C.J.M. and Nijkamp P. 2003. The Ecological Economics of Biodiversity: Methods Values and Policy Applications. Edward Elgar, Cheltenham, UK.

Nunes P.A.L.D. 2002. The Contingent Valuation of Natural Parks: Assessing the Warmglow Propensity Factor, New Horizons in Environmental Economics Series, Edward Elgar, Cheltenham, UK.

Nunes P.A.L.D and van den Bergh J.C.J.M. 2001. Economic valuation of biodiversity: sense or nonsense?'. Ecological Economics, 39: 203–222.

Nunes P.A.L.D., van den Bergh J.C.J.M. and Nijkamp P. 2001. Integration of economic and ecological indicators of biodiversity. In: Valuation of Biodiversity Studies: Selected Studies. OECD, Environment Directorate, Paris, France, pp. 153–182.

Salim E., Djalins U. and Suntana A. 1997. 'Trade and timber certification: international setting and Indonesian experience' paper presented at the XI World Forestry Congress, Antalya, Turkey.

Simpson R.D., Sedjo R.A. and Reid J.W. 1996. Valuing biodiversity for use in pharmaceutical research. Journal of Political Economy, 101: 163–185.

van Bellegem T., Beijerman A.M. and Eijs A. 1999. Green investment funds: organic farming in the Netherlands. In: Handbook of Incentive Measures for Biodiversity: Design and Implementation. OECD, Paris, France.

van der Grijp N.M. and den Hond F. 1999. Green Supply Chain Initiatives in the European Food and Retailing Industry, Institute for Environmental Studies, Vrije Universiteit Amsterdam, The Netherlands.

van van Ravenswaay E. 1995. Public Perceptions of Agrochemicals, Ames, IA Council on Agricultural Science and Technology. Task Force Report, No. 123.

van Ravenswaay E. 1996. Emerging Demands on Our Food and Agricultural System: Developments in Ecolabeling, mimeo. Department of Agricultural Economics, Michigan State University, Ann arbor, Michigan.

van Ravenswaay E. and Blend J. 1997. Using Ecolabeling to Encourage Adoption of Innovative Environmental Technologies in Agriculture, mimeo. Department of Agricultural Economics, Michigan State University, Ann arbor, Michigan.

[16]

Designing Ecolabels In Order To Mitigate Market Failures[1]

An Application to Agrofood Products

Douadia Bougherara

National Institute of Agronomic Research (INRA), Rennes, France

Gilles Grolleau

UMR INRA-ENESAD, Dijon, France

ABSTRACT

For the market for ecofriendly characteristics of agrofood products to function effectively, means of mitigating asymmetric information, informational overload and public goods properties are necessary. Ecolabel success requires a design and an implementation capable of mitigating simultaneously these three sources of market failures. Our contribution differs from many to date by (1) introducing and analyzing the informational overload as a source of market failure and (2) considering the ecolabel, not only as a tool to re-establish information symmetry between the producer and consumer but also as a way to overcome informational overload and public goods problems. We analyze how these sources of market failures may be mitigated by providing information perceived as trustworthy, tying credence and public attributes to verifiable and private attributes and designing the ecolabel as a cognitive support for consumers. We provide an exploratory qualitative study of several French ecolabels to stress how they more or less succeed in attenuating the identified sources of market failures. Several implications for policymakers and managers are stressed. We conclude by suggesting several issues requiring further investigations.

Key words: Ecolabeling; Market failures; Agrofood products.

JEL Classification Numbers: D11, D21, L15

Douadia Bougherara, INRA ESR, 4, Allée Adolphe Bobierre CS 61103, 35011, RENNES, CEDEX, France, Tel: + 33 2 23 48 56 03 – Fax: + 33 2 23 48 53 80, Douadia.Bougherara@rennes, inra,fr
Corresponding author: Gilles Grolleau
UMR INRA-ENESAD (CESAER) 26 Bd Dr Petitjean BP 87999, 21079 DIJON CEDEX, France, Tel: + 33 3 80 77 24 43 – Fax: 33 3 80 77 25 71, grolleau@enesad,inra.fr

1. INTRODUCTION

Many consumers express a willingness to pay for agrofood products that they perceive as less harmful for the environment compared to similar products (Table 1) and producers are willing to market ecofriendly products to capture the consumers' willingness to pay.

The "invisible hand" of free markets fails to provide environmental characteristics[2] embedded in agrofood products at a Paretian optimum, notably because of information asymmetry. In the academic literature, ecolabeling is mostly presented as a market based tool for addressing the market failure resulting from asymmetric information. Ecolabels can help reduce the asymmetric information between producers and consumers by conveying to consumers information related to environmental impacts of products. Consequently, ecolabels help transform environmental awareness of consumers into consumption changes. However, ecolabels also confront two additional problems capable of disrupting the market mechanism, i.e., the limited abilities of consumers to process increasing flows of information and the public nature of environmental characteristics. The main insight of this paper is to consider whether the *ecolabel may simultaneously overcome information asymmetry, information overload and public goods problems.*

The paper is organized as follows. Section 2 shows that environmental attributes constitute a source of market failures, which may explain why declarative willingness to pay does not necessarily correspond to effective purchases. Section 3 analyzes how these market failures may be mitigated by several devices such as emphasizing private and verifiable benefits and designing the ecolabel as a trustworthy and cognitive support for consumers. Section 4 provides an exploratory qualitative study of several existing ecolabels on French agrofood products in order to stress how designers take into account and attenuate the previous sources of market failures. Section 5 outlines the relevance of the results for managers and policy makers and concludes.

2. ENVIRONMENTAL ATTRIBUTES OF AGROFOOD PRODUCTS: A SOURCE OF MARKET FAILURES

"Eco-labeling can provide the missing market information about production process attributes and be used as a mechanism revealing consumer valuation of environmental attributes of agricultural commodities" (Moon et al., 2002). "The overall purpose of the environmental label derives from the intention to overcome market failures caused by information asymmetries" (Karl and Orwat, 1999). Ecolabeling is frequently considered as a way to overcome the market failure resulting from asymmetric information between the producer

TABLE 1

Consumers' willingness to pay for ecofriendly food products

Product	Designation	Segment size Willingness to pay a price premium	Method	Country	Source Date
Corn	Integrated pest Management	84% of the respondents willing to pay 10% more	Consumer survey	United States	Anderson et al. (1996)
Fresh vegetables	Environmental friendly vegetables	70% willing to pay 5% more	Consumer survey	Netherlands	Wandel and Bugge (1997)
Fresh vegetables	Integrated pest management	37,8% of the respondents willing to pay at least 10% more	Consumer survey	United States	Govindasamy and Italia (1998)
Apples	Ecolabeled products	With a premium of $.40 over 40% would still buy	Consumer survey	United States	Blend and Ravensawaay (1999)
Pork products	Embedded with environmental attributes	62 % of the participants paid an average premium of 22 %	Experimental auction	United States	Kliebenstein and Hurley (1999)
Coffee	Shade grown coffee	42% willing to pay at least 1$ more per pound	Consumer survey	Canada	CEC (1999)
Coffee	Shade grown coffee	36% willing to pay at least 1$ more per pound	Consumer survey	Mexico	CEC (1999)
Coffee	Shade grown coffee	28% willing to pay at least 1$ US more per pound	Consumer survey	United States	CEC (1999)
Fresh food products	Agromilieukeur	"Consumers are typically paying a small premium (up to 10%) for these products"	Real premium on the market	Netherlands	Granatstein (2000)
Apple juice	"Green" apple juice	"Some consumers are willing to pay up to 40 per cent more"	Consumer survey	Canada	Van Duren and Rollins (2000)
Lamb	Produced in an environmentally friendly way	More than 50% of consumers are willing to pay a premium of 10 to 20%	Consumer survey	France Germany	Rahmann et al. (2001)
Orange	Food Alliance apples	The premium is small, about 5 cents per pound over an initial price of 99 cents	Consumer survey	United States	Loureiro et al. (2002)

414

and the consumer. However this view is restrictive, because marketing environmental characteristics involves other market failures which also determine the success of ecolabeling schemes.

2.1. Asymmetric information between producers and consumers

Environmental characteristics of agrofood products correspond notably to impacts of farming and processing practices on the environment, which are dissociated from product consumption. For consumers, these environmental attributes are frequently credence goods according to the terms coined by Nelson (1970) and Darby and Karni (1973). According to the considered attribute, consumers will undertake different kinds of expenditures to get "informational input" of their decision making process[3]. In the case of search attributes, consumers inspect the product before purchase. For experience attributes, the most cost-effective way to get the information is to consume the product. If product quality remains constant, consumers can base their future purchases on the information acquired during the first consumption. Credence attributes cannot be accurately and efficiently evaluated even after purchase or consumption, because the consumer lacks technical expertise and/or the cost of acquiring sufficient and accurate information is more costly than its expected value and/or there is no tangible link between the expected attributes and consumption of the product. In most cases, credence attributes are known by producers, but remain hidden to consumers[4] (Table 2).

This information asymmetry may lead to adverse selection and moral hazard. Adverse selection is *ex ante* opportunism due to hidden information while moral hazard is an *ex post* opportunism due to hidden action. Because producing environmental characteristics depends mainly on initial investments done once for all, e.g., acquisition of knowledge, skills, materials, we can consider, at first glance, that the moral hazard framework is inaccurate[5]. If consumers are unable to check environmental characteristics, fraudulent producers can market bad products with a green image and free ride on the ecofriendly market. Consumers anticipate producers' temptation to cheat and then reduce their willingness to pay for environmental friendly products. Honest ecofriendly producers cannot cover their higher production cost and exit the market (Akerlof, 1970).

2.2. Information overload: a potential source of market failure

Informational asymmetry differs from informational overload, even if the dividing line is fuzzy. Informational asymmetry corresponds to an

TABLE 2

Classification of attributes of goods according to transaction costs

Attributes	Definition	Transaction costs		Examples	Seminal authors
		Pre-transaction	Post-transaction		
Search attributes	Consumers get information before purchase by inspecting the attribute	Search costs Low	–	Visible color of	Nelson (1970) an apple
Experience attributes	Consumers get information by experimenting the attribute after purchase because this way is the more cost-effective	Experience costs High	Low	Taste of an apple	Nelson (1970)
Credence attributes	Consumers give credence to a "proxy" because getting information by more direct means is prohibitively costly	Credence costs High	High	Apple from an ecofriendly agricultural process	Darby and Karni (1973)

unequal repartition of information between two agents, which allows an opportunistic behavior of the best-informed agent. Informational overload arises because agents have limited cognitive abilities. Even in a world characterized by symmetric information, agents can be overwhelmed by increasing flows of information. If consumers are provided with too much information such that it exceeds their processing limits, overload occurs and may lead to poorer decision making. The consumer notably makes two decisions: (1) how much information to collect and the subsequent likelihood to devote processing abilities and time to useless information and (2) whether to purchase the product with the risk of purchasing a product that does not correspond to his preferences. Under symmetric information, the problem is solely an exogenous (to the consumer) constraint, under the load problem the problem is partially endogenous – the constraint is a consumer-based one and the decision is partially the consumer's (he decides how much information to consume). Note that consumers differ in terms of their cognitive abilities, hence information overload and the optimal level of information is somewhat a consumer specific phenomenon. This obviously complicates labeling design.

More formally, consider a decision maker willing to acquire information. He has to invest more and more resources, e.g., time and money for acquiring additional information. On the one hand, the marginal cost of information assimilation tends to rise exponentially. On the other hand, the marginal benefit of the gathered information, i.e., the extra value to the decision maker of an additional bit of information, rises initially, at a decreasing rate. Up to a certain point, additional information increases the decision maker utility. Beyond an optimal level of information (from the consumer viewpoint), which is frequently lower than the level of symmetric or complete information, the marginal net benefit from processing additional information may be negative because confusion increases. Moreover, information overload may also lead to more (valuable) information being ignored or not used. Note that perceived information overload would be associated with higher satisfaction and lower performance of decision makers than would be perceived information underload (O'Reilly, 1980). The decision procedure described above can be considered as an application of the psychological law of Yerkes-Dodson (1908). The essence of this law is the relationship between performance, e.g., the decision quality and pressure, e.g., the information load (Leibenstein, 1987). If we visualize the decision quality on the ordinate and the information flows on the abscissa, the Yerkes-Dodson law may be represented by a roughly inverted-U curve (Figure 1). As informational load increases, decision quality comes closer to maximization. Beyond some point informational load may be so high that decision makers find it rather difficult to cope.

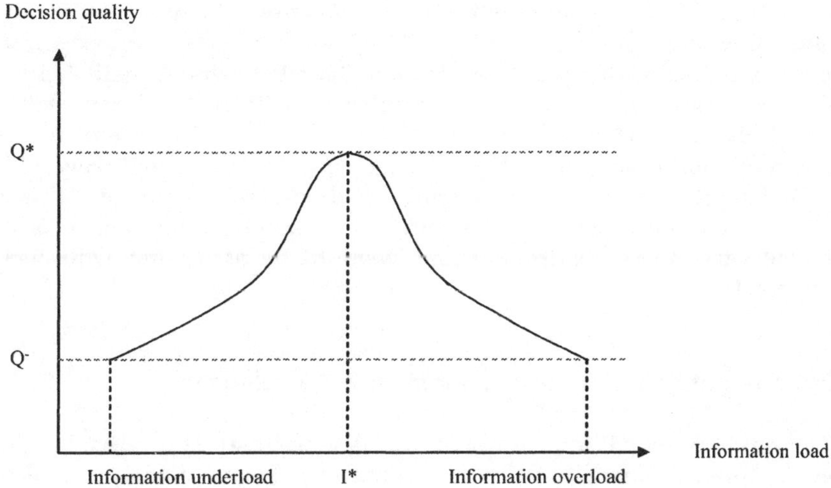

Figure 1. The Yerkes-Dodson law

Consequently, because of limited human processing capacity, agents need to allocate their attention among competitive sources. The consumers' attention becomes the scarcest resource instead of information. Once they devote their attention to given information, they may need to mitigate information asymmetry. Herbert Simon[6] (1995) recognizes that information symmetry is not enough to ensure market efficiency: "What information consumes is rather obvious: it consumes the attention of its recipients. Hence a wealth of information creates a poverty of attention, and a need to allocate that attention efficiently among the overabundance of information sources that might consume it." Indeed, consumers have less time and limited abilities to process increasing information flows. At first glance, competition for consumers' attention occurs in rich informational context, because the label itself provides too much information or competition between several sources capable of capturing consumers' attention. New information and communication technologies reinforce this situation by providing huge amounts of information. In a well-documented study on different ecolabels, Wynne (1994, p. 95) claims that "simply making information available to consumers in no way assures that they will process it". He shows that environmental report cards[7] establish symmetrical but useless information because consumers are overloaded by such amounts of information. One must distinguish between "information provision" and "information impact", because there is no one to one relationship between information provided and the impact, if any, of this information on the recipient". Recently, the German Federal Environmental Agency have stressed the negative effects of competition between eco-claims and other impulses for attention: "The flood of other ecolabels also poses a problem for the first environmental label [Blue Angel]. A great deal of packaging is meanwhile emblazoned with half-a-dozen badges all of them courting the customer's favor. Attracting attention has become more difficult. (...) The average person is confronted daily with 3000 advertising impulses" (German Federal Environmental Agency, 2002[8]). An interesting question, but out of scope of this paper, is to identify the specific attributes that tend to be associated with such overload.

2.3. The public nature of environmental characteristics

The environmental improvement (or degradation) generated by the environmental characteristics of ecofriendly (conventional) products frequently have a public good nature. The characteristics of non-rivalry and non-excludability imply that the purchase of ecofriendly products does not guarantee to the purchaser an exclusive utility from the environmental improvements generated by his purchase. Generally, environmental impacts are global and consumer verification of the impacts is impossible, such as the state of the ozone layer. Other environmental characteristics correspond to long term horizons, exceeding the consumer's life expectancy. Several environmental characteristics result from the total of polluting emissions generated by a product market. Frequently, it is not feasible to exclude those who do not consume ecofriendly products from the benefits generated by environmentally conscious purchasers, e.g., air quality. Environmental characteristics of agrofood products can affect the welfare of future generations who are unable to express their preferences. Moreover, an individual consumer cannot frequently evaluate the real benefits of his contribution. Consequently, the individual does not bear all the consequences of his decision. Public goods lead to a misallocation of scarce resources because the decision making process does not take into account the whole cost[9]. The private production of these environmental characteristics has to mitigate two distinct but interrelated problems: ***(1) The free riding problem:*** Since the environmental improvement is available to everybody, the free riders consume it without purchasing the corresponding ecofriendly product. The presumption of neoclassical economics is that the public good will be under-provided by private and decentralized markets. ***(2) The assurance problem:*** The production of a perceptible environmental improvement can require a minimum level of contributions. Consumers are willing to contribute if they are convinced that an adequate threshold of contributions will be reached, i.e., a sufficient number of contributors will also contribute.

If contributions are insufficient, the environmental improvement will not be produced. Consequently, the individual thinks he will squander his contribution by purchasing an ecofriendly product, corresponding to the "sucker" payoff (Wiener and Doescher, 1991). The assurance problem is especially crucial when environmental quality depends on the environmentally conscious behavior of many consumers.

The three sources of market failures are represented in figure 1, from the most internal problem (inside the individual) to the most external one (the entire context is affected). Individuals may experience an internal conflict between their consumers' preferences and their citizens' preferences. Preference orderings in these two fields are not always compatible[10]. Consumer preferences are about what the consumer wants whereas citizen preferences are about what we should want as a group. Note that Margolis (1982) have proposed a utility function that has more than one domain, $U = f(U_S, U_G)$, where U_S is self-interested preferences and U_G is group-interested preferences. Despite its importance, this question is out of scope of this paper. Moreover, between consumers and sellers, there is a two-sided informational asymmetry. Consumers face a multi-sided situation because of the free riding behavior and the assurance problem among themselves. Lastly, informational overload forces consumers to allocate their attention among different alternatives.

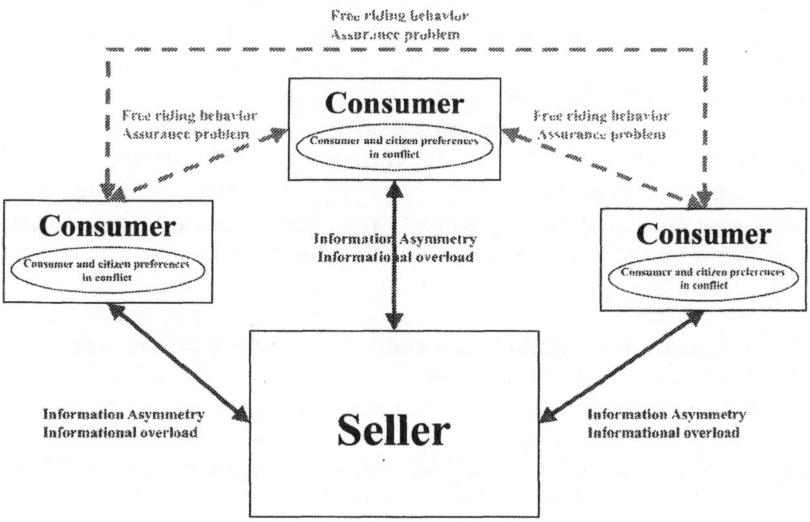

Figure 2. Sources of market failures arising with ecofriendly products

420

3. HOW COULD ECOLABELING CONTRIBUTE TO MITIGATE MARKET FAILURES?

Ecolabeling schemes constitute an explicit institution including two fundamental components, i.e., a system of rules or principles and a system of enforcement. Although ecolabeling schemes were originally considered as voluntary instruments for addressing informational asymmetry, we have shown that this perspective is not sufficient. To ensure market efficiency, an ecolabel has to mitigate simultaneously several sources of market failures. This section stresses how this aim can be reached for each market failure separately and concludes by considering them together.

3.1. Tying the ecolabel sign to a trustworthy third party reputation

To mitigate the adverse selection generated by credence attributes, several authors argue that the participation of third parties with an established reputation for credibility is generally necessary and may constitute a cost effective alternative (Caswell and Modjuszka, 1996; McCluskey, 2000; Sporleder and Goldsmith, 2001). Concretely, the ecolabel sign is frequently tied to the third party certifier – e.g., putting the name or the logo of the certifier – in order to communicate the trustworthy character to consumers. Their role may notably include a participation in (1) the *definition stage* of what is an ecofriendly food product i.e., the criteria selection allowing to use the ecolabel[11], (2) *the monitoring stage* of previously defined criteria to assess product conformity and (3) *the signaling* stage where producers may use a device allowing to distinguish true ecofriendly products from conventional ones and possibly to rank among different levels of ecofriendliness[12]. Each stage is vulnerable to opportunistic behaviors. Consequently the transaction costs implied to achieve a credible procedure may become excessive and swallow up consumers' willingness to pay.

3.2. Designing the ecolabel as a cognitive support for consumers

In an environment overloaded with information, the success of ecolabels depends on its abilities to capture the consumer's attention rather than only providing factual, correct and complete information unprocessable by consumers. Ecolabels design matters. Instead of spending excessive money in direct measurement, consumers search for a cognitive support capable of capturing attention and summarize complex information, e.g., an eco-seal of approval. Teisl and Roe (1998) make the argument that a standardized label is easier to understand and point out that consumers are boundedly rational, which means that they face both time and cognitive constraints, when interpreting environmental labels. Karl and Orwat (1999) argue that "a single sign avoids an 'information overload situation, which can be caused when limited information processing ability is confronted by large amounts of information[13]. Up to a specific level, consumers' ignorance is efficient, because consumers use a proxy which is more cost-effective than a direct measurement of environmental promises (Barzel, 1982). Frequently, the effectiveness of such cognitive supports depends more on the reputation and status of the third party, e.g., environmental associations, than on the precise knowledge of the methods of its intervention. Moreover, the level of information provision is likely to differ according to the product considered. Higher information content may be more suited for decisions about more important purchases e.g., durable goods like cars, washing machines. Consumers may also place more weight on environmental characteristics when buying more frequently purchased products as their high frequency of purchase allows the consumer to perceive themselves as making a larger environmental impact (O'Brien and Teisl, 2004). Frequency of purchase can also lead to more exposure to the eco-label and its message; thus increasing the eco-labels impact (Thøgersen, 2000).

3.3. Reducing free riding and assurance problems by emphasizing private benefits

In the real world, people contribute to public goods at levels that exceed the predictions of the neoclassical theory. Several explanations have been provided, such as the "warm glow" concept (i.e., the increased utility from the act of giving, rather than receiving) (Andreoni, 1990) or the presence of social pressure or private inducements (e.g., the contributor derives more important joint private benefits than others from providing a collective good) or even the "Veblen effect" for which consumption may be conspicuous, thereby having a status value (Ackerman, 1997). The assurance problem can be mitigated by "assurance contracts", i.e., contractual agreements that contribute to a collective good project. These contracts guarantee each party, that his contribution will not be wasted, e.g., by money back guarantees, if the collective good is financially under-supported (Schmidtz, 1991).

In the case of ecolabeled products, public good problems can be alleviated by tying them to private benefits which are emphasized. The initial development of this argument i.e., the private provision of public goods by tying them to goods which have low exclusion cost, can be found in Demsetz (1970). Note that Loureiro et al. (2003)

suggest a similar rationale for organic and eco-labeled products but do not take into account the assurance problem. The association between public environmental attributes and private ones such as taste or health can come from labeling, but may also already exist in consumer's mind, regardless of scientific evidence. It is intuitively convincing that low chemical input production (good for the environment) may induce low pesticide residues in food (good for the consumer's health). Tying public attributes to private ones may reduce free riding and assurance problems by creating excludability. The market switches to a quasi-conventional one with private goods where environmental attributes are provided as joint benefits. The purchase driver remains conventional private benefits and environmental attributes are bundled in the product. People purchase ecofriendly products because they enjoy primarily the private attributes resulting from an environmentally friendly production. For example, Moon et al. (2002) show that "respondents who are more concerned about food safety associated with vegetables are more likely to be willing to pay a premium for the environmental attributes of agricultural products." Moreover, such an analysis explains why products claiming the provision of a local public good are more likely to succeed than one claiming the preservation of a global public good[14].

Let us consider that agro-food products include private attributes like health or taste and public environmental attributes which are interlinked for consumers. Consider a conventional product C on figure 3. Let us assume that product C is now produced with an ecofriendly process. It objectively becomes an ecofriendly product E. Let us now consider consumers' perception of product E. Due to the cognitive problems raised, the consumer's beliefs and marketing efforts, the consumer may perceive product E as having more (or less) environmental attributes than it has in reality. Consumers may perceive product E as having more private attributes than it really has. Consumers may then perceive product E as being product P. In sum, consumers will value environmental attributes and subsequent signaling as a proxy for other attributes that are private.

For example, many consumers expect that agrofood products from an ecofriendly process will have a better taste than conventional products (Deliza et al., 1999; Johansson et al., 1999). Such a perceived better taste allows consumers to (1) mitigate the free riding and the assurance situations by emphasizing private benefits (2) attenuate information asymmetry because consumers use the verifiable attribute as an indicator of the ecofriendly claim credibility, without strictly proving its truthfulness (3) reduce information overload by focusing consumers' attention on a conventional and well-known dimension of food quality. Indeed, taste is then used as a proxy for the existence of related environmental attributes. Consumers may be more familiar with

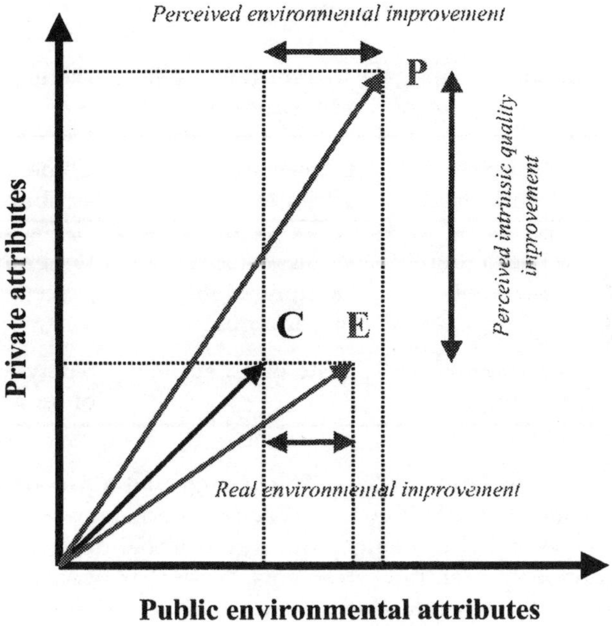

Figure 3. Consumer perception of ecofriendly food products

information about taste than with information on the environmental characteristics. Consequently, they process it more easily.

Table 3 stresses that ecolabel designers want to switch from focusing mainly on public attributes, to emphasizing private attributes, and from credence attributes to verifiable indicators. Several examples of such a strategy are provided in the following section, where we show how several ecolabels attempt to mitigate these potential sources of market failures.

4. HOW FRENCH AGROFOOD ECOLABELS MITIGATE MARKET FAILURES? AN EXPLORATORY ANALYSIS

In France as in most European countries, agrofood products are not included in the governmental ecolabeling schemes. The arguments of this exclusion have been discussed elsewhere (Bougherara et al. 2003). Consequently, French agrofood ecolabels are mainly self-declared claims, which are made without third-party certification, by producers and retailers. To obtain the 'raw data', we have trained about 30 students preparing a master's degree who were in charge of investigating the most important and representative food stores in the metropolitan area

424

TABLE 3

Guidelines to mitigate market failures in marketing ecofriendly products

Types of attributes	Search attributes	Experience attributes	Credence attributes
Public attributes	Packaging material	Waste volume – Compressibility of packaging	Environmental process
Private attributes	Aspect of an apple	Taste of an apple	Safety attributes of an apple

of Dijon (France) in January 2002. They bought all agrofood products bearing an ecolabel. According to a specified framework, a qualitative analysis was conducted to identify the way ecolabel designers mitigate the three previously identified problems. Note that our sample is too small to draw any accurate quantitative analysis. So the percentages given are only indicative (Table 4).

The analysis shows that food ecolabels use especially the association of public and intangible environmental attributes with private and/or verifiable attributes such as health or taste. This may be specific to food products for which the link between the two kinds of attributes is often already in consumer's mind. A more in depth examination of ecolabels allows an interesting insight about the confusion, seemingly sustained by marketers' efforts, that exists between the impact of the production process on the environment and the impact of the environment on food products. For example, on a well-known bottle of mineral water, the producer claims that "the spring [...] benefits from a natural preserved environment"[15]. This water coming from a preserved environment says nothing about the environmental efforts performances of the producer. Rather it is referring to the absence of contamination of the bottled water by the production site. The Mac Cain ecolabel claims that the firm "selects safe fields [...] to plant its potatoes"[16]. The firm focuses on selecting a safe or non polluted place in order to reduce the risk of environmental contamination of potatoes. It does not promise to decrease the environmental impact of production practices. Besides, some French retailers argue that their fresh products are grown far from any source of pollution e.g., a freeway or a dump. They use this argument to substantiate environmental claims. According to the French *Commission Nationale des Labels et des Certificats*[17], the focus of these claims is product safety, not an environmental improvement. Indeed, these practices protect agrofood products from contamination by a polluted

TABLE 4

How type French agrofood ecolabels are designed to mitigate market failures

		Type of market failures and mechanisms attenuating them			
		Public goods problem	Informational asymmetry		Informational overload
		Adding private benefits	Adding experience attributes	Third party certification	Use of a logo considered as easy to process
Product name	Product type				
Bonduelle	Salad				X
Jordans	Cereals	X	X	X	X
Mac Cain	French fries	X	X		
Coopérative Nangica	Potatoes	X	X		
Mas de Nans	Wheat	X	X		
Trilégumes	Potatoes	X	X		
Fermiers de Loué	Turkey				X
Gerblé	Biscuits	X	X		X
Casino	Fresh Vegetables	X	X	X	X
Carrefour	Fresh Vegetables, Meat	X	X	X	X
Auchan	Fresh Vegetables, Meat	X	X	X	X
Cora	Fresh Vegetables, Meat	X	X	X	X
5ème Saison	Salad	X	X	X	
Les Crudettes	Salad	X	X	X	
Florette	Salad	X	X	X	
Candia	Milk				X
Thonon	Mineral water				X
Saupiquet	Tuna				X
Paul	Bread	X	X		X
La Mie Câline	Bread	X	X		
Point Chaud	Bread				
Bret's	Chips				
Peyronnet	Salad				
Nactalia	Milk				
Milka Alp action	Chocolate				X
	Number of products	15 out of 25 (60%)	15 out of 25 (60%)	8 out of 25 (32%)	13 out of 25 (52%)

environment, but they do not produce environmental goods. It is well-known that environmental claims are frequently spurious and generate consumers' confusion between a "preserved area" and an environmental friendly process of production, i.e., the protection of environment. Many claims are narrowly focused and respond more to safety concerns than to environmental ones (Leubuscher et al., 1998). Other mechanisms enabling consumers to trust ecolabels appear through the analyzed ecolabels. Given that numerous products are branded, reputation assets can be transferred on the environmental attributes making them more credible (Sporleder and Goldmsith, 2001). Moreover, some ecolabels refer to sponsoring by organizations such as the World Wide Fund For Nature (WWF) or the French National Forest Agency (ONF) that can generate environmental spillovers capable of reinforcing the credibility of the seller's claims. Such reputation interactions may also lead to adverse effects, such as projecting the green image on other products of the same firm that are not covered by the agreement between the environmental activists and the firm (Grolleau, et al., 2004).

5. SOME POLICY CONSIDERATIONS AND CONCLUDING REMARKS

To mitigate simultaneously the previously identified sources of market failures – asymmetric information, limited cognitive abilities and public goods problems – ecolabel design matters. Ecolabel design determines its overall effectiveness by communicating trustworthiness, constituting a cognitive support and tying public attributes to private ones. Although ecolabeling can be an alternative to mitigate the previous market failures, a careful analysis is necessary to assess that this instrument is not more costly than the market failures it attempts to alleviate and other solutions. Indeed, in some cases we can just shift from a market failure to an ecolabel failure and dissipate money. Consumers' willingnesses to pay may be wasted in high transaction costs rather than used to improve environmental quality. In a coasean framework, the use of ecolabels as an alternative to another policy tool is efficient if the costs of designing, implementing and running ecolabeling schemes are lower (1) than those of the other solutions to mitigate market failures including the laissez faire and (2) than its expected benefits. Policy makers can play important roles by defining the 'rules of the game' which may shape the end result in terms of environmental effectiveness and economic efficiency. These rules include parameters such as participation rules for different interest groups, property rights to use environmental claims, anti-deception laws and level of enforcement. Moreover, consumers' inferences have important managerial implications for producers. They cannot rely on only third party disclosure to ensure the credibility of their claims but have to invest in the production of related attributes, even if the inferences based on them are subjective. There are clearly several issues left to investigate. The success of ecolabeling schemes requires the identification of the optimal level of information provision, which may differ significantly from symmetric or complete information. Lastly, assessing ecolabels overall effects is rather complex and would need further theoretical analysis and empirical studies. The multiple dimensions of ecolabeling has opened a large field of investigation certainly still to be deepened.

NOTES

1 We are indebted to Luc Thiébaut for stimulating conversations and constant encouragement. We greatly appreciate the comments of the referees. A preliminary version of this paper has been presented at the Conference on Ecolabels and the Greening of the Food Market, November, 7-9, 2002, Tufts University, Boston, MA. We thank the French Agency for Environment and Energy Management (ADEME), the Regional Council of Burgundy, the Scientific Committee of the ENESAD and the Inter-Institutional Program for Research in Environmental Economics (PIREE) for their financial support. The usual disclaimer applies.

2 Unlike the economic literature that uses the term "characteristics" for features of a product that are objectively measured and the term "attributes" for features that are perceived by consumers, we use the two terms interchangeably.

3 According to Beales et al. (1981), the demand for attribute information derives from the demand for attributes themselves. The willingness to pay for attribute information is bounded by the willingness to pay for the attributes themselves. This hypothesis enables us not to take into account situations where information disclosure generates a utility that is not linked to the consumption of the attributes themselves, such as environmental activists who are willing to acquire information at a high cost. Indeed, their willingness to pay for information is not derived from the demand for attributes.

4 Some credence attributes do not correspond to an informational asymmetry between seller and purchaser but to, incompleteness, uncertainty and controversies shared by all agents. Scientific knowledge is not established and nobody can probabilize the future, like about health effects of genetically modified organisms (GMO) or global climatic change.

5 For a similar argument, see Rogerson (1983).

6 Herbert Simon, in Scientific American, September, 1995, p. 201.

7 For example, Green Cross is a US environmental report card, which provides detailed and graphical information about performance and environmental impacts of the product, based on a cradle-to-grave study of the product, without value judgement. The report card resembles a nutritional label and aims at allowing the purchaser to compare the environmental burden of one product relative to another.

8 Federal Protection Agency, 2002, The Blue Angel Makes a Fresh Start - New Paths to Public Awareness, Umweltzeichen Newsletter, 5: 1-2.

9 In some cases, certain consumers may want to reduce the environmental impacts of their own consumption and the transaction can be considered as involving private goods. Others care about the whole environmental quality and face a true public good problem. We also admit some environmental characteristics provide private benefits e.g., lower consumption of energy during the consumption phase,

428

longer durability or less packaging. However, we consider that the environmental characteristics of ecofriendly products are mainly public goods.

10 To describe this internal conflict, Sagoff (1988) states "I love my car; I hate the bus. Yet I vote for candidates who promise to tax gasoline to pay for public transportation. I send my dues to the Sierra Club to protect areas in Alaska I shall never visit. [...] I have an "Ecology Now" sticker on a car that drips oil everywhere it's parked."

11 There is a lack of studies devoted to this first stage despite its significant implications. For example, the claim "GMO free" supposes a previous definition of GMO product itself, the choice of a detection threshold and the level of investigation of the food into agrofood chain, e.g., what is about cow's milk bred with GMO cereals? Because of the individual consumer inabilities to define his desired level of environmental quality, the environmental criteria are thus often settled by consulting all the parties e.g., industry, environmental and consumers associations, unions. Some parties may influence the environmental criteria to make them as close as possible to their preferences. Firms attempt to influence products' criteria in order to get the ecolabel without making any environmental efforts or to exclude competitors (Nadaï, 1998). West (1995) states that, "in practice, the needs of industry often take precedence over those of other interest groups in the decision-making process". In the case of the European Ecolabel for washing machines, he also relates that "all German manufacturers were able to meet the criteria from the beginning", or that "in the Netherlands, 60 percent of the tissue paper produced could meet standards without change to manufacturing practices". This situation leads West (1995) to the following statement: "voluntary eco-labeling scheme are rapidly degenerating into a means whereby industry can set the standard it likes".

12 More formally, a signal is effective if it leads to a separating equilibrium which requires the following sufficient conditions: (1) High ability agents can acquire the signal at a lower cost than low ability agents; (2) For high quality agents, the expected profit with a signal minus the cost of acquiring the signal is greater than the expected profit without the signal. (3) For low quality agents, the expected profit without signal is greater than the expected profit minus the cost of acquiring the signal (Macho Stadtler and Castrillo, 1997).

13 Some consumers might not feel sufficiently informed by a single sign, which prevent them from exercising their consumer sovereignty. More in depth research is necessary to compare welfare loss and gains resulting from information reduction.

14 Note that a local public good, e.g., a neighbouring countryside, can also have an advantage because it is more verifiable than a global public good, e.g., the state of the ozone layer.

15 Translated from French.

16 Translated from French.

17 French Agency of Agro-food Labels.

REFERENCES

Ackerman, F. (1997). Consumed in Theory: Alternative Perspectives on the Economics of Consumption. *Journal of Economic Issues*, **31**: 651–664.

Akerlof, G. (1970). The Market for Lemons: Quality Uncertainty and Market Mechanism. *Quarterly Journal of Economics*, **84**: 488–500.

Andreoni, J. (1990). Impure Altruism and Donations to Public Goods: A Theory of Warm-Glow Giving? *Economic Journal*, **100**: 464–477.

Bougherara, D., Grolleau, G. and Thiébaut, L. (2003). Mise en Perspective et Discussion de l'Exclusion des Produits Agro-alimentaires du Dispositif d'Écolabellisation. *Economie Rurale*, **275**: 83–90.

Anderson, M., Hollingsworth, C., Van Zee, V., Coli, W. and Rhodes, M. (1996). Consumer Response to Integrated Pest Management and Certification. *Agriculture, Ecosystems and Environment*, **60**: 97–106.

Barzel, Y. (1982). Measurement Costs and the Organization of Markets. *Journal of Law and Economics*, **25**(1): 27–48.

Blend, J.R. and Van Ravenswaay, E.O. (1999). Measuring Consumer Demand for Ecolabeled Apples. *American Journal of Agricultural Economics*, **81**(5): 1078–1083.

Caswell, J.A. and Modjuszka, E.M. (1996). Using Informational Labelling to Influence the Market for Quality in Food Products. *American Journal of Agricultural Economics*, **78**: 1248–1253.

Commission for Environmental Cooperation (1999). Measuring Consumer Interest in Mexican Shade-grown Coffee: An Assessment of the Canadian. Mexican and US Market. Montreal, Quebec.

Darby, M.R. and Karni, E. (1973). Free Competition and the Optimal Amount of Fraud. *Journal of Law and Economics*, **16**: 67–88.

Demsetz, H. (1970). The Private Production of Public Goods. *Journal of Law and Economics*, **13**(2): 293–306.

Deliza, R., Rosenthal, A., Hedderley, D., MacFie, H.D. and Frewer, L.J. (1999). The Importance of Brand, Product Information and Manufacturing Process in the Development of Novel Environmentally Friendly Vegetable Oils. *Journal of International Food and Agribusiness Marketing*, **10**(3): 67–77.

Govindasamy, R., Italia, J., Thatch, D. and Adelaja, A. (1998). Consumer Response to IPM-grown Produce. *Journal of Extension*, **36**(4). Available at http://www.joe.org/joe/1998august/rb2.html.

Granatstein, D. (2000). Emerging Ecolabels for Food Products. Center for Sustaining Agriculture and Natural Resources Washington State University, Wenatchee, WA.

Grolleau, G., Lakhal, T. and Mzoughi, N. (2004). Does Ethical Activism Lead to Firm Relocation? *Kyklos International Review for Social Sciences*, **3**, 391–406.

Hurley, S.P. and Kliebenstein, J.B. (1999). The Potential For Marketing Pork Products with Embedded Environmental Attributes: Results from an Experimental Study. American Agricultural Economics Association, Annual Meeting, August 8–11, Nashville, Tennesee.

Johansson, L., Haglund, Å., Berglund, L., Lea, P. and Risvik, J.E. (1999). Preference for Tomatoes, Affected by Sensory Attributes and Information about Growth Conditions. *Food Quality and Preference*, **10**(4–5): 289–298.

Karl, H., Orwat, C. (1999) Economic Aspects of Environmental Labelling, in H. Folmer and T. Tietenberg (eds), *Yearbook of Environmental and Resource Economics 1999/2000*, Edward Elgar Cheltenham (UK), p. 107–170.

Leibenstein, H. (1987). *Inside The Firm*, Harvard University Press, Cambridge, MA.

Leubuscher, S., Hager, W., Wattiez, C., Mombrù, J. and Liaska, E. (1998). Study on Verification and Control of Environmental Product Claims. Final Report by Prospect C&S for DG Health and Consumer Protection.

Loureiro, M.L., McCluskey, J.J. and Mittelhammer, R.C. (2002). Will Consumers Pay a Premium for Eco-labeled Apples? *Journal of Consumer Affairs*, **36**(2): 203–219.

Loureiro, M.L., McCluskey, J.J. and Mittelhammer, R.C. (2003). Are Stated Preferences Good Predictors of Market Behavior? *Land Economics*, **79**(1): 44–55.

Margolis, H. (1982). *Selfishness, altruism and rationality*. Cambridge, Cambridge University Press, MA.

Macho-Stadler, I. and Perez Castrillo, D.J. (1997). *Introduction to the Economics of Information: Incentives and Contracts*. Oxford: Oxford University Press.

McCLuskey, J. (2000). A Game Theoretic Approach to Organic Foods: An Analysis of Asymmetric Information and Policy, *Agricultural and Resource Economics*, **29**: 1–9.

Moon, W., Florkowski, W.J., Brückner, B. and Schonhof, I. (2002). Willingness to Pay for Environmental Practices: Implications for Eco-Labeling, *Land Economics*, **78**(1): 88–102.

430

Nadaï, A. (1998). Concurrence dans la Qualification Environnementale des Produits. *Revue d'Economie Industrielle*, **83**: 197–212.

Nelson, P. (1970). Information and Consumer Behaviour. *Journal of Political Economy*, **78**: 311–329.

O'Brien, K.A. and Teisl, M.F. (2004). Eco-information and its effect on consumer values for environmentally certified forest products, *Journal of Forest Economics*, **10**(2): 75–96.

O'Reilly, C.A. (1980). Individuals and Information Overload in Organizations: Is More Necessarily Better?, *Academy of Management Journal*, **23**(4): 685–696.

Rahmann, G., Ashworth, S.W., Boutonnet, J.P., Brunori, G. and Papadopoulos, I. (2001). Opportunities and Barriers for Niche Marketing of Lamb in European LFAs based on consumer attitudes to product quality. *Agrarwirtschaft – German Journal of Agricultural Economics*, **50**: 293–301.

Sagoff, M. (1988). *The Economy of the Earth*, Cambridge University Press, Cambridge.

Schmidtz, D. (1991). *The Limits of Government, An Essay on the Public Goods Argument*, Westview Press.

Sporleder, T.L. and Goldsmith, P.D. (2001). Alternative Firm Strategies for Signaling Quality in the Food System, *Canadian Journal of Agricultural Economics*, **49**(4): 591–604.

Teisl, M.F. and Roe, B. (1998). The Economics of Labeling: An Overview of Issues for Health and Environmental Disclosure, *Agriculture and Resource Economics Review*, **27**: 140–150.

Thøgersen, J. (2002). Promoting "Green" Consumer Behavior with Eco-Labels, in *New Tools for Environmental protection*, edited by T. Dietzand and P. Stern, Washington, DC: National Academy Press Chapter 5: pp. 83–104.

West, K. (1995). Ecolabels: The Industrialization of Environmental Standards. *The Ecologist*, **25**(1): 16–21.

Wiener, J.L. and Doescher, T.A. (1991). A Framework for Promoting Cooperation, *Journal of Marketing*, **55**(2): 38–47.

Wynne, R.D. (1994). The Emperor's New Ecologos? A Critical Review of the Scientific Certification Systems Environmental Report Cards and the Green Seal Certification Mark Programs. *Virginia Environmental Law Journal*, **14**: 51–149.

[17]

Psychological Determinants of Paying Attention to Eco-Labels in Purchase Decisions: Model Development and Multinational Validation

John Thøgersen

ABSTRACT. Environmental labels are useful from an environmental policy perspective only if they are noticed by the consumer in the shopping situation and next – what is more – understood, trusted, and valued as a tool for decision-making. In this paper, a psychological model explaining variations in consumer attention towards environmental labels is developed and its ability to predict attention towards environmental labels in various European countries is tested.

Attention is a major scarce resource
(Herbert A. Simon)

As the abatement of pollution from large industrial sources seems to be within reach, environmental policy in Europe and elsewhere focuses increasingly on reducing the impact of "non-point" sources of pollution (Miljøstyrelsen, 1996), in particular on the pollution (and resource use) associated with private consumption (Geyer-Allély & Eppel, 1997; Norwegian Ministry of Environment, 1994b; OECD, 1997; Sitarz, 1994).

Pollution and the consumption of scarce resources are associated with all phases of the consumption cycle (Pieters, 1991): acquisition of consumer goods and services, their use in the production of utilities in (or outside) the household, and the disposal of possible rest products. Until now, most environmental policy directed at private households has targeted the use phase (e.g., lowering of thermostats, switching off lights when leaving a room) or the disposal phase (e.g., recycling). However, in line with the view that in environmental policy "upstream" solutions (i.e., prevention) are generally preferable to "downstream" ones (i.e., cure) (Eurostat et al., 1995), the focus of this sub-field of environmental policy has shifted towards influencing consumer purchases (e.g., Heiskanen et al., 1998; Miljø- og Energiministeriet, 1995; Miljøstyrelsen, 1996).

As strongly emphasised at the Oslo Symposium on Sustainable

Consumption in 1994, both for democratic and control reasons the needed changes in consumption patterns cannot be achieved by force, but only with "the willing participation of the consumers" (Norwegian Ministry of Environment, 1994a, p. 13). For these (and other) reasons, information policy plays a key role in environmental policy targeting consumers. One of the most promising forms of environmental information policy, in terms of providing timely and relevant information for the consumer, is environmental labelling (Hansen & Kull, 1994; Miljø- og Energiministeriet, 1995; Scammon & Mayer, 1993). For example, a Swedish study credited environmental labelling, and consumer choice of labelled products, for a considerable reduction in the environmental impact caused by the Swedish forest industry (Naturvårdsverket, 1997).

Beginning with the German "Blauer Engel" in 1977, a number of national environmental labelling schemes – and a plethora of industry or industry association schemes of varying seriousness – have developed (e.g., Enger, 1998; EPA, 1998; Forbrugerstyrelsen, 1993; Mayer & Gray-Lee, 1995; Scammon & Mayer, 1993).[1] The U.S. Environmental Protection Agency was able to collect information about 49 "third-party" schemes issued worldwide in 1997, of which only 17 existed in 1989, and the number is still growing (EPA, 1998). In recent years, labels based on cross-national schemes have begun to appear in European supermarkets, notably the "Swan" in the Nordic countries and EU's "flower." Table I shows the status (in the late 1990s) of industry's adoption of some of the most important environmental labels in the industrialised countries.

TABLE I
The Adoption of Environmental Labels by Industry in a Number of Countries

Label	Country(ies)	Year founded	Number of products	Number of product groups
Blue Angel[1]	Germany	1977	4,350	80
Nordic Swan[2]	The Nordic Countries	1989	>1,500	46
EU Flower[2]	EU	1992	200	14
Green Seal[3]	USA	1989	318	19
Environmental Choice[3]	Canada	1988	1,600	48
Eco-Mark[3]	Japan	1989	2,023	71

Sources: [1]The www home page of the Swiss Ministry of Environment (status summer 1997). [2]Regeringskansliet (1999) (status spring 1999). [3]OECD (1997) (status 1996).

Of course, environmental labels are useful from an environmental policy perspective only if consumers use them in their decision-making. However, there are still few published studies of the effectiveness of labelling schemes in this respect (OECD, 1991, 1997).

Most of the published studies focus on consumers' recognition of or knowledge about labels and/or their trust in them (OECD, 1997; for examples, see, e.g., Bekholm & Sejersen, 1997; Tufte & Lavik, 1997), implicitly or explicitly assuming that these are fundamental prerequisites for the use of a label in decision-making. The (implicit or explicit) framework of such studies seems to be a response hierarchy or stage model of decision-making, described in any standard text in consumer behaviour (e.g., Peter, Olson, & Grunert, 1999), the stages including at least knowing and trusting, deciding, and buying (followed by a behavioural outcome stage, which is where the real success of the labelling scheme should be measured).

However, practically all studies are purely descriptive, leaving the question *why* consumers know, notice, and use labels only sporadically answered. With few exceptions (e.g., Verplanken & Weenig, 1993) it is not systematically considered how the decisions that the labels are meant to influence are made and which influences this may have for the functioning and effectiveness of labelling.

For instance, there is plenty of evidence that how, and how much, consumers attend to information in a buying situation depends on their involvement in the decision (e.g., Celsi & Olson, 1988; Herr & Fazio, 1993; Kokkinaki, 1997). In general, one cannot count on information about environmental consequences, in the form of a label or otherwise, producing high involvement in itself. The isolated consequences – environmental as well as personal – of each individual decision are simply too small in most cases (Thøgersen, 1998a). If this is the case, and if other self-relevant information competes for the consumer's attention – sometimes to a degree where the consumer experiences information overload (Jacoby, 1984) – consumers may easily fail to notice relevant labels in the buying situation.

The purpose of this paper is to contribute to a systematic understanding of how eco-labels (and possibly other types of labelling) work. (The terms "environmental label" and the shorter "eco-label," Hansen & Kull, 1994, are used interchangeably in this paper.) The specific objectives are to develop a psychological model explaining when and why consumers pay attention to environmental labels in the buying situation and to demonstrate the model's ability to predict the attention paid to environmental labels in various European countries. The model is tested and cross-validated by means of data from the former West Germany, the European country with the longest tradition for environmental labels, the former East Germany, Great Britain, Ireland, and Italy. But first a (selective) review of previous research with a bearing on whether consumers notice, understand, trust, and value environmental labels is presented.

PREVIOUS RESEARCH

Knowing a label is a prerequisite for using it in decision-making and understanding it is a prerequisite for using it correctly. Understanding a label implies that the person knows it exists, what it looks like, and what it means. For example, a survey in 1997 in the Nordic countries found that 44% of Danish consumers, as opposed to from 5 to 15% of the consumers from other Nordic countries, had no idea what the Swan label meant, and many who thought they knew were wrong (Lindberg, 1998)![2] Another survey from the same year confirmed the low recognition of the Swan label among Danish consumers and found an even lower recognition of EU's flower (Bekholm & Sejersen, 1997), which is still rarely seen in Danish shops. Ninety percent knew the label that is used to indicate that a product is recyclable (particularly used on wine bottles) or is produced from recycled materials (particularly on paper products). A study of the Danish "state controlled organic" label for food products five years after its introduction (in 1990) found that 57% of a broad sample of consumers were not able to identify the correct label among three alternative designs (Thøgersen & Andersen, 1996).

Recognising a label is not the same as understanding the exact meaning of it. For example, one study found that only about 5% of a representative sample of US consumers exhibited a thorough understanding of the terms "recycled" and "recyclable" (Morris, Hastak, & Mazis, 1995).

A consumer will use a label (as intended) in decision-making only if he or she trusts the message it conveys (Hansen & Kull, 1994). A large number of studies have found that consumers are sceptical towards green product claims (see Peattie, 1995). One study cited by Peattie found that 71 percent of British consumers thought that companies were using green issues as an excuse to charge higher prices. However, many studies find that labels and other types of environmental information provided by public and other independent sources are trusted more than information provided by producers or retailers (e.g., Eden, 1994/95; Enger & Lavik, 1995; MacKenzie, 1991; Schlegelmilch, Bohlen, & Diamantopoulos, 1996; Tufte & Lavik, 1997). Unfortunately, and perhaps because the state controlled labels are outnumbered so many times by private labels and other types of environmental information, many consumers are uncertain or hold outright erroneous beliefs about who issues state controlled labels, such as the Nordic Swan. A Norwegian study found that such mistakes reduce the trust in the Nordic Swan (Tufte & Lavik, 1997).

Even consumers who know and trust a relevant environmental label will not use it if they do not notice it due to information overload (Jacoby, 1984) or for other reasons. For instance, in 1992 it was estimated that there were 400–600 private labels, in addition to 36 labelling schemes issued by public authorities, targeted at Danish consumers (Forbrugerstyrelsen, 1993). A study in 1996 found environmental claims on 63% of the packaged goods within 16 product categories in the major supermarkets in Oslo (Enger, 1998). A minority of 8% of the goods carried an environmental label issued by the state or another "third party" organisation.

Besides trust and knowledge, a fundamental prerequisite for paying attention to eco-labels in an information-crowded environment, seldom mentioned in the labelling literature (but see, e.g., Palm & Windahl, 1998), is that the consumer believes that the label helps him or her attain some goal (e.g., Forbrugerstyrelsen, 1993; Nilsson, Nissen, Thøgersen, & Vilby, 1999). Just like unit pricing helps the consumer obtain the goal of value for money and nutrition declarations health-related goals, environmental labelling helps consumers obtain environmental goals. Hence, an informative label about a product's environmental performance will influence decision making only if consumers desire environment-friendly products (unless they believe that other advantages are associated with environmental friendliness, Thøgersen, 1998b). Studies have found that large segments of Western European and North American consumers demand environmentally friendly products in such diverse areas as packaging (Bech-Larsen, 1996; Thøgersen, 1996), food products (Biel & Dahlstrand, 1998; Grunert & Juhl, 1995; Sparks & Shepherd, 1992; Thøgersen, 1998b), paint (Buchtele & Holzmuller, 1990), and heating systems (Berger,

Ratchford, & Haines, Jr., 1994). For example, in 1997 61–71% of representative samples of consumers from the Nordic countries claimed that they "sometimes" or "always" check out the environment-friendliness of the products they buy (Lindberg, 1998).

However, consumers obviously vary in the importance they attach to environmental goals, often referred to as pro-environmental attitudes. It is well documented that pro-environmental attitudes increases consumers' propensity to buy environment-friendly products (e.g., Berger, Ratchford, & Haines Jr., 1994; Biel & Dahlstrand, 1998; Buchtele & Holzmuller, 1990; Sparks & Shepherd, 1992; Thøgersen, 1998b). Less researched in this connection is Fazio's (1986; Roskos-Ewoldsen & Fazio, 1992) proposition that attitudes also influence which information about a product a consumer pays attention to, including information about the product's environment-relevant characteristics (but see Thøgersen, 1999).

THEORETICAL FRAMEWORK

As indicated above, "paying attention to eco-labels" is hardly a goal in itself, but rather a means to a goal: buying (more) environment-friendly products, which is a means to the more abstract goal of protecting the environment (and perhaps to more selfish goals as well). Hence, it is unlikely that a consumer pays attention to an environmental label unless he or she values protecting the environment, perceives buying (more) environmentally friendly products as an effective means to achieve this goal (Ölander & Thøgersen, 1995; Stern, Dietz, Abel, Guagnano, & Kalof, 1999), and perceives the information that the label conveys as useful for this purpose (Hansen & Kull, 1994).

Few products are acquired with the sole (or main) purpose of protecting the environment. Typically, consumers buy goods for the private utility they provide. However, consumers seem to be increasingly aware and concerned that serious environmental impacts may be associated with the production, use, and/or disposal of products they buy. Many consumers are willing to make an effort to diminish the negative environmental impact of their consumption, and environmental labels are welcomed as a tool for this purpose. However, given that environmental attributes – as long as they do not represent any personal threat – are clearly peripheral to what the consumer wants to achieve through the purchase (e.g., Kotler, 1991), usually the issue will not be a high involvement one.

In an environmentally concerned consumer's mental script for buying certain goods, there may be a sequence labelled "choose the most environmentally friendly alternative within the consideration set." Depending on how frequently and recently the consumer has made this particular purchase, the sequence may be more or less automatic. If an eco-label exists for the product category, paying attention to it is almost certainly a part of the sequence, although little consciousness needs to be devoted to the task in the highly rehearsed cases. If no eco-label is available, other sources of information (or cues) about the product alternatives' environmental attributes may be consulted.

If the consumer is inexperienced with the buying of this kind of product, if environmental information about the product has just become available, or if he or she has – for some other reason – not previously been aware that there are potential environmental hazards associated with this type of purchase, they have to go through a somewhat more elaborate procedure in order to be able to make an environment-friendly choice. (1) Find out which types of environmental information – including eco-labels – are available, what an eventual label looks like, where it is placed, and what it means. (2) Evaluate the usefulness and reliability of the information.

Based on the understanding that interested consumers hold or develop scripts for the purchase of environment-friendly products and that they use eco-labels as a means of carrying out such scripts (similar to the way they use a menu in a restaurant), one may specify a series of steps that an individual's decision process must pass through before an environment-labelled product is purchased. All consumers will not pass through all steps every time. Once made, the results of passing through a step are stored in memory and can be activated, if relevant, in subsequent decision situations. Hence, the assumption is only that for an individual to consciously buy an eco-labelled product, at some time prior to the purchase he or she must have come to:

1. form a personal goal of protecting the environment (a pro-environmental attitude),
2a. believe in cautious buying as a suitable strategy for achieving this goal,

2b. know about eco-labels: that they exist, what they look like, what they mean,
2c. trust labels: that the information provided is relevant and important, that it is true.

In the buying situation, he or she must at least:

3. pay attention to eco-labels, and
4. decide to buy the eco-labelled product.

As indicated by this sequence, it is the purchase of environment-friendly products as a goal-directed behaviour, not as the coincidental outcome of other factors, which is interesting when considering the functioning of eco-labels. Further, the sequence refers to cases where the decision to buy an eco-labelled product is made in the shopping situation. This is typical for low-involvement, but less so for high-involvement purchases. Three steps are given the same number 2, followed by a different letter, in order to indicate that these steps can be taken in different sequences with more or less equal *a priori* likelihood.

Other factors, besides the indicated means-goals concerns, may produce variation in the attention paid to eco-labels. Most obviously, it may depend on external conditions and opportunities – in particular the availability of eco-labels in the shopping environment. In addition, variation may be caused by personality-related factors. In this category, researchers dealing with environment-friendly behaviour have particularly emphasised the importance of self-perceptions regarding one's ability to influence the attainment of goals like a cleaner environment (often termed perceived consumer effectiveness, PCE, e.g., Berger & Corbin, 1992; Ellen, Wiener, & Cobb-Walgren, 1991; Kinnear, Taylor & Ahmed, 1974).[3]

PCE, a pro-environmental attitude, a belief in environment-friendly buying, and trust may all be thought of as motivating factors, i.e., determinants of how hard a consumer will look for eco-labels in a buying situation. From their positions in the decision process it follows that pro-environmental attitude is assumed to be a more distal determinant than belief in environment-friendly buying and trust, and that its influence on the subsequent steps is partly or wholly mediated through, these other factors. This assumption is consistent with what one can infer from the specification levels of the three variables. While the latter two constructs share, more or less, the specification level of the next step in the sequence, pro-environmental attitude is a more general concept. The influence of a general attitudinal variable on a specific behavioural criterion is typically found to be more distal than, and partly or wholly mediated through, more specific determinants (Eagly & Chaiken, 1993). Hence, it also follows that the influence of general personality traits, such as PCE, should be expected to be mediated through more specific determinants such as belief in environment-friendly buying and trust.

Causally, knowledge about and availability of labels represent internal and external conditions that, together with motivation, co-determine paying attention to labels.[4] Knowledge and availability of labels are hardly mutually independent either. The eco-labels offered – and the information used to promote them – are among consumers' most important sources of knowledge about eco-labels (Nilsson et al., 1999). It also seems reasonable to expect that these two variables influence paying attention to eco-labels, both directly, as already suggested, and indirectly, through consumers' motivation to look for eco-labels.[5]

From this discussion a causal path model, explaining consumers' propensity to pay attention to eco-labels, can be extracted. Figure 1 outlines this model.

The model illustrated in Figure 1 can, of course, be extended. For example, trust in environmental labels depends not only on the two determinants included in the model, but also on external factors, notably the bodies that provide and/or control the label (e.g., Eden, 1994/95; Enger & Lavik, 1995; Hansen & Kull, 1994; MacKenzie,

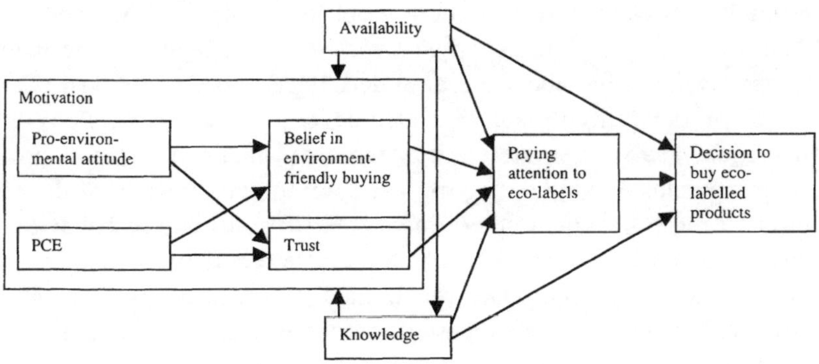

Figure 1. Predicting paying attention to eco-labels and the purchase of labelled products.

1991; Schlegelmilch et al., 1996; Tufte & Lavik, 1997). Most other variables in the model also have additional exogenous antecedents. The model can also be extended forward, adding possible outcomes of the decision (buying behaviour and behavioural outcomes). Further, since purchase decisions involving eco-labels are often recurrent, one could add feedback effects, linking a "downstream" variable (in one decision-making process) with a more "upstream" one (in a later decision-making process). These extensions would produce a more complete, but also a more cluttered model. As in most other cases, clarity and focus are achieved at the expense of completeness.

Theories of decision-making under conditions not conducive to extensive deliberation, such as Fazio's (1986) attitude-to-behaviour theory, fuels the expectation that some of the suggested determinants of paying attention to eco-labels could interact.[6] Fazio suggests that attitudes guide behaviour by creating selective attention towards issues and objects in our surroundings that are consistent with the attitude. For instance, a consumer with a pro-environmental attitude is presumably more attentive than one without to cues that inform about products' environmental qualities, such as eco-labels. Therefore, he or she is also more likely to buy environment-friendly products. This is consistent with the reasoning above.

Fazio makes the important additional point, however, that attitudes have this effect only under certain conditions. Basically, whether or not attitudes guide attention and behaviour depends on their accessibility from memory, or strength. Strong attitudes get activated by mere exposure to an attitude object while the activation of weaker ones depends on stronger stimuli. The strength of an attitude depends on a number of factors such as how much the individual has considered (or "rehearsed") it, experience with, knowledge about, and faith in one's knowledge about the attitude object, and the personal relevance of the attitude and the attitude object (see, e.g., Petty & Krosnick, 1995). In Figure 1 it is indicated that the effect of pro-environmental attitude on paying attention to eco-labels is mediated through more proximal determinants. Fazio's attitude-to-behaviour theory and attitude strength research indicate that the proximal determinants may also moderate the impact of the pro-environmental attitude. It has also been suggested that the effects of motivation and conditions on environment-friendly behaviour are not additive, but that they interact (e.g., Guagnano, Stern & Dietz, 1995; Ölander & Thøgersen, 1995).

THE DATA

The motivation-attention part of the model is tested by means of cross-national survey data collected by the European Consortium for Comparative Social Surveys (COMPASS) in 1993 as part of the so-called REAP programme.[7] The programme applies to five countries, Germany, Britain, Ireland, Italy, and the Netherlands. In Germany, two samples were used in order to be able to control for possible differences between the federal states of Eastern and Western Germany remaining only a couple of years after the reunion. In each country or area, a stratified probability sample of from 957 (Ireland) to 1,261 (Great Britain) people was interviewed. However, due to missing values and the use of listwise deletion, the active sample sizes are reduced up to 15%. The objective was to ask the same questions in the different countries, save for translations. However, some items used in the present study unfortunately differ substantially between the Dutch and the other questionnaires. Therefore, the Netherlands was excluded from the present study, leaving us with two German, a British, an Irish, and an Italian sample.

Still, the data set offers an unusual opportunity to evaluate the international generality of the pattern of determinants of an important social behaviour. Obviously, the survey was not designed to test the model suggested here. This has two important implications. The data set contains no information about availability of and knowledge about eco-labels, meaning that only the motivation-attention part of the model can be tested. Further, even though the data set contains indicators that can be used to represent the concepts in, and to perform at least a preliminary test of, this – crucial – part of the model, they are not necessarily ideal for the purpose in all cases. Both of these limitations should be considered carefully when interpreting the results.

In the survey, paying attention to environmental labels was measured by the question: "When you are choosing a product, how often do you pay attention to any environmental labelling before deciding to buy?" with the response categories always, often, sometimes, and never. Before the question was posed, the issue was introduced by the text "Now a few questions about environmental labelling, that is, information about how a product or its packaging may affect the environment." In the following analyses, the response categories are coded as 4, 3, 2, and 1.

The willingness to carry economic sacrifices, one way or the other, to protect the environment is used as an indicator for the personal importance of environmental protection in the individual's goal hierarchy, or pro-environmental attitude. The construct is represented by three questions: "How willing would you be to pay much higher prices/to pay much higher taxes/to accept cuts in your standard of living in order to protect the environment?" Answers were registered on a five-point scale with the end points "very willing" and "very unwilling."

The trust in eco-labels is registered by the question "If you look at environmental labelling on products, how often do you trust it?" with the response categories always, often, sometimes, and never. In the following analyses, these categories are coded as 4, 3, 2, and 1. Unfortunately, in the REAP survey those that never paid attention to environmental labels were not asked about their trust in such labels. So it is not possible to analyse the importance of distrust for totally ignoring labels. However, the data can be used to investigate the importance of trust for variations in the frequency of paying attention to eco-labels among those who do this at least sometimes.

Perceived consumer effectiveness (PCE) is represented by a single item from the survey: "It is just too difficult for someone like me to do much about the environment," measured on a five-point scale with the end points "strongly agree" and "strongly disagree."

The survey contained no direct measures of consumers' belief in cautious buying as a strategy to protect the environment. However, items focusing on the individual's effort to avoid potential environmental hazards associated with purchases in different situations can be used to construct an indirect measure of such a belief. A strong tradition in attitude theory holds that behavioural indicators such as these reflect and, hence, can be used to represent attitudes (e.g., Eagly & Chaiken, 1993). Arguably, belief in environment-friendly buying is, if not identical to, then at least a large component of the attitude towards environment-friendly buying. So, lacking better alternatives, the extent to which an individual makes an effort to avoid environmental hazards associated with purchases in different situations is used as an indicator of his or her belief in environment-friendly buying. One would have wished to have recourse to a large number of items focusing on efforts of this sort in various types of buying decisions (other than attending to eco-labels). Unfortunately, only two relevant items were included in the REAP survey: "How often do you make a special effort to buy fruits and vegetables grown without pesticides or chemicals?"[8] and "When you are shopping, how often do you pay attention to the amount of wrapping or packaging used on products before you decide to buy something?" The response categories were always, often, sometimes, and never. In the following analyses, these categories are coded as 4, 3, 2, and 1.

The availability of and knowledge about environmental labels were not registered in the REAP survey. Hence, these variables may be sources of structural differences in the international comparison. Knowledge, in particular, may be a source of rest variance within a country.

ANALYSES

Figure 2 shows the frequency of paying attention to eco-labels in the analysed countries. It appears that a large majority of consumers in these countries pays attention to eco-labels, at least sometimes. Only from 8% (Great Britain) to 15% (Ireland) never do that. Among those who pay attention, the Irish seem to be most, and the Germans least, consistent. Measured on a 4-point scale, the mean score varies within a relatively narrow range, from 2.56 (West Germany) to 2.98 (Ireland). The differences between means are statistically significant ($p < 0.01$) except for the difference between East Germany and Italy and that between Great Britain and Ireland. Interestingly, and somewhat surprisingly, it seems that consumers pay least attention to eco-labels in the country with the strongest tradition for such labels, West Germany. However, caution is recommended before drawing such a conclusion since the finding is sensitive to possible selection and response biases and also to national variations in the understanding of the question.

The analysis of relationships between constructs in the model consists of three steps. First, the suggested causal paths are tested on the West German data. Next, the generality of the found direct relationships between motivational antecedents and paying attention is tested by comparing the West German data with the other four data sets. Third, possible interactions between pro-environmental attitudes and other motivational variables are searched for in each of the five data sets.

Based on the polychoric correlation matrices in the Appendix,

Figure 2. Frequency of paying attention to eco-labels by country, N = 5,323.

structural equation analysis, using LISREL version 8.30 with WLS estimation (Jöreskog, Sörbom, Toit, & Toit, 1999), is used in the first two steps. An important advantage of this method is that it makes it possible to separate the part of the variance in an observed variable that reflects a theoretical construct under study from the part that reflects other factors ("error" from the point of view of the analysis) in cases where there are more indicators than one of a theoretical construct (which unfortunately holds for only two of the constructs in this study). In addition, with structural equation analysis, it is possible to estimate the hypothesised complex path model in one run and, hence, control for all possible relationships among variables included in the model. In a comparative study, multiple sample structural equation analysis offers a rigorous test of differences between samples as well as the possibility to distinguish differences with substantive causes from differences due to variation in the interpretation of questions.

Table II presents the structural equation analysis of the motivation-attention part of the model in Figure 1 for West Germany.

The analysis of the measurement model shows that the two latent constructs based on multiple items have acceptable internal reliabilities (ρ_ξ) and also acceptable individual factor loadings (λ) and reliabilities ($1 - \theta$). The highest correlation between latent constructs is 0.63 (between belief in buying strategy and paying attention), indicating discriminant validity as well.[9] The fit indices indicate an acceptable overall fit, thus confirming that the measurement model and the suggested structural relationships give a good representation of the data.

Basically, the structural equation analysis confirms the suggested relationships. Paying attention to environmental labels is correlated with the personal importance of the goal of protecting the environment (pro-environmental attitude) and with the consumer's perceived effectiveness regarding environmental problems. However, the structural model confirms that the influence of both of these broad concepts is indirect (see the non-significant *t*-values for the direct paths from these variables to paying attention) and mediated through the belief in buying as a strategy to protect the environment and through the trust in environmental labels (only pro-environmental attitude). Together, these latter two concepts account for 58% ($100\chi(1 - \zeta)$) of the variation in paying attention to environmental labels. Pro-environmental attitude and PCE account for 25% of the variation in belief in buying

TABLE II

Determinants of Paying Attention to Eco-Labels in West Germany (N = 824)

Measurement Model

	Pro1	Pro2	Pro3	PCE	Orga	Pack	Trust	Atten
λ	0.88	0.82	0.75	1.00	0.63	0.73	1.00	1.00
θ	0.22	0.33	0.43	–	0.60	0.47	–	–

	Pro-environmental attitude	Belief in buying strategy
ρ_ξ	0.86	0.63

Structural Model

Path from:	Belief in buying strategy	Trust	Pro-environmental attitude	PCE	ζ
To:					
Paying attention	0.63 (9.12)*	0.42 (11.02)	–0.08 (–1.62)	–0.03 (–0.84)	0.42
Belief in buying strategy			0.39 (6.98)	–0.21 (–3.98)	0.75
Trust			0.12 (2.64)	–0.00 (–0.00)	0.99

Covariance Matrix of Latent Variables

	Paying attention	Trust	Belief in buying strategy	Pro-environmental attitude	PCE
Paying attention	1.00				
Trust	0.45	1.00			
Belief in buying strategy	0.63	0.05	1.00		
Pro-environmental attitude	0.27	0.12	0.45	1.00	
PCE	–0.23	–0.04	–0.33	–0.31	1.00

Overall fit: GFI = 0.99, CFI = 0.95, RMSEA = 0.072

* *t*-values in parentheses.

strategy, but only for 1% of the variation in trust in eco-labels. As one could imagine, the covariance matrix of latent variables shows that the two exogenous variables are negatively correlated ($p < 0.001$). The negative correlation indicates that the more the individual believes that it is too difficult to do much about the environment (i.e., the lower the PCE), the less he or she is willing to offer to protect the environment (i.e., the less pro-environmental the attitude).

The generality of these results across countries and regions is tested by means of the multiple sample structural equation analysis, reported in Tables III and IV. The five correlation matrices in the Appendix are used as input. A brief visual inspection of the correlation matrices reveals a striking similarity in the overall pattern of correlations, in spite of their origin in three different language areas and an even larger number of different cultures. This overall impression should be kept in mind when interpreting the formal comparative analysis.

The formal test consists of six steps. In Step 1, it is assumed that both structural and measurement parameters are identical across countries and regions. In the following steps, this assumption is relaxed.

The question of the highest theoretical relevance in this context is whether the structure of the motivational antecedents to paying attention to eco-labels differs between countries. This is analysed in Step 5. The change in χ^2 found in this step is not significant. Hence, the structure of the determinants of paying attention to eco-labels found in the analysis of West German data has indeed cross-national generality.[10] This does not mean that there are not variations in these path parameters between countries and regions (see Table IV). However, only one of the estimated path parameters conflicts with the hypothesis that the influence of the two more general motivational antecedents is mediated through the two more specific ones (a significant path from PCE to paying attention in Great Britain). Further, the differences are not statistically significant overall, as seen in Table III.

Returning to Table III, it appears that in spite of the similarities there are a number of significant differences between the country samples. First, in Steps 2 and 3, it is controlled whether the "observed"

TABLE III
Differences in Determination of Paying Attention to Eco-Labels in West Germany, East Germany, Great Britain, Ireland, and Italy (N = 3,624)

Step[#]	χ^2	df	$\Delta\chi^2$	df
1	361.2	157		
2	331.7	145	29.4	12
3	298.9	125	32.8	20
4	231.2	101	67.7	24
5	207.0	85	24.2*	16
6	196.47	81	10.5	4

* Not significant at $p < 0.05$.

TABLE IV
Determinants of Paying Attention to Eco-Labels in West Germany, East Germany, Great Britain, Ireland, and Italy (N = 3,624)

Country/region	Belief in buying strategy	Trust	Pro-environmental attitude	PCE	ζ
West Germany	0.86 (9.97)*	0.35 (7.80)	–0.08 (–1.55)	–0.03 (–0.89)	0.39
East Germany	0.78 (8.41)	0.30 (5.85)	0.07 (1.31)	0.05 (1.29)	0.48
Great Britain	0.89 (10.15)	0.33 (6.27)	0.00 (0.07)	–0.14 (–3.47)	0.27
Ireland	0.83 (7.42)	0.30 (4.35)	0.05 (0.68)	0.02 (0.36)	0.39
Italy	0.66 (6.57)	0.34 (6.50)	0.03 (0.52)	–0.03 (–0.64)	0.56

* *t*-values in parentheses.

variables are perceived identically by individuals in different countries and regions. In Step 2, factor loadings on the latent variables (λ), and in Step 3, the error variances of the individual measures (θ) are allowed to vary. Both steps produce a significant improvement in χ^2, thus rejecting the assumption that the observed variables are perceived totally identically by individuals in different countries and regions. Apparently, individuals in different countries neither hold totally identical perceptions of the basic meaning of the questions (the significant change in χ^2 in Step 2), nor hold these perceptions with equal certainty (the significant change in Step 3). Hence, we are reminded that the interpretation of questions in a questionnaire depends on language, culture, and history. This implies, among other things, that if these effects are not controlled, differences that are found in the assumed causal paths may not have substantive roots, but may just reflect differences in the understanding of questions.

In Steps 4 to 6, the structural parameters, i.e., the covariances of the latent independent variables (Step 4), the paths from independent to dependent variable(s) (Step 5, as already commented), and the structural equation's error variance (Step 6), are allowed to vary between countries and regions. The significant change in χ^2 in Step 4 shows that correlations between motivational antecedents differ between countries and regions. The significant χ^2 change in Step 6 indicates that there are also differences between countries and regions as to how much of the variation in paying attention to eco-labels is explained by motivational antecedents (and, by implication, how much is explained by internal and external restrictions and other non-measured variables). Table IV shows that among these countries and regions, the motivational antecedents explain from 44 to 73% of the variation in paying attention to eco-labels, most in Great Britain and least in Italy.

Possible interactions between pro-environmental attitude and other motivational antecedents of paying attention to eco-labels are detected by means of hierarchical regression analysis (Aiken & West, 1991). This method involves a series of multiple regression analyses, including first only direct effects and then the product(s) of variables assumed to interact. For this purpose, multi-item indicators are summed and all independent variables centred.

According to the analytical framework generating the hypothesis that some motivational antecedents may interact in their influence on paying attention to eco-labels, such interactions are caused by variables moderating the influence of an attitude on attention and behaviour. Hence, in these calculations pro-environmental attitude assumes the role of main determinant and other motivational antecedents the role of moderators. The possible moderator effect of a variable is calculated in two steps. First, the variable is included in the multiple regression analysis as an additional independent variable. Next, the product of this variable and the pro-environmental attitude is included as yet another independent variable. In the first of these steps the main effect, and in the second step the interaction effect of the additional variable with the main independent variable(s), are analysed. The strength of the alleged moderator variable is measured by the change in the overall coefficient of determination (ΔR^2, adjusted) that it leads to and its significance is tested by the F-change test.

The hypothesis that a possible moderator variable moderates the relationship between pro-environmental attitude and paying attention is confirmed if the inclusion of *the interaction term* leads to a significant increase in the coefficient of determination. The results of the hierarchical regression analysis are reported in Table V.

It appears from Table V that there are few interaction effects among the included variables. Only trust in the label acts as a moderator of the relationship between pro-environmental attitude and paying attention towards eco-labels and only in three of the five cases. In all the three cases, the sign of the interaction term is as expected, i.e., the influence of pro-environmental attitude is stronger among consumers who trust the label than among those who do not.[11]

TABLE V

Hierarchical Regression Analysis of Possible Interactions Between Pro-Environmental Attitude and Other Motivational Antecedents of Paying Attention to Eco-Labels

Independent variable(s):	0: p-e attitude	1.1: p-e attitude, belief in buying	1.2: p-e attitude, belief in buying, product term	2.1: p-e attitude, trust	2.2: p-e attitude, trust, product term	3.1: p-e attitude, PCE	3.2: p-e attitude, PCE, product term
West Germany	N =	946		848		932	
Adj. R^2	0.053	0.330	0.332	0.214	0.216	0.090	0.089
ΔR^2		0.277	0.002	0.161	0.002	0.038	–0.001
F-value			3.017		2.411		–0.999
East Germany,	N =	987		879		959	
Adj. R^2	0.042	0.260	0.260	0.183	0.193	0.049	0.048
ΔR^2		0.218	0.000	0.141	0.009	0.007	0.000
F-value			–0.027		10.179*		–0.451
Great Britain,	N =	1158		930		1115	
Adj. R^2	0.110	0.388	0.387	0.279	0.280	0.126	0.126
ΔR^2		0.278	–0.001	0.169	0.001	0.017	–0.001
F-value			–0.942		1.608		–0.940
Ireland,	N =	936		644		932	
Adj. R^2	0.078	0.307	0.307	0.252	0.277	0.103	0.102
ΔR^2		0.229	–0.001	0.174	0.025	0.024	–0.001
F-value			–0.941		22.365**		–0.971
Italy,	N =	944		771		938	
Adj. R^2	0.049	0.163	0.163	0.175	0.186	0.057	0.056
ΔR^2		0.113	0.000	0.125	0.011	0.007	–0.001
F-value			–0.191		10.293*		–0.970

* $p < 0.01$.
** $p < 0.001$.

SUMMARY AND IMPLICATIONS

Based on the conviction that environmental labelling schemes are potentially useful tools in environment policy, but that they are of no use if they are not noticed by the consumer in the shopping situation, this paper deals with the issue of what makes consumers pay attention to eco-labels. It is suggested that "paying attention to eco-labels" is hardly a goal in itself, but rather a means to a goal: buying environment-friendly products, which is a means to a more abstract goal about protecting the environment. Hence, it is unlikely that a consumer pays attention to an environmental label unless he or she values protecting the environment, perceives buying (more) environmentally friendly products as an effective means to achieve this goal, and perceives the information that the label conveys as useful for this purpose. In addition, the availability of eco-labelled products in the shops and the consumer's ability to recognise and understand them undoubtedly influences attention towards this type of labels.

A model of consumer attention towards eco-labels is developed and the motivation-attention part of the model is tested by means of data from a number of European counties. Due to lack of data, the importance of the availability of eco-labelled products in the shops and of the consumer's ability to recognise and understand them is not included in the test. It is a major strength of this study that key findings are replicated in five countries. Its major weaknesses are the missing variables and that the quality of some of the empirical indicators leaves a lot to be desired. The uncertainty produced by these weaknesses should be kept in mind when interpreting the results.

A large majority of the consumers in all the analysed countries pays attention to eco-labels at least sometimes. As predicted, paying attention to eco-labels is strongly influenced by the belief in considerate buying as a means of protecting the environment and by the trust in the labels in all the included countries and regions. The personal importance of environmental protection (pro-environmental attitude) and perceived effectiveness as regards solving environmental problems also influence paying attention to eco-labels, but this influence is mediated through the former two concepts. In three of the five analysed countries and regions there is also an interaction effect between pro-environmental attitude and trust, meaning that the influence of pro-environmental attitude on paying attention is higher when the consumer trusts the label (and the influence of trust higher when the consumer holds pro-environmental attitudes). From 27 to 56% unexplained variance indicates that important determinants are missing from the test, as we are well aware.

The missing variables notwithstanding, the study has important policy implications. Particularly, it documents that the usefulness of eco-labels as a policy tool depends on the credibility of such labels and, with reservations following from our use of a somewhat imprecise measurement instrument, on consumers being convinced that their purchase behaviour has environmental significance. Of course, it only shows that these variables are necessary, not that they are sufficient conditions for achieving environmental policy goals by means of this instrument.

This study gives empirical support to the frequently expressed assumption that consumers pay attention to and use labels in their buying decisions only if they trust them. Hence, it is of utmost importance to supply and use labels that consumers feel they can trust and to promote the labels in a way that builds trust. This study shows that a consumer's trust in an environmental label to some degree is coloured by, and that its effect on paying attention to the label in some cases in influenced by, his or her attitudes. However, none of these effects are strong.

The fundamental requirements for achieving credibility are obviously not covered by this study. According to a large number of studies, the most important requirement is that the label is issued and controlled by a public or independent authority, a so-called "third party" (e.g., Eden, 1994/95; Enger & Lavik, 1995; MacKenzie, 1991; Schlegelmilch et al., 1996; Tufte & Lavik, 1997). Unfortunately, such labels are currently outnumbered many times by producers' private environmental claims (Enger, 1998; Mayer & Gray-Lee, 1995). This situation can hardly be avoided. However, due to previous lack of discipline among producers, consumer trust in producers' private green claims has been undermined (Peattie, 1995), and due to widespread consumer uncertainty about who issues public and independent labels, such as the Swan, the credibility of such labels has also been hurt (Tufte & Lavik, 1997).

There is no doubt that producers now are more cautious about using green claims than they were in the beginning of the nineties (e.g., Mayer & Gray-Lee, 1995), among other things because regulatory authorities in many countries as well as international bodies such as the ICC and the ISO-organisation have now issued guidances on the use of green claims in marketing (Kuhre, 1997). As a result of this and of the greater prevalence of "third party" eco-labels (see Table I), consumer trust in green claims seems to be improving (cf., e.g., Konsumentverket, 1995/96, 1998). However, a recent Norwegian study documents that the credibility of "third party" eco-labels needs to be strengthened and maintained by means of supportive information (Tufte & Lavik, 1997).

Another important contribution of this study is the proposition and empirical documentation that paying attention to and using eco-labels depends on the priority the consumer gives to the goal of environmental protection and on his or her belief in environment-friendly buying as a strategy to achieve this goal. The study shows that consumer belief in responsible purchase behaviour as a means of achieving environmental goals depends partly on pro-environmental attitude and partly on personality traits (perceived effectiveness).

Both of these results make sense and should be acknowledged. They imply that environmental labelling schemes are a supplement to – not a substitute for – general environmental awareness and self-confidence enhancing information and education efforts. But it should also be noticed that together the two variables leave most of the variation in consumer beliefs about this issue unexplained. As discussed in connection with Figure 1, there are other important determinants that were, unfortunately, not measured in the survey. It should be rather self-evident that consumer belief in the environmental significance of responsible purchase behaviour depends on their knowledge about eco-labels. Such knowledge may stem both from consumer education about the issue and from contact with eco-labelled products. Nilsson et al. (1999) suggest that the labels and labelled products themselves and their immediate surroundings in the shops are the most important sources of learning about eco-labels – what a label looks like, what it means, which types of products and environmental problems it is relevant for, etc. Hence, the proper education of consumers about eco-labels first and foremost depends on increasing the prevalence of eco-labelled products in the shops.

The availability of environmentally differentiated (i.e., eco-labelled) products in the retail sector may, in fact, be the most important key to increasing the attention paid to eco-labels. A strong prevalence of eco-labelled products not only makes eco-labels easier to notice by chance and, as just mentioned, facilitate consumer learning about eco-labels. In addition, it is likely that consumer belief in the environmental significance of responsible purchase behaviour is strengthened by a strong prevalence because it makes it more credible that consumers can make a difference by choosing such products. Hence, future empirical studies should address the key issue of the availability of environmental labels and its effects.

APPENDIX: POLYCHORIC CORRELATION MATRICES BY COUNTRY

	Atten	Pro1	Pro2	Pro3	PCE	Orga	Pack	Trust
West Germany								
Atten	1.00							
Pro1	0.21	1.00						
Pro2	0.10	0.75	1.00					
Pro3	0.23	0.62	0.59	1.00				
PCE	–0.23	–0.21	–0.15	–0.32	1.00			
Orga	0.42	0.16	0.14	0.16	–0.24	1.00		
Pack	0.54	0.28	0.20	0.29	–0.20	0.43	1.00	
Trust	0.52	0.07	0.03	0.11	–0.04	0.15	0.20	1.00
East Germany								
Atten	1.00							
Pro1	0.16	1.00						
Pro2	0.19	0.70	1.00					
Pro3	0.20	0.62	0.61	1.00				
PCE	–0.14	–0.22	–0.16	–0.28	1.00			
Orga	0.35	0.03	0.07	0.14	–0.09	1.00		
Pack	0.48	0.16	0.10	0.23	–0.27	0.38	1.00	
Trust	0.51	0.15	0.15	0.13	–0.03	0.20	0.23	1.00
Italy								
Atten	1.00							
Pro1	0.20	1.00						
Pro2	0.17	0.64	1.00					
Pro3	0.20	0.50	0.54	1.00				
PCE	–0.09	–0.10	–0.07	–0.23	1.00			
Orga	0.26	0.09	0.08	0.15	–0.04	1.00		
Pack	0.35	0.21	0.12	0.18	–0.04	0.11	1.00	
Trust	0.47	0.13	0.11	0.16	–0.08	0.22	0.13	1.00
Ireland								
Atten	1.00							
Pro1	0.31	1.00						
Pro2	0.17	0.66	1.00					
Pro3	0.23	0.66	0.65	1.00				
PCE	–0.16	–0.29	–0.11	–0.24	1.00			
Orga	0.33	0.12	0.13	0.12	–0.07	1.00		
Pack	0.49	0.18	0.16	0.20	–0.14	0.26	1.00	
Trust	0.57	0.27	0.16	0.21	–0.12	0.14	0.29	1.00
Great Britain								
Pro1	0.28	1.00						
Pro2	0.27	0.81	1.00					
Pro3	0.33	0.66	0.72	1.00				
PCE	–0.28	–0.27	–0.27	–0.34	1.00			
Orga	0.41	0.21	0.17	0.21	0.00	1.00		
Pack	0.57	0.18	0.17	0.23	–0.11	0.38	1.00	
Trust	0.60	0.17	0.15	0.22	–0.25	0.18	0.30	1.00

Atten = Paying attention to eco-labels. Pro1-Pro3 = Pro-environmental attitude. Orga-Pack = Belief in buying as a means of protecting the environment.

NOTES

[1] Some environmental labels were known even earlier, for instance labels certified by bio-dynamic growers' associations in several countries, hazard labelling, and report cards (EPA, 1998), but the knowledge about them enjoyed a very limited diffusion in any society.

[2] Whereas the Swan label was introduced in the other Nordic countries in 1989, Denmark only became a full member of this labelling scheme at the end of 1997, which undoubtedly explains the difference.

[3] Much more could, of course, be said about the importance of personality factors in this connection.

[4] Framed within a general attitude-behaviour relationship discussion, similar ideas have been suggested by Guagnano, Stern, and Dietz (1995) and Ölander and Thøgersen (1995).

[5] Knowledge about labelling schemes most likely influence both a consumer's belief in cautious buying as a strategy to achieve environmental goals and his or her trust in the labels. Also, by conveying information about opportunities for environment-friendly buying, the availability of labelled products may influence consumers' beliefs in cautious buying as a strategy to achieve environmental goals.

[6] Dual process models of decision-making, such as Fazio's (1990) MODE model, suggest that thorough, or deliberative, decision-making requires that the individual is motivated, able, and has the opportunity to spend time and effort on deliberating. If any of these requirements are missing, a more spontaneous decision-making mode will be applied. This is the case for low-involvement decisions, for instance. Fazio's (1986) attitude-to-behaviour theory is a model of spontaneous decision-making.

[7] COMPASS was founded by the European Union. It comprises five specialist social survey research organisations: EURISKO, Milan, Italy; Instituut voor sociaal-wetenschappelijk Onderzoek (IVA), Tilburg, the Netherlands; Social and Community Planning Research (SCPR), London, UK; Social Science Research Centre (SSRC), Dublin, Ireland; and Zentrum für Umfragen, Methoden und Analysen (ZUMA), Mannheim, Germany. The work programme was called "Research into Environmental Attitudes and Perceptions in Five EU Countries" (REAP). Main topics include energy use in the home, automobile use, household waste management, and environmental labelling. More details about the procedure and design are reported in Witherspoon and Mohler (1995), which can be downloaded from the internet address www.zuma-mannheim.de/cooperation/en/compass/reap.htm.

[8] Labels issued by organic and bio-dynamic growers' organisations have been used in the marketing of pesticide and chemical free vegetables for decades, but at the time of this survey they played a marginal role in the covered countries. Making an effort in this regard would instead involve going to speciality shops or farmers' markets.

[9] Notice that the size of the correlation may be attenuated in cases where one or both of the latent variables are measured with only one item. Hence, the evidence supporting discriminant validity is tentative in these cases.

[10] Only the equation analysing the direct paths from the four antecedents to paying attention is included in the multiple sample analysis.

[11] In the interest of saving space the parameters of the equations are not shown, but they can be acquired from the author on request.

REFERENCES

Aiken, L. S., & West, S. G. (1991). *Multiple regression: Testing and interpreting interactions*. Newbury Park, CA: Sage.

Bech-Larsen, T. (1996). Danish consumers' attitudes to the functional and environmental characteristics of food packaging. *Journal of Consumer Policy, 19*, 339–363.

Bekholm, M., & Sejersen, B. (1997). *Mærkeanvendelse og associationer. Personlig omnibus 44. Udarbejdet for Forbrugerstyrelsen. Tabel- og kommentarrapport.* (Use of and associations provided by labels). København: AIM Nielsen.

Berger, I. E., & Corbin, R. M. (1992). Perceived consumer effectiveness and faith in others as moderators of environmentally responsible behaviors. *Journal of Public Policy & Marketing, 11*, 79–89.

Berger, I. E., Ratchford, B. T., & Haines, G. H., Jr. (1994). Subjective product knowledge as a moderator of the relationship between attitudes and purchase intentions for a durable product. *Journal of Economic Psychology, 15*, 301–314.

Biel, A., & Dahlstrand, U. (1998). *Habits and the establishment of ecological purchase behavior*. Gothenburg: Göteborg University, Department of Psychology.

Buchtele, F., & Holzmuller, H. H. (1990). Die Bedeutung der Umweltvertraglichkeit von Produkten für die Kaufpräferenz. *Jahrbuch der Absatz- und Verbrauchsforschung, 36*(1), 86–103.

Celsi, R. L., & Olson, J. C. (1988). The role of involvement in attention and comprehension processes. *Journal of Consumer Research, 15*, 210–224.

Eagly, A. H., & Chaiken, S. (1993). *The psychology of attitudes*. Fort Worth: Harcourt Brace Jovanovich.

Eden, S. (1994/95). Business, trust and environmental information: Perceptions from consumer and retailers. *Business Strategy and the Environment, 3*(4), 1–7.

Ellen, P. S., Wiener, J. L., & Cobb-Walgren, C. (1991). The role of perceived consumer effectiveness in motivating environmentally conscious behaviors. *Journal of Public Policy & Marketing, 10*, 102–117.

Enger, A. (1998). *Miljøargumentasjon i markedsføring* (Environmental arguments in marketing). Lysaker (Norway): National Institute for Consumer Research. SIFO-Rapport nr. 1 1998.

Enger, A., & Lavik, R. (1995). Eco-labelling in Norway: Consumer knowledge and attitudes. In: E. Stø (Ed.), *Sustainable consumption – Report from the International Conference on Sustainable Consumption*, pp. 479–502. Lysaker (Norway): National Institute for Consumer Research (SIFO).

EPA (1998). *Environmental labeling issues, policies, and practices worldwide*. Washington DC: U.S. Environmental Protection Agency.

Eurostat et al. (1995). *Europe's environment: Statistical compendium for the Dobris assessment*. Luxembourg: Office for Official Publications of the European Communities.

Fazio, R. H. (1986). How do attitudes guide behavior? In: R. M. Sorrentino & E. T. Higgins (Eds.), *The handbook of motivation and cognition: Foundations of social behavior*, pp. 204–243. New York: Guilford Press.

Fazio, R. H. (1990). Multiple processes by which attitudes guide behavior: The MODE model as an integrative framework. In M. P. Zanna (Ed.), *Advances in experimental social psychology (Vol. 23)*, pp. 75–109. New York: Academic Press.

Forbrugerstyrelsen (1993). *Mærkning rettet til forbrugerne* (Labeling directed at consumers). Copenhagen: The Ministry of Trade and Industry.

Geyer-Allély, E., & Eppel, J. (1997). Consumption and production patterns: Making the change. In: *Sustainable Development: OECD Policy Approaches for the 21st century*. Prepared by the OECD Task Force on Sustainable Development. Paris: OECD.

Grunert, S. C., & Juhl, H. J. (1995). Values, environmental attitudes, and buying of organic foods. *Journal of Economic Psychology, 16*, 39–62.

Guagnano, G. A., Stern, P. C., & Dietz, T. (1995). Influences on attitude-behavior relationships. A natural experiment with curbside recycling. *Environment and Behavior, 27*, 699–718.

Hansen, U., & Kull, S. (1994). Öko-Label als umweltbezogenes Informationsinstrument: Begründungszusammenhänge und Interessen. *Marketing, 4*, 265–273.

Heiskanen, E., Kärnä, A., Niva, M., Timonen, P., Rosenschöld, E. M., Pripp, L., & Thidell, Å. (1998). *Environmental improvement in product chains*. Copenhagen: Nordic Council/Nordic Council of Ministers. TemaNord 1998: 546.

Herr, P. M., & Fazio, R. H. (1993). The attitude-to-behavior process: Implications for consumer behavior. In: A. A. Mitchell (Ed.), *Advertising exposure, memory, and choice. Advertising and consumer psychology*, pp. 119–140. Hillsdale: Lawrence Erlbaum Associates.

Jacoby, J. (1984). Perspectives on information overload. *Journal of Consumer Research, 11*, 569–573.

Jöreskog, K. G., Sörbom, D., de Toit, S., & de Toit, M. (1999). *LISREL8: New statistical features*. Chicago: Scientific Software International.

Kinnear, T., Taylor, J. R., & Ahmed, S. (1974). Ecologically concerned consumers: Who are they? *Journal of Marketing, 38*(2), 20–24.

Kokkinaki, F. (1997). Involvement as a determinant of the process through which attitudes guide behaviour. In: *Proceedings from the IAREP XXII Conference*, pp. 52–67. Valencia, Spain: Corpas, C.B.

Konsumentverket (1995/96). *The consumer and the environment: Results of a survey into awareness of the environment amongst Swedish consumers*. Stockholm: The Swedish Consumer Agency. Report 1995/96:54.

Konsumentverket (1998). *Allmänhetens kunskaper, attityder och agerande i miljøfrågor* (Knowledge, attitudes and behaviour with regard to the environment). Stockholm: The Swedish Consumer Agency. Rapport 1998:7.

Kotler, P. (1991). *Marketing management. Analysis, planning, implementation, and control* (7th ed.). Englewood Cliffs, NJ: Prentice-Hall.

Kuhre, W. L. (1997). *ISO 14020s: Environmental labelling-marketing*. Upper Saddle River, NJ: Prentice-Hall PTR.

Lindberg, K.-E. (1998). *Nordisk omnibus svanemerket. Utarbeidet for Nordisk Miljömärkning*. Tabelrapport (A report on the Nordic Swan). Oslo: Markeds- og Mediainstituttet.

MacKenzie, D. (1991). The rise of the green consumer. *Consumer Policy Review, 1*(2), 68–75.

Mayer, R. W., & Gray-Lee, J. (1995). Environmental marketing claims in the U.S.A.: Trends since issurance of the FTC guides. In: E. Stø (Ed.), *Sustainable Consumption – Report from the International Conference on Sustainable Consumption*, pp. 407–426. Lysaker (Norway): National Institute for Consumer Research (SIFO).

Miljø- og Energiministeriet (1995). *Natur- og miljøpolitisk redegørelse 1995* (Nature and environment in Denmark). Copenhagen: The Ministry of Environment and Energy.

Miljøstyrelsen (1996). *En styrket produktorienteret miljøindsats. Et debatoplæg* (Strengthening the environment by improving products. A discussion paper). Copenhagen: The Danish Environmental Protection Agency.

Morris, L. A., Hastak, M., & Mazis, M. B. (1995). Consumer comprehension of environmental advertising and labeling claims. *Journal of Consumer Affairs, 29*, 328–350.

Naturvårdsverket (1997). *En studie ar hur olika styrmedel påverkat skogsindustrin.* (A study of the impact of different means of regulating on the forest industry). Stockholm: The Swedish Environmental Protection Agency. Rapport 4724.

Nilsson, O. S., Nissen, N. P., Thøgersen, J., & Vilby, K. (1999). Rapport fra "forbrugergruppen" under Erhvervsministeriets mærkningsudvalg (A report from the consumer study group to the Labels Committee). In: *Mærkningsudvalgets redegørelse*, pp. 152–199. Copenhagen: The Ministry of Trade and Industry, The Consumer Agency.

Norwegian Ministry of Environment (1994a). Summary of issues raised at the symposium on sustainable consumption. In: *Symposium: Sustainable consumption. 19–20 January 1994, Oslo, Norway*, pp. 22–25. Oslo: The Norwegian Ministry of Environment.

Norwegian Ministry of Environment (1994b). *Symposium: Sustainable consumption. 19–20 January 1994, Oslo, Norway*. Oslo: The Norwegian Ministry of Environment.

OECD (1991). *Environmental labelling in OECD countries*. Paris: OECD.

OECD (1997). *Sustainable consumption and production*. Paris: OECD.

Ölander, F., & Thøgersen, J. (1995). Understanding of consumer behaviour as a prerequisite for environmental protection. *Journal of Consumer Policy, 18*, 317–357.

Palm, L., & Windahl, S. (1998). *How Swedish consumers interpret and use environmental information: A study of quantitative environmental product declarations.* Stockholm: The Swedish Consumer Agency. Report No. 1998:28.

Peattie, K. (1995). *Environmental marketing management: Meeting the green challenge*. London: Pitman Publishing.

Peter, J. P., Olson, J. C., & Grunert, K. G. (1999), *Consumer behaviour and marketing strategy. European edition*. London: McGraw Hill.

Petty, R. E., & Krosnick, J. A. (Eds.) (1995). *Attitude strength: Antecedents and consequences*. Mahwah, NJ: Erlbaum.

Pieters, R. G. M. (1991). Changing garbage disposal patterns of consumers: Motivation, ability, and performance. *Journal of Public Policy & Marketing, 10*, 59–76.

Regeringskansliet (1999). *Märk Väl! Betänkande av Utredningen gällande konsumentinformation om dagligvaror* (White Paper on how to provide consumers with information about non-durable goods). Stockholm: Swedish Government Official Reports 1999:7.

Roskos-Ewoldsen, D. R., & Fazio, R. H. (1992). On the orienting value of attitudes: Attitude accessibility as a determinant of an object's attraction of visual attention. *Journal of Personality and Social Psychology, 63*, 198–211.

Scammon, D., & Mayer, R. N. (1993). Environmental labelling and advertising claims: International action and policy issues. In: *European advances in consumer research, Vol. 1*, pp. 338–344. Provo, UT: Association for Consumer Research.

Schlegelmilch, B. B., Bohlen, G. M., & Diamantopoulos, A. (1996). The link between green purchasing decisions and measures of environmental consciousness. *European Journal of Marketing, 30*(5), 35–55.

Sitarz, D. (Ed.) (1994). *Agenda 21: The earth summit strategy to save our planet*. Boulder: EarthPress.

Sparks, P., & Shepherd, R. (1992). Self-identity and the theory of planned behavior: Assessing the role of identification with "green consumerism". *Social Psychology Quarterly, 55*, 388–399.

Stern, P. C., Dietz, T., Abel, T., Guagnano, G. A., & Kalof, L. (1999). A value-belief-norm theory of support for social movements: The case of environmentalism. *Human Ecology Review, 6*, 81–97.

Thøgersen, J. (1996). *The demand for environmentally friendly packaging in Germany*. Aarhus: The Aarhus School of Business, Department of Marketing. MAPP Working Paper 30.

Thøgersen, J. (1998a). *The norm-attitude-behavior relationship. Theory and applications in the environmental domain*. Aarhus: The Aarhus School of Business, Department of Marketing. Working Paper 98-3.

Thøgersen, J. (1998b). *Understanding behaviours with mixed motives. An application of a modified theory of reasoned action on consumer purchase of organic food products*. Aarhus: The Aarhus School of Business, Department of Marketing. Working Paper 98-2.

Thøgersen, J. (1999). The ethical consumer. Moral norms and packaging choice. *Journal of Consumer Policy, 22*, 439–460.

Thøgersen, J., & Andersen, A. K. (1996). Environmentally friendly consumer behavior: The interplay of moral attitudes, private costs, and facilitating conditions. In: R. P. Hill & C. R. Taylor (Eds.), *Marketing and Public Policy Conference Proceedings, Vol. 6*, pp. 80–96. Chicago: American Marketing Association.

Tufte, P. A., & Lavik, R. (1997). *Helse- og miljøinformasjon. Forbrukernes behov for informasjon om skadelige stoffer i produkter* (Information about health and environment. Consumers' need for information on dangerous substances in products). Lysaker (Norway): Statens Institutt for Forbruksforskning (SIFO). Rapport 4 – 1997.

Verplanken, B., & Weenig, M. W. H. (1993). Graphical energy labels and consumers' decisions about home appliances: A process tracing approach. *Journal of Economic Psychology, 14*, 739–752.

Witherspoon, S., & Mohler, P. P. (1995). *Research into environmental attitudes and perceptions. Final report*. Mannheim: Compass c/o Zuma.

THE AUTHOR

John Thøgersen is an Associate Professor in the Department of Marketing, The Aarhus School of Business, Haslegaardsvej 10, DK-8210 Aarhus V, Denmark. Fax: +45 86 15 39 88; e-mail: John.Thogersen@asb.dk.

The first draft of this paper was written during a Visiting Professorship at Zentrum für Umfragen, Methoden und Analysen (ZUMA), Mannheim, Germany, in July 1997.

The author is grateful for constructive comments from three anonymous reviewers and the Editors.

[18]

What We May Have Is a Failure to Communicate*: Labeling Environmentally Certified Forest Products

Mario F. Teisl

ABSTRACT. To develop an understanding of the performance of different eco-labels for certified forest products, we administered a mail survey to a nationally representative sample of U.S. residents. In the survey, respondents viewed different eco-labels and were asked to perform a series of tasks designed to measure the labels' communication performance. Results indicate that relatively detailed labels are more beneficial for consumers (and environmentally sensitive forest product manufacturers) than simpler eco-seals. Eco-seals are the least credible type of label and, in general, do not allow consumers an adequate basis for product differentiation. However, we also find evidence that marginal changes can significantly improve the performance of simple eco-seals. The results suggest that U.S. consumers value the environmental benefits created from more environmentally benign forestry. Thus, consumer-driven purchases could potentially support a future of environmentally benign forest management practices with less reliance on other policy alternatives. However, the results also suggest that the current state of forest product labeling, where firms can (voluntarily) label their products with simple eco-seals from a variety of certification organizations, is not the most effective labeling approach. Parties who are interested in the long-run success of these programs need to consider altering current labeling approaches. FOR. SCI. 49(5):668–680.

Key Words: Forest management, credibility, market experiment, information, policy.

SPURRED BY ORGANIZATIONS like the Forest Stewardship Council (FSC)[1] and the American Forest and Paper Association (AFPA), environmental certification programs for forests and forest products are rapidly being implemented. *Forest* certification is a process in which a party measures current forest management practices against some environmental management standards. *Forest product* certification requires a party to perform a chain-of-custody audit to confirm that wood from certified forests is being used in product lines. Currently, there are hundreds of U.S. companies selling a diverse range of certified forest products (FSC 1998). In addition, large purchasers of wood (Home Depot, IKEA International) are committed to stocking environmentally certified wood products.

Sellers of certified forest products are able to use environmental labeling approaches to designate their product comes from an environmentally managed forest. (It is common for individuals in the forest certification policy community to use the terms "label" and "labeling" to designate the underlying forest management standards or

* Paraphrase of a statement made by Strother Martin in the film *Cool Hand Luke*, 1967.

Mario F. Teisl, Associate Professor, Department of Resource Economics and Policy, 5782 Winslow Hall, University of Maine, Orono, ME 04469—207-581-3162; Fax: 207-581-4278; E-mail: TEISL@maine.edu.

Acknowledgments: This project was financed in part by a U.S. Department of Agriculture National Research Initiative Grant (#99-35400-7851), and the Maine Agricultural and Forest Experiment Station. Although the research described in this article has been funded in part by the USDA, it has not been subject to the Agency's required peer and policy review and therefore does not necessarily reflect the views of USDA, and no official endorsement should be inferred. All errors and omissions are the sole responsibility of the author. Maine Agricultural and Forest Experiment Station Publication No. 2605.

production practices resulting from certification. We do not use this convention. Here these terms mean the information devices placed on a forest product.) The function of this labeling is to improve the flow of information to consumers (Mitra and Lynch 1995) who, in turn, alter their information search and/or product purchase behaviors. These changes in behaviors may lead to changes in producer behaviors (see Moorman 2001, p. 106–116, for a list of studies looking at firms' responses to changes in information policies).[2] By doing so, labeling may allow for the correction of market failures in forest product production. The market failure here is that a consumer with preferences for (or against) specific forest management outputs lacks the information to reward (or punish) a forest product manufacturer through their market decisions (see Haener and Luckert 1998 for a nice presentation of this issue). Firms that are better able to take advantage of labeling will be rewarded with a comparative advantage, gain market share and maximize any value-added rents (Mitnick 1981, Moorman and Slotegraaf 1999).

Proponents of eco-labeling certified forest products perceive them to be an effective method of altering consumer and producer behavior. Indeed, demand-side research indicates a potential mainstream market for environmentally certified forest products (Forsyth et al. 1999, Ozanne and Vlosky 1997, 1998) and, hence, a derived demand for certification services and altered forestry practices. From a supply and demand point of view, the problem appears simple. Some consumers may be willing to pay more for products from forests managed in an environmentally acceptable fashion while some forest owners and managers are willing to alter practices in order to capture these additional premiums. Environmental labeling may be able to organize and match those who demand and those who provide such improved forestry services.

However, even if consumers hold preferences for specific forest management outputs, several communication issues may make the exchange of premiums for altered forestry practices difficult and could delay or derail the potential benefits of labeling. Because the promise of improved forestry practices is impossible for most consumers to verify, the success of the labeling aspect of a forest product certification program uniquely hinges on forest product companies being able to credibly communicate to the consumer that forestry practices have been altered. Notably, it is an open question whether the current state of labeling certified forest products will lead to increased purchases of products from environmentally managed forests.

Many articles have presented "production side" critiques and comparisons of forest and forest product certification (e.g., debate about the principles, criteria, and implications of sustainable forest management). For a nice review of some of these production side issues see AFPA 2002, Cumbo and Smith 2001, Meridian Institute 2001, NRCM 2001, Peterson 2001, Stone 2001, SCS 1995, Vogt et al. 2000.[3] However, there exists a paucity of research looking at how consumers view and react to environmental labels on forest products. Our purpose here is to illustrate some of the communication issues and to highlight potential questions surrounding the labeling of certified forest products.

Characteristics of Labeling Policies

There is limited evidence available concerning consumer acceptance and trust of environmental labels on forest products. Ozanne and Vlosky (1998) is the only study we found that uses consumer research to address any part of the information communications process as it relates to credible certification programs. They find that forest product certification programs implemented by nongovernmental environmental organizations would garner the most consumer credibility, followed by private certification companies, federal government agencies, and wood product industry organizations. Beyond this study, little public domain research exists concerning consumer understanding and acceptance of forest product eco-labeling.

The ability of labeling to communicate the environmental attributes of a forest product to the consumer can vary along three major continua: compulsoriness, standardization and explicitness.[4] Here, *compulsoriness* denotes the degree to which firms are required to provide product information. At one extreme, labeling restrictions are mandatory; certain pieces of information are required to be displayed on the product. At the other extreme, labeling restrictions are voluntary; currently forest product eco-labeling falls into the voluntary category.[5] A great deal of attention has been devoted to the process by which consumers infer a value for missing information (possible under a voluntary labeling approach) and how this affects choice (Lee and Olshavsky 1997). This literature suggests that consumers may turn to information from other brands (Jaccard and Wood 1998, Ross and Creyer 1992), or turn to information about other attributes of the same product (Ford and Smith 1996, Johnson and Levin 1985, Levin et al. 1984) to "fill in" for the missing information. Others suggest that consumers do not infer missing values at all but merely pay less attention to a product with missing information (Simmons and Lynch 1991).

The second major component of labeling policies is *standardization*: the degree to which regulations require the information be provided in a standardized fashion across products. At one extreme, a labeling policy can make presentation format requirements quite explicit, where the firm has no discretion over the presentation. For example, warning labels typically have wording, font size, typeface, and message location prescribed by regulation. Alternatively, the content of the information may be regulated but the firm has some discretion over the presentation of the information. Studies suggest that standardized displays provide the largest benefit to consumers (Kleinmutz and Schkade 1993, Schkade and Kleinmutz 1994) because they increase the number of products or attributes considered during choice, allowing for more accurate choice decisions (Coupey 1994, Winneg et al. 1998, p. XIII–XIV).

A potential problem with the current state of labeling certified forest products is that the information disclosed to consumers is not in a standardized form. The use of different methods of presenting environmental certification information to consumers may actually increase consumer confusion and erode consumer confidence in environmental labeling. Lyke (1996) notes that at least seven independent programs have been certifying forests across the world. When multiple certification seals or labels are introduced to the market, consumers may come to believe that every firm has found a way to claim their product is environmentally friendly. For example, its questionable whether a consumer will be able to differentiate forest products when one product exhibits a Green Seal logo[6], another product exhibits a FSC logo and a third product states that it comes from forests certified as sustainable by the Rainforest Alliance's "SmartWood" program (note SmartWood is an approved FSC certifier).[7]

The third major component of labelling policies is *explicitness*: the degree of information detail presented to consumers. For simplicity we will discuss only two types of labels that are differentiated by the level of information detail. Eco-seals, such as seals of approval issued by certification programs, communicate little detail concerning attribute values. Only those that are intimately familiar with the certification agency and its standards understand the full meaning of the symbol. At the other extreme are full disclosure labels that provide detailed information about several of the product's attributes (e.g., nutrition labels on food) and the disclosure typically involves continuous or categorical information about each element (e.g., grams of fat). These more detailed labels are generally considered the most objective of the label categories, while seals of approval are often considered the most normative. Currently, it is unknown whether consumers demand just a general assurance of environmental improvement (potentially conveyed by a simple label) or demand specific changes in forest management practices (potentially conveyed by a more detailed environmental label).[8]

Consumer scientists have long understood that more information is not always better due to the possibility of information overload (Scammon 1977) and potential distraction away from more relevant information (Roe, Levy and Derby 1999). Recently, Bei and Widdows (1999) explored how simple (i.e., summary ratings) and complex (i.e., attribute-level ratings) information affected consumers with different levels of experience with the product. They found that more experienced consumers obtained a relatively larger benefit from the more complex information. The implication is that as a consumer becomes more experienced with a product category they become more tolerant of increased information (i.e., they are less likely to experience information overload problems).

Methods

To help evaluate the factors that are most important in designing an environmental labeling policy, we administered a mail survey that contained several experiments specifically designed to elicit respondent reactions to alternative labeling approaches.

Sampling

We obtained a sample of 3,290 prerecruited U.S. adult residents from International Communications Research of Media, Pennsylvania. They conducted a telephone screening survey, using random-digit dialing (RDD), during the spring of 2000 to identify potential mail survey respondents. The sample design consisted of a nationally representative group of adults with an additional oversample of New England and Maine residents. Except for the oversampling, the randomness of the dialing process should produce a sample similar to one drawn through the use of probability sampling if there is no telephone noncoverage bias in the area under study and there is no nonresponse bias.[9] However, one should note that the screening procedure might have identified a sample that is more tolerant of answering surveys and sifting through information.

Survey Administration

During the summer of 2000, we conducted the mail survey of the prerecruited respondents. The survey was administered in three waves; a $5.00 incentive (paid when individuals returned their survey) was provided to increase response. In total 1,948 individuals responded to the mail survey, and 36 were returned as undeliverable, for a response rate of 60% (1,948/3,290 – 36).[10] In general, our survey respondents are relatively representative of the characteristics of the U.S. adult population.

Survey Design

The survey questionnaire featured two experimental tasks designed to elicit reactions from respondents to alternative labeling approaches. Given that one purpose of eco-labeling is to reduce a potential market failure (i.e., it provides information about product attributes that are unknown to the consumer but which are important to the consumer's utility), we used focus group research to help identify the forest management attributes most important to consumers (subsequently used on the more detailed labels). In addition, the design of the experiments and of the information treatments was also based on this earlier focus group research (Teisl et al. 2000).

Experiment I.—In this experiment, respondents were presented a single environmentally labeled generic wood product (Figure 1). The amount and type of information presented on the label was experimentally manipulated. Specifically, the labels varied by the explicitness of the information. The label could present four different levels of information: (1) an eco-seal logo [LOGO], (2) an eco-seal with contact information (phone number and website) [CONTACT], (3) an eco-seal with contact information and a summary score (with accompanying text) [SUMMARY], or (4) an eco-seal with contact information, summary score, and a detailed table listing the five components of the environmental summary score along with associated scores for each component [TABLE].

The labels also differed across respondents by the organization certifying the environmental information (i.e., the organization represented by the eco-seal logo). The organizations included the U.S. Environmental Protection Agency (EPA), the Forest Stewardship Council (FSC) or the Sierra Club (SC). There was an additional variant: some individuals viewed the FSC logo along with a "Made in Maine" logo from the Maine Wood Products Association (ME). Note that of the four variants, only the FSC and Made in Maine labels are

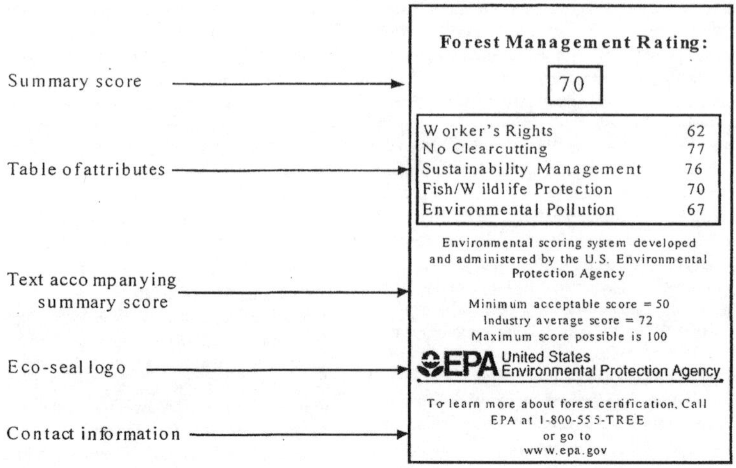

Figure 1. Example of information variations on label.

actually used in the market. The EPA and SC seals are actual organizational logos but are currently not used as part of a forest certification program; so their use here is hypothetical. Finally, there are some organizational seals that are, or could soon be, used as part of a forest product label (e.g., Scientific Certification Systems, or AFPA's Sustainable Forestry Initiative – SFI - logo[11]) but were not used as part of the experiment. It was our interest in this experiment to present organizations that differed in levels of familiarity and by type of organization (government organization, independent certifier, and environmental group). The earlier focus group research indicated that the organizations we used would be adequate for the purposes of the experiment.

Environmental scores were also manipulated. When summary scores were displayed without the corresponding table of scores, then the environmental summary scores were: FSC = 75, ME = 91, EPA = 67 and SC = 82. When summary scores were displayed with the corresponding table of scores, then the environmental summary scores were: FSC = 91, ME = 72, EPA = 82, and SC = 67. Note that the individual scores displayed in the table were perfectly correlated so that a unit movement in the summary score corresponds to a unit movement in each of the individual scores displayed in the tables.

After viewing the label, respondents were asked to: (1) rate the credibility of the label, (2) rate the environmental friendliness of the product, (3) rate their level of satisfaction with the amount of information presented, and (4) rate the likelihood that they would buy the product if the price and quality were the same as the brand of wood product they currently purchase. All four questions used Likert-type ratings scales. For the credibility question, the scale runs from 1, which denotes the label was "not credible," to 5, which denotes the label was "very credible." For the environmental ratings question, the scale ran from 1, "not eco-friendly," to 5, "very eco-friendly." In the information load question 1 denotes "not enough information," 3 denotes to "just enough information," and 5 denotes "too much information." In the likelihood-to-buy question 1 denotes "highly unlikely, " 3 denotes "no opinion either way," and 5 denotes "highly likely."

Each survey respondent viewed and performed two label evaluations. To eliminate any order or sequencing effects, the presentation order of label treatments was randomized across surveys. For the analysis, responses to both label evaluations are pooled.

Experiment II.—In the second experiment, respondents were asked to picture themselves in a store looking to purchase a wood product and finding three different brands available: A, B, and C (Figure 2). Respondents were asked to assume that all three brands were exactly the same except for the price of the product and the information presented on the environmental labels. They were instructed that sometimes the environmental labels might be blank (i.e., there was no environmental information available for these brands). They were also told that products with blank labels may, or may not be, better for the environment than the other (labeled) products.

The price and environmental attributes for the three products were correlated so that C was the low price, least environmentally sound product; B was the high price, most environmentally sound product; and A was midrange for both attributes. Prices and environmental attributes were held constant across respondents. The price of the product was always disclosed; however, the disclosure of environmental information was experimentally manipulated. After viewing the products, respondents were asked which product they would buy.

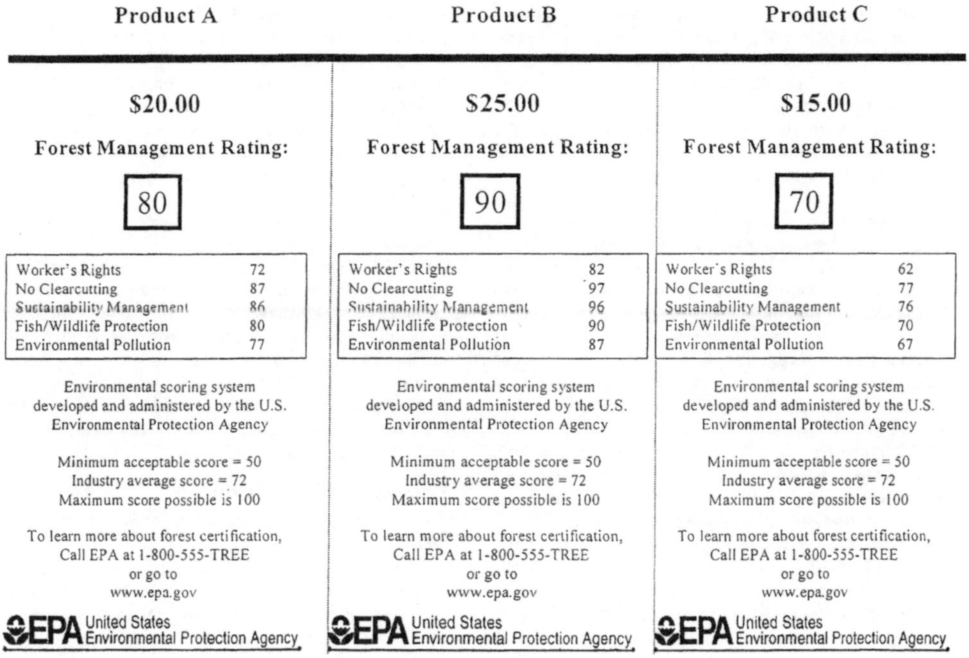

Figure 2. Example of Experiment II label treatment—compulsory, detailed labels, standard certifier.

We manipulated three key elements of the product information. First, we manipulated the *explicitness* of the information disclosure. The regime could take on three different levels: (1) no disclosure [BLANK], (2) an eco-seal with contact information [ECOSEAL], or (3) an eco-seal with contact information, a summary score, and a detailed table listing the five components of the environmental summary score along with associated scores for each component [DETAILED LABEL]. Second, we manipulated the *compulsoriness* of the disclosure. Regimes could assume three levels: no environmental disclosure, products A and B disclose environmental information [VOLUNTARY] or all three products disclose environmental information [COMPULSORY]. Finally, we manipulated the level of *standardization* in the organizations used to certify the environmental information (e.g., STANDARD indicates the respondent views the same certifier across labels; NON-STANDARD indicates the respondent views different certifiers across labels).

Data Analysis

Experiment I.—Here we are interested in estimating the association between individuals' perceptions of the label and/or product across the different information treatments. In turn we estimated four different equations, each with similar sets of independent variables (reflecting the 16 different information treatments) but with different dependent measures. The general form of each equation is:

$$DEP = \alpha_1 INT_1 + \alpha_2 INT_2 + \alpha_3 INT_3 + \beta_{FSC} FSCSEAL + \beta_{ME} MESEAL + \beta_{EPA} EPASEAL + \beta_{SC} SCSEAL + \delta_{FSC} FSCCONT + \delta_{ME} MECONT + \delta_{EPA} EPACONT + \delta_{SC} SCCONT + \phi_{FSC} FSCSUMY + \phi_{ME} MESUMY + \phi_{EPA} EPASUMY + \phi_{SC} SCSUMY + \gamma_{FSC} FSCTBLE + \gamma_{ME} METBLE + \gamma_{EPA} EPATBLE + \gamma_{SC} SCTBLE + e$$

where *DEP* denotes the dependent variable of one of the four equations. The dependent variable is a response to one of the four Likert-ratings questions measuring either the: (1) label's credibility, (2) product's eco-friendliness, (3) respondent's satisfaction with the amount of information presented, or (4) respondent's likelihood to buy the product. INT_1, INT_2 and INT_3 denote the three intercepts. *FSCSEAL, MESEAL, EPASEAL* and *SCSEAL* is a set of four dummy variables denoting whether the label displayed a particular certifying organization's eco-seal (either the Forest Stewardship Council (FSC), Forest Stewardship Council with a Made-in-Maine label (ME), Environmental Protection Agency (EPA) or the Sierra Club (SC), 0 otherwise. *FSCCONT, MECONT, EPACONT* and *SCCONT* is a set of four dummy variables denoting whether the label displayed the organizations' contact information along with the eco-seal, 0 otherwise. *FSCSUMY, MESUMY, EPASUMY* and *SCSUMY* is a set of four dummy variables denoting whether the label displayed an environmental summary score along with the organization's contact information and eco-seal, 0 otherwise. *FSCTBLE, METBLE, EPATBLE* and *SCTBLE* is a set of four dummy variables denoting whether the label displayed a table of environmental scores along with the environmental summary score, organization's contact information and eco-seal, 0 otherwise. Given that the dependent variables are ordered, we use ordered-logit techniques.

Given the manner in which the variables are coded, the β's provide baseline information of how respondents react to labels and products bearing only an organization's eco-seal. The δ's provide information on how the reactions of respondents who saw the eco-seal with contact information are different from those who only saw an eco-seal (i.e., the δ's measure the impact of adding the contact information). The ϕ's provide information on how the reactions of respondents who saw the eco-seal with contact and summary score information are different from those who only saw the eco-seal with contact information (i.e., the ϕ's measure the impact of adding the summary information). Finally, the γ's provide information on how the reactions of respondents who saw the eco-seal with contact, summary score, and table information are different from those who only saw the eco-seal with contact and summary score information (i.e., the γ's measure the impact of adding the table information).

We can test whether adding information alters respondents' perceptions of the label and/or product by testing the equivalence of individual pairs of parameters sharing the same organizational subscript (e.g., $\beta_{EPA} = \delta_{EPA}$ tests whether adding contact information to the EPA eco-seal significantly alters respondent perceptions). We can also test whether the certifying organization matters by testing the equivalence of individual pairs of parameters sharing the same Greek letter (e.g., $\beta_{EPA} = \beta_{FSC}$ tests whether responses to the eco-seals differ whether the certifying organization is the EPA or the FSC).

Because of the difficulty in directly relating equation parameters to changes in the dependent variables (i.e., the coefficients in an ordered-logit equation do not represent marginal probabilities[12]), we use the estimated equations with appropriate variable and intercept coding to provide estimates of the "average" score for each information treatment. Average scores are calculated as: $AVERAGE = \Sigma_j Prob (DEP = j) * j$; where j denotes the integer range (1, 2, 3, 4, 5) used in the Likert-scale.

Experiment II.—We use contingency tables to test whether the distributions of choice responses are affected by the various information treatments. We first test all of the information treatments against a base case of no environmental labeling to determine if environmental labeling affects respondents''choices of wood products. We then test various pairings of information treatments to determine if there are significant differences between: less explicit eco-seal approaches versus more detailed environmental disclosure; compulsory versus voluntary labeling; and standardized versus nonstandardized labeling.

Results and Discussion

Experiment I—Tests Across Information Treatments

Credibility.—Altering the amount of information on the label significantly altered the credibility of the label (Table 1).[13] In all cases, adding contact information to the seal significantly increased the credibility rating. This supports the findings from earlier focus group research (Teisl et al. 2000, p. 11, 14–15, 17, 23) in which participants stated that the credibility of an eco-seal could be greatly enhanced by the inclusion of contact information.

Table 1. Average ratings across levels of information.[a]

	Certifying organization			
	FSC	ME	EPA	Sierra Club
Credibility[b]				
Eco-seal only	3.20 A	3.21 A	3.27 A	3.40 A
Eco-seal with contact information	3.47 B	3.54 B	3.46 B	3.62 B
Eco-seal with contact information and summary score	3.15 C	3.49 B	3.05 C	3.51 B
Eco-seal with contact information, summary score and table	3.66 D	3.46 B	3.46 B	3.58 B
Environmental[c]				
Eco-seal only	3.36 A	3.40 A	3.28 A	3.59 A
Eco-seal with contact information	3.61 B	3.60 B	3.34 A	3.72 A
Eco-seal with contact information and summary score	3.18 C	3.79 B	2.66 B	3.62 A
Eco-seal with contact information, summary score and table	3.95 D	3.26 C	3.33 A	3.23 B
Satisfaction with information[d]				
Eco-seal only	2.35 A	2.21 A	2.32 A	2.54 A
Eco-seal with contact information	2.61 B	2.61 B	2.48 B	2.78 B
Eco-seal with contact information and summary score	2.55 B	2.76 B	2.34 B	2.94 C
Eco-seal with contact information, summary score and table	3.25 C	3.07 C	3.00 C	3.26 D
Likelihood-to-buy[e]				
Eco-seal only	3.82 A	3.93 A	3.67 A	3.88 A, B
Eco-seal with contact information	3.94 A	3.97 A	3.90 B	3.88 A, B
Eco-seal with contact information and summary score	3.66 B	3.92 A	3.32 C	4.03 A
Eco-seal with contact information, summary score and table	4.11 C	3.71 B	3.72 A	3.77 B

[a] Results sharing a letter across information treatments (down a column) are not significantly different from each other.
[b] Rating scale runs from 1 to 5 where 1 denotes not credible and 5 denotes very credible.
[c] Rating scale runs from 1 to 5 where 1 denotes not eco-friendly and 5 denotes very eco-friendly.
[d] Rating scale runs from 1 to 5 where 1 denotes not enough information, 3 denotes just enough information and 5 denotes too much information.
[e] Rating scale runs from 1 to 5 where 1 denotes highly unlikely, 3 denotes no opinion either way and 5 denotes highly likely.

Adding summary score information *decreased* the labels' credibility, even when the environmental scores are high (although this effect is only significant for the FSC and EPA labels). Seemingly, adding the summary score increased the respondents' level of skepticism about the overall information. This type of response has been observed in other contexts; simplified, positive information can *negatively* affect consumer perceptions of products if the information is incongruous with the consumer's prior perceptions of the product (Levy, Derby, and Roe 1997, p. 2, 34; Teisl, Roe, and Levy 1999). Similarly, it may be that respondents here distrust the presentation of quantitative information that is accompanied by only vague information of how that quantitative information was derived. This reaction, in which consumers seem to penalize products for displaying information appearing incongruent or overkill, may occur when consumers invoke a "schemers' schema" (Wright 1986) where the information causes the consumer to shift cognitive gears from processing product communications (advertising, labeling) to second-guessing the motivations behind these communications.

In general, adding the table of environmental scores increased the labels' credibility, although this effect is only significant for the FSC and EPA labels. In all cases, the most detailed label is seen as more credible than the simple eco-seals. However, in three of the four cases, the most detailed label is not significantly more credible than the eco-seal with contact information. For the FSC label, the table greatly increased the label's credibility, probably due to the fact that the FSC table disclosed a relatively high environmental score and the credibility for the FSC baseline is relatively low. This seems to indicate that unless a firm has a relatively high environmental score there may be no additional credibility benefit in providing more detailed information (over the eco-seal with contact information).

Environmental Ratings .—One potential measure of the effectiveness of an information policy is if consumers can adequately rank competing products by key attributes. Such attribute-level ranking can be an important intermediate input into the consumer choice process (Lee and Geistfeld 1998). For all four certifiers, altering the amount of information on the label significantly altered the environmental rating of the product. Adding contact information to the eco-seal increased the perceived "eco-friendliness" of the product; however, this effect was only significant for the FSC and ME cases. Adding relatively low summary scores (the FSC and EPA cases) significantly decreased the perceived "eco-friendliness." In all four cases, adding the table of environmental scores significantly altered respondents' ratings of the products in the correct direction from the summary score situation. That is, respondents viewing a table of higher (lower) environmental scores correctly provided a relatively higher (lower) environmental rating.

One goal of economic and consumer science research is to understand how consumers evaluate products when faced with incomplete or imperfect information (see Lee and Olshavsky, 1997, for a recent review of this literature). In this experiment, the eco-seal with contact information does not provide any explicit environmental score. However, we presume that respondents form some *a priori* expectation of what the eco-seal means in terms of such a score. Analyzing significant differences in respondent reactions between the eco-seal case, and the cases when environmental scores are provided, may provide some insights about how respondents' perceive the eco-seal in terms of an environmental score.

Respondents' environmental ratings of the eco-seals (with contact information) are significantly higher than the ratings given by respondents when scores of 75 (FSC), 72 (ME) and 67 (EPA and SC) are presented. In addition, respondents' ratings of the eco-seals are significantly lower than the ratings given by respondents when a score of 91 (FSC) is presented. The rating for the eco-seal is not significantly different than the ratings given when scores of 91 (ME) and 82 (EPA and SC) are presented. Thus, it appears that respondents only viewing an eco-seal have an expectation that the eco-sealed product has obtained a score centered in the mid-eighties.

Satisfaction with the Level of Information .—For all four certifiers, altering the amount of information on the label significantly impacted respondents' level of satisfaction with the level of information. In all cases, adding contact information to the eco-seal increased respondent satisfaction with the level of information. Except for the SC case, summary scores did not significantly alter respondent satisfaction with the level of information. However, in all four cases, adding the table of environmental scores significantly increased respondent satisfaction relative to the summary score situation. When we compare the most detailed labels with all the alternative information treatments we find that respondents were most satisfied with the most detailed labels, irrespective of certifier.

Likelihood-to-Buy.—For all four certifiers, altering the amount of information on the label significantly altered the respondents' stated likelihood-to-buy. However, adding contact information to the eco-seal significantly increased the likelihood-to-buy only for the EPA case. Adding relatively low summary scores (the FSC and EPA cases) significantly decreased the likelihood-to-buy. In all cases, adding the table of environmental scores significantly altered the respondents' likelihood-to-buy from the summary score situation. Products with relatively low environmental scores (FSC and SC cases) had lower likelihood-to-buy ratings compared to the summary score case; products exhibiting relatively high scores (ME and EPA cases) had higher likelihood-to-buy ratings.

Experiment I—Tests Across Certifiers

Credibility.—There are significant differences in the credibility scores across certification organizations for the simple eco-seal (Table 2). In general, the SC label is seen as the most credible and the FSC label as the least credible. This effect may be due to differences in respondent familiarity with the organizations. Psychology research indicates that the credibility of an information source is greatly influenced by a person's familiarity with the source (Brown et al. 2002); this familiarity effect was also observed in earlier focus group

Table 2. Average ratings across certifiers[a].

	Certifying organization			
	FSC	ME	EPA	Sierra Club
Credibility[b]				
Eco-seal only	3.20 A	3.21 A	3.27 A, B	3.40 B
Eco-seal with contact information	3.47 A	3.54 A	3.46 A	3.62 A
Eco-seal with contact information and summary score	3.15 A	3.49 B	3.05 A	3.51 B
Eco-seal with contact information, summary score and table	3.66 A	3.46 B	3.46 B	3.58 A, B
Environmental[c]				
Eco-seal only	3.36 A	3.40 A, C	3.28 B	3.59 C
Eco-seal with contact information	3.61 A	3.60 A	3.34 B	3.72 A
Eco-seal with contact information and summary score	3.18 A	3.79 B	2.66 C	3.62 D
Eco-seal with contact information, summary score and table	3.95 A	3.26 B	3.33 B	3.23 B
Satisfaction with information[d]				
Eco-seal only	2.35 A	2.21 A	2.32 A	2.54 B
Eco-seal with contact information	2.61 A	2.61 A, C	2.48 B	2.78 C
Eco-seal with contact information and summary score	2.55 A	2.76 B	2.34 C	2.94 D
Eco-seal with contact information, summary score and table	3.25 A	3.07 B	3.00 B	3.26 A
Likelihood-to-buy[e]				
Eco-seal only	3.82 A	3.93 A	3.67 B	3.88 A
Eco-seal with contact information	3.94 A	3.97 A	3.90 A	3.88 A
Eco-seal with contact information and summary score	3.66 A	3.92 B	3.32 C	4.03 B
Eco-seal with contact information, summary score and table	4.11 A	3.71 B	3.72 B	3.77 B

[a] Results sharing a letter across certifiers (along a row) are not significantly different from each other.
[b] Rating scale runs from 1 to 5 where 1 denotes not credible and 5 denotes very credible.
[c] Rating scale runs from 1 to 5 where 1 denotes not eco-friendly and 5 denotes very eco-friendly.
[d] Rating scale runs from 1 to 5 where 1 denotes not enough information, 3 denotes just enough information and 5 denotes too much information.
[e] Rating scale runs from 1 to 5 where 1 denotes highly unlikely, 3 denotes no opinion either way and 5 denotes highly likely.

research (Teisl et al. 2000, p. 4, 11, 13–14, 17, 23). Importantly, there are no differences in credibility across certifiers for the eco-seal with contact information case. Thus, it seems that any differences in credibility due to this familiarity effect can be effectively mitigated by the presence of contact information.

The credibility of the labels are different across certifiers in the presence of more specific environmental information such as summary scores or more detailed tables of scores. In general, when summary scores are displayed, then the labels that display higher environmental scores are seen as significantly more credible than those that exhibit lower scores. Specifically, the ME (score = 91) and SC (score = 82) labels are viewed as more credible than either the FSC (score = 75) or EPA (score = 67) labels. However, the two certifiers with the lowest credibility ratings in the summary score case (FSC and EPA) also have the lowest credibility ratings under the contact-information-only scenario (when no environmental scores are provided). Thus, the positive association between the environmental score and the credibility of the label may be spurious. When making a similar comparison with the detailed environmental tables, we find that the product with the highest credibility rating (FSC) *is* associated with the highest environmental score (score = 91). However, this credibility rating is not significantly different than the SC product with the lowest environmental score (67).

Environmental Ratings .—There are significant differences in respondents' environmental perceptions of the product across certification organizations for each level of information disclosure. Looking at the results for "eco-seal only" we find that the SC certified product is seen as environmentally better than either the EPA or FSC certified products, and that the FSC and ME certified products are seen as environmentally better than the EPA product. Note that this is different than the familiarity effect found earlier, where we found that EPA was rated relatively high in terms of credibility. It appears that respondents view EPA as a credible source of information, but may not view the agency as necessarily ensuring the most environmentally sound product. When looking at the results for eco-seal with contact information, we find that the EPA certified product was still viewed as the least eco-friendly product but that all the other certifiers were seen as equivalent. Adding the contact information mitigated the observed differences across the other certifiers found under the eco-seal baseline case. In general, when summary scores or more detailed tables are displayed, then products that display higher environmental scores are rated as significantly better in terms of the environment. Specifically, the products are correctly ranked by their environmental scores. Hence, respondents' were able to correctly use the more detailed environmental information (summary and detailed tables of scores) to rank products with respect to their environmental profiles.

Satisfaction with the Level of Information.—There are significant differences in respondents' satisfaction with the level of information across certification organizations for each level of information disclosure. Looking at the results for "eco-seal only," we find that respondents were most satisfied with the level of information provided by the Sierra Club (significantly higher than the other three certifiers). This is interesting given that the actual amount of information on the eco-seals was the same. When looking at the results for "eco-seal with contact information" we find that respondents were least satisfied with the amount of information on the EPA label and most satisfied with the Sierra Club label. In general, when summary or table scores are displayed, then respondents are most satisfied with the Sierra Club labels and the products that display a higher environmental score.

Likelihood-to-Buy.—There are significant differences in respondents' environmental perceptions of the product across certification organizations for the simple eco-seal treatment, for the summary score treatment, and for the detailed table of scores treatment. However, there are no differences across certifiers for the eco-seal with contact information treatment. Again, comparing the eco-seal with contact information treatment against the simple eco-seal treatment seems to indicate that adding contact information mitigates any initial differences in respondent ratings of products.

In general, when summary scores or more detailed tables are displayed, then products that display higher environmental scores, or are certified by the Sierra Club, are more likely to be bought. For the more detailed labels, the likelihood-to-buy decision seems to reflect a joint effect of the label's perceived credibility and the perceived environmental rating. For example, except for the Sierra Club label, the likelihood-to-buy score generally reflects the environmental ranking of the products. With the Sierra Club label, the likelihood-to-buy score is higher even though the EPA labeled product actually displayed a higher environmental score; this reaction may reflect the relatively higher perceived credibility of the Sierra Club label.[14]

Experiment II

Comparing Choice with and without Eco-Labeling.—In all but one case, we find that presenting environmental information to respondents significantly affects their choice of product (Table 3). When no environmental information is presented (i.e., treatment 1—only prices are given) then most respondents choose the low price product. Except for the case when Products A and B both exhibit an FSC eco-seal, presenting environmental information reduces respondent choice of the low-price, least environmental product. The above comparisons also indicate that the effect of an eco-seal may be contingent upon the certifying organization. The EPA and SC eco-seals had an impact on choice where the FSC eco-seal did not. This may be due to the EPA and Sierra Club being either more familiar to respondents (both the EPA and Sierra Club are more familiar than the FSC) or seen as more credible (Experiment I indicates that the Sierra Club is viewed as more credible).

Comparing Choice across Alternative Eco-Labeling Treatments.—When we compare the voluntary standard eco-seal treatments (treatments 2–4) with the equivalent nonstandard treatments (treatments 5–7) we find that having standard certifiers increased the purchases of Product A but did *not* alter purchases of Product B (Table 4). However, having nonstandard certifiers increased consumer choice of Product B. Apparently, when respondents view multiple products bearing a common environmental seal and different prices they assume that the environmental characteristics of the products are similar and are not willing to pay a price differential between the two certified products. However, when respondents view a similar situation with competing environmental seals they assume that the environmental characteristics of the higher priced product are better and, at least some of these individuals, are willing to pay the price differential between the two certified products. When we make a similar comparison among more detailed label [voluntary nonstandard (treatments 8–10) versus standard (treatments 11–13)] treatments we do not find a similar result. The increased level of detail counters any differential impact across standard versus nonstandard certifier.

This impact of increased detail is also evident when making comparisons within the voluntary standard and non-standard treatments but across level of detail. That is, when we compare voluntary nonstandard eco-seal versus detailed label treatments (5 with 11, 6 with 12, and 7 with 13) and standard eco-seal versus detailed label treatments (2 with 8,

Table 3. Effect of different labeling treatments on respondents' choice.

		Percent choosing product			Tests of equivalence[a]	
Number	Information treatment	A	B	C	χ^2	P
1	No environmental labels	37	4	59	NA[b]	NA
Voluntary eco-seals, standard certifier						
2	A = EPA; B = EPA; C = Blank	70	7	23	19.8	0.00
3	A = FSC; B = FSC; C = Blank	49	7	43	3.2	0.20
4	A = SC; B = SC; C = Blank	61	9	30	10.4	0.00
Voluntary eco-seals, non-standard certifier						
5	A = EPA; B = FSC; C = Blank	44	28	28	20.4	0.00
6	A = EPA; B = SC; C = Blank	39	30	31	19.3	0.00
7	A = FSC; B = SC; C = Blank	43	29	28	21.1	0.00
Voluntary detailed labels, standard certifier						
8	A = EPA; B = EPA; C = Blank	38	34	29	22.5	0.00
9	A = FSC; B = FSC; C = Blank	35	44	22	35.6	0.00
10	A = SC; B = SC; C = Blank	40	39	21	31.1	0.00
Voluntary detailed labels, non-standard certifier						
11	A = EPA; B = FSC; C = Blank	36	36	28	24.8	0.00
12	A = EPA; B = SC; C = Blank	44	35	21	27.4	0.00
13	A = FSC; B = SC; C = Blank	30	47	23	35.0	0.00
Compulsory detailed labels, standard certifier						
14	A = EPA; B = EPA; C = EPA	37	41	22	31.3	0.00
15	A = FSC; B = FSC; C = FSC	28	46	26	32.3	0.00
16	A = SC; B = SC; C = SC	37	34	30	21.8	0.00

[a] Tests of equivalence are between the alternative labeling treatments and the no label control (Treatment 1).
[b] NA denotes not applicable.

Table 4. Tests of equivalence between various labeling treatments.

	χ^2	P
Standard versus nonstandard certifier—Voluntary, eco-seal		
Treatment 2 versus treatment 5	11.46	0.00
Treatment 2 versus treatment 6	17.79	0.00
Treatment 3 versus treatment 5	11.49	0.00
Treatment 3 versus treatment 7	10.52	0.00
Treatment 4 versus treatment 6	12.26	0.00
Treatment 4 versus treatment 7	7.99	0.01
Standard versus nonstandard certifier—Voluntary, detailed label		
Treatment 8 versus treatment 11	0.05	0.97
Treatment 8 versus treatment 12	1.08	0.58
Treatment 9 versus treatment 11	1.18	0.55
Treatment 9 versus treatment 13	0.27	0.87
Treatment 10 versus treatment 12	0.16	0.92
Treatment 10 versus treatment 13	2.23	0.33
Eco-seal versus detailed label—Voluntary, nonstandard certifier		
Treatment 5 versus treatment 11	3.37	0.18
Treatment 6 versus treatment 12	0.67	0.71
Treatment 7 versus treatment 13	7.43	0.02
Eco-seal versus detailed label—Voluntary, standard certifier		
Treatment 2 versus treatment 8	22.59	0.00
Treatment 3 versus treatment 9	27.81	0.00
Treatment 4 versus treatment 10	14.84	0.00
Compulsory versus voluntary—Standard certifier, detailed label		
Treatment 8 versus treatment 14	0.68	0.79
Treatment 9 versus treatment 15	0.58	0.74
Treatment 10 versus treatment 16	1.30	0.52

3 with 9, and 4 with 10) we find that detail is less important with nonstandardized certifiers but is important when certifiers are standardized. In this latter case, increased label detail allows for greater differentiation in product choice; more detailed disclosure; and increased consumer choice of the high price, environmentally better product. The bottom line is that more detailed disclosures seem to assist consumers significantly better than simple eco-seals; the eco-seal format simply did not allow identification of the most environmentally benign product except when certifiers and prices are different and positively correlated.

When we compare similar voluntary and compulsory label treatments (8 with 14, 9 with 15, and 10 with 16) we find that the move from voluntary to compulsory labeling did not significantly alter choice. Apparently, respondents were able to correctly infer that the lack of environmental information provided on Product C was a signal that this product performed relatively poorly on this characteristic. This would indicate that detailed voluntary labeling regimes might work as well as a similar compulsory labeling policy. However, we did not consider scenarios in which the voluntary disclosure of detailed information did not perfectly correlate with the actual environmental profile of the product. Hence, we do not know how consumers would fare under voluntary programs in the presence of vague or misleading environmental marketing claims. The use of misleading environmental claims is not trivial; Kangun et al. (1991) found that approximately 50% of environmental advertising was misleading or deceptive.

Market Implications .—Many of the political maneuvers surrounding information disclosure regulations may be driven by firms' views of how such disclosure might impact market share (Roe et al. 2001). We provide some initial exploration of this question by analyzing how market shares of the three firms in this hypothetical market perform under different disclosure regimes. We do preface this discussion with the obvious point that market dynamics and strategy will depend greatly on the number of firms and the relative strengths of each firm engaged in the market. The hypothetical market under study here has only three firms, and each firm has a unique strength. If more than one firm competed for each market niche (e.g., lower in price and better for the environment) the impact of disclosure regimes on market share could greatly change. Nonetheless, we feel the analysis below is enlightening as an example of the interplay between market share and information regulation.

In terms of firm-level strategies, the firm marketing product A, the mid-priced, mid-environmental product experiences greatest market share when the environmental characteristics of the product are disclosed only through a standardized eco-seal. Under this regime, Product A can be differentiated from Product C, the low-price, least environmentally benign product but not differentiated from Product B, the high-price, most environmentally benign product. A medium priced, environmentally sound firm's market share seems to benefit from environmental labeling as long as the disclosure is standardized and relatively less explicit.

In contrast, the firm marketing product B experiences greatest market share when the environmental characteristics of the product are directly comparable via detailed disclosure. Apparently the environmental characteristics of the product are important enough that a strategy of disclosing B's relatively strong environmental record would be rewarded. A high priced firm with a strong environmental record would likely prefer regimes with detailed disclosure of standardized environmental information. At a minimum, they would advocate the development of nonstandardized eco-seals.

Finally, the firm marketing product C experiences greatest market share when the environmental characteristics of the product are not disclosed at all. The strategic implication for the firm marketing the low-priced product is simple: avoid compulsory environmental disclosure and focus solely on price. Moving from a state of no disclosure to any policy of environmental disclosure erodes product C's market share. No doubt this result will be sensitive to the relative price of the products and consumers' willingness to pay for environmental attributes.

Conclusions

In the debate surrounding forest certification, some have argued that the current lack of consumer response to forest products which are labeled as environmentally certified may indicate that consumers don't really care about, or at least are not willing to pay more for, these types of products. Although the above explanation is potentially valid, it is not necessarily true. One alternative explanation is that consumers do care about, and are willing to pay for, environmentally certified forest products, but the current state of labeling these products is slowing the development of this market. Research in other markets, using actual market data, has indicated that well-designed health (Ippolito and Mathios 1990, 1996, p. 119–159; Teisl, Bockstael, and Levy 2001) and environmental (Teisl, Roe, and Hicks 2002) labeling can significantly alter consumer and producer behavior.

The results here suggest that U.S. consumers do value the environmental benefits created from more environmentally benign forest management practices. Thus, consumer-driven purchases could potentially support a future of environmentally benign forest management practices with less reliance upon other policy alternatives. However, the results also suggest that the current state of forest product labeling, where firms can (voluntarily) label their products with simple eco-seals from a variety of certification organizations, may not be the most effective method of communicating the environmental attributes of forest products to consumers. Parties who are interested in the long-run success of these programs (e.g., environmental organizations, forest products manufacturers and environmental certifiers) need to consider altering current labeling approaches.

The results indicate that labels that display detailed information about environmental attributes would be more beneficial for consumers (and environmentally sensitive forest product manufacturers) than simple eco-seal policies. Among the labeling approaches studied here, eco-seals are the least credible, provide individuals with the lowest level of satisfaction and, in general, do not allow consumers an adequate basis for product differentiation. However, we also find evidence that seemingly marginal changes (e.g., adding contact information) can significantly improve the performance of simple eco-seals across several different measures (e.g., credibility).

An alternative approach to providing more detail on labels is for certifiers to administer an informational campaign to educate consumers as to the goals and means of the certification program. Currently, most respondents are not familiar with eco-labeled wood products and are either unfamiliar or just do not prefer the two biggest certifying organizations.[15] An informational campaign has the potential to increase the number of individuals familiar with the organization (thus increasing the credibility of a simpler label). The campaign may also increase Americans' desires to put a specific certifier in charge of forest product certification programs. Given consumers' low level of familiarity with the current certifiers of forest products, their low level of understanding of forest management concepts, and the complex nature of forest management and its many possible consequences, it is likely that any such information campaign would need to be significant.

Our results also suggest that some voluntary labeling regimes can yield benefits similar to compulsory disclosure; however, we fail to consider scenarios in which the voluntary disclosure of information by a firm did not perfectly correlate with the actual environmental profiles of the products. Hence, we do not know how consumers would fare under voluntary labeling in the presence of vague marketing claims. It is unlikely that individual firms or certifiers would ever find it privately beneficial to organize and display product information in a standardized format that would also benefit consumers.[16] One might envision a broad set of consumer reactions to missing disclosure data in the presence of vague and contradictory marketing claims; this would be an excellent avenue of future research. One should note that an additional benefit of more detailed environmental labels is that, unlike eco-seals, they may help restrict the seller's ability to make such misleading and false environmental claims and may help improve the credibility of firms' privately sponsored communications.

One should also be mindful of the hypothetical nature of the experiments. First, using the mail survey approach may have allowed respondents to evaluate the labels more fully, and with potentially fewer distractions, than they would in an actual purchase setting. Second, externally validated experiments indicate that when respondents do not face a real budget constraint, they are not as sensitive to price differences as they are in real markets. While progress has been made in calibrating environmental intentions stated in surveys to real world behavior (Champ and Bishop 2001, Byrnes et al. 1999, Bala et al. 1998), this survey has no real-world validity comparison. Further caution is warranted because hypothetical biases may be exacerbated when the respondent has little experience with the product in question and, to date, respondents have had little real-world experience in choosing among certified forest products.

Endnotes

[1] FSC is not a certification agency but an accreditor of certification agencies; each FSC-approved certifier must adhere to FSC's Principles and Criteria.

[2] Importantly, not all consumers in a market need to be affected by the information program to alter markets; only a subset of consumers need to respond to the information to affect producer behaviors (Moorman 1998, Dunn and Ray 1980, Capon and Lutz 1979).

[3] Many of the production-side questions are similar, or intimately linked, to the labeling (communication) issues. For example, questions about what information is presented to the consumer are related to the standards and criteria used in the certification.

[4] We note that the optimal form of an information policy may differ across consumers and products (Teisl, Bockstael, and Levy 1996, p. 171-182).

[5] We use the term "voluntary" in the regulatory sense; we do note that other pressures (market pressures from consumers or retailers, or nonmarket pressures from nongovernmental organizations) can make a voluntary program work similarly to a mandatory one.

[6] Green Seal is a nonprofit environmental organization that awards a "Green Seal of Approval" to products that cause relatively less harm to the environment.

[7] One of the main reasons that the FSC was created was to standardize certification systems. However, just because a forest product retailer is certified by an FSC-approved agency (thus permitting use of the FSC logo) doesn't mean that the FSC logo is actually displayed. For example, in the Early Fall 1999 product catalog from the National Wildlife Federation, they market a "SmartWood Bread Board." Nowhere within the text describing the product or within the text describing the SmartWood program is there any mention of FSC, a display of the FSC logo, or any mention that SmartWood is an approved FSC auditor. So a consumer deciding between two breadboards, one displaying an FSC logo and the other displaying text about the SmartWood program, would not know that these two products represent similar forest management practices. Even though the FSC helps standardize the production issues of certification, the information presented to the consumer is not necessarily standardized.

[8] Many individuals who live within driving distance of large tracts of publicly managed forests hold preferences for specific forest-management practices (Boyle and Teisl 1999, p. 6–10). However, given that most potential forest product customers are not within driving distance of the forests receiving the altered management practices and therefore are unlikely to ever derive direct benefit from visiting the forest, demand for specific practices may be limited.

[9] All analysis is weighted to correct for the oversampling.

[10] The surveys were mailed under a nonprofit organization permit, and thus we were not supposed to receive undeliverable returns (unlike a first class mailing). As a result, the count of 36 undeliverables is likely to be an underestimate of the actual number of undeliverables.

[11] We did not include the SFI program as an alternative because when we designed the experiment there was no indication that AFPA was going to develop a label as part of the SFI program. In addition, previous research had indicated that most consumers did not want an industry-sponsored certification program.

[12] Marginal effects can be calculated to ease comparison of parameters. However, calculated marginals usually hold all other variables in the equation at their means. In our case, all of the variables are dummy variables, and it would be inappropriate to hold all these other variables at their means. One possibility is to calculate the marginals while coding all the other variables to their appropriate values (either 0 or 1); the procedure used here is a variant of this approach.

[13] For simplicity, parameter estimates and tests of parameter equivalence are not reported but are available from the author.

[14] One should not necessarily interpret the relatively low ratings of the EPA labels as implying that respondents generally dislike/disapprove of all government-sponsored certification schemes. In another section of the survey, we find that certification programs sponsored by federal agencies were preferred over alternative organizations (e.g., industry or nongovernmental). However, we also found that most respondents who favored a federal approach preferred that the USDA Forest Service administer the certification scheme rather than the EPA.

[15] Only about 5% of respondents stated that they wanted either the FSC or AFPA to administer and enforce an eco-labeling program for forest products.

[16] Gehrig and Jost (1995) note that, in some cases firms may find it beneficial to impose some type of self-regulation of information provision. However, their results focus on experience attributes; credence and search attributes may have unique aspects that affect self-regulation incentives.

Literature Cited

AFPA (American Forest & Paper Association). 2002. Sustainable Forestry Initiative Program: Program overview, standard, verification/certification principles and procedures, and qualification criteria for verifiers. 44 p.

Bala, M.V., L.L. Wood, S.C. Cates, and S.P. Gambin. 1998. Predicting participation intentions for optional energy services. Resour. Energ. Econ. 20(3):287–301.

Bei, L.T., and R. Widdows. 1999. Product knowledge and product involvement as moderators of the effects of information on purchase decisions: A case study using the perfect information frontier approach. J. Consum. Aff. 33(1):165–186.

Boyle, K.J., and M.F. Teisl. 1999. Public preferences for timber harvesting on private forest land purchased for public ownership in Maine. Misc. Rep. 414. Maine Agric. and For.Exp. Sta. Orono. 18 p.

Brown, A.S., L.A. Brown, and S.L Zoccoli. 2002. Repetition-based credibility enhancement of unfamiliar faces. Am. J. Psychol. 115(2):199–209.

Byrnes, B., C. Jones, and S. Goodman. 1999. Contingent valuation and real economic commitments: Evidence from electric utility green pricing programmes. J. Environ. Plan. Manage. 42(2):149–166.

Capon, N., and R.J. Lutz. 1983. The marketing of consumer information. J. Market. 47(Summer):108–12.

Champ, P.A., and R.C. Bishop. 2001. Donation payment mechanisms and contingent valuation: An empirical study of hypothetical bias. Environ. Resour. Econ. 19:383–402.

Cumbo, D., and R. Smith. 2001. Forest products certification in 2001: Issues and answers. Cent. for For. Prod. Market. and Manage. Dep. of Wood Sci. and For. Prod., Virginia Polytech. Inst. and State Univ. , Blacksburg. 21 p.

Coupey, E. 1994. Restructuring: Constructive processing of information displays in consumer choice. J. Consum. Res. 21(1):83–99.

Dunn, D.A., and M.L. Ray. 1980. A plan for consumer information system development, implementation, and evaluation. Adv. Consum. Res. 7:250–254.

Ford, G.T., and R.A. Smith. 1987. Inferential beliefs in consumer evaluations: An assessment of alternative processing strategies. J. Consum. Res. 14(December):363–371.

Forsyth, K., D. Haley, and R. Kozak. 1999. Will consumers pay more for certified wood products? J. For. 97(2):18–22.

FSC (Forest Stewardship Council). 1998. Reaching the critical mass. FSC U.S. Init. Update 2(2).

Gehrig, T., and P.J. Jost. 1995. Quacks, lemons, and self regulation: A welfare analysis. J. Reg. Econ. 7:309–325.

Haener, M.K., and M.K. Luckert. 1998. Forest certification: Economic issues and welfare implications. Can. Public Pol. 24(Spec. Suppl. 2):S83–94.

Ippolito, P.M., and A.D. Mathios. 1990. The regulation of science-based claims in advertising. J. Consum. Pol. 13:413–445.

Ippolito, P.M., and A.D. Mathios. 1996. Information and advertising policy: A study of fat and cholesterol consumption in the United States, 1977–1990. Bur. of Econ. Staff Rep., Federal Trade Comm. Sept. 257 p.

Jaccard, J., and G. Wood. 1988. The effects of incomplete information on the formation of attitudes toward behavioral alternatives. J. Pers. Soc. Psych. 54(April):580–591.

Johnson, R.D., and I.P. Levin. 1985. More than meets the eye: The effects of missing information on purchasing evaluation. J. Consum. Res. 12(April):169–177.

KANGUN, N., L. CARLSON, AND S.J. GROVE. 1991. Environmental advertising claims: A preliminary investigation. J. Public Pol. Mark. 10(2):45–58.

KLEINMUTZ, D.N., AND D.A. SCHKADE. 1993. Information displays and decision processes. Psych. Sci. 4(4):221–227.

LEE, J., AND L.V. GEISTFELD. 1998. Enhancing consumer choice: Are we making appropriate recommendations? J. Consum. Aff. 32(2):227–251.

LEE, D.H., AND R.W. OLSHAVSKY. 1997. Consumers' use of alternative information sources in inference generation: A replication study. J. Bus. Res. 39(3):257–269.

LEVIN, I.P., R.D. JOHNSON, AND S.P. FARAONE. 1984. Information integration in price-quality tradeoffs: The effect of missing information. Mem. Cognition. 12(January):96–102.

LEVY, A.S., B.M. DERBY, AND B. ROE. 1997. Consumer impacts of health claims: An experimental study. Cent. for Food Safety and Appl. Nutrit., U.S. Food and Drug Admin., Washington DC. January. 36 p.

LYKE, J. 1996. Forest product certification revisited: An update. J. For. 94(10):16–20.

MERIDIAN INSTITUTE. 2001. Comparative analysis of the forest stewardship council and sustainable forestry initiative certification program: volume 1—introduction and consensus statement on salient similarities and differences between the two programs. Washington, DC. October. 64 p.

MITNICK, B.M. 1981. The strategic uses of regulation and deregulation. Bus. Horizons 24(2):71–83.

MITRA, A., AND J.G. LYNCH, JR. 1995. Toward a reconciliation of market power and information theories of advertising effects on price elasticity. J. Consum. Res. 21(March):644–659.

MOORMAN, C. 1998. The market-level impacts of information: competitive responses and consumer dynamics. J. Marketing Res. 34(February):82–98.

MOORMAN, C. 2001. Firm responses to consumer information policy. Handbook of marketing and society, Bloom, P.N., and G.T. Gundlach (eds.). Sage Publications, London. 543 p.

MOORMAN, C., AND R.J. SLOTEGRAAF. 1999. The contingency value of complementary capabilities in product development. J. Market. Res. 36(May):239–257.

NRCM (NATURAL RESOURCES COUNCIL OF MAINE). 2001. A comparison of the American Forest & Paper Association's Sustainable Forestry Initiative and the Forest Stewardship Council's Certification System. June. 8 p.

OZANNE, L.K., AND R.P. VLOSKY. 1997. Willingness to pay for environmentally certified wood products: A consumer perspective. For. Prod. J. 47(6):1–8.

OZANNE, L.K., AND R.P. VLOSKY. 1998. Environmental certification of wood products. *Women Natur. Resour.* 19(3):4–48.

PETERSON, J. 2001. Certification wars—Why the SFI Program is going to win. *Eng. Wood J.* Fall: 38–40.

ROE, B., M.F. TEISL, H. RONG ,AND A.S. LEVY. 2001. Characteristics of successful labeling policies: Experimental evidence from price and environmental disclosure for deregulated electricity services. J. Consum. Aff. 35(1):1–26.

ROE, B., A.S. LEVY, AND B.M. DERBY. 1999. The impact of health claims on consumer search and product evaluation outcomes: Results from FDA experimental data. J. Public Policy Mark. 18(Spring):89–105.

ROSS, W., AND E.H. CREYER. 1992. Making inferences about missing information: The effects of existing information. J. Consum. Res. 19(June):14–25.

SCAMMON, D.L. 1977. Information load and consumers. J. Consum. Res. 4(December):148–155.

SCHKADE, D.A., AND D.N. KLEINMUNTZ. 1994. Information displays and choice processes: Differential effects of organization, form, and sequence. Org. Behav. Hum. Dec. Proc. 57(3):319–337.

SCS (SCIENTIFIC CERTIFICATION SYSTEMS). 1995. The forest conservation program: Program description and operations manual. Oakland, CA. October. 72 p.

SIMMONS, C.J. ,AND J.G. LYNCH, JR. 1991. Inference effects without inference making? Effects of missing information on discounting and use of presented information. J. Consum. Res. 17(March):477–491.

STONE, A. 2001. Private landowner concerns with forest certification. Timber Produc. Oct.: 24-27.

TEISL, M.F., F. NEWMANN, J. BUONO, AND M. HERRMANN. 2000. Designing effective environmental labels for forest products: Results of focus group research. Maine Agric. and For. Exp. Sta. Misc. Rep. No. 420. 42 p.

TEISL, M.F., B. ROE, AND R.L. HICKS. 2002. Can eco-labels tune a market? Evidence from dolphin-safe labeling. J. Environ. Econ. Manage. 43:339–359.

TEISL, M.F., N.E. BOCKSTAEL, AND A.S. LEVY. 1996. Preferences for food labels: A discrete choice approach. *In* Proc., Strategy and policy in the food system: Emerging issues. Annu. Meet. of the North East Reg. Res. Proj. —165. Washington, DC. 247 p.

TEISL, M.F., B. ROE, AND A.S. LEVY. 1999. Eco-certification: Why it may not be a 'field of dreams' Am. J. Agric. Econ. 81(5):1066–1071.

TEISL, M.F., N.E. BOCKSTAEL, AND A.S. LEVY. 2001. Measuring the welfare effects of nutrition information. Am. J. Agric. Econ. 83(1):133–149.

VOGT, K.A., B.C. LARSON, J.C. GORDON, D.J. VOGT, AND A. FANZERES. 2000. Forest certification: Roots, issues, challenges, and benefits. CRC Press, Washington, DC. 374 p.

WINNEG, K., M.J. HERRMANN, A.S. LEVY, AND B. ROE. 1998. Label testing: Results of mall-intercept study. Consum. Inf. Disclosure Ser., The Reg. Assist. Proj. Gardiner, Maine. October. 75 p.

WRIGHT, P. 1986. Schemer schema: Consumers' intuitive theories about marketers' influence tactics. Adv. Consum. Res. 13:1–3.

[19]

Product-related environmental information to guide consumer purchases – a review and analysis of research on perceptions, understanding and use among Nordic consumers

Charlotte Leire*, Åke Thidell

International Institute for Industrial Environmental Economics – Lund University, P.O. Box 196, SE – 221 00 Lund, Sweden

Accepted 20 December 2004

Abstract

Informed product choices are a prerequisite for the greening of consumption. Nordic consumers are subject to an ever-growing amount of product-related environmental information that is available in a number of formats. The degree to which they use information is disputed. This paper reviews Nordic research on consumer perceptions, understanding and use of product-related environmental information. The majority of the publications point to a great number of consumers, who recognise, know and possess sufficient confidence in the main eco-labels and should therefore, be able to use them. However, this is not always the case. How the product-related environmental information works in the market place is dependent upon many determinants related to the purchasing moment. Therefore, in contrast to the majority of research on the topic, studies of the use of product-related environmental information suggests that the focus should be on the purchasing decision and take into account the dynamic context of the diverse purchasing situations.

Keywords: Product-related environmental information; Literature review; Nordic consumers; Purchase

1. Background

1.1. Information as a product policy instrument

The environmental performance of products has gained increased political attention during the last two decades and has resulted in restrictions of certain substances, mandated deposit–refund systems, advance disposal fees, environmental surcharges, voluntary agreements and, in particular, the introduction of informative instruments.

Informative instruments can be voluntary or mandatory, negative, neutral or positive. Mandatory negative information is primarily used for chemical products with hazardous properties. Voluntary neutral or positive schemes, providing product-related environmental information, are considered suitable for stimulating environmentally conscious production and consumption of goods and services due to their flexibility, low administrative and control costs, etc. The possibilities to influence consumers with information are frequently mentioned, sometimes as information in combination with other incentives [43,10]. Information is used to complement economic and regulatory instruments [43,9].

Positive information guides consumers by pointing out the environmentally preferable products among otherwise equal ones, while neutral information gives the consumer an opportunity to judge the products' environmental performance according to their individual priorities. The assumption is that product-related environmental information, in combination with preconditions such as environmental awareness, knowledge

* Corresponding author. Tel.: +46 46 2220281; fax: +46 46 2220240.

E-mail address: charlotte.leire@iiiee.lu.se (C. Leire).

and attitudes, will lead consumers to make informed choices when purchasing products. By promoting the demand for environmentally superior goods and gradually increasing the stringency of the requirements, environmental information should give incentives for the improvement of the performance of the products.

Influencing consumer behaviour is becoming a central priority in European environmental and consumer policy [35]. Consumers influence the entire product life cycle, which is particularly reflected by patterns of purchasing, use and product discard [9]. The European Commission stresses that "*consumers must have easy access to understandable, relevant, credible information either through labelling on the product or from another readily accessible source (e.g. consumer and environmental NGOs, websites, public authorities)*" [10]. Moreover, the lack of adequate information is claimed to be the most important obstacle for environmentally conscious product purchasing [9].

1.2. Approaching consumers in the context of product-related environmental information

Knowledge about how product-related environmental information is received and used by consumers[1] is an important aspect in the shaping of product policies. Consumers in the Nordic countries[2] are generally aware of the fact that products are associated with complex environmental problems. Among other reasons, they know this because they have been exposed to eco-labels and lists of environmentally benign products long before the concept of sustainable development became common knowledge. The increasing coverage of the main eco-labels occurs in parallel with an increase in consumer recognition [45].

The available base for product-related environmental information consists of the many different information schemes initiated by environmental NGOs, industries, industry associations or policy-makers. In the Nordic countries, certified eco-labelling schemes (ISO Type I)[3] offer licensing criteria for a growing number of products and product groups. Moreover, in addition to ISO Type I labels, a range of different environmental product declaration systems has emerged, exemplified by voluntary self-declarations issued by individual producers as well as a Nordic initiative for third-party certified environmental product declarations (ISO Type III). The voluntary self-declarations differ from self-declared claims (ISO Type II) because they provide more extensive information, often with a life cycle perspective, and are mostly intended for business-to-business communication. In many cases, the format of a self-declaration is the result of agreements within an industry sector. Compared to the certified ISO Type III declarations, self-declarations are generally not based on formal life cycle assessments, but provide both qualitative and quantitative data. Producers and retailers also use their own environmentally-oriented brands as a way to differentiate products. Overall, information systems are varied enough to potentially suit most types of consumers. However, despite high recognition and good consumer intentions, studies report the limited ability, interest and willingness of consumers to absorb and act upon the information. As a result, product-related environmental information has been repeatedly scrutinised in regards to its communication capacity.

With continuously increasing consumption and related material flows, there is cause to make product information even more accessible and relevant to consumers. For instance, the Swedish Ministry of Environment recently called for improvements in and the development of existing information systems and heightened attention to the role of product-related environmental information in sustainable consumption [32].

Consumer research, in a restricted sense, can prove inadequate for an appropriate policy formulation. Traditional approaches to study consumer decisions have been based on the assumption of consumers as rational decision-makers with a well-defined and stable system of preferences. These approaches have later been challenged, based on the view that the decision-making models with respect to the purchase of products are often based on heuristics that over-simplify the purchase decision process. Incorporating environmental considerations into these models is a challenging task for environmental policy-makers [35].

Therefore, instead of examining information from the perspective of the consumer as recipients of information, this study takes as its starting point the product-related information that is actually provided to the consumers.

1.3. Purpose

The results presented in this paper summarise key-findings of how consumers in the Nordic markets

---

[1] Consumers refer to end-consumers as well as professional purchasers in the private and public sectors.

[2] The Nordic countries are Denmark, Finland, Iceland, Norway and Sweden.

[3] The International Organization for Standardization (ISO) has established standards for environmental labels and declarations. Type I labelling, which is the traditional third-party certified and independent eco-labelling schemes, are defined in the standard ISO 14024 [56]. The self-declared claims (Type II), that are the short statements, symbols or graphics a producer can be used in order to indicate an environmental aspect of his/her product, component or packaging, are governed by the standard ISO 14023. Finally, the Environmental Declarations (Type III) are built according to certain methodologies, which are described in the technical report ISO/DTR 14025 [57]. It should be stated that most of the Environmental Declarations discussed in the text do not adhere to the standard and thus, are outside the categories defined by ISO.

perceive, understand and use voluntary environmental information on products. The study presents relevant research results and experiences from 1995 to mid-2003. The purpose of the study is to survey well-researched aspects, as well as issues that need further investigation. The results may contribute to the understanding and shaping of policy instruments based on product-related environmental information.

An important rationale for the Nordic focus is the unique, high diffusion of the product-related environmental information in the Nordic markets. Several policy initiatives and information systems have also been developed as joint-Nordic schemes. In addition, several relevant research projects have had a Nordic perspective. A small number of the results have been published in English. Through this article, the authors wish to give interested actors insight and information from research from all five Nordic countries.

1.4. Framework and method

The successful use of product-related environmental information can generally be defined by how the information fulfils its role as a policy instrument. Even if the ultimate role is to reduce the environmental stress, could the role of the information be expressed as changing behaviours, stimulating environmentally sound product development, educating the consumers, etc. The US-EPA developed a framework of specific metrics to measure effectiveness of environmental certification and labelling programmes [54]. In this framework, the goals of several programmes were grouped into five steps that can be used to evaluate the effectiveness of information systems and labelling schemes. The steps are:

- Consumer awareness;
- Consumer acceptance (credibility/comprehension);
- Consumer behaviour change;
- Manufacturer behaviour change; and
- End benefits.

The end benefits address reduced environmental stress or real improvements of the environmental conditions that are due to changed behaviours of consumers and manufacturers, which in turn are triggered by the informative policy instruments. The framework has previously been used for analysing the Nordic Swan scheme [4]. A pilot study has shown three steps of the framework that address consumers to be useful as an analytical framework for this application. These three consumer-related steps are explained below.

- Consumer awareness addresses underlying environmental issues, in general, and relations between products and their environmental impacts, in particular. In addition, this step includes the awareness of specific labelling programmes and their symbols.
- Consumer acceptance covers issues related to credibility of information, knowledge of information systems as well as the environmental attributes of products. The key points are the public's understanding of the relevant issues, understanding of connections between the relevant issues and product choices, accurate and clearly understood presentations of product attributes and what specific actions individuals can take in response to the information provided. Credibility and knowledge are particularly valid issues for labelling programmes.
- Consumer behaviour change deals with how readily consumers change their purchase patterns and actually choose environmentally benign products.

The study is based on a structured literature search covering surveys, scientific reports and journal articles. The following key words, translated into the respective Nordic languages, were used for the searches: consumer, consumption, purchase, environment, declaration, labelling, product, knowledge and understanding.

The literature search was supplemented with interviews of people with valuable and relevant knowledge and experience in order to check that no important research results were missed and to obtain their insights related to the reported research. The interviewees were selected using a snowballing technique.[4] Partners in Finland and Iceland conducted the literature search and interviews, in those two countries, in order to obtain data and information presented in their local languages.

The search revealed 90 applicable publications of which 54 were considered directly relevant.[5] Research on consumer willingness to purchase green goods does not always examine the role of the available product information. In our study, the relevance of the literature findings was determined by whether the scope comprised product-related environmental information and consumers' perception, understanding and use of it. The directly relevant studies were reviewed in regards to the messages conveyed and the methods employed.

2. Review of the literature

A variety of eco-labels for products and organic food, energy labels, environmental product declarations and producers' self-declared claims are common subjects of

[4] In essence, '*snowballing*', is the technique whereby experts are identified from research papers, reports, etc. and when they are interviewed, they are asked for the names of other experts they know who may also have relevant information.

[5] Not all of the 54, directly relevant studies, are cited in the reference list in this paper but are documented in a report [29].

the Nordic studies. To date, the predominant focus is on ISO Type I eco-labels. The Nordic Swan label, together with national labels for organic food (such as the KRAV-label and the Ø-labels[6]), holds a unique position in terms of available research.

The number of studies on environmental product declarations (voluntary self-declarations and ISO Type III) is just a fraction of the number that deals with the Type I eco-labels. The existing certified ISO Type III environmental product declaration scheme has not been the subject of any of the identified studies. The studies of environmental product declarations mainly cover private consumers' perceptions and understanding of proposed declaration formats, as well as professional users' understanding and use of existing self-declarations (without any certification or control system), that primarily focus upon the information and communication technology (ICT) products, textiles and building materials.

2.1. Consumer awareness

2.1.1. Awareness about specific information programmes

The dissemination of eco-labels in the Nordic countries is considered to be a success, as exemplified by a comparison between Norway and selected European countries [44]. The Swan label is reported to be well recognised (>90% of the consumers in Sweden, slightly less in Norway, Finland and Denmark) and is also generally regarded as trustworthy among Nordic consumers [37,25,48,31,46,6]. However, in Iceland, only about 50% of the consumers were reported to be familiar with the Swan label [28,36]. Both the turnover and the number of Swan labelled products are steadily increasing on the Nordic market. The main national labels for organic food and in Sweden the Good Environmental Choice[7] label follow a similar pattern. The EU Flower-label[8] is subject to fewer Nordic investigations; however, a notable increase in research on this labelling scheme has been documented.

Ambivalent findings are reported regarding the consequences of the wide range of labels available on the market. According to some authors, the great number of eco-labels has contributed to consumer confusion [21,16,50,8]. This is seen as a reason for the reduced credibility in all labels, stemming from difficulties in distinguishing certified eco-labelling from other types of labels [42].

2.1.2. Awareness about underlying environmental issues of products

In general, consumers hold a positive attitude towards green products and eco-labels [12]. Some claim that the more environmentally aware consumers make use of and appreciate environmental information to a higher degree [35,49,25]. However, another study claims that the link between environmental literacy and preferences for purchasing environmentally benign products is rather weak [15]. A number of studies report that consumers declare a need for more environmental information, in general, to be able to make greener product choices [1,34]. With regard to the studies that address the environmental awareness and literacy issue, it seems as if consumers, in general, have difficulties in relating environmental problems to products, in making distinctions between green and conventional products and in grasping the environmental benefits of purchasing eco-labelled products.

2.2. Consumer acceptance

2.2.1. The connection between relevant issues and product choices

Even though recognition of the main labels is generally considered high, consumers admit to have a weak comprehension of each label [14,53,18]. Apparently, many consumers who value product-related environmental information, do not differentiate between information on environmental, health and safety issues, but tend to view them holistically as an additional property of the product [47]. It has also been found that health is the main reason for consumers to choose eco-labelled products [33,19]. The proximity of the product to the consumer's body or to her/his own activities is a central factor in their utilisation of environmental information [14,37].

Most studies report that consumers generally trust, to a high degree, the main eco-labels. In the few cases where lower consumer trust has been reported, it is explained by insufficient information on who stands behind the schemes, how the schemes are organised, the environmental relevance of the product and in what ways labelled products are superior to conventional ones [18,5,42]. The studies do not show if the consumers, who express doubts regarding the eco-labelling systems, overlap with those who are less likely to consider environmental performance of products in the purchase decisions, nor if the expressed uncertainties are real barriers for a wider acceptance and use of the information.

Consumer perception on the performance quality of eco-labelled products as compared to their conventional

[6] The KRAV is the organic food label in Sweden, the Ø-label points systems are used for labelling of organic food products in Denmark and Norway.

[7] The Good Environmental Choice or the Falcon is an NGO, eco-label, launched by the Swedish Society for Conservation of Nature. It is similar to a Type I label.

[8] The EU Flower-label is a EU-wide Type I eco-labelling scheme, which is applicable in the Nordic countries in parallel with the Nordic Swan label.

C. Leire, Å. Thidell / Journal of Cleaner Production 13 (2005) 1061–1070 1065

alternatives, varies from equal to lower [24]. Moreover, consumers were repeatedly reported to perceive eco-labelled products as being more expensive [26,2], but claim to be willing to pay a premium price for such products [21,19].

2.2.2. Acceptance of different types of information formats and sources

Findings on consumer preferences in terms of the information format and the level of detail are somewhat inconsistent. Some claim that a majority of consumers appreciate the simplicity of logotypes as information carriers for most everyday products [19,34,46]. A diverging message, found in some literature sources, is that a number of consumers find the simple symbols insufficient and would like to obtain additional information [15,50]. As a way to meet the declared demand for more information, these studies propose expanded textual information, simple declarations explaining the significant environmental features of the products or a combination of labels and explanatory texts.

Environmental product declarations provide detailed information and often consider a product's whole life cycle. Consumers seem to perceive the opportunity to prioritise among environmental aspects as positive. Furthermore, comprehensive and quantitative information is reported to trigger positive judgements about the reliability of the information [38,41]. On the other hand, both private and professional consumers have insufficient understanding of the information provided by environmental product declarations [20,52,51] and need assistance in the form of user guides or training [41,11,38,23]. A reason for confusion is that multiple formats of environmental product declarations are in use for comparable products.

Findings on mandatory environmental product labels are less elaborate. Although not substantially dealt with in the reviewed literature, the main finding is that consumers are positive towards energy labels. Apparently, the reason is that consumers find the grading system helpful [15,22].

Studies of producers' self-declared environmental claims are limited, but thus far the findings seem to indicate a low degree of trust in such information. It is reported that consumers react negatively to the information that is issued by producers and similarly to information that is not certified by third parties [19,40]. In general, consumers prefer independently verified information [22].

2.3. Consumer behaviour

According to several segmentation studies [48,24,12], 20–50% of the consumers give priority to environmentally related properties of the products they purchase. Although, many studies show that consumers claim they have good intentions to buy environmentally benign products, for instance, through using product-related environmental information, the reports repeatedly conclude that there is a discrepancy between consumers' intentions and their behaviour. Consumers are often found to overestimate their use of product-related environmental information [27,3,25,5]. A significant portion of the literature was designed to identify the reasons why consumers do not use eco-labels. The most influential factors were found to be that consumers prioritise other aspects, such as price and quality, and that purchases are guided by habits [12]. Both private and professional consumers are reported to lack the time and financial resources for searching and examining environmental information during the purchase situation [17,33].

3. Analysis

3.1. Consumer awareness

According to the framework used for the review in this study, one important determinant for the successful use of product-related environmental information is consumer motivation to choose environmentally sound alternatives among otherwise equal products. An observation from the literature is that public awareness of the link between products, consumption and the caused environmental impacts is an area that requires further attention. A great number of individuals are still not perceptive to the environmental impacts of particular products and thus, are not likely to be interested in product-related environmental information.

Another determinant is the awareness and recognition of specific labelling programmes. Eco-labelling was, by far, the subject to the most research on the recognition of product-related environmental information. The main eco-labels were found to be very well recognised in the Nordic countries. Therefore, further research within this area may be redundant.

3.2. Consumer acceptance

A third determinant for consumers to use product-related environmental information is that the information is understood and accepted as important to them. On one hand, the literature findings point to a high recognition and knowledge of the main eco-labels and conversely to a confusion of what the eco-labels stand for among some consumers. However, the authors' interpretation is that the confusion caused by the numerous labels is overstated. The reason for this is the significantly higher recognition and knowledge of the main eco-labels when compared to the other labels. Consequently, confusion does not necessarily occur in the daily purchase situations. However, it would be valuable to investigate potential correlations between consumers who emphasis their confusion and distrust in the eco-labelling schemes and those consumers who are not likely to prioritise green products anyway. Based on the literature, this seems to be an important research area.

Credibility is mentioned as the fourth important determinant for product-related environmental information programmes. Two main factors seem to influence consumer trust. Firstly, the reliability of the information sources, which relates to consumers' perception and knowledge of the labelling organisation. Secondly, consumers' knowledge about the environmental attributes of products can influence their understanding of the environmental relevance of their own actions. The literature findings on this topic are that consumers generally perceive the main eco-labels as highly trustworthy, but lack knowledge about the sources of the information, i.e. the organisation behind the eco-labels. The knowledge of correlation between this determinant and the expressed trust is therefore, concluded to be weak. Moreover, consumers' knowledge of the environmental attributes of eco-labelled products was also found to be low. Diverse means to increase such knowledge among consumers need to be explored. We suggest further research in these areas in order to explore methods to build additional trustworthiness into the schemes.

Ultimately, the schemes would probably be in a stronger position if indications of impacts could be displayed so that they do not have to rely entirely on consumers' general environmental awareness or altruistic feelings for the environment. We believe that from this perspective, a potential approach could be to develop mechanisms providing feedback on shifts on market shares for eco-labelled products as well as estimations of environmental benefits to consumers. Positive feedback is also encouraged, based on the results of some of the studies [26,30].

Environmental marketing and in particular self-declared claims (ISO Type II) of producers are occasionally mentioned as complementary to the eco-labels [38] or as an alternative for product groups in which labelling is considered less suitable. Therefore, it is possible that we will see an increasing use of these claims in the future. However, the use of self-declared claims in environmental marketing is a poorly researched area. Data on current use and trends in the use of claims is notably missing in the literature. Moreover, as the relevance of these claims is regulated by both an international standard [55] and consumer protection legislation, it is important to gain a better understanding of how these control mechanisms work in practice and how producers justify their claims; which may influence the credibility of this information.

The type of format in which the information is presented affects consumer acceptance. Simple information carriers, such as eco-labels, seem to be desired by most consumers and are sufficient for decision-making. In contrast, the few studies on environmental product declarations clearly indicate that the level of detailed information provided by these declarations is perceived as far too complex to be useful in making purchase decisions. One immediate issue is that consumers (private and professional) face severe difficulties in interpreting and making use of the information those declarations provide. Although, the idea of declarations is that consumers should evaluate and compare alternatives based on their own preferences, most of these studies point to a need for enhanced utilisation of transparent guidelines and interpretation tools in order to assist consumers to make more informed choices. Curiously, even when consumers cannot interpret the detailed information of extensive declarations, such information is perceived as trustworthy and does build environmental credence toward the products.

The different kinds of environmental product declarations are, to a large extent, still a virgin area for research. In general, there is a need to better understand the role of both certified and uncertified environmental product declarations as policy instruments. Overall, environmental product declarations may be useful in the supply chain, especially for buyers who are obliged to consider the environmental aspects of products for aggregation purposes, more complex products or in product groups without eco-labelling criteria. There are also cases of producers adopting ISO Type III declarations as a substitute for other information systems, such as eco-labels [7]. Existing studies on the knowledge concerning declarations need to be supplemented with more specific studies on how consumers and professional buyers perceive, understand and use such declarations.

3.3. Consumer behaviour

The understanding is still weak regarding how to increase the use of existing product-related environmental information. The discrepancy between consumers' stated purchases of environmentally benign products and their actual behaviour, as well as the limited impact of product-related environmental information outside the green consumer segments, indicate the inherent limitations of information as a policy instrument.

We believe it is valuable to examine the role of guidance from sales personnel in order to stimulate the use of the eco-labels and to influence actual consumer behaviour. A limited number of studies [27,39] cover the role of retailers and sales staff in the process of supporting environmentally conscious consumers. These actors appear to serve as ecological gatekeepers. The findings indicate that generally, sales personnel have limited knowledge about eco-labels, but that they could potentially, help to guide consumers towards using product-related environmental information. This requires that sales personnel should be provided with appropriate knowledge about eco-labels and environmental performance of products. Similarly, the display of eco-labels and information about environmentally sounder products in shops may play a role for product choices and, as such, may function as a leverage point for this type of information. No studies were found that investigated these areas.

We suggest that comparisons or benchmarking studies of different information displays in combination with training of sales personnel could help identify approaches to guide consumers to more fully utilise the information in their purchase decision-making processes.

A main remark is that there is information in the market place that consumers are aware of but do not use to the extent they claim. More product-related environmental information is certainly needed for additional products and product groups, but currently the information is not the limiting factor for consumers to choose green products. The existing information is not fully utilised.

4. Discussion on the research methodologies

A substantial portion of the reviewed literature reaches the point of concluding that there is an inconsistency between consumers' intentions and actual behaviour. Most studies do not go further in their analysis. The method chosen by a particular study seems to play a vital role in furthering the knowledge on Nordic consumers in action.

In the reviewed literature, both quantitative and qualitative methodologies have been used and have generated different kinds of data. Examples of methodologies employed in the studies reviewed are questionnaires, qualitative methods such as personal interviews, group discussions, protocol interviews, telephone surveys and consumer panels. Among the 54, directly relevant studies, approximately a third had adopted an approach involving several steps and methods. Table 1 lists the methodologies and the knowledge generated that was considered relevant for their study. The most common approach employed was a survey or a combination of methods typically involving a survey. Approximately one tenth of the studies were based on in situ interviews.

One aspect of the methodologies employed in the reviewed literature was that the majority of up-to-date consumer surveys seemed to capture the consumer knowledge of certain information systems, perceived credibility, claimed usage, etc., from citizen-interviewees rather than from consumer-interviewees. Mostly, the

Table 1
Research methodologies utilised within the literature reviewed for this article

Methodology	Relevant focus area identified
Combination of survey and interviews or survey and focus groups or discussions, number of publications: 16	Acquisition of information Attitudes Demand General environmental awareness Knowledge Need for additional information Perception Recognition Trust Understanding Use Use in particular markets
Focus groups, number of publications: 4	Knowledge Level of information complexity Need for additional information Perception Preferences Understanding
In situ, number of publications: 6	Attitudes In situ factors that influence use Knowledge Perception Understanding Use
Interview, number of publications: 8	Attitude Knowledge Level of information complexity Need for environmental information Need for guidance Perception Preference Use
Literature review (information on food products), number of publications: 2	Interpretation Knowledge Understanding Use
Survey, number of publications: 18	Attitudes Environmental literacy Knowledge Level of consumer attention Need for additional information Perception Preferences Recognition Sufficiency of information Trust Understanding Use

interviews were performed in locations outside of the actual purchase situations and therefore, may not provide an accurate picture of the real decision-making processes. Furthermore, these studies have a tendency to aggregate the answers statistically, implying that consumers follow static and regular behavioural patterns. We believe that there is a potential to be realised in the existing product-related environmental information schemes through understanding the factors that, according to the segmentation model, would imply a mobility of individuals between consumer segments. That is, consumers are more dynamic and consider product-related environmental information differently in different purchase situations.

A suggestion on how to explore this potential is to adopt study methods that bring the investigation closer to the consumers, in action and to study how they use product-related environmental information in different purchasing situations and contexts. The answers about when individuals use and do not use product-related environmental information could bring to light different behavioural patterns and actions that vary with changing situations. These situations may be influenced by factors, such as:

- Available time and money;
- External stimuli like campaigns and information;
- The availability of eco-labelled products;
- Trust in the information provided; and
- Motivation and understanding of the relevance of different product choices.

An in situ approach was proposed for a broader study on information on food products [13], where the contextual conditions (high recognition and knowledge about the labels/information systems, trust, etc.) is similar to cases that are valid for product-related environmental information. The authors of this paper conducted pilot studies (protocol interviews), which produced interesting results. They suggested that further research should be carried out as close to the purchase situation as possible and should also include attitudinal factors. The in situ approach (exit-interviews and background survey) was later used in a full-scale survey [12] with the objective of coming closer to an understanding of the factors that determine the extent to which the information is effective. This kind of research on available product-related environmental information is a task for which a number of research disciplines will need to cooperate.

5. Conclusions

Eco-labels and especially the Nordic Swan label have a predominant place in Nordic research on consumers' perceptions, understanding and use of product-related environmental information. With regards to the analytical framework used in this study, the literature reviewed indicates that most of the determinants for a successful use of the information are in place in the Nordic market.

Overall, consumers express their preferences for environmentally benign products, which reveal that they have an awareness of the issues. In terms of eco-labels, the main ones are highly recognised and are well-known, and consumers perceive them as rather trustworthy. Yet, a number of reports conclude that consumers tend to overemphasise their purchases of eco-labelled products, which implies a limited consumer behaviour change.

Our conclusion is that, despite the fact that product-related environmental information is lacking for a range of products, available information has a potential to be further used for consumer guidance. In general, the current problem is not insufficient information. Other potential reasons seem to be that consumers consider the eco-label but decide not to choose the product, or that consumers do not use the information to the extent they state in surveys. The current knowledge gaps in Nordic research identified from using the analytical framework highlight the need for more insight into determinants for consumer acceptance of the information, such as trust and motivation, and for factors that result in consumer behaviour changes.

Consumer knowledge of environmental attributes of the products and the relevance of the labelling criteria for certain product groups seem to be relatively weak. Our conclusion is that further research is needed on how enhanced knowledge on the environmental consequences of products and consumption would influence consumers' actual use of the information. A deeper understanding may reinforce the trust of the labelling schemes as well as making consumers motivated to choose environmentally sound products.

Also, a deeper understanding of environmental attributes is essential if the information in environmental product declarations should be used for more than the symbolic value or building credence attributes into a product.

Furthermore, consumer motivation may be enhanced if product-related environmental information schemes more adequately document that they result in environmental improvements. Developing environmental evaluation models for environmental information schemes is indeed a research area separate from consumers' perceptions and understanding of the information.

Finally, our main conclusion relates to a need for employing research methodologies that capture consumers' considerations and actual use of the product-related environmental information, in the purchase situation. Such in situ studies may generate suggestions on how the information should be displayed, explained and delivered to generate a higher use among consumers.

C. Leire, Å. Thidell / Journal of Cleaner Production 13 (2005) 1061–1070 1069

References

[1] Armila M, Kähkönen N. The purchase decision making process of an environmentally conscious consumer – information search and alternative evaluation. Finland: Helsinki School of Economics and Business Administration; 1997.

[2] Arvola A, Biel A, Grankvist G, Koivisto-Hursti U-K, Larsson M, Magnusson M, et al. Ekologiska livsmedel – konsumenternas attityder, vanor och värderingar. Mat21 Research Program. Sweden; 2000.

[3] Asikainen J. Environmental labelling as means to affect consumer choices – case of nordic swan. Helsinki School of Economics and Business Administration, Department of Marketing/Logistics; 2000.

[4] Backman M, Lindhqvist T, Thidell Å. The Nordic white swan: issues concerning some key problems in environmental labelling. In: Stø, editor. Sustainable consumption. Report from the international conference on sustainable consumption. Lillehammer 1995. Working report no. 2-1995. Norway: SIFO; 1995.

[5] Björk P. The effect on green labels on consumer decision making. Svenska Handelshögskolan, Swedish School of Economics and Business Administration; 1997.

[6] Björner T, Hansen G, Russell C, Olsen T. The effect of the nordic swan label on consumers' choice. Denmark: AKF förlaget; 2002.

[7] Edlund S, Leire C, Thidell Å. Svanens roll i förhållande till andra miljöinformationssystem och miljöledning. TemaNord 2002;517. Nordic Council of Ministers.

[8] Erhvervsministeriet. Mærkning – Mærkningsudvalgets redegørelse. Forbrugerstyrelsen. Denmark; 1999.

[9] European Commission: DGXI. Integrated product policy. A study analysing national and international developments with regard to integrated product policy in the environment field and providing elements for an EC policy in this area. Executive summary from the final report. Available from: http://europa.eu.int/comm/environment/ipp/pdf/ippsum.pdf; 1998.

[10] European Commission. Green paper on integrated product policy. Brussels 07/02/2001, COM 68 final; 2001.

[11] Fallenius F, Sjöstedt C, Solér C. Rekommendationer för kommunikation av miljömärkning Typ III inom ramen för ISO 14000. Gothenburg Research Institute; 1997.

[12] Hansen PJ, Knudsen PE, Holm A. Forbrugernes krav til fodevarumärkning og vareinformation. A Pan-Nordic study of consumer behavior and attitudes towards food labelling. TemaNord 2001;573. Nordic Council of Ministers.

[13] Hansen PJ, Mörch Marckmann B. Forbrugernes brug af födevaremärkning. Del 1. Litteratur- og pilotstudie. TemaNord 2001;525. Nordic Council of Ministers.

[14] Heiskanen E, Timonen P. Consumption, knowledge, attitudes and environmental policy. Working papers 18. Finland: National Consumer Research Centre; 1995.

[15] Heiskanen E, Timonen P. Environmental information and consumer decisions: final report: consumers' needs for environmental information. Finland: National Consumer Research Centre; 1996. Publication 9.

[16] Himmelstrup Dahl E, Hjort T. Indragelse af miljøhensyn ved indkøb og design af tekstilprodukter – interviews af designere og indkøbere. Denmark: Miljøstyrelsen; 2002.

[17] Holm og Kildevang, 1996 in Unger, T. Gennemskuelighed og tillid mellem landmænd og forbrugere i Landbrugslauget – Delprojekt 1. Karakteristika for forbrugerandelshaverne. Denmark: Direktoratet for FødevareErhverv; 2002.

[18] Holmberg H-E. Konsumentundersökning om ekologiska produkter/krav. Sweden: KRAV and LRF; 1999.

[19] Jensen ML, Sörensen SY, Schmidt A, Grinderslev M, Hjelmar U. Forbrugernes muligheder og interesse for køb af grøn elektronik. Denmark: Miljøstyrelsen; 2002.

[20] Jönsson K. Communicating the environmental characteristics of products – the use of environmental product declarations in the building, energy and automotive industries. IIIEE dissertation. Lund University; 2000. p. 148.

[21] Juhl HJ, Poulsen CS. Den forbrugerbaserede værdi af en märkning – med fokus på svanemärket. Denmark: MAPP centret, Handelshøjskolen; 2002.

[22] Kiljunen P. Environmental grading of electricity. Study on public opinion. Finnish Energy Industries Federation – FINERGY. Yhdyskuntatutkimus Oy [Social research company]. Study report no. 6. Finland; 1998.

[23] Kogg B, Thidell Å. Utvärdering av system för egendeklarationer av farliga kemiska ämnen i varor – Exempel från Byggsektorn och Textilindustrin. Stockholm, Sweden: National Chemicals Inspectorate; 2003.

[24] Konsumentverket. Konsumenten och miljön – Resultat från en undersökning av svensk konsumenters miljömedvetenhet. Rapport 1995/96: 13. Stockholm, Sweden; 1996.

[25] Konsumentverket. Allmänhetens kunskap, attityder och agerande i miljöfrågor. Rapport 1998: 7. Stockholm, Sweden; 1998.

[26] Konsumentverket. Mitt hem är ingen sopstation... och andra tankar konsumtion och miljö. Rapport 2001: 11. Stockholm, Sweden; 2001.

[27] Kuusela H. Factors affecting the acquisition of domestic appliances and energy saving. Linkki. Research program on consumer habits and energy conservation. Publication 13. Finland; 1996.

[28] Landsvirkjun (The National Power Company). Umhverfisvitund – Viðhorfsrannsókn [Environmental awareness – attitude survey]. Performed by Gallup. Iceland; 2001.

[29] Leire C, Thidell Å. Consumer perceptions, understanding and use of product related environmental information. A literature review of the Nordic knowledge base. TemaNord 2004;539. Nordic Council of Ministers.

[30] Methi N. Tenker ikke miljø når vi handler. Forbrukerrådet. [Online]. Available from: http://forbrukerportalen.no/Artikler/forbrukerrapporten/2000/1023903397.32.05.12.2000; 2000. [11 Nov. 2003].

[31] Methi N, Stø E, Throne-Holst H, Vittersø G. Consumption and environment in Fredrikstad. Norwegian national report for the ToolSust program. Lysaker, Norway: SIFO; 2001.

[32] Miljödepartementet – press release. Informationen om produkters miljöpåverkan utvecklas. [Online]. Available from: http://www.regeringen.se/sb/d/670/a/34274; 2004 [26 Nov. 2004].

[33] Neergaard P, Andersen M, Bech L. Environmental-friendly textiles. Miljöstyrelsen, Denmark: Copenhagen Business School; 2002.

[34] Niva M, Heiskanen E, Timonen P. Environmental information in consumer decision making. Finland: National Consumer Research Centre; 1996. Publication 11.

[35] Niva M, Timonen P. The role of consumers in product-oriented environmental policy: can the consumer be the driving forces for environmental improvements? Article in the International Journal of Consumer Studies December 2001;25(4):331–8.

[36] Nordisk Miljömärkning. The Nordic Swan Label. Conducted by Taloustutkimus Oy. Finland; 2002.

[37] Palm L, Jarlbro G. Nordiska konsumenter om Svanen – livsstil, kännedom, attityd och förtroende. TemaNord 1999;592. Nordic Council of Ministers.

[38] Palm L, Windahl S. How Swedish consumers interpret and use environmental information. Rapport 1998: 28, Konsumentverket. Stockholm, Sweden; 1998.

[39] Puttonen J, Wennerström P. Kuluttajien energiansäästötutkimuksen 2001 yhteenveto. Motiva Oy Taloustutkimus Oy; 2001.

[40] Smallbone T, Page P. Green claims – environmental claims on products and packaging in the shop. An international study. Consumers International; 1999.

[41] Solér C. Communication of product related environmental information. User requirement studies of Environmental Product Declaration, EPD, systems. Report from the NIMBUS project. CPM 2001: 4; 2001.
[42] Statens offentliga utredningar. Märk väl! Betänkandet av utredningen gällande konsumentinformation om dagligvaror [Nota bene! Report of the investigation applicable to consumer information about everyday commodities], SOU 1999:7, Statens offentliga utredningar. Stockholm, Sweden; 1999.
[43] Stern P. Information, incentives, and proenvironmental consumer behavior. Journal of Consumer Policy 1999;22:461–78.
[44] Stø E, Strandbakken P, Strand M. Consumers and environmental information – a survey of consumers' use of and attitudes towards environmental product information in Germany, Italy, Norway and Spain. Professional report no. 3. Norway: SIFO; 2002.
[45] Strandbakken P. Environmental product information schemes (EPIS) in Norway. Norway: Statens institutt for forbruksforskning (SIFO); 2001.
[46] Strandbakken P, Stø E. Advantages and limitations of eco-labels as consumer and environmental political instruments. National report from the European DEEP project. Norway: SIFO; 2002.
[47] Szatek A. Vägen till Marknaden. LUI for LRF and Ekologiska Lantbrukarna. Sweden; 2001.
[48] Szatek A. Den svenska modellen – En studie bland svenska konsumenter. Sweden: LRF, TEMO; 2003.
[49] Thøgersen J. Psychological determinants of paying attention to eco-labels in purchase decisions: model development and multinational validation. Denmark: The Aarhus School of Business; 2000. Article published in Journal of Consumer Policy.
[50] Timonen P, Heiskanen E, Kärnä A, Niva M. Improving the environmental quality of products: the views of actors in the product chain. Finland: National Consumer Research Centre; 1998.
[51] Tufte P-A. Helse- og Miljømerking av produkter. En studie av forbrukernes oppfatninger om produktmerking av miljövarefakta. Arbeidsrapport nr. 6. Norway: SIFO; 1997.
[52] Tufte P-A, Lavik R. Helse- og Miljøinformasion. Rapport nr. 4. Norway: SIFO; 1997.
[53] Vanninen M, Viinikainen H. The Nordic environmental label. Consumer research study. Finland: Finnish Consumer Agency. Mikkeli Provincial State office, Consumer Project. Mikkeli Polytechnic, Institute of Business; 1995.
[54] Winters Lynch J. Determinants of effectiveness for environmental certification and labeling programs. Office of Pollution Prevention and Toxics. US EPA; 1994.

ISO-standards

[55] ISO 14021. Environmental labels and declarations – self-declared environmental claims (type II environmental labelling). Geneva: International Organization for Standardization; 1999.
[56] ISO 14024. Environmental labels and declarations – type I environmental labelling – principles and procedures. Geneva: International Organization for Standardization; 1999.
[57] ISO/DTR 14025. Environmental labels and declarations – type III environmental declarations. Geneva: International Organization for Standardization; 2000.

[20]

Eco-labeling of electricity—strategies and tradeoffs in the definition of environmental standards

Bernhard Truffer[a],*, Jochen Markard[a], Rolf Wüstenhagen[b,1]

[a]Swiss Federal Institute for Environmental Science and Technology (EAWAG), Limnological Research Centre, CH-6047 Kastanienbaum/Switzerland
[b]Institute for Economy and the Environment, University of St. Gallen (IWOe-HSG), Tigerbergstrasse 2, CH-9000 St. Gallen/Switzerland

Received 4 December 2000

Abstract

The liberalization of electricity markets opens up new opportunities for electricity providers to differentiate their products along environmental characteristics. Due to considerable search and evaluation costs for the consumer, the success of green power products depends to a large degree on a professional communication strategy. Besides professional marketing, instruments like environmental labeling have attracted considerable attention. The present paper analyzes the fundamentals of voluntary, third-party eco-labeling of electricity products and compares different labeling approaches, which have been developed in Europe and in the US recently. We distinguish six major dimensions along which labeling schemes can differ and identify two main strategies to deal with the basic tradeoffs, correspondingly. Labeling concepts may either be simple, require low or medium environmental standards and strive for the support of at least one stakeholder group. Alternatively, eco-labels may judge a broad range of environmental impacts and integrate several stakeholder groups with opposing interests. Experiences show that labels of the first type can be set up quickly and have the advantage to set a standard early in the market. In the medium and long term, however, public awareness and potential criticisms may rise and rather simple eco-labeling schemes will have to be adapted. The paper finally discusses the prospects for a common European eco-labeling strategy.

Keywords: Environmental labeling; Green power; Electricity Marketing

1. Introduction

The structure of electricity markets is undergoing profound changes in many countries of the world. Liberalization leads to increased competition among producers (Loskow, 1998) and consumers may now choose among a range of differentiated products. The net effect of these transformations on the environment, in general, and renewable energy, in particular, is yet to be determined (Schelling, 1998). Increased cost pressure—particularly in a situation where external costs of conventional power generation are not sufficiently internalized—makes it more difficult for emerging renewable technologies to compete in the market. However, recent experience shows that many customers are willing to pay more for renewable energy. Green power products, i.e. electricity offerings, which are based on environmentally preferable energy sources, are gaining importance in electricity markets worldwide. They show that customer demand for green electricity may be an important driving factor for supporting environmental sustainability in a liberalized market. The success of green power products depends to an important degree on a professional communication strategy. Besides establishing specific marketing strategies for green power products, transparency-enhancing instruments like environmental labeling may be decisive.

The present paper focuses, both theoretically and empirically, on the analysis of third party eco-labels for electricity. Third-party environmental labeling programs are carried out by an independent body, whereas first-party labeling is performed by marketers on their own behalf (EPA, 1998). In the following, we concentrate on voluntary certification schemes and we will use the short form eco-labeling mainly. What seems, at first hand, a simple problem of putting a tag on an electricity bill, leads quickly to a host of highly complex issues: which

* Corresponding author. Tel.: + 41-41-349-2111; fax: + 41-41-349-2162.

E-mail addresses: truffer@eawag.ch (B. Truffer), markard@eawag.ch (J. Markard), rolf.wuestenhagen@unisg.ch (R. Wüstenhagen).

[1] Tel.: + 41-71-224-2739; fax: + 41-71-224-2722. http//www.oekostrom.eawag.ch. http://www.iwoe.unisg.ch/institut/rwu.htm.

environmental standard should be chosen as a benchmark? How many and which environmental impacts should be considered? How will the ultimate environmental benefits of a green power purchase be assessed and signaled to the customers? By analyzing and comparing some major green electricity labels, which have recently been launched, different strategies for responding to these questions can be identified. The aim of the paper is (i) to provide a fundamental assessment of developing ecological product standards in the electricity market and (ii) to draw future perspectives for green electricity labeling.

We will, however, not discuss the more fundamental issues like (i) the role of third-party eco-labeling in the concert of environmental policy instruments, (ii) the ecological effectiveness of eco-labeling (Salzman, 1997) or (iii) strengths and weaknesses of labeling in general (Morris, 1997). We state that the green power market is one important means to promote renewable energy but will not be sufficient to bring about a sustainable electricity future by its own (Jegen and Wüstenhagen, 2000). From this point of view, eco-labeling can be a promising tool to support green consumerism but always has to be regarded in the concert of energy policy instruments.

In the following section, we will describe first the preconditions of a professional product differentiation strategy both from a producer's and a consumer's point of view. Section 3 discusses different means to reduce transaction costs for the market actors. Special emphasis will be laid on the role of third-party eco-labeling. In Section 4, the fundamental functions of an eco-label, namely the consideration of costs, the definition of what might be "environmentally preferable" and the transfer of "credibility" to the labeled products will be analyzed. In Section 5 this leads to the comparison of different eco-labeling approaches, in Europe and in the U.S. We will conclude with an outlook on the future development of labeling initiatives in Europe and world wide.

2. Differentiating electricity along environmental dimensions

From a competitive strategy point of view, companies in a liberalized electricity market can basically choose between two of Porter's (1980) generic strategies, namely cost leadership or differentiation. To improve their competitive position, they can either try to achieve lower costs than their competitors or establish a unique selling proposition, which allows them to charge customers a premium and hence increase revenues.

2.1. Marketing challenges for electricity providers

To differentiate from its competitors, a company has to identify characteristics of its product, which make a relevant difference to the consumer, and take this as a foundation for its marketing strategy. This is a relatively new approach in the electricity industry, but it is of course everyday business in almost any other industry. Global brands like McDonalds, Coca Cola, Heineken, American Express, Evian or Nescafe have demonstrated for a long time how differentiation can increase returns in otherwise homogeneous markets. In the case of electricity there are several features that may serve as a basis for differentiation:

- technical features (e.g. availability at peak load times, reliability),
- customer segments (e.g. offerings for small and medium enterprises like bakeries),
- additional services (e.g. energy efficiency advice, online billing) or
- environmental characteristics (e.g. renewable, low air emissions, nuclear free).

The latter seems particularly promising from an energy policy point of view, as it might be a means to get customer preferences in line with societal needs. There are some indications that green power offerings can be a viable option for electricity providers. One is that opinion polls and *market research* in many countries show that customers state clear preferences and an associated willingness to pay for renewable energy sources. As a general rule these studies indicate that about 20% of the households are willing to pay a price premium of 10–20% on their electricity bill for environmentally preferable electricity (Farhar, 1999). Empirical data from liberalized markets show that the actual customer response to green power offerings is initially lower than these surveys propose (Wüstenhagen, 2000, Fouquet, 1998), with market shares often below 1%

The second indication is the *experience of other industries*. For example, recent years have shown a significant success in green marketing in the food market. The No. 2 Swiss food retailer, Coop, is one of these success stories (Villiger et al., 2000).[2] Having embraced an ambitious green marketing strategy in 1993, Coop's sales in organic food products have grown within 6 years from zero to CHF 500 million (approx. 300 Mio Euro) in 1999, and the company's stated goal is to double this figure over the coming years. The comprehensive green marketing strategy has also helped the company to significantly increase its share in the saturated Swiss food market against its prime competitor, Migros.

The experience of food marketers as well as early empirical evidence from the green power market indicate

[2] Successful green marketing strategies can be identified in other countries as well, such as the organic product range of Sainsbury's in the UK or Wholefoods Market, a retailer specialized on organic food, in the US.

that some particularities should be taken into account when designing a green power marketing strategy (Wüstenhagen, 2000; Holt, 1997a). In order to capture not only the small niche of highly environmentally aware customers (the "Trueblue Greens" as referred to by Roper Starch Worldwide, 1998), the marketing mix has to be carefully balanced between environmental performance, economic constraints and the customers' preferences. Customers want to purchase a product that is environmentally preferable and has a good public reputation at the same time. Furthermore, green power products have to be convenient, i.e. readily available and easy to purchase with regard to metering, billing or the change of supplier, and the green supply contracts should be likely to cover 100% of the customer's electricity demand.[3]

The pricing of green power products should be in a reasonable range compared to conventional products, e.g. 10–30% above the price for conventional products. A range of products with differentiated prices helps to meet the varying willingness to pay of different customer segments. Communication is another essential element of the green power marketing mix. Professional promotion usually means higher budgets, but beyond mere advertising, personal communication with the customer via call centers or the internet is important, too.

Marketing theory providers the key to deal with most of the mentioned success factors. Currently marketing is a rather new field in electricity markets and product differentiation for environmentally preferable goods is an uncommon field of activity. Improving the professional quality of marketing activities may therefore considerably enhance the prospects of green power markets (Wüstenhagen, 2000).

Apart from that, providers of green goods have to face another major challenge, which is related to the credibility of their offerings. Unlike short-term marketing targets like increasing brand awareness, building up credibility is much less under the control of the individual firm and depends on the interplay between different market actors and stakeholders.

2.2. *Information problems for green power consumers*

The issue of credibility is closely related to the customer's purchasing motives for environmentally friendly products. Until very recently, electricity had not been perceived as a commodity and there had not been any opportunity to express specific preferences either.

From an individual consumer's perspective the decision in favor or against a green electricity product is ridden with some major difficulties. Environmental and social benefits have the particularity that they accrue to the society at large, whereas the costs for the products fall on the green consumer. Why should a rational individual have an increased willingness to pay for these products? Some consumers may be motivated to improve their individual environmental impact record by consuming "clean" electricity. Business customers will probably consider this effect in terms of improving their environmental image. Some motives of households may be identified as forms of impure altruism (e.g. feelings of warm glow, cf. Oberholzer-Gee, 2000). Beyond this, more socially relevant motives may exist: either there might be the conviction that someone has to take the lead for compensating for market failures (De Jasay, 1989) or that immature technologies need an initial push to become competitive because of substantial technological advances.

Given a certain distribution of the people in the population, which express an increased willingness to pay based on the above-mentioned motives, the resulting market potential might still not be realized because of high *transaction costs* the consumer is confronted with. Transaction costs may occur with regard to the purchasing decision, as environmentally friendly goods are usually difficult to evaluate from a customer's point of view (Haberer, 1996). In the case of green electricity, customers would have to invest considerable effort and time in order to

- obtain credible information about a product's environmental characteristics,
- evaluate a product's ecological impacts and
- compare different products.

To avoid transaction costs, customers may also choose to rely on the environmental marketing claims of green power providers. This option will highly depend on the credibility of a particular offering and the image of the supplier, respectively. Given the difficulty to verify the marketing claims, this may not prevent the customer from picking a low-quality product. In this situation of quality uncertainty, choosing a low-cost product may be the best way for the customer to avoid risk. The information asymmetries between the green power producer and the customer are likely to cause withdrawal from more expensive, high-quality products. Thus environmentally preferable products may eventually be crowded out of the market (cf. the phenomenon of adverse selection, Akerlof, 1970).

3. Enhancing transparency in green electricity markets

Several means for reducing information problems have been described in the literature (see Russel, 1998; EPA, 1998; Holt, 1997b). For the green electricity market, the

---

[3] Some green pricing programs just raise premiums on the regular electricity bill thereby contributing into a green electricity fund. Although customers support environmental friendly energy sources in this case the electricity delivery contract is still based on conventional power (e.g. see Future Energy, 1999).

following strategies merit some attention: screening, signaling, disclosure standards and third-party eco-labeling.

Screening is a solution strategy for the less informed economic actor, i.e. the individual consumer. He tries to get a better understanding about the electricity offerings by using secondary, aggregated information. The information may stem from consumer organizations, specialist journals or from experienced friends and relatives. *Signaling* relates to the better informed actor, i.e. the producer who can emit signals for product quality in order to convince the customer. Costlier signals tend to be more credible because they are more difficult to be imitated by competitors who offer low-quality products. In the case of green power, the installation of new, costly generation plants like wind turbines may serve as an example.[4] Moreover, companies can invest into their public image as a trustworthy and reliable promoter of environmentally and socially responsible products through sponsoring, advertising and the like. However, such a strategy is likely to be a long-term endeavor.

Alternatively, legitimacy and credibility can be transferred from other actors, which are considered as trustworthy by the relevant consumer segments. One possibility is the implementation of a *disclosure standard*, i.e. a declaration of the content of the respective product. In the case of electricity the declaration may include a disclosure of power source and fuel mix, air emissions, price and contractual terms. Uniform and trustworthy disclosure can be seen as a basic prerequisite for green electricity markets to develop. Consumers are, in principle at least, empowered to make their purchasing decisions based on their preferences (Wiser, 1999; NCCEI, 1998; REA, 1998). A disclosure standard is most valuable if it is mandatory for all products in the market and not only for green power offerings.

A full disclosure of power sources is a valuable tool for consumers to get information about the available products. However, a large group of consumers might not be willing and/or able to read through all the details of a product declaration. Additionally, they may not be informed about the environmental impacts of different power sources. In order to enable low involved consumers to make informed decisions in the messy environment of a deregulated market, there is a need for higher-aggregated information and guidance. One way of providing such a guiding tool is eco-labeling.

Third-party Eco-Labels are attributed to products after having been certified to fulfill specific criteria (EPA, 1998). Compared to the disclosure of electricity products, labels normally include a specific value judgement thus implicating an explicit purchase recommendation for the labeled product. Eco-labels have been established in different product sectors (Salzman, 1997). The most popular eco-labels are of a generic kind, i.e. they apply to a wide variety of products like toilet paper, varnish or cat litter. Well-known examples in Europe are the German Blauer Engel or the European Flower. Blauer Engel has been set up 20 years ago and products from about 1000 companies have been labeled since (Umweltbundesamt, 1998). Generic labels seem to be well suited for product sectors where criteria are rather easily defined and where no controversial political conflicts exist. For more complex products, sector-specific labels have been developed. As a rule, they are tailor-made for the specific problem at hand and have been initiated by specific stakeholders in the area. The Forest Stewardship Council label for sustainable forest management is a well-known example for the latter kind (FSC, 2000). Before we describe and assess the strengths and weaknesses of different labeling approaches, we will have to analyze the structure and function of eco-labels for green electricity in more detail. Fig. 1 shows some logos of eco-labels for green electricity worldwide.

Sources: Center for Resource Solutions, Energy Savings Trust, SEDA, Environment Canada, Swedish Society for Nature Conservation, Naturemade, Grüner Strom Label e.V., TÜV Süddeutschland

Fig. 1. Examples of green electricity labels (USA, UK, Australia, Canada, Sweden, Switzerland and Germany).

4. Principles of green electricity labeling

As electricity supply is a rather sophisticated issue in terms of technical, economical and ecological constraints, green electricity labeling becomes a highly complex undertaking. Experiences have shown that there is a large variety of possibilities to set up a certification scheme (Holt, 1997b). In the following, we will identify the key dimensions along which different eco-labeling approaches may be distinguished.

We mentioned before that enhancing customer information is the key function of eco-labeling. This role depends on the simplicity and market penetration of the label, which lowers the recognition effort for the consumer (cf. Section 4.1). Furthermore, the accuracy of the criteria and claims of the label have to correspond to the expectations of the consumers (cf. Section 4.2). Finally, the trustworthiness of the supporting body and the objectivity of the quality control procedures, which

[4] For a more detailed description of screening and signaling with regard to the green power market cf. Markard et al. (2000a). For a general description see Haberer (1996).

are implemented by the eco-labeling institution have to be guaranteed (cf. Section 4.3).

4.1. Reducing transaction costs for consumers and producers

The potential of eco-labeling to reduce transaction costs depends on whether the label is easily recognized as a trustworthy indicator by the relevant consumer segments. Single level labels will, as a rule, be easier to understand than labels with two or even more levels. Labels, which are based on simpler criteria, are easier to communicate than more complex ones. However, a label will not be accepted if it is not able to represent the environmental quality of a product in an appropriate way. So, the label is bound to deal adequately with the underlying complexity of the environmental impacts.

Recognition by consumers will, furthermore, greatly be enhanced, if the label is accepted as a standard, not only by the consumers but also by the different competitors, which offer green power products. The more the market grows, the more products will be sold with the label, the easier it will be for consumers to recognize the label and therefore to base their purchasing decision on it. Market penetration will therefore also depend on the applicability of the label for producers and traders.

Acceptance by producers is dependent on whether the benefits of using the label exceed its costs. The benefits are related to an increased willingness to pay for green products and a reduction in communication efforts because of the labels' high credibility. The costs for obtaining the label will mainly depend on the cost of gathering information in order to decide about the criteria and the measures, which have to be taken in order to fulfill the criteria. Furthermore, license fees typically have to be paid to the labeling organization.

Of considerable importance is the impact of a label on communication costs. This depends on the complexity of the label, but also to an important degree on whether or not other parties will support or contest the label. An eco-label, which will be opposed by specific environmental organizations, may eventually increase—instead of decrease—the need for communication. Alternatively, if environmental organizations support the label and advertise it to their members, a positive image transfer on the respective products may be reached at low cost (Mayer et al., 1999).

4.2. What is "environmentally preferable"?

Product differentiation will ultimately be based on environmental characteristics of the products. A key question every green electricity label has to answer is, therefore, how "environmentally preferable" electricity products look like. At a first glance, the answer seems simple: The consumption of electricity by a green power consumer should produce low environmental impact. The translation of this claim into operational criteria, however, proves to be far more complex.

4.2.1. Environmental standards and completeness of the assessment

In general, electricity products can be evaluated with regard to a selection of environmental impacts, which are associated with their generation. Their score may then be compared to either the average score of a national (or in the case of Europe UCPTE) electricity mix, or to the score of a marginal power system, which would currently be built, if additional capacity was needed (e.g. combined cycle plants). The final result will depend crucially on the completeness of the assessment, i.e. the environmental impacts selected for comparison and the applied minimum standards.

For the sake of simplicity, the assessment of environmental impacts may be reduced to some salient environmental issues. Many labels limit the eligible energy systems to *renewable energy sources* (RES, i.e. photovoltaics, wind, geothermal, biomass and hydropower). Electricity from RES, so goes the argument, is green because it does not use up finite resources. An alternative single criterion approach is represented by the "*climate friendliness*" of the products. This may be further simplified by assessing and comparing direct CO_2 emissions from different power sources.

Both shortcut criteria—renewable or climate friendly—have the advantage that they are easily communicated and understood by the customers. From an environmental point of view, however, a focus on mere fuel-based criteria seems incomplete. A more thorough approach has to assess emissions on the basis of an encompassing *life cycle assessment (LCA)*. Some labels therefore benchmark their products against the normal electricity mix by using integrated life cycle methods, like the Eco-Indicator (Frischknecht and Jungbluth, 2000; Goedkoop, 1995).

Even though life cycle assessment aims at considering the entirety of the environmental impacts of a product, some issues are difficult to handle. This is true for the majority of those impacts, which vary considerably from plant to plant and which are strongly related to the *location specific* characteristics of a plant. Impacts on landscape or ecosystems fall under this category. A way of dealing with these problems is to accept only plants, which fulfill a minimum standard at the local level.[5]

[5] These problems are especially relevant in the case of hydropower but also for some types of power plants using biomass or wind energy. As a consequence, some eco-labeling schemes have set up particular, locally adapted assessment procedures for hydropower or, in the case of biomass, are restricted to organically grown crops or FSC-certified firewood. Another example are wind farms, which should not be built in passage corridors for migrating birds (for further details see Section 5).

Finally, one might argue that given the broader set of goals implied by *sustainable development* an even more encompassing approach would be needed. To assess the sustainability of energy products, a label should incorporate social and economic impacts as well. Such an attitude is represented in the claim that the most sustainable kWh of electricity would be the one that did not have to be produced by any generating capacity (the "negawatt", cf. Lovins et al., 1999). According to this, green electricity products should only be labeled if they are part of a more encompassing energy savings service package.

The mentioned approaches to define green electricity have the fact in common that they endeavor to assess environmental product quality in a more or less objective way. However, in order to be understood and accepted in the market, a label must also respond to the subjective perception of power systems by the customers. For example, solar power is perceived as environmentally cleaner than combined cycle plants by a majority of customers, despite the fact that inefficiencies in today's PV manufacturing processes might make this technology score lower in a life cycle assessment than combined cycle plants. Such differences in scientific assessment versus lay man's perception give rise to tradeoffs that have to be carefully dealt with. Neither sticking to scientific assessments only, nor opportunistically following customer perceptions will provide a satisfying answer. For this (and other) reasons, critical voices fundamentally reject the concept of eco-labeling as such (cf. Morris, 1997).

The choice of evaluation criteria is a matter of costs too. A comprehensive and complex assessment procedure will increase the development costs of a label as well as expenses for certification later on. Moreover, it may lead to an increased need for communication in order to make the customers understand the label. On the other hand, if a reduced criteria set is chosen, the label may be accused to promote environmentally dubious forms of power plants. The problem of labeling hydropower may be illustrative in this context. Although hydropower scores extremely high on most LCA measures, it is—in some countries at least—strongly opposed by local environmental groups and a further expansion of power plants would be perceived as severely damaging to the environment. Several labels have tried to solve this problem by either accepting only small or already existing hydropower plants. However, from an environmental point of view this strategy proves to be essentially meaningless and does not solve the problem of the criticisms (Markard et al., 2000b).

A final tradeoff lies between the market penetration of the label and its credibility. This is reflected in the decision about the level of the minimum environmental quality, a product has to reach in order to be eligible for the label. If this level is relatively low, labeled products may spread in the market more easily. If the level is relatively high, then the label might have higher credibility, but its application will be restricted to a smaller number of products. One possible solutions to this tradeoff is a multi-level label with a differentiation in the environmental quality of the products.

4.2.2. Promotional effects

Given the complexity of assessing the environmental impacts, one might argue that customers are ultimately not very much interested in a thorough evaluation of existing plants. Rather, they want to improve the environmental situation associated with their electricity consumption (Markard, 1998). Therefore, the label has to guarantee that additional green generation capacity is connected to the grid and that less desirable plants will be disconnected as a consequence. Today, the existing power supply from RES already exceeds the actual demand for green power products in most electricity markets by far.[6] The introduction of an eco-label will have little or no effect on the environment under such circumstances, because a certification of existing power plants would just result in a "virtual" attribution of some green electricity to specific user segments. Customers would improve their individual environmental record, but only in proportion to a decline of the environmental record of other households or firms.

As consumers state a clear preference for making a contribution to an improved electricity system, additional criteria for *promoting new capacity* have been defined by many labeling initiatives. Most of them demand minimum quota for new plants to be built in the context of marketing green power products. Another possibility to engender environmental improvements is an ecological upgrading of existing RES power plants, e.g. in the case of hydropower (see Naturemade, 2000).

How the ultimate effect of green power purchases may be determined depends crucially on the kind of assumptions and scenarios one chooses for the development of certain technologies. An expansion of new capacity will be all the more effective, if the supported technologies are able to reap learning economies as a consequence of the expansion of the green power market. The increased willingness to pay of green power customers would then operate as an innovation promotion tool and will speed-up competitiveness of RES. The amount of actual innovation support, will depend on the governmental policies in place. In a country with a strong promotion policy, the role of the green power market

[6] In Germany, for example, the 1999 green electricity sales were about 90 GWh in comparison to power generation from wind energy (5500 GWh), small hydropower (5600 GWh), biomass (1100 GWh) and photovoltaics (20 GWh) (cf. Markard et al., 2000b).

will probably be weaker than in a country without such a policy.

4.3. Support and quality control

All the intricate details of different criteria sets, probably will not be penetrable for the majority of the customers. Instead of personally deciding about the optimal criteria list, they will be happy to delegate the information seeking, valuing and assessment decisions to a trustworthy institution (Haberer, 1996). The crucial resource of such a body is its *credibility*. In the domain of environmental standards, environmental NGOs profit from a high trustworthiness. Alternatively, science-based institutes, consumer organizations or technical test laboratories may be accepted as intermediaries by the consumer, as well. In general, intermediaries will be more accepted if they are perceived as being independent from the immediate interests of the marketers. Moreover, a broad involvement of different stakeholders may further increase credibility and lower the risk of fundamental opposition against a labeling-concept.

Standards have, however, not only to be accepted by the consumers, but also by the producers. A pure NGO-based label will have less success in the market place, if it is not accepted by the electric utilities. In complex products sectors like sustainable wood product, marine fishery, organic food, labels have been most successful, which were developed in a joint effort uniting representatives of the environmentally motivated firms and environmental NGOs.

Credibility of the label will also depend on the ability of the labeling organization to guarantee for the application of the criteria to the products according to a transparent and objective procedure. *Quality control* is delegated to professional auditors, which certify the compliance to the demanded criteria. Although the quality control of the label is often a highly sophisticated procedure, transparency should be maintained wherever possible. Customers should, in principle at least, be able to understand and even verify the criteria. In the context of marketing activities, customers who are interested should be able to visit the power plants for which they have signed their electricity contracts.

Transparency and credibility are key for the success of green electricity markets. Nevertheless, some tradeoffs have to be considered here, as well. The inclusion of different stakeholders in the development process of criteria may prove to be extremely time consuming and cumbersome. Compared to a label, which is supported by a single interest group, value judgements have to be made explicit and a common understanding has to be developed. This procedure is important, but carries high risks. At the level of quality control, the definition of an explicit auditing procedure by a professional auditor will, in general, be more costly.

5. Analyzing eco-labeling in practice

Having developed the dimensions in which labels may differ, we will now pass to an empirical analysis of six certification schemes from Europe. Additionally, we have included the US-American Green-e label. At first, a short description of each initiative as well as a tabular overview is provided showing that each label represents a specific approach to the issue (Table 1).

One of the first initiatives in Europe was the Swedish "Brå Miljøval" eco-label. It is owned and operated by the Swedish Society for Nature Conservation (SSNC), a Swedish environmental NGO. Brå Miljøval first set up criteria for green power in 1996 (SSNC, 1995). In 1999, the sales of certified green electricity were 8.9 TWh which is about 6% of the Swedish national electricity demand (Eiderstrom, 2000). For 2000, the label acknowledged 70 licenses, including all major utilities. Since 1998, the label has been expanded to Norway and Finland.

In Germany, competition between different stakeholder groups led to the launching of three different labeling initiatives. The first in the certification of green power products was the association of German TÜVs (VdTÜV, 1998) in late 1998. The different regional TÜV-organizations are traditional auditing and technical supervision companies in Germany. Meanwhile, they have developed five different criteria catalogues for end-user products, two other for generation plants and one for certifying green power customers. In 2000, the TÜV companies certified more than 30 products, including hydropower products from Austria and Switzerland. Besides the TÜV, a private association "Grüner Strom Label" had been founded by environmental organizations and associations for the promotion of RES in 1999 (Eurosolar, 1999). The Grüner Strom Label certified two products so far, including an umbrella organization of 18 municipal utilities. Finally, an institute for applied research on ecology developed a certification scheme and launched a green power label (Öko-Institut, 1999). Öko-Institut, together with a consumer organization and the German World Wide Fund for Nature (WWF) founded a private organization "EnergieVision". In a pilot-phase four products have been labeled.

Green power products in the UK can apply for an accreditation within the "Future Energy" scheme, run by the Energy Savings Trust, which is supported by the British Government (EST, 1999). The Future Energy scheme distinguishes two types of green power offerings: renewable energy supply and renewable energy funds. By mid-1999, accreditation has been awarded to 13 products from 12 suppliers (Elliott, 2000).

Table 1
Description of the analyzed eco-labels electricity

	Brå Miljøval Sweden, Norway, Finnland	TÜV Germany, Austria, Switzerland	Grüner Strom Label Germany	EnergieVision Germany	Future Energy United Kingdom	Naturemade Switzerland	Green-e California Pennsylvania
Simplicity	Single level	Several levels	Two levels (Gold, Silver)	Two levels (Supply, Fund)	Two levels (Supply, Fund)	Two levels (Basic, Star)	Single level
Environmental standard	RES	RES Qualified COGEN[a]	Qualified RES, Qualified COGEN	Qualified RES, Qualified COGEN	RES, Waste combustion	RES, Qualified RES	RES (> 50%), Conventional fuels
Completeness of assessment	List of eligible sources, Local criteria	List of eligible Sources	List of eligible sources, local criteria	LCA	List of eligible sources	List of eligible sources (Basic) LCA, local criteria (star)	List of eligible sources
Promotional effects	Not required	Not required, capacity enlargement[b]	Capacity enlargement	Capacity enlargement	Not required (Supply), inherent (Fund)	Capacity enlargement, improvement for existing hydro	Capacity enlargement
Support by stake-holder groups	Environmental organization	Testing and auditing organization(s)	Environmental and consumer organizations, RES support associations	Research Institute, WWF and consumer organization	Government and WWF	Environmental and consumer organizations, RES support associations, utilities	Consumer organizations, RES support association
Acceptance/ Criticism	No major criticism	Criticized by environmental organizations	Criticized by electricity industry and environmental organizations	Not yet available	Criticized by environmental organizations	Not yet available	No major criticism

[a]COGEN = Co-generation.
[b]One TÜV guideline (VdTÜV 1303) requires capacity enlargement. TÜV EE02 (see Table 2) does not require any promotional effects.

In Switzerland a two-level labeling scheme for green electricity has been launched following a broad stakeholder process: "naturemade basic" and "naturemade star". In a pilot-phase the evaluation procedure has been applied to more than 10 marketers and the first certificates have been delivered in November 2000 (Naturemade, 2000). The label is supported by electricity producers and suppliers, major environmental organizations and consumer representatives.

The Green-e label has been introduced in California in 1997 and has meanwhile certified 23 green power products of 14 power marketing companies (Wingate et al., 2000). A geographical rollout of the program is foreseen, e.g. to the Mid-Atlantic and New England states. First state of adoption was Pennsylvania in 1998.

5.1. Comparison of eco-labels

On the one hand, the described labeling schemes have main goals and principles in common like enhancing market transparency or strengthening customer demand for green electricity. All initiatives but two in our sample are specific labels, which have been explicitly founded for green electricity. The exceptions are Brå Miljøval, which is a label for green products in general and the German TÜV, who uses its seal for technical certification of machines and appliances also. Furthermore, the majority of eco-labels rely on professional, external auditors to manage the quality control of the certified products. Only the TÜV-concept is based on internal auditing, i.e. criteria setup, certification and control are executed by the same organization.

On the other hand, the individual labeling initiatives turn out to have fundamental differences, which will be compared according to the dimensions described in the previous section (see Table 1 for the details)

First, the *simplicity* and the *environmental standards* differ between the labels. In some eco-labels different environmental quality standards are applied to the certification levels introduced. Grüner Strom Label, for example, distinguishes pure RES products (Gold) and RES including co-generation (Silver). Naturemade certifies RES, in general, with its basic level and reserves the star level for ecological leaders among RES. EnergieVision and Future Energy separate products, which are based on a green power supply from products like eco-funds, which rely on conventional power supply but include promotional effects for RES. Green-e and Brå Miljøval, finally, opted for a single level eco-labeling approach. The TÜV has a simple seal, which represents a bewildering variety of guidelines.

With regard to the *completeness of the assessment* of potential environmental impacts, most eco-labels just use a list of eligible sources. Only EnergieVision and Naturemade star have defined their criteria on the basis of a life cycle assessment. Naturemade star is even more encompassing as it combines LCA with a set of minimum criteria for local environmental impacts.[7]

The *promotional effects* of green power products have been dealt with differently by the individual labeling initiatives. Grüner Strom Label, EnergieVision, Naturemade and Green-e set minimum shares for new power plants. Naturemade also accepts the improvement of the eco-performance of existing hydropower plants as a promotional effect. Future Energy does not explicitly include requirements for promotional effects. Still, renewable energy funds, which are approved under the label include investments in new capacity by definition. According to the TÜV concept green power suppliers can choose whether they want to integrate promotional effects or not. Brå Miljøval does not require any capacity enlargement or promotion at all.

The *support by stakeholder groups,* i.e. the kind of constituency of a labelling initiative is a decisive factor for generating credibility for the labeled products.[8] Most of the existing labels are either supported (Grüner Strom Label, EnergieVision and Naturemade) or directly operated by environmental organizations (Brå Miljøval). Moreover, Grüner Strom Label and especially Naturemade seek for a broader stakeholder support including RES producer associations or consumer organizations. Only the Energy Savings Trust, the holding organization of the British Future Energy scheme depends on public funding.

Another key issue is the *acceptance, respectively, criticism* of the label by other stakeholder groups than those officially supporting the labeling initiative. In the case of Future Energy for instance, the declaration of waste incineration as a renewable energy source caused criticism by Friends of the Earth, a major UK environmental NGO (Elliott, 2000). Also, the TÜV is explicitly opposed by environmental organizations. They claim that the TÜV-label seems to promise environmental benefits, which most certified products cannot keep. The concept of Grüner Strom Label is criticized by environmental organizations and the electricity industry for excluding suppliers who are engaged in the conventional power business as well.

[7] The different approaches are especially visible in the case of hydropower: Naturemade star is based on a comprehensive evaluation procedure. Other schemes rely on single criterion approaches like "smaller than 10 MW" or "built before 1995", for further details see: Markard et al. (2000b). The US label Green-e started with a short-hand criterion (30 MW for run of river plants). However, due to negative experiences with the approach they are about to adopt an assessment procedure, which is based on an encompassing local criteria list also (LIHI, 2000).

[8] For a description of the different interests and stakeholder positions see Markard et al. (2000a).

5.2. Identification of general labeling strategies

In order to compare the seven examples in a qualitative way, we have divided each of the six dimensions into three ordinal categories [1–3], which range from low to high contribution to the core functions of the label. The dimensions have been set up as follows:

- *Simplicity* of the label: More than two levels differentiated [1], two levels [2], only single level [3].
- The *environmental standard* of labeled products: The products are just better than average [1], are restricted to RES incl. qualified forms of co-generation [2], or are only the "best" RES included [3].
- The *completeness of the assessment*: The criteria are given as a list of eligible sources [1], further criteria are applied for selected energy sources [2], an LCA (perhaps even including local criteria) is included [3].
- The *promotional effects:* No criteria for promotion required [1], indirect promotion [2], explicit promotion [3].
- The *support of stakeholder groups:* The eco-label is supported by one group[1], by two[2], or more than two different interest groups [3].
- The *acceptance/criticism* of the label: The label is criticized by two or more different stakeholder groups [1], by one kind of interest groups [2], or faces no major criticism, i.e. it is accepted by the major actors in the electricity market [3].

The scores of the seven different labels are represented in Table 2.

Before we may proceed to analyze the table, some methodological qualifications have to be made: The dimensions Simplicity, Support and Acceptance apply to the label as a whole. The other dimensions have to be differentiated according to the levels of each label. In the following argument, we have chosen some extreme cases (esp. TÜV-level "EE02" and Naturemade level "star"). This selection may seem unfair in some cases. Our purpose is, however, not to distinguish "good" from "bad" labels, but to illustrate the fundamental approaches and their associated tradeoffs.

We have grouped the individual dimensions into two categories. The environmental standard, completeness of assessment, promotion and support add up to a measure of overall credibility (column 6 of Table 2). Simplicity and acceptance, on the other hand, are summarized and attributed to a category representing the costs of a label (column 9 of Table 2). We have to admit, however, that our measure for cost is somewhat incomplete as no evaluation of the costs for obtaining the label (assessment of criteria, license fees, etc.) could be included, due to a lack of data.

As an overall measure for the quality of a label, we finally define the sum of all single scores (cf. last column of Table 2). According to this measure, three labels score high (Naturemade, EnergieVision and Grüner Strom Label), two labels range in the middle (Green-e and Brå Miljøval) and two labels score rather low (TÜV, Future Energy).

This simple scoring exercise may also be discussed with regard to credibility and costs. The credibility index distinguishes quite sharply between the three high scoring labels (Naturemade, EnergieVision and Grüner Strom Label) and the rest. In theses labels, consumers are perceived as wanting to be assured that no major environmental impacts are associated with the products they buy. They also represent approaches, which have been developed in multi-stakeholder processes and are therefore likely to encounter less criticism by other stakeholder groups.

However, credibility comes at a cost. The three high-credibility labels score only middle with regard to costs. The two first mover labels in the US and Europe (Green-e

Table 2
Scorers of the analyzed eco-labels for electricity

	Credibility					Costs			Sum
	Environmental standard	Complete-ness of the assessment	Promotional effects	Supported by stakeholder groups	Total	Simplicity	Acceptance criticism	Total	Total
TÜV EE02	1	1	1	1	4	1	2	3	7
Future Energy	1	1	2	1	5	2	2	4	9
Brå Miljøval	2	2	1	1	6	3	3	6	12
Green-e	1	1	3	2	7	3	3	6	13
Grüner Strom Label Gold	3	2	3	3	11	2	2	4	15
Energie Vision Supply	3	3	3	2	11	2	(2)[a]	4	15
Naturemade star	3	3	3	3	12	2	(2)[a]	4	16

[a] Acceptance not yet available, average value chosen.

and Brå Miljøval) have very favourable cost characteristics. They had been designed as simple labels, which encounter high acceptance in their respective home markets. Finally, the two remaining labels (TÜV and Future Energy) score medium to low on the cost dimension as they encounter substantial criticism in their home markets and do not score high with regard to simplicity.[9]

The cost of the label for the producer is directly linked to the complexity of the criteria and the associated procedure to test the criteria. But also indirectly, complexity of the label results in high development costs, which are likely to increase license fees. This argument is particularly important in an early phase of market penetration. The one label which is able to enter the market first is likely to "set the standard" for all laggards. So, there have been quite high opportunity costs for more complex approaches especially in countries with competing labeling initiatives. As a matter of fact, simple labels had been able to position themselves in the market early on (Green-e, Brå Miljøval, TÜV). With time, however, these initiatives realized shortcomings of their approaches and considered more complex and complete criteria lists: Green-e, for instance, is about to introduce a local criteria list for hydropower plants (LIHI, 2000) and Brå Miljøval wants to implement a more thorough evaluation of hydropower too. Future Energy currently reviews its criteria with regard to the criticism of environmentalists about the inclusion of energy from waste technologies. The TÜV follows a strategy of continuously creating new guidelines to cover newly emerging needs.

Summarizing, the individual labeling approaches have found specific ways to deal with the tradeoffs, we discussed in Section four. The resulting concepts depend on:

- How are customers' expectations represented in the criteria setup?
- Which interest groups are involved in the development of the criteria and how are the respective value judgements made?
- What role is attributed to the label in the context of the specific national energy policy?

Depending on the three questions above, our empirical cases seem to follow one of two basic strategies: Eco-labels are either designed to differentiate conventional electricity offerings and products being just better than average. The focus of those labels is rather "to assure that customers get what they paid for" (i.e. that the product meets the marketer's claims) than "to identify and support products which are environmental leaders". As a matter of fact, the label can rely on short-cut evaluation criteria and a concept, which is simple and easy to understand. In most cases, stakeholder involvement is kept to a minimum. The second strategy is to integrate stakeholder groups with divergent interests.

In the beginnings, the first strategy may be successful because it helps setting standards early in the market and keep costs of the label low. Furthermore, the supply of labeled products can grow rapidly as long as many products and generation technologies are eligible. Due to shortcomings concerning the environmental assessment or weak stakeholder involvement, the concept risks to encounter criticisms from consumers sooner or later.

The broadly based and balanced strategy, on the other hand, has considerable advantages in the medium and long term, when public awareness rises and consumers ask for more encompassing information about the environmental consequences of their product. Such eco-labels may be well suited as long-lasting environmental standards for the market as a whole. The simple concepts represent more of a "test and adapt" strategy in the early phases of market development.

6. Conclusion and outlook

Eco-labeling of electricity is an important means to empower consumers to freely express their preferences by reducing search and evaluation costs. We argued in the present paper that the substantiation of this goal leads into a host of complex issues, which reflect the complexity of electricity generation and consumption as a whole. Key dimensions for the development of an eco-label for electricity as well as fundamental tradeoffs could be identified: the simplicity of the labeling design, the completeness of the assessment procedure, the environmental standard chosen, the promotional effects demanded and the way different stakeholder groups are included. These dimensions could be verified quite well at the level of the individual labeling approaches. Much more difficult was the assessment of cost terms. The market has not developed enough to compare the different approaches accurately. Our analysis showed, that two basic strategies had been chosen to deal with the above-mentioned dimensions. Either a label which is simple and has rather low or medium environmental standards or an eco-label which aims at representing the entirety of environmental impacts associated with a product. In an early phase, the quicker labels have a considerable advantage to set the standard in the market. In the medium and long term, however, public awareness will rise and consumers will ask for encompassing information about the environmental consequences of their product.

This analysis may be of interest to other regions in the world where electricity markets are about to be liberalized and where quality assurance for green power products

[9] Although the TÜV seems to present a simple label to the consumers, the concept does not reduce complexity at all because of its high number of different criteria guidelines.

will become an issue of public interest. In Europe the question will be (i) what many and which of the existing labels will survive in the long run and (ii) what an eco-label should look like for the common European electricity market. As the market for green power grows and EU wide trade in electricity expands, transparency and credibility of environmental quality claims will increase in importance.

Based on the analysis of our paper, we may describe a rough road map for the development of a European label:

- First, promoters of a European label should define common goals and a common conceptual approach. The integration of different actors representing different interests will turn out to be a crucial precondition for success.
- In the second step, basic definitions and value judgements would have to be agreed on.
- Finally and probably most importantly, different national backgrounds will have to be taken into consideration: the existing national power generation mix, the structure of the electricity industry, current use and future potential of RES in the member states, prevalent environmental impacts, social and political preferences and so on.

A possible strategy for resolving the third problem may be to agree on basic principles and a common concept, while applying the subsidiary principle to regulate the details of certification in different national contexts. (Markard and Truffer, 1999). Electricity traded under the European label would also have to be recognized by other schemes, which join the system. Furthermore, there is a considerable potential to enhance efficiency and speed up the development process if the European label would be based on (at least some of) the existing approaches. A European label could function as a common platform, which could coordinate the decentralized initiatives and knowledge pools. European policy could help here to start such stakeholder processes in order to arrive at a broadly shared standard for what might count as environmentally preferable electricity products.

Acknowledgements

The authors would like to thank the anonymous reviewer for his valuable comments. The responsibility for remaining errors is with the authors.

References

Akerlof, G.A., 1970. The market for 'lemons': quality uncertainty and the market mechanism. Quarterly Journal of Economics 84, 488–500.

De Jasay, A., 1989. Social contract, free ride: a study of the public goods problem. Oxford University Press, Oxford.

Eiderstrom, E., 2000. Swedish society for nature conservation—good environmental choice electricity. Paper Presented at the NGO Workshop on European Green Electricity Label, Brussels, March 15–16.

Elliott, D., 2000. Prospects for the green power market in the UK. Proceedings of the World Renewable Energy Congress VI, Brighton, pp. 1428–1432.

EPA, 1998. Environmental labeling. Issues, policies, and practices worldwide. United States Environmental Protection Agency (EPA), EPA 742-R-98-009, Washington DC.

EST, 1999. Future energy—a scheme for accrediting renewable energy. Scheme description and criteria. Issue 1, Energy Saving Trust (EST), London.

EUROSOLAR, 1999. Kennzeichnungskriterien für das Label als Grüner Stromanbieter, Der Markt für Grünen Strom, Eurosolar-Verlag, Bochum.

Farhar, B.C., 1999. Willingness to pay for electricity from renewable resources: a review of utility market research. National Renewable Energy Laboratory, NREL/TP.550.26148, Golden, July 1999.

Fouquet, R., 1998. The United Kingdom demand for renewable electricity in a liberalised market. Energy Policy 26, 4, 281–193.

Frischknecht, R., Jungbluth, N., 2000. Globale Umweltkriterien für Ökostrom, ESU-services, Zürich.

FSC, 2000. Forest Stewardship Council (FSC) United States. See http://www.foreststewardship.com.

Future Energy, 1999. A scheme for accrediting renewable electricity. Scheme description and criteria. Issue 1, Energy Savings Trust, London.

Goedkoop, M., 1995. The eco-indicator 95: weighting method for environmental effects that damage ecosystems or human health on a European scale. PRé Consultants, Amersfoort, NL.

Haberer, A., 1996. Umweltbezogene Informationsasymmetrien und transparenzschaffende Institutionen. Metropolis-Verlag, Marburg.

Holt, E., 1997a. Green pricing resource guide. The regulatory assistance project, Gardiner, Maine.

Holt, E., 1997b. Disclosure and certification: truth and labeling for electric power. Issue Briefs No. 5, Renewable Energy Policy Project, University of Maryland, College Park.

Jegen, M., Wüstenhagen, R., 2000. Modernise it, sustainabilise it! Swiss energy policy on the eve of electricity market liberalisation. Energy Policy 29 (1), 45–54.

LIHI, 2000. Low impact Hydropower Institute. See http://www.lowimpacthydro.org/.

Loskow, P.L., 1998. Electricity sectors in transition. The Energy Journal 19 (2), 25–54.

Lovins, A.B., Lovins, H.L., Hawken, P., 1999. A road map for natural capitalism. Harvard Business Review, 77, 145–158.

Markard, J., 1998. Green-pricing – Welchen Beitrag können freiwillige Zahlungen von Stromkunden zur Förderung regenerativer Energien leisten?, Werkstattreihe Nr. 104, Öko-Institut-Verlag, Freiburg.

Markard, J., Rothenberger, D., Truffer, B., 2000a. Analysing market mechanisms of third party eco-labeling: the case of green power certification. Paper Presented at ESEE conference, Vienna, May 3–6.

Markard, J., Truffer, B., 1999. Der lange Weg zu einem Euro-Label für Strom. Energiewirtschaftliche Tagesfragen 49 (11), 724–728.

Markard, J., Truffer, B., Bratrich, C., 2000b. Ecological and competitive green electricity from certified hydropower plants in liberalized electricity markets. Hydro 2000 — Making Hydro more Competitive, Conference Proceedings, The International Journal on Hydropower & Dams, Sutton, pp. 717–725.

Mayer, R., Blank, E., Swezey, B., 1999. The grassroots are greener: a community-based approach to marketing green power. Research Report No. 8, Renewable Energy Policy Project, University of Maryland, College Park.

Morris, J., 1997. Green Goods? Consumers, product labels and the environment. IEA Studies on the Environment No. 8, The Institute of Economic Affairs, London.

Naturemade, 2000. Informationen zum Qualitätszeichen naturemade basic. Version 2.0, Verein für unweltgerechte Elektrizität, Zurich.

NCCEI, 1998. Synthesis Report: A Summary of Research on Information Disclosure. Draft National Council on Competition and the Electricity Industry (NCCEI). The Regulatory Assistance Project, Gardiner, Maine.

Oberholzer-Gee, F., 2000. Your contribution counts! an empirical analysis of the decision to support solar energy. Paper Presented at the Wharton Impact Conference, Philadelphia, January.

Öko-Institut, 1999. Entwicklung eines Zertifizierungsverfahrens für "Grünen Strom", Endbericht, Öko-Institut, Darmstadt Freiburg Berlin.

Porter, M.E., 1980. Competitive Strategy. The Free Press, New York.

REA, 1998. Disclosure: A Vital Tool for Consumers. Renewable Energy Alliance (REA), Oakland. See also http://www.realliance.org.

Roper Starch Worldwide, 1998. The 1998 Green Gauge Reports, Executive Report. Roper Starch Worldwide, New York.

Russel, T., 1998. Greener purchasing. Opportunities and innovations. Greenleaf Publishing, Sheffield, UK.

Salzman, J., 1997. Informing the green consumer. The debate over the use and abuse of environmental labels. Journal of Industrial Ecology 1 (2), 11–21.

Schelling, T., 1998. The environmental challenges of power generation. The Energy Journal 19 (2), 115–124.

SSNC, 1995. Environmental criteria for electricity. Brå Miljøval. Swedish Society for Nature Conservation (SSNC), Göteborg.

Umweltbundesamt, 1998. Erfolgskontrolle Umweltzeichen. Texte NR. 61/98, German Federal Environmental Agency, Berlin.

VdTÜV, 1998. Bereitstellung von Strom aus Erneuerbaren Energien. Vergaberichtlinie für ein TÜV-Zertifikat. Energie- und Gebäudetechnik 1303. Verband der technischen Überwachungsvereine (VdTÜV), Essen.

Villiger, A., Wüstenhagen, R., Meyer, A., 2000. Jenseits der Öko-Nische. Birkhäuser Verlag, Basel Boston Berlin.

Wingate, M., et al., 2000. Green Power certification — environmental and consumer protection benefits of the Green-e programme. Renewable Energy World, 3(3), 63–73.

Wiser, R., 1999. The role of public policy in emerging green power markets: an analysis of marketer preferences, LBNL Report No. 44178. Lawrence Berkeley National Laboratory, Berkeley.

Wüstenhagen, R., 2000. Ökostrom — von der Nische zum Massenmarkt, vdf Hochschulverlag AG, Zurich.

[21]

The use of eco-labeling like initiatives on food products to promote quality assurance—is there enough credibility?

Helen Nilsson *, Burcu Tunçer, Åke Thidell

International Institute for Industrial Environmental Economics (IIIEE), Lund University, Tegnérsplatsen 4, Box 196, SE- 221 00, Lund, Sweden

Received 26 August 2002; accepted 19 June 2003

Abstract

The development of this paper was inspired by the increase in the number of different eco-label like quality assurance sche established in the European Union in the past few years. These schemes are flooding the market and the researchers raise question, is there enough credibility in these schemes? This piece of research is based on 58 eco-labeling like food schemes dissects them into the building blocks of the different factors that makes a scheme credible and analyses the structure of tl blocks as well as a literature survey on the perceptions of both the consumers and producers to these schemes. The sort of credib structure that the building blocks of the schemes (ownership, stakeholder dialogue, traceability, transparency, etc.) have cre was discussed and it was questioned if they meet the expectations of the consumers. The main findings of the research indi that although small groups of consumers may be satisfied with a number of the different schemes, the majority of them fall s of providing a credible quality assurance scheme. Improvements were suggested such as involving a more diverse range of st holders and expanding the responsibility of processors and retailers. An alternative to the current form of labeling is also discus

Keywords: Quality Assurance Schemes; EU; Credibility; Food labeling

1. Introduction

1.1. Background

The state of the farming and food processing industry has, not least in Europe, has been put under the spotlight in recent years due to reoccurring food scares and disease outbreaks on farms. The spread of BSE, Salmonella, the recent outbreak of foot and mouth, not to mention the pesticide residues in our drinking water, additives put into livestock's fodder and the mass transportation of thousands of animals every year in cramped conditions, throughout the whole of Europe, has raised concerns with both consumers and politicians alike [1,2].

There has been a development, in parallel to these scares, in the declaration of the quality and environmental credentials of a product or its production metho competition increases on the European food prod marketplace. A number of initiatives have been c structed to correctly communicate the efforts of ducers and their products' features to consumers improve their environmental performance. These init ives have been developed by a number of diffe actors; such as primary producers, consumer organ tions, producers, retailers and farmers associations. I however, worth noticing that quality and environme aspects include a number of dimensions, for exam animal welfare, health, ethics and taste.

Food labeling can be seen as one of the comr routes to deliver the message about the quality of food product to consumers. However, with the multit of labeling schemes now on the market, it must be as if it is possible to be sure that the situation has b improved and if there is any substance behind the dif ent schemes. The criteria set up by the labels are fo to go in many different directions. For example, s monitor the presence of pesticide residues, others additives and preservatives, and other schemes st

* Corresponding author. Tel.: +46-46-222-0200; fax: +46-46-222-0210.

E-mail address: helen.nilsson@iiiee.lu.se (H. Nilsson).

other characteristics, while they are all claiming the products are environmentally friendly [3].

The market share of labeled food products promoting higher quality or better environmental performance is still rather small in most European countries; typically no more than 10%. Research shows that consumers claim to be willing to pay for improvements in the quality of the product and the environmental performance of the production method. Researchers have also considered how to assure a scheme's credibility for the consumers. It was proposed by Salzhauer that a credible, nationally recognized eco-labeling programme would be a suitable tool to assist consumers in identifying credible schemes, but she does not propose how we can guarantee the credibility of the programme [4].

With the idea of credibility in mind it was found that there appears to be little agreement as to the message that the labeling schemes wish to convey to the consumer. The labels' focus, instead, seem to address existing quality assurance schemes and if they actually improve the performance in identified areas like animal welfare, environment, health etc rather than an analysis of consumer preferences.

This paper is partly based on research conducted during 2001 and reported in Tunçer [5] on European Quality Assurance Schemes in general. Without claiming completeness, 107 different schemes were found and examined in EU countries. However, the research for this paper is based on the part of the research that focused on the 58 European single message schemes found. The residual schemes either used other modes for communicating the content or addressed other target groups. Here this background material has been further scrutinized and analyzed and compared with outcomes from a literature survey examining consumers and producers understanding of the credibility of these schemes.

The purpose of this paper is not to propose a new definition of the term credibility, as this research was not seen as a theory paper, instead a definition will be used that was quoted in Tunçer. This states that the least formal way to manage the relationships to the suppliers or customers of an organization is to develop a reputation i.e. [...] "state in which an organization is held in high regard and trusted by other parties because of its fair and honest business practices" [5]. Credibility tools, in that regard, are needed to build a positive reputation on the quality assurance aspect of the initiative. An understanding of the term has also been gleaned from a survey of the available literature and interviews with stakeholders.

1.2. Methodology

This paper consists of two separate parts, one investigating the building blocks of existing European single message food quality/environmental performance labeling schemes and one survey of producers' and consumers' perceptions of credibility of particular schemes (Fig. 1). The majority of the research was conducted as a desk-based study.

The first part of the methodology, section two on the building blocks, is based on Tunçer. A survey was conducted during 2001 on European Quality Assurance Schemes. Through Internet searches, expert interviews, questionnaires to environmental expert networks, literature and newsprint search, 107 different such schemes were found and described in EU countries. The survey does not, however, claiming completeness. Additional interviews were conducted with key stakeholders operating these schemes. This was done so that the labels used in this study could be identified and their properties understood. Once the information was collected the analysis was based on the building blocks that contribute to the build up of the credibility of the labeling schemes. Out of the 107 schemes found 58 were included in the group for single message labels; and were used as the base for the research in this paper. References to all of the 107 schemes are given at the end of this paper.

The second part, based on Internet and literature search for relevant literature and interviews with large consumer organizations and key stakeholders of some of the major European schemes, follows a structure that illustrates what consumers expect from the labeling schemes. However, as these references are not following a homogenous structure and indicate the credibility issue in different formats, it was not fruitful to sort the information in a fixed format. Instead, the researchers searched the texts and statements for relevant information and interpreted these in a credibility context.

2. Building blocks of credibility in a labeling scheme

Comprehension of the build up of credibility for 58 single message labeling schemes that consist of:

- 20 organic labeling schemes;
- 4 integrated production (IP) schemes;
- 10 farm quality assurance schemes;
- 13 retailer or processor led schemes;
- 11 regional schemes.

The building blocks or assets of each scheme are those that add value to it and lead to schemes standing out from each other and giving them credibility. Ownership structure, stakeholder coverage, quality assurance traceability, marketing system and transparency are accepted as the main properties of the labeling schemes studied. These building blocks were established after a review of the literature published by experts in the field of food quality and policy found these components to be the most important elements when considering the credibility of a labeling scheme. An explanation of the differ-

H. Nilsson et al. / Journal of Cleaner Production 12 (2004) 517–526 519

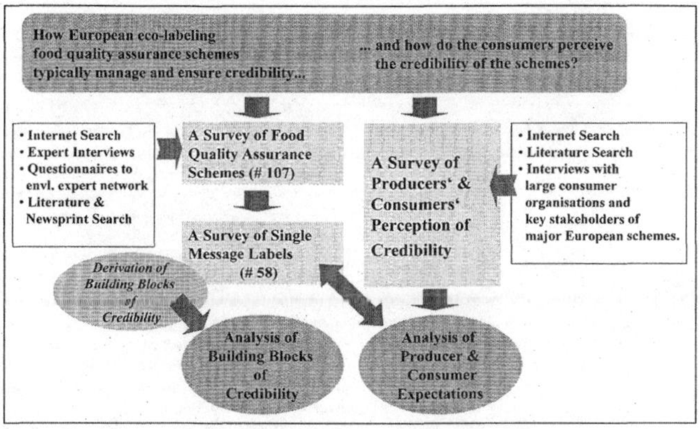

Fig. 1. An overview of the methodology followed in the paper.

ent building blocks and examples of their application can be found in Table 1.

3. Review of consumer and producer perceptions

There are very few studies concerning the messages and specific quality aspects that the actors, from primary producers to actors further down in the food production chain, wish to convey to the end-consumers. However, some articles in newsprints and sector journals point at the producer's conservative attitude and unwillingness to change the production methods. There is more of a defensive attitude stating that the foodstuffs already have good quality produced under controlled conditions. Therefore, the discussion instead seems to revolve around issues, such as, if the quality assurance schemes actually improve the performance in identified areas like animal welfare, environment, health etc., in bulk production rather than focusing on analyses of consumer preferences and market differentiation.

This paper does not pretend to cover all European food eco-labeling schemes or all available producer and consumer credibility surveys. Though from the research that was done it can be said that there are not that many available studies on how the receivers of the messages—the consumer—actually perceive the messages or labeling schemes and why they find them credible or not. However, the limited number of studies found, illustrated below, give an indication of the situation.

3.1. Consumers

The UK consumer is well represented on the public forum with a number of different governmental and non-governmental organizations focusing on protection of the consumer rights. When interviewing these major consumer organizations, there appears to be, however, very little work done to establish a deeper understanding of the consumers' perceptions of trust and belief in quality assurance schemes and labels. One explanation may be the fact that many of these schemes are still so new.

In the survey on consumer attitudes, published by the UK National Food Standards Agency, 35% of the respondents raised concerns about the accuracy of information found on labels. It was, however, also found that 10% of those questioned looked regularly for ethical information when purchasing food products, within this category the survey includes production methods, animal welfare and free trade [6].

A few Swedish studies have, over the past decade, investigated consumer perceptions of environmental issues, eco-labels and food issues. The most common scheme communicated through a label is, in Sweden, the KRAV-label for organic products. The Swedish Consumer Agency commissioned three interview surveys during the 1990s on, among other issues, consumer recognition of the various food labels. These surveys indicate that recognition is relatively high, typically exceeding 50% each year, and still increasing. The organic label and retailers' private brands are generally connected to issues like "decreased environmental impacts" or "organic production". They are typically seen as very or fairly trustworthy by 50–80% of the public [7].

The consumers in the survey were also asked what labeling schemes they know are related to food products. More than 10% could not point at any scheme and about 40% suggested the Nordic eco-label (the Swan) for goods and services. However, the KRAV-label was the most known food product label while the knowledge of other schemes was far lower.

Table 1
The different building blocks, their definition and examples of their application

Building block	Definition	Example of application
Ownership	The organization or stakeholders that has initiated and stands behind the scheme, manages it and, in a sense, also guarantees the trustworthiness.	Organic labeling schemes—the ownership structure was found to have a wide spectrum of structures; from state ownership such as the Bio-Siegel (Bio Seal) in Germany or the Stats Kontrolleret Økologisk (State Controlled Organic) in Denmark to private companies such as the Ecocert in France. However, farmer associations, multi-stakeholder bodies, non-profit organizations, regional promotion organizations and trust companies are the other types of organizations behind organic labeling observed in the European market.
Stakeholder dialogue	The means of communication between the institutions, organizations, groups or people involved with or attached to the labeling schemes in order to reach a common ground and acceptance for the criteria.	Integrated Production and Organic Farming—they consult stakeholders like consumer groups, NGOs and governmental institutions when they develop the criteria and requirements for the scheme. National conferences or yearly meetings are forums for such dialogues. As most of these individual schemes are members of international accreditation bodies, they also work to influence them. There are, however, no examples found of consumer or producer polls preceding the criteria development.
Quality assurance scheme	A framework of standards for production and performance; Verification for compliance with these.	Organic Farming—Most organic labels are established in line with an international primary production concept. The quality assurance criteria are therefore developed in accordance with internationally or nationally well-accepted frameworks by the certification bodies. For example, in the case of 20 organic agriculture certification labels found in the European market, the production criteria are based on guideline standards or regulations developed either by International Federation of Organic Agriculture Movements (IFOAM) or Council Regulation (EC) No. 2092/91 or Codex Alimentarius (9). This wide range of separate requirements is one of the major reasons for the boom in alternative certification organizations in the Western European countries.
Traceability	The systems used to describe the production history of the product and any further processes that the food may undergo along the way from the primary producer to the consumer. The most common traceability tool used by the labeling schemes is found to be documentation follow-up.	Organic Farming and Integrated Production—have to present the invoices and certificates of the raw materials or the final product to the scheme inspectors. Farm Quality Assurance—traceability in the meat supply chain is achieved with the use of ear tags and a passport system. Information and Communication Technology (ICT) based applications such as use of Internet via product code inputs are also emerging.
Marketing	Marketing efforts refer to the target markets addressed by the schemes and the promotion efforts performed by the labeling scheme owners in order to enhance their market acceptance.	Producer and retailer managed labels—get their message to the private consumers via their corporate marketing campaigns. Retailers, for instance, usually have their own special marketing groups for their own brands, which introduce general guidelines for promotions of the schemes at the individual store level. Regional labels—Farmers' markets, which offer their products from the defined region or even from a single farm directly to the private consumers via their shops or market areas.
Transparency	The openness for internal and externals parts to observe, comment and influence the criteria development and verification processes. It refers to the opportunities of these stakeholders who do not regularly participate in the set stakeholder dialogues to understand what priorities and decisions that form the background to the schemes and finally result in requirements, criteria etc. The transparency is judged depending on level and ambition of openness, extent of stakeholder involvement and communication and follow-up.	Organic farming and Integrated Production—high level of involvement of stakeholders for the criteria development is found with these labels found to be important. These interest groups are mainly consumers, scientists, traders and ordinary citizens. However, there is a need for initiatives from these stakeholders. The schemes publish, not least online, substantial information but when individuals want additional information they have to either phone or write to information departments. As most of the quality assurance schemes were found to be open and willing to share information.

H. Nilsson et al. / Journal of Cleaner Production 12 (2004) 517–526 521

Confidence reasons like "trust the products/I know what I get" were mentioned by 5% of the frequent buyers as main argument to buy labeled products. Both frequent and infrequent buyers of organic foodstuff pointed out environmental concerns, content of chemicals, and animal welfare, safety, quality and health aspects as main advantages of labeled products.

A broader group interview survey from 2001 commissioned by the Swedish Consumer Agency [8] confirms the picture. The purpose of the survey was to study the focus group participants' awareness and interest in environmentally related consumption issues and according concrete actions. Furthermore, the purpose was to map out what is needed to promote environmental conscious behaviour.

Animal welfare proved to be of significant importance in the participants' environmental view and they therefore claimed to prefer Swedish meat products due to perceived higher ethical standards. Some consumers also purchased locally produced foodstuffs when possible in order to assure the origin, or products marketed with various claims like "eggs from free-range hens" and "happy pigs" etc. However, an impression is that the connection between ethical values and organic food production is rather weak.

In a recent study, panel members expressed uncertainties about the eco-labeled products' superior environmental performance and questioned if they are better, and if so to what extent? In contrast however, many consider the KRAV label (for organic food) as trustworthy while some were suspicious of the concept of organic production and questioned the difference between these products and conventional and believed that organic foodstuff was a market gimmick. This indicates the issue of trust: the consumers in the panel do not, despite the awareness of what the organic food stands for, have sufficient belief in the claimed benefits of the scheme.

3.2. Producers

Despite thorough research it was again found that little research has been done in this area, this has lead to much of this section being anecdotal rather than based on a conclusive literature review.

Why are the producers motivated to label their products and establish new schemes? One of the main reasons is the assurance of consumer trust in their brands and products. Brom argues that there has been an increase in recent years in consumer concerns regarding food safety and the production methods being used. The producers feel that there is a growing demand from some consumer groups for assurances that the products they buy are of a credible quality and that they also are willing to pay a price premium for that [9].

The Soil Association, an organic certification body in the UK, has good contact with both producers and consumers when it is setting its standards. Producers have to be able to display certificates in order to assure consumers that the products are organic [10]. The need for labels is very much consumer driven and producers are finding that they have to listen to what the consumers want. There are also other reasons that motivate the producers to place some sort of label on their product or join an accreditation scheme. One of the most significant is to gain a competitive advantage over their competitors and differentiate their products from the others on the market. There has also been a change in policy with support being taken away from farming so that they have to compete on a world stage, and therefore feel that they have to differentiate and offer something else to the consumer Ayland suggested that there are three ways for the producers to assure the consumers of the credibility of the scheme [11]:

- Have support of people who have the highest public trust on environmental issues, e.g. WWF,
- Have a totally credible and transparent process—by providing as much information as possible for the consumer,
- Have the support of the retailers, who may already have a degree of consumer trust, and get them to stand behind the schemes and support them and promote them in their stores.

Despite the relatively well spread knowledge about food eco-labeling initiatives, only 10% of the agricultural land in Sweden is currently converted to organic production and the market share for organic products is still small (2% of the Swedish food market in 2001) but it is growing by 30% every year [12]. In contrast, the newly established Swedish seal scheme, a farm quality assurance system promoted by the farmers' organization built on elements from integrated production, is hardly recognized or seen by the consumers covers about 25% of the cereals production for human consumption [13]. Other schemes, for instance pure Integrated Production, have smaller shares.

4. Analysis

The first part of this section discusses the main findings from section two and the implications the different 'building blocks' have for the creation of a scheme's credibility. The second part then discusses the elements that were found in section three when considering the labels through the eye of the consumer and if there is a gap between what the producers provide and what the consumers want.

4.1. What sort of credibility structure have the building blocks built?

4.1.1. Ownership

Among the variety of management structures found, primary producer and the retailer owned schemes are found to be very frequent, whilst multi-stakeholder ownerships are found to be quite rare. It is apparent that the retailer chains and food processing industries are backing up their labeling schemes with the credibility of their own brand names. The primary producer governed schemes either try to boost the credibility by pointing out domestic production as more trustworthy and appealing to connections to rural values or by alliances with more trustworthy bodies like NGOs, authorities etc. It appears that producer organizations that speak for all members also own and thereby set criteria, control and assure the specific qualities for labeling schemes.

For success in the build up of credibility, a reverse picture, i.e. involvement of many stakeholders that can put their good names behind the scheme was seen as a necessity. The involvement of consumer representatives especially in scheme boards prevents disputes in criteria set-up and also enhances the acceptance of the scheme in the market.

4.1.2. Stakeholder dialogue

Even if very few examples of multi-stakeholder ownership were found, most schemes do apparently recognize the need for stakeholder dialogue. At least in the criteria/requirements setting process most schemes do consult national leading players and opinion formers among consumers, researchers, government institutions and environmental NGOs and actor representatives along the supply chains. These groups often also participate on the boards of the schemes. This is seen to enhance the trustworthiness of the schemes. The stakeholder involvement may gain a better support for the schemes in the public debate when several actors feel involved and are able to influence the requirements.

The lack of full food chain actor coverage indicates that most schemes focus on a few selected quality aspects controlled by the participating actors. This implies that a number of quality aspects are excluded from the schemes but also that the superior quality can not be guaranteed in other aspects than those regulated.

4.1.3. Quality assurance schemes

Quality assurance criteria can be seen as the major property of the schemes and are well-defined for primary production while elimination of health and safety risks becomes the focal point for the activities of actors down the chain. The strictness of the criteria obviously depends on the quality concerns of the labeling schemes such as whether animal welfare is emphasized or lessening general impacts on environment is set to be the major objective. However, more importantly, the study found that the criteria strictness differs according to the level of intensity of both interaction with the food supply chain actors and rising of stakeholders' voice.

Still, the level of strictness was not observed as the major factor for credibility built-up. Rather, the process of criteria development together with the level of openness of this process gains importance. Furthermore, third-party verifications or use of independent auditors are found to be a common route used to enhance trustworthiness and impartiality. However, the thoroughness of the control systems is less transparent and audit reports are rarely public. Instead, the consumers seem to accept and trust the mere presence of control systems.

4.1.4. Traceability

The investigated schemes apparently prioritize the traceability issue for the sake of not mixing quality assured products with others. The systems are constructed in order to facilitate follow-ups for the controlling bodies, which also seem to have the appropriate tools to trace raw materials and products. The traceability systems would most likely gain importance for credibility build-up if they also allow the information collected to be passed on to the consumers. It would make it possible for them to both follow the production chain and get some kind of knowledge about the producers. The traceability systems of the current schemes do not allow that service.

4.1.5. Marketing

It was found in the case of retailer or producer owned schemes; the advantage of passing on the message of the label more effectively has been realized. These promotion methods allow the consumers to test the product physically via point of sale materials or life-style promotion efforts. This eliminates the sole need to make the consumers only believe in quality aspects performed by the actors in the beginning of the chain i.e. primary producers.

In relation to the above point, aggressive marketing is identified as a significant tool to aid the consumers identify themselves with a scheme or a brand which then leads to trustworthiness built-up.

4.1.6. Transparency

The transparency of the scheme is important in communication with producers who are to be recruited to the scheme. If consumers should be convinced to purchase labeled products, they should ultimately also know what quality aspects that are superior and what the schemes are doing to maintain and guarantee the levels. As mentioned in section two, the schemes show open attitudes though there is mainly a need for some initial efforts to contact scheme representatives, which may be seen as a barrier. In parallel to traceability, ICT could serve as a tool to make the information and communication easier accessible.

4.2. Consumer expectations and gains from the labeling schemes

4.2.1. Information needs of the consumer

The majority of labeling and quality assurance schemes recognize the importance of providing a credible label for the consumers that is attractive to the producers as well. The development of the Internet in recent years has allowed enormous amounts of information to be accessible for a wider audience at the push of a button. Many schemes have taken advantage of this technology and have made their standards and certification procedures available for viewing on their homepages. Although a large number of consumers are uncertain when it comes to who owns and manages the schemes. The control and ownership of the schemes actually constitutes one of the main differences between the food labels and traditional type I eco-labels, which shall be independent from specific groups of stakeholders. Most of the food labels are in fact controlled and managed by producers or retailers.

When asked about schemes in surveys a reoccurring comment by consumers was that they would like more information. This is obviously an area where the labeling bodies have to work harder to communicate their purposes, priorities and approaches to consumers. In surveys on consumers' purchase habits of organic food, the majority had faith in the label, but were actually uncertain to what the actual concept was all about and had difficulties in separate production practices and supposed environmental and quality benefits. This sort of information is mainly available on websites etc but labeling organizations may have to develop other methods to reach the consumers.

4.2.2. Motivation

The increasing distance between consumer and producer, both geographically and mentally is one reoccurring reasons for a lack of trust that appears in the literature. Most consumers no longer see where the food is produced, or who is producing it. In parallel, the farmer is no longer aware of the needs and wants of the consumer. In contrast to this is one of Rudolf Steiner's guiding principles behind the organic movement, the emphasis on local production and consumption of resources. To many this vision now seems distant. This trend towards isolation of the consumer and producer may partially explain the success of the farmers' markets initiative, which is spreading across Europe, though this initiative is still only on a small scale. Here consumers and producers can meet each other and discuss issues that are important to them, as well as buying farm fresh produce.

4.2.3. Consumers' past experiences

In recent surveys made by the Swedish organic certification body, KRAV, consumers who purchase organic food on a regular basis show a loyalty to the label and repeat their purchasing. In fact, even though consumers feel that they lack information, they still have faith in the label they see and are happy to purchase and repurchase products with organic labels.

4.2.4. The environmental potential of the label

Most schemes provide rationale in general terms for consumers to buy products labeled with their logos. A large percentage of consumers apparently believe that one of the main benefits of organic labels is their positive effect on the environment [14]. Whether the consumers are really aware of the impacts or merely presume them is unknown. Yet no efforts have been detected to connect the purchase of labeled products to environmental improvements or providing positive feedback to the consumers. Thereby the consumers have to trust the statements that the products and production methods are preferable but are not provided with any evidence that their actions actually make a difference. One would ask if this is a condition that holds in the long run without jeopardizing the trustworthiness of the schemes. As mentioned above, there have been scientific studies made on the environmental benefits of organic farming and other production methods. However, the interpretations of the results are still under debate.

5. Conclusions

5.1. Summary of the main findings

This paper attempted to answer the question of how European eco-labeling food quality assurance schemes typically manage and ensure credibility and how the consumers perceive the credibility of the schemes. The overarching question, "is there enough credibility?" though deserves some further elaboration. Quality assured foodstuffs still constitute rather small niche markets. They have, though, reached small but loyal consumer groups that probably find the quality assurance schemes sufficiently credible today. The reasoning behind the limited market penetration is not known, though willingness to pay, lack of knowledge or lack of credibility of the quality assurance could be considered as possible reasons.

The overall conclusion reached was that, despite the attempts of the different schemes to create credibility, the attempt to provide credibility fell short. To substantiate this opinion, the following points are a summary of the main findings of the research.

- No independent third party that formulates objectives, sets criteria, and follows-up compliance. When one or a few strong actors own and manage a scheme, others may either be excluded or have difficulties making their voices heard. The credibility of the system is then more related to their good names rather than the content.

- Unclear underlying values and objectives. The quality issues regarding foodstuffs are apparently mostly multidimensional. The distinction between different environmental aspects, ethical values, health issues, rural pattern development and maintenance is rarely explicit, which may be confusing and difficult for the consumers to know in what way the choice of the product makes a difference or what additional qualities they get. The few studies reported in the literature that have investigated perceptions of the objectives of the existing schemes show that consumers in some countries have a fairly good general understanding of them but appear to know less about their specific qualities.
- Criteria based and excluding. Referring to the multitude of quality aspects mentioned above, the prioritization of the most significant aspects to be controlled through criteria becomes substantially more difficult than, for instance, criteria setting in traditional eco-labeling schemes for products and services. This difficulty is perhaps accelerated by the unclear quality aspects of the schemes. A non-labeled product may perform better according to consumers' priorities than a product that complies with the criteria.
- Limited food supply chain coverage. Most of the current schemes set criteria for selected actors in the chain and ignore several of the other actors or aspects that may be of significant importance for the overall objective the consumer prioritize.

Throughout this article we have referred to the various schemes as eco-label like initiatives or schemes in comparison to true eco-labels. According to ISO, eco-labels can be divided into three different categories; third party eco-labeling schemes where an independent body develops criteria, based on impacts over the life-cycle, which a label has to meet to be approved (type I); Statements from Industry, which have to be justifiable (type II); and finally there are those based on life cycle analysis that follow ISO standards (14 040 series) (type III). Where the schemes covered in this research differ from the eco-labels mentioned above is that the life cycle approach is not taken, only selected elements of the environmental impacts of the production processes are taken into account, instead of the whole chain.

5.2. Observations for future improvements in labeling like schemes

5.2.1. Improvement of the information chain

Many Swedish consumers can identify quality assured labels even if labeled products only constitute a niche market, they would like to receive more information about the labels. Although pieces of the information are already available for the consumer, often on the Internet, they do not seem to be aware of this source. Instead one option would be to enlist the assistance of the retailer to play a more active role in the promotion and marketing of the schemes, as can already be seen in some supermarkets. This endorsement is an additional proof of credibility for the consumer, as the retailer is often seen as being in a position where they put the well-being of the consumers first. One of the Swedish consumer surveys shows that the majority of infrequent buyers of organic foodstuff have a fair understanding of what the label stands for, but reject the products due to the higher price. This could either be interpreted as low demand for/rejection of the offered added values, distrust in the quality assurance scheme or a vague understanding/little recognition of the added values. At least the latter explanation calls for improved information to the consumers if conscious decisions are desired.

5.2.2. Enhancement of marketing efforts

It is certainly not enough to just launch a label on the market. The scheme management has to back it up by professional promotion, public relations and education of the target market. For such purposes, separate sections for strategy development, as already done for audits, can be established. These subdivisions can also enhance collaboration between the scheme participants and improve the understanding of the consumers' needs through market research.

5.2.3. Expansion of the responsibilities of processors and retailers

The role of processors and retailers in a scheme has to be altered. They could be assigned not only to cope with the production standards but also implement strategies to make the consumer grasp the essentials of a label. This would also assist to reestablish the contact between producer and consumer that has been lost in recent years due to the centralization of the distribution channels.

5.2.4. Involvement of a diverse range of stakeholders

Particularly for the choice of ownership and criteria development, involvement of as many stakeholders as possible among international authorities, national leading players, opinion formers among consumers, researchers, government institutions and environmental and consumer NGO communities shall be intended, while verification for compliance shall be through third part audits.

5.3. Alternative to label like initiatives

The existing eco-label like initiatives cover various stages in the production chain, but certain sections are not covered, for example transport and packaging, which apparently are features considered important by some consumers. There is no labeling system that covers the entire food production chain which could install eco-efficiency in the production chain. A traditional eco-labeling structure according to ISO 14 020–24 could be an alternative. However, in the case of food production and products, where the interpretations of environmental aspects and priorities vary to the extent they do, such a system would probably show significant disadvantages. Abandoning the labeling principle as the means for communication would be an alternative measure to inform chain actors and consumers. For instance, quality declaration schemes would give producers the opportunity to differentiate their products and convey their messages and openly tell who is responsible for the products and their added qualities. It would also move the schemes away from the implicit neutral impression an eco-label gives.

The disadvantage would be a slightly more complex scheme but the advantages are numerous. It makes the scheme transparent in respect to whom the sender of the message is, what quality aspects and objectives that actors want to offer the consumer and how well the products perform. An indicator system would be of use when it comes to communicate the quality aspects to them purchasing the food products. For most quality dimensions consumers cannot ascertain the quality before or during the purchase. In order to make purchase decisions, consumers, therefore, have to form quality expectations. They do this by inferring expected quality from the information at hand at the time of purchase [15].

This alternative approach could measure appointed quality aspects in indicators for the whole food product chain and report them to interested parties and consumers. All products can be measured, i.e. no producer or product is excluded and the consumer could then, based on his or her preferences, choose products that fulfill desired quality aspects in relation to the cost of the products. Additionally such a system would be educative as it also provides the background information to the indicator measurements However, no system of this kind is currently in use and has therefore to be developed through research and development on existing food product chains.

It is the belief of the authors that there is a need to change the traditional thinking in terms of production methods and scale of the production processes. The arguments for economies of scale have begun to show deficits; there are signs of consumers and policy makers that demand traceability and accountability from food producers, and the industry is finding that there is a need to concentrate on producing quality products rather than producing large quantities where the safety or quality of the product cannot be guaranteed.

6. Homepage addresses and interviewees contacted for section two of the paper

(These references are not cited throughout the paper):

For organic labeling schemes;

AB (Agriculture Biologique). (2001) http://www.agriculture.gouv.fr/alim/sign/agri/welcome.html
ALMAVERDE BIO. (2001) http://www.apofruit.it/eng/almabio/
BIO SUISSE. (2001) http://www.bio-suisse.ch/e—index.html
Bioagricert. (2001) http://www.bioagricert.org/
Bioland. (2001) http://www.bioland.de/index.htm
BIO-Zeichen. (2001) http://www.ama.at/AMA-Marketing/Aktivitaeten/bio.html
Das Bio-Siegel. (2002) http://www.bio-siegel.de/intro-49.htm
Demeter. (2001) http://www.bio-siegel.de/intro-49.htm
ECOCERT. (2001) http://www.ecocert.fr/dyn/?MIval=PageAccueil&lg=en
EKO. (2001) http://www.skal.com/english/—skal/frameset1.htm
Ernte für das Leben (Harvest for Life). (2001) http://www.bio-ernte.at
Eva Mattson. Standards Manager *KRAV* . Telephone Interview (2001, August 1)
Gunilla Anderson. *Ekologisktmarknadscentrum.* Personnel Interview Lund (2001, June 5)
Gunnar Rundgren. General Manager *GroLink AB*. Telephone Interview (2001, July 16)
Kagfreiland. (2001) http://www.kagfreiland.ch/
KRAV. (2001) http://www.krav.se/
Naturland. (2001) http://www.naturland.de/englisch/frame—defs/framedef.html
Ø-merket. (2001) http://www.debio.no/diverse/deb—eng.htm
Rasmus Kjeldahl, Sektor for økologi Sektorchef, *Plantedirektoratet—Ministry of Food, Agriculture and Fisheries*, Telephone Interview (2001, July 23)
Soil Association. (2001) http://www.soilassociation.org/
SÖL. (2001). Organic Agriculture in Greece by Nicolette van der Smissen. http://www.organic-europe.net/country—reports/greece/default.asp
SÖL. (2001). Organic Agriculture in Spain by Picazos, J, Parra, A. http://www.organic-europe.net/country—reports/spain/default.asp.
Stats Kontrolleret Økologisk. (2001) http://www.foedevaredirektoratet.dk/java—enab/f—uk.html

For integrated production schemes;

AMA Gutesiegel. (2001) http://www.ama.at/
Assured Produce Scheme (AP). (2001) http://www.littleredtractor.org.uk/assured—veg.asp
Flandria. (2001) http://www.vlam.be/sitemap/index—en.html
Milieukeur. (2001) http://www.milieukeur.nl/

For farm quality assurance schemes;

Assured British Meat. (2001) http://www.abm.org.uk/
Assured British Meat. Scheme Standards and Inspection Protocols. http://www.abm.org.uk/farmers/main3.htm [2001, August 8]
Assured Chicken Production (ACP). (2001) http://www.abm.org.uk/farmers/main3.htm.
Assured Combinable Crops. About ACCS. http://www.assuredcrops.co.uk/ACCS/accs002.asp [2001, August 1]
British Farm Standard (The Little Red Tractor). (2001) http://www.littleredtractor.org.uk/
Cerafel. The Growers' Organisation. http://www.cerafel.com/en/organisation.html [2001, August 1]
Peter's Farm. (n.d.) http://www.petersfarm.com/menu.asp: [2001, August 1]
Rainbow Growers. (2001) [Online]. Available http://www.rainbow-growers.nl/
Scottish Pig Industry Initiative (SPII). (2001) http://www.scottishqualitypork.com/
SQF. (2001) http://www.sqf.wa.gov.au/
The Green Nature Group. (2001) http://www.green-nature-group.nl/
For retailer or processor led schemes;

Änglamark. (2001) http://www.anglamark.com/
Baby Organix. (2001) http://www.babyorganix.co.uk/
Bell Natura. (2001) http://www.admin.ch/ch/d/sr/c910—132—5.html.
CoopNATURAplan. (2001) http://naturaplan.coop.ch
FDB. (2001) http://www.fdb.dk/
Hanegal Organic Meats. (2001) http://www.hanegal.dk/
Hipp. (2001) http://www.hipp.de/
Hruschka, Thomas. (thomas.hruschka@17und4.at). (2001, June 13). Re: Uber Ja Naturlich! E-mail to Burcu Tuncer (burcu.tuncer.118@student.lu.se)
Ingemar Hjelm, *Affarsomrade Gröna Konsum Miljosamordnare* (*Environmental Coordinator at Gröna Konsum*). Personnel Interview Göteborg (2001, June 28)
Ja Naturlich. (2001) http://www.janatuerlich.at/janatuerlich/default.asp
Migros 7 Punkte Garantie (Migros 7 Points Guarantee). (2001)
http://www.miosphere.ch/d/produkte/fleisch1.php3
Migros BIO. (2001) http://www.miosphere.ch/d/produkte/labels.php3#Anchor-Migros-Bio-Produktion-49575
Nestlé Alete. (2001) http://www.alete.de/
Tengelmann's Naturkind. (2001) http://www.tengelmann.de/deutsch/umweltschutz/super.htm
URTEKRAM Denmark. (2000). About the company. http://www.urtekram.dk/uk/Page100.htm. [2001, August 1]
For regional schemes;

Bäuerlicher Rastmarkt Aurach. (2001) http://www.rastmarkt.de/index.htm
Chris Day. *Vellinge Municipality. Environmental Manager.* Personal Interview (2001, July 25)
Comté Cheese. (2001) http://www.comte.com/
Food From Britain. (2001) http://www.foodfrombritain.com/speciality—food.html
Gåsamärkt. (2001) http://www.vellinge.se/turism/gasen.htm
Hofmarke. (2001) http://www.hofmarke.at
Hofpfisterei. (2001) http://www.hofpfisterei.de/
Natürlich Aargau. (2001) http://www.ruebli.ch/home.htm
Nautilus. (2001) http://www.nautilusorganic.nl/
Rolf Axel Nordström. *Ängavallens Gård Manager.* Onsite Interview. (2001, July 26)
Somerset Levels and Moors. (2001) http://www.somersetorganics.co.uk/
Tastes of Anglia. (2001) http://www.tastesofanglia.com/
Ängavallens Picknick Park. (2001) http://www.angavallen.se/shopbox/angavallen/a—startsida.htm

References

[1] Emerson T. Where's the beef? Newsweek 26-02-2001.
[2] Palm B, Rydén D. Kunder vill ha makt over maten. Sydsvenskan 04-03-2001.
[3] Pretty J. The living land. London: EarthScan Publications Ltd, 1998.
[4] Salzhauer AL. Obstacles and opportunities for a consumer ecolabel. Environment November 1991
[5] Jones GR. Organizational theory text and cases. New Jersey: Prentice Hall, 2000.
[6] Food Standards Agency. Consumer attitudes to food standards—UK report. London: Food Standards Agency and COI Communications, 2002.
[7] Konsumentsverket. Konsumenten och miljön-Resultat från en undersökning av svenska konsumenters miljömedvetenhet. Konsumentverkets rapport 1995/1996:13. (The consumer and the environment - results from a survey of the Swedish consumers environmental awareness).
[8] Konsumentverket. Mitt hem är ingen sopstation Konsumentverket rapport 2001:11 (My home is not a refuse dump).
[9] Brom FWA. Food, consumer concerns, and trust: food ethics for a globalising market. Journal of Agricultural and Environmental Ethics 2000;12:127–39.
[10] Jones, Lisa (ljones@soilassociation.org) (2002-03-14) Research into consumer confidence in organic labeling Email to Helen Nilsson (Helen.nilsson@iiiee.lu.se)
[11] Ayland R. What's in a label? Green Futures 2001;September/October:41.
[12] KRAV. Fakta om KRAV (the facts about KRAV) 2002 at http://www.krav.se/allmana.html.
[13] Svenskt Sigill (Swedish Seal) (2002) at http://www.krav.se/allmana.html.
[14] Holmberg Hans-E, (1999)
[15] Gruner K. Perceptions of food quality—expectations, inferences and process characteristics. In: Food Chain 2001—Safe, Sustainable, Ethical. SLU, Uppsala, Sweden, 14-16 March 2001, 29-31.

[22]

The Impact of Environmental Labelling on Consumer Preference: Negative vs. Positive Labels

Gunne Grankvist, Ulf Dahlstrand and Anders Biel

ABSTRACT. Eco-labels in use today signify environmentally benign outcomes: "Choose this product, it is better for the environment than the average product." Another strategy would be to indicate negative outcomes with the purpose of trying to persuade consumers to avoid a product: "Do not choose this product, it is worse for the environment than the average product." In a computer-based experiment, it was investigated how these two types of labels affected preference for some everyday products. Individuals who had a weak or no interest in environmental issues were unaffected by either kind of label. Individuals with an intermediate interest in environmental issues were more affected by a negative label than by a positive label. Individuals with a strong interest in environmental protection were equally affected by the two kinds of labels.

Imagine that you are facing a choice between two alternatives (A and B) and that you receive a new piece of information that is either positive for alternative A or negative for alternative B. Which kind of information would be most likely to move your preference towards alternative A, that is, would the pull force of the positive information be stronger or weaker than the push force of the negative information? To make the example more tangible, imagine that you are entering a grocery store with the intention of buying tomatoes. In the fruit and vegetable section you find just two kinds of tomatoes, both looking fresh. However, one kind is cheaper than the other one and you go for the cheaper. Just as you are about to pick up the tomatoes, you notice that they have a red eco-label (meaning worse than average for the environment), while the more expensive ones have a yellow eco-label (indicating about average in the impact on the environment). For most people that piece of information would move their preference away from the cheaper, but not necessarily all the way to a reversed choice. The push or avoidance force of the red label can be compared with a pull or approach force of a green label (signifying better than average impact on the environment) in the case where the more expensive tomatoes have a green label and the cheaper a yellow label.

The main question in this study is concerned with the importance of a positive as compared to a negative piece of environmentally related information for a consumer's product preference in a choice situation. In our view, this question cannot be answered in a simple way; its ramifications will be discussed later. Before presenting the theoretical background, we will summarize current research on eco-labels and their effect on consumer decision-making.

ECO-LABELS AND CONSUMER CHOICE

The intention behind the introduction of eco-labels and organic food labels is to make it possible for the consumers to distinguish products less harmful to the environment from other products. Furthermore, consumers' preference for eco-labelled products could give producers of relatively environmentally friendly products a competitive advantage, allowing them to gradually push less environmentally friendly products out of the market. Such a competitive advantage could give companies an incentive to develop new, and more environmentally benign, products (Thøgersen, 2002).

Results from a number of studies suggest that two of the major reasons why consumers choose eco-labelled food products are consideration for the environment and/or for their own health (Chinnici, D'Amico, & Pecorino, 2002; Davies, Titterington, & Cochrane, 1995; Fotopoulos & Krystallis, 2002; Harper & Makatouni, 2002; Nordlund, Garvill, & Marell, 2001; Tregear, Dent, & McGregor, 1994; Wandel, 1994; Wandel & Bugge, 1997; Wier & Calverley, 2002).

However, there is no straight and simple relationship between attitudes and pro-environmental behaviour. Magnusson, Arvola, Koivisto Hursti, Åberg, and Sjödén (2001) found in a study conducted in Sweden that women, university graduates, and young people (18–25 years) had more positive attitudes toward eco-labelled variants of some everyday food products. The differences were statistically significant, but rather small in absolute values. The gender and age differences found in attitudes were not found in relative purchase frequency of eco-labelled alternatives of everyday food products.

Nevertheless, some consumers may be more attuned to eco-labels than others. In a survey conducted in Norway (Torjusen, Lieblein, Wandel, & Francis, 2001) it was found that traditional food quality aspects such as appearance, freshness, and taste were important to all consumers. In addition, those who purchased eco-labelled or organic foods were more affected by ethical, environmental, and health issues. Torjusen et al. (2001) defined traditional food quality aspects as "observational" traits and the second category, characterized by ethical, environmental, and health concerns, as "reflection" traits. The "reflected" consumer was concerned about many parts of the food system: how food is produced, processed, and handled, and how these circumstances affect people, animals, and nature. Eco-labels could provide a signal and activate such reflections. The majority of consumers may not be easily swayed though.

A prerequisite for the use of an eco-label is knowledge of its existence. In a survey conducted in the Nordic countries, Palm and Jarlbro (1999) found that in Denmark 31% of the respondents mentioned the national organic food label (the ø-label) unaided. In Sweden the national organic food label was mentioned – unaided – by 16% of the respondents. Corresponding figures were in Finland 5% and in Norway 1.4%. The unaided recall of the EU eco-label, the Flower label, was below 2% in all four countries.

The recognition of a label is, however, not the same as understanding even the approximate meaning of it. Several studies have found that consumers often have a hard time in understanding what labels are intended to communicate, and uncertainty about what a label means could be associated with mistrust (see Thøgersen, 2002). Thøgersen showed that consumers pay attention to and use environmental labels in their buying decisions only if they trust such labels. Attitudes towards eco-labels were positively correlated with consumers' trust in and, as a consequence, the degree of attention paid to eco-labels. Furthermore, attention to and use of eco-labels depends on how highly the consumer prioritizes environmental protection, and how strong his or her beliefs are in the purchase of eco-labelled products as a strategy towards achieving this goal (Thøgersen, 2000). However, to highly value environmental protection is not tantamount to choosing eco-labelled alternatives. Many studies have reported weak correlations (i.e., weak linear relationships) between environmental concern and the choice of eco-labelled products (see, for example, Magnusson et al., 2001).[1]

Multilevel Label Systems

It seems likely that consumers with some basic knowledge of eco-labels associate eco-labelled products with the implicit message, "Choose this product, it is less negative or 'better' for the environment than the average product." A further development of the labelling system in use today would be a system with several levels, including labels associated with neutral and negative outcomes. The purpose of negative environmental outcome labels would be to get consumers to avoid certain products. Such labels would indicate: "Do not choose this product, it is worse for the environment than the average product."

The EU Energy Label (see, e.g., Saveenergy, 2003) is an example of such a multilevel label system. A scale with seven levels, ranging from A to G, with "A" as the most efficient and "G" as the least efficient, is used. Label A is coloured in dark green, and B and C are in lighter green colours. Label D holds a middle position and is yellow. Labels E and F are in light red colours, while label G is coloured in dark red. The red labels (E, F, and G) could be interpreted as indicating negative environmental outcomes and the yellow label a neutral outcome.

The EU Energy Label system was introduced in 1992 by an EU Council Directive (European Union On-Line, 2003), and the Member States are obliged to adopt the Energy Label system as a national law. The introduction of this multilevel energy label system, embracing levels associated with disapproved outcomes (high energy consumption), shows that a label system that includes negative outcomes can be introduced by legislation. Below, the theoretical underpinnings of the expected effects of information about positive and negative outcomes are presented along with our hypotheses about consumer reactions to positive and negative eco-labels.

THEORETICAL BACKGROUND

Positive vs. Negative Information

One well-known description of the relative importance of gains and losses (positive and negative information) is the value-function curve proposed by Kahneman and Tversky (1984). The value function has three essential characteristics: "*Reference dependence*: the carriers of value are gains and losses defined relative to a reference point. *Loss aversion*: the function is steeper in the negative than in the positive domain: losses loom larger than corresponding gains. *Diminishing sensitivity*: the marginal value of both gains and losses decreases with their size" (Tversky & Kahneman, 2000, p. 143). The first two characteristics have implications for the present study. Loss aversion implies that the negative (red) label would move the preference more than the positive (green) label. However, reference dependence signifies that individuals will react differently, depending on the person's present reference point. This point will be elucidated in the section on prevention versus promotion focus.

There are many studies in different areas showing a greater effect for negative than for positive events. For example, negative traits are given greater weight than positive traits in forming an impression of a person (Anderson, 1974). Furthermore, negative events evoke stronger physiological responses than positive events. In studies of attitudes where subjects have been presented with opinions they agree or disagree with, one has found a greater arousal in the case of disagreement (Schwartz, 1990). Further support for the greater impact of negative information comes from the domain of attention research. People attend to negative information more often and for a longer period of time than they do to positive information (Fiske, 1980). In crowds, threatening faces "pop out" more often than happy faces (Hansen & Hansen, 1988). Wegner and Bargh (1998) concluded that negative social information has permanent access to a judgmental process, provided that the information is relevant and applicable (see also Higgins, 1996). The evidence for a greater effect of negative as compared with positive information dominates, but there are studies showing the opposite pattern (e.g., Skowronski & Carlston, 1987).

Environmental Information and Prevention vs. Promotion Focus

According to self-regulatory theories, whether positive or negative information has the greatest impact on people's judgments depends on the current goals of the individual. In this research domain, importance is attached to the valence of a *reference* or *regulation value* that can be either positive or negative (Carver & Scheier, 1990). Consequently, the direction of a motivated movement may be to approach a positive regulation value or to avoid a negative one, that is, regulating oneself may be an attempt to reduce a discrepancy between a present self-state and a desired self-state or an attempt to amplify a discrepancy to an undesired self-state. According to self-discrepancy theory (Higgins, 1987), people have representations of *ideal* selves, which consist of the attributes that one would like to possess (i.e., a representation of one's hopes, aspirations, or wishes); and people have representations of *ought* selves, which consist of the attributes that one believes one should or ought to possess (i.e., a representation of one's sense of duty, obligations, or responsibilities). As has been shown by Higgins and others (Higgins, Roney, Crowe, & Hymes, 1994; Higgins & Tykocinski, 1992), regulation in relation to an ideal self-guide involves a greater sensitivity to the presence and absence of positive outcomes, whereas regulation in relation to an ought self-guide makes a person more sensitive to the presence and absence of negative outcomes. Hence, a person can primarily have access to a *promotion focus*, related to an ideal self, or to a *prevention focus*, related to an ought self. Accessibility may in turn be affected by more permanent or personal factors, or by situational factors such as priming.

We hypothesize that people with a strong environmental concern tend to have a promotion focus with respect to environmentally relevant behaviour, while people with a less strong concern tend to have a prevention focus in this respect. The hypothesis is based partly on the finding that there is a substantial correlation between environmental concern and the rated gravity of environmental problems (Dahlstrand & Biel, 1997). Thus, the greater the problems seem to be, the greater one's engagement to improve the state of affairs should be. In case one does not regard the environmental problems as grave, and consequently, to a larger extent is prepared to accept the present state of affairs, one should be more concerned with the prevention of things getting worse than with their improvement. Accordingly, people with a prevention focus will act in line with Kahneman and Tversky's value function, while people with a promotion focus will not.

In the present context, a green eco-label may be seen as representing the presence of a positive outcome whereas the yellow label represents a neutral outcome in the sense that no change in the present state of the environment is likely. Likewise, a red eco-label may represent the presence of a negative outcome, that is, one which makes the present state of the environment worse.

HYPOTHESES

The application of Higgins' self-discrepancy theory and the assumption that environmentally strongly concerned people have an ideal self-guide as regards environmentally friendly behaviour lead to the hypothesis that this group of people will be affected at least as much by positive environmental information as by negative information. However, it should be noted that this group is assumed to have improvement as a regulation value, rather than "status quo," with the implication that both a yellow and a red eco-label represent a deviation from the ideal, but to different degrees. Thus, for this group a yellow label, when compared to a red label, may be as important as a green label when compared to a yellow label.

Furthermore, when making the assumption that people with an intermediate environmental concern are to a higher extent regulating their environmentally relevant behaviour in relation to an ought self-guide, one arrives at the hypothesis that this group is more affected by negative environmental information than by positive information. Thus, the difference between these two groups in terms of self-regulation is that the latter group tends to avoid things that are environmentally bad, whereas the former group reaches for things that are environmentally good. Assuming that the latter group more or less accepts the present state of the environment, the difference between a green eco-label and a yellow one should be negligible for its members, while the difference between the yellow and the red one should be large as for them, the red eco-label represents the regulation value that should be avoided. Obviously, people who are not environmentally concerned are assumed to be not at all affected by environmentally relevant information.

The typical environmental information that meets the customer is an eco-label. This label, e.g., "the Nordic Swan," "Good Environmental Choice," or "the German Blue Angel," signals that, with regard to environmental qualities, the product is to be preferred to a non-labelled product. Hence, the information is of a positive nature. To test the interaction effect put forth above, the present study introduced, along with the positive label, a label signalling negative environmental consequences. Based on Higgins' proposals, applied to the environmental domain, two hypotheses were derived.

Hypothesis 1. Individuals with a stronger environmental concern will be more affected by information about both positive and negative environmental outcomes than those with a weaker environmental concern

The second hypothesis addresses the suggestion that the effect of negative and positive information will not be equally strong for all individuals:

Hypothesis 2. Individuals with an intermediate environmental concern will be more affected by information about negative than by information about positive environmental outcomes.

METHOD

Participants

Forty undergraduates at Göteborg University in Sweden, 25 women and 15 men, participated in return for one cinema ticket at a value of 80 SEK (about 8 US$). Their mean age was 26.2 years (SD = 6.7) within a range of 19 to 56 years of age.

Design and Procedure

Upon arrival at the laboratory, participants were told that all instructions would be displayed on a computer screen. Participants were seated in private booths facing a personal computer. All instructions and other information were presented in the Swedish language. Software QuestionAsker 2000 (DirSoft©) was used for all applications in this experiment.

Initially participants were informed that a number of pairs of products would be presented and that their task would be to indicate product preference within each pair. Next, a three-level eco-label system was introduced. This system was adopted from the design of a traffic light. Accordingly, the colour of the top signal or label was red and defined as: "Compared to other products of the same category, the environmental consequences of this product are much worse than average." Environmental consequences associated with the yellow signal were described as "average," while environmental consequences defined by the green label were "much better than for the average product."

Sixteen pairs of products, eight food (bread, potatoes, milk, meat, apples, tomatoes, coffee, and pasta) and eight non-food commodities (washing-up detergent, soap, washing powder, toilet paper, battery, T-shirt, a light bulb, and writing paper), were selected as target stimuli. Two pictures, one for each product in the pair, were presented simultaneously and described on three to four attributes such as price, quality, country of origin, and expiry date. In addition, each commodity carried an eco-label. Two versions of each pair were created. In one version, the control condition, each alternative was labelled with a yellow label.[2] In the other version, the experimental condition, one of the alternatives was depicted with a yellow label while the other alternative carried a red label for half of the cases, and a green label for the remaining cases. The order of the experimental and control conditions was counterbalanced between subjects. All pairs were presented in two trials, and half of the trials were controls, i.e., both alternatives carried a yellow label. The order of presentation within trials was individually randomized. The second trial was preceded by a one-minute pause to avoid subjects remembering, and adjusting to, previously made ratings.

A session lasted for approximately 30 minutes. All participants were thoroughly debriefed and paid.

Measures

For each pair of products, one alternative designated A and the other B, respondents were asked to indicate their product preference on a 21-point scale anchored on "I would definitely choose product A" (1)/ "I would definitely choose product B" (21). The midpoint of the scale (11) was defined as "I could just as well choose product A as product B". Furthermore, subjects were further asked to rate "the importance attached to environmental consequences when purchasing everyday commodities." Ratings were made on a 7-point scale anchored on "not at all important" (1) / "very important" (7). Finally, subjects were asked to provide information about their age and gender.

RESULTS

For each participant and pair of products the difference in preference ratings between the control (that is, where both alternatives had a yellow label) and the experimental condition (that is, where one product had a yellow label and the other product a red or a green label) was calculated. Next, for each participant one mean preference change score was calculated across the sixteen pairs of products.

On average, participants changed their preference ratings 2.89 points in the experimental condition as compared to the control condition. This difference in preference ratings was significant; $t\ (39) = 5.28$, $p < 0.001$. Very few participants changed product preference in an unexpected direction.[3]

The first hypothesis stated that the stronger their environmental concern was, the more participants would be affected by information about environmental consequences. The corresponding correlation between preference change in the experimental condition, as compared to the control condition, and the importance attached to the purchase criterion "environmental consequences" was 0.55, $p < 0.001$. That is, a significant positive correlation was obtained in support of the first hypothesis.

Next, for each individual the difference between the control condition and the mean change in preference associated with the positive and the negative labels, respectively, was calculated. In Figure 1, the effects of the positive and the negative label are presented as a function of environmental concern. While individuals with low or no environmental concern were unaffected by environmental information, negative or positive, those with a strong environmental concern were affected the most. This group was affected as much by positive as by negative information. Individuals with an intermediate concern displayed a differentiated pattern in that they were affected more by negative than by positive information.

These results are in line with the second hypothesis that individuals with an intermediate environmental concern would become more affected by information about negative than by information about positive environmental outcomes. According to Hypotheses 1 and 2, an interaction between type of environmental information (positive or negative) and environmental concern was expected. To test this interaction effect, a new dependent variable was calculated. For each individual the difference between mean preference change associated with the positive and negative labels was calculated.[4] This measure was used as the dependent variable in a non-linear (quadratic) regression analysis with environmental concern as the independent variable. Support for the two hypotheses would result in a difference value close to zero for weak or strong environmental concern and a negative difference value for an intermediate environmental concern;

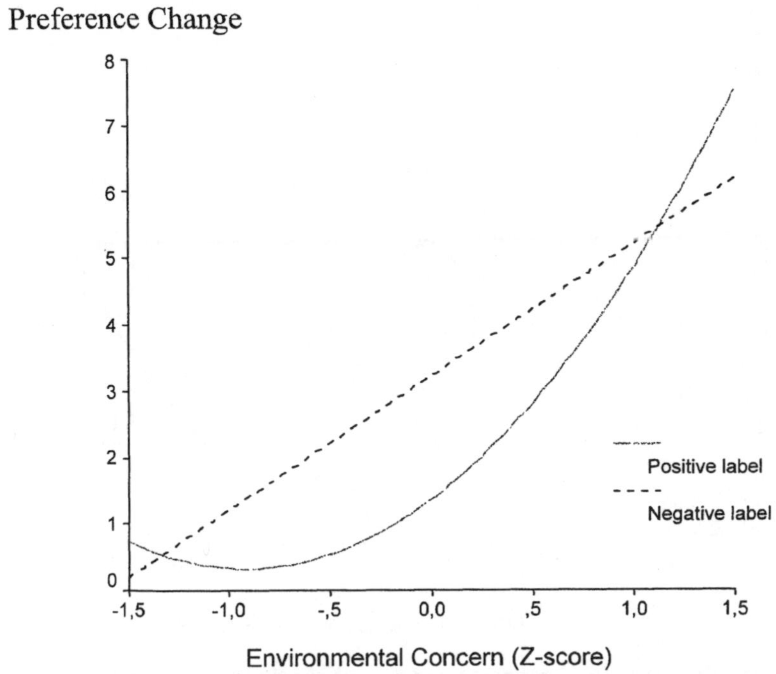

Note: The regression function associated with the effect of the positive eco-label was $f(X) = 1.36 + 2.26X + 1.23X^2$. The corresponding function for the negative eco-label effect was $f(X) = 3.21 + 2.00X$.

Figure 1. Mean change in product preference due to positive and negative eco-labels as a function of environmental concern.

that is, a U-shaped relationship between environmental concern and the difference variable.

In Figure 2 the regression line associated with the quadratic regression analysis is shown. The shape of this line, and the fact that a significant quadratic trend factor was found ($p < 0.02$), support the second hypothesis.

Finally, the analyses presented above were conducted on the subsets of food and non-food products. For both subsets, the tendencies found were in line with Hypotheses 1 and 2. However, some of these tendencies were not strong enough to reach statistical significance in each subset. A possible explanation is that since these measures, that is changes in preference ratings, were based on only half as many products as in the aggregate analysis, they became less reliable.

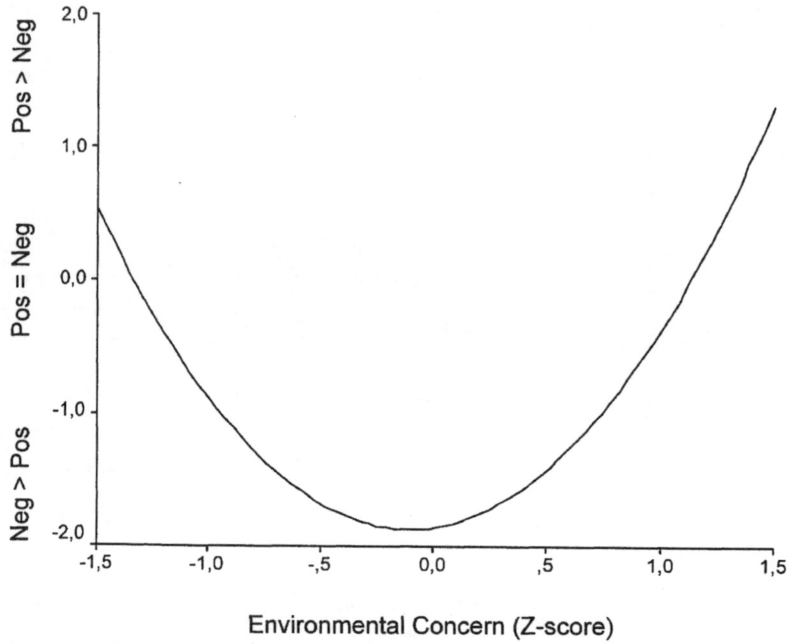

Note: The regression function associated with the difference variable was $f(X) = -1.86 + 0.26X + 1.25X^2$. $R^2 = 0.16$, $R^2_{adj} = 0.12$.

Figure 2. Relative preference change, the effect of negative eco labels subtracted from the effect of positive labels, as a function of environmental concern.

DISCUSSION

As predicted, information about environmental outcomes provided by eco-labels did influence product preference. Furthermore, participants who attached high importance to the purchase criterion "environmental consequences" were more affected by the labels than participants who valued environmental consequences less.

Consistent with Higgins' self-discrepancy theory (Higgins, 1996), participants with a strong environmental concern, indicating an ideal self-guide, were sensitive to information about positive environmental consequences. Participants with a weaker concern, indicating an ought self-guide, were primarily attuned to labels signalling negative environmental consequences. That respondents lacking environmental concern did not change their preferences in face of either positive or negative environmental information was also to be expected. Since environmental concern is not embodied in their self-guide, be it ideal or ought, they will be insensitive to environmental information. Taken together, these are promising results. Before any firm conclusions are drawn, though, a stricter test of the theory is called for. Such a test would include independent measures of self-guides rather than a proxy as in the present study. Furthermore, other domains than the environmental should be used.

A proviso on the interpretation of ought in the present context should be made explicit. The ought self-guide notion has been used in a very general sense. However, it is clear from Higgins (1987) that the ought self refers to a person's representation of attributes that the person him- or herself, *or* another significant person, believes the person ought to possess. This distinction relates to findings in the attitude and environment literature.[5] A common finding here is that personal norms predict environmentally responsible behaviour (see, e.g., Harland, Staats, & Wilke, 1999; Thøgersen, 1999). Moral obligations contribute to the striving for positive outcomes when they reside in the person him- or herself. Another aspect of moral obligation implies a social norm, where sanctions (and rewards) are administered by other people. Our interpretation is that the distinction between a promotion and a prevention focus can be linked to the distinction between personal and social norms in that a personal norm tends to orient an individual towards a promotion focus, and a social norm towards a prevention focus. Future research may elucidate this relation.

Eco-labels in use today signal positive environmental outcomes. When the terms "eco-labelled," "ecological," or "organic" are used in everyday language it is implicit that one is referring to a product with environmentally benign outcomes. As shown above a change in product preference due to a positive eco-label was associated with a strong concern for environmental protection.

As mentioned above, weak correlations between environmental concern and the choice of eco-labelled products have been found in a number of studies (for example Magnusson et al., 2001). However, a weak relationship between environmental concern or attitudes and behaviour must not imply that consumers lack interest in eco-labelled products. Some consumers probably do pay some attention to eco-labels when choosing between products, but their attention is not strong enough to make them choose eco-labelled alternatives. These consumers are most likely guided by ought, rather than ideal, environmental self-guides. Consequently, their product preference is only partially swayed by information about positive environmental outcomes, as signalled by positive eco-labels. However, being more sensitive to negative environmental outcomes these consumers would probably not choose a product with a negative eco-label. The weak correlation between environmental concern and the influence of positive eco-labels is thus not expected for negative eco-labels. On the contrary, a strong relationship between environmental concern and the tendency not to choose environmentally harmful versions of a product is expected.

That it takes a strong environmental concern to choose products with positive eco-labels may partially explain the weak correlations between environmental attitudes and purchase behaviour. There are of course a number of supplementary explanations. Several studies have reported a conflict between concern for environmental protection and a desire to cut down one's own expenses (e.g., Miles & Frewer, 2001; Torjusen et al., 2001; Tregear et al., 1994). Thus, that eco-labelled variants often are purchased at a higher price could be one additional factor. In other studies consumers with negative beliefs about eco-labelled products have reported a conflict between environmental protection and demand for high quality products (e.g., Davies et al., 1995; Wandel & Bugge, 1997). Such clashes between concern for the environment and other motives will result in an attenuated relationship between environmental attitudes and, for example, how often an individual purchases eco-labelled food products.

The purchase of many everyday products, such as bread, potatoes, milk, and meat is a frequent behaviour (often several times per week), performed in a stable context (e.g., the familiar supermarket), and with rewarding consequences (one keeps on purchasing gratifying products). These are three factors that have been shown to increase the probability of developing a habitual behaviour (Ouellette & Wood, 1998). Furthermore, individuals with strong habits have been shown to attend less to contextual information, to display less appreciation of choice options and attributes, and to show less integration of information into judgments (Verplanken, Aarts, & van Knippenberg, 1997). When such is the case, negative labels could be more effective in drawing attention to environmental aspects of a product among consumers with only an intermediate environmental concern.

In line with the results above, could eco-labels that signal negative environmental consequences be of practical use and contribute to a different purchase pattern? Today independent eco-labelling organizations examine products. If standards are satisfied, a product may become certified and eco-labelled. Similarly, organizations may sometimes introduce their own labelling system. An information system based on colour signals has recently been introduced in Sweden by the World Wide Fund for Nature (WWF) to guide consumers in their choice of fish. A green light indicates "Bon appetite," a yellow light "reflect," while a red light signals "abstain!" (WWF:s recommendation about fish and the environment, 2002). Note, however, that currently eco-labelling is done on a voluntary basis and implies a financial cost for the producer. It is not likely that a producer would voluntarily pay an independent organization to have a product labelled as "worse than average for the environment." To implement a system that not only includes positive, but also neutral and negative eco-labels, a regulation that prescribes that all products should be classified into one of these three categories will be needed. The EU Council Directive concerning the Energy Label (European Union On-Line, 2003), a system which is based on order of rank, from most to least energy efficient, shows that legislation can be used to introduce a label system that not only indicates positive but also neutral and negative outcomes.

If a three-level eco-label system were to be introduced, this could affect both consumers and producers. Consumers with an intermediate environmental concern would perhaps not choose products with a positive eco-label, but they would sort out products with negative labels. This in turn could foster product development in a more environmentally benign direction. A regulatory system with negative labels may also drive products out of the market. If it becomes common knowledge that unless certain standards are met, a negative label has to be attached to the product, such products may be withdrawn.

NOTES

1 Magnusson et al. (2001) found that between 46% and 67% agreed that it was quite or very good, wise, and important to purchase eco-labelled (organic) food products. However, only between 16% and 8% stated that they quite often, very often, or always chose an eco-labelled (organic) version of some everyday food products.

2 As judged by an independent sample, when both products of a pair were provided with a yellow (neutral) eco-label (control condition), the two alternatives were perceived as equally attractive.

3 Only one individual showed a stronger preference for products with a negative rather than a neutral eco-label (a mean deviation of more than one unit), while three

individuals rated products as less attractive when the eco-label was positive rather than neutral.

[4] A positive difference indicates a stronger influence by the positive labels and vice versa.

[5] We thank Folke Ölander for drawing our attention to this parallel.

REFERENCES

Anderson, N. H. (1974). Cognitive algebra. In: L. Berkowitz (Ed.), *Advances in Experimental Social Psychology*, Vol. 7, pp. 1–101. San Diego: Academic Press.

Carver, C. S., & Scheier, M. F. (1990). Origins and functions of positive and negative affect: A control-process view. *Psychological Review*, *97*, 19–35.

Chinnici, G., D'Amico, M., & Pecorino, B. (2002). A multivariate statistical analysis on the consumers of organic products. *British Food Journal*, *104*, 187–199.

Dahlstrand, U., & Biel, A. (1997). Pro-environmental habits: Propensity levels in behavioural change. *Journal of Applied Social Psychology*, *27*, 588–601.

Davies, A., Titterington, A. J., & Cochrane, C. (1995). Who buys organic food? A profile of purchasers of organic food in northern Ireland, *British Food Journal*, *97*, 17–23.

European Union On-Line (2003). *Council Directive 92/75/EEC (31992L0075) of 22 September 1992*. Retrieved June 18, 2003, from http://www.europa.eu.int

Fiske, S. T. (1980). Attention and weight in person perception: The impact of negative and extreme behaviour. *Journal of Personality and Social Psychology*, *38*, 889–906.

Fotopoulos, C., & Krystallis, A. (2002). Purchasing motives and profile of the Greek organic consumer: A countrywide survey. *British Food Journal*, *104*, 730–765.

Hansen, C. H., & Hansen, R. D. (1988). Finding the face in the crowd: An anger superiority effect. *Journal of Personality and Social Psychology*, *54*, 917–924.

Harland, P., Staats, H., & Wilke, H. A. M. (1999). Explaining pro-environmental intention and behaviour by personal norms and the theory of planned behaviour. *Journal of Applied Social Psychology*, *29*, 2505–2528.

Harper, G. C., & Makatouni, A. (2002). Consumer perception of organic food perception and farm animal welfare. *British Food Journal*, *104*, 287–299.

Higgins, E. T. (1987). Self-discrepancy: A theory relating self and affect. *Psychological Review*, *94*, 319–340.

Higgins, E. T. (1996). Knowledge activation: Accessibility, applicability, and salience. In: E. T. Higgins & A. Kruglanski (Eds.), *Social Psychology: A Handbook of Basic Principles*, pp. 133–168. New York: Guilford Press.

Higgins, E. T., Roney, C. J. R., Crowe, E., & Hymes, C. (1994). Ideal versus ought predilections for approach and avoidance: Distinct self-regulatory systems. *Journal of Personality and Social Psychology*, *66*, 276–286.

Higgins, E. T., & Tykocinski, O. (1992). Self-discrepancies and biographical memory: Personality and cognition at the level of the psychological situation. *Personality and Social Psychology Bulletin*, *18*, 527–535.

Kahneman, D., & Tversky, A. (1984). Choices, values, and frames. *American Psychologist*, *39*, 341–350.

Magnusson, M. K., Arvola, A., Koivisto Hursti, U.-K., Åberg. L., & Sjödén, P.-O. (2001). Attitudes towards organic foods among Swedish consumers. *British Food Journal*, *103*, 209–227.

Miles, S., & Frewer, L. Y. (2001). Investigating specific concerns about different food hazards. *Food Quality and Preference, 12*, 47–61.

Nordlund, A., Garvill, J., & Marell, A. (2000). Motives for pro-environmental behavior. In: E. Hölzl (Ed.), *Fairness & cooperation. Proceedings IAREP/SABE 2000, Vienna/Austria*, pp. 309–313.

Ouellette, J. A., & Wood, W. (1998). Habit and intention in everyday life: The multiple processes by which past behavior predicts future behavior. *Psychological Bulletin, 124*, 54–74.

Palm, L., & Jarlbro, G. (1999). *Nordiska konsumenter om Svanen – livsstil, kännedom, attityd och förtroende* (Nordic consumers about the Swan – life style, knowledge, attitude and trust). Copenhagen: Nordic Council of Ministers. TemaNord 1999:592.

Saveenergy (2003). *EU Energy Label.* Retrieved June 17, 2003, from http://www.saveenergy.co.uk/whatto/choosing/label.cfm

Schwartz, N. (1990). Feelings as information: Informational and motivational functions of affective states. In E. T. Higgins & R. M. Sorrentino (Eds.), *Handbook of motivation and cognition*, Vol. 2, pp. 527–561. New York: Guilford.

Skowronski, J. J., & Carlston, D. E. (1987). Social judgment and social memory: The role of cue diagnosticity in negativity, positivity, and extremity biases. *Journal of Personality and Social Psychology, 52*, 689–699.

Thøgersen, J. (1999). The ethical consumer. Moral norms and packaging choice. *Journal of Consumer Policy, 22*, 439–460.

Thøgersen, J. (2000). Psychological determinants of paying attention to eco-labels in purchase decisions: Model development and multinational validation. *Journal of Consumer Policy, 23*, 285–313.

Thøgersen, J. (2002). Promoting "green" consumer behavior with eco-labels. In: T. Dietz & P. C. Stern (Eds.), *New Tools for Environmental Protection*, pp. 83–104. Washington, DC: National Academy Press.

Torjusen, H., Lieblein, G., Wandel, M., & Francis, C. A. (2001). Food system orientation and quality perception among consumers and producers of organic food in Hedmark county, Norway. *Food Quality and Preference, 12*, 207–216.

Tregear, A., Dent, J. B., & McGregor, M. J. (1994). The demand for organically grown produce. *British Food Journal, 96*, 21–25.

Tversky, A., & Kahneman, D. (2000). Loss aversion in riskless choice: A reference-dependent model. In: D. Kahneman & A. Tversky (Eds.), *Choices, Values, and Frames*, pp. 143–158. New York: Cambridge University Press.

Verplanken, B., Aarts, H., & van Knippenberg, A. (1997). Habit, information acquisition, and the process of making travel mode choices. *European Journal of Social Psychology, 27*, 539–560.

Wandel, M. (1994). Understanding consumer concern about food-related health risks. *British Food Journal, 96*, 35–40.

Wandel. M., & Bugge, A. (1997). Environmental concern in consumer evaluation of food quality. *Food Quality and Preference, 8*, 19–26.

Wegner, D. M., & Bargh, J. A. (1998). Control and automaticity in social life. In: D. T. Gilbert, S. T. Fiske, & G. Lindzey (Eds.), *Handbook of Social Psychology*, Vol. 1, pp. 446–496. New York: McGraw-Hill.

Wier, M., & Calverly, C. (2002). Market potential for organic foods in Europe. *British Food Journal, 104*, 45–62.

WWF:s recommendation on fish and the environment (2002). Second ed. Retrieved November 20, 2003, from http://konsumentverket.se.

THE AUTHORS

Gunne Grankvist is an Assistant Professor in the Department for Studies of the Individual and Society, University of Trollhättan/Uddevalla, P.O. Box 1236, SE 462 28 Vänersborg, Sweden; Fax + 46 521 26 40 99; E-mail: Gunne.Grankvist@htu.se. Ulf Dahlstrand is a Lecturer and Anders Biel an Associate Professor in the Department of Psychology, Göteborg University, P.O. Box 500, SE 405 30 Göteborg, Sweden; Fax: +46 31 773 46 28. E-mail: Ulf.Dahlstrand@psy.gu.se; Anders.Biel@psy.gu.se.

This research was financially supported by grants from Foundation for Strategic Environmental Research to the research programme Food 21 and the third author. We thank two anonymous reviewers for helpful comments on an earlier draft of this paper.

[23]

Measuring Consumer Preferences for Ecolabeled Seafood: An International Comparison

Robert J. Johnston, Cathy R. Wessells, Holger Donath, and Frank Asche

An analysis of consumer preferences for seafood labeled with information about environmental production attributes is introduced into the food labeling literature. International seafood ecolabeling programs have been proposed to create market-based incentives for fisheries managers to promote sustainable fisheries. We investigate differences in consumer preferences for ecolabeled seafood across the United States and Norway. Using a contingent-choice telephone survey of random households in each nation, a wide range of factors is found to influence consumers' likelihood of purchasing ecolabeled seafood. Consumer preferences differ by price premium, species, consumer group, and certifying agency. The effect of these factors often differs between the United States and Norway, suggesting heterogeneity in international reactions to seafood ecolabels.

Key words: consumer preferences, contingent choice, ecolabeling, food labeling, seafood

Introduction

Concern over the status of natural resource stocks, combined with well-known limitations of command-and-control management mechanisms, has led to a variety of ecolabeling initiatives in resource-based industries (Swallow and Sedjo; Wessells, Donath, and Johnston). In general, such programs evaluate the production process with regard to established environmental standards set by an independent third party. If the process meets these standards, the producer or marketer may buy a license to display a specific ecolabel on their product. In effect, the label conveys to the consumer otherwise unobservable information concerning a product's environmental impact. The increasing use of such programs for many consumer products notwithstanding, ecolabeling initiatives worldwide have met with varying degrees of consumer acceptance [Organization for Economic Cooperation and Development (OECD)].

Within the seafood industry, both industry and non-industry groups have proposed ecolabeling as a means to identify seafood harvested under management regimes that demonstrably prevent over-exploitation of natural stocks. Presumably, consumers who value sustainable harvest as an attribute of seafood products will reveal a greater demand for labeled seafood products than for unlabeled products, thereby creating a market incentive for "sustainable" management.

Robert J. Johnston is assistant professor, Cathy R. Wessells is professor and department chair, and Holger Donath is research assistant, all in the Department of Environmental and Natural Resource Economics, University of Rhode Island; Frank Asche is professor in the Stavanger University College and Centre for Fisheries Economics, Norwegian School of Economics and Business Administration. Funding for this project was provided by the Rhode Island Sea Grant Program, the Rhode Island Agricultural Experiment Station, and the Norwegian Research Council.

Review coordinated by Gary D. Thompson.

For example, the Marine Stewardship Council (MSC) was created in 1996 through a cooperative effort of the World Wildlife Fund (WWF) and Unilever, a multi-national corporation (McHale; MSC). The goal of the MSC is to provide a standardized mechanism for certifying and labeling seafood products worldwide.[1] The MSC ecolabel may only be applied to a product after certification by an independent certification firm that the fishery is well managed. Several large supermarket chains in the United States and Europe have become MSC partners, pledging to promote and buy only certified seafood from sustainable sources, once the certification program is well established. In contrast to the international program promoted by the MSC, several regional governmental labeling programs are being promoted (e.g., through the Nordic Council), as well as programs promoted by other environmental groups (e.g., the Audobon Society), and programs promoted by the seafood industry.

With a few exceptions (Wessells, Donath, and Johnston; and Tiesl, Roe, and Hicks), the economic literature regarding the potential impact of seafood ecolabels is sparse. However, work in food product labeling establishes that the ultimate economic impact of labels depends on consumer acceptance (e.g., Caswell and Mojduszka; Caswell). Moreover, as certification programs often involve some degree of unavoidable cost, the extent to which consumers are willing to pay a price premium may have a significant impact on program success (Gudmundsson and Wessells). Although past research indicates most consumers prefer ecolabeled products to nonlabeled products, ceteris paribus (Blend and van Ravenswaay; Forsyth, Haley, and Kozak), and many consumers are willing to pay a premium for such products (Nimon and Beghin), consumer acceptance is likely to differ across product classes, demographics, and consumer preferences.

Consumer acceptance may also be influenced by other factors: the credibility of the agency providing the ecolabel, perceptions of the links between product choices and environmental impact, and understanding of the label's meaning [U.S. Environmental Protection Agency (EPA)]. Finally, cultural and other differences may lead otherwise similar consumers in separate countries to have varied reactions to international product certification programs (OECD; U.S. EPA). International ecolabeling or certification programs based solely on research conducted in a single country may therefore face unanticipated obstacles or even outright failure, if consumers in all nations do not react similarly to the labeling program.

This study evaluates factors that may influence consumers' acceptance of an ecolabeling program for seafood products. To evaluate potential differences in consumers' acceptance of seafood ecolabels in different countries, we compare the results of parallel consumer preference research conducted in both the United States and Norway. The primary goal is to assess: (*a*) whether consumers prefer ecolabeled seafood, (*b*) what factors influence those choices, and (*c*) whether these determinants or their impacts differ across countries.

As no large-scale market for ecolabeled seafood currently exists,[2] this study relies on the results of a multi-attribute contingent-choice survey (Opaluch et al.). Although often

[1] Additional information on the MSC is provided through its web site at www.msc.org.

[2] Consumers may purchase dolphin-safe tuna. However, the purpose of ecolabeling of tuna is to protect dolphins; it does nothing to prevent overfishing of tuna. In addition, because there are no unlabeled canned tuna products in the U.S. market, consumers have no choice but to buy ecolabeled canned tuna. In other words, we have no data available to measure consumers' preferences for ecolabeled canned tuna. In a 1997 study, Tiesl, Roe, and Hicks measured demand changes for canned tuna between the periods of pre- and post-labeling.

used to assess tradeoffs in the design of multi-attribute public policies, such analysis is also useful in the consideration of new or proposed market products for which there is little available data concerning consumer demand (Anderson and Bettencourt). Previous analyses of food markets (Blend and van Ravenswaay), and in particular seafood markets, have employed contingent-choice techniques (Holland and Wessells).

The Contingent-Choice Model

The contingent-choice survey format asks respondents to make a discrete choice among multiple product alternatives. By analyzing preferences for a variety of potential products differing according to a chosen set of variables, researchers can estimate the relative importance of particular variables in determining respondents' choices. In the present case, the contingent-choice approach is applied to consumers' choices of ecolabeled (certified) versus nonlabeled (uncertified) seafood. An underlying assumption of the chosen model is that for a given seafood species, consumers will choose either certified or uncertified seafood on any single purchasing occasion, but will not buy a combination of the two products.

Utility from a seafood product is assumed to be a function of the attributes of the product including certification, the cost of the product to consumers, and the characteristics of the consumer. We assume the principal shopper of the household has previously made a selection of the desired seafood species, based on the characteristics of that species relative to substitute products. The consumer must then choose between certified (ecolabeled) and uncertified (nonlabeled) products for that species. Specifically, we only model the choice of certified versus uncertified products within particular species groups; we do not model the choice among different seafood species.[3]

To model discrete-choice behavior, the contingent-choice method relies on the random utility model, in which individual utility is divided into observable and unobservable components (Hanemann). Within this framework, the utility derived from seafood product i is assumed to be a function of physical characteristics of the product (a vector $\mathbf{X}_i$); environmental characteristics of the product, represented by the presence or absence of an ecolabel (L_i, set equal to 0 for uncertified product and 1 for certified product); the agency backing or guaranteeing the ecolabel (G); the consumer's demographic attributes (a vector $\mathbf{D}$); household income (Y) minus the price of product i (P_i), and a vector of other goods ($\mathbf{S}$). This may be formally represented as:

$$(1) \qquad U_i(\mathbf{X}_i, L_i, G_i, \mathbf{D}, Y - P_i, \mathbf{S}) = v_i(\mathbf{X}_i, L_i, G_i, \mathbf{D}, Y - P_i, \mathbf{S}) + \varepsilon_i \quad \text{for } i = \{A, B\},$$

where $U_i(\cdot)$ is the total utility related to the seafood purchase, $v_i(\cdot)$ is a function representing the empirically measurable component of utility, and ε_i is a term representing the unobservable component. Note that (1) represents the utility of a representative household food shopper, assuming a single person is the primary shopper for each household. Accordingly, the vector of relevant demographic attributes ($\mathbf{D}$) may include

[3] It is possible the presence of a certified seafood product could alter consumers' choices among different seafood species. However, the goal of this study is to address consumers' choices of certified versus noncertified products for a given species, not to address inter-species choices.

demographic household attributes (e.g., the size of the household) together with individual consumer attributes (e.g., age, education).

The consumer compares the utility derived from a certified seafood product, identified by the subscript A, and an otherwise identical noncertified seafood product, identified by the subscript B. Thus, for the certified product, $L_i = L_A = 1$, and for the noncertified product, $L_i = L_B = 0$. Also note that the agency backing the ecolabel (G_i) only influences utility if the certified product is chosen; hence, $G_i = 0$ for the uncertified product.

The premium for the certified product may be positive, zero, or negative, with the difference between the price of the certified and noncertified product given by:

$$P_A = P_B + P_L, \tag{2}$$

where P_B represents the per unit "base price" for the unlabeled product, and P_L denotes the premium paid over P_B for the certified product. We assume the quantity of seafood to be purchased is fixed in the short run (i.e., fixed at the amount of seafood needed to feed the household), where we define the short run as a single shopping trip. This assumption is based on the results of focus groups and survey pretests with seafood consumers. (The rationale for this assumption is discussed further in the following section.)

The difference in utility (dU) resulting from the purchase of the certified versus uncertified product is specified as:

$$\begin{aligned} dU &= U_A(\mathbf{X}_A, L_A, G_A, \mathbf{D}, Y - (P_B + P_L), \mathbf{S}) \\ &\quad - U_B(\mathbf{X}_B, 0, 0, \mathbf{D}, Y - P_B, \mathbf{S}) \\ &= v_A(\mathbf{X}_A, L_A, G, \mathbf{D}, Y - (P_B + P_L), \mathbf{S}) \\ &\quad - v_B(\mathbf{X}_B, 0, 0, \mathbf{D}, Y - P_B, \mathbf{S}) - [\varepsilon_B - \varepsilon_A] \\ &= dv - \theta, \end{aligned} \tag{3}$$

where $U_i(\cdot)$, $v_i(\cdot)$, and ε_i are as defined above. Accordingly, dv represents the difference in observable utility between the two competing products, and θ represents the difference in the unobservable or stochastic component.

Although dU is unobservable to the researcher, respondents' choices may be observed between the certified and uncertified products. That is, the respondent compares the two products, assesses the difference in utility between the two products, and indicates the sign of the utility difference (dU) by choosing either the certified product or the uncertified product. If one assumes θ follows a logistic distribution, the probability of choosing a particular product (e.g., the certified product) can then be modeled using the standard conditional logit model (Maddala). As this model does not allow for a "status quo" response in which survey respondents may choose to purchase neither the certified nor the uncertified product (Adamowicz et al.), model findings should be interpreted as *conditional* on the choice to purchase the specified seafood product. Hence, the model addresses factors influencing the choice of certified versus noncertified products, given a prior decision to purchase a particular type of seafood. Although this assumption constrains the interpretation of model results, it matches common purchase behavior revealed in focus groups.

The Econometric Specification

The logit model specifies the probability of choosing certified seafood as follows:

$$\text{Pr}(\textit{certified}) = \frac{1}{1 + e^{-dv}}, \tag{4}$$

where dv represents the utility-difference function shown in (3). In order to test for differences in preferences between Norwegian and U.S. consumers, we supplement the linear specification of dv with a set of dummy variables allowing for differences in intercept and partial slopes between the two groups. The resulting econometric specification is given by:

$$\begin{aligned} dv = {} & \beta_0(NR) + \beta_1(X)(NR) + \beta_2(\mathbf{D})(NR) + \beta_3(\mathbf{S})(NR) + \beta_4(P_L)(NR) \\ & + \beta_5(G)(NR) + \lambda_0(US) + \lambda_1(X)(US) + \lambda_2(\mathbf{D})(US) + \lambda_3(\mathbf{S})(US) \\ & + \lambda_4(P_L)(US) + \lambda_5(G)(US), \end{aligned} \tag{5}$$

where X is a dummy variable identifying the seafood species under consideration (cod or shrimp), $\mathbf{D}$ is a vector of demographic variables including household income, and $\mathbf{S}$ is a vector characterizing the household's general seafood shopping patterns and budget. The latter variables serve as a proxy for information regarding other seafood goods purchased by the consumer. P_L is the premium paid for the certified product, and G is a variable characterizing consumer trust in the particular certifying agency under consideration. US is a dummy variable identifying observations from United States consumers, where $US = 1$ for U.S. consumers, and $US = 0$ for Norwegian consumers. NR is a dummy variable identifying observations from Norwegian consumers, where $NR = 1$ for Norwegian consumers, and $NR = 0$ for U.S. consumers. Finally, $\beta_0, \ldots, \beta_5$ and $\lambda_0, \ldots, \lambda_5$ are parameters (or conforming vectors of parameters) to be estimated.

The coefficients corresponding to variables multiplied by NR ($\beta_0, \ldots, \beta_5$) measure effects on dv for Norwegian consumers. Effects for United States consumers are given by coefficients corresponding to variables multiplied by US ($\lambda_0, \ldots, \lambda_5$). Table 1 describes the set of variables included in the final model.

Rationale Underlying Quantity Assumptions

As noted above, we assume the quantity of seafood to be purchased is fixed in the short run, where the short run is defined as a single shopping trip. This assumption, based on evidence from focus groups, suggests household shoppers generally buy the quantity of seafood they perceive to be sufficient to feed household members. More specifically, seafood purchases of any given product, on a given shopping trip, may be viewed as a discrete choice. Depending on the price and quality attributes of the product, the consumer will either buy an approximately fixed quantity of the product, or will forego the product entirely and substitute another seafood, meat, or other food product.

Second, consumers in many cases use rules of thumb such as the "size of the fillet" to measure seafood quantity—i.e., when purchasing seafood, consumers often base their selection on the apparent size of the piece of fish, rather than according to a known weight measure. Indeed, a number of those interviewed indicated they did not really know the number of pounds of fish they typically bought, but they almost always bought

Table 1. Model Variables and Their Definitions

Variable Name (U.S., Norway)	Definition	Mean Value U.S.	Mean Value Norway
US_INTERCEPT *NR_INTERCEPT*	Model intercept	1.00	1.00
US_PERCENT *NR_PERCENT*	Certification premium as a percentage of the price for uncertified product	0.27	0.22
US_COD *NR_COD*	Dummy variable with a value of 1 if species in observation is cod	0.49	0.51
US_TRUST *NR_TRUST*	Dummy variable with a value of 1 if respondent's most trusted agency is the reported certification agency for purposes of discrete-choice questions	0.23	0.25
US_ANTI-ECO *NR_ANTI-ECO*	Factor score indicating latent aspects of ecological purchasing behavior (see main text and tables 2 and 3)	0.38	-0.33
US_NO_PURCH *NR_NO_PURCH*	Factor score indicating latent aspects of ecological purchasing behavior (see main text and tables 2 and 3)	0.09	-0.08
US_NO_CHANGE *NR_NO_CHANGE*	Factor score indicating latent aspects of ecological purchasing behavior (see main text and tables 2 and 3)	-0.14	0.11
US_ENVIR *NR_ENVIR*	Dummy variable with a value of 1 if respondent is a member of an environmental organization	0.15	0.05
US_FRESH *NR_FRESH*	Dummy variable with a value of 1 if respondent most often purchases fresh seafood	0.58	0.43
US_OFTEN *NR_OFTEN*	Dummy variable with a value of 1 if respondent consumes seafood at least once a week	0.34	0.87
US_LOWB *NR_LOWB*	Dummy variable with a value of 1 if respondent's weekly seafood budget is less than U.S. $10 or 80 NOK	0.66	0.40
US_HIEDU *NR_HIEDU*	Dummy variable with a value of 1 if respondent has at least a 4-year college degree	0.46	0.45
US_FEMALE *NR_FEMALE*	Dummy variable with a value of 1 if respondent is female	0.65	0.50
US_OLD *NR_OLD*	Dummy variable with a value of 1 if respondent's age is at least 45	0.49	0.47
US_INCOME *NR_INCOME*	Dummy variable with a value of 1 if respondent's income is greater than U.S. $75,000 or 200,000 NOK	0.21	0.47

the same size fillet or steak. Consequently, introducing an explicit weight measure into choice questions would have added a form of methodological misspecification, as it would have presented choice scenarios different from those commonly understood by respondents.[4]

Primary Data

A telephone survey format was chosen for data collection, allowing random nationwide sampling in both Norway and the United States. The survey was administered in the United States in the summer of 1998, and in Norway during spring 1999. Telephone

[4] Households purchasing greater (or lesser) quantities of seafood per shopping trip possibly would be either more or less willing to purchase certified seafood at any given premium. However, statistical evidence suggests such patterns do not influence respondents' choices in this case. For example, household size, which one would expect to be highly correlated with quantity of seafood purchased, could not be shown to be statistically significant in any model specification.

numbers to be called were generated by random digit dialing within regions of each country, where the number of surveys completed within each region was determined by regional population relative to national population. Thus, more heavily populated regions generated proportionately more respondents. Regions were defined as states in the United States, and *fylke* within Norway (roughly the equivalent of a U.S. county). The U.S. sample includes 1,640 completed surveys; the Norwegian sample includes 2,039. Surveys were completed by the "principal shopper" in seafood-consuming households; households that did not consume seafood were excluded from the survey. The final data therefore represent a stratified random sample of seafood-consuming households in each country.[5]

Survey development included background research, interviews with those involved in seafood ecolabeling initiatives, interviews and focus groups with seafood consumers, and extensive pretesting. Focus groups and pretests emphasized both the need to provide respondents with sufficient information to make informed product choices, and a requirement that product descriptions and survey language be kept straightforward and succinct. Hence, the number of attributes of each seafood species considered by respondents was minimized to include only those central to the choice of certified versus uncertified seafood.

Respondents considered the choice of certified versus uncertified seafood for two different species: cod and shrimp. The order in which the species were considered was randomized, to prevent question-order bias (Mitchell and Carson). For each species, certified seafood was described simply as being "caught under strict controls that prevent too much fishing." The survey emphasized the fact that both certified and uncertified seafood were of equal quality, texture, and freshness. Prior to the presentation of discrete-choice questions, respondents were provided with background information detailing the meaning of certified seafood, and reminding respondents of their budget constraint.

For each choice instance, respondents were provided with both the price of the uncertified product and the price of the certified product (per pound in the United States, and per kilogram in Norway), where the premium is defined as the difference between the two prices. These premiums ranged from −20 Norwegian kroner (NOK) to +50 NOK in Norway, and from −$2 to +$5 in the United States.[6] Although the surveys expressed this premium in either dollars or kroner, as appropriate, the data analysis converts price premium to a percentage of the uncertified price. This convention was adopted as a means to compare premiums across countries.

In addition to species, price, and premium, each survey listed a specific "certifying agency" for each set of questions for any one respondent, maintaining the same agency for each respondent. These certifying agencies included the World Wildlife Fund (WWF) and the Marine Stewardship Council (MSC) in both countries, as well as the National Marine Fisheries Service (NMFS) in the United States and the Norwegian Fisheries Directorate (NFD) in Norway.

The three certification organizations for each country can be regarded as representatives from: (*a*) a national governmental body, (*b*) a well-known environmental organization, and (*c*) an unknown initiative. Fractional factorial design was used to construct the

[5] There are no data on the population of seafood consumers; thus it is impossible to make a general comparison between our samples and this population. We can compare our samples to the overall population of each country, which indicates we have oversampled households with high income and education levels.

[6] At the time of this writing, the exchange rate for Norwegian kroner was approximately 9.0 per U.S. dollar.

range of discrete-choice question attributes (Addelman and Kempthorne), resulting in 54 unique contingent-choice questions, divided among 18 survey versions in each country.

In addition to the discrete-choice questions described above, the survey incorporated questions addressing a number of factors including: (*a*) seafood consumption behavior and budget, (*b*) trust in potential certification agencies, (*c*) demographic characteristics, (*d*) the extent to which environmental concerns influenced past purchasing behavior, and (*e*) perceptions regarding the status of particular seafood stocks.

Characterizing Respondents' Ecological Purchase Behavior

Respondents often display heterogeneous preferences for environmental goods or attributes (Swallow et al.). In some cases, differences among respondents' preferences may be explained through the inclusion of demographic variables in the utility difference function. However, in other cases, heterogeneity in responses may be due to unobservable, latent factors which influence behavior (Bollen). These factors are often estimated through analysis of Likert-scale responses to multiple questions linked to a set of underlying concepts (e.g., Variyam, Jorday, and Epperson). Details of the various methods of factor and principal component analysis are presented by numerous authors (see, for example, Harman; Reyment and Jöreskog).

In an attempt to better model preference heterogeneity, each survey included a set of 10 questions designed to characterize the extent to which environmental concerns influenced respondents' purchasing behavior. These questions were selected from the standardized ecologically conscious consumer behavior (ECCB) scale (Roberts), which asks respondents to rate the veracity of various statements regarding their purchase behavior with respect to environmental product attributes. Responses to these questions are summarized in table 2.

Following Variyam, Jorday, and Epperson, factor analysis is conducted to estimate a small number of underlying constructs which together account for a large percentage of the observed variation in responses. Responses are analyzed using principal-component factor analysis of the response correlation matrix, with three factors retained and rotated using the VARIMAX method (Kaiser). Retained factors were chosen based on a threshold eigenvalue of one (Variyam, Jorday, and Epperson). Rotated VARIMAX factor loadings are illustrated in table 3.

Factor 1 is characterized by high factor loading (with values appearing in bolded italics in table 3) on questions 4, 7, 8, 9, and 10, where high loadings indicate respondents considered the statements to be less true. These statements tend to reflect a consumer's willingness to forego desired products for abstract environmental reasons—no personal economic gain is involved and no specific purchases are described. High scores on these questions reflect a lack of willingness to give up products solely for abstract environmental reasons. This factor is accordingly characterized as "anti-environmental" (denoted *ANTI-ECO*).

Factor 2 is characterized by high loading on questions 1, 2, and 3 (table 3). These questions concern the likelihood of purchasing specific types of environmentally-friendly products, often with potential long-term economic benefits to the purchaser. These include products with low energy costs or reduced packaging. Hence, this factor is characterized as representing the degree to which a consumer has made expenditures for

Table 2. Ecologically Conscious Consumer Behavior (ECCB) Questions and Response Statistics

Quest. No.	Syntax	U.S. Mean [a] (Std. Dev.)	Norway Mean [a] (Std. Dev.)	Difference of Means [b]
1	I have purchased a household appliance because it uses less electricity than other brands.	2.91 (1.11)	3.01 (1.53)	0.11**
2	I have purchased light bulbs that are more expensive but saved energy.	2.88 (1.12)	2.25 (1.46)	-0.62***
3	I will not buy products that have excessive packaging.	2.95 (1.04)	2.74 (1.34)	-0.21***
4	If I understand the potential damage to the environment that some products can cause, I do not purchase these products.	2.32 (0.94)	1.76 (1.14)	-0.56***
5	I have switched products for ecological reasons.	2.85 (1.01)	2.79 (1.57)	-0.06
6	I have convinced members of my family or friends not to buy some products that are harmful to the environment.	3.24 (1.13)	3.44 (1.60)	0.20***
7	Whenever possible, I buy products packaged in reusable containers.	2.53 (0.98)	1.82 (1.21)	-0.70***
8	When I have a choice between two equal products, I always purchase the one less harmful to other people and the environment.	2.24 (0.93)	1.83 (1.13)	-0.42***
9	I will not buy a product if the company that sells it is ecologically irresponsible.	2.34 (0.96)	1.94 (1.16)	-0.39***
10	I do not buy household products that harm the environment.	2.40 (0.87)	1.86 (1.07)	-0.55***

Note: Questions were selected from Roberts' 1996 ECCB Scale.

[a] Scoring scale: always true = 1, mostly true = 2, sometimes true = 3, rarely true = 4, and never true = 5. Note that *higher* numbers represent a *lower* probability of engaging in the particular behavior.

[b] Double and triple asterisks (*) denote significance at $p < 0.05$ and $p < 0.01$, respectively.

Table 3. Rotated VARIMAX Factor Loadings: Responses to ECCB Questions

Quest. No.[a]	Factor 1	Factor 2	Factor 3	Quest. No.[a]	Factor 1	Factor 2	Factor 3
1	-0.024	***0.778***	0.154	6	0.081	0.133	***0.838***
2	0.236	***0.705***	-0.039	7	***0.537***	0.305	0.219
3	0.254	***0.535***	0.226	8	***0.654***	0.196	0.147
4	***0.731***	0.056	0.071	9	***0.753***	0.091	0.123
5	0.266	0.112	***0.757***	10	***0.771***	0.118	0.104

[a] Refer to table 2 for the syntax corresponding to questions 1–10.

specific energy- or packaging-related environmental benefits. Because high scores for this factor indicate an unwillingness to undertake this sort of specific behavior, we characterize this factor (denoted *NO_PURCH*) as "no specific energy-packaging purchases."

Factor 3 (table 3) is characterized by high loading on questions 5 and 6—questions addressing an active change in behavior in response to environmental information. Those with high scores on this factor will not change products in response to ecological information, nor will they convince others to do so. Therefore, this factor is characterized as indicating an "unwillingness to change" (denoted *NO_CHANGE*) purchase patterns for environmental reasons.

The three factors are included in the logit model as standardized factor scores—i.e., the original factors are transformed so as to have a mean of zero and standard deviation of one. This procedure simplifies interpretation of estimated logit parameters, as the scores indicate the extent to which a factor score for a particular respondent differs from that of the sample mean (Kline).

Model Results

Logit results for the choice of certified versus uncertified seafood are reported in table 4. After deletion of observations with missing data for variables, the final model includes 6,220 observations. Of these, 3,100 are U.S. observations and 3,120 are Norwegian observations.[7] A log-likelihood ratio test ($-2\text{LnL} = 1{,}838.354$, df = 30) shows the model is significant at $p < 0.0001$. Of 28 non-intercept variables in the model, 18 are statistically significant. The model predicts 76% of in-sample observations correctly.

A log-likelihood test of the unrestricted model versus a restricted model in which U.S. and Norwegian effects were constrained to be equal indicates the restrictions have a statistically significant impact on the model at $p < 0.0001$ ($\chi^2 = 340.915$, df = 15). We conclude that statistical differences exist between the choices of Norwegian and U.S. respondents, although the pairwise statistical equivalence of U.S. and Norwegian parameter estimates cannot be rejected for all variables.

Table 5 provides results of log-likelihood χ^2 tests of pairwise equality between U.S. and Norwegian parameter estimates. The null hypothesis of pairwise equality may be rejected at $p < 0.10$ for eight of the 15 model variables considered, including the intercepts. In the discussion below, we examine implications of these results, as well as the implication of the signs and magnitudes of particular parameter estimates. Following the variable groupings of equation (5), we focus the following discussion on the effects of price premium (P_L), species (X), agency trust (G), environmental purchase and seafood consumption patterns ($\mathbf{S}$), and demographics ($\mathbf{D}$).

Price Premium

A log-likelihood test ($\chi^2 = 662.465$, df = 2) clearly indicates joint significance of variables associated with the price premium P_L (*US_PERCENT*, *NR_PERCENT*) ($p < 0.0001$), confirming the results of individual tests of statistical significance. As expected, the price premium measured in percentage terms has a negative impact on consumers' likelihood of choosing certified seafood. Although the negative effect holds in both countries, it is more pronounced in Norway: the variable *NR_PERCENT* is greater in absolute value than *US_PERCENT*, and the difference is statistically significant at $p < 0.0001$ (table 5).

---

[7] The 1,640 U.S. surveys generated 3,280 observations (one observation each for cod and shrimp per survey). However, 180 observations were dropped due to missing data. The 2,039 Norway surveys generated 4,078 observations (again, one observation each for cod and shrimp per survey). Of these, 958 observations were dropped due to missing data. Data most often missing include demographic data such as income and education.

Table 4. Logit Model Results for Choice of Certified versus Uncertified Seafood

Variable	Parameter Estimates	Standard Error	Wald χ^2	Prob > χ^2
US_INTERCEPT	1.67	0.15	119.44	0.0001***
NR_INTERCEPT	0.47	0.16	9.23	0.0024***
US_PERCENT	-2.45	0.16	231.92	0.0001***
NR_PERCENT	-3.57	0.19	350.94	0.0001***
US_COD	0.47	0.10	22.79	0.0001***
NR_COD	0.40	0.08	23.05	0.0001***
US_TRUST	0.08	0.11	0.62	0.4305
NR_TRUST	0.47	0.09	24.87	0.0001***
US_ANTI-ECO	-0.35	0.05	43.11	0.0001***
NR_ANTI-ECO	-0.27	0.04	42.75	0.0001***
US_NO_PURCH	-0.13	0.05	6.50	0.0108**
NR_NO_PURCH	-0.01	0.04	0.11	0.7458
US_NO_CHANGE	-0.11	0.06	3.53	0.0604*
NR_NO_CHANGE	-0.14	0.04	15.85	0.0001***
US_ENVIR	0.32	0.14	5.66	0.0174**
NR_ENVIR	0.11	0.18	0.34	0.5608
US_FRESH	0.33	0.09	13.52	0.0002***
NR_FRESH	0.07	0.08	0.64	0.4228
US_OFTEN	-0.15	0.10	2.43	0.1191
NR_OFTEN	-0.02	0.12	0.03	0.8662
US_LOWB	-0.46	0.11	18.48	0.0001***
NR_LOWB	-0.19	0.09	5.19	0.0228**
US_HIEDU	0.09	0.09	0.96	0.3265
NR_HIEDU	-0.44	0.08	27.52	0.0001***
US_FEMALE	0.13	0.09	2.06	0.1509
NR_FEMALE	0.61	0.08	55.67	0.0001***
US_OLD	0.10	0.09	1.35	0.2457
NR_OLD	0.19	0.08	5.07	0.0243**
US_INCOME	0.10	0.12	0.66	0.4177
NR_INCOME	0.20	0.09	5.45	0.0195**
No. Observations	= 6,220			
-2LnL	= 6,784.397			
-2LnL χ^2	= 1,838.354 (df = 30)			0.0001***

Note: Single, double, and triple asterisks (*) denote significance at $p < 0.10$, $p < 0.05$, and $p < 0.01$, respectively.

The greater price sensitivity of Norwegian consumers may appear counterintuitive, given Norwegians are in general more willing to change purchase behavior based on environmental concerns (table 2). However, Norwegians' greater attention to environmental attributes, combined with Europeans' more significant experience with ecolabeling programs in general (U.S. EPA), may generate a greater awareness of tradeoffs associated with premiums paid for labeled products. More specifically, Norwegians' greater experience with ecolabels and greater seafood consumption may provide a better defined sense of when a price premium is "too high" relative to premiums for other ecolabeled products, leading to the potential for greater price sensitivity.

Table 5. Results of Log-Likelihood Test of Equality Between U.S. and Norway Parameter Estimates

Null Hypothesis	Wald χ^2	Prob > χ^2
US_INTERCEPT = NR_INTERCEPT	30.04	0.0001***
US_PERCENT = NR_PERCENT	20.22	0.0001***
US_COD = NR_COD	0.36	0.5493
US_TRUST = NR_TRUST	7.50	0.0062***
US_ANTI-ECO = NR_ANTI-ECO	1.56	0.2124
US_NO_PURCH = NR_NO_PURCH	3.52	0.0605*
US_NO_CHANGE = NR_NO_CHANGE	0.16	0.6923
US_ENVIR = NR_ENVIR	0.94	0.3330
US_FRESH = NR_FRESH	4.78	0.0288**
US_OFTEN = NR_OFTEN	0.71	0.3983
US_LOWB = NR_LOWB	3.67	0.0553*
US_HIEDU = NR_HIEDU	18.13	0.0001***
US_FEMALE = NR_FEMALE	15.03	0.0001***
US_OLD = NR_OLD	0.50	0.4802
US_INCOME = NR_INCOME	0.44	0.5071

Note: Single, double, and triple asterisks (*) denote significance at $p < 0.10$, $p < 0.05$, and $p < 0.01$, respectively.

Species

To evaluate the relative importance of species on respondents' choices, the model includes dummy variables (*US_COD*, *NR_COD*) identifying observations associated with cod rather than shrimp for U.S. and Norwegian observations. These variables are significant and positive ($p < 0.0001$), indicating both U.S. and Norwegian consumers are relatively more likely to choose certified cod than certified shrimp. Because there is no statistical difference between the parameter estimates for the U.S. and Norway (table 5), we conclude that the same choice patterns hold for U.S. and Norwegian respondents. Hence, certification appears to have a stronger influence on purchase behavior for cod, compared to shrimp. The existence of such effects suggests the success of ecolabeling programs will likely differ across species.

Although one might conclude the preference for certified cod is due to frequent media reports regarding the depletion of cod stocks (Cramer), in fact only a low percentage of respondents stated they believed cod to be "severely overfished." Moreover, preliminary models could establish no significant correlation between a belief that cod stocks were overfished and the likelihood of choosing certified cod; this belief variable was subsequently deleted from the model.

Agency Trust

Prior to discrete-choice questions, the survey asked respondents to indicate which of a list of agencies would be most trusted to guarantee certification. U.S. respondents could choose among the World Wildlife Fund (WWF), the Marine Stewardship Council (MSC), and the National Marine Fisheries Service (NMFS). Norwegian respondents were given a choice among the WWF, the MSC, and the Norwegian Fisheries Directorate (NFD). Respondents were not provided additional information regarding these agencies, to mimic actual buying conditions in which consumers would likely not have on-site access to additional information regarding a certifying agency.

Of the agencies considered by U.S. respondents, the NMFS garnered the highest trust ratings, with 49% of respondents indicating they would trust this agency most to provide certification. The WWF was chosen by 23% of respondents, and the MSC by 5%. The remaining 23% were unsure of their most trusted agency. Approximately 81% of Norwegian respondents reported they would trust the government agency most to provide certification. The WWF was chosen by 16% of Norwegian respondents, and the MSC by 3%. Private (e.g., seafood industry) certifying organizations were not considered, as prior research regarding seafood safety has shown seafood consumers place relatively little value on guarantees offered solely by industry groups (Wessells and Anderson).

While trust in the certification agency might be expected to influence the probability of selecting certified products, model results support this conclusion only for Norwegian consumers. More specifically, in cases where the discrete-choice selection of certification agency for guaranteeing certification was also reported as the respondent's "most trusted" agency, Norwegian respondents were more likely to choose the certified product (*NR_TRUST* is significant and positive). However, this effect did not hold for U.S. consumers (*US_TRUST* is not statistically significant).

Based on these findings, choice of certifying agency may be an important element in the success of a certification program. Certifications offered by little-known organizations (such as the MSC) may in some cases result in a lower probability that consumers will choose a certified product, compared to certifications offered by better known government agencies. Although consumer trust in such agencies may be bolstered by public-relations or educational campaigns, the efficacy of such programs will likely vary across nations, just as the relative importance of trust in the certification agency differs. In nations where significant changes in consumer acceptance are associated with agency trust (e.g., Norway), costly promotional and educational campaigns may be particularly advantageous; such activities may be less effective in countries where agency trust has little or no demonstrable impact on consumer behavior.

Environmental Purchase Patterns

An initial log-likelihood test was employed to assess the role of heterogeneity in respondents' ecological purchase behavior, as measured by the 10 ECCB-scale questions (table 2) characterizing the extent to which environmental concerns influenced respondents' purchasing behavior (Roberts). Six model variables are constructed from these questions, with *NR_ANTI-ECO*, *NR_NO_PURCH*, and *NR_NO_CHANGE* representing factor scores for Norwegian observations, and *US_ANTI-ECO*, *US_NO_PURCH*, and *US_NO_CHANGE* representing the corresponding scores for U.S. observations (table 1). Hypothesis test results ($\chi^2 = 117.381$, df = 6) document the joint significance of these variables at $p < 0.0001$, and confirm that the factor loadings successfully capture underlying preferences which influence respondents' hypothetical choices for certified seafood.

The individual latent factors derived from ECCB responses can help identify types of consumers who may be particularly likely or unlikely to respond to seafood certification programs. For example, consumers who score highly on the first factor measuring abstract anti-environmentalism (*NR_ANTI-ECO* and *US_ANTI-ECO*) in both Norway and the United States are less likely to select certified seafood than those with lower scores for this factor. The impact of this variable cannot be shown to be different across the two countries (table 5). Similarly, the third factor (*NR_NO_CHANGE*, *US_NO_CHANGE*) is significant for both countries and negative, indicating respondents who score highly on the "*unwillingness to change*" factor are less likely to choose certified seafood. In contrast, the second factor (*NR_NO_PURCH* and *US_NO_PURCH*, denoting no specific energy-packaging purchases) has no apparent influence on the behavior of Norwegian respondents. This may be due to the relatively high profile of energy supply issues in Norway (Kalgraf, Owens, and Raaholt); such issues may be viewed as distinct from other environmental concerns such as overfishing. However, for U.S. respondents, high scores for this factor are associated with a lower probability of selecting certified seafood ($p < 0.02$).

Summarizing these results, anti-ecological purchasing tendencies as measured by high scores on the three latent factors are associated with a significant decrease in the estimated probability of selecting an ecolabeled seafood product. Moreover, the effects appear largely the same across the two countries, notwithstanding differences associated with the second factor. These results suggest that active targeting of marketing and information toward consumers with identifiable tendencies related to ecological purchasing behavior may be an important determinant of the success of seafood ecolabeling programs.

One might also seek to characterize heterogeneity in respondents' environmental attitudes and behavior using directly observable variables, such as those indicating membership in various environmental organizations (Swallow et al.). Unlike the latent variables discussed above, which have largely similar impacts on Norwegian and U.S. choices, the effects of environmental organization membership appear to differ. Holding all else constant, membership in environmental organizations (*NR_ENVIR* = 1) cannot be shown to have a statistically significant impact on the likelihood of choosing certified seafood for Norwegian respondents.[8] For U.S. respondents, environmental organization membership (*US_ENVIR* = 1) is associated with a significant increase in the likelihood of choosing certified seafood. However, despite the difference in statistical significance, we cannot reject the equality of the two parameter estimates at $p < 0.10$ (table 5). Hence, we cannot appropriately claim the effect differs between the two countries.

Seafood Consumption Patterns

From the model results, Norwegians who most often purchase fresh seafood products rather than frozen (*NR_FRESH*) are no more likely to choose certified seafood. However, U.S. respondents who most often purchase fresh seafood (*US_FRESH*) are more

[8] Although it is possible that degrading multicollinearity between ecological purchasing behavior and membership in environmental organizations could contribute to the insignificance of the latter variable, these variables are not, in fact, highly correlated. For example, the Pearson correlation coefficients between *NR_ENVIR* and the three factor scores for ecological purchasing behavior are in all cases less than 0.093 in absolute value. Moreover, assessing the condition indices of the data matrix reveals that the largest condition index is 25.90.

likely to select certified seafood. Such patterns could be of importance in the development of certification programs, as Unilever, a co-founder of the MSC and a primary private-sector proponent of seafood ecolabeling, is a large-volume seller of frozen seafood to both Europe and the U.S. under the Gorton's® and Bird's Eye® brands (McHale).

Frequency of seafood consumption does not influence consumers' choices of certified seafood in either country, despite substantial differences in the average rate of seafood consumption in the two countries surveyed. Neither *NR_OFTEN* nor *US_OFTEN* can be shown to be statistically significant. However, model results do support the hypothesis that those with low weekly seafood budgets are less likely to purchase certified products. Norwegian consumers with relatively low seafood budgets (see table 1) are less likely to choose certified products ($p < 0.0228$). The variable *US_LOWB* is also negative and significant ($p < 0.0001$), revealing less likelihood for U.S. consumers with a relatively low seafood budget to choose certified products. Hence, although the frequency of seafood consumption is not a key indicator of preferences for certified seafood, the average weekly budget for seafood products is a significant factor.

Demographics

Demographic factors also influence respondents' preferences for certified seafood products. However, the impacts of these variables are not consistent across the two countries surveyed. For example, the independent variables *US_HIEDU* and *NR_HIEDU* identify respondents with at least a four-year college degree. For Norwegian respondents, this variable is associated with a lower probability of selecting certified seafood; for U.S. respondents, it cannot be shown to have a statistically significant effect. The coefficients of gender (*US_FEMALE, NR_FEMALE*), age (*US_OLD, NR_OLD*), and income (*US_INCOME, NR_INCOME*) have the same sign across both countries, but differ in statistical significance, magnitude, or both. For example, a greater likelihood of selecting ecolabeled seafood is associated with females in Norway, but not with females in the United States. Those identified as being in a high-income group (cf. table 1) are more likely to select certified seafood in Norway, but this effect is not statistically significant in the United States. The results suggest significant heterogeneity across countries with regard to the role of specific demographic indicators.

Forecasting the Probability of Selecting Ecolabeled Seafood

Although the significance and magnitude of specific effects may differ across countries, this in itself does not guarantee the average probability of selecting ecolabeled seafood will differ to a statistically significant degree. To assess potential differences in the overall probability across the two sampled countries, table 6 forecasts the probability that an "average" consumer will choose certified seafood, at different premium levels. To offset potential effects related to differences in the demographic characteristics of the two samples, consumer characteristics are fixed at sample means for both countries combined. Table 6 displays the probabilities of an identical consumer in Norway and the United States. Forecasts are generated for price premium percentages of 0%, 24.2% (the premium sample mean across both countries), 50%, and 75%, with all except the 75% premium representing in-sample forecasts. Probabilities may also be calculated using

Table 6. Estimated Probability of Selecting Certified Seafood: U.S. versus Norway

Price Premium (*PERCENT*)	U.S. Estimated Prob. (Std. Error)[c]	Norway Estimated Prob. (Std. Error)[d]	Probability Difference (Std. Error)	*t*-Statistic (H_0: Prob. Diff. = 0)
0%	0.880 (0.009)	0.735 (0.014)	0.142 (0.016)	8.87***
24.2%[a]	0.802 (0.010)	0.539 (0.014)	0.263 (0.017)	15.47***
50%	0.682 (0.014)	0.317 (0.016)	0.364 (0.021)	17.33***
75%[b]	0.538 (0.021)	0.160 (0.015)	0.378 (0.027)	14.00***

Notes: Asterisks (***) denote significance at $p < 0.01$. All variables except *PERCENT* are held constant at mean values.
[a] The 24.2% price premium is the sample mean across both countries.
[b] Out-of-sample prediction.
[c] If U.S. probabilities are calculated using means for U.S. observations only, they become (moving down the column) 0.868, 0.784, 0.658, and 0.511.
[d] If Norwegian probabilities are calculated using means for Norway observations only, they become (moving down the column) 0.749, 0.556, 0.333, and 0.17. Differences from U.S. probabilities remain significant.

the mean values for each country's sample—i.e., U.S. sample means used to calculate U.S. probabilities and Norwegian sample means used to calculate Norway probabilities. These latter results are shown in footnotes (c) and (d) to table 6.

Estimated probabilities are calculated directly from estimated model parameters and mean values for associated variables, based on the logistic function (4). Following Poe, Welsh, and Champ, and Krinsky and Robb, standard errors for the estimated probability are generated using a bootstrap of the estimated variance-covariance matrix. This simulation method may be used to establish the empirical distribution of any estimator which is a nonlinear function of estimated parameters, and explicitly accounts for both the variability associated with estimated coefficients and interactions among coefficients (Park, Loomis, and Creel). We randomly draw 10,000 sets of coefficient estimates from the maximum-likelihood estimates and accompanying variance-covariance matrix. Probability estimates are calculated for each of the 10,000 draws, resulting in an empirical distribution of probability for each scenario (Poe, Welsh, and Champ). This distribution is used to calculate standard errors for the probability of choosing certified seafood. Finally, these estimated standard errors are used to generate *t*-statistics for the null hypothesis that the difference between estimated Norwegian and U.S. probabilities is equal to zero.

In all cases shown in table 6, the estimated probability difference—for otherwise identical consumers at identical premium levels—is statistically different from zero at $p < 0.05$. For example, at a 0% price premium, the estimated probability of a Norwegian consumer choosing certified seafood is approximately 74%. The equivalent estimated probability for a U.S. consumer is 88%. This probability difference is significant at $p < 0.01$. Moreover, Norwegian respondents are apparently more sensitive to increases in the price premium. For example, from table 6, a change in price premium from 0% to 50% of the uncertified price leads to a 0.42 percentage point decrease in the probability of selecting certified seafood for Norwegians, but only a 0.20 percentage point decrease for U.S. respondents. Similar changes occur between all in- and out-of-sample premium levels, with Norwegian consumers always less likely to choose certified products.[9] These results are particularly notable, given that Norwegian respondents report a greater importance of ecological attributes in general purchase behavior (cf. table 2).

These results contrast with results found in past international studies of consumer willingness to purchase "environmentally friendly" products. The few existing international comparisons of ecological purchasing behavior often indicate U.S. residents are *less* willing to pay premiums for improvements in environmental or ecological product attributes. For example, a recent international study concluded U.S. residents were more reluctant to pay a premium for Mexican shade-grown coffee—a production method offering environmental and other benefits—than were Canadian or Mexican consumers (Commission for Environmental Cooperation). In the case of seafood, the ubiquity of seafood consumption in Norway may lead to a greater awareness of price distributions for various seafood products, and hence a greater awareness of cases in which a premium would place the price of a particular seafood product far out of line with the prices of competing seafood products.

Implications

The organizational blueprint for the Marine Stewardship Council was developed by the consulting firm of Coopers & Lybrand (McHale). According to this firm, "the power of [this eco-seal] is in how customers, voting with their pocket, actually exert economic leverage all the way back up the value chain" (R. Cooke, quoted in McHale, p. 38). Consumer reactions may influence the impact of an ongoing labeling program on seafood harvest patterns, as well as the very existence and economic viability of labeling, given that the revenue-generating capacity of an ecolabel may or may not provide funds sufficient to support required monitoring and management activities (McHale). The results of our analysis suggest that the long-term viability and impact of a seafood ecolabel will depend on a wide range of factors, including the characteristics of consumers, the country in which the label is used, popular trust in the labeling agency, and the species chosen for labeling.

Because the market for seafood is global, with large volumes traded among countries, popular acceptance of an ecolabel in a single nation or group of nations (e.g., the European Union) may be insufficient to prevent overfishing of valued stocks, particularly for migratory species or stocks which straddle national boundaries. The necessity of international acceptance, combined with heterogeneity in consumer behavior across nations, highlights the importance of flexibility in the implementation of labeling programs across national borders. The marketing of a successful ecolabeling program for seafood products cannot follow a simple "one-size-fits-all" approach.

Based on our model results, differences among countries may have significant implications for the success of such programs. For example, differences in price sensitivity

[9] Equivalent results are found for various alternative methods of estimating the "average" probability of selecting certified seafood. Equivalent results are found if one: (*a*) calculates probabilities according to the different means for each country, or (*b*) calculates the probability of purchasing certified seafood for each observation, and then assesses the mean across all observations.

between the United States and Norway may lessen the ability of the seafood industry to charge a constant price premium in some countries sufficient to cover costs associated with labeling programs.[10] To cover costs across the global market, price discrimination by international corporations across different countries may be required.

Model results also have potential implications for the targeting and marketing of seafood ecolabeling programs. For example, programs targeted at consumers of fresh seafood may meet with relatively greater success in the United States, where those who generally purchase fresh seafood are more likely to select certified seafood, compared to those who generally purchase frozen seafood. However, such patterns are not apparent in the Norwegian sample, suggesting this type of targeting may be less effective in Norway. Unless labeling programs incorporate the flexibility necessary to account for these and other differences in consumer behavior across national borders, unilateral international ecolabeling programs may lead only to a reallocation of trade patterns, without any long-term benefit to fish stocks.

In summary, model results highlight the need for thorough analyses of consumer preferences for ecolabeled seafood, particularly given that seafood ecolabels must compete with other valued attributes of fish—safety, quality, price, brand, etc.—to attract consumer purchases. Model results also highlight the significance of cultural and other differences across nations; we find substantial divergence between Norway and the United States concerning the impact of a wide range of variables. A comparison between consumers in industrialized versus developing nations would likely yield even more interesting results, but limited funding temporarily prevents such a study.

Despite the insights provided by comparisons of Norwegian and U.S. survey responses, this research has important limitations that may be addressed by future research. First, the lack of an actual, large-scale market for ecolabeled seafood necessitated a stated-preference approach, which may result in upwardly biased estimates of consumers' actual willingness to pay to obtain ecolabeled products or probability of selecting such products at any given premium (Arrow et al.). Second, restrictions on quantity purchased, although a direct result of consumer behavior in focus groups, limit the welfare information which may be estimated from the random utility model. Finally, the model addresses the choice of ecolabeled seafood contingent on the prior choice to purchase a particular seafood species. It does not address the impact of labels on consumers' choices among different seafood species, or among seafood and other food products. Research which addresses these limitations may provide yet additional information relevant to the design of international seafood ecolabeling programs.

[*Received November 2000; final revision received April 2001.*]

[10] There are four costs of bringing ecolabeled fish to market. First, there are the costs of certification, which vary according to complexity of the fishery being certified. For three fisheries certified, one was paid for entirely by the fishing industry (Thames River herring, UK), one by a combination of industry and government (Western Australian rock lobster), and the third entirely borne by the Alaska state government (Alaskan salmon). Second, there are costs associated with licensing the use of the logo (label). These costs are based on total sales of the firm. Third, chain-of-custody certification is required at several levels of the market to ensure certified fish products are not mixed with noncertified products. Finally, there are costs of achieving a sustainable fishery, which may require better data collection and better management systems. Such costs are highly variable, and depend on the characteristics of the fishery. These costs are typically borne by the government.

References

Adamowicz, W., P. Boxall, M. Williams, and J. Louviere. "Stated Preference Approaches for Measuring Passive Use Values: Choice Experiments and Contingent Valuation." *Amer. J. Agr. Econ.* 80(1998): 64–75.

Addelman, S., and O. Kempthorne. "Orthogonal Main Effect Plans." ARL Tech. Rep. No. 79, U.S. Air Force Office of Aerospace Research, Aeronautical Research Library, Dayton OH, 1961.

Anderson, J. L., and S. Bettencourt. "A Conjoint Approach to Model Product Preferences: The New England Market for Fresh and Frozen Salmon." *Marine Resour. Econ.* 8(1993):31–49.

Arrow, K., R. Solow, E. Leamer, P. Portney, R. Rander, and H. Schuman. "Report of the NOAA Panel on Contingent Valuation." *Federal Register* 58(1993):4602–14.

Blend, J. R., and E. O. van Ravenswaay. "Measuring Consumer Demand for Ecolabeled Apples." *Amer. J. Agr. Econ.* 81(1999):1072–77.

Bollen, K. *Structural Equations with Latent Variables.* New York: John Wiley and Sons, 1989.

Caswell, J. A. "How Labeling of Safety and Process Attributes Affects Markets for Food." *Agr. and Resour. Econ. Rev.* 2(1998):151–58.

Caswell, J. A., and E. M. Mojduszka. "Using Informational Labeling to Influence the Market for Quality in Food Products." *Amer. J. Agr. Econ.* 78(1996):1248–53.

Commission for Environmental Cooperation. "Measuring Consumer Interest in Mexican Shade-Grown Coffee: An Assessment of the Canadian, Mexican, and U.S. Markets." CEC, Communications and Public Outreach Department, Montreal, Canada, 1999.

Cramer, D. "Troubled Waters." *Atlantic Monthly* 275(1995):22–26.

Forsyth, K., D. Haley, and R. Kozak. "Will Consumers Pay More for Certified Wood Products?" *J. Forestry* 97(1999):18–22.

Gudmundsson, E., and C. R. Wessells. "Ecolabeling Seafood for Sustainable Production: Implications for Fisheries Management." *Marine Resour. Econ.* 15(2000):97–113.

Hanemann, W. M. "Welfare Evaluations in Contingent Valuation Experiments with Discrete Responses." *Amer. J. Agr. Econ.* 66(1984):332–41.

Harman, H. *Modern Factor Analysis.* Chicago: University of Chicago Press, 1976.

Holland, D., and C. R. Wessells. "Predicting Consumer Preferences for Fresh Salmon: The Influence of Safety Inspection and Production Method Attributes." *Agr. and Resour. Econ. Rev.* 27(1998):1–14.

Kaiser, H. "The Varimax Criterion for Varimax Rotation in Factor Analysis." *Psychometrica* 23(1958): 187–200.

Kalgraf, K., J. Owens, and M. H. Raaholt. "A Practical Plan for Stabilizing Norway's Energy Consumption." *Energy Policy* 14,6(1986):500–15.

Kline, J. D. "Measuring Public Preferences to Improve the Design of Farmland Preservation Programs." Unpub. Ph.D. diss., Dept. of Environ. and Nat. Resour. Econ., University of Rhode Island, 1996.

Krinsky, I., and R. Robb. "On Approximating the Statistical Properties of Elasticities." *Rev. Econ. and Statis.* 68(1986):715–19.

Maddala, G. S. *Limited Dependent and Qualitative Variables in Econometrics.* Cambridge: Cambridge University Press, 1983.

Marine Stewardship Council (MSC). Information on certifying and labeling of seafood products worldwide, 2000. Website. Online at www.msc.org.

McHale, T. "What's in a Label?" *Pacific Fishing* (February 1997):36–39.

Mitchell, R. C., and R. T. Carson. *Using Surveys to Value Public Goods: The Contingent Valuation Method.* Washington DC: Resources for the Future, 1989.

Nimon, W., and J. Beghin. "Are Ecolabels Valuable? Evidence from the Apparel Industry." *Amer. J. Agr. Econ.* 81(1999):801–11.

Opaluch, J. J., S. K. Swallow, T. F. Weaver, C. W. Wessells, and D. Wichelns. "Evaluating Impacts from Noxious Facilities: Including Public Preferences in Current Siting Mechanisms." *J. Environ. Econ. and Manage.* 24(1993):41–59.

Organization for Economic Cooperation and Development (OECD). *Towards Sustainable Consumption Patterns: A Progress Report on Member Country Initiatives.* Paris, France, 1998.

Park, T., J. B. Loomis, and M. Creel. "Confidence Intervals for Evaluating Benefits Estimates from Dichotomous Choice Contingent Valuation Studies." *Land Econ.* 67(1991):64–73.

Poe, G. L., M. P. Welsh, and P. A. Champ. "Measuring the Difference in Mean Willingness to Pay When Dichotomous Choice Contingent Valuation Responses Are Not Independent." *Land Econ.* 73(1997): 255–67.

Reyment, R., and K. Jöreskog. *Applied Factor Analysis in the Natural Sciences.* Cambridge, UK: Cambridge University Press, 1996.

Roberts, J. A. "Green Consumers in the 1990s: Profile and Implications for Advertising." *J. Bus. Res.* 36(1996):217–31.

Swallow, S. K., and R. A. Sedjo. "Eco-Labeling Consequences in General Equilibrium: A Graphic Assessment." *Land Econ.* 76(2000):28–36.

Swallow, S. K., T. Weaver, J. J. Opaluch, and T. S. Michelman. "Heterogeneous Preferences and Aggregation in Environmental Policy Analysis: A Landfill Siting Case." *Amer. J. Agr. Econ.* 76(1994): 431–43.

Tiesl, M., B. Roe, and R. L. Hicks. "Can Ecolabels Tune a Market? Evidence from Dolphin-Safe Labeling." Paper presented at annual meetings of the American Agricultural Economics Association, Toronto, Canada, 27–30 July 1997.

U.S. Environmental Protection Agency, Office of Pollution Prevention and Toxics. "Environmental Labeling: Issues, Policies, and Practices Worldwide." EPA Pub. No. 742-R-98-009, Office of Prevention, Pesticides and Toxic Substances, Washington DC, December 1998.

Variyam, J. N., J. L. Jorday, and J. E. Epperson. "Preferences of Citizens for Agricultural Policies: Evidence from a National Survey." *Amer. J. Agr. Econ.* 72(1990):257–67.

Wessells, C. R., and J. G. Anderson. "Consumer Willingness to Pay for Seafood Safety Assurances." *J. Consumer Affairs* 29(1995):85–107.

Wessells, C. R., H. Donath, and R. Johnston. "Assessing Consumer Preferences for Ecolabeled Seafood: The Influence of Species, Certifier, and Household Attributes." *Amer. J. Agr. Econ.* 81(1999):1084–89.

[24]

Forest certification (eco-labeling) programs and their policy-making authority: explaining divergence among North American and European case studies

Benjamin Cashore[a,*], Graeme Auld[b], Deanna Newsom[c]

[a]*Chair, Program on Forest Certification, Global Institute of Sustainable Forestry, Yale School of Forestry and Environmental Studies, 360 Prospect Street, New Haven, CT 06511-2189, USA*
[b]*Program on Forest Certification, Global Institute of Sustainable Forestry, Yale University, School of Forestry and Environmental Studies, 205 Prospect St., New Haven, CT 06511, USA*
[c]*TREES Program, Rainforest Alliance, Richmond, VT, USA*

Received 8 November 2001; accepted 28 March 2002

Abstract

In recent years, transnational and domestic non-governmental organizations have created private standard setting bodies whose purpose is to recognize officially companies and landowners practicing 'sustainable forest management'. Eschewing traditional state processes and state authority, these certification programs have turned to the market to create incentives and force compliance to their rules. This paper compares the emergence of this non-state market driven (NSMD) phenomenon in the forest sector in eight regions in North Am40erica and Europe. We specifically seek to understand the role of forest companies and landowners in granting competing forest certification programs 'legitimacy' to create the rules. We identify distinct legitimation dynamics in each of our cases, and then develop seven hypotheses to explain differences in support for forest certification.

Keywords: Forest certification; Market instruments; Voluntary initiatives; Privatization of governance; Eco-labelling; Forest sustainability; Sustainable forest management; Environmental governance; United States; Canada; Europe; Forest sector; Market access; Competitiveness

1. Introduction

In recent years, transnational and domestic non-governmental organizations have created private standard setting bodies whose purpose is to recognize officially companies and landowners practicing 'sustainable forest management' (SFM) (Meidinger, 1997). Eschewing traditional state processes and state authority, these organizations have turned to the market, soliciting the support of customers of forest products in order to achieve compliance with their rules (Fletcher and Hansen, 1999).

This trend toward using non-state market driven (NSMD) governance to address matters of concern

*Corresponding author. Tel.: +1-203-432-4001; fax: +1-203-432-3809.
E-mail address: benjamin.cashore@yale.edu (B. Cashore).

to global civil society has its roots in increasing economic globalization (Berger and Dore, 1996) and corresponding 'internationalization' (Bernstein and Cashore, 2000), where non-governmental organizations and international institutions have often acted to reverse the 'downward' effects of globalization on environmental, social and labor standards (Vogel, 1995).

Despite these forces for convergence, differences have emerged among domestic political struggles over which non-governmental certification program gets the right to create sustainable forest management rules. In particular, the support of forest land owners and forest land managers—those companies, individuals, and organizations who must actually implement the certification rules—has varied dramatically. In some regions, these 'supply side' audiences have committed to pursuing the detailed, prescriptive performance standards of the international Forest Stewardship Council (FSC) certification program, while in the other jurisdictions, supply side forest companies and landowners have coalesced around the more flexible industry and landowner initiated programs that were created to compete with the FSC for rule-making authority. In other cases, a hybrid has developed.

Given very similar global pressures for forest certification, why have supply side actors in some regions chosen to support the FSC, while those in other regions have maintained firm support for competitor programs? The answer to this question is important for the implications for forest management in different regions, but also sheds light on the conditions under which NSMD governance may or may not gain long-term rule-making authority.

This paper addresses the divergence question in three steps: First, it reviews the historical context behind the development of certification programs and briefly describes the 'legitimacy framework' we have employed in our comparative study. Second, it briefly reviews the methodological hurdles in comparing certification legitimation dynamics that move across regional, national, and international bodies. Third, using a qualitative case study approach, we use the classification system to offer a nuanced description of how forest owners and forest land managers accept, or fail to accept, the rules of a particular certification program in Canada (specifically British Columbia and the Maritimes), the United States (specifically the pacific coast region, the south-east and north-east), the United Kingdom, and Germany.

Fourth, we develop working hypothesis to explain key divergence among the cases. The central argument that emerges from this approach is that four related factors, together, explain the degree of support the FSC and its competitor programs achieved from supply audiences: a region's degree of market globalization, the degree to which forestry practices in a region are seen as a 'problem' on the political agenda, the structure of a region's forest industry (in terms of land ownership fragmentation and industry concentration), and the characteristics of the FSC's competitor programs.

2. Historical context and emergence of forest certification as a non-state marked driven governance system

The emergence and interest in non-state market-driven forest certification programs can be traced to the economic and political trends in the last 10 years in which market-oriented policy instruments have been given increasing salience domestically and internationally (Harrison, 1999; Rosenbaum, 1995; Tollefson, 1998). Originally research on economic globalization found that increased capital mobility, international trade, and foreign direct investment appeared to reduce or constrain domestic policy choices, sometimes leading to downward protection in environmental and social standards (Berger, 1996: 12). However, other scholars noted that a parallel process was taking place in which domestic policy arenas were facing increasing scrutiny by transnational actors, international rules, and norms (see Risse-Kappen, 1995; Keck and Sikkink, 1998), sometimes leading to a reversal of the 'downward' effect of globalization (Vogel, 1995), a process Bernstein and Cashore (2000) refer to as internationalization.[1] Indeed, 'interna-

[1] A similar definition was first offered by Doern et al. (1996).

tionalization' processes often get their policy influence by using market-based boycott campaigns to force upward governmental and firm-level environmental protection. Internationalization provided lessons to environmental non-governmental organizations (ENGOs) about the power in using market forces to shape policy responses where efforts to change policy were often easier than attempting to influence domestic and international business dominated policy networks. This recognition increased the salience of market-manipulation campaigns generally, but also of forestry specifically (Stanbury et al., 1995).

Non-governmental organizations' interest in market mechanisms in forestry also increased following the failure of the Earth Summit in 1992 to sign a global forest convention (Bernstein and Cashore, 1999, 2000). ENGOs felt that they had spent significant effort and resources on state-sanctioned international fora with no discernible policy gains. As a result, transnational groups, led by the World Wide Fund for Nature (WWF)[2] turned to the market for influence by certifying forest landowners and forest companies, who practiced WWF's and other interest groups' definition of 'sustainable forestry', expanding the traditional boycott campaign's stick approach by offering carrots as well. These international trends were mirrored by increasing interest at domestic levels as well (Harrison, 1999; Rosenbaum, 1995; Tollefson, 1998).

The Forest Stewardship Council (FSC) itself can be traced back to 1990, when, following widespread scrutiny of tropical deforestation, timber retailers and distributors began looking for fiber from well-managed forests (Meidinger, 2000: 4). Following 3 years of discussions, the FSC was created in Toronto, Canada in 1993 and became legally registered in 1995 in Oaxaca, Mexico. The FSC originally created nine 'principles' and more detailed 'criteria' that are performance based, broad in scope, including tenure and use rights, community relations, worker's rights, environmental impact, management plans, monitoring and preservation of old growth forests (see Moffat, 1998: 44; Forest Stewardship Council, 1999). Importantly, the FSC program required the development of regional or national working groups given the task of developing specific indicators and verifiers with which to apply the principles and criteria. This institutional dynamic is crucial because it directed many political struggles over certification to the territories designated as a national/regional working group. The FSC can be identified with a conception of NSMD governance that diverged dramatically from traditional clientele public policy processes, as it institutionalized decision making processes that limited business influence, forbade government involvement in rule making, and developed wide ranging policy initiatives.

In all the cases under review in this paper, introduction of the FSC program resulted in industry and/or landowner organizations developing their own NSMD programs to compete with FSC governance. These programs contained a very different conception of governance, highlighted by a belief in business dominating rule making, a narrow policy realm, and broad, flexible policy options. Our task in this paper is to document the levels of supply side support for FSC and alternative programs at two points in time: 'initial' support, when the FSC first emerged, and 'current' support, which, in our project, ranges from autumn 2000 to the autumn of 2001, and to explain changes and regional differences.

In order to facilitate comparison across an array of national and regional settings, our study developed a comparative heuristic that categorizes both the types of audiences from which certification programs seek authority, and the different levels of support these audiences give. To accomplish this task, we use Cashore (2002) application of Suchman's (1995) seminal article on organizational legitimacy, a subset of organization sociology literature. Suchman (1995: 574) defines legitimacy as:

> ...a generalized perception or assumption that the actions of an entity are desirable, proper, or appropriate within some socially constructed system of norms, values, beliefs and definitions.

[2] Although the acronym has not changed, The World Wide Fund for Nature changed its name from the World Wildlife Fund in 1995 but the Canadian and US sections opted to keep their original names.

Two sub-categories of legitimacy are central to this paper. *Pragmatic legitimacy* rests on the 'self-interested calculations of an organization's most immediate audiences' (Suchman, 1995: 578), in which the material 'well being' of the legitimacy grantor is enhanced (Suchman, 1995: 589).[3] In contrast *moral legitimacy* reflects a ''positive normative evaluation of the organization and its activities. It rests not on judgments about whether a given activity promotes the goals of the evaluator, but rather on judgments about whether the activity is 'the right thing to do...''' (Suchman 1995: 579).[4] Our cases below reveal that, arguably owing to the nascent nature of forest certification, most firms evaluations fell into the pragmatic legitimacy category, with the choice over which certification program to support, if any, a central theme. On the other hand, moral legitimacy granted to the Forest Stewardship Council by environmental groups and other 'core audiences', played a key role in understanding how far the FSC could go in their efforts to achieve pragmatic legitimacy from supply side interests.

3. Methodological hurdles

Comparing NSMD legitimation dynamics in regions with different types of political boundaries (national and sub-national) comes with certain difficulties. Likewise, the certification programs that we are examining are also designed to act at different levels: international vs. national vs. sub-national. The FSC, on the one hand, is international in scope, but asserts the need for certification based on specific standards, which are developed at the national/regional level. Conversely, competitor programs have gone in the opposite direction: while they generally have domestic origins, subsequent developments saw them moving to a regional, or even international, scale. While these issues raise methodological constraints—how does one compare a country with a sub-national region?—complex comparisons often yield more nuanced findings precisely because of this complexity. We address this in our comparison by shifting our research lens in and out, depending on the region and certification program being examined.

The cases that are investigated have two key components that make them relevant to this study. First, they are all actively involved in the production and consumption of industrial wood and paper products. In some cases, the United Kingdom (UK) and Germany, higher reliance on imports exists, while in another, British Columbia (BC), exports are the driver of local production. Since NSMD is based around market transactions and the process of production, distribution and end product consumption, this variety of market conditions provides an important continuum of supply side audience types. Second, given the emerging nature of NSMD, these cases are important, as they were all early participants in forest certification, which gives us the opportunity to identify key factors that lead to differing paths of development.

4. The cases

This section of the paper describes the initial and current levels of support given to the FSC and competitor programs in each region. Our inductive approach revealed that, in contrast to the relatively stable ENGO support for the FSC, support by forest companies and landowners for FSC varied among cases. Although the supply side audience initially gave legitimacy to the alternative program in every region, we observed a shift in BC and the UK, where the supply side has increasingly legitimized the FSC. See Table 1 for a summary

[3] The definition of pragmatic legitimacy falls outside existing political science international relations work on legitimacy (for a review see Bernstein, 2001) which sees legitimacy as entailing a 'logic of appropriateness' supported by moral or cognitive dimensions, and which stand in contrast to rational self-interest support for governance structures (March and Olsen, 1998).

[4] Suchman and our broader study identified a third category of *cognitive legitimacy*, which is based neither on interests nor on moral motivations, but rather on 'comprehensibility' or 'taken for grantedness'. In the former case legitimacy is given because the actions of an organization are understandable, in the latter case legitimacy is given because 'for things to be otherwise is literally unthinkable' (Suchman, 1995: 583). We found no examples of this type of legitimacy in our current research, and thus leave out this distinction for purposes of this paper.

Table 1
Type of legitimacy given by forest companies and non-industrial private forest (NIPF) landowners to the FSC and the FSC's competitor program in our study regions. Bolded text indicates where changes from initial to current legitimacy have occurred

Case	Time	Forest companies		NIPF landowners	
		FSC	FSC's competitor program[a]	FSC	FSC's competitor program
British Columbia (Canada)					
	Initial	None	**Pragmatic Moral**	N/a	N/a
	Current	**Pragmatic**	**Pragmatic Moral**	N/a	N/a
Maritimes (Canada)					
	Initial	None	None	None	None
	Current	None	None	None	None
Pacific Coast (US)					
	Initial	None	**Moral**	None	Moral
	Current	None	**Moral**	None	Moral
Northeast (US)					
	Initial	None	**Pragmatic Moral**	None	None
	Current	None	**Pragmatic Moral**	None	None
Southeast (US)					
	Initial	None	**Pragmatic**	None	None
	Current	None	**Pragmatic**	None	None
United Kingdom					
	Initial	None	**Moral**	None	**Moral**
	Current	**Pragmatic**	**Moral**	None	**Moral**
Germany					
	Initial	N/a	N/a	None	**Pragmatic**
	Current	N/a	N/a	None	**Pragmatic**

[a] The alternative program is the CSA in Canada, the SFI in the US, the PEFC in the UK, and the PEFC in Germany.

of initial and current legitimacy for the FSC and alternative programs in each region.

4.1. Canada

The case of forest certification in Canada illustrates the way international, national and regional level struggles influence legitimation dynamics and support for FSC and its competitor programs. The decision by FSC to create regional standards bodies in Canada, and the Canadian Standards Association (CSA) to stick with one national certification system in which individual companies created specific implementation rules, has meant that only a regional/national analysis can capture the complex ways supply side groups have acted to support the systems.

Once the FSC emerged in Canada in 1996, a two-pronged approach was taken. A national office was established to coordinate the FSC in Canada, while regional working groups in the Maritimes, the Great Lakes-Saint Lawrence and British Columbia, took on the responsibility of developing standards. Most of the political struggles have occurred within these regional processes, however, the presence of the FSC spurred the creation of the CSA forest certification program in which industry attempted to frame the certification debate as a national-level issue. Led by the Canadian Pulp and Paper Association (CPPA), 23 industry associations joined ranks to form the Canadian Sustainable Forestry Certification Coalition. This group approached the CSA in 1994 to develop an industry initiated Canadian forest certification program (Elliott, 1999). The process was completed by 1996, with the first successful certification occurring in 1999 (Canadian Standards Association Sustainable Forest Management System, 2000).

The major role offered to the Canadian forest sector in the development of the CSA program reflects the close relationship between the government and the forest industry. In the National Forest Strategy, the federal government proposed the following partnership: 'Industry and governments will work cooperatively to pursue joint technical discussions aimed at internationalizing product standards, codes and certification procedures' (National Forest Strategy Coalition 1997: 4–22). The portrayal of the CSA certification process as a partnership between government and industry, and its emphasis on using the CSA as a way to secure market access (Elliott, 1999), led to most Canadian forest companies granting it early pragmatic legitimacy. Environmental groups never appeared to give any type of legitimacy. Though there were some signs that ENGOs were considering granting pragmatic legitimacy, they quickly moved away from such explorations.[5]

While these national developments were key to CSA legitimacy dynamics, FSC dynamics were largely regional, and it is at this level where divergence in the acceptance of forest certification programs by supply side members is most clearly demonstrated by our two regional Canadian case studies. While ENGO support for the FSC has remained relatively constant throughout the development of certification in both British Columbia and the Maritimes, we have observed a shift in the attitudes of supply side members in BC, who have gone from strictly supporting the CSA program to the reluctant acceptance of the FSC program as what one industry official described as 'a cost of doing business' (personal interview). This shift did not occur in the Canadian Maritimes, where, on the contrary, the early support of FSC by the JD Irving company was later revoked.

4.1.1. British Columbia

Initially, forest certification developments in British Columbia showed striking similarities to developments in other regions: local and international ENGOs were quick to give the FSC regional process moral legitimacy (i.e. they indicated that the FSC was consistent with their ideals), while forest companies played a central role in the creation of a competitor to the FSC, the CSA certification program. Environmental groups based in BC and international groups concerned over BC forest management were clearly excited about the prospects of increasing their influence on forest management policy in the province. The supply side, on the other hand, supported the national developments of the CSA (mentioned above) as a morally appropriate and pragmatically appealing body to develop a workable certification program (Lapointe, 1998; Abusow and Rotherham, 1998). At this point the BC forest industry believed that it could handle the forest certification agenda 'in-house' using the CSA process (Paget and Morton, 1999: 7), and that the CSA program would meet the requirements of its buyers, avert boycott threats and ensure international customers that 'Canada is working towards sustainability in its forests (Forest Alliance of British Columbia, 1996).' Many companies announced early on that they were pursuing CSA certification and expressed skepticism about the FSC process (Forest Alliance of British Columbia, 1997, 1998; Lapointe, 1998). Similarly governmental actors expressed strong concern about the FSC (personal interviews).

The initial patterns of industry support for certification programs changed dramatically in 1998: the exclusive support of supply side members for CSA quickly fell apart, as individual forest companies targeted by ENGO's expressed interest in the FSC (Stanbury, 2000). While not the first to announce intentions to pursue FSC certification, the moves of MacMillan Bloedel to adopt a new 'green' strategy, in which Greenpeace campaigners, at a highly reported press conference, presented then president Tom Stephens with champagne (MacMillian Bloedel Ltd., 1998a,b) are telling of the dramatic shifts that occurred within the supply side. Overall, by 2001, many of BC's large supply side companies have expressed interest in the FSC (Stanbury, 2000; Hogben, 1998; Canfor Corporation, 1999). Even the Forest Alliance, often criticized by environmental groups (Forest Alliance of British Columbia, 1997), actually joined the FSC (Jordan, 1999), and the pro-timber industry IWA applied for membership in the FSC. It has also been reported that companies have provided direct financial support to the re-kindled BC FSC working group and its standards development team (personal interview). All these efforts appeared to fall under the 'pragmatic legitimacy' category. Companies still supported the CSA, and even later an expanded SFI (American Forest and Paper Association, 2000b), but many recognized that it was now necessary to shape FSC regional standards and show some kind of support (Hogben, 1998; Hayward, 1998; Hamilton, 1998), lest they be targeted as environmentally unfriendly. Some governmental agencies appeared to alter their position on the FSC. For example, initial opposition changing, with some agencies/divisions showing

[5] Although the first CSA technical committee contained a broad membership encompassing many interest groups, ENGO support was minimal from the start, perhaps related to the unauthorized listing of two prominent Canadian ENGO executives on the CSA technical committee, and their subsequent withdrawal from the process (Elliott, 1999).

signs of change. For example, the province's Small Business Forest Enterprise Program contracted PricewaterhouseCoopers to provide a gap analysis to determine how feasible it would be to become certified, not just under CSA rules, but also under the more stringent FSC approach (PricewaterhouseCoopers, 1999).

Interestingly, the increased involvement of the BC supply side in the development of FSC BC regional standards has been met with skepticism by some BC ENGOs, who fear that the rigor of the standards may be in jeopardy. Although most ENGOs are still behind the FSC, moral legitimacy may be withdrawn if the FSC's regional standards no longer resonate with the goals of ENGOs. Overall these pronounced shifts in industry support and potential shifts in ENGO support have set BC apart from many other regions in North America where the FSC is still pursuing support from the supply side while enjoying relatively unwavering support from ENGOs.

4.1.2. Maritimes

Maritimes standards development began early in FSC history, with little national or centralized regional guidance. Research conducted by Lawson and Cashore (2001) found that local environmental groups, themselves products of anti-pesticide campaigns in the 1970s and 1980s, quickly gave FSC moral legitimacy, hoping to influence certification rules to embrace concerns they were unable to raise within traditional public policy processes. Aboriginal groups, also an important actor as a social interest with specific and evolving rights to forested areas, gave the FSC pragmatic legitimacy and were active in developing its rules (Lawson and Cashore, 2001).

On the other hand, non-industrial private forest (NIPF) landowners, who comprise roughly one-half of the forest land base in this region, were divided in their support for FSC. A highly active 'anti-industrial forestry' group gave FSC moral legitimacy and, like the ENGOs in the region, sought to create FSC regional standard rules that would fundamentally suit forest management in the region. However, the vast majority of landowners and other associations took a more typical approach, expressing strong concerns about the influence of the FSC on their rights to manage their property.

Indeed, by 1996 many NIPF owners gave pragmatic legitimacy to the CSA standards, resulting in many joining a certification market project sponsored by the CSA (Lawson and Cashore, 2001; Moffat, 1998).

On the other hand, and mirroring many of the cases under review, large and medium-sized industrial forest companies either favored no certification at all, or if they had to, gave reluctant pragmatic legitimacy to the CSA. Government agencies at both the provincial and federal levels were disinclined to legitimate the FSC or the CSA at all.

The one exception to these general patterns was found with the region's most significant industrial forest company, JD Irving Ltd, who actively sought, and obtained, FSC certification in the region for some of its holding in the provinces (Elliott, 1999). Evidence indicates that this support was pragmatic in nature, as Irving felt it could gain market access and recognition as a green leader by being an industry leader and innovator (Lawson and Cashore, 2001). However, by the summer of 1999, Irving's initial support changed dramatically, removing its pragmatic support of FSC. It gave back its FSC certificate and withdrew from participation in the standards development process, which changed significantly as the politics over FSC regional standards became more intense and polarized.

The FSC case reveals fairly consistent positions on the part of the key immediate audiences, illustrating again that the supply side seems to be the most significant area of change in support for certification programs. The 'Irving exception' to these general patterns is also unique among our cases because in this case a large vertically integrated company supported FSC relatively quickly, but later removed its support—thus moving in the opposite direction from the way industry moved in British Columbia.

4.2. United States

Similar to Canada, forest certification in the US is dominated by two programs: the FSC and the

American Forest and Paper Association's (AF&PA) Sustainable Forestry Initiative (SFI) program. And like Canada, part of the certification issue is a complex intersection of national and regional dynamics in which the SFI frames the issue as a national one in which companies are given flexibility in implementation of the rules, and the FSC, in which regional processes create region–specific certification requirements that are generally more prescriptive and wider in scope. The regional–national intersection is important because, as we will see, there are striking similarities among the level of supply side support for certification programs in our three US cases.

What is striking is that in all of the US cases, nearly all members of the forest industry, organized as the AF&PA, have rejected the FSC certification process: companies have resisted FSC certification on their own lands and have not participated in the FSC standards development process in their regions. NIPF landowners in all three regions are either uninformed about certification or support landowner-run programs like the Tree Farm program (Coulombe and Brown, 1999). ENGO support for the FSC has generally remained constant in all regions.

Although the origin of SFI is actually earlier than that of the FSC, the SFI's 'certification' role has been categorized as a direct response to the creation of the FSC program (personal interviews). The SFI program was originally created to address the polarized atmosphere that had developed in the US over forestry issues in the late 1980s and early 1990s.[6] For analytical purposes, we treat the SFI re-creation in 1994 as its forest certification origins.

Nationally, the SFI program initially enjoyed the support of industry as a program to address public concerns about forest management (Wallinger, 1995) and to increase the industry's lobbying presence in Washington DC. The program's principles were designed to focus on a combination of traditional measures of 'good' forest management (e.g. prompt and successful regeneration)[7] and some broader environmental issues (e.g. clean water) (American Forest and Paper Association, 2000a,d). The requirements set were viewed as quite similar to existing state regulations, so the program was deemed an appealing public relations tool for regaining public faith.

Since the inception of the two programs, we have seen the FSC gain pragmatic legitimacy from the demand side audience while remaining unsupported by the majority of the supply side. The creation of the North American Certified Forest Products Council (CFPC) has been a major activity of the FSC at the national level (World Wildlife Fund United Kingdom, 2000). Membership in this includes high-profile companies like the Home Depot, Lowe's, and Kaufman and Board Home Corp (Caulfield, 1999; Carlton, 2000). This support appears merely to be pragmatic at this stage, as many of these companies have instituted purchasing policies, yet their implementation has been characterized as poor by some ENGOs (Forest Ethics, 2001). Whether this demand side support will increase supply side support for the FSC is not yet discernible.

While the FSC has focused on regional standards detailed below, the FSC national office has been undertaking increasing policy influence and legitimation achievement strategies, moving beyond coordinator to become a key source of expertise. Indeed, by the end of 2000, the national office was developing national guidelines in an effort to head off industry criticisms that its stan-

[6] The SFI was formally created after an AF&PA commissioned study revealed different perceptions on the part of civil society and industry about whether sustainable forestry was being practiced (Wallinger, 1995), cited in Hansen and Juslin (1999).

[7] In the 1999 SFI annual report the AF&PA state, 'it is not enough to reforest harvested areas—it must be done quickly. SFI participants have successfully reforested, on average, 99% of the acres harvested under the SFI program within 5 years. In 1998, 59% of sites schedules for planting were replanted within 1 year; 98% were planted within 2 years (American Forest and Paper Association, 1999)'. While this is sold as an accomplishment, its ecological value could be contested given that late seral and early seral stages are categorized as the most beneficial for wildlife habitat. In particular, the brush stage that may occur in stands that naturally regenerate after disturbance are removed from the succession process when industrial planting techniques are employed. So, while reforestation is important, regenerating trees in the most expedited manner is not necessarily the most ecologically beneficial management strategy.

dards varied among regions not for ecological reasons, but for political ones.

4.2.1. Pacific Coast

Initial legitimation dynamics in the US Pacific Coast generally converged with most other cases in this paper: ENGOs gave FSC moral legitimacy. ENGOs appeared eager to use FSC to force changes on private and state owned forest lands, which came on the heels of their successful efforts to significantly reduce harvesting on federally owned forest lands in the region (Cashore, 1997, 1999; Hoberg, 2000).[8] On the other hand, large vertically integrated forest companies in the region gave pragmatic legitimacy to the SFI program, which, as noted above, was designed to compete with the FSC in the US. Likewise, most NIPF landowners in the region either failed to support any certification efforts, or were attracted to the more flexible Tree Farm program. Indeed, members of AF&PA operating in the region failed to become involved with the FSC regional standards working group. Participation was limited to certain smaller companies, and landowners such as Collins Pine and the Hoopa Valley Native American tribe (Forest Stewardship Council United States, 1998). Their participation stood in stark contrast to the otherwise startling homogeneity of firm level responses.

However, unlike legitimation dynamics north of the border in BC, members of the AF&PA have to date maintained their support for SFI, failing to give the FSC any type of legitimacy. Indeed, the creation of the CFPC in 1997 (World Wildlife Fund United Kingdom, 2000), and well publicized commitments of large retailers and developers in the US (mentioned above) to purchase certified wood conforming to FSC standards, has led only a few companies and landowners to consider FSC certification. The program has yet to gain inroads with the regions larger operators.

Both the integrated forest companies and NIPF landowners were openly critical of the FSC regional standards setting process, despite their own boycott of these discussions. They argued that the initial draft standards surrounding reduced clear cuts and selection harvesting systems were designed for drier interior regions, rather than for those operating in the temperate forest in the western part of these states.[9] This illustrates the national–regional dynamic, as there is evidence that concerns over these clear cut rules were a catalyst for the creation of the FSC regional standards harmonization process created at the national level. The supply side audience explains their lack of support for FSC by pointing to uncertainty surrounding how national harmonization will affect these regional standards and the ambiguous support demand side purchasing policies are offering the FSC. The latter being the concerns for ENGOs as companies have not been quick to implement policies favoring FSC wood (Forest Ethics, 2001).

Other immediate audiences have played a less direct role at this point. Governmental support for the FSC in the region has been mixed. Some US national forest service officials have expressed concern that its lands have been exempted from the certification process, while others have raised concerns about FSC rules and procedures, mirroring critiques by professional organizations such as the Society of American Foresters. At the state governmental level there is a split between forest management agencies whose procedures follow a more flexible approach (at least in Oregon and Washington), and those who lead more conservation-focused agencies who have expressed an interest in having state-owned land FSC certified (Price, 2000).

In the US Pacific Coast we observed clear legitimation dynamics, with FSC maintaining moral support from ENGOs, pragmatic from national retailers and some lumber dealers, and SFI maintaining pragmatic and arguably moral support from large vertically integrated forest companies. Landowners either failed to support certification at all, or opted for an FSC competitor program geared toward the NIPF landowner.

[8] Indeed, an important subtext is that many environmental groups in the area did not want FSC to apply on national forest lands, as they feared that this might allow increased harvesting in these regions (MacCleery, 1999).

[9] This inequity was reported as a point of debate among participants in the regional standards process: there were even discussions of splitting the region to better capture the different ecosystems in drier southern interior forests and the wetter northern coastal forests.

4.2.2. North-east

US north-east legitimation dynamics largely played out in similar ways to experiences in the Pacific Coast and the south-east. In the early stages, many large industrial firms, especially pulp and paper producers, did not see any potential for certification at all, and feared the FSC's disclosure requirements and prescriptive rules process. Once the idea of certification was seen to be taking hold, most AF&PA members supported making SFI a condition of their membership as a way of countering the FSC. Thus, most large forest companies gave pragmatic legitimacy to the SFI, while ENGOs gave moral legitimacy to the FSC. On the other hand, some small-to-medium-sized timber brokers and specialty wood-products manufacturers joined the FSC process, seeing both pragmatic reasons to participate in niche markets by differentiating themselves from larger suppliers (Lawson and Cashore, 2001).

There were notable exceptions to the overall support of industrial forest companies for the SFI, for instance a Maine forest products company, Seven Islands sought and received FSC certification for its operation. And as in the Maritimes, the most marked initial support for the FSC came from JD Irving—and unlike its experience in the Maritimes, Irving has maintained its FSC commitment in this region, owing to more industry friendly FSC regional standards (Lawson and Cashore, 2001).

Few companies have joined the Irving and Seven Islands approaches. Instead, industry has worked towards enhancing the credibility of the SFI in the region. For example, the Maine SFI committee worked with the Maine Forest Products Council to expand the role of the SFI, increasing third-party SFI certifications, setting minimum criteria under such certifications, and created an expert panel at the local level that would report on such certifications to the legislature (Lawson and Cashore, 2001). Partly as a result, increased ENGO efforts to expand the FSC in this region have failed, mirroring broad national trends.

4.2.3. South-east

Ownership and management of forest lands in the US South are generally conducted by two distinct groups, with vastly different levels of knowledge about and involvement in forest certification: industrial forest companies, who manage a large proportion of Southern forest lands and virtually unanimously support the SFI program, and NIPF landowners, who generally know very little about forest certification (Newsom et al., 2002). Industrial forestry operations in the south-east rely on non-industrial private landowners for 60% of their raw materials (Hyde and Stuart, 1998)

Due to the national nature of the SFI and the regional nature of the FSC, the two programs developed in the south-east with relatively little influence on one another. Both the SFI program and FSC US south-east working group were developed in 1996 (Wallinger, 1995; The Forest Management Trust, 1999). Although all members of the AF&PA are obligated to participate in the SFI program, the degree to which southern AF&PA members gave moral legitimacy to the SFI appears ambiguous. Interviews conducted in the US south-east with key US forest company officials indicate that, although an element of moral legitimacy in the basic approach taken by the SFI may have existed, it appears that most SFI members in the region saw the program as a pragmatic marketing and educational tool from its very inception. NIPF landowners have generally not yet entered the certification debate in the south-east.

The FSC south-east working group gained moral legitimacy from regional environmental groups through its focus on minimizing the spread of plantations and other industrial forestry practices, but failed to get participation by members of the forest industry (Humphries, 1999). Interviews indicate that the refusal of virtually all south-eastern forest industry members to participate in the FSC standards-development process was a conscious decision: internal sanctions within the industry were used against one company that expressed interest in participating (personal interviews). The company in question eventually withdrew from the FSC regional process. In the absence of forest industry participation, the FSC standards setting team recruited the help of professional foresters to bring a 'realistic touch' (personal interview) to the process and act as a proxy for the voice of the forest industry.

The FSC still faces entrenched opposition to its program in the south-east, mainly from the large industrial supply side actors. Those NIPF landowners that are aware of the certification issue consider local or state-level initiatives most appropriate, and give legitimacy to initiatives like the Tree Farm program, which is specifically run by and tailored to NIPF landowners (Newsom et al., 2002). Some NIPFs appear to give pragmatic legitimacy to the certification concept as potential protection against more stringent public policy (personal interview). Despite reports of having felt pressure from some national-level buyers group members (personal interviews), most large industrial members of supply side, with the exception of the Anderson-Tully Company (FSC US, 2001), are currently maintaining a strong resistance to the FSC, as are their counterparts in other regions of the US.

4.3. Europe

Two programs dominate forest certification politics in Europe: the FSC and the Pan-European Forest Certification (PEFC) program. When the FSC made headway in Europe in the mid 1990s, supply side interests in the UK and Germany, supported by their respective state agencies, proposed labels of origin as a means of countering the FSC approach. The 'local grown' certificate was called the FICGB Woodmark in the UK (Forestry Industry Council of Great Britain, 2000; Kiekens, 1997) and the *Herkunftszeichen,* or 'label of origin,' in Germany (Klins, 2000). Generally there was agreement between the government and the forest sector that FSC certification was unnecessary given the extensive and rigorous set of state regulations (Forestry Commission, 1998) (Personal interviews). Their view of the FSC was that it had an environmental bias, a top down decision-making structure, lacked accountability, and was totally unnecessary in countries where government regulations were, they felt, more than sufficient.

The PEFC program was conceived in 1998 by forest companies and private forest owners in six European countries, including Germany (Hansen et al., 1999). Although the PEFC has become a Europe-wide initiative with 15 members (including the UK) and an overarching set of principles, based on the Helsinki Criteria, developments within this program are still occurring mainly at the national, not European, level. Each member country proposes its own set of standards, which are then evaluated by the European PEFC office, and, if they conform to the requirements, are accepted. The member's countries have their own national office, and much autonomy in terms of the development of national auditing system. Like the FSC national working groups, each national PEFC office is not constrained by its international framework when it comes to national-level political or strategic decision-making. This national focus allows us to examine each of our European case studies separately, without the problems of scale associated with our other cases.

4.3.1. United Kingdom

Two opposing forces similar to those present in Germany and in North America shaped initial legitimacy dynamics in the UK. On the one side, a local industry, composed of NIPF owners and processors in concert with the government, stood behind existing forest regulations, defending their history of ensuring 'good' forest management. Representing the interests of local producers, the FICGB and the government's Forestry Commission established the FICGB Woodmark (mentioned above) (Kiekens, 1997). With the addition of a new government felling license, this program was able to ensure that 'wood (products were) derived from British grown timber and that the tree was felled in accordance with official government regulations (Forestry Industry Council of Great Britain, 2000).'

The opposing force was a coupling of environmental groups and members of the demand side audience who gave moral support to the international FSC program. The WWF and the Do-It-Yourself (DIY) retailer chain, B&Q, were two important supporters that saw the management of tropical forests as a key issue of concern (Viana et al., 1996). With the addition of a number of locally focused environmental groups, the FSC regional standards process was supported as an important starting point for attaining global forest sustainability (Tickel and World Wildlife Fund for Nature, 2000). These groups were not satisfied with 'local grown' labels, since tropical timber imports often came with dubious sustainability claims (Global Witness, 1999). Only an international process independent of government was perceived as a credible and morally appropriate option to ensure the procurement of sustainable forest products.

Subsequent events provoked a significant change in the polarized views on certification; arguably, certification has become an accepted reality and has been deemed morally appropriate by most of the immediate audience (Tickel and World Wildlife Fund for Nature, 2000). However, the legitimacy dynamics are complex. The key element of this shift was the UK, Woodland Assurance Scheme (UKWAS), which was a government-facilitated process independent of the FSC that successfully merged the draft FSC Great Britain standard with the government's UK Forestry Standard (Goodall, 2000). UKWAS represents a compromise that achieved moral support from all members of the immediate audience. Yet, it is merely a standard rather than a certification program (UKWAS Steering Group, 2000) and its moral legitimacy has not been automatically transferred to the FSC. Only the environmental community and the government are giving moral legitimacy to the FSC. The WWF 95+ group has increased its membership, however, not all members are purely specifying FSC wood, and even B&Q, an original participant in the creation of the FSC, has begun to consider alternative sources of certified wood (Knight, 2000).

The processing sector has framed the debate in pragmatic terms. At present, FSC is the best alternative available; however, manufacturers and managers have expressed openness to other alternatives if their clients demand them. NIPF landowners are the least supportive of certification and their persistent mistrust for the FSC has provoked interest in the landowner-initiated PEFC program in Europe. The recently incorporated PEFC UK Limited hopes to gain access to the UKWAS standard (Pan European Forest Certification United Kingdom, 2001); however, this has been prevented through the veto of certain members of the UKWAS steering group (Yull, 2000, 2001).

4.3.2. Germany

Although the roots of forest certification were already present in Germany in the late 1980s, when boycotts of tropical wood were promoted by German environmental groups, certification would not directly impact domestic forest landowners and managers in Germany until nearly a decade later, in 1996. At that time, pressure for certification had spread from tropical to temperate ecosystems, and forest certification programs such as *Naturland* and eco-timber had been created by German environmental groups, including Greenpeace and NABU (Klins, 2000). The German Forestry Association (*Deutscher Forstwirtschaftsrat*), (made up of state forestry agencies, forest products associations, private landowner associations and academics), and private landowners rejected those labels, saying that they were inappropriate for German land ownership patterns, which saw 45% of forest land owned by private individuals or enterprises (more than 75% of which own less that 5 ha), and 55% of forest lands managed directly by state and municipal agencies (Bundesministerium für Ernährung Landwirtschaft und Forsten, 2000). Instead, the forest sector chose to create its own labeling program, the *Herkunftszeichen*, or 'label of origin', in 1996, which was available to all German forest owners (Klins, 2000).

When the German FSC working group was created in 1997, the forest sector reacted by working with other landowners and managers throughout Europe to create its own label, the PEFC, which was created in June 1998 and based on the Helsinki Criteria of the 2nd Forest Ministers Conference in 1993 (Hofmann et al., 2000). While the FSC had early buy-in from all major ENGOs, forestry labor unions, and a handful of forest landowners, state agencies and municipalities, the PEFC, with its systems-based standards and strong voice for private landowners, had broad support from nearly all private and state landowners. Although this support has a moral element, given the PEFC's recognition of the long tradition of sustained yield forestry in Germany, the PEFC is mainly given pragmatic legitimacy, because forest land managers want to maintain their market access, and came to believe that in the absence of a competitor program, the market might force them to adopt FSC. Thus, the PEFC is almost seen as the lesser of two evils: one PEFC supporter even called the PEFC 'the certification program that nobody wants' (personal interview). This case reveals the importance of understanding how legitimacy granting for one type of NSMD certification program may affect the legitimacy of a competitor program.

Since the PEFC's creation, the forest sector's support for the PEFC has not changed significantly; if anything, early patterns of support have become more ingrained and two distinct and increasingly polarized factions—the FSC with its ENGO supporters and the PEFC with its landowner and state supporters—have emerged. Recent attempts at mutual recognition were unsuccessful and formal communication between the two programs was recently ended by the FSC (FSC Germany, 2001).

Table 1 summarizes the developments described in the previous seven case studies by showing the initial and current type of legitimacy given by forest companies and landowners to the FSC and competitor programs. Again, significant areas of divergence become visible, which are shown in bold type: the level of forest company support for the FSC changed from 'none' to 'pragmatic' in both the BC and UK cases. In all other regions, legitimacy has remained stable. Why do we observe change in BC and the UK, and stability elsewhere? Which independent or intersecting factors propelled these supply side audiences to embrace the prescriptive, broad ranging rules of the FSC? Our attention now turns to offering an explanation for the divergence observed in Table 1.

5. Explanatory variables

This section seeks to account for the differences in supply side audience decisions across the regions—namely why some forest companies and landowners would maintain resolute opposition to FSC certification programs (usually giving FSC competitor programs strong pragmatic and in some cases moral legitimacy), while in other regions companies and landowners have begun to accept FSC authority and started the process of seeking certification with this program.

We draw on existing globalization, public policy literature, as well as our own observations of the cases to develop seven working hypotheses that fall under four interrelated factors: (1) the degree of a region's dependence on foreign markets; (2) the extent to which forest management is seen as a 'problem' on the political agenda; (3) the structure of the supply side; and (4) the character of the FSC's competitor programs. These factors, through both their independent and intersecting effects, help us to understand why the FSC has gained differing levels of supply side legitimacy in each region.

6. Degree of dependence on foreign markets

Much of the research on globalization and policy convergence has pointed to the susceptibility of a region to global and transnational pressures when it relies on foreign markets for the source of its products (Berger and Dore, 1996). One of the central ideas behind this argument is that it is easier for transnational actors to raise concern about a firm's behavior when that firm is outside of the political system in which it operates. There can be no charges of breaching territorial 'sovereignty' (Bernstein and Cashore, 2000), nor do those exerting pressure have to be wary of domestic consequences of their actions. These ideas lead to our first preliminary hypothesis to explain divergence:

Hypothesis 1. Supply side companies in a region that sells a high proportion of its forest products to foreign markets will be more susceptible to demand side manipulations by the FSC and its supporters.

This hypothesis helps account for divergence among our cases because it addresses those heavily export oriented regions in which ENGO efforts to force change have been directed toward non-domestic purchases of the products. If this hypothesis has merit, it should most certainly help explain legitimation dynamics in the province of BC, since it fits the description of an export oriented region in which transnational groups have gone to non-domestic purchasers of their products. We see both indirect and direct support for this hypothesis.

Indirect support comes from dynamics that occurred before certification but followed the same logic. From the beginning of the 1990s supply side companies in BC, particularly coastal producers, were under market pressure, first from boycott campaigns (Greenpeace International, 1993) and then later from demands for FSC certified wood (Greenpeace UK, 1998). In both instances, ENGOs pressured demand side companies in Europe and the US to terminate their contracts with companies operating in BC (Vertinsky et al., 1997; Stanbury, 2000). It was evident that BC coastal producers felt the heaviest impacts of these pressures, which were made worse by the limited US market access, owing to the 1996 Canada–US Softwood Lumber Agreement, and Asian economic woes peaking in 1998 (Taylor and van Leeuwen, 2000; Stanbury, 2000). This made Europe—a relatively small market for BC products—appear greatly important, as other alternative markets were inaccessible. As Bernstein and Cashore (2000) noted, going international allowed demand side members that had been operating in Canada, or more specifically, BC, to avoid national political backlash.

The BC forest industry attempted to counteract the demands of their international buyers, but with little success. BC's main forest labor union, the IWA, even mounted counter boycotts against companies that proposed terminating contracts with BC firms (Stanbury, 2000). However, with no stake in the domestic political ramifications of a buying policy specifying FSC wood, these international demand side companies appear to have been quite malleable to the demands of European ENGOs.

Direct support for this hypothesis is found in ENGO efforts to have retailers such as Home Depot in the US announce policies for FSC wood. This announcement had a discernible impact on BC, which imports most of its forest products to the US: many companies in BC announced their intention to become FSC certified following the Home Depot announcement. In line with this hypothesis, the announcement had much more of a direct effect on forest companies in Canada than in the US. The Home Depot announcement in the US simply appeared to galvanize and reinvigorate support for the SFI, whose supporters believed that they could alter Home Depot's approach through a highly strategy domestic campaign.

While more research needs to be done into other similar cases, preliminary evidence indicates that companies operating in export-oriented markets and are the target of transnational campaigns to have purchasers adopt FSC wood are more likely to grant FSC pragmatic legitimacy.

Hypothesis 2. Supply side members in a region that imports a large proportion of its forest products will be more susceptible to demand side FSC manipulation strategies within its own borders.

When the supply side in a region is small and cannot produce the volume of forest products required to meet local demand, it becomes susceptible to competition from FSC imports. This was observed in the UK where the supply side responded to the combination of domestic demand for FSC, and increasing international competition. In 1999, the UK was the least self-sufficient EU country in terms of wood production capacity and, internationally, placed second behind Japan in terms of net value of wood products imported (Forestry Industry Council of Great Britain, 2000). When B&Q issued an ultimatum to it suppliers that by the end of 1999 it intended to purchase only FSC certified wood, local processors were not exempted (Stanbury, 2000; National Home Center News, 1998; DIY, 1998). And because competition from FSC-certified in suppliers Sweden was increasing and the potential that countries in the Baltic States, such as Latvia and Estonia, would follow suit (Tickel and World Wildlife Fund for Nature, 2000; Hansen and Juslin, 1999), local producers recognized their market share was in jeopardy. This solidified the pragmatic appeal of the FSC, as it was a logical way to meet customer demands and maintain market share. The fear of increased competition from international suppliers of FSC wood also exists among the supply side members in Germany, who sell most of their products domestically; however, these suppliers have not yet felt the combined effects demands for FSC wood and increasing international competition (personal interviews).

In sum, the use of the international market appears to have been salient for the FSC's success in gaining BC and UK supply side pragmatic legitimacy, which highlights the importance *dependence* plays in determining the success or failure of market manipulation strategies. Overall, the lack of market diversity—either as an exporter or domestic supplier—and openness to the international markets appeared to be key intervening factors that motivated the shift in supply side legitimacy in both these regions.

7. The extent to which forest management issues are seen as a 'problem' on the policy agenda

In the traditional public 'policy cycle', a problem is brought to the government's attention, policy options are formulated, and a particular course of action is put into effect and monitored (Howlett and Ramesh, 1995). When controversial forest management practices or issues have reached a region's policy agenda, supply side members may choose to rely exclusively on the government's solutions, or may choose to take independent action to avoid potential controversy and address the issue. We would expect to observe this type of action especially in an internationalized sector such as forestry, where trans-national environmental groups have the resources and expertise to publicize controversial issues quickly. We propose the following hypothesis:

Hypothesis 3. Supply side companies are more likely to give the FSC pragmatic legitimacy when forestry practices are perceived to be a major problem on the governmental policy agenda. The strength of this relationship is heightened when governmental initiatives have failed to remove the issue as a perceived problem.

In each of our study regions, the appropriateness of certain forest management practices, to some extent, were an issue on the political agenda. Yet in particular instances these issues reached a heightened prominence that forced supply side members to find solutions to controversies themselves, even after governments had sought to develop public policy solutions. To the extent that the perceived problem is a threat to profits or a loss of 'social license', we expect that supply side audiences will give pragmatic legitimacy to the FSC.

The BC case provides support for this hypothesis. There, both domestic and international civil societies were well aware of the global significance of BC's few remaining coastal temperate rainforests. Documented evidence pointed to a history of mismanagement (Tripp et al., 1992; Tripp, 1994), which was closely linked to criticism of existing public policy (Wilson, 1998; Marchak et al., 1999). Initially the provincial government acted independently to develop an internationally defendable Forest Practices Code (British Columbia. Ministry of Forests, 1996) and associated land use planning process (British Columbia. Integrated Resources Planning Committee and Ministry of Forests, 1993; Wilson, 1998) and the federal government acted, in concert with industry, to back the CSA (Elliott, 1999). However, the salience of the ecological crisis increased in fervor as ENGOs in North America and Europe continued to press for changes (Greenpeace UK, 1998; Stanbury, 2000). Finally, when MacMillan Bloedel, one of the hardest hit forest companies, released its Forest Plan, certification was one component of a series of reforms tailored to address deep-seated concerns within civil society (MacMillian Bloedel Ltd., 1998a,b); as a solution, this move went beyond what was required by public policies. The persistence of the perceived 'ecological crisis' can be viewed as a key element of this shift, which has been absent from the other cases we have studied.

In the UK and Germany, supply side buy-in to the FSC was not associated with a specific domestic ecological issue but rather concerns about tropical and temperate rainforests. Both countries import large amounts of wood from these types of ecosystems (Forestry Industry Council of Great

Britain, 2000; European Forest Institute, 2001).[10] Hence, the perceived ecological crisis was the deforestation and the associated loss of biodiversity occurring in these forests. Morally implicated by their dependence on imports, ENGOs in both regions responded with campaigns for the adoption of a certification standard domestically, to ensure international parity in their demands: in the UK, the standard became the FSC for reasons discussed in the previous section, while in Germany, the PEFC was seen as adequate. In the US south-east, the proliferation of pine plantations and the resulting loss in biodiversity is considered a problem by many regional ENGOs: the number of acres in southern pine plantations increased from less than two million acres in 1952 to over 24 million acres in 1992, and is expected to continue increasing (Hyde and Stuart, 1998). However, this problem has not reached a crisis level among the general public, and has not, therefore, had market or social license implications for supply side members there.

It appears that the persistence of an ecological problem and demands that it be addressed heighten a company's search for solutions, which in the BC case, involved FSC certification. By conforming to the moral ideals of domestic and international ENGOs the FSC has been able to offer a 'way out' for forest companies that were under campaign scrutiny. Unlike the traditional state-centered policy responses to 'ecological crises', the FSC offered a set of standards endorsed by both domestic and international ENGOs that indicated to civil society that a solution was being pursued.

8. Structure of the supply side in a region

By understanding the way that forest lands are distributed, and the avenues through which the managers of those lands relate to each other, we are able to build an understanding of the motivations companies might have for participating in a certain forest certification program. The following two hypotheses are an exploration of this relationship.

Hypothesis 4. Supply side members in a region where land ownership is fragmented will be less likely to grant the FSC pragmatic legitimacy, owing to high transaction and implementation costs.

The roots of this hypothesis derive more from traditional economic theory than public policy literature. Nevertheless, it appears to be an important explanation in supply side legitimation dynamics. If this hypothesis holds, regions with a large number of small land holdings will show a low acceptance of FSC certification owing to the diseconomies of scale associated with certifying small forest tracts. We found both direct and indirect evidence in our cases that support this hypothesis.

The most illustrative support exists in the US south-east. Most forestland is owned by NIPFs in small tracts, which together supply the majority of the region's domestic wood fiber (United States Forest Service, 2000; DeCoster, 1998). Wood processors in the region require a continuous fiber supply in order to feed their highly specialized capital-intensive mills (Hyde and Stuart, 1998: 24); consequently industrial forest companies consider the logistical problems associated with the FSC's chain of custody requirements overwhelming given their large number of small fiber suppliers (personal interviews). The FSC's attempts to address this practical barrier have met with little success. The regional working group has promoted group certification and resource manager certification as options for the many small landowners in the Southeast, but neither industrial managers nor NIPFs themselves have taken much interest in them.

Conversely, we observed cases in which there were a small number of large landowners who enjoyed economies of scale in the costs of FSC certification. In BC, land is concentrated in the hands of a few large companies (Marchak et al., 1999) (and ultimately in the hands of one landowner—the provincial government), and in the UK, land is held by the government (Forestry Industry Council of Great Britain, 2000). Although

[10] Six percent of UK softwood sawnwood and 15% of its pulp imports in 1997 were of Canadian origin (Forestry Industry Council of Great Britain, 2000), and of its tropical imports 90% were reported to come from Brazil, Indonesia, and Malaysia (Matthew, 2001).

economies of scale and cost issues were not mentioned explicitly during interviews in BC, they were cited during UK interviews as an important reason for supply side support for FSC there.

While chain of custody requirements were a hurdle for UK processors, two factors made a resolution possible there. First, the FSC took a conforming strategy to meet the substantive needs of these companies through a change to the threshold percent-requirements for an FSC product to carry a label (Forest Stewardship Council, 2000).[11] Second, the successful FSC certification of large areas of government lands meant that processors had sufficient FSC wood supply even without the certification of any large portion of the private growers (personal interview). Consequently, where certification has been stifled by a complex wood basket in the US south-east, the presence of one key landowner open to certification has made possible broader acceptance of the FSC in the UK.

The different structural and economic limitations on owners of small and large land holdings appear to have explanatory power when exploring the divergent levels of supply side support for FSC in our cases.

Hypothesis 5. Members of the supply side in a region where companies and landowners are well represented and unified are less likely, everything else being equal, to grant the FSC pragmatic legitimacy.

Inhibiting the production of FSC certified products in a region, and therefore blocking FSC supporters' market strategies by keeping supply of FSC certified wood low, would seem to be an effective way to prevent the FSC from gaining headway. However, any such unified action runs the risk of being sabotaged by defectors—in the case of certification, the defectors would be companies or landowners who take advantage of a relatively high demand for FSC certified products that is not being met. The ability of a region's forest sector to keep supply side members from defecting should be a measure of its success in denying the FSC legitimacy.

This hypothesis appears to be an important explanation for FSC's lack of success in gaining legitimacy from any large portion of the US supply side. Up until the early 1990s every element of the US forest sector was noticeably fragmented (Lertzman et al., 1996). However, with the threat of declining access to timber from National Forests in the Pacific Northwest the industry as a whole moved to increase their national presence.[12] The result was the AF&PA that amalgamated a number of national associations (Cashore, 1997).

When certification emerged, the AF&PA was well resourced and situated to develop an industry wide response to the demand side and environmental group pressures. Throughout the US the SFI has rapidly developed tougher standards and third-party procedures (American Forest and Paper Association, 2000a), in an attempt to win ENGO support or at least to neutralize general societal recognition for ENGO policy (personal communications). The strength of the industry's cohesion is central to its resistance of the FSC in the US. Not only has the surprising consistency in the industry's 'boycotting' of the FSC regional processes been important, but also the AF&PA has agreed to recognize the existing American Forest Foundation's Tree Farm Program as the most appropriate approach to certifying NIPF landowners (American Forest and Paper Association, 2000c), demonstrating that the industrial portion of the supply side has united with one of the only nationally, broadly supported landowner associations. Their joint stance on certification represents unprecedented unity in a country that has traditionally been characterized as extremely politically fragmented (Grant, 1989).

By contrast, the Canadian forest sector was represented by a fragmented and regionally focused associational system (Coleman, 1987, 1988), which inhibited the adoption of a unified

[11] In fact, when the changes occurred, many of the UK mills were immediately able to offer FSC certified panel products to the market.

[12] The achievements of the environmental movement in the north-west had resulted from court action seeking proper implementation of federal statues by the certain agencies. The industry felt it lacked the ability to lobby the federal government to ensure future legislation did not give the environmental movement this kind of power.

stance on certification issues, and helps explain the pragmatic legitimacy the supply side in BC has given to the FSC. Lacking a logical association in which to house an industry response, the Canadian forest sector drew on the expertise and resources of the CSA to create its FSC response (Abusow and Rotherham, 1998; Lapointe, 1998). Within BC, unified supply side support for the CSA was further restrained given dynamics created with the emergence of the Forest Alliance of BC (FABC). Its mandate was to bolster the public image of the BC forest industry. However, when the FSC became a serious issue, neither the already established Council of Forest Industries (COFI) nor the FABC had a clear mandate to develop a response. It seems that the FABC's diverse membership limited its ability to act purely on the industries behalf. Once a few companies showed an interest in pursuing FSC, 'the dam broke' and other companies followed suit (personal communication).

9. The nature of the FSC's competitor programs

Hypothesis 6. Supply side members are more likely to give the FSC pragmatic legitimacy when the alternative programs have rigorous standards, even when alternative programs might provide market access.

Hypothesis 7. Supply side members are more likely to give the FSC pragmatic legitimacy when the alternative programs have lax standards accompanied with poor market access.

While seemingly contradictory, our inductive research seems to indicate that in order for FSC competitor programs to maintain support from companies and landowners (or at least, not have to share legitimacy with the FSC), they need to develop a program that walks a fine line between being too prescriptive (and thus the costs of FSC do not seem more difficult) and too weak (thus the market will not accept the program).

A central element of pragmatic legitimacy is an evaluation on the part of the supply side member of the 'costs' and 'benefits' of certification under different programs. In these hypotheses we consider how the costs of different programs, in terms of compliance with the program's standards, and benefits, in terms of market acceptance and expected economic gains, have affected supply side support for the FSC. Empirical evidence seems to indicate that supply side members' decisions to support the FSC (or not) are influenced strongly by how well the competitor program balances costs and benefits.

In some cases, the standards of an FSC competitor program were perceived as being too rigorous to be acceptable by the supply side. Canadian supply side companies initially endorsed the CSA; however, the costs, level of public scrutiny, and time commitments of many aspects of this program eventually met or exceeded those of the FSC (personal interview), yet could not offer the potential market benefits of the FSC. Consequently what had been a program that was meant to limit FSC success, became one among a number of certification alternatives (Forest Alliance of British Columbia, 2000). In its attempts to gain support from the ENGO community by conforming to existing Canadian public policy norms about public consultation, it appears that the program lost some of its pragmatic appeal with the industry.

Likewise, a competing certification program may have less onerous standards than the FSC. This may be pragmatically appealing to the supply side, but may be a disincentive for participation if the system has no market acceptance. However, we did not observe this situation in any of our cases: rather, it appeared that when the alternative program had weaker standards and low market acceptance, it provided other, less tangible benefits that mitigated against the lack of market support. For example, in the case of Germany, no retailers or wood processors are demanding PEFC certified wood; however, the PEFC is associated with the avoidance of *Fremdbestimmung*, or the 'ruling from the outside' of domestic forest practices, which is seen by supply side members as the major threat of the three-chamber FSC system. Similar feelings persist even in the UK, where the private growing community is seeking recognition of the UKWAS, a standard recognized as equivalent to the nationally endorsed UK FSC standard, to the

PEFC for mutual recognition under its framework (Yull, 2000, 2001).

In the US, SFI certified wood has not been accepted by members of FSC buyers groups in the US or elsewhere, but the SFI offers supply side members an opportunity to maintain a cohesive front and 'starve' the US demand for FSC certified products. Furthermore, the SFI, being initially created as a public relations program to regain public trust, has continued to offer its members other benefits. For instance, the SFI continually uses informing strategies directed at general society that point to the many beneficial things the program is achieving (American Forest and Paper Association, 2000b,c,d). While linked to its competition with the FSC, these give tangible benefits to participant companies that are, in a sense, independent of market acceptance.

Overall, the cases highlight the struggle that the FSC's competitor programs face to keep the level of certification standards acceptable to supply side members while simultaneously getting acceptance in the marketplace. So far, no competitor program has met this challenge; however, some programs have managed to offer the supply side other important incentives. It appears that, in many of our cases, these non-market benefits tend to be reason enough for supply side support.

10. Conclusion

Supply side response to the creation of the FSC certification program and its competitors began with remarkable homogeneity in all of our study regions: with few exceptions, supply-side forest companies and landowners gave pragmatic and sometimes moral legitimacy to the competitor programs. However, since 1998 a dramatic shift has occurred, which saw the supply side audience in BC and the UK moved to give the FSC pragmatic legitimacy. In this paper we explored this divergence and offered a first attempt at applying a heuristic developed by Cashore (2002) that builds on the seminal work of Suchman (1995) on organizational legitimacy. Its importance hinges on the way in which it allows us to investigate firm response to rules generated outside the purview of the state, and governed by interactions in the market. Our research found that emerging NSMD governance programs, while theoretically able to garner pragmatic, moral and cognitive support from supply-side members, are presently focused on gaining pragmatic and moral legitimacy, the least durable categories.

Empirical evidence showed that forest companies' and landowners' decisions to support a certain certification program are constrained and influenced by a complex group of factors. Supply-side members in regions with a high reliance on foreign markets appear to be more likely to support the FSC program, since these members' international buyers can make (sometimes controversial) demands for FSC certified wood without risking the political backlash that domestic companies might experience. Supply side members in regions where forest management practices have reached the status of 'problem' on the policy agenda are more likely to support the FSC as a way to expedite problem resolution and avoid controversy, or as a way to gain 'social license'. The structure of the forest sector was also shown to be closely related to diverging levels of FSC acceptance: regions with fragmented land ownership faced more resistance to the FSC, due to hurdles such as increased transaction costs and decreased economies of scale. A high level of forest sector cohesion, on the other hand, was closely related to the ability of the supply side to resist FSC certification. The inability of competitor programs to simultaneously minimize the substantive costs of certification yet maximize market acceptance appears to be related to a higher level of supply side support for FSC in some cases; however, in other cases, competitor programs have secured supply side support through the provision of other non-market incentives, despite being unable to promise market acceptance.

While the methodological hurdles of such a study are significant, the importance of a new governance system like forest certification means that more research needs to be done to understand just what is emerging, who is giving it support, and how durable this support might be. This paper is an initial effort to do this by focusing on issues of divergence, which may shed light on where certification is headed in other regions and other sectors. It also serves as an example of another policy realm that has come under intense pressure for convergence, but has nonetheless experienced important divergence.

References

Abusow, K., Rotherham, T., 1998. Canadian progress toward SFM system certification. Forestry Chronicle 74 (3), 405–408.

American Forest and Paper Association, 1999. SFI: 4th Annual Progress Report. Washington DC: American Forest and Paper Association's SFI program.

American Forest and Paper Association, 2000a. Fifth Annual Progress Report. Washington, DC: SFI Program.

American Forest and Paper Association, 2000b. Fraser 3rd Party Certifies all its Forestlands: Lands in Canada First Outside U.S. to be Independently Audited under SFIsm. Washington DC: American Forest and Paper Association's SFI Program.

American Forest and Paper Association, 2000c. SFISM AFF's American Tree Farm System® and AF&PA's Sustainable Forestry Initiative (SFI)SM Program Collaborate to Expand the Practice of Sustainable Forestry. Washington DC: American Forest and Paper Association's SFI Program.

American Forest and Paper Association, 2000d. Sustainable Forestry Initiative Standard: Principles and Objectives. Washington DC: American Forest and Paper Association's SFI Program.

Berger, S., 1996. Introduction. In: Berger, S., Dore, R. (Eds.), National Diversity and Global Capitalism. Cornell University Press, Ithaca and London.

Berger, S., Dore, R., 1996. National Diversity and Global Capitalism. In: Katzenstein, P.J. (Ed.), Cornell Studies in Political Economy. Cornell University Press, Ithaca and London.

Bernstein, S., 2001. Legitimacy in Global Governance: Three Conceptions. Paper prepared for delivery at the annual meeting of the American Political Science Association, August 29–September 1, San Francisco.

Bernstein, S., Cashore, B., 1999. World trends and Canadian forest policy: trade, international institutions, consumers and transnational environmentalism. Forestry Chronicle 75 (1), 34–38.

Bernstein, S., Cashore, B., 2000. Globalization, four paths of internationalization and domestic policy change: the case of eco-forestry policy change in British Columbia, Canada. Canadian Journal of Political Science 33 (1), 67–99.

British Columbia. Integrated Resources Planning Committee, and Ministry of Forests, 1993. Land and Resource Management Planning: A Statement of Principles and Process. British Columbia.

British Columbia. Ministry of Forests, 1996. B.C.'s Forest Practices Code: An International Comparison. Victoria: Province of British Columbia. Ministry of Forests.

Bundesministerium für Ernährung Landwirtschaft und Forsten, 2000. Land- und Forstwirtschaft in Deutschland: Daten und Fakten.

Cashore, B., 2002. Legitimacy and the privitization of environmental governance: how Non-State Market-Driver (NSMD) Governance systems gain rule-making authority. Governance 15 (4), 503–529.

Canadian Standards Association, Sustainable Forest Management System, 2000. Certification Status in Canada: Shaping the Future of Canadian Forests [Internet]. CSA/SFMS, December 10, 2000 [cited December 19 2000].

Canfor Corporation, 1999. Annual Information Form. Vancouver.

Carlton, J., 2000. Home Builders Centex and Kaufman Agree Not to Buy Endangered Wood. Wall Street Journal, March 31.

Cashore, B., 1997. Governing Forestry: Environmental Group Influence in British Columbia and the US Pacific Northwest. PhD, Political Science, University of Toronto, Toronto.

Cashore, B., 1999. Chapter Three: US Pacific Northwest. In: Wilson, B., Kooten, K.V., Vertinsky, I., Arthur, L. (Eds.), Forest Policy: International Case Studies. CABI Publications, Oxon, UK.

Caulfield, John. 1999. Selling 'safe' wood gets a major advocate. National Home Center News, October 11.

Coleman, W.D., 1987. Federalism and Interest Group Organization. In: Bakvis, H., Chandler, W.M. (Eds.), Federalism and the Role of the State. University of Toronto Press, Toronto.

Coleman, W.D., 1988. Business and Politics: A Study of Collective Action. McGill-Queen's University Press, Kingston.

Coulombe, M., Brown, M., 1999. The Society of American Foresters Task Force on Forest Management Certification Programs: The Society of American Foresters.

DeCoster, L., 1998. The Boom in Forest Owners—A Bust for Forestry?.

DIY, 1998. Hodkinson tells World Bank of B&Q's FSC plans. DIY, January 23, 1998.

Doern, G.B., Pal, L.A., Tomlin, B.W., 1996. The Internationalization of Canadian Public Policy. Border Crossings: The Internationalization of Canadian Public Policy. Oxford University Press, Toronto.

Elliott, C., 1999. Forest Certification: Analysis from a Policy Network Perspective. PhD, Departement de Genie Rural, Ecole Polytechnique Federale de Lausanne, Lausanne, Switzerland.

European Forest Institute, 2001. Country Reports—Germany. 2001 [cited 2001]. Available from http://www.efi.fi/cis/english/creports/germany.html.

Fletcher, R., Hansen, E., 1999. Forest Certification Trends in North America and Europe, New Zealand.

Forest Alliance of British Columbia, 1996. Sustainability—defining the goal. In: Choices: Issues and Options for B.C. Forests. Vancouver: Forest Alliance of British Columbia.

Forest Alliance of British Columbia, 2001. Greenpeace boycott campaigns hits BC [News release]. Forest Alliance of British

Columbia 1997 [cited February 14 2001]. Available from http://www.forest.org/frameset.cfm?pagerequest=mediacentre.

Forest Alliance of British Columbia, 1998. FSC should get the axe. Vancouver: Forest Alliance of British Columbia.

Forest Alliance of British Columbia, 2000. Forest Alliance pleased with provincial government commitment to certification. Vancouver, BC: Forest Alliance of British Columbia.

Forestry Commission, 1998. Forestry Briefing, Forestry Commission of Great Britain, Secretariat Divison, Edinburgh.

Forestry Industry Council of Great Britain, 2000. A reference for the forestry industry. London: Forest Industry Council of Great Britain.

Forest Stewardship Council, 1999. FSC Principles and Criteria: Forest Stewardship Council.

Forest Stewardship Council 2000. FSC Policy of Percentage Based Claims. Oaxaca, Mexico: FSC International.

Forest Stewardship Council United States, 2001. Pacific Coast Working Group, Draft Standards [Archived item]. Forest Stewardship Council United States 1998 [cited April 8 2001]. Available from http://fscus.org/html/standards_policies/us_regional_standards/archives/pacificnorthwest.html#p1.

Forest Ethics, 2001. Employment Opportunities: Job Posting: Corporate Advisor—Traction Consultant [online job information]. Forest Ethics 2001 [cited March 20 2001]. Available from http://www.forestethics.org/html/eng/939-AA.shtml.

FSC Germany, 2001Annäherung der Zertifizierungssysteme nicht möglich: Keine weiteren Gespräche auf Vorstandsebene zwischen FSC and PEFC. Forest Stewardship Council Arbeitsgruppe Deutschland e.V.

FSC US, 2001. Anderson-Tully Company First Southeastern U.S. Timberland Company To Earn Forest Stewardship Council Certification For All Its Lands. Washington DC: Forest Stewardship Council United States.

Global Witness, 1999. Made in Vietnam—Cut in Cambodia: How the Garden Furniture Trade is Destroying Rainforests. London: Global Witness Ltd., in association with Friends of the Earth.

Goodall, S., 2000. Forest Certification and the UK Woodland Assurance Scheme. Quarterly Journal of Forestry 94 (3), 239–244.

Grant, W., 1989. Government and Industry: A Comparative Analysis of the U.S., Canada and the U.K. Aldershot Hants, England: Edward Elgar Publishing Limited.

Greenpeace International, 1993. Greenpeace Annual Report 1992–1993: Greenpeace International.

Greenpeace UK, 1998. Spring Campaign Report. London, UK: Greenpeace, UK.

Hamilton, G., 1998. MacBlo Decides to Abandon Pro-Logging Forest Alliance. The Vancouver Sun, August 18, Internet posting.

Hansen, E., Forsyth, K., Juslin, H., 1999. A Forest Certification Update for the ECE Region, Summer 1999. Geneva, Switzerland: United Nations Economic Commission for Europe and Food and Agriculture Organization of the United Nations, Timber Section.

Hansen, E., Juslin, H., 1999. The Status of Forest Certification in the ECE Region. New York and Geneva: United Nations, Timber Section, Trade Division, UN-Economic Commission for Europe.

Harrison, K., 1999. Talking with the donkey: cooperative approaches to environmental protection. Journal of Industrial Ecology 2 (3), 51–72.

Hayward, J., 1998. Certifying industrial forestry in B.C.: European market drives big timber to the FSC. Understory 8.

Hoberg, G., 2000. How the way we make policy governs the policy we choose. In: Salazar, D., Alper, D. (Eds.), Forging Truces in the War in the Woods: Sustaining the Forests of the Pacific Coast. UBC Press, Vancouver.

Hofmann, F., Kill, J., Meder, R., Plachter, H., Volz, K.-R., 2000. Waldnutzung in Deutschland: Bestandsaufnahme, Handlungsbedarf und Maßnahmen zur Umsetzung des Leitbildes einer nachhaltigen Entwicklung. D.R. v. S.f. Umweltfragen. vol. 35: Metzler-Poeschel Stuttgart.

Hogben, D., 1998. Western seeks certification to satisfy European buyers. The Vancouver Sun, June 5,1998, F2.

Howlett, M., Ramesh, M., 1995. Studying Public Policy Policy Cycles and Policy Subsystems. Oxford University Press, Toronto.

Humphries, S., 1999. Forest Certification Handbook: For the Southeastern United States. The Forest Management Trust, Gainesville, FL.

Hyde, B., Stuart, W.B., 1998. Chapter Two: US South. In: Wilson, B., Kooten, G.C.v., Vertinsky, I., Arthur, L. (Eds.), Forest Policy: International Comparisons. CAB International, Oxon, UK.

Jordan, D., 1999. Forest Alliance to Join the Forest Stewardship Council. Business in Vancouver, June 15, 1999, 3.

Keck, M.E., Sikkink, K., 1998. Activists Beyond Borders: Advocacy Networks in International Politics. Cornell University Press, Ithaca and London.

Kiekens, J.-P., 1997. Certification: International Trends and Forestry and Trade Implications. Environmental Strategies Europe, Bruxelles.

Klins, U., 2000. Die Zertifizierung von Wald und Hoizprodukten in Deutschland: Eine Forstpolitische Analyse. Ph.D, Forstwissenschaftliche Fakultaet, Technische Universitaet Muenchen, Muenchen.

Knight, A., 2000. Seeing the Customer and the Trees: A Proposed Revised Buying Policy for B&Q for Beyond 2000 for Consultation. B&Q, Chandlers Ford, Eastleigh, Hampshire.

Lapointe, G., 1998. Sustainable forest management certification: The Canadian Programme. Forestry Chronicle 74 (2), 227–230.

Lawson, J., Cashore, B., 2001. Forest Certification Experiences in Maine and the Canadian Maritimes: A Comparative Institutional Study. Paper presented to the Auburn Forest Policy Center's Conference on Globalization and Private Forestry, March 25–27, Atlanta, Georgia.

Lertzman, K., Wilson, J., Rayner, J., 1996. Learning and change in the B.C. forest policy sector: a consideration of Sabatier's advocacy coalition framework. Canadian Journal of Political Science 29 (1), 112–133.

MacCleery, D., 1999. NFS Certification Issues. USDA Forest Service, Washington, DC.

MacMillian Bloedel Ltd., 1998a. MacMillan Bloedel to Phase Out Clearcutting; Old-Growth Conservation is Key Goal; Customers to Be Offered Certified Products. Vancouver, BC: MacMillan Bloedel.

MacMillian Bloedel Ltd. 1998b. MB Plans to Stop Clearcutting. MB Journal 18 (7).

March, J.G., Olsen, J.P., 1998. Institutional dynamics of international political orders. International Organization 52 (4), 943–969.

Marchak, M.P., Scott, L.A., Herbert, D.M., 1999. Falldown: Forest Policy in British Columbia. David Suzuki Foundation and Ecotrust Canada, Vancouver, BC.

Matthew, E., 2001. Import of Illegal Tropical Timber into UK. Friends of the Earth, Living World Campaign, London.

Meidinger, E., 1997. Look who's making the rules: international environmental standard setting by non-governmental organizations. Human Ecology Review 4 (1), 52–54.

Meidinger, E., 2000. 'Private' Environmental Regulation, Human Rights, and Community. [Online Paper], [cited March 2001] Available: http://www.law.buffalo.edu/homepage/eemeid/scholarship/hrec.pdf.

Moffat, A.C., 1998. Forest Certification: An Examination of the Compatibility of the Canadian Standards Association and Forest Stewardship Council Systems in the Maritime Region. MES, Environmental Studies, Dalhousie University, Halifax, Nova Scotia.

National Forest Strategy Coalition. 1997. National Forest Strategy 'Sustainable Forests: A Canadian Commitment' Final Evaluation Report. Hull, Quebec: National Forest Strategy Coalition.

National Home Center News. 1998. B&Q pledges to work towards forest certification. National Home Center News 24, 9.

Newsom, D., Cashore, B., Auld, G., Granskog, J., 2002. In Press. Forest Certification in the Heart of Dixie: A Survey of Alabama Landowners. In: Teeter, L., Cashore, B., Zhang, D. (Eds.), Forest Policy for Private Forestry: Global and Regional Challenges. Wallingford, UK: CABI Publishing.

Paget, G., Morton, B., 1999. Forest Certification and 'Radical', Forest Stewardship in British Columbia, Canada: the Influence of Corporate Environmental Procurement. Paper presented at Greening of Industry Network Conference—1999, November, Kenan-Flagler Business School, University of North Carolina, Chapel Hill.

Pan European Forest Certification United Kingdom, 2001. New company will promote certified timber production in Britain [Press Release]. Pan European Forest Certification UK Ltd., September 15, 2000 [cited February 17 2001]. Available from www.pefe.co.uk/page4.html.

Price, W., 2000. Forest Certification Pilot Projects Expanded on State and Tribal Lands. Pinchot Institute for Conservation, Washington DC.

PricewaterhouseCoopers, 1999. B.C. Ministry of Forests Small Business Forest Enterprise Program (SBEEP) Certification Pre-Assessment. Vancouver: PricewaterhouseCoopers.

Risse-Kappen, T., (Ed.) 1995. Bringing Transnational Relations Back In: Non-state Actors, Domestic Structures and International Institutions. Cambridge: Cambridge University Press.

Rosenbaum, W.A., 1995. Environmental Politics and Policy. 3rd ed. CQ Press, Washington.

Stanbury, W.T. 2000. Environmental Groups and the International Conflict Over the Forest of British Columbia 1990 to 2000. Vancouver: SFU-UBC Centre for the Study of Government and Business.

Stanbury, W.T., Vertinsky, I.B., Wilson, B., 1995. The Challenge to Canadian Forest Products in Europe: Managing a Complex Environmental Issue. Forest Economics and Policy Analysis Research Unit, University of British Columbia, Vancouver.

Suchman, M.C., 1995. Managing legitimacy: strategic and institutional approaches. Academic Management Review 20 (3), 571–610.

Taylor, R.E., van Leeuwen, G., 2000. Outlook for Hemlock Prices & Volumes in Japan and Western Red Cedar Prices & Volumes in the USA. R.E. Taylor & Associates Ltd, Vancouver, BC.
Produced for Coastal Forest & Lumber Association.

The Forest Management Trust, 1999. Forest Certification Handbook: for the Southeastern United States: The Forest Management Trust.

Tickel, O., World Wildlife Fund for Nature, 2000. Certification: A Future for the World's Forests. Godalming, Surrey, UK: WWF Forests for Life Campaign.

Tollefson, C., (Ed.) 1998. The Wealth of Forests: Markets, Regulation, and Sustainable Forestry. Vancouver: UBC Press.

Tripp, D., Nixon, A., Dunlop, R., 1992. The Application and Effectiveness of the Coastal Fisheries Forestry Guidelines in Selected Cut Blocks on Vancouver Island. D Tripp Biological Consultants Ltd, Nanaimo, B.C.
Prepared for the Ministry of environment, Lands and Parks.

Tripp, D., 1994. The Use and Effectiveness of the Coastal Fisheries Forestry Guidelines in Selected Forest Districts of Coastal British Columbia. Victoria.

UKWAS Steering Group, 2000. Introduction to the UK Woodland Assurance Scheme, UKWAS Steering Group, Edinburgh.

United States Forest Service, 2000. 1997 RPA Assessment: The United States Forest Resources Current Situation (Final Statistics). Washington, DC.: United States Department of Agriculture.

Vertinsky, I.B., Stanbury, W.T., 1997. Boycotts in conflicts over forestry issues: the case of clayoquot sound. Commonwealth Forestry Review 76, 1.

Viana, V., Ervin, J., Donovan, R., Elliott, C., Gholz, H., 1996. Certification of Forest Products: Issues and Perspectives. Island Press.

Vogel, D., 1995. Trading Up: Consumer and Environmental Regulation in a Global Economy. Harvard University Press, Cambridge, Mass.

Wallinger, S., 1995. A commitment to the future: AF&PA's sustainable forestry initiative. Journal of Forestry 93 (1), 16–19.

Wilson, J., 1998. Talk and Log: Wilderness Politics in British Columbia. University of British Columbia Press, Vancouver.

World Wildlife Fund United Kingdom, 2000. WWF 95+ Group. Surrey, UK: World Wildlife Fund for Nature.

Yull, L., 2000. Comment and analysis: pressing engagements. Timber Growers Autumn 2000 (156), 12–13.

Yull, L., 2001. Comment and analysis: certification progress steady. Timber Growers Spring 2001 (158), 10.

[25]

Eco-labeling for energy efficiency and sustainability: a meta-evaluation of US programs

Abhijit Banerjee[a,*], Barry D. Solomon[b]

[a] *Center for Energy and Environmental Policy, University of Delaware, Newark, DE 19716-7301, USA*
[b] *Department of Social Sciences, Michigan Technological University, Houghton, MI 49931-1295, USA*

Abstract

Eco-labeling is a promising market-based approach for improving the environmental performance of products through consumer choice. While eco-labeling itself is not new, eco-labeling to promote energy efficiency or sustainability is a more recent phenomenon. Five such energy-labeling programs in the US are evaluated: Green Seal, Scientific Certification Systems, Energy Guide, Energy Star, and Green-e. Of these, the first four certify energy-efficient appliances while the last one certifies renewable electricity. While Energy Guide and Energy Star are government-run programs, the rest are privately administered. Two evaluation criteria were used—consumer response and manufacturer/marketer response. Very few studies have been done on the private programs making comparative analysis more difficult. It was found that government programs, in general, and Energy Star, in particular, were much more successful than the private programs. For appliance energy labeling, the private programs were found to have an almost insignificant effect on the market. Government support proved to be crucial in determining a program's credibility, financial stability, and long-term viability. Further, targeting of specific product categories, legislative mandates, incentives, and partnerships were found to be other important factors for program success. Simple seal-of-approval logos and labels have generally affected consumer behavior more than the complex information-disclosure labels

Keywords: Environmental labeling; Green power; Electricity marketing

1. Introduction

Eco-labeling may be defined as making relevant environmental information about a product available to the appropriate consumers through the product label to promote an environmental goal, cause or objective through consumer choice (Truffer et al., 2001). Products may be labeled based on a wide range of environmental considerations, such as recycled content, biodegradability, toxic emissions, waste generation, harm to wildlife, etc. By contributing to the decision-making process inherent in product selection, purchasing, use and disposal, eco-labeling has wide implications for consumers, businesses and government.

In response to an apparent consumer demand for "green" products, the number of products in US markets making environmental claims started mushrooming in the late 1980s. Products making claims such as recyclable, biodegradable, ozone-safe or eco-friendly were being introduced at a rate 20–30 times greater than that of other goods (EPA, 1991), and the percentage of new products making environmental claims grew from 1.1 in 1986 to 11.4 in 1990 (Rhodes and Brown, 1997). Such explosion of claims led to widespread confusion and distrust among consumers. The growing number of environmental claims led the Federal Trade Commission (FTC) in 1992 to issue guidelines for responsible green marketing (FTC, 1992).

By the late 1990s, largely due to the FTC guidelines, misleading environmental claims had nearly disappeared and third-party certification programs gained greater popularity and influence. Today, there are over 20 eco-labeling programs in the US covering a wide range of product categories. Examples include both government and private initiatives: pesticide labeling under the Federal Insecticide, Fungicide, and Rodenticide Act (FIFRA), "Dolphin-Safe" labeling of fish products by the Marine Stewardship Council, "Smartwood" labeling of wood products by the Forest

*Corresponding author. Tel.: +1-302-831-6044; fax: +1-302-831-3098.

E-mail addresses: abanerje@udel.edu (A. Banerjee), bdsolomo@mtu.edu (B.D. Solomon).

Stewardship Council, and energy-efficient appliance labeling by the US Environmental Protection Agency (EPA) and the US Department of Energy (DOE).

The recent trend toward partial deregulation in the electric utility industry is seen by many as a renewed opportunity for energy-efficiency programs. For the first time, utilities will not only have to invest in energy-efficiency measures to cut costs and remain competitive, but will also face direct competition with renewable-energy providers. Though much of the public will naturally choose the cheapest provider available, evidence suggests that there is a distinct market niche for renewable energy even at a price premium (Lamarre, 1997). At this juncture, third-party eco-labels certifying that the energy being marketed as green is indeed renewable and sustainable will go a long way toward instilling customer confidence in the credibility of the suppliers' claims.

Amory Lovins coined the term "negawatts" to denote the electricity saved from energy-efficient technologies and reducing waste that could be used for other purposes (Lovins, 1996). This innovative concept finds widespread application in the electric power industry, where utilities, facing rising costs, mounting opposition for siting new power plants, and directives from public utility commissions, started investing heavily in demand-side management (DSM) programs in the 1980s and early 1990s. With more states deregulating the electric power industry, however, utility spending on DSM has greatly decreased. Eco-labels certifying energy efficiency can thus play a critical role by helping to promote the direct use of energy-efficient products, though low energy prices may still hinder program effectiveness. Since investments in energy efficiency often have a high rate of return, with proper publicity campaigns, energy-efficiency labels can affect consumer behavior leading to higher market penetration of energy-efficient products.

In this paper, the five major energy-labeling programs in the US are reviewed and evaluated. Evaluation criteria considered include consumer response and manufacturer/marketer response. Seven major determinants of program success are identified. Several recommendations for policy-makers conclude the paper.

2. Energy-labeling programs

2.1. Green Seal

Green Seal is an independent, non-profit organization involved in promoting the manufacture and use of environmentally responsible consumer products through eco-labeling. Established in 1989, Green Seal issues a third-party, seal-of-approval label to products that "cause less harm to the environment than other similar products" (Green Seal, 2001). Energy-efficient appliances (refrigerators, heaters/coolers, washers, ovens, lights, etc.) are a small part of the wide range of products that Green Seal certifies. An annual certification fee of $5000–10,000 and a slightly smaller monitoring fee are charged to manufacturers. All certified products display the Green Seal-of-Approval logo (Fig. 1).

Fig. 1. The Green Seal-of-Approval logo. Reprinted with permission of Green Seal.

In addition to the labeling program, Green Seal established the Environmental Partners Program in 1994. Businesses, government agencies, universities, schools or other organizations may join the program by agreeing to the Environmental Partners Pledge, thereby committing to choose environmentally preferable products and services as part of their procurement policies. In exchange, Green Seal provides its partners with guidance in eco-friendly purchasing in the form of advice and reports.

Once product categories are chosen, they are then assessed based on a life-cycle analysis (LCA) of several products on the market. The environmental impacts during various stages of a product's life cycle are taken into account from raw material extraction, manufacturing, transportation and distribution, to product use and disposal. The environmental impacts that are taken into consideration include toxins generation, energy consumption, resource consumption, air and water pollution, and impacts on wildlife. During the LCA, Green Seal depends on previous studies done on that product category as well as independent testing and studies by their own experts. Products within a category must also comply with minimum performance standards (Green Seal, 2001).

2.2. Scientific Certification Systems

Scientific Certification Systems (SCS) is an independent organization that primarily labels for food safety and quality issues (organic, pesticide-free, etc.). Started in 1984, they have diversified into a wide range of certification, labeling and consulting services since then (SCS, 2001). Their Environmental Claims Certification Program is a single-attribute certification developed in

Fig. 2. SCS Environmentally preferable logo. Reprinted with permission of Scientific Certification Systems.

1990 to independently verify the accuracy of environmental claims on consumer products (e.g. biodegradable, recycled, etc.). Since environmental claims historically were unregulated, it has been difficult for customers to know which claims to believe (Rhodes and Brown, 1997). Under this program, SCS conducts detailed investigations to determine whether a claim can be verified. Once a product has been certified by SCS, its packaging may display an authorized certification emblem accompanied by a description of the verified claim. A one-time certification fee that varies from $1500–8000 as well as a quarterly certification update are charged to firms. The SCS Certified Eco-Profile Labeling Program is a neutral information disclosure about the environmental performance of a product, which leaves the decision making to the consumer based on their informed judgement (Fig. 2).

The Eco-Profile evolved from the SCS Environmental Report Card that was introduced in 1993. Both approaches use LCA studies that cover all relevant impacts for each of the product's life-cycle stages. But while the Report Card provided a quantified inventory of the resources used and emissions associated with a product, the Eco-Profile uses a more advanced form of LCA, known as life-cycle stressor-effects assessment (LCSEA). LCSEA takes into account the specific environments in which the extraction and releases occur to more accurately reflect its actual environmental effects. SCS claims that this approach is more scientific and compatible with international procedures, and charges $15,000–50,000 to conduct an LCSEA. If a product or service performs better than average in its market category on 15–20 core indicators, then it earns the right to use the label Certified Environmentally Preferable, shown in Fig. 3 (SCS, 2001).

2.3. Energy Guide

The Congress passed the Energy Policy and Conservation Act in 1975 (EPCA), the primary purpose of which was to conserve energy by enabling consumers to use energy-efficiency criteria while purchasing appliances. The EPCA required that mandatory Energy Guide labels should be placed on certain new home appliances such as refrigerators, freezers, clothes washers, air conditioners, etc. (FTC, 1997). The National Appliance Energy Conservation Act of 1987 also established minimum efficiency standards for all EPCA product categories and the Energy Policy Act of

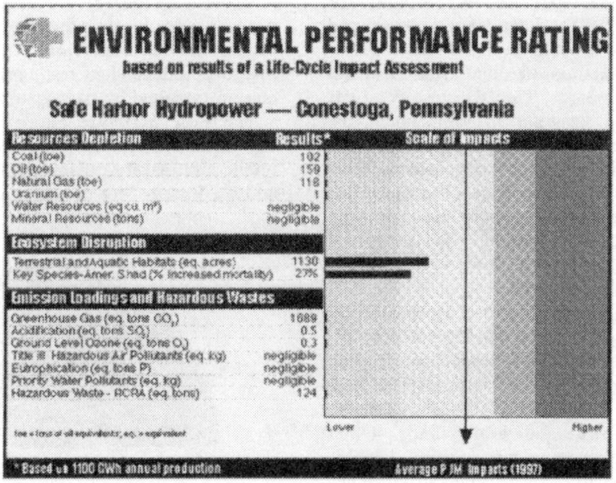

Fig. 3. SCS Eco-Profile Label scientific certification systems. Reprinted with permission of SCS.

112 A. Banerjee, B.D. Solomon / Energy Policy 31 (2003) 109–123

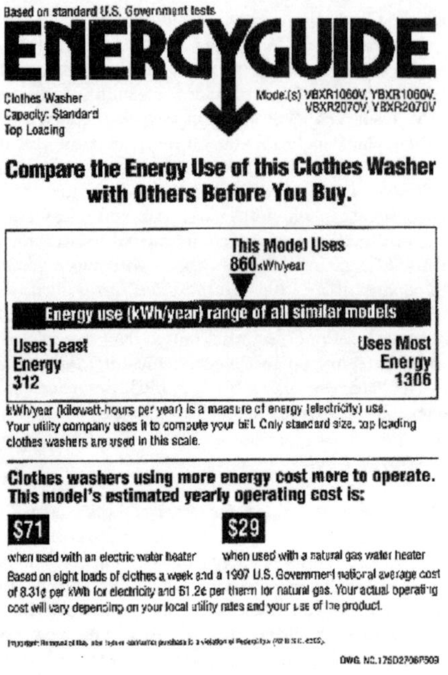

Fig. 4. Energy Guide label.

1992 (EPAct) added several other categories to the initial EPCA list.

Under the EPCA, the FTC has responsibility for establishing the format of the Energy Guide labels and ensuring implementation, while the DOE has responsibility for promulgating standardized test procedures and minimum-efficiency standards. The DOE also conducts a consumer education program to complement the labeling program (EPA, 1998). The Energy Guide label is a neutral, information-disclosure label leaving decision making to the consumers based on informed judgement (Fig. 4). It is mandatory under law, requiring all manufacturers of appliances in the categories specified by EPCA to display the label and prohibits the removal of the label before sale.

Under the EPCA, appliances selected for mandatory Energy Guide labeling have a significant potential for reducing energy use due to altered consumer purchasing behavior. These categories include refrigerators, freezers, water heaters, dishwashers, clothes washers, furnaces, air-conditioners, and heat pumps. Other appliances considered were omitted from labeling requirements when the difference in energy efficiency among models was insignificant. Lights, toilets, and showerheads also require labeling for energy or water consumption under EPAct but do not require Energy Guide labels (FTC, 1997).

The original label designed by the FTC in 1979 required disclosure of the appliance's estimated annual operating cost as well as a comparison with similar models. But critics soon pointed out that expressing energy use in dollars could be confusing since the cost of energy varies over time and between regions. In response, the FTC introduced a new format in 1994, which it claimed to be more user-friendly. In the new format, the primary information in the center of the label is units of energy consumed per year (in kilowatt-hours, therms or gallons). A range of comparability showing how the appliance compares with similar models on energy consumption is also provided. Estimated annual operating cost is still provided in some cases, but it appears as a smaller figure at the bottom and is accompanied by the energy price used in its calculation, thus avoiding the problem of cost fluctuations (FTC, 1997).

2.4. Energy Star

Energy Star is an umbrella of voluntary programs aimed at reducing air pollution and climate change by promoting the development and use of energy-efficient products. It was started by the US EPA in 1992, but since 1996, it has been jointly administered by the EPA and DOE (Energy Star, 2001). The Energy Star label is a single-attribute certification that promotes energy efficiency through its seal-of-approval logo. When a manufacturer meets the energy-efficiency criteria set by the EPA and DOE for a particular product category, it is allowed to display the Energy Star logo on that product and in its advertising (Fig. 5). For setting product standards, previous literature about the product category, independent testing and auditing, and information provided by participating producers are used.

The EPA promotes Energy Star Partnerships with businesses, government agencies, schools, and other public and private organizations who volunteer to buy and use Energy Star products. Both the EPA and DOE

Money Isn't All You're Saving

Fig. 5. Energy Star logo.

actively promote Energy Star through the Internet, television ads, public information campaigns as well as through partnerships. The Energy Star web-site features a complete range of certified products including information on where and how to get them (Energy Star, 2001). The EPA estimates that through all its partnership programs, US consumers will save nearly 50 billion kilowatt-hours of energy over the next 10 years, which translates to a net savings of $3 billion a year (EPA, 2000). In July 2001, an agreement was reached to bring Energy Star products to Canada (EPA, 2001).

There are several Energy Star Programs. The first and best known program is for office equipment, which expanded to cover wide areas including consumer electronics, household appliances, heating and cooling equipment, lights and light fixtures, exit signs and transformers. Energy Star computers, monitors, copiers, fax machines and printers drop down to a sleep mode when not in use, representing a 50–75% reduction compared to normal power draw.

The Energy Star Homes Program was established in 1995 as a partnership between EPA and homebuilders and developers. Builders and developers who join the program agree to build energy-efficient homes incorporating energy-efficient lighting systems, ventilation, heating and cooling systems, and extra insulation. When a newly built home meets all the criteria, it is certified as an Energy Star Home and can be advertised as such in real estate listings. Homeowners may also inquire about energy-efficiency upgrades to their existing homes. The EPA maintains a detailed nationwide database for buying (or upgrading) Energy Star Homes on a region-by-region basis (Energy Star, 2001).

The Green Lights program was launched by the EPA in 1991 and aimed at encouraging the widespread use of energy-efficient lighting. Partners agree to install energy-efficient lighting where profitable, as long as lighting quality is maintained or improved (EPA, 1998). In 1996, the Green Lights program was expanded into the Energy Star Buildings program, primarily aimed at commercial buildings where energy wastage was widespread. In order to become a partner, commercial building owners agree to upgrade their building to become more efficient. This may involve installing energy-efficient lighting, completing building tuneups, performing load reductions, undertaking insulation and ventilation upgrades, and upgrading heating and cooling systems. In 1999, the EPA introduced a new performance-rating system for commercial buildings in which buildings, in the top 25% in energy efficiency within their own category, would be awarded the Energy Star Building label. According to the EPA, this has increased response to the program and 90 buildings have already been certified with several hundred more in the evaluation process (EPA, 2000).

2.5. *Green-e*

As electric utility markets began to deregulate in the late 1990s, considerable consumer confusion has arisen about the nature and credibility of the green claims of the electricity service providers. To remove such confusion, the non-profit Center for Resource Solutions (CRS) started a voluntary third-party certification program in 1997 known as Green-e. The Green-e is a seal-of-approval label that certifies that power marketed by an electricity provider has been generated from renewable sources (Fig. 6).

Originally started in California, the Green-e program has spread to 10 eastern states. Though Green-e suffered a severe market setback in California in 2000–2001 due to the electricity market crisis, not only are results from the other states as encouraging as before, but Green-e is close to expanding to several new states (CRS, 2000). Green-e is also actively involved in promoting the program through public education campaigns by partnering with other organizations.

The Green-e program is different from the other labeling programs in one important respect. While the other programs promote energy efficiency through energy savings, Green-e promotes sustainability through renewable-energy resource development and use, but at a price premium. In that sense, the scope of the Green-e program may be limited to a small segment of the market, i.e. people who are environmentally concerned enough to be willing and able to spend more for the public benefit.

The Green-e program is administered by the Green Power Board set up by the CRS, consisting of stakeholders such as environmental advocates, consumer advocates, technical experts, and public officials. However, for the development of standards and criteria, input is taken from all stakeholders including power-marketing companies. Since the regulatory and sometimes economic climate may vary considerably from state to state, standards have to be developed for each state, or at least regionally (Rabago et al., 1998).

Green-e accepts five categories of sources as renewable: geothermal, biomass, solar, wind, and small

Fig. 6. The Green-e label. Reprinted with permission of Center for Resource Solutions.

hydroelectric. In each case, the environmental impacts of power generation must be minimal. A wide range of products are certified by Green-e, containing a mix of some or all the above categories and sometimes even non-renewable. Nuclear energy, however, is prohibited from the mix, and other non-renewables have to meet stringent environmental performance standards that are higher than the market average in that category. The lowest level of renewable product currently accepted by Green-e is 50% (Green-e, 2001). The supplier decides the mix that it is willing to market based on the demand and generation options in its area. Green-e also has a rule that any certified electricity provider has to supply at least 5% new renewables in their second year of operation and 10% thereafter (CRS, 2000). A $5000 annual application fee is charged by CRS for wholesale and retail electricity providers or a $3500 annual application fee for providers reselling an existing Green-e certified product. To ensure that a product has the energy mix that it claims, Green-e verifies whether the provider produced or bought the right mix and whether such generation met all relevant standards. Marketers must provide all the information necessary for verification. Each product is verified in the first year in which it is introduced and subsequently spot-checked, and is required to undergo an annual process audit. Marketers must meet additional requirements before they can use the Green-e logo: fuel source disclosure, contract information disclosure, and adhere to a professional code of conduct (Rabago et al., 1998).

3. A meta-evaluation

3.1. Evaluation criteria

Two criteria were used for evaluating the energy-labeling programs—consumer response and manufacturer/marketer response:

- *Consumer response:* A favorable and widespread consumer response is the primary goal of an eco-labeling program since good consumer response is essential to create a market for environmentally preferable products. As such, consumer response is an important criterion for evaluating the public acceptance of a program and has three dimensions: awareness, understanding, and behavior. Awareness indicates to what extent consumers know of the program's existence. Understanding indicates how far a consumer can interpret the connection between the environmental issue, the label's meaning, and action needed to elicit results. Behavior indicates how far consumers' awareness and understanding are translated into the desired action, i.e. purchase of the labeled products. Awareness and understanding are easy to measure through consumer surveys, though such results are often overstated due to inherent survey biases (EPA, 1994). On the other hand, behavior is usually more difficult to measure since a consumer's purchase decision is based on numerous factors besides the label. All three dimensions of consumer response will be considered as far as data permit.
- *Manufacturer/marketer response:* Manufacturer/marketer response represents the supply side of an eco-labeling program and is the key to market transformation. The more manufacturers sign up to get their products certified, and the more marketers agree to sell such certified products, the more successful a labeling program will be. For "green" power, quite often marketers rather than manufacturers apply for certification. That is the primary reason for including marketers for evaluation in the same category with manufacturers. Manufacturer/ marketer data are obtainable directly from the labeling program administrators and are fairly comparable among programs.

3.2. Data sources and gaps

Since this study is a meta-analysis, the primary data source used was published studies and reports. These comprise a wide range of sources including the program administrators themselves, government research institutions, non-profit research organizations, state agencies, electric utilities, and other researchers. Data for the manufacturer/ marketer response criterion, however, were almost entirely obtained directly from the program administrators, either by personal communication or from their annual reports.

While many studies exist that evaluate the government-run programs, none were publicly available for the privately run programs. This can be explained by the enormous differences in funding between the private and government-run programs (see Table 3). Consequently, anecdotal evidence, where available, had to be used to evaluate the private programs. Naturally gaps remain in these data, making comparisons among different programs more difficult.

3.3. Evaluation based on consumer response

Concern about environmental and energy issues, in general, is very high among the public according to numerous opinion polls. There is overwhelming support, in principle, for policies to promote energy efficiency and renewable energy. But such high claims about awareness and understanding are seldom borne out by tests of behavior. While surveys suggest that a majority of consumers may be willing to pay more for energy-efficient appliances or renewable energy, in practice only a few do so, with most people considering price, quality, or brand name before energy efficiency. Although most people indicate a strong preference for a credible, third-party labeling system and claim to use it, very few actually use the labels for purchasing decisions. Even their perception of the connections between different issues, policies, and actions often lacks basic understanding. Such overstated responses leading to survey bias are thought to be due to the "warm glow" effect, which often casts doubt on the validity of such results. This has led many researchers to believe that a particular segment of the population (better educated, higher income) is responsible for sustained demand for energy efficiency and renewables, and many marketing and campaign strategies are developed to target this niche (Egan and Brown, 2001).

3.3.1. Green Seal

Market studies on consumer response to Green Seal have not been done. However, in a study conducted during their preliminary market research in 1991, Green Seal reported that four out of five consumers said that they would choose a product with the Green Seal logo over a product without it, quality and price being equal (EPA, 1994). This was just the stated willingness to purchase and no later studies were conducted to gauge actual consumer awareness or behavior.

Since Green Seal has a limited advertising budget, it tries to increase its presence by targeting institutional buyers who agree to buy certified "green" products through its Environmental Partners program. To date, more than 500 organizations have joined the program as partners, which is an indirect measure of consumer response (Petruzzi, 2001).

3.3.2. Scientific Certification Systems

As in the case of Green Seal, only limited and old data are available for the SCS Report Card in the form of a focus group study conducted by Gerstman & Meyers, Inc., (1993). This study gauged consumer understanding of the label and was not meant to determine awareness or behavior–response levels. The report found that on the whole, respondents perceived the Report Card as legitimate and trustworthy, in part due to its "scientific presentation". However, respondents also told the researchers that they would be willing to use the Report Card as a decision-making tool for small household goods, but would be less likely to use it when purchasing larger items (e.g. appliances and cars), where other factors such as safety may take precedence over environmental criteria. The report also suggested that consumers might erroneously perceive the Report Card as endorsing a product when it is only aimed at information disclosure.

The SCS Eco-Profile is generally regarded as the most complex comparison label now in use. It provides consumers with detailed, quantitative information about a product's life-cycle environmental burden by quantifying 18 product characteristics in the form of both numerical values and a bar graph. The label has been criticized as too complex for the average consumer to use (Harris and Casey-McCabe, 1996). Like Green Seal, SCS also targets institutional buyers to rapidly increase their market presence. Further, it has been noted that the greatest demand for the Eco-Profile was in industry (EPA, 1998), though estimates of the number of organizations involved were not available.

3.3.3. Energy Guide

A 1999 Wisconsin survey of 2263 people showed that nearly 70% claimed that they had seen or heard of the Energy Guide label. Those who had shopped for or purchased an appliance in the previous year were significantly more aware of it than those who had not. But when the survey went on to ask the respondents about the information provided by the label, only 49% were able to accurately describe some aspect of the pertinent information (Opinion Dynamics Corp., 2000). Another Wisconsin study found that about two-thirds of Wisconsin respondents claimed to be aware of the Energy Guide label, but when asked to describe the information it provides, only half of the respondents could do so correctly (Feldman and Tannenbaum, 2000).

An American Council for an Energy-Efficient Economy (ACEEE)-sponsored study also found very low levels of understanding of the label. The specific test of awareness presented respondents with five possible Energy Guide label designs. All had the standard yellow color and heading now in use, but the energy information format was changed in each. The respondents were asked to identify as to which one was in current use. Only 12% could correctly do so. This finding vastly undermines the claim that the Energy Guide label is familiar to most Americans, suggesting that when one controls for the most striking visual elements (color and headline), the actual informational elements of the label are in fact unfamiliar. This brings into serious question the label's actual use and the extent to which it is understood (Shugoll Research, 2000).

Comprehension of the Energy Guide label has been debated. For example, du Pont found that consumers cannot necessarily tell as to which product is actually more energy-efficient based on the label (du Pont, 1998). More recently, ACEEE conducted research that indicated that for some consumers, comprehension is problematic, but that a more overarching problem is that consumers perceive the current label to be "boring, overly technical and unmotivating" (Egan et al., 2000). Egan also found that some people were confusing the Energy Guide informational label with a Energy Star-type seal-of-approval for energy efficiency (Egan, 2000). Various alternative formats for the label are being tested by the ACEEE and, depending upon the findings, the Energy Guide label may be altered in the near future.

3.3.4. Energy Star

In the first five years of the Energy Star program, since office equipment was the major focus, most studies reported that logo recognition by consumers was almost always linked to computers (Weissman, 1999). But recent studies have found more extensive awareness levels. In May 2000, an EPA-contracted study conducted 250 intercept interviews at shopping malls in five cities. In an aided test of awareness where respondents were shown the logo and asked whether or not they had ever seen the symbol, 48% claimed to have seen the logo. Of those who had seen it, 70% could correctly identify some aspect of what it stood for. Of those who had not seen it previously, 34% could correctly guess its actual meaning. This indicates that the Energy Star logo effectively communicates even upon initial sighting, a very promising indicator for the future of Energy Star recognition (Egan and Brown, 2001, p. 37).

In a New England states study, 46% of the survey respondents claimed to be at least somewhat familiar with the Energy Star label. Of those, more than half could correctly state its purpose. The same study also found that awareness levels were higher among residents who had shopped for or purchased an appliance in the last year than those who had not, indicating that exposure in the retail setting was enhancing awareness (Ulrich Research Services, 2000).

Evidence on the actual effect of the label on purchasing decisions is, however, not always so unequivocal. A Wisconsin phone survey asked respondents who displayed valid awareness and who had purchased appliances in the last 12 months the extent to which the label was influential in their purchase decision. A total of 54% said that it was somewhat or very influential (HB&SFMC, 2000). However, another Wisconsin study of refrigerator shoppers found that approximately three purchasers in 10 reported having noticed the logo when making the purchase. Among those who did notice the logo, one in two reported to have been at least somewhat influenced by it in their decision. Thus, the joint effect of these two measures, the proportion of purchasers who both recall the logo and report that it affected their decision, is one purchaser in six. This seems to indicate that the overall influence of the logo is still modest (Feldman and Tannenbaum, 2000).

Finally, a recent nationwide survey by the Consortium for Energy Efficiency (CEE) found that nationwide, 41% of households reported having seen the Energy Star label. Over one-half of all households nationwide, including those who had previously seen the label and those who were reacting to it for the first time, had at least a general understanding of the label's message. Nationwide, about 50% of those who reported that they bought an Energy Star-labeled product also reported that they were influenced by the label to buy the product. The study also found that label awareness was much higher in the high-publicity areas than in the low-publicity areas: 52% versus 37%. Not only was awareness higher, but the proportion of consumers influenced by the label in purchasing qualifying products also was significantly higher in the high-publicity areas compared with the low-publicity areas: 58% versus 35% (CEE, 2000).

The EPA also reported that in 1999 alone, Americans bought more than 100 million Energy Star products, all categories taken together. A qualitative study on the Energy Star Homes program found that though consumers generally associated the Energy Star logo with energy efficiency, they had difficulty in understanding different energy-efficiency features or designs. Moreover, they tended to perceive energy efficiency as an expensive sacrifice. The study also showed that consumer acceptance could be significantly increased if Energy Star Homes were marketed as money-saving in the long run, with higher resale value and greater comfort. Financial incentives, such as favorable mortgage terms, also was found to boost the demand for the homes (Werling et al., 1998).

3.3.5. Green-e

Though consumer attitudes about green pricing for renewable energy have been extensively studied (cf. Holt, 1997; Egan and Brown, 2001), there has been no systematic effort to study consumer response to the Green-e program either by the program administrators or by independent researchers. As such, fragmentary data have been used as a proxy in this discussion.

According to the CRS, over 128,000 households in California (approximately 350,000 people) and over 16,700 households in Pennsylvania (approximately 50,000 people) received Green-e certified electricity in 1999. There also were about 26,500 commercial accounts in California and 1233 in Pennsylvania served by Green-e certified power. Such accounts include companies such as Kinko's and The Gap, the City of Santa Monica, California, and the EPA laboratory in Richmond, California. The report also showed that at the end of 1999, Green-e certified offerings represented 80% of the direct access residential electricity market in California. In other words, 80% of the people who switched their power providers in California switched to a Green-e certified product. In Pennsylvania, the corresponding figure was only 4% since, unlike California, Pennsylvania does not provide subsidies in the form of a "Customer Credit Sub-account", which enables electricity providers to offer renewable electricity at a discount (CRS, 1999).

Though in 2000–2001 Green-e suffered a setback in California, losing substantial market share due to skyrocketing electricity prices, evidence suggests that the program is steadily increasing in other states. In 2000, Green-e also certified the green pricing programs of three public utilities—Tennessee Valley Authority, Wisconsin Electric and Madison Gas & Electric, which have the potential to provide green electricity to 3 million customers (CRS, 2000).

Since there is strong public support for disclosure and labeling of fuel sources by electricity providers in a deregulated market (Holt, 1997), one study conducted by the National Council on Competition and the Electric Industry examined consumer attitudes toward mandatory disclosure versus voluntary certification. Consumers were shown hypothetical fuel and environmental burdens disclosure labels as well as the Green-e logo (as an example of voluntary certification) and asked for their preferences. The results were mixed. Certification did not appear to be a viable substitute for disclosure. On the other hand, certification appeared to be useful as a supplement to the label (Winneg et al., 1998). This finding appears to be an important consideration for Green-e program design.

Green-e's image has been challenged recently as some consumer and citizen groups have charged it with "green-washing". Several of its certified providers have been blamed for passing off not-so-green energy as green (such as waste incineration as biomass, large hydro as low-impact hydro, etc.) instead of investing in clean renewables like wind and solar, and Green-e has shared the blame for being their certifier. This problem has been particularly acute in the Northeast, where renewable generation capacity is smaller than in California. Critics have called for governmental regulatory oversight of the standard setting and certification process (Gimein, 1999; Ewall, 2001). Green-e, however, maintains that both their standards and verification procedures are defensible and foolproof enough to ensure credibility (Rabago et al., 1998).

3.4. Evaluation based on manufacturer/marketer response

A label's effectiveness also can be measured as a function of its ability to change manufacturer behavior to meet program criteria. Regardless of the reasons that a manufacturer changes its behavior (e.g., increasing market share of new product, avoid losing market share of existing product to labeled competitor or to improve stockholder and employee relations), a labeling program that induces the manufacturer to meet the criteria of its label has been successful in reducing environmental burdens inasmuch as there are significant differences between products that do and do not meet the award criteria. Manufacturer response is thus a central criterion for evaluation representing the supply side performance of a labeling program.

This section examines the extent of manufacturer participation in a labeling program as a measure of its success. Data on marketer participation, where available, also has been used to judge greater (or lesser) market penetration. In case of Green-e, companies applying for certification are often just marketers rather than manufacturers (producers), and thus have been treated as one and the same. Manufacturer-participation data were readily available from program administrators, making this criterion the most complete and hence comparable among programs. Sales data, however, which also may be useful in judging a labeling program's success, are considered proprietary and not usually released.

3.4.1. Green Seal

Communication with Green Seal revealed that when all their labeling activities are taken together, so far they have developed standards for 90 product categories. Among these, 300 individual products already have been certified representing a total of 14 companies. When just electrical appliances are considered, standards have been developed for 19 categories. About 50–75 models of appliances have been certified representing seven companies. Green Seal certification for appliances had been going well before the government-run Energy Star program was launched. Since then, market interest in the Green Seal for energy appliances has largely evaporated (Petruzzi, 2001).

3.4.2. Scientific Certification Systems

The SCS Environmental Claims Certification program currently certifies 10 different types of claims. Overall over 3000 individual claims have been certified, representing about 500 companies. The Eco-Profile has been developed for 10 different product categories; Eco-Profiles have been certified for 30 individual products representing 15 companies (Brown, 2001).

For both electrical appliances and electricity generation only the Eco-Profile (and not Claims Certification) is available. Both of these categories are relatively new and have been in existence for about 3 years. So far standards have been developed for only one category of electrical appliances, and only one appliance has been certified for one company. For electricity generation, six products have so far been certified for four companies (Brown, 2001).

3.4.3. Energy Guide

The Energy Guide label was developed only for those categories of household appliances (such as refrigerators, freezers, clothes washers, heaters, air-conditioners, etc.) for which significant differences in energy consumption exists across different models. Other appliances (such as television sets, kitchen ranges and ovens, clothes dryers, etc.), for which energy consumption was largely similar across different models, were left out of the program. Since the label is required under federal law as discussed earlier, all manufacturers making appliances in the nearly 20 categories covered by the program have to display a label. In a mandatory labeling program, market penetration of the label is automatically 100% by definition, and program success therefore cannot be judged by the criterion of manufacturer participation. However, some relevant data on manufacturer or marketer response on the effectiveness of the label were obtained from the literature and is discussed below.

A Bonneville Power Administration study of retail outlets found that overall, 77% of salespersons at appliance stores understood the labels well enough to explain them intelligently. However, only 13% of all dealers said that they used the Energy Guide label in their sales pitches (BPA, 1988). In recent ACEEE-sponsored interviews, both manufacturers and marketers agreed that the aim of the labeling program was justified but they thought that the label was not having its desired results. This was because they thought that energy efficiency was a low priority for consumers and that the information provided in the label was not necessarily understandable to the average consumer. Most sales staff agreed that only a small subset of shoppers, particularly those looking for refrigerators and air-conditioners, ever asked about energy-efficiency criteria. They also indicated that the most common use of the label was in situations, where energy efficiency was a deciding variable between two otherwise equivalent products. It was even noted that for many installed appliances, such as furnaces and boilers, the consumers typically do not see the label until after the purchase (Egan, 2000).

3.4.4. Energy Star

Extensive data are available on manufacturer participation and market penetration for Energy Star products. The EPA claims that more than 1200 manufacturers were producing 7000 Energy Star-labeled product models in 31 consumer product categories by the end of 1999 (EPA, 2000). The most successful Energy Star category has been for office equipment, in which the EPA introduced the low-power sleep mode. In less than a decade, the program accomplished a virtual 100% market transformation with all the major manufacturers today using the logo. Other categories of appliances, though progressing steadily, have not attained such success rates. The market penetration rates for household appliances are anywhere from 5-20%, depending upon the appliance and the year it was brought under Energy Star (Feldman and Tannenbaum, 2000).

Over 8000 new homes qualified as Energy Star Homes in 1999, an increase of more than 50% over 1998, bringing the total number of labeled homes to 15,000. In 1999, the threshold of 1000 builder partners also was crossed, and the manufactured-housing sector, representing about 30% of all new housing began participating in the program. More than 800 industry allies, including 40 utilities, have signed agreements to promote Energy Star Homes. In the Energy Star Buildings program more than 5500 organizations have partnered with EPA to improve their energy performance, committing over 10 billion square feet or 15% of the total commercial, public and industrial building market. Nearly 1.6 billion square feet of investor-owned office properties also joined Energy Star, representing over 70% of the office properties market (EPA, 2000).

Twenty-five retail partners with more than 4600 storefronts in the US have so far agreed to promote and sell Energy Star products. The number of utilities and state administrators actively promoting Energy Star grew from 29 in 1998 to 62 in 1999, servicing more than 60 million households across the nation (EPA, 2000). A Wisconsin study also reported similarly high levels of participation when they found 70–80% of the retailers in the state participating in Energy Star partnerships. The same study also reported that both awareness and participation was significantly lower in neighboring Iowa, a state without an Energy Star promotion program (Feldman and Tannenbaum, 2000).

3.4.5. Green-e

In 1999, Green-e branched out from California to include Pennsylvania and New Jersey. Nationwide in 1999, four wholesalers and seven retailers participated in the Green-e program, offering 18 renewable products. All of them were at least 50% renewable, with most either containing 75% or 100% renewables. The total load served by Green-e electricity products in 1999 was approximately 1,227,000 MW h. An estimated 62% of this load was sold to residential customers and the remaining 38% to commercial customers (CRS, 1999).

In 2000, just before the electricity crisis, there were 13 companies selling 23 Green-e certified products in California, all of which were 100% renewable. This meant that almost all of the wholesale and retail power marketers had at least one product certified by Green-e. Though Green-e suffered a temporary setback in California due to the state electricity crisis, evidence shows that they are steadily increasing market-share elsewhere as the program now operates in 11 states.

3.5. Overall assessment

Table 1 is based on manufacturer-response data taken from the previous section to facilitate easy comparison. For Green Seal and SCS, which have a wide range of product categories, only the energy-related product categories were chosen since this analysis considers

Table 1
Comparison of different eco-labeling programs based on manufacturer response

Program	Number of product categories for which criteria have been developed	Number of products actually certified	Number of manufacturers with certified products
Green seal	19	50–75	7
SCS	2	7	5
Energy Guide	20	All models	All manufacturers
Energy Star	31	7000	1200
Green-e	1	23	13

Table 2
Overall ranking of eco-labeling programs based on market influence

Market influence	Program
High	Energy Star
Modest	Energy Guide
Low but significant in a few states	Green-e
Insignificant	Green Seal
	SCS

energy labeling. For SCS, data on both their appliance and electricity generation categories have been combined together in the table.

From Table 1 one fact is easily noticeable. The government-run programs have a much higher market penetration, as evidenced from the number of certified products and manufacturers as compared to the private ones. However, it should be noted that Energy Guide has a 100% market penetration because it is mandated by law. But when compared to the private programs, Energy Star is several orders of magnitude ahead in terms of manufacturer response. It can be argued that the more successful or reputable a labeling program is, the more manufacturers will try to get certified in their own interest. In that sense then, the table showing manufacturer response can be said to be a rough indicator of the success of a labeling program.

Based upon Table 1 and our earlier discussion, the five labeling programs have been ranked according to their market influence (Table 2). Clearly, the government-run programs have a higher market influence than the private ones. The success of Energy Star is beyond question both in terms of consumer and manufacturer response. Though the market penetration (and recognition) of the Energy Guide is higher than that of Energy Star, its usefulness to consumers has been seriously questioned and its market influence is much more limited. Though the Green-e program has not been around as long as Green Seal or SCS, it has certified more products and manufacturers than either of the other two, though it certifies just one product category (viz. renewable energy). The principle reasons for these findings will now be discussed.

4. Determinants of success

Based on our evaluation, seven determinants of the success of energy-labeling programs in the US can be identified:

(1) *Government support and credibility*: Government run programs have been far more successful than the private ones. Government support to a labeling program not only increases its credibility and recognition, but also improves financial stability, legal protection and long-term viability. In the US, private labeling programs have enjoyed limited success in niche markets with participation, primarily restricted to smaller companies interested in increasing their market share. This is chiefly because the private programs have not yet satisfactorily established their credibility and long-term viability, making larger corporations reluctant to join them (Paulos, 1998). Particularly, after Energy Star became widely recognized and accepted, manufacturer interest in energy-efficiency labeling by the private programs has largely disappeared.

(2) *Budget*: A labeling program's scale of operations is a direct function of its budget, with a limited budget affecting both its demand and supply side operations (Table 3). Though higher budgets are usually a result of government involvement, it has been treated as a separate factor because private programs have access to additional resources through donations or grants.

Clearly, the government-run programs have much larger budgets than the private ones. In particular, Energy Star's budget is 10–100 times that of the private programs. For Green Seal and SCS, which are not exclusively energy labels like Energy Guide and Energy Star, their already small budgets are further divided for each of the wide range of product categories they certify, making the amount of money available to their energy-labeling efforts even smaller.

(3) *Publicity and partnerships*: Publicity is not only vitally important for increasing consumer awareness and recognition, but also helps in getting the attention of manufacturers. An early BPA (1988) study e.g. found that the influence of the Energy Guide label on consumer purchasing behavior was much greater with promotional materials than with just exposure to the label.

With a multimillion-dollar advertising budget, Energy Star ensures high recognition of its label by continually reaching millions of consumers through its extensive promotional campaign, which includes advertisements in print media, television commercials, and retail

Table 3
Scale of operations of different eco-labeling programs in the US

Program	2000 Budget (in US dollars)	Employees (in 2000)
Green Seal[a]	0.5 million	10
SCS[b]	5 million	35
Energy Guide[c]	8 million[d]	12[d]
Energy Star[e]	50 million[d]	100[d]
Green-e[f]	1 million	17

[a] Weissman (2001).
[b] Brown (2001).
[c] Pollock (2001) and Newsome (2001).
[d] Estimates.
[e] Sullivan (2001).
[f] CRS (2000).

promotions. Through December 1999, the cumulative number of impressions made in all advertising media for Energy Star exceeded 2.5 billion, reaching a cumulative audience of over 1 billion people (Egan and Brown, 2001). In contrast, private programs with more limited budgets have to rely upon free publicity, such as public service announcements and when manufacturers advertise their certified products.

Partnerships among stakeholders are another important way to greatly increase a program's visibility and recognition as the more partners a program has, the more it is publicized. Partners include a wide range of organizations including state governments, other government agencies, utilities, small businesses, corporations and non-profits.

(4) *Label clarity*: Seal-of-approval labels are usually better understood by consumers than information-disclosure labels. While seal-of-approval labels may be oversimplified and judgmental, experience has shown that the proportion of informed consumers who are willing and able to use technical information effectively is low. Also, consumers often confuse disclosure labels as a seal-of-approval thus defeating the label's purpose. In surveys, consumers have repeatedly complained about their difficulty in understanding both the SCS Eco-Profile and the Energy Guide. Thus, seal-of-approval labels are generally more useful to consumers, which is the main reason why most of the successful European eco-labels are seal-of-approval types (Truffer et al., 2001). This explains as to why even though the visibility and recognition of the Energy Guide is much higher than that of Energy Star, its influence on consumer behavior is much lower.

(5) *Targeted product category*: It is most effective to target one specific product category at a time. These should be selected based on two criteria: first, there should be large room for improvement in product standards; and second, considerable improvement should be achieved with relatively little effort. This was the logic behind both the Green Lights and Energy Star Office Products programs. Since the programs were designed for minimal, if any, initial investment on the part of manufacturers, and a quick payback and eventual profitability was guaranteed, manufacturers (of office products) and commercial-building owners (for Green Lights) quickly agreed to participate to gain wide recognition. This resulted in high and rapid levels of market transformation, making these programs the most successful under the Energy Star umbrella (Geller and Nadel, 1994). Even Green-e, exploiting a similar situation in California, achieved considerable recognition in two years.

(6) *Legislative mandates*: Besides direct government involvement, other government actions may help an eco-labeling program. For example, federal law has made the Energy Guide label mandatory on several household appliances. Thus, by forcing manufacturers to disclose energy-efficiency data, such laws ensure a high visibility of the Energy Guide label. Similarly, former President Clinton's Executive Order 12873 in 1994, which directed all federal agencies to adopt environmentally preferable purchasing policies, created an automatic market for Energy Star products, particularly office products. Manufacturers thus were forced to participate since they could not ignore the biggest customer in the nation.

(7) *Incentives*: Incentive-based promotions have a remarkable effect on both consumer and manufacturer behavior. Utilities such as the BPA, which as part of their DSM programs rebated consumers who bought Energy Star appliances from retailers, reported higher consumer response from dealers offering rebates (EPA, 1994). Similarly, the EPA has maintained an extensive system of rewards and recognition for partners in its Green Lights and Energy Star Buildings programs. For instance, buildings qualifying for the Energy Star logo are not only mentioned in EPA publications and the web-site, but also receive awards in public ceremonies such as Earth Day. Also, consumers are given favorable mortgage terms if they agree to buy Energy Star Homes (Werling et al., 1998). Such incentives translate into higher participation rates.

5. Policy recommendations

From the preceding discussion, clear lessons emerge on as to why some labeling programs are more successful than the others. Based on these lessons, seven policy recommendations are provided to improve the effectiveness of energy-labeling programs that may be useful to the environmental and energy policy community. While energy labeling alone will not improve the environment, experience has shown that a properly designed labeling program can be a significant stimulus for market transformation toward environmentally preferable products.

- *Increase government involvement:* As shown in the preceding section, government support is the most critical factor for the success of a labeling program. Though European experience has shown that a government-administered broad-spectrum national eco-label is the most successful, this is unlikely to happen in the US market in the present political climate. However, the EPA, DOE or the FTC should not only continue, improve, and expand their reasonably successful programs, but also should provide strategic support to the private labeling efforts. Such support may take the form of funding, technical assistance, promotion, oversight or representation on the program boards. Government support may do wonders to the struggling private programs' credibility and long-term viability and thus will attract greater manufacturer participation. For example, in the rapidly changing electricity market, government support to Green-e (backed by oversight) is likely to cast it into national prominence and provide a further boost for renewable energy development. At a minimum, government agencies should establish and coordinate an ongoing dialogue between existing programs to better harmonize the divergent efforts.
- *Avoid confusion:* More than one label in the market for the same product category not only creates consumer confusion about which one to trust, but also confuses manufacturers about which program to join. Having two or more programs labeling the same product categories works at cross-purposes and one will inevitably prosper at the expense of the other. This is exactly the case today with Energy Star, Green Seal, and SCS for the household appliance-labeling category and should be avoided. Through an ongoing dialogue representing all stakeholders, responsibilities for different product categories should be distributed among the different programs according to experience and expertise so that no more than one label is used for one product category. For example, under the same Energy Star umbrella, the EPA can certify homes and buildings, DOE can certify appliances, and Green Seal and SCS can certify residential and commercial lighting between themselves. This will ensure that the different programs complement each other instead of compete. Otherwise, the EPA may act as a "certifier-of-certifiers" much like the Forest Stewardship Council, which certifies other certifiers for their sustainable forestry certification program (EPA, 1998).
- *Target-specific product categories:* Since a broad-spectrum national eco-label is out of question in the present political climate, preliminary feasibility analyses should be conducted to narrow down specific product categories. These target categories should be chosen based on two criteria—whether there is ample room for improvement in product categories and whether considerable improvement can be achieved with relatively little effort. The Energy Star Office Equipment and Green Lights designations were selected in this way and both programs have been highly successful (Geller and Nadel, 1994).
- *Improve label clarity:* Repeated research has shown that seal-of-approval labels are more useful to the average consumer than information-disclosure types, because few consumers can understand the "technical" information or are not willing to spend time doing so. Moreover, they often tend to interpret disclosure labels as a seal-of-approval, thereby completely defeating the label's purpose (Egan and Brown, 2001). To reconcile the relative advantages and disadvantages of the two kinds of labels, two compromise options can be suggested. In the first option, the seal-of-approval format could be retained but the standard setting and verification should be based on a broader life-cycle analysis rather than a single attribute such as energy efficiency. In the second, the information-disclosure format could be retained but the seal-of-approval should be added to products performing above a certain level. This would satisfy both the "info-phobic" and "info-philic" consumer. Actually, the ACEEE is currently researching such an option for combining the Energy Guide and Energy Star labels for appliances (Egan, 2000).
- *Increase promotions and partnerships*: The importance of education campaigns cannot be stressed enough since research has consistently shown that participation in an eco-labeling program rises with education (EPA, 1994). Advertising budgets should be increased as far as possible and efforts should be made to cover as many kinds of media as possible. Research also should be performed so that ads are designed in a way that is both appealing and understandable to consumers. At the same time, every effort should be made to increase the number of partnerships, targeting as wide an audience as possible including state governments, municipal governments, small businesses, corporations, universities, school districts, among other organizations. Expanding partnerships increases both recognition and credibility by positive feedback, and does so at no extra cost.
- *Support with legislative mandates:* As shown in the previous section, carefully planned legislative mandates go a long way toward improving the effectiveness of a labeling program. Practicable legislative mandate options to reinforce existing labeling initiatives should be investigated by governments. For example, a mandatory Renewable Portfolio Standard (RPS), which entails that a certain portion of any power provider's electricity mix must come from renewables, would be a shot in the arm for a green power certifying label. Such RPS requirements

already exist in at least 10 states but a federal mandate would be more desirable.

- *Improve incentives:* Whether applying for certification or buying certified products, labeling usually involves extra investment and effort for both manufacturer and consumer. Since only a limited few companies or people will normally go that length, incentives are an important tool to expand participation in a labeling program. Manufacturer incentives could take the form of tax breaks, awards, or recognition; and consumer incentives could take the form of rebates, prizes, or assured protection against future price increases for green power customers. The EPA and its partners already have an elaborate system of incentives that could be improved and expanded further. Money for such incentives could come from shared savings from the avoided power.

References

BPA, 1988. The evaluation of phase II of the BPA appliance energy efficiency promotion campaign: promotion of energy efficient refrigerators and freezers. Bonneville Power Administration. Office of Conservation, Portland, OR.

Brown, L., 2001. Personal communication. Vice-President, Scientific Certification Systems, July.

CEE, 2000. CEE Energy Star Household Survey Report (2000): Part I—Results. Consortium for Energy Efficiency, Boston, MA.

CRS, 1999. Green-e Verification Results 1999. Center for Resource Solutions, San Francisco, CA.

CRS, 2000. Center for Resource Solutions: 2000 Annual Report. Center for Resource Solutions, San Francisco, CA.

du Pont, P., 1998. Energy policy and consumer reality: the role of energy in the purchase of household appliances in the US and Thailand. Ph.D. Dissertation, University of Delaware, Newark, DE.

Egan, C., 2000. An evaluation of the Federal Trade Commission's energy guide appliance label: an interim summary of findings. American Council for an Energy-Efficient Economy. Report No. A003, Washington, DC.

Egan, C., Brown, E., 2001. An analysis of public opinion and communication campaign research on energy efficiency and related topics. American Council for an Energy-Efficient Economy. Report No. A013, Washington, DC.

Egan, C., Payne, C., Thorne, J., 2000. Consumer Interview Analysis For The ACEEE Appliance Labeling Project. American Council for an Energy-Efficient Economy, Washington, DC.

Energy Star, 2001. US EPA Energy Star Homepage. http://www.e-nergystar.gov.

EPA, 1991. Assessing the environmental consumer market. EPA-21P-1003. Prepared by Abt Associates, Inc. for the US Environmental Protection Agency, Washington, DC.

EPA, 1994. Determinants of effectiveness for environmental certification and labeling programs. EPA-742-R-94-001. Abt Associates, Inc., Washington, DC.

EPA, 1998. Environmental labeling: Issues, policies, and practices worldwide. EPA-742-R-98-009. Abt Associates, Inc., Washington, DC.

EPA, 2000. 1999 Energy Star annual report. Washington, DC.

EPA, 2001. US Energy Star heads north for Canadian debut: Countries announce agreement to promote energy efficiency. US Environmental Protection Agency press release, July 19, 2001, Washington, DC.

Ewall, M., 2001. Pennsylvania Environmental Network, Green Energy Leadership Team. http://www.penweb.org/issues/energy

Feldman, S., Tannenbaum, B., 2000. Swoosh!? Awareness and effects of the energy star brand in Wisconsin appliance labeling programs. Proceedings of the 2000 ACEEE Summer Study on Energy Efficiency in Buildings. American Council for an Energy-Efficient Economy, Washington, DC.

FTC, 1992. Guides for the use of environmental marketing claims. Federal Trade Commission, Washington, DC.

FTC, 1997. Rule concerning disclosures regarding energy consumption and water use of certain home appliances and other products required under the Energy Policy Act (the appliance labeling rule). Code of Federal Regulations 16 CFR 305, 1 January 1997.

Geller, H., Nadel, S., 1994. Market transformation strategies to promote end-use efficiency. Annual Review of Energy and the Environment 19, 301–346.

Gerstman & Meyers, Inc., 1993. Scientific Certification Systems environmental report card consumer research evaluation report. Oakland, CA.

Gimein, M., 1999. Green power next for huge IPO. The Industry Standard. April 26.

Green-e, 2001. http://www.green-e.org.

Green Seal, 2001. http://www.greenseal.org.

Harris, J.P., Casey-McCabe, N., 1996. Energy efficient product labeling: market impact on buyers and sellers. Proceedings of the 1996 ACEEE Summer Study on Energy Efficiency in Buildings, American Council for an Energy-Efficient Economy, Washington, DC.

HB&SFMC, 2000. The Energy Star products program, draft interim report. Hagler Bailly & Shel Feldman Management Consulting, Madison, WI, cited in Egon and Brown 2001.

Holt, E., 1997. Disclosure and certification: truth and labeling for electric power. Issues Brief Number 5, Renewable Energy Policy Project, University of Maryland, College Park, MD.

Lamarre, L., 1997. Utility customers go for the green. EPRI Journal 22 (2), 6–15.

Lovins, A.B., 1996. Negawatts: twelve transitions, eight improvements and one distraction. Energy Policy 24 (4), 331–343.

Newsome, H., 2001. Personal communication. Federal Trade Commission, July.

Opinion Dynamics Corp., 2000. Appliance sales tracking: 1999 residential survey. Report 195–1. Energy Center of Wisconsin, Madison, WI.

Paulos, B., 1998. Green power in perspective: lessons from green mar-keting of consumer goods. The Electricity Journal 11 (1), 46–55.

Petruzzi, M., 2001. Personal communication. Director of Communication, Green Seal, July.

Pollock, E., 2001. Personal communication. US Department of Energy, August.

Rabago, K., Wiser, R., Hamrin, J., 1998. The green-e program: an opportunity for customers. The Electricity Journal 11 (1), 37–45.

Rhodes, S., Brown, L., 1997. Consumers look for the ecolabel. Forum for Applied Research and Public Policy 12 (1), 109–115.

SCS, 2001. Scientific Certification Systems homepage. http://www.scs1.com.

Shugoll Research, 2000. A focus group study to assess consumer reaction to proposed alternatives to the FTC energy guide label. Prepared for the American Council for an Energy-Efficient Economy. Shugoll Research, Bethesda, MD, cited in Egon and Brown 2001, p. 39.

Sullivan, J., 2001. Personal communication. US Environmental Protection Agency, August.

Truffer, B., Markard, J., Wustenhagen, W., 2001. Eco-labeling of electricity—strategies and tradeoffs in the definition of environmental standards. Energy Policy 29 (11), 885–897.

Ulrich Research Services, Inc., 2000. Northeast homeowner opinion survey. Orange Park, FL, cited in Egon and Brown 2001.

Weissman, A., 2001. Personal communication. President, Green Seal, July.

Weissman, R., 1999. Lessons learned from the Alliance to Save Energy's public awareness campaign. Presented at ACEEE's Workshop on Strategies for Increasing Consumer Interest in Energy Efficiency, November 30, 1999, Easton, MD.

Werling, E., Hall, J., Meisegeier, D., Rashkin, S., Collison, B., Chinery, G., 1998. Lessons learned in the Energy Star Homes program. Proceedings of the 1998 ACEEE Summer Study on Energy Efficiency in Buildings, American Council for an Energy-Efficient Economy, Washington, DC.

Winneg, K., Herrmann, M., Levy, A., Roe, B., 1998. Label testing: results of mall intercept study. The National Council on Competition and the Electric Industry. The Consumer Information Disclosure Series.

Part IV
International Dimensions of Environmental Labelling

[26]

Eco-labelling and the Trade-Environment Debate

Daniel Melser and Peter E. Robertson

University of New South Wales, Australia

1. INTRODUCTION

THE purpose of this paper is to consider the potential global environmental benefits of eco-labelling programmes, paying attention in particular, to internationally traded commodities. This is important for two reasons. First, we shall argue that in some instances the impetus behind the introduction of eco-labels is that they are seen as an alternative to more trade-restrictive environmental policies, such as import bans or tariffs on goods with harmful environmental effects. In such cases, however, it is desirable to consider the extent to which eco-labelling programmes and trade restrictions might be substitutable with respect to their impact on the environment. Second, eco-labelling requirements for traded goods may be quite costly due to the potential for abuse of labelling requirements as technical trade barriers, and the costs of compliance, such as monitoring foreign production methods. In view of these costs, there is a presumption that eco-labelling requirements for imported goods should yield significant environmental benefits.

The paper is organised as follows. In the next section we characterise some features of eco-labelling programmes and show how these contrast with the more common health and safety labels. Section 3 then considers the political economy of eco-labelling programmes – specifically the relationship between eco-labelling and trade-environment disputes. In Section 4 we discuss the incentive problems inherent in eco-labelling policies, and identify their appropriateness as a tool of environmental policy. Section 5 considers these issues in the context of the debate over GM organisms, and discusses some of the special issues that arise in that case, before concluding in Section 6.

The authors would like to thank two anonymous referees who provided helpful comments on an earlier draft of the paper. They are also grateful to Wei-Fang Lin, who provided very helpful research assistance.

2. CHARACTERISTICS OF ECO-LABELLING PROGRAMMES

Eco-labelling is a relatively recent phenomenon having grown in popularity world-wide since Germany's *Blue Angel* eco-label was introduced in 1977.[1] By 1992, eco-labelling was explicitly endorsed by the 156 countries who were signatories to *Agenda 21*, at the Earth Summit (United Nations Conference on Environment and Development, Rio de Janeiro, 1992); Appleton (1997). These eco-labelling programmes are aimed at conserving environmental resources through various channels such as promoting consumer awareness, creating markets for green goods and providing consumers with the ability to make informed choices.[2]

In some respects the practice of eco-labelling is very similar to the more widespread practice of health and safety labelling. Clearly both require producers to comply with labelling requirements that help consumers make informed choices. With respect to international trade policy, they pose similar problems in that both may require an importing country to insist that foreign exporters meet labelling requirements, thus having the potential to be misused as technical trade barriers.[3] A significant amount of attention has focused on the potential for health and safety labels to be misused as a technical barrier to trade, and this concern remains for eco-labels (OECD, 1997, 2003; and WTO, 1998).

Nevertheless there are several aspects of eco-labelling policies that make them different and more controversial than health and safety labels. First, whereas health and safety issues tend to be relatively homogeneous across countries, attitudes toward the environment may differ widely. Hence there is often scope for disagreement between countries over the validity of a specific requirement. Moreover, since demand for environmental quality tends to be income elastic, these differences in attitude are often greatest between developed and less developed countries. In this sense, the motive for eco-labelling requirements on internationally traded commodities may be viewed with more suspicion by developing-country exporters than other types of labels. This concern is heightened for mandatory eco-labelling requirements describing harmful environmental characteristics of the good, but may also hold for voluntary programmes.[4]

[1] The *Blue Angel* was followed by over 30 national eco-labelling programmes with the USA alone having over 20 eco-labelling schemes in a wide range of industries (WTO, 2000).

[2] For a broader discussion of eco-labelling see Vossenaar (1997), Childs and Whiting (1998), Dosi and Moretto (1998), Shams (1995), Mason (2001) and Mattoo and Singh (1997).

[3] The Technical Barriers to Trade (TBT) Agreement explicitly mentions labelling requirements. As stated in the preamble, the aim of the TBT Agreement is, '. . . to ensure that technical regulations and standards, including packaging, marking and labelling requirements, and procedures for assessment of conformity with technical regulations and standards do not create unnecessary obstacles to international trade'.

[4] In a voluntary programme the organising body, usually a government agency, sets a standard, and products meeting these standards are permitted to use the eco-label. The programmes are typically therefore voluntary, in the sense that producers can choose whether or not they want to meet the

Second, since eco-labels contain information regarding potential environmental impacts, they are designed to inform consumers about costs that occur at the production and disposal stages of a product's life-cycle. This contrasts with health and safety labels that are typically concerned with costs that are incurred during consumption. This difference is important from the perspective of international trade, since WTO rules exclude trade policy measures based solely on different process and production methods (PPMs). Eco-labelling information, however, will often refer to precisely these criteria, and this has implications for the political economy of eco-labelling discussed in Section 3 below.

The third factor is the difference in incentives. Health and safety labels convey information about various attributes and effects of the product on the *consumer*. Examples include the possible allergenicity of a food product or the cancer-causing effects of cigarettes. The effectiveness of health and safety labelling therefore depends on a consumer's willingness to pay a premium to protect their own well-being. Eco-labelling, however, depends on consumers' willingness to protect the environment. Thus eco-labelling depends implicitly on consumers' desires to contribute to the wellbeing of others as the environment is a shared resource. Clearly the incentives facing the consumers in each case are quite different. We will argue in Section 4 that even a perfectly monitored eco-labelling scheme cannot be relied upon to prevent excess environmental damage, due to the 'tragedy of the commons'. However, we defer further discussion of this issue and first consider the role of eco-labelling in international trade disputes.

3. BARRIERS TO TRADE, THE WTO AND ECO-LABELS

a. Trade and Environment Disputes

Recent environmental disputes brought before the WTO have highlighted the conflict faced by governments who wish to pursue domestic environmental policies and also comply with WTO rules.[5] First, as mentioned above, the WTO principles of non-discrimination require that *like* products are treated equally irrespective of the country of origin.[6] This poses a problem for the environmental

standard and use the eco-label. In some cases, however, the eco-label may be a mandatory warning about negative environmental effects.

[5] For a useful and contemporary discussion of WTO-environment disputes see Kelly (2003).

[6] This principle appears throughout the GATT Agreement; for example, consider Article III, Part 4, which reads, 'The products of the territory of any contracting party imported into the territory of any other contracting party shall be accorded treatment no less favourable than that accorded to like products of national origin in respect of all laws, regulations and requirements affecting their internal sale, offering for sale, purchase, transportation, distribution or use.'

lobby since it is often the production method that differentiates an environmentally friendly commodity, from one that is not.

The problem became apparent with the GATT 'Tuna-Dolphin' dispute, where the harvesting of tuna was associated with the incidental killing of dolphins by tuna fishing fleets using purse-seine nets. Under the terms of the Marine Mammal Protection Act, the USA placed import restrictions on countries with less stringent tuna-fishing rules than its own. This led the EU to file a claim against the USA's ban on imported tuna under the GATT.

The USA case for a ban on tuna caught with purse-seine nets, rested on GATT Article XX (b) and (g), which provide exemptions in the case of trade policy measures, '. . . necessary to protect human, animal or plant life or health' and 'relating to the conservation of exhaustible natural resources . . .'. Despite these exceptions, the GATT panel found that the restrictions were unjustified.

In the more recent 'Shrimp-Turtle' case, the USA required that local shrimp trawlers use nets fitted with trapdoors, known as turtle excluder devices (TEDs), to prevent excessive mortality of endangered sea-turtles.[7] The USA attempted to ban imports of shrimp from South and East Asia which is harvested without the use of TEDs. In this case, the USA was eventually able to have the restrictions upheld, but this decision was reached after a protracted six-year dispute where both the original WTO panel and appellate body ruled against the USA.[8]

b. Eco-labelling and Trade-environment Disputes

A legacy of these landmark disputes is a sense of uncertainty surrounding how environmental and trade laws are to be reconciled. Over the last decade or more, governments have thus sought to adopt environmental policies that might satisfy WTO guidelines. It appears that eco-labels may be suitable policy instruments. Currently the WTO Committee on Trade and Environment (CTE) has been considering the issues surrounding eco-labelling after urging from member states in recent submissions (WTO, 2002; and WTO, 2003a). Part of the Doha Ministerial Declaration, prepared by the CTE (WTO, 2003b), summarises members' opinions on this issue:

> Most Members agreed that voluntary, participatory, market-based and transparent environmental labelling schemes were potentially *efficient economic instruments* in order to inform consumers about environmentally friendly products. As such they could help move consumption on to a

[7] The TEDs cost between US$75 and US$500. It was estimated that shrimp catches fell by around two per cent with its introduction and turtle mortality by around 97 per cent; Ranné (1999).

[8] Somewhat ironically, the resolution of this dispute in favour of the US environmental movement has been followed by anti-dumping duties of around 100 per cent on Chinese and Vietnamese shrimp, and similar duties are likely to be imposed on other East Asian countries; *The Economist* (2004b).

more sustainable footing. Moreover, they tended, generally, to be less trade restrictive than other instruments. [original emphasis] (WTO, 2003b, p. 8).[9]

Thus, whereas WTO rules explicitly try to prevent mandated discrimination on the basis of production methods, eco-labels provide an opportunity for consumers to discriminate. In view of this, eco-labelling policies, and in particular voluntary programmes, potentially offer some room for compromise between the demands of the environmental lobby and WTO principles.

For example, in the Forestry industry, eco-labelling, in the form of forestry certification arose in response to numerous calls for trade bans and consumer boycotts of the international trade in tropical timber during the 1980s. Timber accounts for a large fraction of developing countries' output, and most of this is exported. Hence international trade and growth is seen as a catalyst for deforestation.[10] In response to these concerns, in 1992 Austria raised tariffs on unsustainably managed timber by 70 per cent and also introduced a compulsory eco-labelling scheme. Shortly after, however, the tariff was dropped because of concern about its GATT legality and the eco-labelling scheme became voluntary; Crossley (1996).[11]

In a similar vein, eco-labelling programmes were introduced as a direct result of the USA failure to impose trade measures on imported Mexican tuna; WTO (1998). Likewise they were also called for as a resolution to the Shrimp-Turtle dispute, and were introduced to a limited extent. These examples suggest, therefore, that eco-labelling programmes have been used as a politically expedient policy option for governments caught between pressure from national environmental lobbies and pressure from exporters and foreign governments to comply with WTO rules.

The difficulty is that there is very little discussion over what the expected benefits of eco-labelling are. The concern must then arise as to whether eco-labelling policies might be adopted in place of more mandatory, but less WTO-compliant policies, without evaluation of the environmental consequences. In the next section we discuss the consequences of using eco-labelling as an environmental policy tool.

---

[9] It is interesting to note that the CTE added, 'One Member questioned the assumption that comprehensive labelling schemes had a positive effect on achieving good environmental outcomes' (WTO, 2003b, p. 8). This is the point that we hope to develop in this paper.

[10] In 1990 Indonesia (60 per cent) and Malaysia (75 per cent) both exported a large proportion of their timber. Moreover, Indonesia and Malaysia account for 65 per cent of world tropical timber exports. The major importers are Japan and the EU, together accounting for 48 per cent of world imports in 1987. Similarly Japan, the EU and the USA account for over 70 per cent of wood furniture imports. Thus, although domestic consumption is an important factor in explaining deforestation, in particular for fuel, the role of foreign demand is also significant; Crossley et al. (1997).

[11] Crossley et al. (1997) report that there were seven national certification schemes in operation in 1994, although their coverage is very small relative to total timber trade. Swallow and Sedjo (2000) discuss the commercial impact of timber certification on the forestry industry and on developing country exports.

4. THE ENVIRONMENTAL EFFECTS OF ECO-LABELLING

a. Eco-labels as a Tool to Create Markets for Green Goods

The potential benefit of eco-labelling policies is put succinctly by *The Economist*:

> The right way to deal with people's desire to eat turtle-friendly shrimp, hormone-free beef or whatever is not to impose one country's values on another: it is to label the products appropriately. Consumers, not governments, can then choose what they eat, and trade can remain free. (*The Economist*, 7 October, 1999).

To make this argument concrete, consider a consumer's choice over two brands of shrimp, say *Brown* and *Green*. *Brown* involves the incidental killing of sea turtles during harvesting and *Green* is the eco-labelled product which is harvested with nets that have TEDs (which, for simplicity, we suppose eliminates turtle deaths). We naturally assume that the price of *Green* exceeds *Brown*, otherwise there would be no demand for *Brown*.

For the purpose of this discussion it does not matter whether the label is a mandatory requirement on *Brown*, or a voluntary label on *Green*, as long as the information on the label is credible. The choice facing a consumer is depicted in Figure 1. Here the demand curve of the consumer for *Brown* and *Green* is depicted by *dd′*, which is the same for both goods as from a consumption perspective these goods are indistinguishable. The consumer's marginal costs of consuming *Green* is simply equal to the price, p_G. The marginal costs of consuming *Brown*, however, is the sum of the price of *Brown* ($p_B < p_G$) and the private disutility the consumer derives from environmental damage, and is given by the schedule *s*. There are two possible reasons why this schedule lies above p_B: (i) the consumer may correctly perceive that their consumption decision has a small, but non-zero, effect on the environment which generates some loss of utility, and (ii) the consumer may face some psychic costs of knowingly causing environmental damage, such as guilt or peer pressure.

The horizontal axis in Figure 1 shows the quantity purchased by the consumer, and hence also the damage caused by this consumer. In choosing whether to purchase *Green* or *Brown* the consumer compares the surplus from each commodity. For *Green* this is given by the triangle dfp_G and for *Brown* it is dep_B. The difference in the two regions is given by the difference between the two shaded regions. As drawn in Figure 1, the consumer will choose *Brown*. It can be seen that the decision depends on the position of the marginal cost schedule *s*. This in turn depends on the price of *Brown*, and the consumer's dislike for killing sea turtles. Environmental preservation thus depends on enough consumers having a strong-enough preference for the particular environmental service, and on the cost of the *Green* eco-labelled substitute not being too much greater than *Brown*.

FIGURE 1
Eco-labels and Consumer Choice

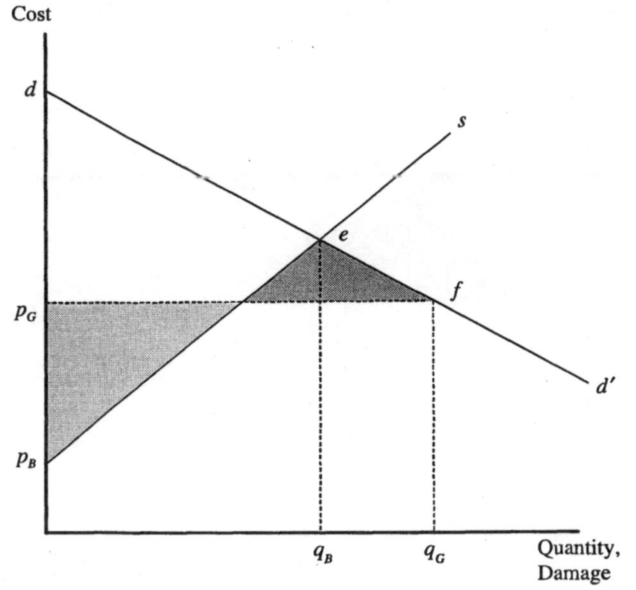

b. Eco-labels and External Environmental Costs

Despite the beneficial effect of eco-labels in creating markets for green goods, there remains a concern whether the self-interest motive provides a strong-enough impetus for consumers to act to reduce environmental costs. The preceding argument implicitly assumes that the production or consumption of the commodities in question, create no external costs. In the shrimp example, however, the main point of the USA opposition to incidental killing of sea-turtles is that they are an endangered species, with all but one species of sea turtle currently threatened with extinction; Liebig (1999). In general it has been argued that species' extinction generates significant external costs, through the loss of biodiversity and the loss of part of the 'global heritage of humanity'; Ferrantino (1997).[12]

[12] Perhaps the best example of the use of eco-labelling for a product where the information concerns potentially large external costs, is in Forestry Certification. Specifically this was designed to reduce deforestation of tropical hardwoods that are associated with two important trans-national externalities: the effect on the carbon cycle and climate change, and bio-diversity; Simula (1997). In general, however, the significance of these implied external costs is open to debate, and needs to be assessed on a case-by-case basis.

FIGURE 2
The Environmental Effects of Eco-labelling

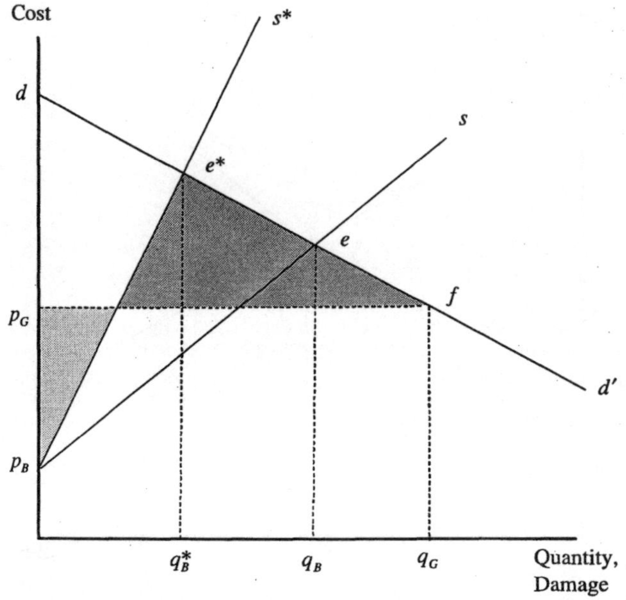

Viewed in this way it is apparent that the hormone-free-beef and turtle-friendly-shrimp problems are actually quite different. Eco-labels in the latter case provide consumers with information about the costs or benefits of production that are external to the consumer. Since, almost by definition, environmental services are public goods, this is a general feature of eco-labels.

To see this more formally, suppose that the *Brown* good has an individual effect on the environment as perceived by the consumer, and reflected in s, and also an effect on the environment which is borne by all consumers. This is illustrated in Figure 2 where the social costs of the purchase of the *Brown* good are given by s^* and lie above s.

The analysis is similar to the standard public good model, since purchasing the *Green* good is equivalent to paying a subscription to the maintenance of a public good, in this case the environment. However, because individuals only consider the benefits of such a subscription on themselves there will be too few consumers of the *Green* good and too many of *Brown*.[13] Thus it can be seen in Figure 2 that

[13] Robertson (2002) contains a formal analysis of free riding behaviour and finds that the number of countries signing onto international eco-labelling agreements may have little or no effect on the environment.

the social surplus from the consumption of *Brown* is smaller than the private surplus, shown in Figure 1. The reduction in the social surplus of *Brown* relative to *Green* implies that in the free market equilibrium there are too few consumers of *Green* and too many of *Brown*. Indeed as depicted in Figure 2 the difference between the social surplus from consuming the *Green* and *Brown* goods is now in *Green's* favour in contrast to Figure 1.

c. *Policy and Political Implications*

Despite the limitations of eco-labelling programmes, they appear to be increasing in popularity. As suggested above, this may be due to the political expedience of eco-labelling policies. In particular, when an environmental problem is international in scope, the usual policy tools for internalising external costs don't exist. A country is limited to imposing trade measures aimed at minimising environmental damage, or to international cooperation, through multilateral environmental agreements. The first avenue is likely to violate WTO rules and the second is likely to be very costly with a highly uncertain outcome. Thus the introduction of eco-labelling regimes that apply to internationally traded goods, may simply be a policy of last resort, that avoids WTO conflict and makes some concessions to the environmental lobby.

Our preceding analysis suggests that whether or not a given eco-labelling regime is an appropriate policy, or an inappropriate but perhaps politically expedient one, depends on the extent of the divergence between social and private environmental costs, s^* and s, associated with the commodity being subjected to eco-labelling requirements. For example, an eco-label on tuna, indicating whether it is dolphin friendly, may be a reasonable policy if it is accepted that dolphin mortality has no particular external environmental costs, perhaps since they are not endangered. Such a label would increase the utility of dolphin-squeamish consumers. In this case, the introduction of eco-labelling may be an appropriate response to the trade dispute. Nevertheless, since there are no significant external environmental costs, this 'eco-label' is more akin to a personal health warning. In cases where externalities exist, however, it would be desirable to attempt to quantify these social costs before determining whether or not eco-labelling is an appropriate policy. If the external costs are indeed significant, then eco-labelling at best should be part of a broader policy package.

5. ECO-LABELS AND GM FOOD

The preceding analysis also has implications for the current debate over trade in GM food products and products derived from GM crops. The production of GM crops has increased rapidly over the last decade, particularly in the USA,

Canada and Argentina.[14] This has been associated with fears about the effects of GM products, resulting in trade bans by the EU in particular. Thus, aspects of this debate are strongly reminiscent of the forestry and fishing examples we cite above. Nevertheless, there are also some important differences which we explore in this section.

a. Background: GMOs and the Trade Conflict Between the USA and the EU

In 1999 the EU introduced a moratorium on the importation of new GMOs into its territories. In a formal complaint with the WTO in May 2003, the USA, Canada and Argentina cited two objections to the EU policy. Firstly, they claim there is a lack of scientific evidence that GM foods are harmful to people or the environment, which means in turn that the moratorium cannot be justified under the Agreement on Sanitary and Phytosanitary Measures (SPS) or Article XX. Secondly, the moratorium violates the PPM rules. According to the USA, GM foods and non-GM foods are like products, or 'substantially equivalent', and the EU moratorium discriminates according to PPMs.[15]

As in our preceding examples, the EU responded to this political pressure by introducing a requirement that all products which contain an ingredient which includes more than 0.9 per cent GMOs, be labelled.[16] With this extended labelling policy in place since late 2003, the moratorium was then allowed to expire in May 2004 (Greenpeace, 2003; and *The Economist*, 2004a). Thus in terms of the political pressure for labelling, the GM case is very similar to the shrimp, tuna and forestry examples.

b. Aspects of the GM Debate

There are, however, several peculiarities with the current GM debate. First, the labelling programme is a mandatory policy – all producers of GM foods must comply with the label. This is contentious since it means the policy places the burden of compliance on foreign producers. Arguably, therefore, it has greater potential to be used as a technical trade barrier than, say, a voluntary 'GM Free' label would. Nevertheless, from the perspective of the GM-producing countries, it is more desirable than the moratorium. In this sense the policy can therefore

[14] In 1995 there were virtually no GM crops. In 2002, there were 59 million hectares of transgenic crops grown globally. Most of this was grown in only four countries: USA (66 per cent), Argentina (23 per cent), Canada (6 per cent) and China (4 per cent) and is dominated by just three crops, soybean (62 per cent), maize (21 per cent) and cotton (12 per cent); James (2002).

[15] For further discussion of the issues see Sheldon (2002).

[16] In fact labelling of GM products is widespread with 26 countries either having labelling regimes or planning to introduce them; Phillips and McNeil (2000).

still be viewed as an attempted political compromise between the interests of GM producers and the environmental lobby.

The second aspect of this debate concerns the factors that are driving consumer demands for labelling. Though the consumers' 'right to know' might be cited as the direct reason, the question remains as to what extent the demand for information is being driven by fears for personal safety, and to what extent they are driven by concerns for the environment.[17] The extensive consultation undertaken by the UK's GM Science Review Panel (2003) suggests that the principle demand for labelling of GM products is due to a concern for personal health. Thus, though these concerns were strongly rejected by the Panel, it appears that, for many consumers, the labelling of GM products is akin to a health and safety warning.[18]

Nevertheless, there is also considerable discussion of the possible negative effects on the environment of GM crops. These include (i) the potential dispersal of the genetically modified organisms in the environment – for example through invasiveness; (ii) the potential transfer of the inserted genetic material to other organisms – for example through cross-pollination; (iii) potential impacts on non-target species; and (iv) potential impacts on soil bacteria and the nitrogen cycle (European Environment Agency, 2002; and Mackenzie et al., 2003). Similar concerns about the irreversible effects of cross-contamination have been voiced by groups such as Greenpeace, Friends of the Earth and GeneWatch UK (Greenpeace, 2003; Friends of the Earth, 2004; and GeneWatch UK, 2001).[19] Though the GM Science Review Panel downplays such fears, it also warns that environmental risk factors of GM crops may prove to be more challenging in future, as more GM crops are developed; GM Science Review Panel (2003).

In conclusion, therefore, though personal safety issues, such as exposure to unknown allergens, remain perhaps the principal source of consumer concern over GM foods, the environmental factors cannot be ignored. This means that labels on GM foods serve two purposes – a potential health and safety label as well as an eco-label.

---

[17] One striking example of the depth of consumer concern can be seen in Europe where a Eurobarometer (2001) survey of 16,029 people in the 15 member states reported that 94.6 per cent of people believed they should have the right to choose with regard to GM foods – implying that these foods should be labelled.

[18] The GM Science Review Panel (2003, p. 22) noted that, '. . . for human health to date there is no evidence currently commercialised GM crop varieties or foods made from them, are toxic, allergenic or nutritionally deleterious'.

[19] According to Greenpeace (2003), 'GM crops pose irreversible long-term risks to the environment. It is possible for GM crops to transfer modified genes to closely related plant species . . . This is particularly worrying because GM oilseed rape and sugar beet have wild relatives in the UK – any contamination of these wild species would be irreversible.'

c. An Appropriate Policy Regime for GM Products

Though the moratorium on GM foods in the EU has been lifted, decisions on individual GM crops are now made on a case-by-case basis by the European Commission; *The Economist* (2004a). Thus the EU still remains a long way from 'free trade' in GM products. The analysis above suggests that labelling requirements on GM products are an appropriate response for the health and safety concerns of consumers, but not for the environmental concerns expressed by lobby groups. In principle, therefore, an informed discussion on these issues should consider the environmental and personal health concerns separately, and form policies based on the perceived attributes of each product. If the concerns over a GM product extend only as far as personal health issues associated with its consumption or disposal, then labelling policies may be all that is required. If concerns extend to local or global environmental costs, then alternative regulations need to be considered.

6. CONCLUSION

Understanding the potential benefits of eco-labelling policies is particularly important for internationally traded goods, where the potential exists for the misuse of labelling as a technical trade barrier. Moreover, in several high-profile cases, eco-labelling was adopted as a substitute for more trade-restrictive measures. Thus we have argued that eco-labelling policies may, in some cases, be simply a politically expedient way of avoiding the difficulties that have arisen between trade and environmental policy objectives. This is a concern, however, since we have also argued that there are grounds to be pessimistic about the ability of eco-labels to solve environmental problems.

Specifically we showed that, while eco-labels can benefit consumers, the market for the labelled good will be characterised by damage levels that exceed the social optimum, because eco-labels only provide information about costs and benefits that are external to each consumer. In some cases, such as the GM food debate, labelling information about environmental costs may be only part of the objective of a labelling programme. Nevertheless, if there are external costs associated with the decision to consume the product, even a perfectly monitored labelling scheme will not force consumers to internalise these costs. Hence eco-labelling should only be seen as a part of a range of policies for environmental management.

REFERENCES

Appleton, A. E. (ed.) (1997), *Environmental Labelling Programmes: International Trade Law Implications. International Environmental Law and Policy Series* (Boston, MA: Kluwer Law International).

Childs, C. and S. Whiting (1998), 'Eco-labelling and the Green Consumer', Department of Environmental Science (University of Bradford).

Crossley, R. (1996), 'A Review of Global Management Certification Initiatives: Political and Institution Aspects', Paper for the Conference on Economic, Social and Political Issues in Certification of Forest Management (Malaysia).

Crossley, R., C. P. Braga and P. Varangis (1997), 'Is there a Commercial Case for Tropical Timber Certification', in R. Vossenaar, S. Zarrilli and V. Jha (eds.), *Eco-labelling and International Trade* (Basingstoke: Macmillan Press).

Dosi, C. and M. Moretto (1998), 'Is Eco-labelling a Reliable Environmental Policy Measure?' (University of Padova, mimeo).

European Environment Agency (2002). *Genetically Modified Organisms: The Significance of Gene Flow Through Pollen Transfer*, Environmental Issue Report, 28.

Eurobarometer (2001), 'Europeans, Science and Technology: Main Results of Eurobarometer', *Eurobarometer*, **55**, 2 (December).

Ferrantino, M. J. (1997), 'International Trade, Environmental Quality and Public Policy', *The World Economy*, **20**, 1, 43–72.

Friends of the Earth (2004), 'Genetically Modified Crops: A Decade of Failure (1994–2004)', (Malaysia: FOE International).

GeneWatch UK (2001), 'Comments on European Commission's Proposed Regulations on GM Good and Feed and the Traceability and Labelling of GMOs' (available at http://www.genewatch.org/cropsandfood/labelling/fsa_evidence.rtf).

GM Science Review Panel (2003), *GM Science Review: First Report*.

Greenpeace (2003), 'The European Union's New Labelling Rules for Genetically Engineered Food and Feed: Implications for the Market of GMO and Non-GMO Products' (Greenpeace European Unit).

James, C. (2002), 'Preview, Global Status of Commercialized Transgenic Crops: 2002', *ISAAA Briefs*.

Kelly, T. (2003), 'The WTO, the Environment and Health and Safety Standards', *The World Economy*, **26**, 2, 131–51.

Liebig, K. (1999), 'The WTO and the Trade-Environment Conflict', *Intereconomics: Review of International Trade and Development*, **34**, 2, 83–90.

Mackenzie, R., F. Burhenne-Guilmin, A. G. M. La Viña and J. D. Werksman (2003), 'An Explanatory Guide to the Cartagena Protocol on Biosafety', *IUCN Environmental Policy and Law Paper No. 46* (Gland, Switzerland, and Cambridge: IUCN).

Mason, C. (2001). 'On the Economics of Eco-labelling', Paper presented at the EAERE Conference (Southampton).

Mattoo, A. and H. V. Singh (1997), 'Eco-labelling, the Environment and International Trade', in R. Vossenaar, S. Zarrilli and V. Jha (eds.), *Eco-labelling and International Trade* (Basingstoke: Macmillan Press).

Organisation for Economic Co-operation and Development (OECD) (1997), 'Eco-labelling: Actual Effects of Selected Programmes' (OCDE/GD(97)105, 30 May).

Organisation for Economic Co-operation and Development (OECD) (2003), 'Analysis of Non-tariff Measures, The Case of Labelling: Overview and Analysis of WTO Data' (TD/TC/WP(203) 40, 13 November).

Phillips, P. W. B. and H. McNeill (2000), 'A Survey of National Labelling Policies for GM Foods', *AgBioForum*, **3**, 4.

Ranné, O. (1999), 'More Leeway for Unilateral Trade Measures? The Report of the Appellate Body in the Shrimp-Turtle Case', *Intereconomics: Review of International Trade and Development*, **34**, 2, 72–83.

Robertson, P. E. (2002), 'Eco-labels and Free-riding' (School of Economics, University of New South Wales, mimeo).

Shams, R. (1995), 'Eco-labelling and Environmental Policy Efforts in Developing Countries', *Intereconomics: Review of International Trade and Development*, **30**, 1, 143–49.

Sheldon, I. M. (2002), 'Regulation of Biotechnology: Will We Ever Freely Trade GMOs?', *European Review of Agricultural Economics*, **29**, 1, 155–76.

Simula, M. (1997), 'Timber Certification Initiatives and their Implications for Developing Countries', in R. Vossenaar, S. Zarrilli and V. Jha (eds.), *Eco-labelling and International Trade* (Basingstoke: Macmillan Press).

Swallow, S. K. and R. A. Sedjo (2000), 'Eco-labelling Consequences in General Equilibrium: A Graphical Assessment', *Land Economics*, **76**, 1, 28–36.

The Economist (1999), 'Why Greens Should Love Trade' (7 October).

The Economist (2004a), 'Another Gene Genie out of the Bottle' (19 May).

The Economist (2004b), 'Shrimp Wars' (10 July).

Vossenaar, R. (1997), Eco-labelling and International Trade: The Main Issues', in R. Vossenaar, S. Zarrilli and V. Jha (eds.), *Eco-Labelling and International Trade* (Basingstoke: Macmillan Press).

World Trade Organisation (WTO) (1998), *Trade and Environment News Bulletin* (TE/023, 14 May).

World Trade Organisation (WTO) (2000), 'Information Relevant to the Consideration of the Market Access Effects of Eco-labelling Schemes', Committee on Trade and the Environment (WT/CTE/W/150).

World Trade Organisation (WTO) (2002), 'Labelling for Environment Purposes (Submission by Switzerland)', Committee on Trade and the Environment (WT/CTE/W/219).

World Trade Organisation (WTO) (2003a), 'Labelling and Requirements of the Agreement on Technical Barriers to Trade (TBT): Framework for Informal, Structured Discussions (Communication from Canada)', Committee on Trade and the Environment (WT/CTE/W/229).

World Trade Organisation (WTO) (2003b), 'Report to the 5th Session of the WTO Ministerial Conference in Cancún', Committee on Trade and the Environment (WT/CTE/8).

[27]

Eco-Labeling and Stages of Development

*Arnab K. Basu, Nancy H. Chau, and Ulrike Grote**

Abstract

The paper examines the effectiveness of eco-labeling in providing a market-based solution to the under-consumption of eco-friendly products in developing and developed countries. The authors show that whether labeling is an effective device in solving the problem of asymmetric information between sellers and buyers, or whether false labeling severs the link between willingness to pay and environmental conscious production choices, depends crucially on how monitoring intensities respond endogenously to economic growth, openness to trade, and technology transfers. In particular, by accounting for endogenous policy responses to economic growth, it is shown that an inverted-U relationship exists between consumer spending on eco-unfriendly products and national income. In addition, while international trade unambiguously benefits the environment in the presence of eco-labeling with perfect enforcement, trade openness may nevertheless delay the turning point of the growth and environment relationship, when the cost of enforcement falls disproportionately on developing countries, and when environmental policies are employed to reap terms-of-trade gains.

1. Introduction

"Eco-labeling," defined as a practice of providing information to consumers about a product which is characterized by improved environmental performance and efficiency compared with similar products, has gained increasing popularity in recent years. Germany's "Blue Angel" label, launched in 1978 to denote "environmentally friendly" products, was a pioneering eco-labeling effort. Since then, products with labels promoting attributes like "recyclable," "degradable" or "ozone-friendly" have flooded the market. More recently, products with eco-labels promoting environmentally friendly process and production methods (PPMs) like "dolphin-safe tuna," furniture made from wood harvested from sustainably managed forests, and fish from sustainable fisheries have also entered the market. Consequently, as Table 1 shows, there has been a notable increase in developing country participation in initiating voluntary eco-labeling programs since 1991. In the meantime, consumer awareness regarding eco-labeled products has also grown substantially. Results of surveys and opinion polls conducted in Japan, Singapore, Canada, and Norway, for example, show that the awareness of national eco-labels had increased by more than 50% only a few years after their introduction; while surveys in developing countries like India, South Korea, Chile, and Mexico have found that consumers would be willing to accept higher prices for environmentally friendlier products (Shams, 1995).

---

*Basu: Department of Economics, College of William & Mary, Williamsburg, VA 23187, USA. E-mail: akbasu@wm.edu. Chau: Department of Applied Economics and Management, Cornell University, Ithaca, NY 14853, USA. E-mail: hyc3@cornell.edu. Grote: Center for Development Research (ZEF), University of Bonn, Walter-Flex-Str. 3, D-53113, Bonn, Germany. E-mail: u.grote@uni-bonn.de. We thank Kym Anderson, Joachim von Braun, Ramon López, Bodo Steiner, Susanna Wolf, seminar participants at the XXIV–IAAE meetings at Berlin and at ZEF for their suggestions and Melina Raffin for able research assistance. We are particularly grateful to an anonymous referee for comments on an earlier version. The usual disclaimer applies.

Table 1. Voluntary Eco-labeling of Retail Products in Selected Countries

	Seal	*Year founded*	*No. of product categories*
Austria	Austrian Eco-Label	1991	35
Brazil	ABNT	1993	2
Canada	Environmental Choice	1988	49
China	Ecomark	1994	12
Croatia	Croatia's Environmental Label	1993	33
Czech Rep.	—	1994	17
Nordic countries	Nordic Swan	1989	42
EU	European Union Ecolabel Award Scheme	1992	11
France	NF-Environment	1992	6
Germany	Blue Angel	1977	88
Germany	Green Dot	1990	7
India	Ecomark	1991	16
Indonesia	BAPEDAL	1995	n.a.
Japan	Ecomark	1989	69
Korea	Ecomark	1992	36
Malaysia	Product Certification Program	1996	1
Netherlands	Stichting Mileukeur	1992	32
New Zealand	Environmental Choice	1990	17
Singapore	Green Label	1992	21
Thailand	Thai Green Label	1993	6
US	Energy Star	1992	26

Source: WTO (2000).

The popularity of eco-labeled products is primarily based on the market-driven approach to achieving environmental goals. The increased concern of consumers for the environment and food safety, reflected by their willingness to pay relatively higher prices for products that has been produced in an environmentally friendly manner, provides a positive incentive for producers to choose techniques that minimize the adverse effects on the environment and improve the quality of final products. However, this popularity is not without controversy. Many developing countries are afraid of losing market access for their products that often do not meet existing PPMs in developed countries since information about production methods, contained in labels, can shift consumers' demand away from a country's product that has not qualified for the eco-label. In view of the resulting adverse terms-of-trade consequences for developing countries, eco-labeling has become a much debated topic within the context of the World Trade Organization (Nordström and Vaughan (1999), United Nations (1997), WTO (2000)).

The growing popularity of national efforts to promote eco-labeling notwithstanding, important questions remain as to how voluntary labeling programs can be credibly designed. While the primary purpose of labeling is to assist consumption decisions by providing product information that consumers value as a private good, market responses to eco-labeling, along with the question of whether deliberate policy action would be required to guide both producers and consumers to make efficient decisions regarding labeling, remains largely unanswered. The theoretical literature on eco-labeling is sparse and no clear policy conclusions can be drawn from those that exist. Notable are studies by Srinivasan (1995) and Brown et al. (1996). Both these studies consider cases of domestic market failure in developing countries that lead to ecologically unfriendly production, and conclude that methods like tax/subsidy schemes imposed by governments can help mitigate the problem. Srinivasan's analysis is particularly relevant since it allows for moral considerations in the consumers' utility functions that allows them to express their environmental concerns through a greater willingness to pay for goods that are eco-friendly. Mattoo and Singh (1997), on the other hand, consider the adverse effects of labeling. Since labeling transforms previously homogenous products into differentiated ones (eco-friendly and unfriendly), demand for both eco-friendly and unfriendly products can rise thereby leading to the possibility that the increase in production of eco-unfriendly goods outweighs the corresponding increase in eco-friendly products in the developing countries.

The obvious difficulty associated with providing credible monitoring of eco-labeling programs raises the additional issue of false-labeling induced by the very green premium that is presumed to entice producers to adopt eco-friendly production processes. In this respect, Bureau et al. (1999) and Marette et al. (1999) adopt the notion of *credence goods* in the sense of Nelson (1970) and Akerlof (1970), and consider cases where adverse selection arising from imperfect information necessitates quality signaling. Bureau et al. show that trade liberalization that allows for potential entry of lower-quality goods induces high-quality producers to test and signal their true quality, thereby leading to an improvement in consumer welfare.[1] However, since these studies take the distribution of firms within each quality type (high and low) as exogenous, the question of costly monitoring and the possibility of false labeling in the scenario where firms endogenously choose quality levels remains to be addressed. In addition, the proliferation of eco-labeling schemes in agriculture, forestry, and other resource-based sectors also implies that oligopolistic market conduct may not apply.

In this context, in sections 2 and 3 we present a basic North–South model where consumers in the developed country (North) value the eco-friendless of imports from the developing country (South). In the absence of eco-labeling, consumers are unable to decipher products that are made via eco-unfriendly methods from eco-friendly ones. The basic setup thus allows for an understanding of how the green premium endogenously emerges depending on the consumers' willingness to pay, the developing economy's openness to trade, the stage of economic development of the country in question, and the degree of monitoring that producers who adopt eco-labeling are subject to. We provide formal arguments which show, in contrast to a situation with asymmetric consumer information in the absence of labeling, that eco-labeling with perfect enforcement guides both consumers and producers to the efficiency frontier of the developed and developing countries. Interestingly, the endogenous price premium thus generated implies that eco-friendly products are normal while eco-unfriendly products are inferior goods in general equilibrium. As a result, income gains achieved via international trade can be beneficial only to the environment as consumers shift demand away from the eco-unfriendly product consequent upon an increase in income.

In sections 4 and 5, these results are contrasted against a more realistic scenario with costly enforcement, to be borne by the developing country. We show that unless monitoring efforts keep pace with either indigenous economic growth or income gains via international trade, eco-unfriendly products become normal goods in general equilibrium precisely as false labeling rises and falls with consumer income. In this context, we find that the incentives on the part of developing economies to credibly enforce national eco-labeling schemes have a threshold property—costly monitoring strictly benefits a developing economy only when national income exceeds a critical threshold. This finding, regarding the role of the critical income threshold in determining the effectiveness of eco-labels in reducing eco-unfriendly products, explains why eco-labeling programs have not gained momentum in the majority of developing countries and provides a theoretical foundation for the empirical findings of Grossman and Krueger (1995) who show that economic growth brings an initial phase of environmental deterioration followed by a subsequent phase of improvement. Grossman and Krueger estimate that this turning point comes, in most cases, before a country reaches a per-capita income of US $8000.

It is of interest to note that our results regarding the effectiveness of eco-labeling in promoting the supply of eco-friendly products under imperfect monitoring can be contrasted with the findings of Copeland and Taylor (1994) regarding the impact of income growth and trade on the incidence of pollution, and hence on the environment. In particular, by decomposing the effects of income growth into scale, technical, and composition effects, Copeland and Taylor show that economic growth in the developed countries (North) increases the technological gap between the North and the developing countries (South), and has the effect of increasing the incidence of world pollution as dirty industries get concentrated in the South. Income growth in the South, on the other hand, reduces the technological gap between the trading partners and reduces world pollution—a finding that is confirmed by Antweiler et al. (2001).

In contrast, we show that eco-labeling allows for the endogenous emergence of a green premium that increases in response to income growth in either the North or the South. However, this increase in the green premium may not be sufficient to ensure that producers choose eco-friendly production techniques when monitoring of labels is imperfect. We show that below a critical Southern threshold income, Southern welfare is decreasing in the monitoring intensity on products labeled eco-friendly originating from the South. Further, the monitoring intensity is endogenous rather than determined through deliberate government action and depends on the economy's openness to trade, the cost-effectiveness of monitoring technologies, and the incentives for the developing country to suppress environmental objectives in order to achieve terms of trade gains. Consequently, in this earlier phase of economic growth, the South will not devote any resources to monitoring. Thus income growth, either indigenous or trade-induced, has the effect of increasing the supply of eco-unfriendly products in consonance with rising consumer expenditure.

On the other hand, Southern welfare improves with the monitoring intensity on eco-labeled products as Southern income exceeds the critical threshold. In this latter phase, if the fall in supply of eco-unfriendly products in response to increased monitoring outweighs the increased consumer expenditure on eco-unfriendly products, eco-labeling serves the objective of increasing the supply of eco-friendly products in general equilibrium—a result akin to Copeland and Taylor's findings on the environmental impact of growth in Southern income. These observations suggest that wholesale promotion of eco-labeling as a device to improve environmental performance should be exercised with caution.

2. A Model of Eco-Labeling

Consumers

The developing country (South) has a large number of consumers, N. The representative consumer is characterized by his preferences, and his disposable income I derived either from the ownership of productive input resources, or from direct production activities, net of any tax on income T.

Each consumer allocates his disposable income (I) between a *numéraire* commodity, y, which has no adverse environmental consequences, and a good x made in the developing country. The demand for x, $d_i^x (i = f, u)$, suitably subscripted, denotes the quantity of eco-friendly and eco-unfriendly products consumed by the representative consumer. More specifically, let the utility function of a representative consumer be

$$U(d_f^x, d_u^x, d^y) = \alpha \log D(d_f^x, d_u^x) + (1-\alpha)\log d^y.$$

$\alpha > 0$ denotes the weight allotted to the consumption of good x. In addition, $D(d_f^x, d_u^x)$ is a utility index which reflects consumers' taste for eco-friendly production methods,[2] with $D(d_f^x, d_u^x) = a_f d_f^x + a_u d_u^x$. In particular, $a_i > 0$ is a preference parameter that captures the private valuation of *one* unit of good d_i^x. Assume, in addition, that $a_f > a_u$ for all consumers. The interpretation of a_i depends on the particular commodity category in which eco-labeling is applied. For instance, in the case dolphin-safe tuna, the ratio $a \equiv a_f/a_u > 1$ (marginal rate of substitution) reflects consumers' willingness to exchange more than one unit of tuna produced via unsafe means for one unit of dolphin-safe tuna. Also let p_l (p_n) be the market price of a labeled (unlabeled) product.[3]

We use subscripts f and u to denote eco-friendly and eco-unfriendly products while the subscripts l and u refers to labeled and unlabeled products. In the present context with perfect enforcement, labeled products are synonymous with eco-friendly products while unlabeled products are synonymous with eco-unfriendly ones. By standard arguments (Dixit and Stiglitz, 1977), the utility-maximizing solution of the consumer's decision problem is given by the expenditure shares α and $1 - \alpha$, which consumers devote to the consumption of goods x and y, along with the two demand functions $d_i^x(p_l, p_n)$ $i = f, u$:

$$d_f^x(p_l, p_n) = \begin{cases} \dfrac{\alpha I}{p_l}, & \text{if } \dfrac{a_f}{p_l} > \dfrac{a_u}{p_n}; \\ 0 & \text{if } \dfrac{a_f}{p_l} < \dfrac{a_u}{p_n}, \end{cases} \tag{1}$$

$$d_u^x(p_l, p_n) = \begin{cases} \dfrac{\alpha I}{p_n}, & \text{if } \dfrac{a_f}{p_l} < \dfrac{a_u}{p_n}; \\ 0 & \text{if } \dfrac{a_f}{p_l} > \dfrac{a_u}{p_n}. \end{cases} \tag{2}$$

Thus, demand is positive for both labeled and unlabeled products if and only if

$$\frac{p_l}{p_n} = \frac{a_f}{a_u} > 1. \tag{3}$$

It follows, therefore, that the introduction of eco-labeling gives rise to consumption demand for both eco-friendly and eco-unfriendly products that are nondecreasing in income (equations (1) and (2)).[4] In addition, demand curves are negatively sloped as long as $d_i^x > 0$.

Northern consumers are assumed to have identical preferences as consumers in the South and hence the utility-maximization problem of a representative Northern consumer is identical to that of their Southern counterparts.

Production

The two sectors y and x constitute the production side of the developing country. Production of the composite *numéraire* y uses as input, a composite resource L_y, with $y = \beta L_y$, where $\beta > 0$ denotes both the average and the marginal productivity of the composite resource. The economy-wide return to the composite input resource is thus given by β.

A large number of producers, M, are engaged in the production of good x in the developing economy. Each individual producer is endowed with an equal amount of capital $\overline{K} > 0$ which can be invested in either the eco-friendly method of production of x or in the eco-unfriendly method. Production of either the eco-friendly or unfriendly product thus requires the use of two inputs: capital and the composite resource. To focus on the choice of production technique by producers, the production function of the eco-friendly product is given simply by

$$x_f = \min\left\{\overline{K}, \frac{\ell_f}{(1+b)}\right\},$$

where ℓ_f is the amount of resource input employed in the production of the eco-friendly product. $b > 0$ denotes a firm-specific technological parameter that captures the cost-effective eco-friendly production to which the producer has access. The production of the eco-unfriendly (unlabeled) product is

$$x_u = \min\{\overline{K}, \ell_u\}.$$

With $b > 0$, the specification of the production functions above requires that eco-friendly production entails higher resource costs. In addition, the technological parameter $b \in [\underline{b}, \infty]$ should be interpreted as a firm-specific production cost parameter: since unit cost, $\beta(1 + b)$, is strictly increasing in b, a higher b indicates a higher per-unit cost of eco-friendly production. The relative cost effectiveness of eco-friendly production among the M producers in the developing country is characterized by a cumulative distribution function $\mu(b, \gamma)$. $\mu(\cdot)$ is taken to be continuously differentiable over the relevant range with a strictly positive density function $\mu_b > 0$. In addition, a decrease in γ represents first-order stochastic dominance—a rightward shift of the cost distribution.

The decision problem of producers involves the choice between the two techniques of production to maximize profits at given prices, p_l and p_u, and at a given per-unit cost of applying an eco-label, k. Specifically, eco-friendly production is profit-maximizing if and only if

$$p_l\overline{K} - \beta(1+b)\overline{K} - k\overline{K} \geq p_u\overline{K} - \beta\overline{K} \Leftrightarrow b \leq \frac{p_l - p_u - k}{\beta}.$$

Thus, individual firms in the developing country self-select into two groups, according to their choice of production technique that is either eco-friendly or unfriendly. In particular, as long as eco-friendly and eco-unfriendly products coexist, we can define

$$\tilde{b}(p_l, p_u, \beta) \equiv \max\left\{\frac{p_l - p_u - k}{\beta}, \underline{b}\right\}. \tag{4}$$

It follows that if $b < \tilde{b}(p_l, p_u, \beta)$, or equivalently, if the price premium of labeled products is sufficiently higher than the associated costs, producers should optimally elect eco-friendly methods of production.[5]

The supply of the eco-friendly product is therefore $x_l = \mu(\tilde{b}, \gamma)M\overline{K}$ while that of the unlabeled product is $x_u = [1 - \mu(\tilde{b}, \gamma)]M\overline{K}$. Since μ is increasing in p_l and decreasing in p_n and β, (i) both eco-friendly and eco-unfriendly products have upwardly sloping (own-price) supply curves; (ii) an increase in wage cost β biases producer incentives away from eco-friendly methods, and (iii) a decrease in γ which shifts the distribution of production costs $\mu(b, \gamma)$ to the right has a similar adverse impact on the supply of eco-friendly products.

Also let $\pi(p_l, p_n, \gamma)$ denote aggregate producer profits in the production of x, where

$$\pi(p_l, p_n, \gamma) = p_l x_l + p_n x_u - \left[\int_{\underline{b}}^{\tilde{b}} (\beta(1+b)+k)d\mu(b,\gamma) + \int_{\tilde{b}}^{\infty} \beta d\mu(b,\gamma)\right]M\overline{K}.$$

By utilizing the definition of green premium, $p_l - p_n = p_n(a - 1)$, we have

$$\pi(p_l, p_n, \gamma) = \left[p_n(1+\mu(\tilde{b},\gamma)(a-1)) - \int_{\underline{b}}^{\tilde{b}} (\beta(1+b)+k)d\mu(b,\gamma) - \int_{\tilde{b}}^{\infty} \beta d\mu(b,\gamma)\right]M\overline{K}. \quad (5)$$

Thus, taking prices p_l and p_n as given, we have

$$\begin{aligned}\frac{\partial \pi(p_l, p_n, \gamma)}{\partial \tilde{b}} &= (p_n(a-1) - \beta(1+\tilde{b}) - k + \beta)\mu_b(\tilde{b},\gamma)M\overline{K} \\ &= (p_l - p_n - k - \beta\tilde{b})\mu_b(\tilde{b},\gamma)M\overline{K} = 0,\end{aligned}$$

where the last equality follows from (3). It follows, therefore, that since eco-labeling implies that relatively low-cost producers are favorably selected in the pool of eco-friendly producers, aggregate producer profits are maximized when the marginal firm has a cost parameter b that is given exactly by $\tilde{b}$. Note that the second-order condition is also satisfied, since $(\partial^2\pi/\partial\tilde{b}) = -\beta\mu_b < 0$.

To complete the characterization of the production side of the two-country world, we assume that the developed economy is specialized in the production of the composite commodity.[6] Let an asterisk (*) denote variables pertaining to the developed country and since $L_y^* = N^*e^*$, total Northern output of y is given as $y^* = \beta^* N^* e^*$.

General Equilibrium

We are now in a position to characterize the determination of the equilibrium prices for labeled and unlabeled products. Since the government recovers $k\mu(\tilde{b}, \gamma)M\overline{K}$ from the license fees collected from producers, government budget balance requires that lump-sum transfers to consumers ($-NT$) be equal to total receipts from license fees:

$$k\mu(\tilde{b},\gamma)M\overline{K} + NT = 0.$$

Thus, total disposable consumer expenditure (including income from resource ownership, producer profits, and lump-sum transfers) on labeled and unlabeled products in the two countries is equal to total revenue of producers of good x if

$$\alpha(\beta eN + \pi(p_l, p_n, \gamma) + k\mu(\tilde{b},\gamma)M\overline{K}) + \beta^* e^* N^* = p_n(1+\mu(\tilde{b},\gamma)(\alpha-1))M\overline{K}. \quad (6)$$

Note that by standard arguments, the market for the composite commodity clears, and can thus be ignored.

3. Market Responses to Eco-labeling under Perfect Monitoring

With the simple analytical setup discussed above,[7] we arrive at a number of conclusions regarding the role of eco-labeling in preventing environmental abuse.

Eco-labeling as a Market-Based Solution

To see that eco-labeling constitutes a market-based solution to the problem of under-consumption of eco-friendly products, note that if labeled and unlabeled products coexist, then from equation (3): (i) the price of labeled products strictly exceeds that of unlabeled products, and (ii) the relative price, p_l/p_n, reflects precisely consumers' relative willingness to pay for (the marginal rate of substitution between) eco-friendly and eco-unfriendly products. In particular, the marginal rate of substitution is defined as

$$\frac{\partial U/\partial d_f^x}{\partial U/\partial d_u^x} = \frac{a_f D(d_f^x, d_u^x)}{a_u D(d_f^x, d_u^x)} = \frac{a_f}{a_u} = \frac{p_l}{p_n},$$

where the last equality follows from equation (3). Thus, consumer's relative willingness to pay for eco-friendly products is equal to the relative market prices of the labeled and unlabeled products.

In contrast, absent labeling, consumers have no means of distinguishing between eco-friendly and eco-unfriendly products. As such, regardless of the production method undertaken by producers, a uniform price p prevails for products produced via either means. It follows, therefore, that the willingness of consumers to pay for eco-friendly products (the marginal rate of substitution, $a_f/a_u > 1$) is strictly greater than the prevailing market price of eco-friendly products relative to eco-unfriendly ones.

Indeed since $p_l - p_n = 0$ in the absence of labeling

$$\tilde{b} = \max\left\{\frac{p_l - p_n - k}{\beta}, \underline{b}\right\} = \underline{b}.$$

As should be expected, producers in this scenario fail to elect eco-friendly methods of production in the absence of a green premium since they have no means of internalizing the environmental consequences of their choice of production methods. To make the case for eco-labeling even stronger, it can be readily shown that self-selection among producers in response to eco-labeling guides production and consumption decisions in the two countries to their efficiency frontier. To see this, consider the problem of a planner whose objective is to maximize $U(d_f^x, d_u^x, d^y)$ by choice of consumption allocation between the two countries, $\{\hat{d}_i^x, \hat{d}_i^{x*}\}, i = f, u$, and $\{\hat{d}^y, \hat{d}^{y*}\}$ and the allocation of producers in the developing country between eco-friendly and eco-unfriendly production processes, $\hat{b}$; subject to the following constraints:

$$d_f^x + d_f^{x*} \le \mu(\tilde{b},\gamma)M\overline{K}, \tag{7}$$

$$d_u^x + d_u^{x*} \le (1-\mu(\tilde{b},\gamma))M\overline{K}, \tag{8}$$

$$d^y + d^{y*} \le y + y^* = \beta e N + \beta^* e^* N^* - \left(\int_{\underline{b}}^{\tilde{b}} (\beta(1+b)+k)d\mu(b,\gamma) + \int_{\tilde{b}}^{\infty} \beta d\mu(b,\gamma)\right)M\overline{K}, \tag{9}$$

$$U(d_f^{x*}, d_u^{x*}, d^{y*}) \ge \overline{U}^*. \tag{10}$$

Maximizing $U(d_f^x, d_u^x, d^y)$ subject to expressions (7)–(10) yields the familiar conditions on consumption allocations:

$$\frac{\partial U/\partial d_f^x}{\partial U/\partial d_u^x} = \frac{\partial U^*/\partial d_f^{x*}}{\partial U^*/\partial d_u^{x*}}, \quad \frac{\partial U/\partial d_f^x}{\partial U/\partial d^y} = \frac{\partial U^*/\partial d_f^{x*}}{\partial U^*/\partial d^{y*}},$$

which may be achieved, under decentralization, via free trade between the developed and developing countries. In addition, ruling out cases in which a is sufficiently small, so that $\tilde{b}$ is an interior solution, efficient allocation of producers in the developing country, $(\hat{b})$, requires that

$$\frac{1}{\beta}\left[\left(\frac{\alpha}{1-\alpha}\right)\frac{\hat{d}^y}{D(\hat{d}_f^x, \hat{d}_u^x)}(a_f - a_u) - k\right] = \frac{1}{\beta}\left[\left(\frac{\alpha}{1-\alpha}\right)\frac{\hat{d}^{y*}}{D(\hat{d}_f^{x*}, \hat{d}_u^{x*})}(a_f - a_u) - k\right] = \hat{b}.$$

To confirm that $\hat{b}$ indeed coincides with $\tilde{b}$ in equation (4), note that

$$\left(\frac{\alpha}{1-\alpha}\right)\frac{d^y}{D(d_f^x, d_u^x)}(a_f - a_u) = \left(\frac{\alpha}{1-\alpha}\right)\left(\frac{(1-\alpha)I}{\alpha I a_u / p_n}\right)(a_f - a_u) = p_l - p_n,$$

where the second equality follows from equation (3), since

$$D(d_f^x, d_u^x) = a_f d_f^x + a_u d_u^x = a_f d_f^x + a_u\left(\frac{\alpha I}{p_n} - \frac{p_f d_f^x}{p_n}\right) = \frac{\alpha I a_u}{p_n}.$$

Thus, $\tilde{b} = \hat{b}$. The intuition as to why eco-labeling, under perfect monitoring, allows countries to reach their efficiency frontier lies in the observation that the absence of labeling gives rise to a consumption distortion as the relative price of eco-friendly products is strictly less than the consumers' relative valuation of eco-friendly products (the marginal rate of substitution between eco-friendly and eco-unfriendly products). The introduction of labeling corrects this distortion by allowing for a positive green premium that enables countries to reach their efficiency frontiers.

Environmental Consequences of Economic Growth

As a first link between the use of eco-labeling and the oft-noted growth–environment tradeoff, note from equation (6), along with the definition of $\tilde{b}$ in equation (4), that

$$\frac{\partial p_n}{\partial e^*} = \frac{\alpha N^* \beta^*}{\Delta} > 0, \quad \frac{\partial p_l}{\partial e^*} = a\frac{\partial p_n}{\partial e^*} > \frac{\partial p_n}{\partial e^*},$$
$$\frac{\partial \tilde{b}}{\partial e^*} = \frac{(a-1)}{\beta}\frac{\partial(p_l - p_n)}{\partial e^*} > 0, \tag{11}$$

where $\Delta = (1 - \alpha)(x_u + ax_l) + [(p_n(a - 1) - \alpha k)/\beta]\mu_b(\tilde{b}, \gamma)(a - 1)M\overline{K} > 0$, since $p_n(a - 1) - \alpha k > p_n(a - 1) - k = \tilde{b} \geq 0$. Likewise, similar manipulations yield

$$\frac{\partial \tilde{b}}{\partial e} = \frac{(a-1)\alpha N\beta}{\Delta} > 0.$$

Thus, growth-driven increase in aggregate demand, originating either from the developed or the developing country, raises the market price of labeled products by proportionately more than that of unlabeled products. Also, the widening gap between the price of eco-friendly and eco-unfriendly products, in turn, translates to a larger equilibrium fraction of producers in the developing country who employ eco-friendly production methods, since $(p_l - p_n - k)/\beta = \tilde{b}$.

In particular, define $\varepsilon = -d\log(1 - \mu(\tilde{b}, \gamma))/d\log(p_l - p_n)$ as the elasticity of supply of the eco-unfriendly output with respect to the green premium. It follows that if and only if $\varepsilon > 1$:

$$\frac{\partial p_n(1-\mu(\tilde{b},\gamma))}{\partial e} < 0 \Leftrightarrow 1 - \mu(\tilde{b}) - \frac{p_n}{B}\mu_b(\tilde{b},\gamma)(a-1) = (1-\mu(\tilde{b},\gamma))(1-\varepsilon) < 0;$$

and hence, *eco-unfriendly products are inferior goods in general equilibrium in the presence of eco-labeling*, in the sense that consumer income $p_n(1 - \mu(\tilde{b}, \gamma))K\overline{M}$ spent on eco-unfriendly products *declines* during the process of economic growth. The intuition as to why eco-unfriendly products are inferior goods in general equilibrium can be seen from the fact that the institution of eco-labeling leads to a relatively higher increase in the price of the labeled products (since consumers' valuation of eco-labeled products is higher). The green-premium thus generated makes the production of eco-friendly products relatively more attractive to the producers. The ensuing increase in the supply of eco-friendly products ensures that the market-clearing output of eco-unfriendly products is lower than without eco-labeling.

In contrast, the absence of eco-labeling implies that consumers' private valuation for environmentally friendly products fails to be reflected in market prices. As such, a growth-driven increase in consumer demand has precisely the opposite impact, as rising consumer income implies a similar increase in consumer spending on eco-unfriendly products as consumers fail to distinguish between eco-friendly and eco-unfriendly products. The absence of a green premium further dissuades producers from undertaking eco-friendly production which translates into an increase in the supply of eco-unfriendly products.[8] It follows, therefore, that *eco-unfriendly products are normal goods in general equilibrium in the absence of labeling.*

Environmental Consequences of International Trade

By similar arguments, international trade with the developed country leads not only to traditional gains from trade via terms-of-trade improvements, but also an increase in demand originating from developed nations has environmental impacts that are similar to indigenous economic growth. In particular, international trade in the presence of eco-labeling has a pro-environment bias, precisely as producers in the developing country respond to the widening gap between the international price of labeled and unlabeled products by switching towards eco-friendly production methods.

The proof of this result is obtained from equation (11) above. With eco-labeling, the price of eco-friendly and eco-unfriendly products in the developing country in the *absence* of international trade, p_l^a and p_n^a, are given by

$$\alpha(\beta eN + \pi(p_l^a, p_n^a, \gamma) + k\mu(\tilde{b}, \gamma)M\overline{K}) = p_n^a\left[1 + \mu\left(\frac{p_l^a - p_n^a - k}{\beta}, \gamma\right)(a-1)\right]M\overline{K},$$

$$p_l^a = \frac{a_f}{a_u} p_n^a.$$

Clearly, $p_i^a < p_i$, $i = l, n$, and $p_l^a - p_n^a < p_l - p_n$ from equation (11), since the green premium, along with the individual prices p_l and p_n, are increasing in e^*. Put differently, commercial policies that restrict the flow of goods between the two countries and adversely affect consumer spending from the developed country on exports from the developing country also *decrease* the number of producers in the developing country who adopt eco-friendly production methods.

Environmental Consequences of Technological Change

We turn now to a characterization of how technologies interact with equilibrium investment in eco-friendly production, both via its direct effect on the cost of production, as well as its general equilibrium effect on the green premium. In particular, note that by definition, an increase in γ shifts the cost distribution to the left. Thus, $\mu_\gamma > 0$. In addition, we show in the Appendix that

$$\frac{\partial p_n}{\partial \gamma} = -\frac{\left[(p_n(a-1)-\alpha k)\mu_\gamma(\tilde{b},\gamma)(a-1)-\alpha\int_{\underline{b}}^{\tilde{b}}\beta\mu_\gamma(b,\gamma)db\right]M\overline{K}}{\Delta},$$

$$\frac{\partial \tilde{b}}{\partial \gamma} = \frac{a-1}{\beta}\frac{\partial p_n}{\partial \gamma},$$

$$\frac{\partial \mu(\tilde{b},\gamma)}{\partial \gamma} = \frac{\mu_b(\tilde{b},\gamma)(a-1)}{\beta}\frac{\partial p_n}{\partial \gamma} + \mu_\gamma(\tilde{b},\gamma)$$

$$= \frac{\left[(1+\mu(\tilde{b},\gamma)(a-1))(1-\alpha)\mu_\gamma(\tilde{b},\gamma)+\left(\alpha\mu_b(\tilde{b},\gamma)(a-1)\int_{\underline{b}}^{\tilde{b}}\mu_\gamma(b,\gamma)db\right)\right]M\overline{K}}{\Delta} > 0.$$

As should be expected, technological change that improves the overall cost-effectiveness of eco-friendly production raises the supply of eco-friendly products.

4. Enforcement and Monitoring under Imperfect Information

Key to the promise that eco-labeling holds for promotion of eco-friendly production methods is the ability on the part of consumers in differentiating between goods made with varying environmental consequences. Consequently, consumers' dis-utility of environmental abuse is reflected in markets through the price premium attached to the consumption of labeled products. Nevertheless, it is easy to see that the same price premium which encourages some producers to shift towards environmentally friendly production methods also has the potential of inducing others to apply eco-labels to mask their true (eco-unfriendly) production means. In particular, absent effective monitoring and enforcement of labeling standards, producers that apply false eco-labels can still capture the price premium that consumers are willing to pay, but at a cost that corresponds to eco-unfriendly production methods. In essence, the absence of effective enforcement and monitoring of producers opens up the possibility that consumers pay a price premium only to obtain an eco-unfriendly product.

The basic model developed above is therefore extended to account for the presence of imperfect and costly monitoring, along with the associated problem of false eco-labeling. In this context, a number of questions arise. First, to what extent can eco-labeling be expected to provide a solution to the problem of under-consumption of eco-friendly products via markets? Second, are there tradeoffs between economic progress and environmental conservation when the success of eco-labeling depends on effective policy intervention via the monitoring of eco-labels? In addition, what are the environmental consequences of international trade particularly when the decision-making of developing countries now involves the interplay between gains from trade and the costs of monitoring eco-labels?

Recall, therefore, our basic two-country setup, and consider (i) consumers who *rationally* account for the probability of consuming a labeled product produced via an eco-unfriendly method, along with (ii) producers who may falsely label depending on the intensity of monitoring. Let λ be the probability that a labeled product is eco-friendly. The analogue of equation (3), which relates the relative prices of labeled and unlabeled goods to consumers' willingness to pay, is simply

$$\frac{p_l}{p_n} = \frac{\lambda a_f + (1-\lambda)a_u}{a_u} = \lambda a + (1-\lambda) \leq a \tag{12}$$

whenever $\lambda \leq 1$. As should be expected, the presence of false labeling decreases the premium that consumers are willing to pay for labeled goods.

On the production side, let ϕ be the probability that false labeling is discovered by a monitoring agency and c the per-unit penalty, upon discovery, for applying false labels. Given that k is the per-unit license fee for a label, producers of eco-unfriendly products are better off by using false eco-labels if and only if their expected profit from false labeling exceeds that from producing unlabeled products, or

$$[(1-\phi)p_l + \phi p_n - \beta - k - \phi c]\overline{K} \geq [p_n - \beta]\overline{K} \Leftrightarrow p_l - p_n \geq \frac{k+\phi c}{1-\phi}. \tag{13}$$

Expression (13) is the *false-labeling* constraint. Thus, (i) the larger the price premium attached to labeled products, and (ii) the lower the monitoring intensity ϕ, the more likely that false labels prevail. On the other hand, the larger the license fee for a label (k) or the penalty for false labeling (c), the lower the likelihood of false labeling.

Expressions (12) and (13) imply that, in equilibrium, the marginal producer who is just indifferent between eco-friendly and eco-unfriendly production is given by

$$\overline{b} = \frac{p_l - p_n - k}{\beta} = \frac{\phi(k+c)}{\beta(1-\phi)}. \tag{14}$$

Simply put, imperfect enforcement implies that aggregate supply of eco-friendly products depends on the monitoring intensity ϕ and the penalty c, precisely as the price premium reflects consumers' rational expectation regarding the fraction of truly eco-friendly products in the pool of labeled exports from the developing country.

Thus, total supply of labeled products (whether eco friendly or not) that are sold at price p_l is given by $x_l = [\mu(\overline{b}, \gamma)/\lambda]M\overline{K}$. Total supply of unlabeled products, inclusive of those that are caught as falsely labeled products, and therefore sold only at p_n, is given by $x_n = (1 - [\mu(\overline{b}, \gamma)/\lambda])M\overline{K}$. Aggregate producer profits are

$$\pi(p_l, p_n, \phi, \gamma) = p_l x_l - \left[k\mu(\overline{b},\gamma) + (k+\phi c)\left(\frac{1/\lambda - 1}{1-\phi}\right)\mu(\overline{b},\gamma) + \int_{\underline{b}}^{\overline{b}} \beta(1+b)d\mu(b,\gamma)\right]M\overline{K}$$
$$+ p_n x_u - \int_{\overline{b}}^{\infty} \beta d\mu(b,\gamma)M\overline{K},$$

where $k\mu(\overline{b}, \gamma)$ is the cost of applying a label for the true labelers while

$$(k+\phi c)\left(\frac{1/\lambda - 1}{1-\phi}\right)\mu(\overline{b},\gamma)$$

is the cost of applying a label for the false labelers.[9] Note in addition, that since $p_n\lambda(a-1) = p_l - p_n = (k + \phi c)/(1-\phi)$ from (13), we have

$$\pi(p_l, p_n, \phi, \gamma) = p_n(1 + \lambda\mu(\overline{b},\gamma)(a-1))M\overline{K}$$
$$-\left[\int_{\underline{b}}^{\overline{b}} (\beta(1+b)+k)d\mu(b,\gamma) + \int_{\overline{b}}^{\infty} \beta d\mu(b,\gamma)\right]M\overline{K}, \tag{15}$$

where the equality follows from (12). Thus, in comparison to the case with perfect monitoring wherein aggregate profits are shown to be maximized via producer self-selection alone, aggregate profits and producer self-selection under imperfect monitoring can be affected, at constant prices p_l and p_n, through a change in the monitoring intensity:

$$\frac{\partial \pi(p_l, p_n, \phi, \gamma)}{\partial \phi} = \left[(p_n\lambda(a-1) - \beta(1+\bar{b}) - k + \beta)\mu_\phi(\bar{b}, \gamma) + p_n(a-1)\mu(\bar{b}, \gamma)\frac{\partial \lambda}{\partial \phi}\right]M\bar{K}$$

$$= \frac{(k+c)}{(1-\phi)^2}\mu(\bar{b}, \gamma)M\bar{K} > 0, \tag{16}$$

where the last equality follows from the definition of $\bar{b}$ (equation (14)) and by substituting for $\lambda = (k + \phi c)/[p_n(a - 1)(1 - \phi)]$. Thus, as long as there exists some degree of false labeling $\lambda < 1$, aggregate producer profits can be increased by simply increasing the fraction of eco-friendly producers through an increase in the monitoring intensity. The result follows since each act of false labeling creates a negative externality on the other producers in the developing country, as consumers rationally account for the presence of a mixed pool of products with differing environmental consequences. As such, consumers' relative valuation for eco-friendly products $p_n(a-1)$ is strictly greater than the increase in unit cost that the marginal firm $\bar{b}$ incurs ($p_l - p_n = p_n\lambda(a - 1)$) whenever $\lambda < 1$.

Finally, let the cost required to monitor eco-labeling programs, in units of the *numéraire*, be given by $\tau\phi$. Given that the government recovers

$$\left[k + (k+\phi c)\left(\frac{1/\lambda - 1}{1-\phi}\right)\right]\mu(\bar{b}, \gamma)M\bar{K}$$

from the penalty, it follows that the government budget constraint is

$$NT = \tau\phi - \left[k + (k+\phi c)\frac{(1/\lambda - 1)}{(1-\phi)}\right]\mu(\bar{b}, \gamma)M\bar{K},$$

where T gives the lump-sum tax levied on each one of the N consumers. It follows that the analogue of the market-clearing relationship in equation (6) is just

$$\alpha(\beta e N - NT + \pi(p_l, p_n, \phi, \gamma) + \beta^* e^* N^*) = p_n\left[1 + \mu(\bar{b}, \gamma)\frac{p_l - p_n}{\lambda p_n}\right]M\bar{K}$$

$$\Leftrightarrow \alpha\left\{\beta e N - \tau\phi + \left[k + (k+\phi c)\left(\frac{1/\lambda - 1}{1-\phi}\right)\right]\mu(\bar{b}, \gamma)M\bar{K} + \pi(p_l, p_n, \phi, \gamma) + \beta^* e^* N^*\right\}$$

$$= p_n(1 + \mu(\bar{b}, \gamma)(a-1))M\bar{K}. \tag{17}$$

Thus, increasing the intensity of monitoring efforts ϕ, raises the fraction of eco-friendly producers $\mu(\bar{b}, \gamma) = \mu(\phi(k + c)/\beta(1 - \phi), \gamma)$, and puts downward pressure on the price p_n of eco-unfriendly products. Specifically:

$$\frac{\partial p_n}{\partial \phi} = -\frac{\tau\alpha + \left(p_n(a-1)(1-\alpha) + \frac{\phi(k+c)}{1-\phi}\right)\mu b(\bar{b}, \gamma)M\bar{K})\frac{k+c}{\beta(1-\phi)^2}}{(1+\mu(\bar{b}, \gamma)(a-1))(1-\alpha)M\bar{K}} < 0. \tag{18}$$

Thus, there are two effects in play. First, raising monitoring intensities increases the fraction of eco-friendly producers, and diverts consumer income away from eco-unfriendly products. Meanwhile, since costly monitoring decreases consumer income via lump-sum taxation T, the negative impact of an increase in ϕ on p_n is further reinforced via this negative income effect.

5. Market Responses to Eco-labeling under Imperfect Monitoring

Eco-labeling as a Market-Based Solution

At this point, it should be apparent that the effectiveness of eco-labeling as a market-based solution to the problem of under-consumption of eco-friendly products is hampered in the presence of imperfect enforcement. In particular, so long as there is pooling of false eco-labels with products that are produced via eco-friendly methods, rational consumers discount their willingness to pay for labeled products. Consequently, as shown in expression (12), the relative market price of labeled goods *underestimates* the utility gains on the part of consumers for eco-friendly products, with

$$\frac{p_l}{p_n} = a - (1-\lambda)(a-1) \leq a.$$

Perhaps more importantly, the presence of imperfect information biases producers' incentives against the use of eco-friendly methods (as compared to the case of perfect monitoring) since the price premium can be captured, depending on monitoring intensity and the penalty, even when production is undertaken by means of eco-unfriendly methods and products are sold via false labeling. Specifically, as long as eco-friendly and eco-unfriendly products co-exist

$$\bar{b} = \frac{p_l - p_n - k}{\beta} = \frac{p_n\lambda(a-1) - k}{\beta}$$

is strictly decreasing in the share of false labels $1 - \lambda$. As a second round consequence of imperfect monitoring, therefore, producers of goods with false labels confer a *negative* externality on all producers of eco-friendly products, precisely since the prevalence of false labels drives consumers' valuation of labeled goods, eco-friendly or otherwise, downwards. Thus, comparing the definition of $\bar{b}$ above with $\tilde{b}$ which is on the efficiency frontier (equation (4)), it should be clear that Pareto optimality is no longer guaranteed.

Environmental Consequences of Economic Growth

In the absence of credible monitoring efforts, the relationship between economic growth and environmental abuse becomes much more complex. Specifically, unless monitoring intensity keeps pace with economic growth, an increase in aggregate demand can have no impact on the incentives of producers to switch in favor of eco-friendly means of production. In order to see why, first consider the market clearance condition

$$\alpha\left\{\beta e N - \tau\phi + \left[k + (k+\phi c)\left(\frac{1/\lambda - 1}{1-\phi}\right)\right]\mu(\bar{b}, \gamma) M\bar{K} + \pi(p_l, p_n, \gamma, \phi) + \beta^* e^* N^*\right\}$$
$$= p_n\left[1 + \mu\left(\frac{\phi(k+c)}{\beta(1-\phi)}, \gamma\right)(a-1)\right] M\bar{K}.$$

Clearly, as long as the monitoring intensity ϕ remains unchanged, total consumer expenditure on eco-unfriendly products

$$p_n\left[1 - \mu\left(\frac{\phi(k+c)}{\beta(1-\phi)}, \gamma\right)\right] M\bar{K}$$

and on the labeled products (eco-friendly or not) is monotonically increasing in e and e^*. However, the price premium $p_l - p_n = (k + \phi c)/(1 - \phi)$, as determined by the false-labeling constraint (13), is independent of the growth-induced changes in aggregate demand and is determined solely by the monitoring intensity, license fee, and the penalty associated with false labeling. Therefore, the increases in p_l and p_n consequent upon an increase in income must be proportionate to one another in order to satisfy the false-labeling constraint.

Thus, an increase in consumer income, in the absence of increased monitoring, has the effect of increasing the supply of both unlabeled (eco-unfriendly) and labeled products. But does an increase in the supply of labeled products imply an increase in the supply of the eco-friendly products? From expressions (12) and (13) we have

$$p_l - p_n = p_n\lambda(a-1) \Leftrightarrow \lambda = \frac{k+\phi c}{1-\phi}\frac{a-1}{p_n}. \tag{19}$$

The fraction of true labels (λ) is thus inversely related to consumer income e^* and e, since p_n is strictly increasing respectively in e^* and e. It follows that the incidence of false labeling $(1 - \lambda)$ rises and falls with consumer income. Accordingly, an increase in the price of labeled products, consequent upon an increase in consumer income, has the effect of increasing the supply of false-labeled products as producers mask their true production methods, rather than switching in favor of eco-friendly ones. Meanwhile, consumer expenditure on products with false labels,[10]

$$p_l\left(\frac{1}{\lambda}-1\right)\mu(\bar{b},\gamma)M\bar{K} = \left(p_n + \frac{k+\phi c}{1-\phi}\right)\left(\frac{1-\lambda}{\lambda}\right)\mu(\bar{b},\gamma)M\bar{K}$$

also rises and falls with consumer income since $\partial p_n/\partial e > 0$ from equation (17). Thus, *eco-unfriendly (falsely labeled and unlabeled) products are normal goods in general equilibrium, when monitoring fails to keep pace with economic growth.* Contrary to the original intent of eco-labeling, therefore, economic progress under imperfect enforcement and information imperfection on the part of consumers is harmful to the environment.

A question that remains, however, is whether the monitoring intensity chosen optimally by the government of a developing country may respond to exogenous economic growth in a predictable fashion. In this context, consider a planner whose objective is to maximize the welfare of the developing country, taking into account the associated net cost of monitoring. The indirect utility of the developing economy is

$$V(\phi) = \log W + \alpha\log\left(\frac{a_u}{p_n}\right) + \log(\alpha)^{\alpha}(1-\alpha)^{1-a},$$

where W denotes gross national product of the developing country, with

$$W = \left\{\beta eN - \tau\phi + \left[k + (k+\phi c)\frac{(1/\lambda - 1)}{(1-\phi)}\right]\mu(\bar{b},\gamma)M\bar{K} + \pi(p_l, p_n, \phi, \gamma)\right\}.$$

Thus:

$$\frac{\partial V(\phi)}{\partial\phi} = \left[p_n(a-1) - \frac{\phi(k+c)}{1-\phi}\right]\frac{\mu_b(\bar{b},\gamma)}{\beta(1-\phi)^2}M\bar{K} - \tau + \frac{\alpha\beta^* e^* N^*}{p_n}\frac{\partial p_n}{\partial_\phi} \geq (<)0.$$

Denoting η as the elasticity of p_n with respect to monitoring intensity $d\log p_n/d\log(1-\phi) > 0\left(\frac{\partial p_n}{\partial\phi} < 0\right)$ above, we get

$$\frac{\partial V(\phi)}{\partial \phi}=\left[p_n(a-1)-\frac{\phi(k+c)}{1-\phi}\right]\frac{\mu_b(\bar{b},\gamma)}{\beta(1-\phi)^2}M\overline{K}-\tau-\alpha\beta^*e^*N^*(1-\phi)\eta\geq(<)0. \quad (20)$$

From equation (20), three effects determine whether credible monitoring should be in place. First, since the marginal cost of producing the eco-friendly product is $\beta\bar{b} = \phi(k + c)/(1 - \phi)$ from equation (14), $p_n(a - 1) - \phi(k + c)/(1 - \phi)$ represents the utility gains from an increase in monitoring intensity, and via an increase in the profits of Southern firms. Clearly, the higher the market valuation of a unit of eco-friendly products $p_n(a - 1)$, the higher will be the direct benefits of an increase in monitoring intensity. Second, τ is the marginal resource cost incurred in monitoring activities; hence, the lower is the marginal resource cost, the higher the gains from increased monitoring. Third, an increase in ϕ embodies a terms-of-trade effect that adversely affects the welfare of the exporting developing country as $\partial p_n/\partial\phi$ is negative from equation (18). Therefore, the smaller the value of η the higher the gains from increased monitoring.

Thus, we can show that there exists an inverted-U relationship between economic growth and consumer expenditure on eco-unfriendly products originating from the developing country. As shown in the Appendix, evaluating the change in welfare at $\phi = 0$

$$\left.\frac{\partial V(\phi)}{\partial \phi}\right|_{\phi=0}\geq 0 \Leftrightarrow W\geq\frac{\tau\beta}{\alpha\mu_b(\underline{b},\gamma)(k+c)(a-1)}\left(1+\frac{\alpha s^*}{1-\alpha}\right)\equiv\overline{W}, \quad (21)$$

where $s^* = \beta^*e^*N^*/(W + \beta^*eN^*)$. Thus, $\overline{W}$ denotes the critical income threshold, such that whenever $W > \overline{W}$, the developing country is strictly better off devoting a positive amount of domestic resources to monitoring efforts. Therefore, economic growth *prior* to the realization of the income threshold can only raise consumer spending on the eco-unfriendly output, as ϕ and hence the fraction of eco-friendly producers $\mu(\bar{b}, \gamma)$ are both equal to zero. This gives the upward-sloping portion of the inverted-U.

Meanwhile, as income exceeds the critical income threshold, welfare of the developing country is strictly increasing with respect to the monitoring intensity and the developing country will be willing to devote a positive amount of resources to monitoring. In order to determine the downward-sloping portion of the inverted-U curve —decreasing consumption expenditure on eco-unfriendly products with respect to rising income—note that consumer expenditure on eco-unfriendly products is given simply by:

$$[p_n(1+\mu(\bar{b},\gamma)(a-1))-p_l\mu(\bar{b},\gamma)]M\overline{K}=\left[p_n(a-1)-\left(\frac{k+\phi c}{1-\phi}\right)\right]\mu(\bar{b},\gamma)M\overline{K}. \quad (22)$$

Thus, whether or not the right-hand side of the above equation is decreasing in the developing country income (e), depends on the effect of a change in e on (i) the price of the unlabeled products, p_n; (ii) the fraction of firms that choose to produce the eco-friendly product, $\mu(\bar{b}, \gamma)$; and (iii) the welfare optimal level of monitoring, $\bar{\phi}$.

First, note that the effect of a change in e on the welfare optimal level of monitoring $\partial\bar{\phi}/\partial e$ can be obtained from equation (20), with

$$\frac{\partial\bar{\phi}}{\partial e}=-\frac{\partial^2 V}{\partial\phi\partial e}\bigg/\frac{\partial^2 V}{\partial\phi^2}.$$

Thus, so long as the objective function $V(\phi)$ is strictly concave, with $\partial^2V/\partial\phi^2 < 0$, the welfare optimal level of monitoring intensity is strictly increasing in e if and only if the numerator $\partial^2V/\partial\phi\partial e$ is strictly positive. As can be readily confirmed using equation

(20), the marginal benefits of increasing monitoring intensity is strictly increasing in Southern consumer income ($\partial^2 V/\partial\phi\partial e > 0$), which implies accordingly that $\bar{\phi}$ is strictly increasing in e if and only if

$$\left[(a-1)M\bar{K}\frac{\partial p_n}{\partial e}\right] - \alpha\beta^* e^* N^*(1-\phi)\frac{\partial \eta}{\partial e} > 0.$$

Two effects thus dictate whether economic growth encourages a strengthening of monitoring. First, since rising consumption demand via an economic growth in the developing country (e) implies a corresponding increase in the valuation for eco-friendly products $p_n(a-1)$ (equation (17)), the first term in square brackets in the equation above is strictly positive. Second, if the size of the terms-of-trade effect due to monitoring—as measured by the elasticity η—responds negatively to economic growth ($(\partial\eta/\partial e) < 0$), or is otherwise sufficiently small, the overall impact of economic growth on the marginal benefits of raising monitoring intensity is strictly positive.

These observations imply that the downward-sloping segment of the inverted-U shape relationship linking economic growth and the consumption of eco-unfriendly products is indeed possible. In particular, since p_n is strictly increasing in e from equation (17), and

$$\mu(\bar{b},\gamma) = \mu\left[\frac{\phi(k+c)}{\beta(1-\phi)},\gamma\right]$$

is increasing in the monitoring intensity ϕ, it can be readily verified that a sufficient condition for consumption expenditure on eco-unfriendly products to decline with economic growth (for the right-hand side of equation (22) to be declining in e) is that the elasticity of eco-unfriendly output supply $\varepsilon = -d\log(1-\mu)/d\log(p_l - p_n)$ be sufficiently large, so that the negative output effect of an increase in monitoring prompted by government response to economic growth dominates the tendency for the price of falsely labeled eco-unfriendly products to increase with growth.

Also of interest at this point is whether the threshold income level is susceptible to change, and should, therefore, be expected to differ across similarly poor developing countries, and across different kinds of environmental abuse. In terms of our model, note from equation (21) that $\bar{W}$ is strictly increasing in τ (decreasing in $\mu_b(\underline{b},\gamma)$, and α). It follows that a decrease in the cost of monitoring, an increase in producer access to cost efficient production methods (an increase in $\mu_b(\underline{b},\gamma)$), and an increase in the share of income devoted to good x all contribute to a *reduction* in the threshold level of income.

Environmental Consequences of International Trade

In a similar fashion, gains from trade that are not accompanied by an increase in monitoring efforts confer damages to the environment since producers only respond to the increase in world demand by means of false labeling, as $\mu(\bar{b},\gamma) = \mu(\phi(k+c)/\beta(1-\phi),\gamma)$ is independent of e or e^*. Perhaps more interestingly, we find that the choice of monitoring intensity on the part of developing economies suffers from a problem that is typical of nonexcludable public goods. In particular, we find that even if consumers view eco-friendly products as private goods, developing countries lack incentives to pay for monitoring since the utility gains from costly monitoring efforts is shared between the developing exporter and the developed importer. In fact, note from equation (21) that $\bar{W}$ is strictly increasing in s^*. Thus, setting $s^* = 0$, which applies in the absence of trade, the developing country benefits from initiating monitoring efforts whenever

$$W \geq \frac{\tau\beta}{\alpha\mu_b(\underline{b},\gamma)(k+c)(a-1)} \equiv \overline{W}^a < \overline{W}.$$

In effect, international trade can shift the inverted-U curve to the right, and delay the turning point of the economic growth and environmental relationship. To see the intuition behind this result, it will be useful to recall from (21) that national welfare change in response to an increase in monitoring intensity ϕ depends on

$$\frac{\partial V(\phi)}{\partial \phi} = 0 \Leftrightarrow \left[p_n(a-1) - \frac{\phi(k+c)}{1-\phi}\right]\mu_b(\overline{b},\gamma)\frac{(k+c)}{\beta(1-\phi)^2}M\overline{K} - \tau + \frac{\alpha\beta^* e^* N^*}{p_n}\frac{\partial p_n}{\partial \phi} = 0.$$

In addition to the direct welfare benefits $(1-\alpha)\mu_b(\overline{b},\gamma)$, net of the costs associated with increased monitoring, economies open to trade also face a *negative* terms-of-trade effect since an increase in ϕ lowers the price of eco-unfriendly exports. Thus, that $\overline{W}^a$ is less than $\overline{W}$ can be attributed to the terms-of-trade losses that are associated with monitoring. In particular, the larger is the share of imports of the developed country in world consumption s^*, the larger the terms-of-trade impact.[11]

6. Conclusion

We have shown that if consumers view labeled products as private goods then labeling does help in altering the incentives of producers in developing countries to shift to eco-friendly methods of production. Indeed, we show that eco-labeling with perfect enforcement guides both consumers and producers to the efficiency frontier of the developed and developing economies. In addition, the endogenous price premium allows choice of production technologies to reflect consumers' relative preference for eco-friendly production, and as such, we find that eco-friendly products are normal goods in general equilibrium. Consequently, income gains achieved via international trade can only be beneficial to the environment.

However, by extending the basic setup to incorporate the more realistic scenario with costly enforcement, our findings suggest that wholesale promotion of eco-labeling to improve environmental performance should be exercised with caution. In particular, we show that unless monitoring efforts keep pace with indigenous economic growth, the presence of false labeling that goes unchecked in turn implies that eco-unfriendly products are normal goods in general equilibrium. Interestingly, the incentive for developing economies to credibly enforce national eco-labeling schemes has a threshold property. In particular, the income threshold varies endogenously with the developing economy's openness to trade, the cost-effectiveness of monitoring technologies, and the terms-of-trade effects.

Appendix

Market Responses to Technological Change under Perfect Enforcement

To derive the results displayed in section 3, note that total disposable consumption income in the South can be written as

$$I = \left[p_n(1+\mu(\tilde{b},\gamma)(a-1)) - \beta(1+\tilde{b}\mu(\tilde{b},\gamma)) + \int_{\underline{b}}^{\tilde{b}} \beta\mu(b,\gamma)db\right]M\overline{K} \tag{A1}$$

by integrating by parts. Thus, making use of market clearance in equation (6), routine differentiation yields

$$\frac{\partial p_n}{\partial \gamma} = -\frac{(p_n(a-1) - \alpha k)\mu_\gamma(\bar{b},\gamma) - \alpha\int_{\underline{b}}^{\bar{b}} \beta\mu_\gamma(b,\gamma)db)M\overline{K}}{\Delta} \tag{A2}$$

as in the text.

Market Responses to Economic Growth under Imperfect Enforcement

To arrive at the income threshold as shown in section 5, make use of (18) to obtain

$$\left.\frac{\partial p_n}{\partial \phi}\right|_{\phi=0} = -\frac{p_n\mu_b(\underline{b},\gamma)(a-1)M\overline{K}(k+c)/\beta + \alpha\tau}{(1-\alpha)M\overline{K}}. \tag{A3}$$

In addition, from equation (20), we have

$$\left.\frac{\partial V(\phi)}{\partial \phi}\right|_{\phi=0} = p_n(a-1)\mu_b(\underline{b},\gamma)(k+c)M\overline{K}/\beta - \tau + \frac{\alpha\beta^* e^* N^*}{p_n}\left.\frac{\partial p_n}{\partial \phi}\right|_{\phi=0}, \tag{A4}$$

since $\phi = 0$ in the absence of monitoring efforts, and in addition $p_n x_u = p_n M\overline{K} = \alpha W + \alpha\beta^* e^* N^*$. The income threshold $\overline{W}$ as in the text follows from substituting (A3) into (A4).

References

Akerlof, G. A., "The Market for 'Lemons': Quality Uncertainty and the Market Mechanism," *Quarterly Journal of Economics* 84 (1970):488–500.

Antweiler, W., B. Copeland, and M. Scott Taylor, "Is Free Trade Good for the Environment?" *American Economic Review* 91 (2001):877–908.

Basu, A. K. and Nancy H. Chau, "Asymmetric Country-of-Origin Effects on Intraindustry Trade and International Quality Race," *Review of Development Economics* 2 (1998):140–66.

———, "Market Access Rivalry and Eco-Labeling Standards: Are Eco-Labels Non-Tariff Barriers in Disguise?" working paper 2001-15, Department of Applied Economics and Management, Cornell University (2001).

Brown, D. K., A. V. Deardorff, and R. M. Stern, "International Labor Standards and Trade: a Theoretical Analysis," in J. Bhagwati and R. Hudec (eds.), *Harmonization and Fair Trade: Prerequisites for Free Trade?*, Cambridge, MA: MIT Press (1996).

Bureau, C., E. Gozlan, and S. Marette, "Quality Signaling and International Trade in Food Products," mimeo, INRA-Economie, BP01, 78850, Grignon, France (1999).

Copeland, B. and M. Scott Taylor, "North–South Trade and the Environment," *Quarterly Journal of Economics* 109 (1994):755–85.

Dixit, A. and J. E. Stiglitz, "Monopolistic Competition and Optimum Product Diversity," *American Economic Review* 67 (1977):297–308.

Grossman, G. and Alan B. Krueger, "Economic Growth and the Environment," *Quarterly Journal of Economics* 110 (1995):353–77.

Marette, S., J. Crespi, and A. Schiavina, "The Role of Common Labelling in the Context of Asymmetric Information," *European Review of Agricultural Economics* 26(2) (1999):167–78.

Mattoo, A. and H. Singh, "Eco-Labelling, the Environment and International Trade," in S. Zarilli, V. Jha, and R. Vossenaar (eds.), *Eco-Labelling and International Trade*, MacMillan, London (1997).

Nelson, P., "Information and Consumer Behavior," *Journal of Political Economy* 78 (1970): 311–29.

Nordström, Hakan and Scott Vaughan, "Trade and Environment," WTO Special Studies 4 (2000).

Shams, R., "Eco-Labelling and Environmental Policy Efforts in Developing Countries," *Intereconomics*, 30(3) May/June (1995):143–49.

Srinivasan, T. N., "International Trade and Labour Standards," in P. van Dyck and G. Faher (eds.), *Challenges to the New World Trade Organization*, Amsterdam: Martinus Nijhoff/Kluwer (1995).

United Nations, "Trade Effects of Eco-Labelling," *Studies in Trade and Investment 27*, United Nations, NY (1997).

World Trade Organization Committee on Trade and Environment (WTO-CTE), "Information Relevant to the Consideration of the Market Access Effects of Eco-labelling Schemes," note by the Secretariat (WT/CTE/W/150) (2000).

Notes

1. See also Basu and Chau (1998) for a study on trade under asymmetric quality information with country-of-origin as signals for quality.

2. See Dixit and Stiglitz (1977) for a discussion of the use of similar utility indexes when product differentiation is of central concern.

3. Note that we assume here that all labeled products are of the eco-friendly variety. The possibility of false labeling will be considered later.

4. Without a full description of the production side of the economy, and the mechanism through which the two prices are determined in general equilibrium, that demand for both goods are increasing in income alone does not necessarily imply that both labeled and unlabeled products are *normal goods* in general equilibrium. We deal with this issue later.

5. Of course, if willingness to pay a is small, so that the green premium is too low to justify costs (k), we have $\tilde{b} = \underline{b}$, and the fraction of producers who apply eco-labels is equal to zero. Thus, whether an eco-labeling program is in place is a matter of indifference to both producers and consumers. In what follows, we assume that a is large enough, so that a positive fraction of producers employ eco-friendly production methods so long as consumers are allowed to make informed choices.

6. Alternatively, let the Armington assumption apply, so that consumers in the developed country perceive imports orginating from the developing country as imperfect substitutes for domestic outputs, as embodied in the production function y^*.

7. Note that market clearance in equation (6) determines the price of the eco-unfriendly (unlabeled) products. Accordingly, the price of labeled products can be had using equation (3). Finally, the fraction of producers in the developing country that elects to produce via eco-friendly means is given once the absolute price premium $p_l - p_n$, and hence $\tilde{b}$, is determined.

8. To see this, note that $\tilde{b} = \underline{b}$ and $p_n x_u = \alpha(\beta^* e^* N^* + \beta e N + \pi)$, where x_u is given simply by $M\overline{K}$.

9. In particular, since $\mu(\overline{b}, \gamma)M$ of the $\mu(\overline{b}, \gamma)(1/\lambda)M$ number of producers of labeled products offer eco-friendly products, the total number of labelers who are nevertheless eco-unfriendly is simply $\mu(\overline{b}, \gamma)(1/\lambda - 1)M\overline{K}$. In addition, since $1 - \phi$ is the fraction of eco-unfriendly producers of labeled products remain undetected, the total number of false labelers must be equal to $\mu(\overline{b}, \gamma)(1/\lambda - 1)M/(1 - \phi)$, and accordingly, the total number of false labelers who are caught is given by $\phi\mu(\overline{b}, \gamma)(1/\lambda - 1)M/(1 - \phi)$.

10. Note that since λ is the fraction of eco-unfriendly outputs in the pool of labeled products, and $\mu(\overline{b}, \gamma)M\overline{K}$ is the total supply of eco-friendly products, $[(1 - \lambda)/\lambda]\mu(\overline{b}, \gamma)$ thus represents the total supply of falsely labeled products.

11. Basu and Chau (2001) show that when the developing and the developed country strategically compete in eco-labeling standards, then the lack of eco-labeling has a pro-trade bias while market access rivalry in the presence of eco-labeling exhibits a tariff-like outcome, with net exporters supplying too much eco-friendly products while net importers supplying too little.

[28]

On export rivalry and the greening of agriculture—the role of eco-labels

Arnab K. Basu[a,1], Nancy H. Chau[b,2], Ulrike Grote[c,*]

[a]*Department of Economics, College of William and Mary, Williamsburg, VA 23187, USA*
[b]*Department of Applied Economics and Management, Cornell University, Ithaca, NY 14853, USA*
[c]*Center for Development Research (ZEF), University of Bonn, Walter-Flex-Str. 3, D-53113, Bonn, Germany*

Abstract

Why do some countries have eco-labelling programs and others do not? In this paper, we provide both theoretical arguments and empirical evidence suggesting that the answer to this question can shed new light on three questions that have taken center-stage in the trade and environment debate: (i) does trade exacerbate the exploitation of the environment; (ii) are countries competing in export markets engaged in a race to the bottom in environmental performance; and (iii) do market-based environmental instruments benefit the rich and hurt the poor?

JEL classification: F1; O1; O3

Keywords: Eco-labelling in agriculture; Export rivalry; Strategic complementarity

1. Introduction

Eco-labelled products provide consumers with information about the environmental externalities associated with the production and consumption processes. The adoption of eco-labels has rapidly grown worldwide, and the prevalence of the practice of eco-labelling is becoming evident in both industrial and agricultural sectors.[3] This paper focuses on the adoption of labelling in the agricultural sector for two reasons. To begin with, labelling initiatives in the agricultural sector clearly preceded similar efforts in the industrial sector.[4] Second, and in the context of the

* Corresponding author.
E-mail addresses: akbasu@wm.edu (A.K. Basu), hyc3@cornell.edu (N.H. Chau), u.grote@uni-bonn.de (U. Grote).
[1] Tel.: +1 757 221 1318.
[2] Tel.: +1 607 255 4463.

[3] The German Blue Angel label, the Canadian Environmental Choice Program, the EcoMark in Japan, the White Swan label in the Nordic Countries, and the Green Seal of the United States, are examples in industrial production.

[4] For example, the earliest agricultural labelling initiatives were established as early as 1924 in Western Europe.

development consequences of market-based environmental initiatives, eco-labelling in agriculture has become a contentious issue. Indeed, a potential conflict of interests between developed and developing countries arises, at least in part, because of the differential degrees of dependence on agriculture for the purposes of export earnings and job creation in these two groups of countries. As such, the argument goes that the market access consequence of eco-labelling may have a disproportionate impact on producers in developing countries.

The debate surrounding eco-labelling in the context of international trade can be more succinctly summarised under three main themes. First, eco-labelling represents a potential market-based policy response to the threat of environmental exploitation in the face of increasing international trade (OECD, 1997a, 1997b; WTO–CTE, 1997; Basu and Chau, 2001). A question that naturally arises is whether countries and/or producers that practice green production methods are indeed favourably selected in the set of countries that have adopted eco-labelling. Relatedly, trade competition and the need to cut cost have led to concerns regarding the threat of a 'race to the bottom' (Basu et al., 2003; Chau and Kanbur, 2000). In this context, eco-labelling provides a potential market-based alternative to regulatory standards aimed at levelling the playing field. Precisely since the adoption of eco-labelling is voluntary, however, an important question remains as to whether observed adoption patterns reveal indication of a 'race'—in the sense that adoption decisions across countries are mutually dependent—and whether observed adoption time patterns in fact reveal a move towards vicious interdependence.

A third theme in the eco-labelling debate involves the North–South dialogue regarding the threat of increasing disparities between the rich and the poor (Basu and Chau, 1998; Basu et al., 2003). By its very nature, eco-labelling takes advantage not just of a country's cost advantage, but potentially also of its 'reputational' comparative advantage in a segmented market where the credibility of labelling programs is key to the determination of the green premium, and hence, the terms of trade. Therefore, the question that arises here is whether eco-labelling can exacerbate income disparities between developed and developing countries via one or more of the following routes: (i) the lack of technological know-how and the inability to shoulder the costs involved in implementing eco-labelling programs; and (ii) the inherent disadvantage that developing countries may be confronted with when industries in developed and developing countries are faced with different green premia.

These observations accordingly serve as the motivation of this paper, and our focus on tracing out the adoption pattern of eco-labelling programs. Our approaches to understand the role of export rivalry in the adoption of green agricultural production methods in this paper are two-fold. We begin by proposing an analytical framework to examine the economic incentives to initiate eco-labelling. The framework entertains export rivalry in a multicountry setting, and yields a set of theoretical implications in a subgame perfect Nash equilibrium, highlighting: (i) the selection criteria of countries that adopt eco-labelling; and (ii) the endogeneity of labelling incentives and the welfare consequences of observed labelling initiatives.

The second approach undertaken in this paper is closely related to the first, and presents an empirical investigation of the determinations of the time to adopt eco-labelling. The objectives are to empirically verify both the relative importance and significance of non-trade and trade related factors associated with the decision to adopt eco-labelling. As the analysis unfolds, we will also present the findings of our empirical estimation of whether systematic divergences in the industry-level green premium between developed and developing countries may be inferred from the time pattern of adoption decisions.

2. The basic model

We consider a setup in which producers in N countries are engaged in the production of two goods:[5] a homogeneous numeraire Y^j, $j = 1 \ldots N$; and an agricultural output X^j. The numeraire commodity employs a composite input, L_y^j with $Y^j = \omega^j L_y^j$, where

[5] A more detailed description of the model is available from the authors upon request.

ω^j denotes the marginal and average product of input L_y^j.

2.1. Production

Production of the agricultural output can be accomplished via either (i) an environmentally sound production technology, X_e^j; or (ii) a baseline production technology, X_o^j. More specifically, let there be M^j number of competitive agricultural producers in country j. Agricultural production also employs the composite input, L_x^j as input; and final agricultural outputs produced via the two techniques are given by:

$$X_e^j = \left(\frac{L_x^j}{a}\right)^\alpha, \qquad X_o^j = (L_x^j)^\alpha,$$

where $\alpha \in (1, \infty)$ is producer-specific, and parameterises the cost of adopting the environmentally friendly production technique. Also, let the cumulative distribution function $F^j(a')$ denote the fraction of producers in country j with $a \leq a'$, and with $a, a' \in (1, \infty)$.

The decision problem of producers is a simple two-stage problem. In the first-stage, individual producers decide to employ one of the two production techniques. Conditional on the choice of production techniques, producers employ inputs to maximise profits in the second-stage.

2.2. Consumption and the green premium

The utility of a representative consumer in country j, $U^j(D_x^j, d_y^j)$, takes as arguments the consumption of the homogeneous numeraire d_y^j, along with an index of effective units of good x consumed, D_x^j, with $\log U^j(D_x^j, d_y^j) = \beta^j \log D_x^j + (1-\beta^j)\log d_y^j$, where $\beta^j \in (0, 1)$ denotes the share of consumer expenditure devoted to the consumption of the agricultural output. Let d_e^{ji} be the quantity demand for the eco-friendly output originating from country i, and d_o^j be the quantity demand for the baseline output, effective consumption D_x^j is given by:

$$D_x^j = \sum_{i=1}^{N}(1+g^i)d_e^{ji} + \sum_{i=1}^{N} d_o^j.$$

Clearly, for demand to be positive for both labelled and unlabelled products originating from each country i, it must be the case that:

$$\frac{p_\ell^i}{p_\ell^k} = \frac{1+g^i}{1+g^k}, \qquad \frac{p_\ell^i}{p_u} = 1+g^i, \quad i,k = 1,\ldots,N. \quad (1)$$

The green premium as defined in Eq. (1) above can be used to show that aggregate agricultural producer revenue in the presence of eco-labelling in country j depends on the green premium in a predictable way:

$$Q_e^j = \left(p_u M^j \left(\frac{\alpha}{\omega^j}\right)^\alpha\right)^{1/1-\alpha} \times \left(1 + (1+g^j)^{1/1-\alpha} \int_1^{(1+g^j)^{1/\alpha}} \times \left[\left(\frac{1}{a}\right)^{1/1-\alpha} - \left(\frac{1}{1+g^j}\right)^{1/1-\alpha}\right] \mathrm{d}F^j(a)\right)$$
$$\equiv (p_u)^{1/1-\alpha}\gamma^j(1+G^j).$$

In addition, revenue in the absence of labelling is given by $Q_o^j(p_u) = (p_u)^{1/(1-\alpha)}\gamma^j$, where $\gamma^j \equiv M^j(\alpha/\omega^j)^{\alpha/(1-\alpha)}$ parameterises the production cost of country j. On the other hand, $p_u G^j$ is an industry-level green premium, and represents the increase in industry-level revenue, holding p_u constant, that may be expected subsequent to eco-labelling. Note, in particular, that the industry-level green premium G^j depends on a demand-side and a supply-side effect. On the demand-side, the higher the country-specific unit green premium $1+g^j$, the higher will be the revenue gains in the presence of eco-labelling. Indeed, it can be readily verified that $G^j > 0$ if and only if $g^j > 0$. Meanwhile, the supply-side response depends directly on the cost distribution among producers in country j, F^j. Specifically, countries in which producers are more concentrated along the lower end of the cost distribution are more likely to benefit from eco-labelling.

2.3. General equilibrium and the incentives to adopt eco-labelling

Let I be the set of all countries in which an eco-labelling program is in place, and I_{-i} be the set of all countries in I but country i. With consumer income (aggregate earnings of composite input owners) equal to $\omega^j L^j$ in country j, aggregate world demand for the agricultural output is equal to total producer revenue if and only if:

$$p_u(I) = \left(\frac{\sum_{j=1}^{N} \beta^j \omega^j L^j}{\sum_{j \in I} \gamma^j G^j + \sum_{j=1}^{N} \gamma^j} \right)^{1-\alpha}.$$

By inspection, the price of eco-unfriendly products is strictly decreasing in the number of countries that have instituted an eco-labelling program, as long as $G^j > 0$ for all $i \in 1$.

In addition, let W be the sum total of consumer expenditure on agriculture in the N countries with $W \equiv \sum_{j=1}^{N} \beta^j \omega^j L^j$, and as c^j is the fixed cost required to put in place a credible eco-labelling program, aggregate producer profits rise with market-based voluntary green production via eco-labelling, taking as given the adoption decisions of the rest of the $N-1$ countries, if and only if:

$$\begin{aligned} \log G^j \geq\; & \log \left(\frac{c/(1-\alpha)}{Q_o^j(p_u(I_{-j}))} \right) \\ & - \log \left(1 - \frac{c/(1-\alpha)}{W - Q_o^j(p_u(I_{-j}))} \right) \\ & + \log \left(1 + \frac{\gamma^j}{\sum_{i \in I_{-j}} \gamma^i G^i + \sum_{i \neq j} \gamma^i} \right). \end{aligned} \tag{2}$$

As such, the decision to implement an eco-labelling program reflects a number of factors that are simultaneously in play. To begin with, the larger the industry-level green premium G^j, the more likely it is that the inequality in Eq. (2) is satisfied. In addition, aggregate output level of country j, $Q_o^j(p_u(I_{-j}))$, also plays a key role in the determination of labelling incentives. First, the larger the output level in the absence of an eco-labelling program in country j, $Q_o^j(p_u(I_{-j}))$, the more able are producers in country j in shouldering the fixed cost of labelling. However, and contrary to the first effect, a country with market share $(Q_o^j(p_u(I_{-j}))/W)$ that is sufficiently close to one to begin with may also have little left to gain from market share rivalry via eco-labelling.

The third term in Eq. (2) denotes the magnitude and the nature of peer effects between the N countries. To this end, linearise this last term of Eq. (2), $\log\left(1 + \gamma^j/\left(\sum_{i \in I_{-j}} \gamma^i G^i + \sum_{i \neq j} \gamma^i\right)\right)$, so we have;

$$\begin{aligned} & \log \left(1 + \frac{\gamma^j}{\sum_{i \in I_{-j}} \gamma^i G^i + \sum_{i \neq j} \gamma^i} \right) \\ & \approx \log \left(1 + \frac{\gamma^j}{\sum_{i \neq j} \gamma^i} \right) \\ & \quad - \frac{\gamma^j}{\sum_{i \neq j} \gamma^i} \left(\sum_{i \in I_{-j}} G^i \left(\frac{\gamma^i}{\sum_{i \neq j} \gamma^i} \right) \right). \end{aligned}$$

It follows, therefore, that the strategic incentive for country j to adopt an eco-labelling program depends on the interplay of three effects. In particular, adoption is more likely: (i) the larger the production cost advantage of country j in agricultural production $(\gamma^j / \sum_{i \neq j} \gamma^i)$; (ii) the higher the cumulative number of countries that have currently adopted a labelling program I_{-j}, so long as $G^i > 0$, and finally; (iii) the higher the industry-level green premium of those countries $\gamma^i G^i$ that already have a labelling program in place. Specifically, the higher the industry-level green premium, G^i, the larger will be the added incentives for country j to adopt. In the context of our analysis, therefore, Eq. (2) also opens up a way of testing whether these allegations apply, by examining whether developed and developing countries exert differential influence on the adoption behaviour of countries that have yet to adopt eco-labelling.[6]

3. Empirical analysis

In this section, we present the empirical approach to answer the three questions enumerated at the outset of this paper. A key issue is how observed incidences of eco-labelling may reveal information on producers' perception of the size of the industry green premium.

[6] Eq. (2) can be used to determine the Nash equilibrium welfare implications of eco-labelling. These findings are available upon request and show in particular, that the adoption of eco-labelling in a Nash equilibrium is a necessary, though not sufficient condition for producer welfare to improve, relative to a regime where eco-labelling is not a viable option for all countries engaged in export rivalry.

To this end, we refer to the right hand side of Eq. (2), which suggests the inclusion of regressors to capture and control for (i) the cost of eco-labelling; (ii) scale effects; (iii) production cost; and (iv) peer effects.

The dataset that we assembled accordingly contains macro-economic, environmental, eco-labelling, output, bilateral and aggregate trade information on the food industry[7] of 66 countries between 1976 and 1999. To begin with, the data makes use of the implementation year of food industry related eco-labelling programs, if an eco-labelling program has been put in place prior to or during 1999 (Table 1). The eco-labelling information is obtained via an internet search and through secondary literature.[8]

Since our macro-economic and environmental data are national in scope, and our trade and production data track food industry performance for each country, we pay particular attention to three issues while collecting the eco-labelling data. First and foremost, we scrutinize the product coverage of each national labelling program. For example, the *Hungarian Eco-labelling Program* covers a whole range of products and commodities but specifically excludes food products. Similarly, *Ecolabelling Scheme Indonesia* is concerned with the sustainability of tropical timber. Both of these eco-labelling schemes, while national in scope, has little to do with agriculture or the food industry and are thus excluded from the data.

Second, we focus solely on labelling programs that are national in scope, and do not include labelling programs that are carried out by international or sub-national organisations. For example, *Better Banana Project* label, earlier also known as the *Eco-OK* label, invites the participation of banana growers throughout Central America including Costa Rica, Ecuador, El Salvador, Guatemala, Honduras, Panama and Peru. Certification is administered collectively by the *Rainforest Alliance*. Meanwhile, sub-national labelling programs also prevail in countries, such as the United States. The *Core Values Program for Northeast Apple Growers*, *Organic Growers of Michigan*, and the *IPM label* in New York State are but a few of the many examples.

Finally, we focus on labelling programs that encompass a whole range of agricultural and/or food industry commodities. For example, the *Chinese Greenfood Standard* has certified over 1000 different varieties of products to date, ranging from raw meat to processed food products and beverages.

Figs. 1 and 2 respectively plot the Kaplan–Meier estimates of the survival likelihood—the likelihood that a country has not yet adopted eco-labelling as a function of time elapsed since 1976. As may be expected, developing countries are uniformly more likely to be late adopters of eco-labelling. Towards the end of the sample period, the survival probabilities for developing and developed countries are respectively about 0.75 and 0.20. Fig. 2 shows that net exporters as a group essentially overtook net importing countries in their likelihood of adoption.[9] This is despite the fact that a significant number of net importing countries have already adopted eco-labelling in agriculture at the beginning of the sample period.

To capture the fixed (administrative) cost of eco-labelling, c^j is taken to depend on: (i) the stage of development of an economy—real gross domestic product per capita, 1976–1999 (World Bank, 2001a); (ii) the existing level of food industry environmental damage—average food industry water pollution (share of total BOD emission) (World Bank, 2001a); and/or (iii) the existence of national commitments via multilateral environmental agreements, such as the Kyoto protocol and Montreal protocol.

To capture the scale effect, our data contain average pre-labelling food industry total output share at time t, taken from Trade and Production Database (World Bank, 2001b). The comparative cost advan-

[7] Food industry level output, along with bilateral and aggregate food industry trade follows ISIC 3-digit classification code 311, and includes slaughtering, preparing, and preserving meat; manufacture of dairy products; canning, and preserving of fruits and vegetables; canning, preserving, and processing of fish crustacea and similar foods; manufacture of vegetable, and animal oils, and fats, grain mill products; manufacture of bakery products, sugar factories, and refineries; manufacture of cocoa chocolate, and sugar confectionery; manufacture of food products not elsewhere classified.

[8] This information allows us to ascertain whether each of the 66 countries either initiated their own labelling scheme or certified their export products by external certification agencies. For the 30 countries with eco-labelling programs, we further identified the year indicating when labelling schemes or certification were first adopted in the agricultural sector.

[9] Net exporting (importing) countries are defined as countries that sustained an annual average of over (less than) 50% of food industry exports in total food industry trade (imports plus exports) during the 1976–1999 period.

Table 1
Environmental initiatives, stage of development and regional distribution by export orientation

Net exporters						Net importers					
Country name	Eco-label	Montreal protocol	Kyoto protocol	Developing country	Region[a]	Country name	Eco-label	Montreal protocol	Kyoto protocol	Developing country	Region[a]
Argentina	X	X	X	X	lac	Armenia	–	X	X	X	eca
Australia	X	X	X	–	eap	Austria	X	X	X	–	wena
Bulgaria	–	X	X	X	eca	Bangladesh	–	X	X	X	sa
Canada	X	X	X	–	wena	Bolivia	–	X	X	X	lac
Chile	X	X	X	X	lac	Cameroon	–	X	X	X	ssa
China	X	X	X	X	eap	Cyprus	–	X	X	–	wena
Colombia	–	X	X	X	lac	Egypt, Arab Rep.	X	X	X	X	mena
Costa Rica	–	X	X	X	lac	Finland	X	X	X	–	wena
Denmark	X	X	X	–	wena	France	X	X	X	–	wena
Ecuador	–	X	X	X	lac	Germany	X	X	X	–	wena
Ethiopia	–	–	X	X	ssa	Greece	X	X	X	–	mena
Guatemala	–	X	X	X	lac	Hong Kong, China	–	–	–	–	eap
Honduras	–	X	X	X	lac	Iran, Islamic Rep.	–	X	X	X	mena
Hungary	–	X	X	X	eca	Italy	X	X	X	–	wena
India	X	X	X	X	sa	Japan	X	X	X	–	eap
Indonesia	–	X	X	X	eap	Jordan	–	X	X	X	mena
Ireland	X	X	X	–	wena	Korea, Rep.	X	X	X	X	eap
Kenya	–	X	X	X	ssa	Kuwait	–	X	X	–	mena
Malawi	–	X	X	X	ssa	Latvia	–	X	X	X	eca
Malaysia	–	X	X	X	eap	Macao, China	–	–	–	–	eap
Moldova	–	X	–	X	eca	Mexico	–	X	X	X	lac
Netherlands	X	X	X	–	wena	Morocco	–	X	X	X	mena
New Zealand	X	X	X	–	eap	Nepal	–	X	X	X	sa
Norway	X	X	X	–	wena	Pakistan	–	X	X	X	sa
Peru	–	X	X	X	lac	Panama	–	X	X	X	lac
Philippines	–	X	X	X	eap	Poland	X	X	X	X	eca
South Africa	X	X	X	X	ssa	Portugal	X	X	X	–	mena
Spain	X	X	X	–	wena	Romania	–	X	X	X	eca
Sri Lanka	X	X	X	X	sa	Singapore	–	X	X	–	eap
Thailand	X	X	X	X	eap	Sweden	X	X	X	–	wena
Turkey	X	X	–	X	eca	Trinidad and Tobago	–	X	X	X	lac
United States	X	X	X	–	wena	United Kingdom	X	X	X	–	wena
Uruguay	–	–	X	X	lac	Venezuela, RB	–	X	X	X	lac
	17(33)	31(33)	31(33)	24(33)			13(33)	31(33)	31(33)	18(33)	

Sources: Author's compilation.

[a] Regions: lac, Latin American and the Caribbean; eap, East Asia and Pacific; wena, Western Europe and North America; ssa, Sub-saharan Africa; sa, South Asia; mena, Middle East and North Africa; and eca, East and Central Asia.

tage of country j is proxied by the export orientation of the economy—the average share of total exports to total food industry trade, 1976–1999 (Trade and Production database, World Bank, 2001b). Tables 2a and 2b display the summary statistics of the variables that capture the stage of development, existing pollution levels, the scale effect and the comparative cost advantage in two groups of countries, pre and post eco-labelling. Consistent with the Kaplan–Meier survival curves, countries that have instituted an eco-labelling program during the period (1976–1999) on average enjoy a higher real income per capita, and are on average net food industry exporters. Countries that instituted an eco-labelling program produce an average of about 3% of the total food industry output among all the countries included in our data—more than ten times higher than countries that do not have an eco-labelling program in place. Despite their export orientation in food industry, countries that institute a labelling program

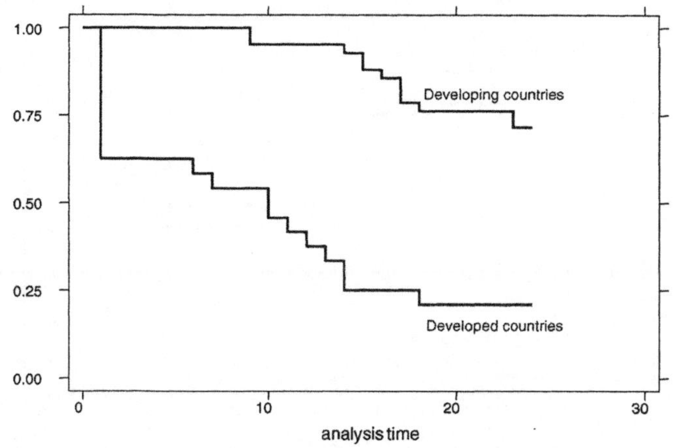

Fig. 1. Kaplan–Meier survival estimates, by stage of development.

appear to have an already relatively low level of food industry share of total water pollution.

Finally, countries are divided into peer groups based on a number of criteria. A first peer grouping is defined based on whether a country is (on average) a net exporter, or a net importer of food industry output, during the period 1976–1999. A second peer grouping takes into account the extent of trade competition at the bilateral trade level. Specifically, each country is given a distinct peer group, which includes the top ten food industry export destinations of the country. To allow for a finer distinction between export destinations, we also single out a sub-peer grouping which includes the top five food industry export destinations of the country. A final peer grouping addresses regional influences, and each country falls into one of a number of regional groups.[10]

Based on these peer groupings, the size of the peer effect for country j at time t is constructed by computing the weighted cumulative number of countries in the peer group other than country j, that have adopted eco-labelling since 1976 till time $t-1$. The weights are taken to be the food industry output $\gamma^j / \sum_{i=1}^{N} \gamma^i$ (Eq. (2)) of country j as a share of the total output of the 66 countries. To compute peer effect based on bilateral export competition, the top 10 export destinations for each country are identified based on food industry bilateral trade data. The weighted cumulative number of these export destinations that have an eco-labelling program in place at each time t gives *wcexdest* at time t. The variable *wcexdest5* is similarly constructed based only on the weighted cumulative number of top 5 export destinations that have instituted an eco-labelling program. The variables *wcreg* and *wcexdum* respectively denote peer effects when peer groupings are based on regional differences, and whether a country is a net importer or a net exporter.

Estimation results

In order to uncover the empirical determinants of the likelihood of adoption, we work with an empirical framework that allows us to analyse the data on food industry eco-labelling illustrated above. The interest here is to empirically ascertain the likelihood of eco-labelling. We take the approach of estimating a proportional hazard model. More specifically, let x_{jt} be a vector of time-varying explanatory variables, where $t = 1976 \ldots T_j$, when country j implements an eco-labelling program.

[10] These include lac (Latin American and the Caribbean), eap (East Asia and Pacific), wena (Western Europe and North America), ssa (Sub-saharan Africa), sa (South Asia), mena (Middle East and North Africa), and eca (East and Central Asia).

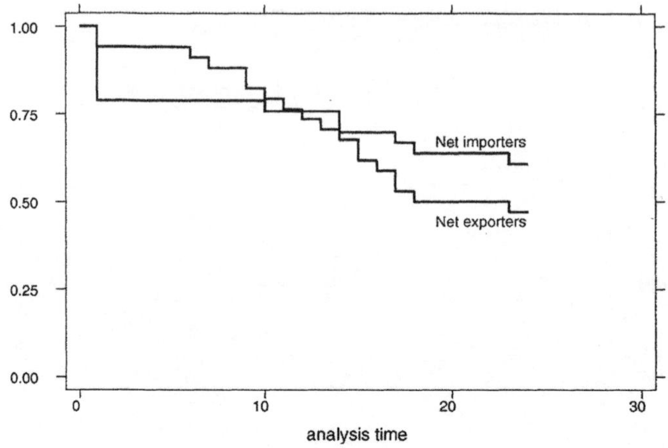

Fig. 2. Kaplan–Meier survival estimates, by export orientation.

Denote the hazard rate at t_j—the probability of adoption when t_j years have passed, given that adoption has not yet taken place—as $h(t_j|x_{jt})$. We assume a model with proportional hazard (Cox (1972)), and specify in addition that each of the K time-varying covariates enter into the determinant of the hazard rate as follows:

$$h(t_j|x_{jt}) = \hat{h}(t_j)\exp\left(\sum_{k=1}^{K}\beta_k x_{jkt}\right) \tag{3}$$

where $\hat{h}(\cdot)$ denotes the baseline hazard function. The hazard ratio for a unit change in x_{jkt} is thus simply $\exp(\beta_k) \geq (<) 1$. Parameter estimates of β_k of the Cox proportional hazard model are obtained by maximising a partial log-likelihood function (Kalfleisch and Prentice, 1980), and has the virtue that the estimation procedure places no restrictions on the unknown functional form of the baseline hazard function.

Empirical estimates of $\exp(\beta_k)$ are presented in three tables.[11] We first turn to a discussion of a number of similarities that emerge in all of our estimations as summarised in Tables 3–5.[12] To begin with, a higher real per capita GDP is associated with a higher likelihood of eco-labelling, as the estimated coefficients are strictly greater than one, and significant at the 1 or 5% level. Based on our theoretical discussion in Section 4, there are a number of possible interpretations of this finding. These include the possibilities that higher income countries (i) are more capable of bearing the fixed cost of eco-labelling; (ii) have in place technologies that are more in line with the production criteria on which eco-labelling programs are typically based; and (iii) have higher industry-level green premia based on perceived or actual credibility of the eco-labelling programs, once one is put into place. Clearly, the first two of these possibilities are relevant to fixed costs considerations that may naturally put developing countries at a disadvantage. The third of these possibilities imply a potentially more contentious problem, that cannot be solved simply via technological transfers. We will return to this question in our discussion of the last part of our estimation results.

[11] Note that the number of incidences of eco-labelling ultimately employed in our estimation is 21 as comparable pre-labelling data on output share, trade orientation are not available to us. These nine countries include Austria, Finland, France, Germany, Italy, New Zealand, Norway, Sweden and the UK, each having established an eco-labelling program prior to 1976. This leaves a total 57 countries that are included in our estimation.

[12] Each table also reports the log likelihood and Wald Chi-squared statistics of the estimation. The hypothesis that all of the estimated coefficients are all equal to one is rejected in all of our estimations, at significance levels of less than 1%.

Table 2a
Trade links and output pre and post eco-labelling

Variable	Pre[a]	S.D.	Post[b]	S.D.	NA[c]	S.D.
Real per capita income (US$ 1995 const.)	10299.750	9400.600	13505.990	12103.420	3487.938	4639.820
Food industry export orientation (% export to total food industry exports and imports)	60.157	21.953	56.469	22.628	48.426	25.143
Food industry export share to US, WE, and JPN (% export to US, WE, JPN to total export)	53.378	21.001	56.622	20.593	47.567	24.983
Food industry output share (% total world output)	3.234	6.927	3.602	6.997	0.222	0.250
Food industry water pollution (% total BOD emission)	44.203	10.298	43.707	9.955	51.046	16.056

[a] Mean country annual averages during the pre-labelling periods for countries that instituted a labelling program after 1976.
[b] Mean country annual averages during the post-labelling periods for countries that instituted a labelling program after 1976.
[c] Mean country annual averages from 1976 to 1999 for countries that never instituted a labelling program.

Table 2b
Export and environmental performance growth

Variable	Labelling[a]		NA[b]	
	Mean	S.D.	Mean	S.D.
Real per capita Income (% change)	43.083	35.215	14.654	28.955
Food industry export orientation (% change)	−4.777	25.925	13.116	48.184
Food industry export share to US, WE, and JPN (% change)	22.861	90.487	−7.716	45.399
Food industry output share (% change)	25.654	35.105	12.853	51.012
Food industry water pollution (% change)	−0.655	8.502	4.432	18.202

[a] Computed based on pre- and post- labelling averages.
[b] Computed based on 1976–1987 and 1988–1999 averages.

A second common feature in our estimations is that of the existing level of food industry related water pollution. In particular, all of the estimated coefficients associated with the existing level of pollution are strictly less than one, and significant at the 5 or 10% level. Interpreting the degree of food industry, water pollution as a proxy for the level of monitoring that may be expected to become necessary once a labelling program is in place, our findings here are consistent with the fact the fixed cost of labelling may be jointly determined by a number of factors. These include the access to green technologies, along with the existing degree of departure from environmentally friendly production practices.

A third common feature relates to the scale effect, as captured by the food industry output share of a country. A food industry output squared term is also included to capture any non-linearity that is suggested by our theoretical discussion. As should be apparent, the scale of the food industry has a significant and positive impact on the likelihood of eco-labelling, though at a rate that is decreasing with size. Thus, size matters and matters positively, but only up to a point.

The fourth common feature across our estimates is the role of comparative cost advantage, accounted for in our estimations by the share of exports to total trade in the food industry of each of the countries. In Eq. (2), comparative cost advantage enters into the determination of the likelihood of adoption in a positive way, and our estimation results likewise indicate that the higher the export share of food industry trade, the more likely it is that a country adopts eco-labelling. In Tables 3–5, the estimated coefficients are strictly greater than unity, and significant at the 1 or 5% level.

As all of the above effects are controlled for, we now turn to strategic interaction between countries based on the peer groupings discussed above. The first set of results is summarised in Table 3. Of these peer effect measures, however, only two are significant and greater than unity, indicating that increases in the

Table 3
Proportional hazard regression: export destination and export orientation peer effects

	Hazard ratios				
	I	II	III	IV	V
Real per capita Income	1.00007***	1.00004*	1.00007***	1.00008***	1.00006**
	0.00002	0.00002	0.00002	0.00002	0.00003
Food industry output share	1.36124***	1.46222***	1.37131***	1.31104**	1.36176**
	0.16898	0.19017	0.16523	0.18003	0.19584
Food industry output share-squared	0.99256**	0.99036***	0.99230**	0.99443	0.99314
	0.00362	0.00383	0.00353	0.00404	0.00430
Food industry water pollution	0.95269***	0.95544***	0.95300***	0.95714**	0.96030**
	0.01728	0.01734	0.01721	0.01900	0.01958
Food industry exports share	1.05236***	1.02343***	1.05383***	1.04847***	1.04459**
	0.01615	0.01613	0.01546	0.01924	0.01969
wcreg		1.02343***			1.01666
		0.00912			0.01050
wcexdum			0.97687		
			0.04180		
wcexdest				0.99896	1.00042
				0.00558	0.00579
wcexdest5				1.06044***	1.04942**
				0.02269	0.02290
Number of observations	1089	1089	1089	1089	1089
Incidences of eco-labelling	21	21	21	21	21
Log likelihood	−63.249	−61.772	−63.184	−60.659	−59.962
Wald Chi-squared	35.220	33.400	46.830	47.840	48.500
Prob > Chi-squared	0.000	0.000	0.000	0.000	0.000

Robust standard errors (Lin and Wei, 1989) in parenthesis.
* Significant at the 10% level.
** Significant at the 5% level.
*** Significant at the 1% level.

weighted cumulative number of countries in these peer groups are associated with a higher likelihood of eco-labelling. Respectively, these are the regional peer effect, *wcreg*, along with the bilateral export destination peer effect for the top five exporting destinations, *wcexdest5*. In addition, when regional and bilateral export destination peer effects are both included (Column V of Table 3), the bilateral export destination peer effect is the only peer effect variable that continues to be significant. Based on our theoretical discussion, these findings may be interpreted as an indication of *strategic complementarity* between countries, and particularly those that are engaged in trade competition. Thus, notwithstanding the argument that international trade tends to encourage the exploitation of the environment, resulting in the possibility of a race to the bottom as countries engage in minimising costs, our findings here suggest the possibility of a reversal of these tendencies in the context of eco-labelling. In addition, rather than a race to the bottom, our findings suggest that a race to the top may be a more appropriate characterisation in the context of food industry eco-labelling initiatives.

Table 4 turns to an examination of the role of multilateral environmental commitments, such as the Kyoto Protocol, and the Montreal Protocol, in determining the likelihood of 'voluntary' mechanisms towards the greening of agriculture, as is the case with eco-labelling. Two sets of regression results are shown, one with all the basic variables and the ratification of the two multilateral environmental agreements, but without the peer effects, and a second set includes the export destination peer effects. The variables *Kyoto* and *Montreal* are both binary variables that take on a value of one (zero) at time t for country j if the corresponding multilateral environmental agreement is (is not) ratified by country j at any time $s < t$. Our findings indicate that neither of

Table 4
Proportional hazard regression: multinational environmental agreements

	Hazard ratios					
	I	II	III	IV	V	VI
Real per capita income	1.00007***	1.00006***	1.00008***	1.00007***	1.00005*	1.00006**
	0.00002	0.00002	0.00002	0.00002	0.00003	0.00003
Food industry output	1.36084***	1.30577*	1.31269*	1.28632*	1.33169*	1.36262**
	0.16971	0.19084	0.18041	0.20233	0.21784	0.19582
Food industry output-squared	0.99257**	0.99380	0.99441	0.99495	0.99375	0.99314
	0.00363	0.00425	0.00404	0.00457	0.00484	0.00430
Food industry water pollution	0.95263***	0.94961***	0.95747**	0.95415**	0.95660**	0.96057**
	0.01718	0.01768	0.01913	0.01913	0.01954	0.01970
Food industry exports share	1.05236***	1.05044***	1.04843***	1.04844***	1.04464**	1.04457**
	0.01616	0.01538	0.01920	0.01857	0.01889	0.01968
wcreg					1.01545	1.01657
					0.01032	0.01053
wcexdesttop10			0.99891	0.99895	1.00035	1.00037
			0.00557	0.00579	0.00595	0.00579
wcexdesttop5			1.06122***	1.05521**	1.04479**	1.04999**
			0.02278	0.02372	0.02370	0.02304
Kyoto	1.08453		0.63127			0.72182
	0.89143		0.52041			0.58616
Montreal		3.54972		2.95939	2.82098	
		3.00581		2.43227	2.21605	
Number of observations	1089	1089	1089	1089	1089	1089
Incidences of Eco-labelling	21	21	21	21	21	21
Log likelihood	−63.248	−61.849	−60.638	−59.713	−59.083	−59.951
Wald Chi-squared	37.950	47.210	47.820	62.750	62.460	48.460
Prob > Chi-squared	0.000	0.000	0.000	0.000	0.000	0.000

Robust standard errors (Lin and Wei, 1989) in parenthesis.
* Significant at the 10% level.
** Significant at the 5% level.
*** Significant at the 1% level.

the two multilateral agreements appears to have significant influence on the likelihood of adoption. Indeed, in the four estimation results displayed in Table 4, none of the estimated coefficients associated with the two Protocols are significant. Comparing these findings with the data presented in Table 1, the result should perhaps not be all that surprising. In particular, while eco-labelling is prevalent mostly in developed countries, and/or in countries that are export oriented, the ratification of the two multilateral environmental conventions is almost universal.

As discussed in Section 4, countries that perceive a higher industry-level green premium are more likely to have a stronger influence on the labelling decision of other countries. This analytical finding suggests one possible way of trying to ascertain whether developed and developing countries may have differential industry-level green premia, as has been popularly argued in the North–South debate surrounding the potential distributional consequences of eco-labelling. To this end, we construct three additional variables based on the peer effect variables established above. *wcdevexdum*, *wcdevexdest*, and *wcdevreg*, respectively denote the weighted cumulative number of developing countries that have adopted a labelling program respectively in the three peer groupings. The inclusion of each of these three additional variables into the estimation thus allows the strategic impact of labelling programs initiated in developing countries to be singled out. Accordingly, if the estimated coefficient of these new variables are less than (greater than) unity, we may infer that the industry-level green premia of developing countries are strictly lower (higher) than that of developed countries.

Table 5
Proportional hazard regression: industry premium differentials

	Hazard ratios		
	I	II	III
Real per capita income	1.00005**	1.00006***	1.00008***
	0.00002	0.00002	0.00002
Food industry output	1.42285***	1.43480***	1.32569**
	0.18607	0.18385	0.17711
Food industry output-squared	0.99099**	0.99112**	0.99420
	0.00389	0.00367	0.00389
Food industry water pollution	0.93737***	0.95553**	0.95482**
	0.01664	0.01842	0.01905
Food industry exports share	1.05461***	1.05781***	1.04927***
	0.01368	0.01461	0.01961
wcreg	1.02342***		
	0.00960		
wcexdum		0.98951	
		0.03944	
wcexdesttop10			1.01314
			0.01246
wcexdesttop5			1.05547**
			0.02263
wcdevreg	0.47443*		
	0.20675		
wcdevexdum		0.85037	
		0.17514	
wcdevexdest			0.93680
			0.04440
Number of observations	1089	1089	1089
Incidences of eco-labelling	21	21	21
Log likelihood	−59.245	−62.923	−60.309
Wald Chi-squared	32.950	50.670	42.510
Prob > Chi-squared	0.000	0.000	0.000

Robust standard errors (Lin and Wei, 1989) in parenthesis.
* Significant at the 10% level.
** Significant at the 5% level.
*** Significant at the 1% level.

The results are presented in Table 5. As should be apparent, all of the estimated coefficients of developing country peer effects are less than one (including *wcdevreg*, *wcdevexdum* and *wcdevexdest*). However, none of the estimated coefficients for developing country peer effects are significant at 5% or less, with the exception of developing country peer effect for the regional peer grouping, which is significant at the 10% level. Returning to our discussion earlier, having to do with the possibility that a country's stage of development may play a role in determining the perceived credibility of eco-labelling programs, our findings here would suggest that at least for the case of food industry labelling, systematic and/or significant evidence suggesting that industry green premia respectively of developed and developing countries diverge in a predictable way has yet to be found.

4. Conclusion

This paper began with the observation that while eco-labelling in agriculture has been longstanding in developed countries in Western Europe, the spread of this practice is beginning to take hold in developing countries worldwide. The data that we assembled for this paper, regarding the time to adoption pattern of eco-labelling programs in 66 countries, reveal further that food industry export orientation appear to be correlated with the speed with which countries implement their own eco-labelling programs.

The theoretical model proposed in this paper suggests two sets of results, having to do with the selection criteria of the adoption of labelling, and the welfare consequences of eco-labelling in a subgame perfect Nash equilibrium. In terms of selection criteria, the model suggests that so long as countries perceive a strictly positive green premium, the speed of adoption depends on: (i) the fixed cost of eco-labelling programs; (ii) a scale effect; and (iii) the comparative cost advantage of the industry in question. Interestingly, the model also suggests the likely prevalence of strategic interactions, so that the decision to adopt by one country depends on the decision to adopt by other countries. Comparing producer welfare with and without competition based on eco-labelling, the theoretical discussion indicates that while countries that find themselves satisfying these selection criteria in a subgame perfect Nash equilibrium are not necessarily made better off, countries that do not satisfy these selection criteria will be worse off.

The empirical findings provide a set of possible answers to the three questions posed at the outset of this paper, and suggest additional research questions. To begin with, while production specialisation induced by international trade may encourage environmental exploitation when the exportable industry is pollution intensive, our findings suggest that the export orientation of an industry can itself be a driving force that makes the practice of eco-labelling an attractive option. Meanwhile, with strategic complementarity in adoption as shown in both the theory and the empirical discussions of this paper, at least insofar as eco-labelling in the food industry is concerned, our findings suggest that quite to the contrary of a 'race to the bottom' may in fact be in play.

Finally, while we find no significant difference between the industry green premia between developed and developing countries based on observed patterns of adoption and strategic interdependence, developing countries are nevertheless late adopters, due possibly to the ability to shoulder the cost of having a credible labelling program. This finding confirms in part the concerns regarding the negative terms of trade consequence of eco-labelling particularly for developing countries.

Acknowledgments

We thank seminar participants at the IAAE 2003 Annual Meeting (Durban, South Africa), the SAET (Society for the Advancement of Economic Theory) 2003 Annual Conference, the University of Hong Kong, and the City University of Hong Kong for useful comments. This paper was completed while Basu and Chau were visiting the Center for Development Research (ZEF), University of Bonn, Germany. The final revision of this paper was undertaken while Basu and Chau were visiting the Department of Economics and Finance at the City University of Hong Kong. The hospitality of the department staff and colleagues is gratefully acknowledged. Finally, Basu thanks the Reves Center at the College of William and Mary for a Faculty Travel Grant.

References

Basu, A.K., Chau, N.H., 1998. Asymmetric country-of-origin effects on intra-industry trade and the international quality race. Rev. Dev. Econ. 2, 140–166.

Basu, A.K., Chau, N.H., 2001. Market Access Rivalry and Eco-labelling Standards: Are Eco-labels Non-tariff Barriers in Disguise? Department of Applied Economics and Management Working Paper, Cornell University, WP, pp. 2001–2015.

Basu, A.K., Chau, N.H., Grote, U., 2003. Eco-labelling and stages of development. Rev. Dev. Econ. 7, 228–247.

Chau, N.H., Kanbur, R., 2000. A Race to the Bottom, from the Bottom. Department of Applied Economics and Management Working Paper, Cornell University, WP, pp. 2000–2016.

Cox, D.R., 1972. Regression models with life-tables. J. R. Stat. Soc. Ser. B 34, 187–220.

Kalfleisch, J.D., Prentice, R.L., 1980. The Statistical Analysis of Failure Time Data. Wiley & Sons, New York.

OECD, 1997a. Eco-labelling: Actual Effects of Selected Programmes. OECD, Paris.

OECD, 1997b. Uses of Food Labelling Regulations, OCDE/GD(97)150. OECD, Paris.

World Bank, 2001a. World Development Indicators, 2001. The World Bank, Washington, DC.

World Bank, 2001b. Trade and Production Database. The World Bank, Washington, DC, http://www.worldbank.org/research/trade.

WTO-CTE, 1997. Trade and Environment; Bulletin No. 18, Press/TE018. WTO, Geneva.

[29]

Eco-labelling scheme, environmental protection, and protectionism

Huilan Tian *Division of Management, University of Toronto at Scarborough*

Abstract. In this paper a model of international duopoly is developed involving competition in both prices and levels of environmental friendliness, and the implications of government policies are studied. It is shown that, contrary to the conventional wisdom, a regulatory increase in the minimum required level of environmental friendliness of imported goods may harm the home firm and may result in a rise in the volume of imported goods. Whether consumers lose or gain from such a regulatory increase depends on consumption spillover effects. We also show that, under certain conditions, the duopoly's equilibrium choice of levels of environmental friendliness is socially optimal. JEL Classification: F00

Eco étiquetage, protection environnementale, et protectionnisme. Ce mémoire développe un modèle de duopole international où la concurrence se fait à la fois par les prix et par les niveaux de sensibilité aux problèmes environnementaux. On étudie les impacts des politiques gouvernementales. Contrairement à ce que suggère la sagesse conventionnelle, il semble qu'un accroissement réglementaire dans le minimum requis de disposition favorable à l'environnement des biens importés peut nuire à l'entreprise locale, et peut résulter en un accroissement du volume de biens importés. Le consommateur perdra ou gagnera en conséquence de cette accroissement de la réglementation selon la nature des effets de retombées. On montre aussi que, sous certaines conditions, le choix des niveaux d'équilibre de sensibilité aux problèmes de l'environnement s'avère socialement optimal.

1. Introduction

While it is widely accepted that international trade is generally gainful for all countries, the increase in trade volume in the past two decades has also been accompanied by environmental problems such as deforestation, increased use

Email: tian@utsc.utoronto.ca

of chemicals in agriculture, depletion of scarce resources. In fact, the growth in trade volume and trade liberalization has gradually led to a widespread public awareness of environmental issues. Such awareness has found its expression in the formation of environmentally minded pressure groups that seek to put pressure on governments in order to influence policies on trade and the environment. The environmental awareness has also been manifested in the market place, as more and more consumers are becoming increasingly concerned with the environmental and health properties of goods and services.

Consumers, especially in developed countries, are becoming more responsive to product-labelling schemes, and some of them are even willing to pay more for a product if its production process is more 'friendly' to the environment. Not only does this kind of consumer activism affect consumption demands for goods labelled as environmentally friendly, it is also reflected in some investment choices, as exemplified by the 'socially responsible investing' (SRI) principles. As of 3 July 2000 British pension funds have been required by law to disclose whether they will take account of the environmental, ethical, and social impacts of their investments.

Firms have also recognized that environmental awareness can be a source of market opportunities. For example, the Home Depot, a huge and widely admired home improvement retailer, has pledged to give preference to certified wood[1] and to stop buying wood from endangered areas by the end of 2002. More and more firms are offering products labelled 'green products.' These are becoming increasingly popular, not only in western economies, but also in emerging market economies, for example, the 'green food' in China and the 'green cotton' in India.

In several countries, working groups have proposed the adoption of eco-labelling schemes, which either require or encourage the supply of information on the environmental characteristics of a commodity to the public. Even though there remain disagreements on some ethical issues concerning trade and environment and on the effectiveness of such schemes, some attempts at formulating criteria and implementation have been undertaken. A prominent example is the German 'Blue Angel' scheme.[2]

Some limited evidence has lent support to the presumption that positive environmental effects could result from eco-labels. (1) Norway's imports of fine papers originating from Brazil declined significantly after the introduction

1 See *Ford Foundation Report*, 4–8. According to this report, the Forest Stewardship Council (FSC), an international non-profit organization that has national working groups in 16 countries, accredits certifying organizations that evaluate management practices in forests. As of now, about 45 million acres, or about 5% of the world's commercially logged forestland, including about 6.2 million acres in the United States, are certified by FSC. To be certified, a forest operation must show evidence of environmetally friendly management practices.

2 Eco-labelling schemes exist in Germany, Canada, Japan, Norway, United States, Sweden, New Zealand, India, Austria, Australia, Republic of Korea, Singapore, France, Netherlands, and Croatia. At present, eco-labelling schemes are mostly voluntary.

of an eco-label; (2) a few years after the introduction of the eco-labels for oil and gas heating appliances under the German Blue Angel program, emission of sulphur dioxide, carbon monoxide, and nitrogen oxides were reduced by more than 30%; (3) in the United States 'dolphin-safe' tuna products have driven the non-eco-label tuna products out of the supermarket shelves.

However, there are many arguments concerning the eco-labelling scheme. One argument is that it can be used as a technical barrier to trade (TBT). The United States has protested against the European attempts to require eco-labelling of agricultural products containing genetically modified organisms, on the grounds that such labelling would discriminate against American products. In addition, in some cases certification of eco-labels can be quite costly. Several case studies have highlighted some possible impacts of eco-labelling schemes on trade flows. Pedro da Motta Veiga (1997) concluded that the proposed EU labels in relation to textiles, pulp and paper, and footwear will be difficult for some Brazilian firms to acquire because of their small size and other technological constraints. They also questioned the relevancy of the criteria. In their investigation of the Colombian case, Lily Ho et al. (1997) concluded that the costs of complying with EU standards were much higher for Colombian producers than for EU producers because of the way the criteria are framed.

To the extent that less developed countries may find it difficult to satisfy some eco-labelling requirements imposed by developed economies, some interesting questions arise: (1) Would firms in developed economies be better off by exerting pressure on their governments to impose a certain level of environmental friendliness for the imported good? (2) Would governments of developed countries respond to such pressure? (3) Do consumers gain or lose from such government intervention? And, more important, (4) Does the volume of imports necessarily fall as a result of increased environmental requirement? (5) Does an increase in the level of environmental friendliness of products really improve the environmental quality?

In an attempt to provide some answers to the above questions, in this paper we present a model of an international duopoly, consisting of a home firm and a foreign firm that face consumers, in the home country, who are concerned about the environmental impacts of the products. The price that consumers are willing to pay for a good is assumed to be an increasing function of the degree of environmental friendliness of products. The firms therefore compete on their levels of environmental friendliness. However, the more environmentally friendly a product is, the higher is the average variable cost of producing the good. In addition, the firms must incur a set-up cost that is a non-decreasing function of the level of environmental friendliness. The game between the two firms is modelled as a two-stage game. In the first stage, firms choose the level of environmental friendliness of their own product and incur a set-up cost. In the second stage, they compete in prices.

Using this model, we first characterize the equilibrium of the game between the duopolists, then proceed to ask a few questions about government policies. Suppose the government of the home country wishes to maximize its political support function, which is a *weighted* average of the gain in consumers' surplus and the gain to the home firm, under what conditions would the home government impose a minimum requirement of environmental friendliness on the imported goods? Would such an action be beneficial to the home firm? Would it result in an environmental improvement?

It turns out that, under certain conditions, imposing a minimum level of environmental friendliness on the imported goods may result in an increase in import volume and a fall in the profit of the home firm. Thus, contrary to the conventional wisdom, imposing environmental requirements is neither necessarily a barrier to trade nor a protectionist device. Moreover, the government may find, under certain conditions, that the laissez-faire levels of environmental friendliness chosen by the duopolists are, in fact, socially optimal. Our analysis reveals that, in general, one cannot presume that an increase in environmental requirements on imported goods will help the home firm, or the environment. The model points to the important trade-offs that one must look for in environmental and trade policies. It also reminds us that a good intention may result in a bad outcome. As Samuelson once put it, the problem is to go beyond the good intention.

Our model is built on the earlier work on duopoly and trade, as exemplified by papers of Brander and Spencer (1985) and Eaton and Grossman (1986). However, in discussing government policies, we focus on the imposition of a minimum environmental requirement. Our paper is different from the existing models on standards, which largely focus on the issues of compatibility and network externalities; (see, for example, Matutes and Regibeau (1996), Jeanneret and Verdier (1996), Economides (1996), de Palma and Leruth (1996)). In the international trade context, Fischer and Serra (2000) analysed a model of Cournot rivalry in which the home country may impose a minimum standard. They obtained a number of interesting results on the relationship between minimum standard and relative market size. They made, however, the simplifying assumption that the utility function does not have a parameter that would shift the demand curve as standards, or quality levels, are changed. In contrast, in our paper, for given prices, consumers prefer products with higher levels of environmental friendliness: this creates incentives for firms to respond to consumers' preferences. Our paper also differs from models of quality choice, such as Leland (1979), Shapiro (1983), Ronnen (1991), because each consumer in our model cares not only about the number of units he consumes, but also about the total damage to the environment that results from the total consumption by the society. Finally, in postulating a political support function, we follow the footsteps of Hillman (1982, 1989, 1991) and Hillman and Ursprung (1988) in their analysis of protection.

We have chosen the duopoly framework because we believe that this type of model offers a greater scope for the analysis of international rivalry in the choice of environmental characteristics of goods. On the other hand, a perfectly competitive market structure would allow focusing on different issues. For example, Copeland and Taylor (1994, 1995) assume perfect competition and focus on general equilibrium linkage between income, pollution, and trade patterns. Schleich and Orden (2000) also assume that firms are perfectly competitive, and focus on 'contribution schedules' that industry pressure groups offer to the government so as to influence policies.

The rest of the paper is organized as follows. In section 2, we present the basic model and analyse the equilibrium of the two-stage game between the duopolists. In section 3, we study the effects of a policy of imposing a minimum environmental requirement on imported goods, and on home goods. Section 4 offers some concluding remarks.

2. A model of an eco-labelling duopoly

2.1 The basic assumptions

We consider, for simplicity, a product category for which the degree of environmental friendliness can be objectively measured and can be represented by an index θ, a real number, in the real interval $[\theta_L, \theta_H]$. For example, cardboard may contain 75% recycled material (thus its $\theta = 0.75$). We must note two important facts: (i) in the real world, eco-labels, to be credible, need certification by independent certifying organizations that are themselves certified;[3] (ii) even in the continuous variable case, certifying organizations do not provide a continuum of labels.[4] For simplicity, we abstract from the discreteness of certification.

The product group consists of two differentiated goods, denoted by X and Y. Even in the absence of eco-labelling, these goods are already differentiated. Eco-labelling provides another dimension of differentiation. For example, good X is a recyclable paper bag, and good Y is a recyclable plastic bag; eco-labelling may indicate the percentage of recycled material used to make these bags. Good X is produced by a single firm, called firm X, located in the home country, and good Y is produced by firm Y in the foreign country. The two firms compete in the home country.

Let θ_i denote the *actual* degree of environmental friendliness of good i ($i = X, Y$); θ_i^r denote the index *reported* on the label; and θ_i^p denote the

3 For example, two certifying organizations in the United States, SmartWood, and Scientific Certification Systems, have been accredited (i.e., certified), in turn, by the Forest Stewardship Council. See *Ford Foundation Report* (2000, 5).

4 For example, chain-of-custody certification is available for manufacturers of furniture who can demonstrate that any assembled final product consists of *at least* 70% certified wood. See *Ford Foundation Report* (2000, 6).

representative consumer's *perceived* level of environmental friendliness of good i. Each firm i has the choice between (a) practising eco-labelling, that is, printing a value θ_i^r and (b) not practising eco-labelling. We assume that there are heavy penalties for telling lies, so that if the firm chooses action (a), it will print the true value, and consumers, knowing that firms do not dare to lie, believe in the reported value θ^r, that is, $\theta_i^r = \theta_i^p = \theta_i$. If a firm i chooses action (b), then $\theta_i^p = \theta_L$, that is, the consumers will think that θ_i is at its lowest level. In what follows, we assume that cost and demand parameters are such that both firms find it profitable to practise eco-labelling.

We now specify the utility function U of the representative consumer in the home country. Assume that U is quasi-linear: it is linear in the numeraire good, denoted by Z, and non-linear in goods X and Y. The market for good Z is perfectly competitive and its price is $P_Z = 1$ by normalization. Let lowercase letters, such as x, y, and z, with subcript j, denote the quantities consumed by individual j. We denote by X_{-j} and Y_{-j} the quantities of goods X and Y consumed by individuals other than j. We adopt the specific functional form,

$$U(x_j, y_j, z_j) = u(x_j, y_j) + z_j,$$

where

$$u(x_j, y_j) = \left\{ A_0 x_j - \frac{B}{2} x_j^2 + A_0 y_j - \frac{B}{2} y_j^2 - D x_j y_j \right\} \\ - \gamma[\theta_H - \theta_X][x_j + \beta X_{-j}] - \gamma[\theta_H - \theta_Y][y_j + \beta Y_{-j}]$$

with $A_0 > 0$, $B > 0$ and $\gamma > 0$. In addition, we postulate that $B^2 > D^2$ to ensure that u is strictly concave in (x_j, y_j). The terms inside the brackets $\{\ldots\}$ are the familiar quadratic utility that one often encounters in industrial organisation models. The term $[\theta_H - \theta_X]$ is a measure of the shortfall of the actual degree of environmental friendliness of good X from the maximum θ_H. Thus, $[\theta_H - \theta_X]x_j$ is consumer j's perception of the harm done to the environment by his consumption, x_j. The parameter γ is the consumer's marginal evaluation of environmental quality of the product. The parameter β measures the degree of consumer' concern about the environmental impact of consumption, and it can take any value in the real interval $[0, 1]$. There are two polar cases. If $\beta = 0$, this indicates that consumer j is concerned only with the impact of his own consumption on the environment. If $\beta = 1$, then $x_j + \beta X_{-j}$ is society's total consumption of good X, and the term $[\theta_H - \theta_X][x_j + \beta X_{-j}] = [\theta_H - \theta_X]X_j$ may be thought of as representing the environmental damage caused by aggregate consumption of good X. In the intermediate case where $0 < \beta < 1$, the consumer is more concerned about the effect of his own consumption on the environment than the effect of other people's consumption. As long as $\beta \neq 0$, other people's consumption of the product will affect consumer j's utility. As we will show in a later section, the impact of the choice of θ_X and θ_Y by firms on the level of social welfare depends on the value of β. Consequently, optimal government policies, such as minimum requirement on the levels θ_X and θ_Y, also depend on β.

Let M_j denote the income of consumer j. He takes the prices P_X, P_Y, and P_Z and the quantities X_{-j} and Y_{-j} as given, and chooses the quantities x_j, y_j, and z_j to maximize his utility U_j, subject to the budget constraint

$$P_X x_j + P_Y y_j + P_Z z_j = M_j.$$

Assuming that the maximum is an interior one, the individual j's demand functions for goods X and Y are

$$x_j = x(P_X, P_Y, \theta_X, \theta_Y) = \frac{A(B-D) - B[P_X - \gamma\theta_X] + D[P_Y - \gamma\theta_Y]}{B^2 - D^2} \tag{1}$$

and

$$y_j = y(P_X, P_Y, \theta_X, \theta_Y) = \frac{A(B-D) - B[P_Y - \gamma\theta_Y] + D[P_X - \gamma\theta_X]}{B^2 - D^2}, \tag{2}$$

where A is defined by

$$A \equiv A_0 - \gamma\theta_H$$

(we assume $A > 0$).

The demand functions (1) and (2) are intuitively appealing. The term $(P_X - \gamma\theta_X)$ may be interpreted as the 'net price' of X: in consuming a unit of good X, which has the environmental friendliness level θ_X, the consumer incurs a 'monetary cost' of P_X dollars, but receives a 'psychological benefit' valued at $\gamma\theta_X$. A similar interpretation applies to $(P_Y - \gamma\theta_Y)$. Note that the two goods are substitutes if $D > 0$. In what follows, we will focus on this case.

2.2. *The two-stage game between the firms*

2.2.1. The setting

We consider the following two-stage game between the duopolists. In the first stage, each firm decides on the level of environmental friendliness of its product, that is, firm X (respectively, Y) chooses a level θ_X (respectively, θ_Y) in the real interval $[\theta_L, \theta_H]$. This choice involves the choice of a production process. In the second stage, the duopolists compete in the home country, by choosing prices. That is, in the second stage the duopolists are Bertrand rivals in differentiated products. It can be shown that the main results are substantially unchanged if the firms are Cournot rivals.

For firm X, the greater is θ_X, the greater will be the average variable cost, denoted by $f(\theta_X)$, which we assume to be independent of the output level. For

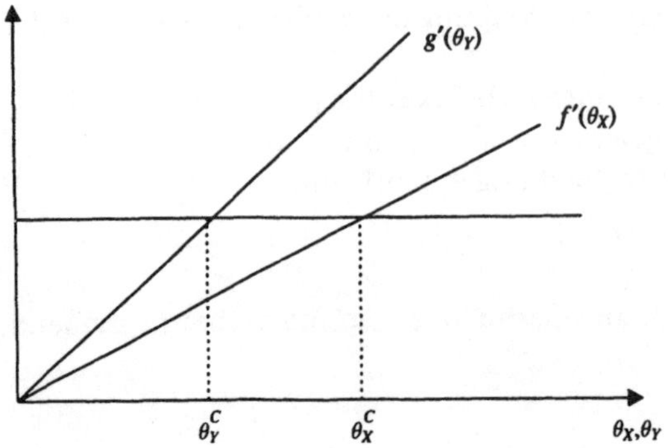

FIGURE 1 Properties of the variable cost functions

firm Y, the average variable cost is denoted by $g(\theta_Y)$. In what follows we assume that the functions $f(.)$ and $g(.)$ are strictly convex and increasing: $f' > 0, f'' > 0$, and $g' > 0$, $g'' > 0$. The functions $f(.)$ and $g(.)$ are not necessarily identical. For example, it is possible that $g(\theta_Y) = b\theta_Y^2$ while $f(\theta_X) = a\theta_X^2$, where $a < b$. For example, firm Y, located in a less-developed country, has a higher cost of achieving any given level of environmental friendliness in its production process.

To avoid corner solutions, we assume that

$$f'(\theta_L) = 0 = g'(\theta_L), \qquad f'(\theta_H) = g'(\theta_H) > \gamma.$$

These assumptions imply (see figure 1) the existence of a unique *critical value* θ_X^C in $[\theta_L, \theta_H]$ and a unique critical value θ_Y^C in $[\theta_L, \theta_H]$ that satisfy

$$f'(\theta_X^C) = \gamma = g'(\theta_Y^C). \tag{3}$$

At the critical value θ_i^C, a marginal increase in the level of environmental friendliness of good i would raise the average variable cost by an amount that is just equal to the consumer's marginal valuation of environmental quality, γ. We will see later that the values θ_X^C and θ_Y^C play an important role in the analysis of market equilibrium and social welfare.

In addition, there is a set-up cost $F(\theta_X)$ for firm X, and $G(\theta_Y)$ for firm Y. The set-up costs are incurred in stage 1, and the variable costs are incurred in stage 2. Since there are N identical consumers in the home country, and since by assumption the firms sell their products only in the home country, the net profit functions of the firms are

$$\pi_X^{net} = [P_X - f(\theta_X)]Nx(P_X, P_Y, \theta_X, \theta_Y) - F(\theta_X)$$
$$\pi_Y^{net} = [P_Y - g(\theta_Y)]Ny(P_X, P_Y, \theta_X, \theta_Y) - G(\theta_Y).$$

Our first step is to find the Bertrand equilibrium in stage 2 for any given pair (θ_X, θ_Y) chosen in stage 1.

2.2.2. Stage 2 equilibrium under Bertrand rivalry
Given (θ_X, θ_Y), firm X chooses P_X to maximize π_X, taking P_Y as given. This gives the first-order condition

$$\frac{\partial \pi_X}{\partial P_X} = Nx(P_X, P_Y, \theta_X, \theta_Y) + N[P_X - f(\theta_X)]\frac{\partial x}{\partial P_X} = 0. \tag{4}$$

Similarly, firm Y's first-order condition is

$$\frac{\partial \pi_Y}{\partial P_Y} = Ny(P_X, P_Y, \theta_X, \theta_Y) + N[P_Y - g(\theta_Y)]\frac{\partial y}{\partial P_Y} = 0. \tag{5}$$

Equations (4) and (5) determine a unique pair of equilibrium prices $P^b_X(\theta_X,\theta_Y)$ and $P^b_Y(\theta_X,\theta_Y)$, where the superscript b indicates this is a Bertrand equilibrium. From (4), we obtain firm X's reaction function as

$$P_X = I_X(\theta_X, \theta_Y) + \frac{D}{2B}P_Y \equiv R_X(P_Y; \theta_X, \theta_Y),$$

where

$$I_X(\theta_X, \theta_Y) = \frac{1}{2B}[A(B - D) + Bf(\theta_X) + \gamma B\theta_X - D\gamma\theta_Y].$$

We can represent this reaction function in a diagram with P_Y along the horizontal axis and P_X on the vertical axis. The graph of the reaction function is a straight line, with positive slope $D/(2B) < 1/2$ (recall that we assumed $B > D > 0$). The vertical intercept of this reaction function is $I_X(\theta_X, \theta_Y)$.

It is useful to note at this stage how changes in these values affect the reaction functions. An increase in θ_Y will shift the vertical intercept $I_X(\theta_X, \theta_Y)$ downward. This means that, *if* P_Y does not change, an increase in θ_Y (meaning good Y becomes more environmental friendly) will cause firm X to lower its price, because it now faces a 'stronger' rival. By contrast, an increase in θ_X will shift vertical intercept of firm X's reaction function upward; hence, for a given P_Y, firm X will raise its price, because good X has become more attractive.

Similarly, from (5), we obtain for firm Y the reaction function

$$P_Y = J_Y(\theta_X, \theta_Y) + \frac{D}{2B}P_X \equiv R_Y(P_X; \theta_X, \theta_Y),$$

where

$$J_Y(\theta_Y) \equiv \frac{1}{2B}[A(B - D) + Bg(\theta_Y) + \gamma B\theta_Y - \gamma D\theta_X].$$

The intersection of the two reaction functions gives the Bertrand equilibrium pair of prices (P^b_X, P^b_Y) for given (θ_X, θ_Y):

$$P^b_X = E + 2\phi B^2[f(\theta_X) + \gamma\theta_X] + \phi BD[g(\theta_Y) - \gamma\theta_Y] + \gamma\phi D^2\theta_X$$
$$P^b_Y = E + 2\phi B^2[g(\theta_Y) + \gamma\theta_Y] + \phi BD[f(\theta_X) - \gamma\theta_X] + \gamma\phi D^2\theta_Y,$$

where

$$\phi \equiv \frac{1}{4B^2 - D^2} \text{ and } E \equiv \phi A(B - D)(2B + D).$$

The Bertrand equilibrium prices (P^b_X, P^b_Y) have the following properties:

$$\frac{\partial P^b_X(\theta_X, \theta_Y)}{\partial\theta_X} = \phi[2B^2 f'(\theta_X) + (2B^2 - D^2)\gamma] > 0 \tag{6}$$

and

$$\frac{\partial P^b_X(\theta_X, \theta_Y)}{\partial\theta_Y} = \phi BD[g'(\theta_Y) - \gamma]. \tag{7}$$

Similarly,

$$\frac{\partial P^b_Y(\theta_X, \theta_Y)}{\partial\theta_Y} = \phi[2B^2 g'(\theta_Y) + (2B^2 - D^2)\gamma] > 0 \tag{8}$$

and

$$\frac{\partial P^b_Y(\theta_X, \theta_Y)}{\partial\theta_X} = \phi BD[f'(\theta_X) - \gamma]. \tag{9}$$

PROPOSITION 1. (Comparative statics on Bertrand equilibrium prices and net prices)

(i) *A marginal increase in θ_i will* always *result in an increase in the Bertrand equilibrium price of good i.*

(ii) *The effect of such increase on its rival's price depends on the initial level of θ_i:* (1) *If $\theta_i < \theta^C_i$, a marginal increase in θ_i will lead to a fall in the Bertrand equilibrium price of good j;* (2) *If $\theta_i > \theta^C_i$, a marginal increase in θ_i will lead to a rise in the Bertrand equilibrium price of good j.*

(iii) *The effect on its own net price also depends on the initial level of θ_i:* (1) *If $\theta_i < \theta^C_i$, then a small increase in θ_i will lead to a fall in the equilibrium* net price *of good i;* (2) *If $\theta_i > \theta^C_i$, then a small increase in θ_i will lead to a rise in the equilibrium* net price *of good i.*

Intuitive explanation. The increase in P^b_X in response to a small increase in θ_X is due to *both* (i) the increase in average variable cost, and (ii) the increase in the attractiveness of good X to the consumers who care about environmental impact of the products. The sign of this equilibrium response is positive, that

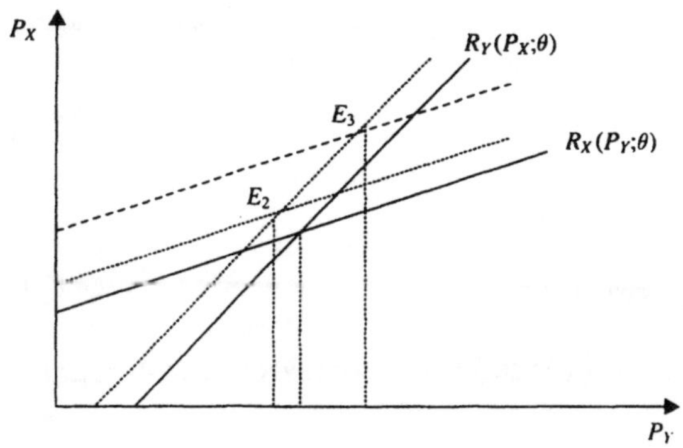

FIGURE 2 Bertrand equilibrium prices

is, it is independent of the initial value θ_X. By contrast, the sign of the effect of a small increase in θ_X on the equilibrium price of the rival product, P^b_Y, depends on the initial level θ_X. A small increase in θ_X causes P^b_Y to fall if θ_X is initially lower than θ^C_X and to rise if θ_X is initially greater than θ^C_X. This is because an increase in θ_X by an amount ϵ shifts the reaction function $R_Y(P_X;\ \theta_X, \theta_Y)$ to the left by the distance $\gamma\varepsilon D/2B$ and at the same time shifts the reaction function $R_X(P_Y;\ \theta_X, \theta_Y)$ upward by (1/2) $[f'(\theta_X)+\gamma]\varepsilon$. This upward shift is greater, the greater is the initial θ_X. The leftward shift in $R_Y(P_X;\ \theta_X, \theta_Y)$ means that good Y becomes relatively less attractive, for given prices, and this by itself *tends* to induce firm Y to reduce its price (point E_2). But the upward shift in $R_X(P_Y;\ \theta_X, \theta_Y)$ means that firm X will increase its price, because of both higher cost and higher demand. If this price hike is largely due to cost rather than demand (and this occurs if the initial θ_X exceeds θ^C_X), then the optimal reaction of firm Y is to increase its price instead (point E_3). (See figure 2.)

We now turn to the equilibrium outputs in stage 2, denoted by $x^b(\theta_X, \theta_Y) \equiv x(P^b_X, P^b_Y, \theta_X, \theta_Y)$ and $y^b(\theta_X, \theta_Y) \equiv y(P^b_X, P^b_Y, \theta_X, \theta_Y)$, which are given by

$$x^b(\theta_X, \theta_Y) = \Omega[A - D) - B^b_X(\theta_X, \theta_Y) - \gamma\theta_X) + D^b_Y(\theta_X,\theta_Y) - \gamma\theta_Y)] \qquad (10)$$

and

$$y^b(\theta_X, \theta_Y) = \Omega[A - D) - B(P^b_Y(\theta_X, \theta_Y) - \gamma\theta_Y) + D^b_X(\theta_X, \theta_Y) - \gamma\theta_X)] \qquad (11)$$

where

$$\Omega \equiv \frac{1}{B^2 - D^2}.$$

Equivalently, and more simply, using the first-order conditions for stage 2 maximization, we can write

$$x^b(\theta_X, \theta_Y) = \Omega B[P_X^b(\theta_X, \theta_Y) - f(\theta_X)] \tag{12}$$

and

$$y^b(\theta_X, \theta_Y) = \Omega B[P_Y^b(\theta_X, \theta_Y) - g(\theta_Y)]. \tag{13}$$

Using either (10) or (12), we obtain the following comparative static results for firm X:

$$\frac{\partial x^b}{\partial \theta_Y} = B^2 D\phi\Omega[g'(\theta_Y) - \gamma] = \Omega B \frac{\partial P_X^b}{\partial \theta_Y} \tag{14}$$

and

$$\begin{aligned} \frac{\partial x^b}{\partial \theta_X} &= \Omega\phi B(2B^2 - D^2)[\gamma - f'(\theta_X)] \\ &= \Omega B\left[\frac{\partial P_X^b}{\partial \theta_X} - f'(\theta_X)\right]. \end{aligned} \tag{15}$$

Thus, the equilibrium quantitity response is in the *same* direction as the equilibrium price response if $\theta_X < \theta_X^C$, but is in the *opposite* direction if θ_X exceeds the critical value θ_X^C.

Similarly, for firm Y,

$$\frac{\partial y^b(\theta_X, \theta_Y)}{\partial \theta_X} = B^2 D\phi\Omega[f'(\theta_X) - \gamma] = \Omega B \frac{\partial P_Y^b}{\partial \theta_X} \tag{16}$$

$$\begin{aligned} \frac{\partial y^b(\theta_X, \theta_Y)}{\partial \theta_Y} &= \Omega\phi B(2B^2 - D^2)[\gamma - g'(\theta_Y)] \\ &= \Omega B\left[\frac{\partial P_Y^b}{\partial \theta_Y} - g'(\theta_Y)\right] \end{aligned} \tag{17}$$

PROPOSITION 2. (Comparative statics on equilibrium quantities)

(i) If θ_i *is initially small, that is, it is below the critical value* θ_i^C, *a marginal increase in* θ_i *will result in an increase in the equilibrium output of good i, and a decrease in the equilibrium output of good j.*

(ii) *If* θ_i *is initially great, that is, it is above the critical value* θ_i^C, *a marginal increase in* θ_i *will result in an decrease in the equilibrium output of good i, and an increase in the equilibrium output of good j.*

Remark. To understand part (i), note that if θ_X is initially small, then, as shown in proposition 1, the effect of a marginal increase in θ_X is to raise P_X^b and to reduce P_Y^b. But the rise in P_X^b is relatively small, because $f'(\theta_X)$ is small (for $\theta_i < \theta_i^C$), compared with the increase in the environmental friendliness of X. In fact, the net price $(P_X^b - \gamma\theta_X)$ will fall, as indicated in proposition 1. Therefore, the equilibrium sale x^b rises. Firm Y responds to the increase in the environmental friendliness of X by reducing its price, but this is not enough to increase its sale. For part (ii), we know from proposition 1 that both P_X^b and P_Y^b rise in response to an increase in θ_X that is initially above θ_X^C. Since the increase in P_X^b in this case is largely due to cost increase, the equilibrium sale x^b falls. Even though firm Y increases its price, its sale y^b still increases.

We are now in a position to provide an answer to the following question: *Does raising environmental requirements really protect the domestic firm and improve the environment?* Consider the case where initially $\theta_Y < \theta_Y^C$. Suppose the government of the home country (whose residents import good Y) requires the foreign firm (firm Y) to *raise* its θ_Y above its initial level θ_Y by a small amount so that θ_Y continues to be below θ_Y^C, and θ_X is kept unchanged. This will result a higher price of good Y, but good Y will become more attractive than before, so more of good Y will be imported. The price of good X will fall (as firm X tries to remain competitive); nonetheless, consumers in the home country will buy less of good X (because good Y has become more attractive.). Therefore, as proposition 1 and proposition 2 predict, a small increase in θ_Y can result in a decrease in both the net profit and the market share of firm X. Thus, such imposition may hurt the domestic firm.

As to the effect on the environment, let the term $\gamma(\theta_H - \theta_X)\ x^b + \gamma(\theta_H - \theta_Y)y^b$ represent the total environmental damaged. Then, the change in the environmental damage when θ_Y is raised to satisfy the environmental requirement is

$$\frac{\partial}{\partial\theta_Y}(\gamma[\theta_H - \theta_X]x^b + \gamma[\theta_H - \theta_Y]y^b)$$
$$= \gamma[\theta_H - \theta_X]\frac{\partial x^b}{\partial\theta_Y} + \gamma[\theta_H - \theta_Y]\frac{\partial y^b}{\partial\theta_Y} - \gamma y^b. \qquad (18)$$

We can classify the RHS of equation (18) into two effects: (1) the first term represents a strategic effect in the sense that any change of θ_Y will affect P_X^b, thus x^b; (2) we can rewrite the second and third terms as

$$\gamma y^b\left(\frac{\partial y^b / y^b}{\partial\theta_Y/(\theta_H - \theta_Y)} - 1\right),$$

and call it own-elasticity effect. If the own-elasticity effect dominates the strategic effect, the RHS of (18) can be positive.

PROPOSITION 3. *A legislated* small *increase in the environmental requirement for the imported goods may hurt the domestice firm and harm the global environment if* $\theta_Y < \theta_Y^C$.

Remark. If D is small, then the right-hand side of equation(18) is positive if the following elasticity condition holds:

$$\left(\frac{\theta_H - \theta_Y}{y^b}\right)\frac{\partial y^b}{\partial \theta_Y} > 1.$$

2.2.3. Equilibrium profits in stage 2

We can now explore the properties of the **equilibrium** *gross* profits in stage two.

$$\pi_X^b = [P_X^b(\theta_X, \theta_Y) - f(\theta_X)]Nx^b(\theta_X, \theta_Y) \quad (19)$$

$$\pi_Y^b = [P_Y^b(\theta_X, \theta_Y) - g(\theta_Y)]Ny^b(\theta_X, \theta_Y). \quad (20)$$

Note that stage 2 equilibrium gross profits are always *non-negative*, because a firm can always set a price such that the demand for its product is zero, and at that price, it will produce zero output, incur zero production costs in stage 2, and earn zero gross profit. From (4), in equilibrium, $P_X - f(\theta_X) = x/(\Omega B)$. Substituting this into (19), we obtain the equilibrium gross profits in stage 2, expressed as functions of the parameters θ_X and θ_Y that were chosen in stage 1:

$$\pi_X^b = N\Omega B[P_X^b(\theta_X, \theta_Y) - f(\theta_X)]^2. \quad (21)$$

Similarly,

$$\pi_Y^b = N\Omega B[P_Y^b(\theta_X, \theta_Y) - g(\theta_Y)]^2. \quad (22)$$

In what follows, we assume that parameter values are such that both firms make strictly positive gross profits. Then we obtain the following results

PROPOSITION 4.

(i) *The stage 2 equilibrium gross profit of firm X is an increasing function of its own level of environmental friendliness* θ_X *for values of* θ_X *smaller than the critical level* θ_X^C *and is a decreasing function of* θ_X *for values of* θ_X *greater than the critical level* θ_X^C.

(ii) *The stage 2 equilibrium gross profit of firm X is a decreasing function of the level of environmental friendliness* θ_Y *of its rival, for values of* θ_Y *smaller than the critical level* θ_Y^C, *and is a increasing function of* θ_Y *for values of* θ_Y *greater than the critical level* θ_Y^C.

Proof.

(i) From (21),

$$\frac{\partial \pi_X^b}{\partial \theta_X} = 2N\Omega B[P_X^b(\theta_X, \theta_Y) - f(\theta_X)]\left[\frac{\partial P_X^b(\theta_X, \theta_Y)}{\partial \theta_X} - f'(\theta_X)\right]$$
$$= 2N\Omega\phi B[2B^2 - D^2][P_X^b(\theta_X, \theta_Y) - f(\theta_X)][\gamma - f'(\theta_X)]. \tag{23}$$

Since the function π_X^b as given by (21) is non-negative and its derivative with respect to its own θ_X, as given by (23), is zero at a unique value θ_X^C and is positive to the left (and positive to the right) of that value, $\pi_X^b(\theta_X, \theta_Y)$ has the single-peaked property with respect to θ_X. In other words, π_X^b is strictly quasi-concave in θ_X. Thus, given θ_Y, gross profit $\pi_X^b(\theta_X, \theta_Y)$ attains its global maximum (with respect to θ_X) when θ_X takes the value θ_X^C.
(ii) From (21)

$$\frac{\partial \pi_X^b}{\partial \theta_Y} = 2N\Omega B[P_X^b(\theta_X, \theta_Y) - f(\theta_X)]\left[\frac{\partial P_X^b(\theta_X, \theta_Y)}{\partial \theta_Y}\right]$$
$$= 2N\Omega\phi B^2 D[P_X^b(\theta_X, \theta_Y) - f(\theta_X)][g'(\theta_Y) - \gamma] \tag{24}$$

and

$$\frac{\partial^2 \pi_X^b}{\partial \theta_Y^2} = 2N\Omega\phi B^2 D[P_X^b(\theta_X, \theta_Y) - f(\theta_X)]g'' + 2N\Omega\phi^2 B^3 D^2[g'(\theta_Y) - \gamma]^2 > 0.$$

Note that $P_X^b(\theta_X, \theta_Y) - f(\theta_X) > 0$ by the assumption that firms make strictly positive profits. It follows that, given θ_X, the equilibrium profit of firm X, when graphed against the possible values of θ_Y chosen by its rival, has the U-shaped property (see figure 3).

2.2.4. Stage 1: the choice of the level of environmental friendliness
Now we are ready to solve the equilibrium in stage 1. The *net* profit functions are

$$\pi_X^{net} = \pi_X^b - F(\theta_X) = N\Omega B[P_X^b(\theta_X, \theta_Y) - f(\theta_X)]^2 - F(\theta_X) \tag{25}$$

and

$$\pi_Y^{net} = \pi_Y^b - G(\theta_Y) = N\Omega B[P_Y^b(\theta_X, \theta_Y) - g(\theta_Y)]^2 - G(\theta_Y). \tag{26}$$

In stage 1, firm X takes θ_Y as given and chooses θ_X to maximize (25). (Note that the function π_X^b attains its maximum at θ_X^C, for any θ_Y.) The first-order condition for an interior maximum is

$$\frac{\partial \pi_X^{net}}{\partial \theta_X} = \frac{\partial \pi_X^b}{\partial \theta_X} - F'(\theta_X) = 0. \tag{27}$$

The second-order condition is assumed to be satisfied; this seems a reasonable assumption if $F(\theta_X)$ is convex.

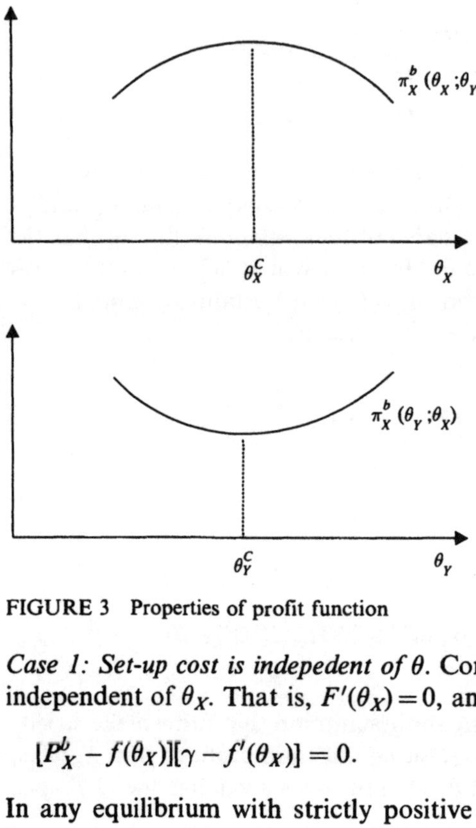

FIGURE 3 Properties of profit function

Case 1: Set-up cost is indepedent of θ. Consider first the case in which $F(\theta_X)$ is independent of θ_X. That is, $F'(\theta_X)=0$, and the above equation gives

$$[P^b_X - f(\theta_X)][\gamma - f'(\theta_X)] = 0. \tag{28}$$

In any equilibrium with strictly positive profits, $P_X - f(\theta_X) > 0$, we conclude from (28) that the equilibrium choice of θ_X in stage 1 satisfies the property

$$f'(\theta^*_X) = \gamma. \tag{29}$$

Thus, $\theta^*_X = \theta^C_X$. Notice that firm X's optimal choice of θ_X is *independent* of firm Y's choice. Firm X's strategy is a *dominant strategy*.

Similarly, for firm Y,

$$g'(\theta^*_Y) = \gamma. \tag{30}$$

The pair $(\theta^*_X, \theta^*_Y) = (\theta^C_X, \theta^C_Y)$ is the free trade equilibrium choice of θ_X and θ_Y, in case 1, where $F'(.) = G'(.) = 0$ identically.

PROPOSITION 5. Assume that the set-up cost is independent of θ_i. In the equilibrium of stage 1, firm i's choice of the environmental friendliness level θ^C_i has the property that a marginal increase in θ_i would raise the average variable cost by an amount which just equals γ (the consumer's marginal evaluation of environmental friendliness). In particular, this equilibrium is a *dominant strategy equilibrium.*

Case 2: Set-up cost is increasing in θ. Turning to the case where $F'(\theta_X) > 0$ and $G'(\theta_Y) > 0$, the necessary conditions are

$$\alpha[P_X^b(\theta_X, \theta_Y) - f(\theta_X)][\gamma - f'(\theta_X)] = F'(\theta_X) > 0 \tag{31}$$

$$\alpha[P_Y^b(\theta_X, \theta_Y) - g(\theta_Y)][\gamma - g'(\theta_Y)] = G'(\theta_Y) > 0, \tag{32}$$

where

$$\alpha \equiv 2N\Omega\phi B[2B^2 - D^2] > 0.$$

Since $P_X^b(\theta_X, \theta_Y) - f(\theta_X) > 0$ in any equilibrium with positive gross profit, it follows from (31) and (32) that the firms' choices of θ_X^{**} and θ_Y^{**} satisfy

$$\gamma - f'(\theta_X^{**}) > 0$$

and

$$\gamma - g'(\theta_Y^{**}) > 0$$

Hence, with $F'(\theta_X) > 0$ and $G'(\theta_Y) > 0$, the Nash equilibrium choice of levels of environmental friendliness θ_X^{**} and θ_Y^{**} are *lower* than the critical levels θ_X^C and θ_Y^C.

PROPOSITION 6. *Assume that $F'(\theta_X) > 0$ and $G'(\theta_Y) > 0$. In the equilibrium of stage 1, firm i's choice of the environmental friendliness level is $\theta_i^{**} < \theta_i^C$. Thus, at the free trade equilibrium, a marginal increase in θ_i would raise the average variable cost by an amount which is* smaller *than γ, the marginal evaluation of environmental quality.*

3. Policy implications

We now consider the case where the government of the home country interferes with the market in order to maximize a certain objective function. For example, the government may wish to maximize a political support function. The view that government policies can be explained in terms of political support motives has been expounded in a number of influential papers; see Hillman (1982, 1989, 1991), Hillman and Ursprung (1988).[5]

5 See also Long and Vousden (1991) and Hillman, Long, and Soubeyran (2001). In a different formulation, Grossman and Helpman (1994) used a common agency model, where each interest group offers to the government a contribution schedule that, promises payments to the government conditional on its acting in favour of the group.

We assume that the political support that the government receives depends on (i) the increase in consumers welfare (relative to the welfare level $\bar{S}$ obtained under the free market regime) and (ii) the increase in the profit of the home firm (relative to its profit level $\bar{\pi}_X^{net}$ under the free market regime) that are generated by government policies. In other words, we suppose that consumers and firms know what their welfare levels would be under the laissez-faire regime, and they use these as benchmarks against which they measure the impact of government policies on their welfare.

Let S denote the consumer's surplus of the representative consumer of the home country and π_X^{net} denote the profit of the home firm. We assume that the political support function is linear in S and π_X^{net}:

$$R = \omega(NS - N\bar{S}) + (1 - \omega)(\pi_X^{net} - \bar{\pi}_X^{net}), \tag{33}$$

where ω is the relative weight of consumer surplus in the political support function $(1 \geq \omega \geq 0)$.

In what follows, the equilibrium levels of environmental friendliness are denoted by $\widehat{\theta}_X$ and $\widehat{\theta}_Y$. Then, in case 1, $\widehat{\theta}_i = \theta_i^* = \theta_i^C$; and in case 2, $\widehat{\theta}_i = \theta_i^{**} < \theta_i^C$.

3.1. Minimum requirement imposed on imported goods

The home government may have many policy instruments at its disposal. In this paper, we focus on only one type of policy instrument: requiring the environmental friendliness of the foreign good or the home good (or both) to satisfy certain mimimum level. We seek to determine conditions under which it is in the interest of the home government to use such instruments, and we analyse the effects of policy decisions on the home firm's profit and on the welfare of consumers.

3.1.1. General considerations

In this subsection, we assume the home government passes a law that requires the product of the *foreign firm* to meet (or exceed) a minimum level θ_m. In the absence of government intervention, the equilibrium choice $\widehat{\theta}_Y$ may be below $\widehat{\theta}_X$, reflecting the fact that the foreign firm, located in a less advanced country, has a cost disadvantage. In such a case, a required minimum level θ_m that exceeds $\widehat{\theta}_Y$ may not be met with strong criticism from international organizations. We now wish to determine whether a small increase in θ_m from the initial value $\widehat{\theta}_Y (= \theta_Y^*$ in case 1, $= \theta_Y^{**}$ in case 2) will generate positive political support for the home government. To do this, we must find the derivative $\partial R / \partial \theta_m$ and evaluate it at $\theta_m = \theta_Y^*$ in case 1 (or θ_Y^{**} in case 2).

In taking this derivative, we must ask what happens to θ_X when the government raises θ_m marginally above the free market equilbrium level. This question may be answered under two alternative scenarios. (a) We assume that θ_X is being held fixed at $\widehat{\theta}_X (= \theta_X^*$ in case 1, $= \theta_X^{**}$ in case 2). (b) We assume that firm X optimally adjusts $\widehat{\theta}_X$ when it learns that firm Y must comply with the government requirement θ_m. As argued in the appendix, under case 1 (i.e., set-up costs are independent of θ_i), scenarios (a) and (b) are *equivalent*, because the home firm's optimal adjustment under case 1 is zero. Under case 2 (i.e., set-up costs are increasing in θ_i), the second scenario is slightly more complicated, but the basic results do not differ from those obtained under scenario (a) and therefore will be relegated to the appendix.

Since $\bar{S}$ and $\bar{\pi}_X^{net}$ are constant, we need only to find $\omega N \partial S/\partial\theta_m + (1-\omega)\partial\pi_X^{net}/\partial\theta_m$, where

$$S = u(x_i, y_i) - P_X^b x_i - P_Y^b y_i,$$

with P_X^b and P_Y^b being stage 2 Bertrand equilibrium prices, given θ_X, and given $\theta_Y = \theta_m$. The quantities $x_i = x^b$ and $y_i = y^b$ are the equilibrium quantities we found in section 2.

Then, given that θ_X is *unchanged* at its initial laissez-faire value,

$$\frac{\partial S}{\partial\theta_m} = \frac{\partial S}{\partial\theta_Y}\frac{\partial\theta_Y}{\partial\theta_m}. \tag{34}$$

That is,

$$\begin{aligned}\frac{\partial S}{\partial\theta_m} ={}& \gamma[y_i + \beta Y_{-i}]\frac{\partial\theta_Y}{\partial\theta_m} + \frac{\partial u}{\partial x_i}\frac{\partial x^b}{\partial\theta_Y}\frac{\partial\theta_Y}{\partial\theta_m} + \frac{\partial u}{\partial y_i}\frac{\partial y^b}{\partial\theta_Y}\frac{\partial\theta_Y}{\partial\theta_m}\\ &- P_X^b\frac{\partial x^b}{\partial\theta_Y}\frac{\partial\theta_Y}{\partial\theta_m} - P_Y^b\frac{\partial y^b}{\partial\theta_Y}\frac{\partial\theta_Y}{\partial\theta_m} - x_i\frac{\partial P_X^b}{\partial\theta_Y}\frac{\partial\theta_Y}{\partial\theta_m} - y_i\frac{\partial P_Y^b}{\partial\theta_Y}\frac{\partial\theta_Y}{\partial\theta_m}\\ &- \beta\gamma[\theta_H - \widehat{\theta}_Y]\frac{\partial Y_{-i}}{\partial\theta_Y}\frac{\partial\theta_Y}{\partial\theta_m} - \beta\gamma[\theta_H - \widehat{\theta}_X]\frac{\partial X_{-i}}{\partial\theta_Y}\frac{\partial\theta_Y}{\partial\theta_m}.\end{aligned} \tag{35}$$

Now, since $\partial u/\partial x_i = P_X^b$ and $\partial u/\partial y_i = P_Y^b$, and since $\partial\theta_Y/\partial\theta_m = 1$, the above expression is simply

$$\begin{aligned}\frac{\partial S}{\partial\theta_m} ={}& \gamma[1 + \beta(N-1)]y^b - \beta(N-1)\gamma[\theta_H - \widehat{\theta}_Y]\frac{\partial y^b}{\partial\theta_Y}\\ &- \beta(N-1)\gamma[\theta_H - \widehat{\theta}_X]\frac{\partial x^b}{\partial\theta_Y} - x^b\frac{\partial P_X^b}{\partial\theta_Y} - y^b\frac{\partial P_Y^b}{\partial\theta_Y}.\end{aligned} \tag{36}$$

This expression deserves some comments. In case 1, $\partial P_X^b/\partial\theta_Y = 0$ but in case 2, $\partial P_X^b/\partial\theta_Y$ is negative because θ_Y^{**} is less than θ_Y^C. So in both cases, the term $-x^b(\partial P_X^b/\partial\theta_Y)$ is greater than or equal to zero, indicating it contributes to a gain in consumers surplus, which is obvious because the price of good X falls. The term $-y^b(\partial P_Y^b/\partial\theta_Y)$ is always negative, because P_Y rises with θ_Y. However, when this term is combined with the positive term γy^b, we have a non-negative expression:

$$\left[\frac{2B^2(\gamma - g'(\theta_Y))y^b}{4B^2 - D^2}\right],$$

which is zero if $\theta_Y = \theta_Y^C$ and positive if $\theta_Y = \theta_Y^{**}$.

It follows that if $\beta = 0$, then a marginal increase in θ_m (starting from the initial laissez-faire equilibrium) will increase the consumer's surplus. If β is positive, then we must evaluate the expression

$$\beta(N-1)\gamma\left[y^b - (\theta_H - \theta_Y)\frac{\partial y^b}{\partial \theta_Y} - (\theta_H - \theta_X)\frac{\partial x^b}{\partial \theta_Y}\right],$$

which measures the effect of an increase in θ_Y on the total environmental damage (see proposition 3).

Turning to the profit of the home firm, we obtain from (21)

$$\frac{\partial \pi_X^{net}}{\partial \theta_m} = 2NB\Omega[P_X^b(\theta_X, \theta_Y) - f(\theta_X)]\frac{\partial P_X^b}{\partial \theta_Y}\frac{\partial \theta_Y}{\partial \theta_m}. \tag{37}$$

Thus, given that the assumption that profits are strictly positive in equilibrium, the sign of this expression is simply the sign of $\partial P_X^b / \partial \theta_Y$ which we know from proposition 4.

3.1.2. Case 1: Set-up costs are independent of environmental friendliness levels

In case 1, since $G' = 0$, we have from section 2, with $\widehat{\theta}_Y = \theta_Y^* = \theta_Y^C$

$$\frac{\partial x^b}{\partial \theta_Y} = 0 = \frac{\partial y^b}{\partial \theta_Y} = \frac{\partial P_X^b}{\partial \theta_Y} \text{ and } \frac{\partial P_Y^b}{\partial \theta_Y} = \gamma.$$

Hence,

$$\frac{\partial R}{\partial \theta_m} = \omega\beta\gamma N(N-1)y^b. \tag{38}$$

Thus, we obtain the following result for case 1.

PROPOSITION 7. *In case 1, if $\beta > 0$ and $\omega > 0$, then a small increase in θ_m will raise the political support for the home government, because consumers' surplus rises and profit of the home firm also rises (though the increase in profit is approximately zero for an infinitesimal increase in θ_m).*

What happens if $\beta = 0$? From (38), an infinitesimal increase in θ_m from the laissez-faire level $\theta_Y^* = \theta_Y^C$ would have no first-order effect on the level of political support. This indicates that R is at a local maximum or a local minimum, or an inflection point at $\theta_Y^* = \theta_Y^C$. To find out, we must look at the second derivatives. From (36), the second derivative of S with respect to θ_m, evaluated at θ_Y^C, and $\beta = 0$, is

$$N\frac{\partial^2 S}{\partial\theta_m^2} = -Nx^b\frac{\partial^2 P_X^b}{\partial\theta_m^2} - Ny^b\frac{\partial^2 P_Y^b}{\partial\theta_m^2}$$
$$= -\phi BDNx^b g'' - 2B^2\phi Ny^b g'', \tag{39}$$

indicating a local maximum for S. From (37), the second derivative of π_X^b with respect to θ_m, evaluated at θ_Y^C, is

$$\frac{\partial^2 \pi_X^b}{\partial\theta_m^2} = 2NB\Omega[P_X^b(\theta_X, \theta_Y) - f(\theta_X)]\frac{\partial^2 P_X^b}{\partial\theta_Y^2}$$
$$= 2Nx^b\phi BDg'', \tag{40}$$

indicating a local minimum for the profit of the home firm.

Combining (39) and (40), we see that the political support function reaches a maximum at $\theta_m = \theta_Y^C$ if the weight given to consumers is sufficiently great, or if D is small. In the special case where $\omega = 0.5$, we can see that, with $\beta = 0$, the political support function, in fact, reaches a local maximum at $\theta_m = \theta_Y^C$ if $y^b \geq x^b$.

PROPOSITION 8. *In case 1, if the spillover parameter is zero ($\beta = 0$), starting from the initial laissez-faire equilibrium, a small increase in θ_m will reduce consumers' surplus but increase the profit of the home firm. If, in addition, $y^b \geq x^b$, then the loss of consumers surplus exceeds the gain in profit. If the weight given to profit is sufficiently great, the home government will have an incentive to increase θ_m.*

3.1.3. Case 2: Set-up costs are increasing in the level of environmental friendliness

Turning to case 2, we use (36), (37), and (14) to obtain

$$\frac{\partial R}{\partial\theta_m} = \omega\beta N(N-1)\gamma\left[y^b - (\theta_H - \theta_Y^{**})\frac{\partial y^b}{\partial\theta_Y} - (\theta_H - \theta_X^{**})\frac{\partial x^b}{\partial\theta_Y}\right]$$
$$+ \omega Ny^b\left[\gamma - \frac{\partial P_Y^b}{\partial\theta_Y}\right] + [2(1-\omega)\phi BDN[g'(\theta_Y^{**}) - \gamma]x^b$$
$$-\omega Nx^b\frac{\partial P_X^b}{\partial\theta_Y}\Bigg]. \tag{41}$$

The right-hand side of (41) consists of three terms. The first term represent the equilibrium effect of an increase in θ_Y on the environmental damage. It is zero if $\beta = 0$. If β is positive, then this term is positive if the environment improves when θ_Y increases. With the help of (8), the second term can be written as

$$\omega Ny^b\left[\frac{2B^2(\gamma - g'(\theta_Y^{**}))}{4B^2 - D^2}\right],$$

which is positive because $\theta_Y^{**} < \theta_Y^C$. Now, since $(P_X^b - f)B\Omega = x^b$ by (4), the third term can be written as

$$\phi BDN[g'(\theta_Y^{**}) - \gamma][2 - 3\omega]x^b.$$

Combining the second and the third terms, we obtain

$$(\gamma - g'(\theta_Y^{**}))NB\phi[2B\omega y^b - (2 - 3\omega)Dx^b].$$

PROPOSITION 9. *In case 2 ($G' > 0$), for all $\beta \geq 0$, a small increase in the environmental requirement imposed on imported goods from the initial value θ_Y^{**} will* reduce *the profit of the home firm and increase the volume of imports. The effects of such a policy on political support are as follows*:

(i) *If $\beta = 0$, consumers' surplus will* unambiguously *increase. Thus, in this case, the increase in the level imposed on imported goods will raise political support if $2\omega By^b \geq (2 - 3\omega)\ Dx^b$, (in the case $\omega = 1/2$, this condition is simply $2By^b > Dx^b$), but it will reduce political support if ω is sufficiently small.*

(ii) *if $\beta > 0$, consumers' surplus may decrease if total environmental damage increases (see proposition 3). In such a case, total political support may fall*[6] *when θ_m is raised.*

3.2. *Minimum requirement for home goods*

Let θ_n be the minimum requirment imposed on the home firm. Again, we assume that initially $\theta_n = \widehat{\theta}_X = \theta_X^*$ in case 1, ($= \theta_Y^{**}$ in case 2.)

Then,

$$\frac{\partial S}{\partial \theta_n} = \gamma[1 + \beta(N - 1)]x^b - \beta(N - 1)\gamma[\theta_H - \widehat{\theta}_X]\frac{\partial x^b}{\partial \theta_X} - \beta(N - 1)\gamma[\theta_H - \widehat{\theta}_Y]\frac{\partial y^b}{\partial \theta_X} - x^b\frac{\partial P_X^b}{\partial \theta_X} - y^b\frac{\partial P_Y^b}{\partial \theta_X}.$$

Turning to profit, we note that $\partial \pi_X^{net}/\partial \theta_n = 0$ when θ_n is evaluated at the initial equilibrium. Hence, we obtain the following propositions. (The proofs are omitted to save space.)

PROPOSITION 10. *In case 1,*

(i) *if the spillover parameter is zero ($\beta = 0$), starting from the initial laissez-faire equilibrium, a small increase in θ_n will reduce consumers' surplus and reduce the profit of the home firm.*

(ii) *if $\beta > 0$, then (starting from the initial laissez-faire equilibrium) a small increase in θ_n may raise consumers' surplus and* reduce *the profit of the*

6 This can occur even if $\omega = 1/2$ and $y^b = x^b$.

home firm. The government will increase θ_n only if the weight ω given to consumers' surplus is sufficiently great.

PROPOSITION 11. *In case 2,*

(i) *with $\beta = 0$, a small increase in θ_n from the initial value θ_X^{**} will raise political support because it will raise consumers' surplus, while the change in profit is zero (as a first-order approximation.)*

(ii) *with $\beta > 0$, a small increase in θ_n from the initial value θ_X^{**} will raise political support if it reduces total environmental damage (or if there is only a very small increase in total environmental damage).*

4. Concluding remarks

We have used a simple model to study the equilibrium choice of levels of environmental friendliness by rival firms, and derived some implications for environmental policy. It is demonstrated that under certain conditions, the market outcome is socially optimal (in the second-best sense). We also show that a government that seeks to maximize political support might want to interfere with the market by setting a minimum level of environmental friendliness on the imported goods or on the home goods. The effects of such intervention on the profit of the home firm and on consumers' surplus vary from case to case. For example, as shown in proposition 9, contrary to expectations derived from the conventional wisdom, an increase in the minimum environmental requirement imposed on foreign goods may *reduce* the profit of the home firm and may increase or reduce the consumers' surplus. When the minimum requirement level is raised, market prices rise but net prices may fall. Moreover, this imposition may increase the total damage done to the environment if the demand is elastic with respect to adjusted prices.

Our analysis indicates that it would be too hasty to conclude that the imposition of a minimum requirement would reduce the volume of imports and would be protectionist in intent or in effect. The model points to some of the possible pitfalls in intuitive reasoning.

To keep the analysis simple and to focus on key issues, we have abstracted from a number of real-world considerations. Among these are (i) consumers are heterogeneous: not all of them care about the environment; (ii) firms may carry out R&D activities to search for least-cost methods of producing environmentally friendly goods; (iii) entry and exit of firms when minimum environmental requirements are imposed. It is no doubt worthwhile to take up these issues.

Appendix

The case in which θ_X adjusts when θ_m is raised marginally above $\widehat{\theta}_Y$:

Case 1: Set-up cost is a constant. In this case firm X finds it optimal not to adjust θ_X away from the equilibrium level $\theta_X^* = \theta_X^C$, because, as we have shown, $\theta_X^* = \theta_X^C$ is the dominant strategy.

Case 2: Set-up cost is not a constant. In this case, applying the implicit function theorem to (27) we get

$$\frac{\partial \theta_X}{\partial \theta_Y} = -\frac{1}{\Delta}\,\frac{\partial^2 \pi_X^{net}}{\partial \theta_Y \partial \theta_X},$$

where

$$\Delta \equiv \frac{\partial^2 \pi_X^{net}}{\partial \theta_X^2},$$

which is negative by the second-order condition, and

$$\frac{\partial^2 \pi_X^{net}}{\partial \theta_Y \partial \theta_X} = 2N\Omega B\phi[2B^2 - D^2][\gamma - f'(\theta_X^{**})]\frac{\partial P_X^b}{\partial \theta_Y} < 0.$$

That is,

$$\frac{\partial^2 \pi_X^{net}}{\partial \theta_Y \partial \theta_X} = \frac{F'(\theta_X^{**})}{P_X^b - f(\theta_X^{**})}[g'(\theta_Y^{**}) - \gamma]\phi BD.$$

Thus,

$$\frac{d\theta_X}{d\theta_Y} = -\left[\frac{1}{\Delta}\right]\left[\frac{F'(\theta_X^{**})}{P_X^b - f(\theta_X^{**})}\right][g'(\theta_Y^{**}) - \gamma]\phi BD < 0.$$

Then,

$$\frac{\partial S}{\partial \theta_m} = \frac{\partial S}{\partial \theta_Y}\,\frac{\partial \theta_Y}{\partial \theta_m} + \frac{\partial S}{\partial \theta_X}\,\frac{\partial \theta_X}{\partial \theta_Y}\,\frac{\partial \theta_Y}{\partial \theta_m},$$

where $\partial\theta_Y/\partial\theta_m = 1$. Thus, we have to add to the right-hand side of (36) the following terms:

$$\gamma[1 + \beta(N-1)]x^b\frac{\partial \theta_X}{\partial \theta_Y} - \beta(N-1)\gamma[\theta_H - \theta_Y]\frac{\partial y^b}{\partial \theta_X}\,\frac{\partial \theta_X}{\partial \theta_Y}$$
$$- \beta(N-1)\gamma[\theta_H - \theta_X]\frac{\partial x^b}{\partial \theta_X}\,\frac{\partial \theta_X}{\partial \theta_Y} - x^b\frac{\partial P_X^b}{\partial \theta_X}\,\frac{\partial \theta_X}{\partial \theta_Y} - y^b\frac{\partial P_Y^b}{\partial \theta_X}\,\frac{\partial \theta_X}{\partial \theta_Y}.$$

Concerning the profit term, we now have

$$\frac{\partial \pi_X^{net}}{\partial \theta_m} = \frac{\partial \pi_X^{net}}{\partial \theta_Y}\frac{\partial \theta_Y}{\partial \theta_m} + \frac{\partial \pi_X^{net}}{\partial \theta_X}\frac{\partial \theta_X}{\partial \theta_Y}\frac{\partial \theta_Y}{\partial \theta_m}.$$

But the second term is zero because $\partial \pi_X^{net}/\partial \theta_X = 0$ at $\theta_X = \widehat{\theta}_X$. It follows that $\partial \pi_X^{net}/\partial \theta_m$ is the same as in the main text.

It is clear that the basic conclusions in the text are qualitatively unchanged when θ_X adjusts in response to a marginal increase in θ_m.

References

Anderson, Kym (1997) 'Environmental standards and international trade,' in *Annual World Bank Conference on Development Economics* (Washington, D.C.: World Bank)

Brander, J.A. and B.J. Spencer (1985) 'Export subsidies and international market share rivalry,' *Journal of International Economics* 18, 83–101

Caswell, J.A. (1998) 'Should use of genetically modified organisms be labeled?' *AgBio-Forum* 1, 22–4

Copeland, Brian, and Scott Taylor (1994) 'North-south trade and the environment,' *Quarterly Journal of Economics* 109, 755–87

—— (1995) 'Trade and transboundary pollution,' *American Economic Review* 85, 716–37

da Motta Veiga, P. (1997) 'Eco-labelling schemes in the EU and their impact on Brazilian exports,' in S. Zarrilli, V. Jha, and R. Vossenaar, eds, *Eco-labelling and International Trade* (UNCTAD)

de Palma, André, and Luc Leruth (1996) 'Variable willingness to pay for network externalities with strategic standardization decisions,' *European Journal of Political Economy* 12, 235–51

Eaton, Jonathan, and Gene M. Grossman (1986) 'Optimal trade and industrial policy under oligopoly,' *Quarterly Journal of Economics* 101, 385–406

Economides, Nicholas (1996) 'Network externalities, complementarities, and invitations to enter,' *European Journal of Political Economy* 12, 211–33

Economist, The (2000) 'Ethical investment: morality plays,' 12, 8–14 July, 78–9

Fischer, Ronald, and Pablo Serra (2000) 'Standards and protection,' *Journal of International Economics* 52, 377–400

Ford Foundation Report, Fall (2000), Volume 31, No. 4. (New York: Fond Foundation)

Grossman, Gene, and Elhanan Helpman (1994) 'Protection for sale,' *American Economic Review* 84, 833–50

Hillman, A.L. (1982) 'Declining industries and political-support protectionist motives,' *American Economic Review* 72, 1180–87, reprinted in J. Bhagwati and P. Rosendorff, eds (1998) *Readings in the Political Economy of Trade Policy* (Cambridge, MA: MIT Press)

—— (1989) *The Political Economy of Protection* (Harwood Academic Publishers, Cambridge, London)

—— (1991) 'Protections, politics and market structure,' in E. Helpman and A. Razin, eds, *International Trade and Trade Policy* (Cambridge, MA: MIT Press)

Hillman A.L. and H.W. Ursprung (1988) 'Domestic politics, foreign interests and international trade policy,' *American Economic Review* 78, 729–45; reprinted in J. Bhagwati and P. Rosendorff, eds (1998) *Readings in the Political Economy of Trade Policy* (Cambridge, MA: MIT Press)

Hillman, A.L., N.V. Long, and A. Soubeyran (2001) 'Protection, lobbying, and market structure,' *Journal of International Economics* 54, 383–409

Ho, Lily, et al. (1997) 'The potential impact of the EU eco-labelling program on Columbian textile exports,' in S. Zarrilli, V. Jha, and R. Vossenaar, eds, *Eco-labelling and International Trade* (UNCTAD)

Jeanneret, Marie-Hélène, and Thierry Verdier (1996) 'Standardization and protection in a vertical differentiation model,' *European Journal of Political Economy* 12, 253–71

Leland, H.E. (1979) 'Quacks, lemons and licensing: a theory of minimum quality standards,' *Journal of Political Economy* 87, 1328–46

Long, Ngo Van, and Neil Vousden (1991) 'Protectionist responses and declining industries,' *Journal of International Economics* 30, 87–104

Long, Ngo Van, and Antoine Soubeyran (1997) 'Cost heterogeneity, industry concentration, and strategic trade policy,' *Journal of International Economics* 43, 207–20

Matutes, Carmen, and Pierre Regibeau (1996) 'A selective review of the economics of standardization: entry deterrence, technological progress and international competition,' *European Journal of Political Economy* 12, 183–209

McClusky, Jill J. (2000a) 'Read the warning: this product may contain GMO,' *Choices, Second Quarter*, 39–42

—— (2000b) 'A game theoretic approach to organic foods: an analysis of asymmetric information and policy,' *Agricultural and Resource Economic Review* 29, 1–9

Moenius, Johannes (1999a) 'Information versus product adaptation: the role of standards in trade,' typescript, Department of Economics, University of California, San Diego

—— (1999b) 'The bilateral standards database: a technical reference manual,' typescript, Department of Economics, University of California, San Diego

Nelson, G.C., A. De Pinto, D. Bullock, et al. (1999) 'The economics and politics of genetically modified organisms in agriculture: implications for WTO 2000,' *Bulletin 809*, University of Illinois, November

Ronnen, U. (1991) 'Minimum quality standards, fixed costs, and competition,' *Rand Journal of Economics* 22, 490–504

Schleich, Joachim, and David Orden (2000) 'Environmental quality and industry protection with non-cooperative domestic and trade policies,' *Review of International Economics* 8, 681–97

Shapiro, C. (1983) 'Premium for higher quality products as returns for reputations,' *Quarterly Journal of Economics* 98, 659–79

Swann, P., P. Temple, and M. Shurmer (1996) 'Standards and trade performance: the UK experience,' *Economic Journal* 106, 1297–313

Name Index